Fluorescent Dye Labels and Stains

Fluorescent Dye Labels and Stains

A Database of Photophysical Properties

Tarso B. Ledur Kist
Federal University of Rio Grande do Sul
Brazil

WILEY

This edition first published 2023

Registered Offices
John Wiley & Sons, Inc., 111 River Street, Hoboken, NJ 07030, USA
John Wiley & Sons Ltd, The Atrium, Southern Gate, Chichester, West Sussex, PO19 8SQ, UK

For details of our global editorial offices, customer services, and more information about Wiley products visit us at www.wiley.com.

Wiley also publishes its books in a variety of electronic formats and by print-on-demand. Some content that appears in standard print versions of this book may not be available in other formats.

Library of Congress Cataloging-in-Publication Data
Names: Ledur Kist, Tarso B., 1962- author. | John Wiley & Sons, publisher.
Title: Fluorescent dye labels and stains : a database of photophysical properties / Tarso B. Ledur Kist.
Description: Hoboken, NJ : John Wiley & Sons, 2023. | Includes bibliographical references and index.
Identifiers: LCCN 2022047668 (print) | LCCN 2022047669 (ebook) | ISBN 9781119835134 (hardback) | ISBN 9781119835141 (pdf) | ISBN 9781119835158 (epub) | ISBN 9781119835165 (ebook)
Subjects: LCSH: Fluorescent labeling. | Fluorescent probes. | Fluorescence microscopy--Technique.
Classification: LCC QH212.F55 L43 2023 (print) | LCC QH212.F55 (ebook) | DDC 570.28/2--dc23/eng/20221117
LC record available at https://lccn.loc.gov/2022047668
LC ebook record available at https://lccn.loc.gov/2022047669

Cover and Author Image: Courtesy of Tarso B. Ledur Kist
Cover Design: Wiley

Set in 9.5/12.5pt STIXTwoText by Integra Software Services Pvt. Ltd, Pondicherry, India

Contents

Preface

This book presents the molecular structures and photophysical properties of over seven hundred fluorescent dyes with medium to high brightness in aqueous solutions. It has been written to be a reliable and useful source of information for users of fluorescent dye labels and stains in general. The specific aqueous solution used to measure the photophysical parameters of each of the seven hundred dyes is clearly provided in this succinct database.

The photophysical parameters presented in this database include: wavelengths of absorption and/or excitation maximum, wavelengths of emission maximum, Stokes shifts, excitation ranges, emission ranges, molar absorption coefficients, fluorescence quantum yields, brightnesses, fluorescence mean-lifetimes, sizes (represented by the number of atoms of the structures), and charges in aqueous solutions at neutral pH. Additional interesting properties of some of the fluorescent dyes have also been included, such as photostability and pK_a (when present at around neutral pH). Scattergrams are systematically used to give panoramic views of the photophysical data that allow users to find desired photophysical properties within a given spectral range.

Previously, this data was only available from varied sources that contained partial lists, were missing data, and, in many cases, omitted descriptions of the liquids in which parameters were measured. It is known that almost all fluorescent dyes that exhibit high molecular fluorescence brightness in aqueous solutions are also bright in organic solvents; however, the opposite is not true. Most dyes that exhibit a high brightness in organic solvents exhibit a lower and, in many cases, a very low brightness in aqueous solutions. Even worse, the great majority of fluorescent dyes available are not soluble enough in aqueous solutions to have their photophysical parameters measured. Hence, the importance of this database, where only dyes whose photophysical parameters were measured in plain water, phosphate-buffered saline, or aqueous buffers (without organic modifiers) are included. Some dyes are no longer commercially available, but are nonetheless kept in the database for completeness and historic record purposes.

The author began recording the spectral and photophysical data of fluorescent dyes in his Master's degree, and has continued throughout his whole academic career. Measurements were collected in practice and from the literature, and now the time has come to share this collection with the whole community in the form of the current printed database. Probably not all bright water-soluble fluorescent dyes are here, as some may have gone unnoticed. Therefore, the author apologizes to the developers of these missing fluorophores for this.

The extensive data collection necessary to make this book a reality would not have been possible without the help of so many bright students and colleagues, as well as the professionalism of many companies (suppliers of fluorescent dye labels, stains, and probes). The author would also like to thank Prof A.Z. Khoury, Prof D. Lüdcke, and Prof A. Manz for their many stimulating conversations; Prof J.N. Picada for the suggestion to write this book; and both Dr W. Nietfeld and Prof H. Lehrach from the Max Planck Institute in Berlin, where many photophysical parameters were measured by the author during a post-Doctorate leave.

Finally, this book is dedicated to all Organic Chemists from all over the world for their talented work that has brought so many interesting fluorescent structures to our daily lives and impressive developments across so many fields. Moreover, these high-tech products are now being developed and synthesized using increasingly sustainable and environmentally-friendly methods.

April 2022

Tarso B. Ledur Kist
Porto Alegre

Acronyms

A	Acetate buffer
ACN	Acetonitrile; ethanenitrile
ADC	Antracene-2,3-dicarboxaldehyde
APD	Avalanche photodiode
B	Borate buffer
BODIPY	4,4-Difluoro-4-bora-3a,4a-diaza-*s*-indacene
C	Carbonate buffer
CCD	Charged coupled device
CBQCA	3-(4-Carboxybenxoyl)quinoline-2-carboxaldehyde
DMF	Dimethylformamide; *N,N*-dimethylmethanamide
DMSO	Dimethylsulfoxide; dimethyl(oxido)sulfur
FRET	Förster resonant energy transfer
HAS	Heteroatom-substituted
HEPES	4-(2-Hydroxyethyl)-1-piperazineethanesulfonic acid
IA	Iodoacetamido
ITC	Isothyocianate
ME	Maleimido
NDA	Naphthalene-2,3-dicarboxaldehyde
NHS	*N*-Hydroxysuccinimidyl ester
NIR	Near-infrared
NMP	*N*-Methyl-2-pyrrolidinone
OPA	*o*-Phthaldialdehyde
P	Phosphate buffer
PBS	Phosphate-buffered saline (usually: 8.0 g/L NaCl, 0.2 g/L KCl, 1.42 g/L Na_2HPO_4, and 0.24 g/L KH_2PO_4)
PD	Photodiode
PC	Propylene carbonate; 4-methyl-1,3-dioxolan-2-one
PMT	Photomultiplier tube
RT	Room temperature
SE	*N*-Hydroxysuccinimidyl ester
THF	Tetrahydrofuran; 1,4-epoxybutane
Tris	Tris(hydroxymethyl)aminomethane; 2-amino-2-(hydroxymethyl)propane-1,3-diol
UV	Ultraviolet
UVA	UV radiation in the 315 and 400 nm range. Informally also known as "black light"
Vis	Visible

Symbols and Conventions

Symbol	Name and definition [Units]
$A(\lambda)$	Absorbance at wavelength λ. $A(\lambda) = \text{Log}_{10}(P^o(\lambda)/P(\lambda)) = \text{Log}_{10}(1/T(\lambda))$. [dimensionless]
b	Cuvette inner length. [cm]
$B(\lambda)$	Molecular fluorescent brightness at wavelength λ. $B(\lambda) = \varepsilon(\lambda)\ \Phi_f$. [$\text{M}^{-1}\text{cm}^{-1}$].
B_{max}	Molecular fluorescent brightness at wavelength $\lambda_{a,max}$. $B_{max} = \varepsilon(\lambda_{x,max})\ \Phi_f = \varepsilon_{max}\ \Phi_f$. [$\text{M}^{-1}\text{cm}^{-1}$].
c	Amount concentration. Expressed in mol dm^{-3} or mol L^{-1}, while the non-SI unit M (small cap M) is used as an abbreviation for mol dm^{-3}. [mol dm^{-3}]
h	Planck constant. $h = 6.626069 \times 10^{-34}$ J s. [Js]
L	Radiance. Defined as the radiant power (P) leaving or passing through a small element of surface (dS) divided by the solid angle of this surface (dΩ) and the orthogonally projected area of this element of surface in a plane normal to the beam direction (dS cosθ). $L = \text{d}^2p/(\text{d}\Omega\ \text{d}S\ \cos\theta)$ [$\text{Js}^{-1}\text{m}^{-2}\text{sr}^{-1}$ or $\text{Wm}^{-2}\text{sr}^{-1}$]. For a parallel beam of radiation the radiance is simple given by $L = \text{d}p/(\text{d}S\ \cos\theta)$. [$\text{Js}^{-1}\text{m}^{-2}$ or Wm^{-2}]
$L(\lambda)$	Spectral radiance. Derivative of radiance, L, with respect to wavelength, λ. [$\text{Js}^{-1}\text{m}^{-2}\text{nm}^{-1}$ or $\text{Wm}^{-2}\text{nm}^{-1}$]
L_p	Photon radiance. Defined as the number of photons leaving or passing through a small element of surface (dS) per second divided by the solid angle of this surface (dΩ) and the orthogonally projected area of this element of surface in a plane normal to the beam direction (dS cosθ). L_p = number of photons per second/(dΩ dS cosθ) [$\text{s}^{-1}\text{m}^{-2}\text{sr}^{-1}$]. For a parallel beam of radiation the photon radiance is simple given by L_p = number of photons per second/(dS cosθ). [$\text{s}^{-1}\text{m}^{-2}$]
$L_p(\lambda)$	Spectral photon radiance. Derivative of photon radiance, L_p, with respect to wavelength, λ. [$\text{s}^{-1}\text{m}^{-2}\text{nm}^{-1}$]
P	Radiant power. Power emitted, transferred, or received as electromagnetic radiation. [Js^{-1} or W]
$P(\lambda)$	Spectral radiant power, at λ. Derivative of radiant power, P, with respect to wavelength, λ. In spectroscopy: $P(\lambda)$ = {total radiant power between [λ –($\Delta\lambda$/2)] and [λ +($\Delta\lambda$/2)] that exits a sample or cuvette filled with a solvent containing the solute}/$\Delta\lambda$. [W nm^{-1}]
$P^o(\lambda)$	Spectral radiant power at λ that exits a blank sample or cuvette "filled with solvent only." $P^o(\lambda)$ = {total radiance, between [λ –($\Delta\lambda$/2)] and [λ +($\Delta\lambda$/2)], that exits a blank sample or cuvette filled with solvent only}/$\Delta\lambda$. [W nm^{-1}]
N_a	Number of atoms of the fluorescent structure (counter-ions of salts are not counted) in its prevalent ionic form in an aqueous solution at a specified pH. [dimensionless]
N_n	Number of negative elementary charges of a fluorescent structure in its prevalent ionic form in an aqueous solution at a specified pH. [dimensionless]

N_p	Number of positive elementary charges of a fluorescent structure in its prevalent ionic form in an aqueous solution at a specified pH. [dimensionless]
q	Net number of elementary charges of a fluorescent structure in its prevalent ionic form in an aqueous solution at a specific pH ($q = N_p - N_n$). [dimensionless]
T	Temperature. [°C and sometimes K]
$T(\lambda)$	Transmittance at wavelength λ. $T(\lambda) = P(\lambda)/P^o(\lambda)$. [dimensionless]
Δ_S	Stokes shift. $\Delta_S = \nu_{x,max} - \nu_{e,max}$. [cm^{-1}]. The following are also used: $\Delta_S = \lambda_{e,max} - \lambda_{x,max}$ and $\Delta_S = \lambda_{e,max} - \lambda_{a,max}$. [nm]
$\varepsilon(\lambda)$	Molar decadic absorption coefficient at wavelength λ. $\varepsilon(\lambda) = A(\lambda)\, b^{-1}c^{-1}$. [M^{-1}cm^{-1}].
ε_{max}	Molar decadic absorption coefficient at the absorption maximum of the longest-wavelength band. $\varepsilon_{max} = A(\lambda_{a,max})\, b^{-1}c^{-1}$. [M^{-1}cm^{-1}].
λ	Vacuum wavelength of the electromagnetic radiation. [nm]
$\lambda_{a,1}$	Wavelength of absorption maximum of the longest wavelength absorption band ($\lambda_{a,1} \equiv \lambda_{a,max}$). [nm]
$\lambda_{a,max}$	Wavelength of absorption maximum of the longest wavelength absorption band ($\lambda_{a,max} \equiv \lambda_{a,1}$). [nm]
$\lambda_{a,-\alpha}$	Wavelength at height α of the left side of the longest wavelength absorption band ($0 < \alpha \leq 1$). [nm]
$\lambda_{a,+\alpha}$	Wavelength at height α of the right side of the longest wavelength absorption band ($0 < \alpha \leq 1$). [nm]
$\lambda_{e,1}$	Wavelength of emission maximum ($\lambda_{e,1} \equiv \lambda_{e,max}$). [nm]
$\lambda_{e,max}$	Wavelength of emission maximum. [nm]
$\lambda_{e,-\alpha}$	Wavelength at height α of the left side of the emission band ($0 < \alpha \leq 1$). [nm]
$\lambda_{e,+\alpha}$	Wavelength at height α of the right side of the emission band ($0 < \alpha \leq 1$). [nm]
$\lambda_{x,max}$	Wavelength of excitation maximum of the longest wavelength band. [nm]
ν	Wavenumber. It is related to wavelength λ [nm] by $\nu = 10^7/\lambda$ [cm^{-1}]
$\nu_{a,max}$	Wavenumber of absorption maximum of the longest wavelength absorption band. [cm^{-1}]
$\nu_{e,max}$	Wavenumber of emission maximum. [cm^{-1}]
$\nu_{x,max}$	Wavenumber of excitation maximum of the longest wavelength band. [cm^{-1}]
$\tau_{1/2}$	Fluorescence half-lifetime. $\tau_{1/2} = \tau_f \ln(2)$. Time required for half the entities, of a sample of initially excited entities, to decay. [ns]
τ_f	Fluorescence mean-lifetime. $\tau_f = \tau_{1/2} / \ln(2)$. [ns]
τ_{pb}	Photobleaching mean-lifetime or mean-lifetime before photobleaching. [s]
$\tau_{pb,r}$	Relative photobleaching mean-lifetime. Relative to a dye exposed to an identical radiance and other conditions. [dimensionless]
Φ_f	Fluorescence quantum yield. Φ_f = number of fluorescent photons emitted/number of photons absorbed. [dimensionless]
Φ_t	Triplet quantum yield. Φ_t = number of triplets created/number of photons absorbed. [dimensionless]
Φ_p	Phosphorescence quantum yield. Φ_p = number of phosphorescent photons emitted/number of photons absorbed. [dimensionless]
Φ_{pb}	Photobleaching quantum yield. Expresses the number of molecules photobleached per absorbed photon. [dimensionless]
M	Unit of amount concentration. [mol dm^{-3} or mol L^{-1}]

1

Introduction

Fluorescent dye labels and stains are used in a great variety of fields such as RNA or DNA detection and analyses (real-time PCR, fluorescence *in situ* hybridization, DNA microarrays, DNA sequencers, and DNA analyzers), proteomics, antibody labeling and immunoassays, flow cytometry, cell sorters, electrophoresis (in platforms capillary, microchip, and slabs), high-performance liquid chromatography, fluorescence microscopy, and fluorescence nanoscopy, to mention a few. Fluorescence is used in so many fields due to its extraordinary sensitivity, high selectivity, simplicity, and wide linear working ranges, sometimes reaching up to six orders of magnitude in concentration. In most cases, fluorescent dyes are used to label or stain a myriad of targets for their separation, detection, quantification, imaging, or tracking in space and/or in time. Measurement of intensity changes is still the predominant mode of usage, although measurements related to changes in polarization, emission lifetimes, and spectra are also used.

High contrasts in imaging, high signal-to-noise ratios, and optimal detection limits in analytical methods are achieved when, together with the use of proper instruments and excitation sources, the best fluorophores are included. However, it is a boring and time demanding task watching dozens of uncomplete spectra viewers, going through catalogs, books, and hundreds of articles to find the best dye for a given application. Each dye has many characteristics that must be confronted with the characteristics of the other hundreds. This book, among the databases provided, gives panoramic views (using scattergrams) of all dyes together.

This present database contains the molecular structures, photophysical properties, and characteristics of over 700 fluorescent dye labels and stains with medium to high brightness in aqueous solutions (plain water, phosphate-buffered saline, or any aqueous buffer containing less than 1% of organic modifiers and without complexing agents) and with absorption maximum in the 300–900 nm range of the spectrum. This database ended up including fluorescent entities from the following main classes: acridines, acridiziniums, acridones, benzothiazoles, benzenes, boron-dipyrromethenes, carbazoles, chromones, coumarins, cyanines, diketopyrrolopyrroles, fluorenes, fluoresceins (and fluorones), hybrid structures, indoles, isoquinolines, mercaptofluoresceins, merocyanines, naphthalenes, naphthalimides, phenazines, phenoxazines, phospholes, pteridines, pyrenes, pyridines, pyrodols, pyrodones, pyronines, quinolines, rhodamines (and rosamines), rhodols (and rosols), squaraines, sulfofluoresceins, sulforhodamines, sulforhodols, and trianguleniums, to mention a few. Hetero-atom substituted (HAS) xanthenes are also included, such as HAS-fluoresceins (and HAS-fluorones), HAS-pyronines, HAS-rhodamines (and HAS-rosamines), and HAS-rhodols (and HAS-rosols).

The photophysical parameters are tabulated and ranked by their importance for the users, such as: the optimal excitation range (including the wavelength of excitation maximum); optimal detection wavelength range (emission range, including the wavelength of emission maximum); Stokes shifts; molar absorption coefficients; fluorescence quantum yields; molecular fluorescent brightness; fluorescence mean-lifetimes; sizes; and net charges of the dyes in aqueous solutions at neutral pH. The photophysical and molecular properties were also examined using scatter plots. Wavelengths of excitation maxima were compared with Stokes shifts, molar absorption coefficients, fluorescence quantum yields, brightness, fluorescence mean-lifetimes, molecular sizes (number of atoms), and charges (at the pH where the photophysical parameters were measured). The photophysical

Fluorescent Dye Labels and Stains: A Database of Photophysical Properties, First Edition. Tarso B. Ledur Kist.

parameters were also compared with each other; for instance, brightness vs. Stokes shift, brightness vs. molecular size, and so on.

The resulting panoramic views provided by the scattergrams reveal many compelling properties for the users, adding efficiency to the work of fluorescent label and stain selection. The resulting panoramic view is also interesting to research groups engaged with the development and synthesis of new dyes as there are many scarcely populated and even empty areas in the scattergrams. There are no fundamental constraints within current theories that prohibit the existence of fluorophores in some of these empty areas. Therefore, these empty or scarcely populated ranges of a desired photophysical or molecular property could be filled with newly synthetized water-soluble organic fluorescent molecules and this, in turn, could lead to new applications in both basic and applied sciences.

Chapter 2 presents the most common light sources, such as lamps, LEDs, and lasers. The most commonly used pieces of hardware for light filtering and dispersion are also presented and explained (filters, polarizers, prisms, and diffraction gratings). Light detectors are fundamental in the field and are therefore also reviewed. Additionally, the properties of light and light beams are of fundamental importance as they allow users to maximize the performance of currently available instruments. Fluorescence imaging and detection set-ups are also briefly presented to show the available options. Finally, the fundamental concepts and definitions of the field, such as excitation and emission spectra, molar absorption coefficient, Stokes shift, fluorescence quantum yield, brightness, and the many factors affecting fluorescence, are presented. Descriptions are based on the most recent theories describing electromagnetic radiation and are presented in a simple, very didactic manner with a large number of illustrations.

In Chapter 3, the techniques of target-fluorophore binding are succinctly presented. It begins with a review of the properties of the following solvents: water, phosphate-buffered saline (PBS), and 29 water mixable organic solvents. Their melting points, boiling points, vapor pressure, and viscosity are also provided. In addition, the evaporation rates of the following solvents are tabulated: methanol, acetone, tetrahydrofuran, acetonitrile, ethanol, 2-propanol, water, PBS, propylene carbonate, *N*,*N*-dimethylformamide, 1-methyl-2-pyrrolidinone, formamide, and dimethylsulfoxide. Fluorogenic reactions (reactions in which only the product of the reaction is fluorescent) are reviewed, focusing on reactions with primary amines, secondary amines, thiols, and α-dicarbonylic compounds. At the end of this chapter, the available labeling reactions are reviewed. A three-columned table contains common functional groups that can be found on target molecules, their respective reactive moiety options for fluorescent labels, and the final functional groups produced by the resulting covalent linkage. Examples of target functional groups are alcohols, aldehydes, amines, carboxylic acids, glycols, ketones, phenols, and thiols.

Chapter 4 presents over 700 fluorescent dyes with medium to high brightness in aqueous solutions, separated into over 40 classes (1 per section). Each section details the molecular structures and photophysical parameters of each fluorescent dye. An additional section contains the names and suppliers of the dyes and stains whose structures have not been disclosed (only 44 of the 777). Molecular structures are presented in the ionic form at which photophysical parameters were measured. For almost all 700 fluorescent dyes, this is within the neutral pH range. Below each molecular structure the following very useful photophysical parameters of the corresponding dye are given: wavelength of excitation maximum (of the longest wavelength band) or wavelength of absorption maximum (if the excitation maximum was not available in the references), Stokes shift, and maximum molecular brightness (expressed in the convenient and practical unit of $\text{mM}^{-1}\text{cm}^{-1}$). Molecular brightness is defined as the product of molar absorption coefficient and fluorescence quantum yield. It expresses both how good a fluorescent dye is at absorbing radiation (absorption cross-section) and the probability of emission of a fluorescent photon upon each photoexcitation. The three photophysical parameters presented are very important from a practical point of view. Therefore, they are shown alongside the aqueous solution in which measurements were taken. The joint presentation of the molecular structure, photophysical parameters, and solution used provides the maximum amount of information in a concise manner. This will be useful for both fluorescent dye users and dye developers (Organic Chemists). Finally, all references from which data was taken are given in the text. Note that there is one entry for each dye, with some rare exceptions when photophysical parameters were measured at two distinct pH values.

Chapter 5 presents scattergrams of the database. These panoramic perspectives give a good overall picture of the dye options available and also reveal many interesting properties that can otherwise be hard to see. For example, the scattergram of molecular absorption coefficient (ε_{max}) vs. wavelength of excitation maximum ($\lambda_{x,max}$) shows how ε_{max} behaves with $\lambda_{x,max}$. Furthermore, different marks and colors are used for each class to make the behavior of each class more visible. Scattergram analysis is also used to examine the behavior of fluorescence quantum yield (Φ_f) vs. $\lambda_{x,max}$. The scattergrams simultaneously give a detailed view of the fluorescence quantum yield of each dye and class along the whole spectrum (300–900 nm).

The product of ε_{max} and Φ_f, which is defined as brightness (B_{max}), provides the answer to some extremely important practical questions: which are the brightest fluorescent dyes available? Which classes contain the brightest structures? Which excitation wavelength ranges have the brightest entities? Which are the brightest fluorophores for a given laser wavelength or light source? Another important scatter plot is Stokes shift vs. $\lambda_{x,max}$. In the present book, Stokes shift is presented as the difference between emission maximum and excitation maximum ($\Delta_S = \lambda_{e,max} - \lambda_{x,max}$). Large Stokes shifts are very important for some applications. This scattergram shows the classes with the largest Stokes shifts and the excitation wavelength ranges containing the entities with the largest Stokes shifts. It also exhibits the typical Stokes shifts of each class. Another interesting scattergram is brightness vs. Stokes shift. Unfortunately, this shows that it is still not possible to have the best of both worlds, as the brightest dyes have medium to small Stokes shifts and the dyes with high Stokes shifts have medium to low brightness. Other graphs worth mentioning here are the scattergram of brightness vs. molecular size, as well as a scattergram presenting brightness vs. dye charges (positive and negative) and $\lambda_{x,max}$ in the same graph! The above-mentioned scattergrams are only a few of the many presented in this chapter.

Chapter 6 discusses brightness values observed in practice, or effective brightness. In reality, laser lines are usually fixed and rarely operate exactly at the $\lambda_{x,max}$ of the fluorescent dyes used. Therefore, effective brightness (termed $B(\lambda)$) in fact goes from zero (0) to B_{max}. The maximum brightness reported is $B_{max} = 112{,}000\ \text{M}^{-1}\text{cm}^{-1}$ or, more conveniently, $112.0\ \text{mM}^{-1}\text{cm}^{-1}$, which was found in the rhodamine class. This chapter also proposes a method to predict excitation ranges (good for excitation) and emission ranges (ideal for detection) in order to obtain predictable effective brightness values from the scarce information available in the literature. The absorption bands and emission bands within a given class somehow have a regular shape if their spectral radiance is plotted against frequency or wavenumber instead of wavelength. This is used to predict, with the aid of an equation which utilizes class information, $\lambda_{x,max}$, and $\lambda_{e,max}$, both the excitation range and detection ranges of fluorescent dyes in aqueous solutions. This is very useful from a practical point of view. Of course, conjugation and protein-rich environments produce some bathochromic shifts that are hard to predict. Nonetheless, band parameters collected from hundreds of excitation and emission bands facilitated the prediction of not only the full width at half maximum, but also the wavelengths at one-quarter of the height of the left side of the excitation bands ($\lambda_{x,-0.25}$), wavelengths at one-half the height of the left side of the excitation bands ($\lambda_{x,-0.5}$), and wavelengths at one-half the height of the right side of the excitation bands ($\lambda_{x,+0.5}$), and so on. The same is predictable from the emission band, i.e., $\lambda_{e,-0.25}$, $\lambda_{e,-0.5}$, $\lambda_{e,+0.5}$, $\lambda_{e,+0.25}$, and $\lambda_{e,+0.1}$.

Chapter 7 presents and discusses the photostability, or photorobustness, of the fluorescent dyes. Usually, fluorescence radiance is measured and compared with other fluorescent dyes under identical illumination conditions. The concepts of photodegradation cross-section, relative photodegradation rates, relative photodegradation mean-lifetimes, or relative photodegradation half-lifetimes are discussed and compared. There is still a lack of precise and reproducible measurements of these photophysical parameters as the results are impacted by the experimental conditions used. Examples of these environmental conditions are: if the aqueous solutions are degassed or not; the illumination conditions (radiance and wavelength used); and additional solutes (ingredients) present in the aqueous solutions. In addition, some dyes are also affected by the pH of the aqueous solution.

Each Appendix presents the fluorescent dyes ranked by a different photophysical parameter. Examples of these parameters include excitation ranges, emission ranges, Stokes shift, brightness, fluorescence mean-lifetime, and dye net charges.

2
Basic Definitions and Fundamentals

2.1 Introduction

This chapter presents some basic tools (light sources, optical components, and light detectors), concepts, and definitions that are important to the field of fluorescence. The definitions of radiant power, radiance, and spectral radiance, for example, are highlighted as they have a precise meaning. The use of the term "intensity" is discouraged as it may refer nonspecifically to photon flux, radiant power, radiance, or spectral radiance [1].

The technologies used to produce, manipulate, and detect electromagnetic radiation, especially light (the visible part of the spectrum), have developed enormously during the last few decades. In the first sections of this chapter, the most important tools such as filters, gratings, and polarizers, which are used to manipulate light in the field of fluorescence imaging and detection, are briefly presented and discussed. The available options of light sources, as well as their pros and cons, are also presented *en passant*. Light detectors are concisely presented in a single section, despite the vastness of the field and the many developments in this area. The interconnected topics of light beams, radiant power, radiance, spectral radiance (emission spectra), microscope objectives, and both imaging and detection set-ups are concisely presented. Finally, the most important definitions related to the field of fluorescent dyes, such as the absorption spectra, excitation spectra, emission spectra, Stokes shift, molar absorption coefficients, fluorescence quantum yields, molecular brightness, and fluorescence mean-lifetimes, are presented.

2.2 Light Sources

The number of available light sources is constantly growing and there is an increasing diversity of light source options for fluorescence excitation. Therefore, there is a suitable light source for every application. **Tungsten filament lamps** (or incandescent lamps) are the oldest and simplest. They emit a continuous spectral radiant power (in both time and frequency) represented by the Planckian curve. Indeed, quantum theory emerged when M. Planck tried to find the equation to model this phenomenon from first principles. The best theory at that time (J.W. Maxwell's Theory of Electromagnetism) could not provide a good model to explain the phenomenon called black body radiation, and this led to the discovery of the granular nature of light. This was an important fact that contributed to the development of quantum mechanics during its early days, the first three decades of the twentieth century. According to quantum theory, electromagnetic radiation is best modeled as packets of energy (photons) traveling in space, instead of a continuous plane wave with an arbitrary amplitude. Moreover, these photons are characterized by the frequency which defines the energy and linear momentum they carry through space, in addition to their states of spin, polarization, and angular momentum (Section 2.5.6).

Fluorescent Dye Labels and Stains: A Database of Photophysical Properties, First Edition. Tarso B. Ledur Kist.

Halogen lamps are a special case of filament lamps, as they are intense light sources (high radiant power) with longer operation lifetimes than common tungsten lamps. In these lamps, a halogen gas is added to a quartz reinforced bulb containing the metal filament. The halogen gas continuously removes the metal deposited on the inner bulb's surface and deposits it back onto the hot filament, thereby increasing the lifespan of filament lamps when they are operated at higher temperatures and high radiant power outputs.

Electrical discharge lamps are another important source of light. In these lamps an electrical discharge passes through a gas and produces light. They use a true gas, at low or high pressure, and sometimes a mixture of gas and metal (e.g., sodium) vapors. Deuterium lamps are an important and widely used source of UV radiation in spectrophotometers, spectrofluorometers, imaging systems, and detection systems in analytical instruments (e.g., chromatography and electrophoresis).

However, all the above-mentioned sources are noisy, bulky, hot, and not efficient at converting electric energy into luminous energy, since a large fraction of the electric energy input is converted into undesired thermal energy (heat, infrared radiation, and microwave radiation). Moreover, the light produced by most of the previous sources has low radiance and is difficult to collimate (into a beam) in space and modulate in time; i.e., it is hard to produce light beams that are both narrow and/or pulsed in time. Pulsed light sources are very important for sensitive detection in many fields. For example, lock-in amplifiers are widely used to extract good signals from noisy baselines. The **LEDs** (Light Emitting Devices) that have been developed in the last few decades are an interesting alternative as they are very compact, more efficient at converting electric energy into luminous energy, exhibit a much longer lifetime, have a more stable luminous power output (radiant power), and the light they produce is easier to collimate and modulate in time.

Finally, **lasers** (an acronym for Light Amplification by Stimulated Emission of Radiation that became a noun) are a very special light source as their fundamental principle of light production is stimulated emission, not spontaneous decay as in all previous examples. Stimulated emission produces light beams with special properties. These beams are monochromatic, collimated, polarized, have a high radiance (L) (or radiant power per unit area, see Section 2.5.1), and are coherent (Section 2.5.5).

2.3 Filtering and Dispersing Light

2.3.1 Absorber Filters

Absorber filters or **color filters** are designed to perform the following tasks: i) to work as bandpass filters; i.e., they let a band of the spectrum pass through while absorbing the neighboring ranges. They are characterized by the transmittance (in percentage) at the center of the band and by the band width (in nm or Hz), which is defined as the width of the band at half height of the band; ii) to work as shortpass filters (they let shorter wavelengths pass through) or, equivalently, highpass filters (they let higher frequencies pass through). Shortpass (or highpass) filters are characterized by a cut-off wavelength (or frequency). They are transparent to wavelengths shorter than the cut-off and opaque to wavelengths above it; iii) longpass filters are the opposite of shortpass filters as they let longer wavelengths pass through. Equivalently, they are lowpass filters as they let lower frequencies pass through. They are also characterized by a cut-off frequency, but the transparent and opaque regions are reversed when compared to shortpass filters (ii).

The working mechanism of these filters relies on the selective absorption of light by the chromophores (pigments) that are usually dissolved in the filter materials (homogeneous solid solutions). A good color filter does not scatter light and does not fluoresce upon illumination, otherwise it would decrease the signal-to-noise ratio of detection systems or the contrast of the images in imaging systems. However, the reflection of ~4% of light at the air–solid interfaces is, in general, unavoidable and must be properly handled to prevent it from entering the detection or imaging sensing area. Inorganic pigments are used more often than organic pigments as they are less prone to photobleaching (Section 2.8) and are more stable over time, especially when strong UV sources are used. Nevertheless, some color filters do exhibit a small undesired fluorescence. This light, in addition to the background stray light, compromises signal-to-noise ratios and contrasts. Finally, it is difficult to produce sharp edges at the cut-off wavelength in the transmitted spectra when using color filters, as the absorption bands always have a smooth change along the spectrum and usually have at least a small tail. This can be solved by using interference filters (Section 2.3.2), which produce sharper edges (Figure 2.1).

Attenuance filters or **neutral-density filters** are very special absorber filters as they are designed to absorb a given fraction (or percentage) of light, no matter the color of the incident light. They are based on the same concept as previously described; i.e., pigments are added. However, what differentiates these filters is that they absorb the same fraction of all colors and their subranges. Therefore, neutral-density filters for the visible always look gray; e.g., 10% of neutral filters are light gray, 50% are medium gray, and 90% are dark gray. They are used to decrease the radiant power (Section 2.5.1) to which samples are exposed and/or to partially block the amount of

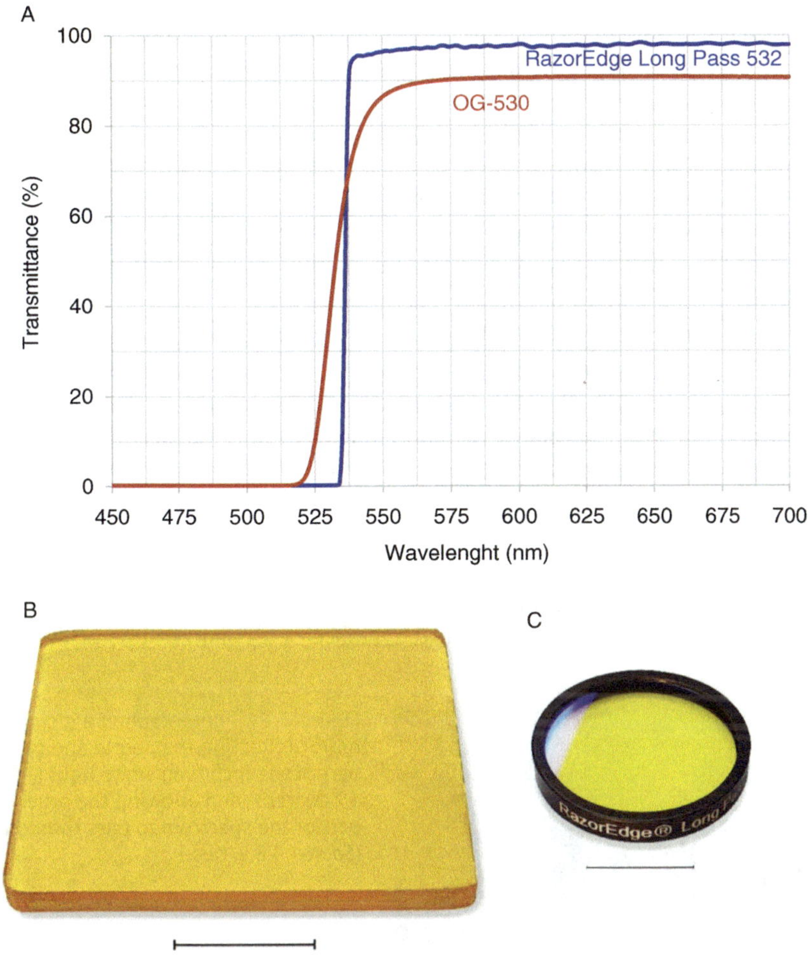

Figure 2.1 (A) Comparison of two longpass filters: a common color filter (OG-530) and a high-performance interference filter (model RazorEdge Long Pass 532 from SemRock, Lake Forest, IL). Photograph of (B) the color filter and (C) interference filter used to produce graph A. This interference filter exerts good suppression at 532 nm and below, but exhibits a high transmittance above 532 nm. This is a good blocking filter to be placed between the target and the sensing device when a DPSS laser with λ = 532 nm is used. (Unpublished results from the author.)

harvested light. In this way, they prevent the saturation of sensing devices such as charged coupled devices (CCDs) cameras, photomultiplier tubes (PMTs), avalanche photodiodes (APDs), silicon photomultipliers (SiPMs), and photodiodes (PDs) (Section 2.4).

2.3.2 Interference Filters

Interference filters are usually made of dielectric-stack Fabry–Pérot interferometers, which sometimes have thin metallic films integrated into them. They are designed to work at a given angle with respect to the incident beam: usually 45 degrees for dichroic mirrors and 90 degrees for all other common interference filters. Light falls on the front surface of the filter, which contains multiple thin layers of dielectric materials which have alternating indexes of refraction. Only part of the spectrum passes through, as the rest is back scattered. This is based on the principle of constructive and destructive interference phenomena on the reflected and transmitted waves.

As with color filters, interference filters can also be bandpass, shortpass, and longpass filters. Differently, interference filters have the advantage of sharper edges (Figure 2.1). Moreover, there are both narrow (~5 nm) and broad (~50 nm) bandpass filters. The former are used to purify laser line beams (remove undesired wavelengths that may mix with sample's fluorescence), while the latter are used to harvest fluorescent light and block undesired wavelengths, especially from the excitation source.

Notch filters are designed to reflect a narrow band and let its neighboring wavelengths pass through. They offer high in-band rejection and high out-of-band transmission, so as to only block light within a small range of wavelengths. Therefore, they are also known as band-stop or band-rejection filters and, in this sense, they are the inverse of bandpass filters. Typically, they are placed between the target area and the photosensor unit to block the scattered laser light. This increases the signal-to-noise ratio in both Raman spectroscopy and fluorescence detection, and increases the contrast in fluorescent and Raman-based imaging systems.

Dichroic mirrors are similar to interference filters (Figure 2.1); however, they are designed to work at an angle of 45 degrees. In typical applications they reflect shorter wavelengths (i.e., purple and blue) and let longer wavelengths (i.e., green and yellow) pass through (Figure 2.2). They are the key part of confocal set-ups (Section 2.6.3) as they allow the excitation light to be inserted between the objective and eye piece (in fluorescence microscopes) or the objective and projector (in confocal scanning microscopes).

2.3.3 Polarizers

Absorptive polarizers do not act like passive absorbing filters and are therefore better described by quantum theory than the theory of classical electromagnetism. When a photon lands on the surface of a polarizer, one of the two

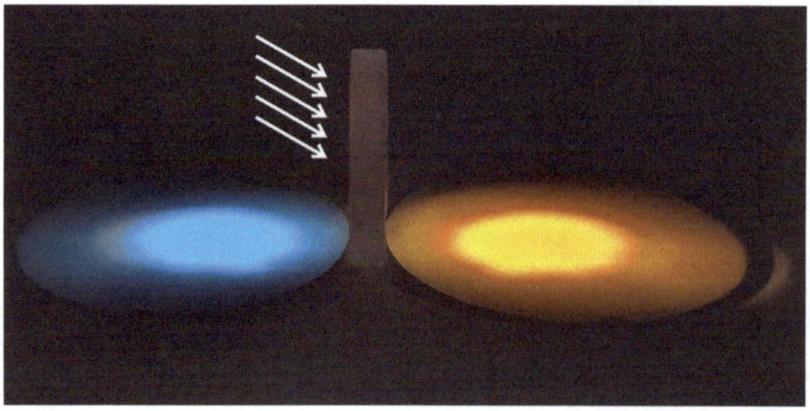

Figure 2.2 Photograph of a dichroic mirror reflecting the purple and blue part of the incoming white light (at 45 degrees) and allowing the other part of the spectrum to pass through. (*Source*: T.B.L. Kist.)

following happens: i) the polarization state of the photon is projected orthogonal to the working direction of the polarizer and is consequently absorbed; or ii) the polarization state of the photon is projected parallel to the working direction of the polarizer and passes through. The probability that one or the other will happen is related to the angular difference between the working direction of the polarizer (polarizer orientation) and the "polarization state" of each incident photon. Figure 2.3 shows four standing polarizers; the working direction of the two polarizers at the center are parallel to each other, the two at the left are orthogonal, and the two at the right are at 45 degrees.

Beam-splitting polarizers act in the same way as absorptive polarizers, with the exception that the photons of the reflected beam have their polarization directions projected parallel to the plane of the beam splitter (dielectric material), and the photons of the transmitted beam have their polarization directions projected orthogonally to the previous one. This occurs on the surface of water and can easily be observed through a handheld polarizer by the naked eye.

The radiant power (P) of a polarized beam that passes through a perfect polarizer is given by **Malus's Law** (named after Étienne-Louis Malus):

$$P = P_0 \cos^2 \alpha, \tag{2.1}$$

where P_0 is the incident radiant power, P is the transmitted radiant power, and α is the angle between the light's initial polarization direction and the working direction of the polarizer.

2.3.4 Prisms

There is a dependence of the refraction index of dielectric materials on the frequency of the radiation. This makes different frequencies of light beams follow different paths when a beam of mixed frequencies hits a transparent dielectric material at an angle other than 90 degrees (between 30 and 60 degrees is most commonly used). This is the basic functioning principle of dispersion by a **prism**. The angle of the prism at the vertices is also important because the dispersing efficiency can be optimized by the combination of a material with a high index of refraction (compared to air) and the correct vertices angle. However, prisms are no longer widely used today as they have a lower angular dispersing power than diffracting gratings.

2.3.5 Grating

Reflection diffracting gratings are made of special dispersing surfaces containing thin and parallel grooves. When a photon hits the surface of a diffracting grating on a trajectory that belongs to a plane that is normal to the

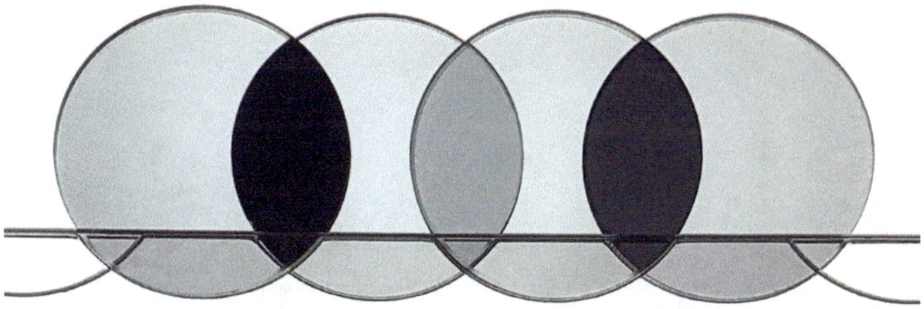

Figure 2.3 Polarizers in action. Photograph of four polarizers, the working directions of the two at the center are parallel to each other, the two at the left are orthogonal to each other, and the two at the right are at 45 degrees to each other. (*Source*: T.B.L. Kist.)

surface and normal to the grooves, it can be scattered in any direction along the plane. However, only a few well-defined directions offer a constructive interference for the wavefunction of the photon when it is simultaneously scattered by two or more adjacent grooves (Figure 2.4). The constructive and destructive interference phenomenon results from the path length difference of two adjacent grooves. Constructive interference only happens when the path length difference is equal to a multiple of the wavelength of the landing photon, and destructive interference occurs in all other cases. **Transmission gratings** work in the same way, except that the scattering of the transmitted light is what counts instead of the back-scattering phenomena.

2.4 Light Detectors

Until the discovery and understanding of the photoelectric effect, at the turn of the mineteenth to the twentieth centuries, there were only a few rustic ways to measure the power of light beams. One such technique was based on the ability of these light beams (luminous energy) to sustain a given heat flow from a target and a cold reservoir. This ability to generate heat was measured in many ways (measuring temperature differences, pressure in a closed gas bulb, voltage of a thermopile, and others). Another technique that became popular after the invention of photography (ca. 1830), was the use of light-sensitive emulsions made of silver salts. However, all of these methods were slow, or not sensitive enough, or too demanding, not practical, or even a combination of all of these issues.

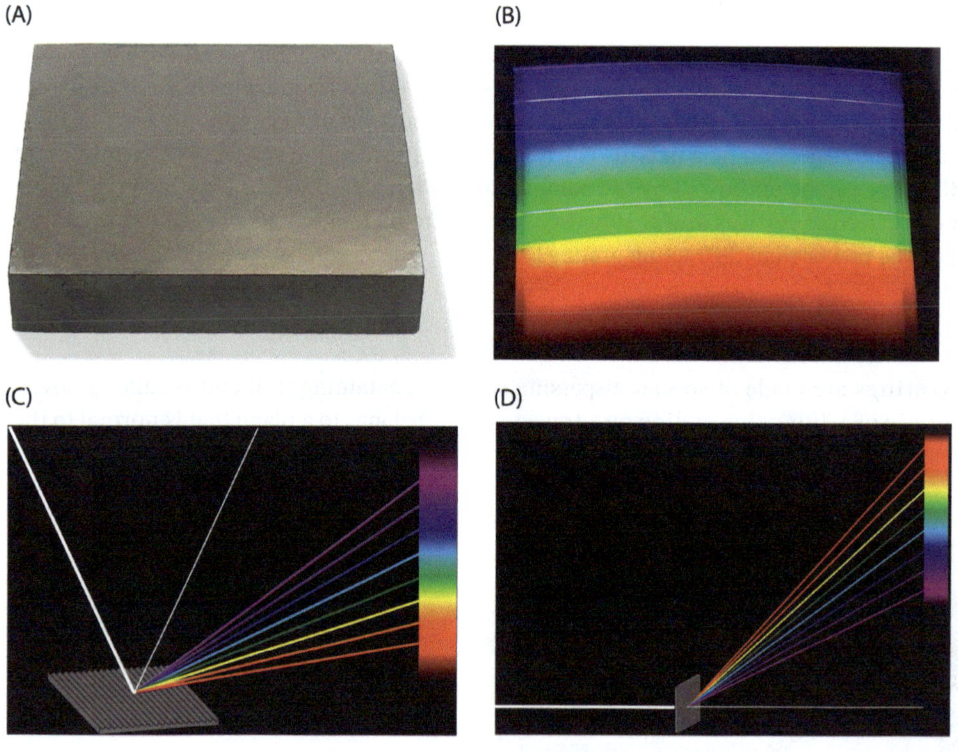

Figure 2.4 (A) Photograph of a reflection diffracting grating showing the surface of the grating when illuminated by a diffuse light and (B) the first-order diffraction of a rod-like fluorescent lamp. (C) The trajectory of light with different frequencies when they hit a reflection diffracting grating and (D) a transmission diffracting grating. Both dispersion mechanisms are based on (C) random back or (D) random forth scattering, in addition to the interference phenomenon. (*Source*: T.B.L. Kist.)

The development of photocells, and later **photomultiplier tubes,** in the first half of the twentieth century, brought great advances to the field of optical detection. Photomultiplier tubes (Figure 2.5A) use a series of solid thin grids or blades that are held a few millimeters apart from each other in a high vacuum tube, among which electrical potential differences (e.g., 100 V dc) are applied. When a photon of sufficient energy strikes the front blade (a photo-sensitive solid, which can be seen behind the glass of Figure 2.5B) that is held at a low potential (e.g., −600 V), the photon rips off an electron through the photo-electron effect. This electron is accelerated by the electrostatic potential difference and collides with the second grid or blade of the series (held at; e.g., −500 V). This collision releases around a dozen electrons, which are accelerated to the third grid or blade (held at −400 V), ripping off more electrons. This continues until the electrons reach the last grid or blade (usually the 6th). In this case, the charge grows exponentially with the number of blades, and a peak of a few nanoAmperes of current is observed at the drain pin connected to the last grid or blade (which may be kept; e.g., at 0 V with respect to first stage or front stage photo-sensitive blade). These spikes of nano currents are very fast and can be amplified. Photomultiplier modules (Figure 2.5B) have everything integrated into them: a high voltage source, voltage divider, and nanocurrent preamplifier. Their combination of high sensitivity (especially in the mid-visible range), high gain, low noise, very fast response, and large light collection area cannot presently be matched by any other detection system. Therefore, they are currently used in a very large range of fields, from astronomy to many medical imaging devices, microscopes, analytical instruments, and clinical analysis instruments.

The initial physical event of light detection in photomultiplier tubes is the photelectric effect. The initial physical event of light detection in **avalanche photodiodes** (APDs) (Fig. 2.5C), **charged coupled devices** (CCDs) (Fig. 2.5D), and even **photodiodes** (Fig. 2.5E), is similar. When a photon with sufficient energy strikes the semiconductor close to a p-n junction it creates an electron-hole pair. This effect is also called the inner photoelectric effect, because both the hole (a missing electron) created in the valence band of a semiconductor material and the electron itself (which is promoted to the conducting band) stay within the solid semiconductor, although they may move. This movement of a charge (electron) and a missing charge (hole) creates the so-called photoinduced current. In CCDs they start to charge a picture element (pixel), whose accumulated charges are repeatedly read out at given time intervals (frequency) and discharged for the next read. In avalanche photodiodes these initial photocurrents are internally amplified by the application of high voltages across the semiconductor, producing an amplified current of a few nanoAmperes whose initial event was a single inner photoelectric effect. These devices also present a constant dark current and the signals; therefore, must be read over this baseline.

The main advantage of CCDs is that they have a spatial resolution. This means that the radiance changes in time and across the sensing area of the detector can be digitalized. Some CCDs have hundreds of thousands of

Figure 2.5 The most commonly used light sensing devices: (A) photomultiplier tube (PMT) (based on a vacuum photocell with internal amplification); (B) photomultiplier module (which includes photomultiplier tube, a high voltage source, high voltage divider, and amplifier circuitry); (C) avalanche photodiodes (APD); (D) charged-couple device (CCD); and (E) photodiode (PD) (monolithic photodiode with on-chip transimpedance amplifier). (*Source*: T.B.L. Kist.)

pixels, which gives an even higher spatial resolution. However, they are slow as they need a few milliseconds to accumulate charge on each pixel (the picture element) and have a limited working dynamic range (each pixel is quickly saturated with only two to three orders of magnitude of charge). On the other hand, the fast response of PMTs can be used to produce spatial resolution. This is performed using a fast-scanning system for the excitation spot of light and the PMT as a fast-recording system. In this way, images with spatial resolution can be reconstructed if the sequence of spatial excitation or illumination addresses are programed or known beforehand.

Finally, there are a few important definitions related to light detectors that are worth mentioning. First is the **spectral responsivity**, which is a graph that expresses the dependence of detector sensitivity on wavelength. For example, photomultiplier tubes are known to exhibit high sensitivity in the mid-visible range. This sensitivity can be extended to the blue and violet region of the spectrum using special photo-sensitive cathodes or, alternatively, to the red and infrared region of the spectra using another class of photo-sensitive cathodes. CCDs and photodiodes, on the other hand, are known to perform well in red and near-infrared regions of the spectrum. Some silicon photodiodes (Si-PD) also operate in some ranges of the UV (e.g., UVC), where they are almost the only option left despite their low sensitivity. A second important definition is **response time**, which expresses the time required by the detector to respond to a pulse of light or photon strike. This is quantitatively expressed as the time constant or the RC time constant. PMTs have the fastest response time, while CCDs have the slowest because they need the charges to accumulate in the micro-capacitor of each pixel before each read. These micro-capacitors are discharged after each read. A third important definition is **sensitivity** or noise-equivalent power. This is expressed as the minimum optical power required to generate a current that is equal to the mean-root-square of the dark current noise. PMTs used to be the most sensitive of all; however, some cooled CCDs and APDs are starting to become competitive in this regard. Nowadays, there are PMTs with single photon sensitivity and high quantum efficiency (over 90% of landing photons are detected). Finally, **dark current** is also an important parameter. It expresses the current (or signal) delivered by the detector in the absence of light. A common source of dark current is background radiation (e.g., gamma rays). Background radiation produces a few thousand counts per minute in any small detector (with a sensing area of a few cm^2) dedicated to detecting gamma rays. Life has adapted to the presence of gamma rays, and so must the optical sensing instruments. Cosmic rays also contribute to dark current and are characterized by occasional spikes. A third source of dark current is thermal energy, which releases electrons that contribute to dark current. Finally, the amplifying electronics of some devices also contribute to dark current.

2.5 Light Beams

2.5.1 Radiant Power and Radiance in Space: Divergent and Collimated Beams

Radiant power (P) or radiant flux is defined as the amount of luminous energy emitted, transferred, or received per unit time. The SI unit is $W \equiv J{\cdot}s^{-1}$. The term **intensity** is also sometimes used; however, this term is vague because it refers to radiant power (W), radiance (Wm^{-2}), or photon flux per unit time (s^{-1}). The IUPAC recommendation is that "intensity" should only be used for qualitative descriptions [2]. **Radiance** (L) of a "parallel beam" (Figure 2.6B) is defined as the "radiant power, P, of all wavelengths leaving or passing through a small element of surface in a given direction from the source divided by the orthogonally projected area of the element in a plane normal to the given direction of the beam" [2, page 1225 of ver. 2.3.3]. If the beam is normal to the surface S, and constant over it, then $L = P/S$, otherwise radiance must be calculated at each point as follows: $L = \mathrm{d}P/(\mathrm{d}S\ \cos\theta)$, where θ is the angle difference between the beam direction and the surface normal.

It should be noted that there are no collimated beams of light that propagate without spreading when considered over many meters. There is always a focal region from which the beam diverges. In good-quality lasers this is only observable after many meters of propagation. In these cases, the focus can be either inside the optical cavity of the laser or far away from it, but the further away the focus is the larger the waist of the beam. The waist

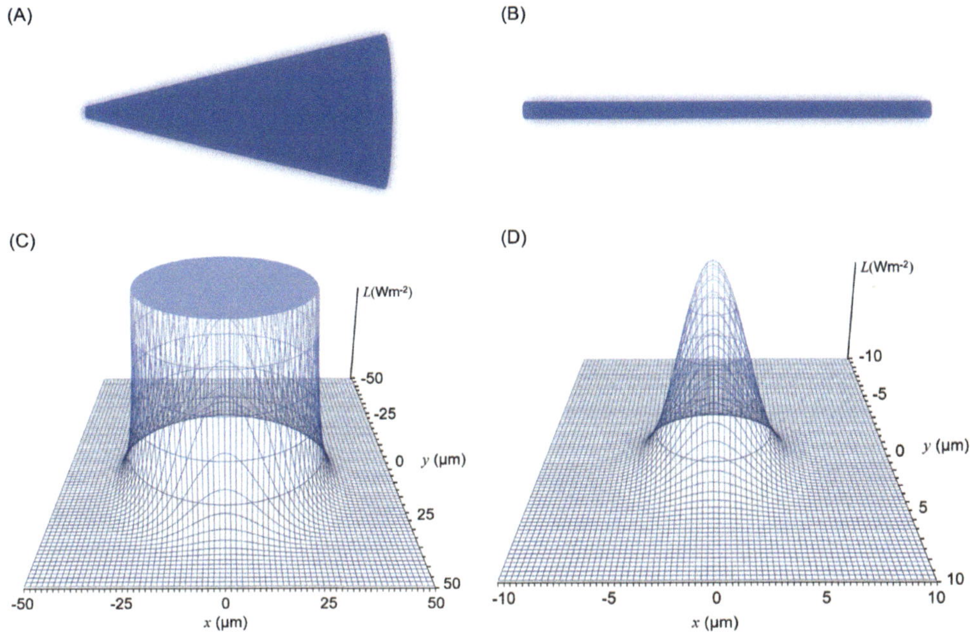

Figure 2.6 Schematic drawing of (A) a divergent light beam and (B) a collimated light beam. (C) Collimated beams with a cylindrical and homogeneous radiance over a large area at the center are sometimes desired, but difficult to produce. (D) Lasers operating at the fundamental transverse electromagnetic mode (TEM_{00}) usually deliver a beam exhibiting a radiance with a 2D Gaussian profile.

of a beam is defined as the width at $1/e$ of the height of Figure 2.6D, where e is the Napier's number or the base of natural logarithms. According to the theory of electromagnetism there is always a lower limit to the waist of the beam, even at the focus. This lower limit cannot be exceeded and is proportional to the wavelength of the radiation of the beam.

2.5.2 Radiant Power and Radiance in Time: Continuous, Modulated, and Pulsed

Continuous wave (cw) light sources are the most commonly used in imaging systems when CCD cameras are used to capture the images. However, **modulated** and **pulsed** light sources (Figure 2.7) are extremely advantageous when fast response detectors are used, as in this case their output signal can be processed by "**lock-in amplifiers**." These schemes, although more complex, have the advantage of being able to recover a weak signal in the midst of a noisy background. Moreover, modulated and pulsed excitation sources are at the heart of time-resolved fluorescence techniques. This is especially interesting with the development of fluorescent dyes with long mean-lifetimes, such as transition metal complexes (e.g., ruthenium (II) coordination complexes with bipyridine and terpyridine). The radiant power of LEDs is easily modulated in time as a square wave (on and off), while the creation of sinusoidal or triangular waveform outputs is more difficult in these cases because the LEDs are designed to operate at a given applied current as the spectral radiant power (Section 2.5.3) changes with the applied current. Incandescent lamps like tungsten filament and halogen lamps are not designed to be modulated electronically, and can only be modulated with the use of mechanical rotating choppers. Some lasers are pulsed by their operating nature (e.g., the nitrogen laser), others may be operated in both the continuous or modulated mode (most of the diode lasers), and some are designed to operate only as a cw.

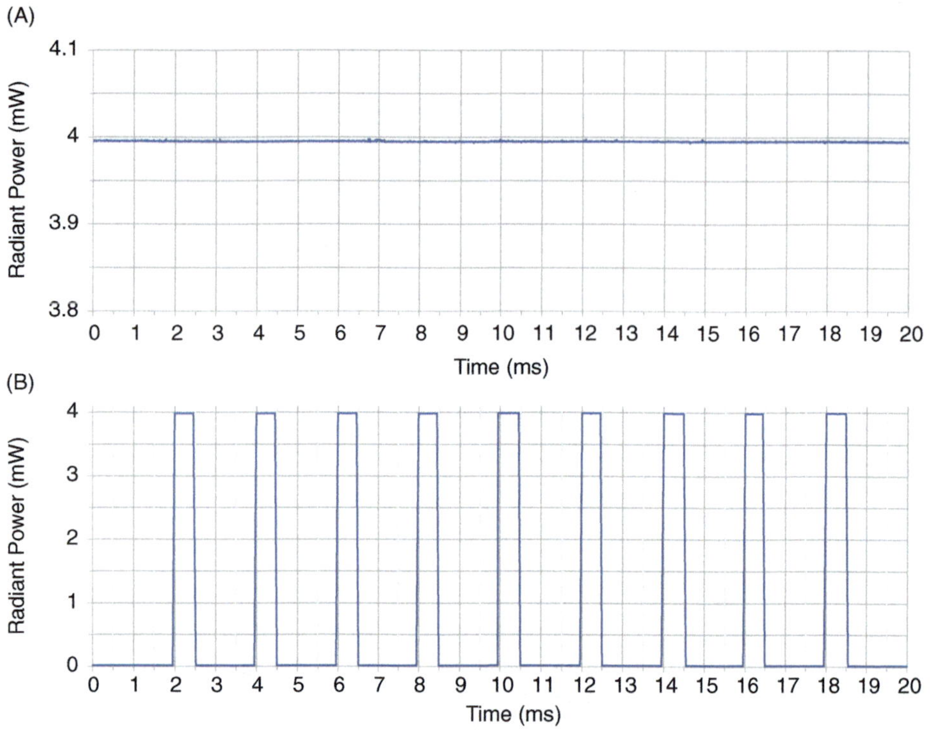

Figure 2.7 Radiant power (*P*), expressed in mW, of a beam of light that has almost parallel propagating photons and is produced by a source that is (A) continuous in time (cw) and (B) modulated in time. The modulated beam was produced by an LED (model B5-435-TL with $\lambda_{e,max}$ = 638 nm and dedicated optics) and powered by a function generator with a square wave output. Detection was achieved by an OPT101 photodiode (shown in Figure 2.5E) and a data acquisition module with analogic-to-digital converter. (Unpublished results from the author.)

2.5.3 Spectral Radiant Power (Emission Spectra) of Lamps, LEDs, and Lasers

The **spectral radiant power** of a beam, denoted as $P(\lambda)$ and commonly known as the emission spectra, varies greatly from source to source. There are many available light sources that emit different spectral radiant powers. Some emit wide bands, and others narrow lines with high radiant power in the NIR, visible, or UV ranges (these are known as monochromatic sources). Some are easily collimated, while others are not. Additionally, some are easily modulated in time (e.g., LEDs), while others are harder to modulate (e.g., halogen and tungsten lamps). Lasers are known to deliver collimated and monochromatic light beams with a high radiance (luminous power per unit area), expressed in Wm^{-2}. Figure 2.8 shows the spectral radiant power of the three commonly used sources: a laser, an LED, and a tungsten filament lamp. Halogen lamps produce spectra that are similar to tungsten filament lamps.

2.5.4 Light Wavelength, Transmittance, and Absorbance

The **wavelength** of a photon changes when it moves from air into liquids and solids, while its frequency (and hence energy) does not. The wavelength of a photon in a given medium (i.e., in water) is given by $\lambda_w = \lambda/n$, where λ is the wavelength of the photon in a vacuum (or vacuum wavelength) and n is the index of refraction of the medium for this wavelength. Note that the index of refraction of air is $n \approx 1$ for visible light at room temperature. Surprisingly, inside water at 20°C ($n \approx$ 1.33 for 532 nm), the 532 nm wavelength of a green laser line is reduced to $\lambda_w \approx$ 532 nm/1.33 = 400 nm! Nevertheless, despite the fact that the photophysical parameters are measured in

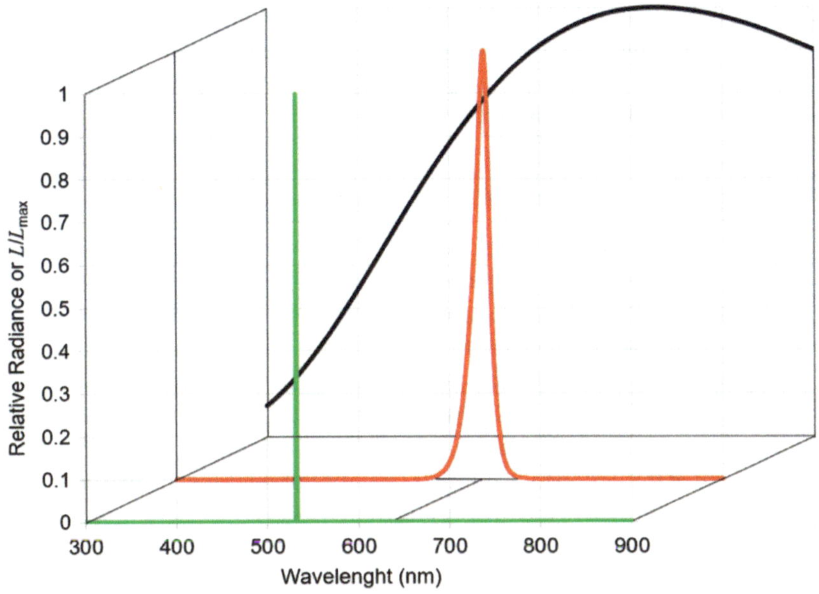

Figure 2.8 Sources with a continuous spectral radiant power output (tungsten filament lamp) (black curve), a band-like output (LED model B5-435-TL with $\lambda_{e,max}$ = 638 nm) (red), and a line output (532 nm from a DPSS laser) (green). (Unpublished results from the author.)

aqueous solutions, the literature only refers to the corresponding "vacuum wavelength" of the radiation used in the measurements, not to the actual wavelength of the radiation inside the aqueous solutions under study (solutions of fluorescent labels and stains). Luckily, the energy of a photon remains the same, whether it is in a vacuum, in air, or in an aqueous solution. The energy is given by $E = hf = hc/\lambda = hv_w/\lambda_w$, where h is Plank's constant, c is the velocity of light in vacuum, and v_w is the velocity of light in water ($v_w = c/n$). In conclusion, while wavelength and velocity change from one medium to another, the frequency and energy do not. Despite these properties, the use of the vacuum wavelength expressed in nanometer (nm) became of common use instead of frequency (f) or wavenumber (ν). Note that f (s^{-1}) and ν (cm^{-1}) have a very simple and linear relationship: $f = 100c\nu$, where 100 (cm/m) is a factor.

Another important aspect is the nature of the **transmittance** and **absorption** spectra itself. It represents the transmittance and absorbance at "each wavelength," following the hypothesis that such a monochromatic source of radiation exists. Let us suppose that a green-emitting laser has a high radiant power output at 532 nm. Even in this case, there will be no one single photon with a λ value that is "exactly" 532 nm produced during the lifespan of any such laser. This is because 532.0000... nm (with infinite zeros) is a single point on the real number axis and the probability of having a single photon with exactly 532 nm is zero. Therefore, the transmittance and absorption spectra are "density probability distributions" because at wavelength λ they are measured using a beam with photons in the range $\lambda - \Delta\lambda$ to $\lambda + \Delta\lambda$. In this case, there may be a bunch of photons reaching the detector per second that can be measured. If $\Delta\lambda$ is too small, then the spectra may be too noisy and it might be difficult for some detectors to measure it. On the other hand, if $\Delta\lambda$ is too large then the absorption spectra may exhibit absurd distortions, such as flattened bands or peaks. Therefore, $\Delta\lambda$ can be increased in order to produce a good spectrum (with low noise), but it cannot be increased too much as this can lead to deformed peaks (flattened tips) and valleys (raised mid points). That said, transmittance at wavelength λ is defined as the ratio between spectral radiant power, at λ, that exits a sample or cuvette "filled with a solvent containing the solute," $P(\lambda)$, and the spectral radiant power at λ that exits a blank sample or cuvette "filled with solvent only," $P^o(\lambda)$. Mathematically, **transmittance** is a dimensionless quantity defined as $T(\lambda) = P(\lambda)/P^o(\lambda)$, where $P(\lambda)$ has units of W nm^{-1} and is defined as $P(\lambda)$ = {total radiant power between [λ −($\Delta\lambda$/2)] and [λ +($\Delta\lambda$/2)] that exits a sample or cuvette filled with a solvent containing the solute, divided by $\Delta\lambda$. This $\Delta\lambda$ is related to the slit width and is usually of the order of 1 nm. $P^o(\lambda)$ also has units of W nm^{-1} and is defined as $P^o(\lambda)$ = {total radiance, between [λ −($\Delta\lambda$/2)] and [λ +($\Delta\lambda$/2)], that exits a blank sample or cuvette filled with solvent only} divided by $\Delta\lambda$. **Absorbance** is a dimensionless quantity and at wavelength λ it is defined as $A(\lambda) = \text{Log}_{10}(P^o(\lambda)/P(\lambda)) = \text{Log}_{10}(1/T(\lambda))$.

2.5.5 Spontaneous Decay and Stimulated Emission in Lasers and STED Nanoscopy

Lasers consist of an active medium (excitable entities) within an optical cavity delimited by two parallel mirrors. One mirror has an ~100% reflection index while the other (the output mirror) typically has a reflection index of ~95–99%. When a valence electron of an atom, an electron of a π-electron system of a molecule, or even a valence band electron of a semiconductor (producing an electron-hole pair) of the active medium is excited to the first excited state by any event, the time following this event can be divided into small time intervals (dt) and the decay event of any of these single systems can be simulated. The probability (dp) of decay during the first time interval dt is given by $dp = \Gamma(1 + n)dt$, where Γ is the average decay rate of the system during this time interval, the number 1 is related to the spontaneous decay rate, and n is the occupancy number (number of photons) of that mode of the electromagnetic field in the optical cavity where the active medium is inserted. This occupancy number is zero at the start. The value of dt must be chosen such that $dp \ll 1$. The time behavior of such single quantum systems can be simulated. A random number (between 0 and 1) must be generated to decide if it will (or will not) emit a photon in this time interval (between $t = 0$ and $t = dt$). This number draw continues, dt after dt, until the excited state decays (and it is said that a quantum jump has occurred) [3, 4]. When the laser is turned on at the beginning ($t = 0$ and $n = 0$), a large number of the atoms, molecules, or the semiconductor (of a diode laser) of the active medium are subjected to excitation (pumping). This produces a large population of excited atoms, excited molecules, or electron-hole pairs in the semiconductor. By chance, one of the **spontaneous decays** emits a photon in the axial direction of the optical cavity of the laser (orthogonal to the mirrors). If the mirror reflects this first photon, then $n = 1$ at this instant, but the back-and-forth reflection stimulates the emission of more photons in the same direction and phase as the first emitted photon (this produces a coherent flux of photons, explaining the **coherence** of laser light beams). In the following fraction of a second the laser cavity is populated by a very large number of photons and reaches a steady state with $n = n_s$. In this state a constant ~5–1% (as 95–99% is reflected) of the cavity radiant power is delivered and the system stays in a steady state caused by the pumping of the active medium and the **stimulated emission** events promoted by the ~n_s photons in that cavity mode of the electromagnetic field. This is where the acronym LASER (light amplification by stimulated emission of radiation) comes from. The term laser is a noun, while LASER is the phenomenon that occurs in lasers.

In stimulated emission depletion (STED) nanoscopy, super-resolution images which have resolutions that are far beyond the limit predicted by the electromagnetic theory of radiation are obtained. The formula expressing the maximum resolution (D) achievable by conventional optical microscopy within the electromagnetic theory of light was deduced by Abbe [5] and is given by $D = \lambda/(2NA)$, where λ is the wavelength used for illumination and NA is the numerical aperture of the microscope objective (defined in Section 2.6.1). **STED nanoscopy** stimulated emission, which is a phenomenon predicted by quantum theory, is promoted by the depletion beam (red arrows in Figure 2.15 and red laser beam in Figure 2.14B). Depletion beams usually has a donut-shaped spatial profile and a longer wavelength than the emission maximum of the dye (red arrows in Figure 2.15 and red laser in Figure 2.14B). Therefore, the use of the high-power depletion beams drastically reduces the volume of the sample where spontaneous emission still occurs (orange spot of Figure 2.14B). This small remaining volume, within which spontaneous decay is still the predominant radiative decay mechanism, is located at the center of the donut-like beam (Figure 2.14B). This happens because $n_s \gg 1$ within the donut volume, which consequently reduces the spontaneous emission within the donut center. Here spontaneous emission also competes with stimulated emission in a 1 to n_s ratio according to the formula $dp = \Gamma(1 + n_s)dt$, demonstrating the importance of using high-power depletion beams. In summary, STED nanoscopic super-resolution images are obtained by the selective deactivation of spontaneous emission of the fluorescent dyes, which consequently minimizes the effective illumination to a very small cross-sectional area around the focal point of the excitation and depletion beam (they are supposed to operate with a coincident focus). The remaining fluorescence from spontaneous emission (in all directions) is then collected by the objective of the scanning confocal microscope to produce the images. The resolution limit predicted by Abbe's formula when using visible light is ~250 nm in conventional microscopy, while the lateral resolution that has already been achieved in practice by STED nanoscopy is ~10 nm.

2.5.6 Energy, Momentum, Polarization, Spin, and Angular Momentum

Photons propagate through space and carry **energy**. The quantity of energy carried is given by $E = hf$, where h is Planck's constant and f is frequency. Besides energy, photons also possess a **linear momentum** that can be calculated using $p = hf/c$, where c is the velocity of light in a vacuum. In addition to energy and linear momentum, photons may also be found in many states of **polarization**. Polarization is the direction along which the electric field oscillates. It can be any direction normal to the propagation direction (from 0 to 180 degrees) or a coherent superposition state of any two orthogonal directions. Finally, photons carry spin and angular momentum. The **spin** of a photon is related to its circular polarization, which can be left or right. Photons are bosons (particles that follow the Bose–Einstein statistics) with integer spins. In this case, the electric field rotates around the axial direction and performs a complete turn per wavelength. This leads to spin values of plus or minus $h/2\pi$, depending on whether the photon is circularly polarized to the right or left, respectively. On the other hand, the **angular momentum** of light beams is associated with helical wavefronts. Light beams with angular momentum are reported to penetrate tissues deeper than light beams without angular momentum [6]. For a didactic explanation of the difference between the spin of a photon and the orbital angular momentum of a light beam, see Padgett and Allen [7], and for a more elaborate mathematical-physics description, see Bliokh and Nori [8].

Circularly polarized luminescence (CPL) based on fluorescent small organic molecules has attracted a lot of attention in recent years [9]. In this phenomenum, dyes with unique spatial structures emit circularly polarized light. For a review of the synthesis of CLP materials, the construction of CPL systems, and working CPL mechanisms, see Chen [10].

2.6 Light Collection Set-Ups

Independent of whether a microscope or an analytical instrument is used, the detection set-ups are generally similar. Some are more commonly used in microscopy (fluorescence microscopy and confocal scanning microscopy), while others in analytical instruments (chromatography, capillary electrophoresis, microchip electrophoresis, DNA analyzers, cell sorters, and flow cytometers). However, there are important differences in nanoscopy, which is a light microscopy with nanometric resolution [11].

2.6.1 Microscope Objectives

In the majority of cases, a good fluorescent light harvesting system collects the maximum fluorescent light and minimal stray light. Hence the importance of the definition of the **angular aperture** (α) and **numerical aperture** (NA) of a **microscope objective**. Numerical aperture is defined as $NA = n \sin(\alpha)$, where α is the angular aperture and n is the refraction index of the medium between the lens and the sample ($n \approx 1.00$ for air, 1.33 for water, and 1.51 for special microscope objective oils). Angular aperture (α) is defined as half the angle of the cone made by the focal point of the lens and the lens perimeter (Figure 2.9B).

In practice, the angular aperture has an upper limit of ~72 degrees ($\theta \leq 72^{\text{o}}$). Therefore, in air, $0 \leq NA \leq 0.95$. This numerical aperture is an indication of both the resolution and light collecting (or harvesting) capability of any particular objective. From this definition it is possible to calculate the maximum amount of light collected by the objective from a punctual source of light in the ideal situation in which no light is reflected at the air–"front surface of the objective lens" interface:

$$\text{Fraction of light collected} = \sin^2\left[\frac{\arcsin\left(\frac{NA}{n}\right)}{2}\right]. \tag{2.2}$$

The results produced by this equation and depicted in Figure 2.9 (ray-paths) are also shown in Figure 2.10 (graph) and 2.11.

2.6.2 Fluorescence Detection Set-Ups

There are many fluorescence-based detection set-ups used in many branches of analytical and bioanalytical chemistry. Innovative set-ups in both illumination (excitation) and fluorescent light collection (harvesting) are still being developed and improved. Most can be found in patents but there are also good review articles; see, for instance, Galievsky et al. (2016) and the references contained therein [12].

There are dozens of **fluorescence detection set-ups** used in high-performance liquid chromatography, capillary electrophoresis, microchip electrophoresis, cell sorters, and flow cytometers, to mention just a few. Figure 2.12 shows three very interesting detection set-ups, all producing good limits of detection with relatively simple instruments: (A) a **confocal set-up**; (B) a **normal set-up** in which fluorescence is induced by a low angle excitation beam; and (C) a **sheath-flow cuvette set-up**. By using confocal set-ups, it is possible to build very compact detection modules that reduce the number of loose cables on the bench. In the second case (B), the hardware is the simplest of all. These set-ups are very effective at achieving optimal detection limits because there is minimal laser light "scattered" into the objective, providing high noise-to-noise ratios. However, an objective with both a high angular aperture and longer focal distance may sometimes be required. Set-up C is more sophisticated. It only works at the outlet of the fused silica microtube or HPLC column, but it is able to detect single molecules.

2.6.3 Fluorescence Imaging Set-Ups

Fluorescence "microscopy" relies on the use of fluorescent labels and stains to produce high resolution images of many kinds of samples. The two most common set-ups used in conventional fluorescence "microscopes" are

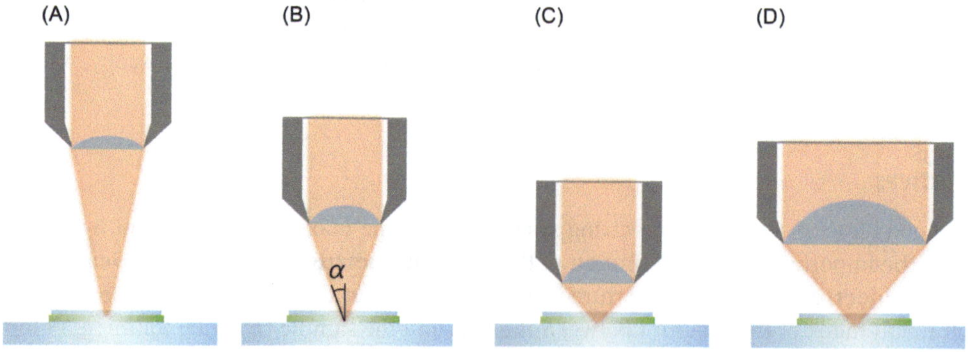

Figure 2.9 Four objectives with angular apertures of (A) 12 degrees; (B) 20 degrees; and (C and D) 40 degrees. The angular aperture (α) is shown in B. Lens D has the same angular aperture as C, but with the advantage of a longer focal length. This is important when the objective cannot get close to the object (for any reason) and a high angular aperture is needed. (*Source*: T.B.L. Kist.)

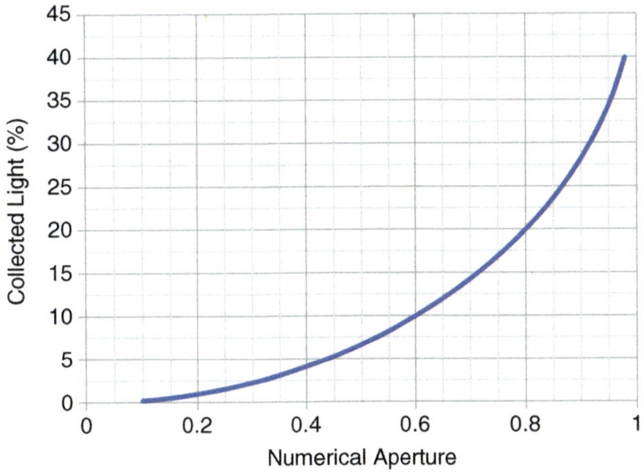

Figure 2.10 Amount of light collected (%) as a function of the microscope objective numerical aperture (*NA*). Note that the fraction of light collected barely reaches 40%. A light collection of 50% would mean that all the light of one hemisphere is collected. This is not possible with lenses for practical reasons, even when using oil-immersed objectives as shown in Figure 2.11.

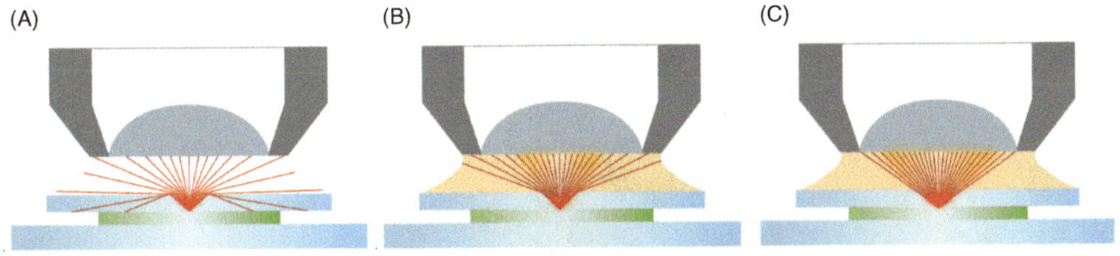

Figure 2.11 The mechanism by which an immersed objective increases the numerical aperture. Sample and objective (A) in air; (B) immersed in a liquid with a refraction index of 1.33 (usually water); and (C) immersed in an oil with a refraction index of 1.50. The internal reflection and the refraction at the cover slide–air interface diminishes the amount of light collected by the objective in A and B. There is a good match between the refraction index of the oil and the glass cover slide in C. (*Source*: T.B.L. Kist.)

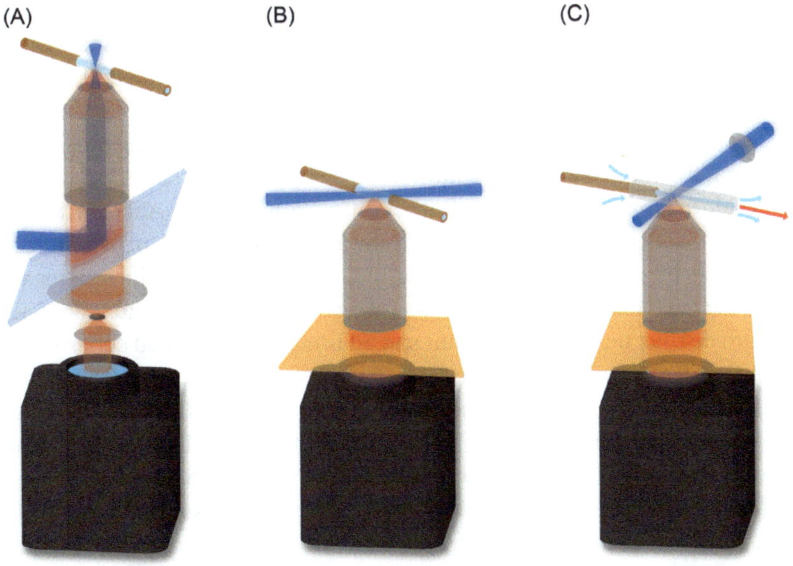

Figure 2.12 Optical set-ups for detection in bioanalysis using (A) a confocal geometry with the aid of a dichroic mirror; (B) a common set-up that uses a low-angle laser for excitation (with the advantage of a simple set-up with low reflection into the detection direction); and (C) a sheath flow cuvette (for ultrasensitive laser induced fluorescence detection in flows). The fused-silica capillaries are shown at the top, above the microscope objectives. In (A), a rectangular dichroic mirror is shown while in (B) and (C) interference or color blocking filters are used. The detectors are at the bottom (black). (*Source*: T.B.L. Kist.)

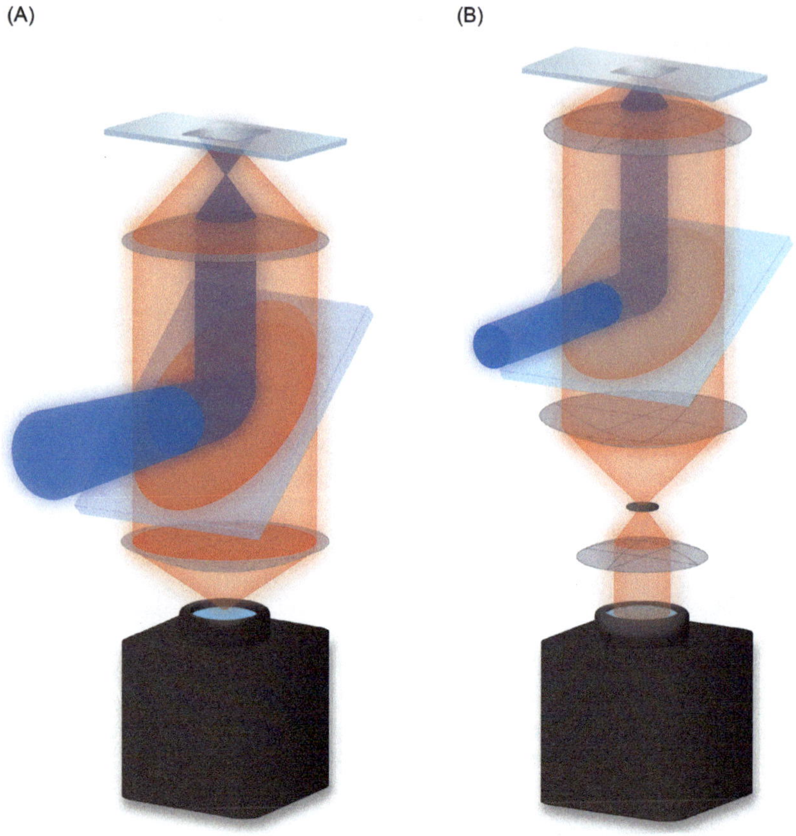

Figure 2.13 The two most common set-ups used in fluorescence "microscopy." (A) Wide-field fluorescence microscope and (B) laser scanning confocal microscope. The glass slides with samples are at the top and light detectors at the bottom. The dichroic mirror is shown in blue (at 45 degrees) and the pinhole (black disk) of the confocal scanning microscope. (*Source*: T.B.L. Kist.)

shown in Figure 2.13: the (A) **wide-field** (common) fluorescence microscope and (B) the **confocal scanning** microscope. The latter scans wide areas (by moving the glass slide with sample and cover slit) with good lateral and in deep resolution up to ~0.25 μm.

The advent of optical nanoscopy (or optical super-resolution microscopy) brought a remarkable development for the field of fluorescence imaging in the first decades of the twenty-first century. There are now dozens of fluorescent techniques which have crossed the optical resolution limit (also known as the "diffraction limit" or Abbe limit [5]) predicted by the theory of electromagnetism. For visible radiation this theoretically achievable maximum resolution is ~250 nm. However, these nanoscopies, which rely on bright fluorescent and photostable dyes, have already achieved resolutions of below 10 nm in both the axial and lateral directions. They are classified into the following four main groups, each possessing two or more distinct techniques: i) point spread function modulated super-resolution nanoscopies; ii) single molecule localization-based super-resolution nanoscopies; iii) fluorescent probe fluctuation based super-resolution nanoscopy; and iv) sample expansion based super-resolution nanoscopy. All of them are very well explained in the elegant reviews of van de Linde et al. [13] and by Liu et al. [14], while the traditional microscopies and some of the above cited groups are also well presented by Valeur and Berberan-Santos [16].

Optical nanoscopes use much more sophisticated arrangements and two schemes are depicted in Figure 2.14. They also require much more sophisticated fluorescent dyes, depending on the type of nanoscopy used. In general, high brightness and high photostability are required. Nanoscopes together with the development of new and state-of-the-art fluorescent labels and stains promoted great developments in the fields of cell biology, molecular biology, genetics, biophysics, and many others relying on lateral and in-depth resolution down to 10 nm.

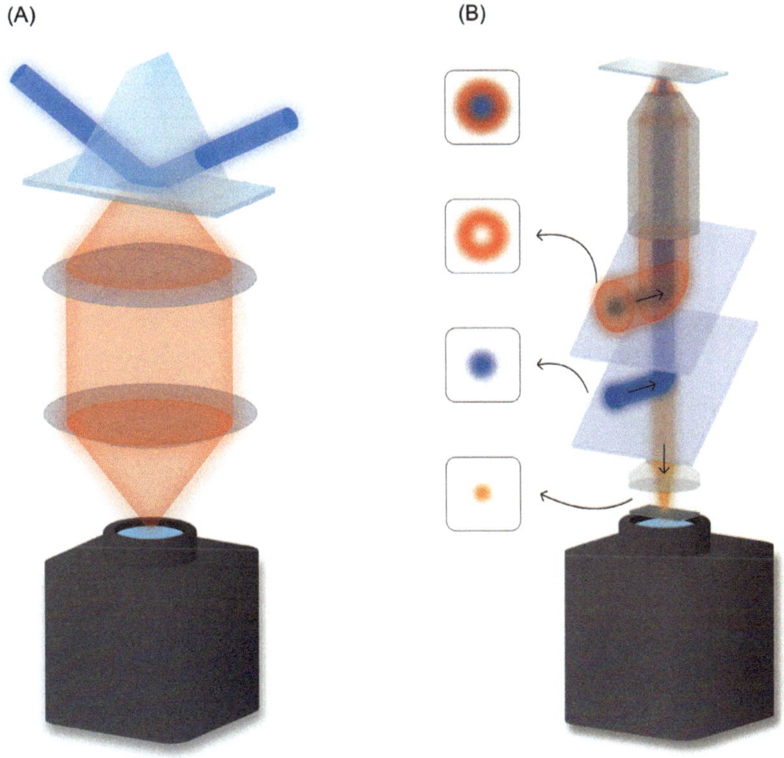

Figure 2.14 This figure shows only two of dozens of imaging set-ups used in fluorescence "nanoscopy." (A) Internal total reflection is used when high sensitivity is sought (e.g., in single molecule tracking schemes); and (B) STED (stimulated emission depletion) set-up which uses a confocal system with an excitation beam (blue) and depletion beam (red). (*Source*: T.B.L. Kist.)

2.7 Fundamentals of Fluorescence

Fluorescence is a fast luminescence phenomenon which occurs during the irradiation of a substance with electromagnetic radiation [1] (according to classical electromagnetism theory) or a flux of photons [15] (according to quantum theory). If light is emitted after a period of time longer than 10 ns and less than 1 μs after excitation, then it is termed "delayed fluorescence." On the other hand, a delay larger than 1 μs is called **phosphorescence** and occurs due to electronic transitions involving changes in spin multiplicity [1]. Therefore, a fluorophore is defined as a molecular entity that emits fluorescence [1]. Small organic fluorescent molecules are also called fluorescent dyes to distinguish them from other organic fluorescent structures (e.g., fluorescent proteins and quantum dots, among others). Figure 2.15 illustrates the photoexcitation and fluorescence phenomena of a fluorescent dye with the aid of a **Jablonski diagram** (after Polish Physicist Aleksander Jabłoński). Nobel laureate J.B. Perrin and his son F. Perrin also made important contributions to the field [17]. Some dyes experience a large dipole moment change upon excitation and upon decay. There is an energy involved with the re-alignment of the water molecules surrounding the fluorescent dye and represented by E_1 and E_2. These energy differences represented by E_1 and E_2 in Figure 2.15 are responsible for one type of extra-large Stokes shifts exhibited by some fluorescent dyes (discussed in Section 2.7.5).

Fluorescence is the method of choice in a great variety of fields due to its extraordinary sensitivity, high selectivity, simplicity, and linear output over a very wide range of target concentrations. Fluorescent dyes are used to label and stain a myriad of targets for their detection, imaging, and tracking in space or in time. Measurement of

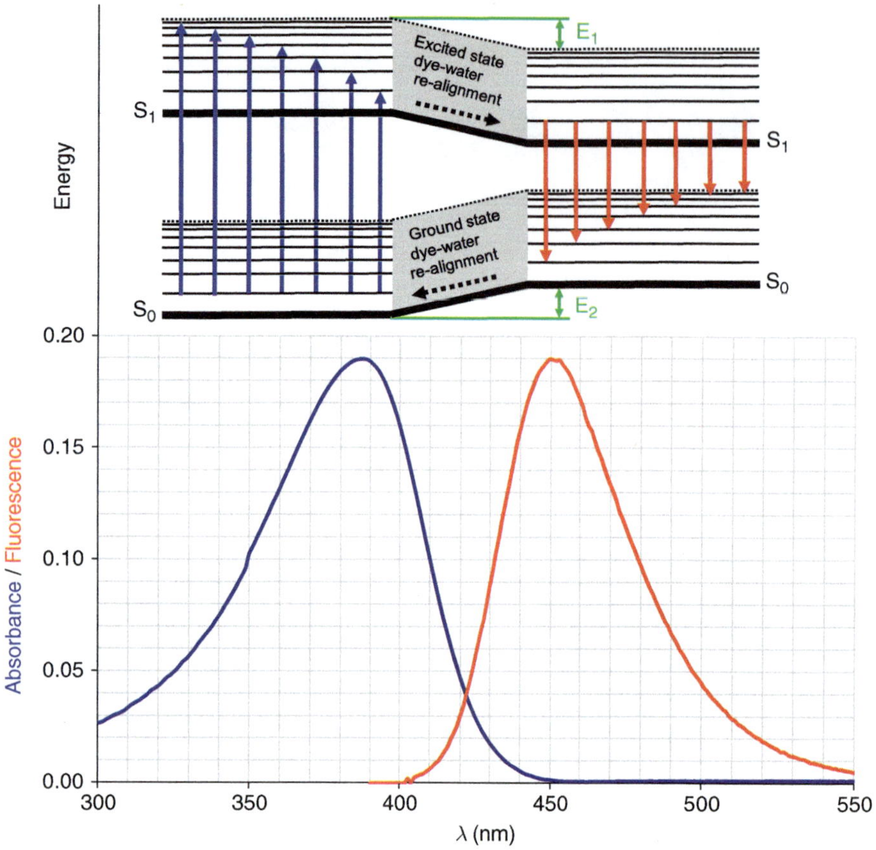

Figure 2.15 The Jablonski diagram illustrating the energy levels and transitions related to absorption and fluorescence of a fluorescent dye with a large Stokes shift. Absorption (blue) and emission (red) spectra of a coumarin (Pacific blue). The thick black lines represent the electronic ground state (S_0) and the first excited singlet electronic state (S_1) and the thin lines the vibrational states. The black dashed arrows indicate the non-radioactive transitions from the vibrational excited states to the vibrational states that are predominant at the temperature of the medium, regardless of whether the electronic state is in the ground or excited state. The blue arrows represent the photoexcitation and the red arrows the fluorescence emission (or decay). (Unpublished results from the author.)

"intensity changes" (indeed "radiant power changes," Section 2.5.1) is still the predominant mode of usage, although measurements related to changes in absorption or emission spectra, polarization (emission anisotropy), and emission mean-lifetimes are also used. Finally, light itself is still being studied and modeled [15], as well as the spontaneous and stimulated quantum jumps of the molecule–photon interaction processes, which has the potential to lead to new developments and applications. All this shows the potential for the further development of fluorescence and its many applications.

2.7.1 Fluorescence: Fields of Application

A large number of applications of fluorescence have emerged during the last decades. Examples of such applications include: DNA sequencing by many of the Sanger's chain termination reaction methods [18]; sequencing by synthesis monitored with fluorescence detection; single molecule real time (SMRT) sequencing; dsDNA sizing (human identification and genotyping); real-time PCR [19]; molecular beacons in a loop-mediated isothermal amplification assay [20]; DNA microarrays [19] with screening of common SNP variations [21]; fluorescence *in-situ* hybridization (FISH) [19]; fluorescence polarization [22]; FRET (Förster Resonant Energy Transfer) acceptors

[23]; antibody labeling; derivatization reactions in HPLC and capillary electrophoresis [24]; flow cytometry [19]; fluorescence microscopy [25]; controlled photoswitching [26]; fluorescence nanoscopy; STED [27, 28]; SIM [28]; SMLM [28]; PALM [29]; STORM [30]; MINSTED [31]; experiments with single molecules; and a myriad of both biosensors and microchips.

Special optical and molecular properties are required for some applications, but in general a maximum signal-to-noise ratio is sought. For this, the optical set-up of the instrument, as well as the fluorophores chosen, is critical. Simply put, the maximum signal (fluorescence light) must be collected with minimal background (auto-fluorescent, reflected, and scattered light) and noise (from excitation radiant power fluctuations and electronics). To achieve this, the following must be considered: i) minimal sample auto-fluorescence, as well as a good match of the excitation light and wavelength of excitation maximum of the fluorescent dyes used; ii) large Stokes shifts, so that the scattered excitation light can easily be filtered out from the collected fluorescence light; iii) high brightness, to achieve a high fluorescence signal with low excitation radiance (Wm^{-2}) and low scattering light; iv) superior photostability [32, 33, 34], to allow long exposition times without flickering or photobleaching; and, in some cases: v) narrow full width at the half maximum of the emission bands, to maximize the number of channels or colors that can be used in multiplexed strategies.

For many applications, the molecular size of the fluorophores is also a relevant parameter to be considered. Small hydrodynamic volumes are required for some applications (derivatization reactions, antibody labeling, fluorescence depolarization schemes, FRET and quenching techniques), whereas large volumes are desired for others (e.g., end-labeled free solution electrophoresis) [35]. The water solubility of the fluorophores is always a matter of concern, as is their sensitivity to temperature and the chemical environment (viscosity and polarity). Additional items to be considered include toxicity, chemical stability, and shelf lifetime. The first excited state fluorescence mean-lifetime is also important for applications based on single molecules, time-resolved spectroscopy, fluorescence depolarization, rotational relaxation measurements, two-photon spectroscopy, and the real-time imaging of fast-moving targets.

2.7.2 Molar Absorption Coefficient

The molar (decadic) absorption coefficient (ε) or molar absorptivity is defined as the decadic absorbance divided by the path-length b and mole concentration c, of the absorbing material [1]. Mathematically, at the wavelength λ, $\varepsilon(\lambda) = A(\lambda)\, b^{-1}c^{-1}$, where $A(\lambda)$ is the absorbance at λ, c is the concentration given in mol L^{-1} (or simply M), and the absorption path length b is given in cm. The condition $A < 0.6$ must be met due to non-linearity in the Lambert–Beer law observed when $A > 0.6$. Absorbance (A) is calculated as the minus logarithm to base ten (decadic) of the transmittance: $A(\lambda) = -\mathrm{Log}_{10}(T(\lambda))$ (Section 2.5.4).

The use of the term "**extinction coefficient**" is discouraged because non-chemical societies reserved the word "extinction" for diffusion of radiation; i.e., the sum of the effects of scattering, absorption, and luminescence [36, p. 3760].

The **molar absorption coefficient** at $\lambda_{a,max}$ is termed ε_{max} and calculated as $\varepsilon_{max} = A(\lambda_{a,max})b^{-1}c^{-1}$, where $\lambda_{a,max}$ is the wavelength of absorption maximum of the longest wavelength absorption band. Expressing the molar absorption coefficients (ε_{max}) in $M^{-1}cm^{-1}$, as is normally done in the literature, has the inconvenience of carrying too many non-significant digits. For example, the bright dye Alexa Fluor 647[1] has $\varepsilon_{max} = 250{,}000\ M^{-1}cm^{-1}$. The last two digits are mathematically meaningless for many reasons: uncertainty of the purity of the reagent used, limited precision of weighing instruments, imprecision in the volumetric dilution steps, limited precision of absorbance measurements, and the temperature variations that occur during both volumetric dilution and absorbance measurements. A simple error propagation calculation shows that $\varepsilon_{max} = 250.0\ mM^{-1}cm^{-1}$ should be used instead of $250{,}000\ M^{-1}cm^{-1}$. For this reason, and also to save space in the table columns, the unit of $mM^{-1}cm^{-1}$ (millimolar decadic absorption coefficient) will be used throughout this book instead of $M^{-1}cm^{-1}$. This $mM^{-1}cm^{-1}$ unit has a

1 A trade mark of Thermo Fisher Scientific (Waltham, MA, USA).

simple meaning: it represents the hypothetic absorbance that would be exhibited by a 1 mM solution when placed in a 1 cm length cuvette if Lambert–Beer's law were linear in very wide ranges of absorbance.

2.7.3 Excitation Spectra

The **excitation spectrum** is the plot of the fluorescence radiance against the wavelength of excitation [1]. Exitance or emittance quantifies how much light is emitted by an object. They can be expressed either as the number of photons per unit time or as energy per unit time (power). For the great majority of the organic fluorescent molecules, the excitation spectrum has a band-like shape with a local maximum at the longest wavelength. This is expressed as $\lambda = \lambda_{x,max}$. In most cases, the longest wavelength band of the excitation spectrum, as well as its maximum ($\lambda_{x,max}$), coincide with the longest wavelength band of the absorption spectrum and its respective maximum ($\lambda_{a,max}$). Therefore, in most cases, $\lambda_{x,max} = \lambda_{a,max}$ and consequently $\varepsilon(\lambda_{x,max}) = \varepsilon(\lambda_{a,max}) \equiv \varepsilon_{max}$. Throughout this book λ_a will be used when λ_x is not available.

A simple example of an excitation spectrum is given in Figure 2.16A. Note that the scan was carried out in an exaggeratedly wide interval, as the excitation wavelength also crossed the interval used to record fluorescence.

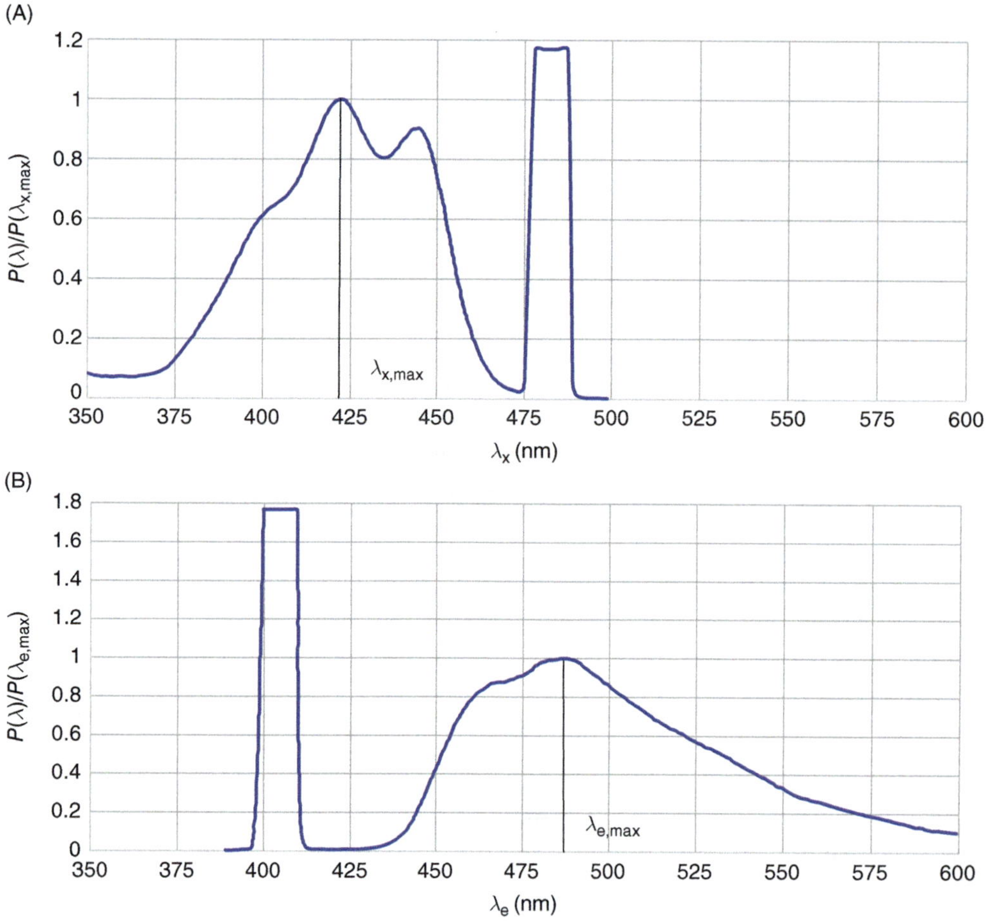

Figure 2.16 (A) Excitation spectrum of lysine-1-cyanobenz[f]isoindole (Section 3.4.1). Fluorescence was detected at 480 nm using a slit of 2.5 nm. (B) Emission spectrum of lysine-1-cyanobenz[f]isoindole (Section 3.4.1). Excitation occurred at 405 nm.

The scattering of the excitation light (due to suspended particles in the dye solution) caused that saturation at 480 nm (the center of the detection range used for this dye).

2.7.4 Emission Spectra

The fluorescence **emission spectrum** or **spectral radiant power** of fluorescence is the plot of the fluorescence radiance against the wavelength or wavenumber (using a fixed excitation band or line). The emission spectrum at λ is calculated by measuring the radiant power carried by the fluorescent photons with a wavelength between $\lambda - \Delta\lambda$ and $\lambda + \Delta\lambda$, which is divided by $2\Delta\lambda$ in the limit in which $\Delta\lambda$ is much smaller than the full width of the emission band at half-maximum. The result is a "density probability distribution," called the spectral radiant power (Section 2.5.3). The maximum of this spectrum occurs at the so-called wavelength of emission maximum, and is denoted by $\lambda_{e,max}$. The typical values of $\Delta\lambda$ used are in the 2–0.2 nm range, which is related to the so-called slit width. Small values of $\Delta\lambda$ must be used for emission bands with sharp peaks, otherwise a rounded or flattened peak with the wrong $\lambda_{e,max}$ value will be recorded. The emission band is always red-shifted with respect to the longest wavelength excitation band, with this shift denoted by $\Delta_S = \lambda_{e,max} - \lambda_{x,max}$ or, in good approximation, by $\Delta_S = \lambda_{e,ma} - \lambda_{a,max}$. This is the so-called Stokes shift [1] and is typically found between 10 nm and 50 nm (Section 2.7.5). For many reasons, some authors express the excitation and emission spectra as a function of wavenumber (ν_a, ν_x, and ν_e in units of cm^{-1}) instead of wavelength (λ_a, λ_x, and λ_e in nm). In the present book the points of the wavenumber axis were first converted to λ (cm) using $\lambda\ (cm) = 1/\nu(cm^{-1})$ and then to λ (nm) using $\lambda\ (nm) = \lambda(cm)\ 10^7\ (nm/cm)$.

Emission spectra should be measured using diluted solutions and/or short path-length cuvettes. Otherwise, the measured emission spectra may appear distorted due to self-absorption, in which part of the radiation is reabsorbed by the dyes before it leaves the cuvette (specially the left side tail in the wavelength plot). This effect is more pronounced with dyes in which there is a significant overlap between the absorption and the emission band.

An example of an emission spectrum is given in Figure 2.16B. Again, the scan was carried out in an exaggeratedly wide interval. It starts at 390 nm, well before the used excitation wavelength (centered at 405 nm), continues past 405 nm (causing saturation due to the scattering of the excitation light by suspended particles in the solution under investigation), and into the range of interest (which is effectively the emission spectrum). Note that the recording stopped at $\lambda_{e,+0.10} = 600$ nm, which is one-tenth of the height at the right side of the emission band maximum.

2.7.5 Stokes Shift

Mirror symmetry between the longest wavelength excitation band and the fluorescence band is observed in typical organic fluorescent molecules in liquid solutions. In these cases, the fluorescence band is always red-shifted with respect to the excitation band. The **Stokes shift** (named after the Irish physicist George Gabriel Stokes) is related to the energy difference between the maximum of the emission band (fluorescence) and the maximum of the excitation band. Therefore, the difference between these maxima can be expressed in electron-Volts (eV), which is an energy quantity, or in frequency (Hz) or wavenumber (cm^{-1}), both are linearly related to energy. However, for practical reasons, the Stokes shift (Δ_S) is usually expressed in wavelength (nm), which is inversely related to frequency and energy.

Since the electronic configuration of a molecule changes upon excitation, the nuclei of the atoms belonging to the π-electron system must move to adjust to the new electronic configuration, which instantaneously sets up a molecular vibration change (some covalently bound pair of atoms of the fluorophore briefly occupy vibrational excited states, also known as nonequilibrium Franck–Condon states) [25]. In other words, electronic excitations

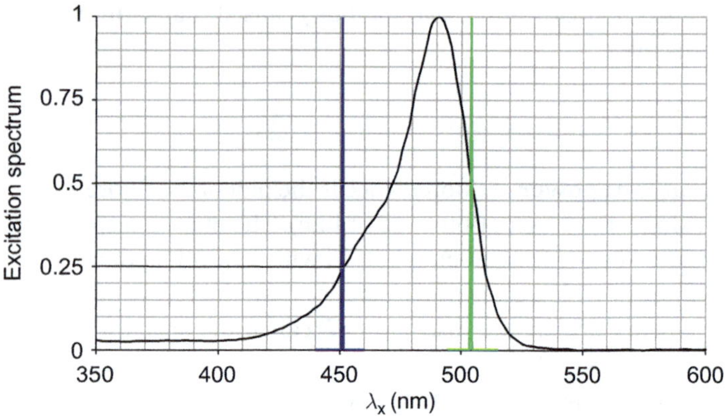

Figure 2.17 (A) A laser line located at one-quarter of the maximum height of the left side of the excitation band ($\lambda_{a,-0.25}$) and (B) a second laser line located at half of the maximum of the right side of the absorption band ($\lambda_{a,+0.5}$).

are accompanied by vibrational excitations. In liquids, these vibrational excited states quickly relax to the vibrational states lying just above the "lowest lying excited electronic state" (S_1), depending on the temperature of the sample (Kasha's Rule) [25]. Therefore, photon emission (fluorescence) with an appreciable yield will predominantly occur from this lowest lying vibrational states, just above the excited electronic state, to the ground electronic state (Figure 2.15). However, here again, the electronic transition or quantum jump (de-excitation) is accompanied by vibrational excitations (now they will be located above the electronic ground state, S_0). Again, these vibrational excited states quickly relax to the vibrational states lying just above the "ground electronic state" (S_0), depending on the temperature of the sample. The applicability of the Franck–Condon principle in both absorption and emission, along with Kasha's Rule, leads to the mirror symmetry between the excitation and the fluorescence spectra of typical organic molecules.

Owing to the loss of vibrational energy during the excitation/emission cycle, fluorescence emission always gives spectrally red-shifted photons with lower energy. This is the basis of the so-called fluorescence Stokes shift, $\Delta_S = \lambda_{e,max} - \lambda_{x,max}$, which should not be confused with the Raman Stokes or Raman anti-Stokes scattering phenomena. These inelastic light scattering events, together with Rayleigh scattering (an elastic scattering phenomenon), are undesired sources of noise (background) in most fluorescence detection and imaging schemes. Fortunately, it is possible to filter them out by several means. Since the Stokes shift of fluorescence is much larger than that of Raman Stokes and anti-Stokes, color filters, interference filters, diffracting gratings, and other dispersion devices may be used to isolate fluorescent light from Raman and Rayleigh scattering. Alternatively, because fluorescence polarization is not preserved by small organic molecules in liquids with low viscosity, Raman and Rayleigh scattering can also be removed by polarizers, allowing depolarized fluorescence light to pass through. Large Stokes shifts are preferred ($\Delta\lambda_S > 50$ nm) as the scattered excitation light is more easily removed using color filters, interference filters, or dispersing devices. This leads to higher signal-to-noise ratios. Fortunately, additional new ways to increase the Stokes shift have been found and are still being developed [37, 38].

Stokes shifts should also not be confused with the so-called pseudo-Stokes shift observed in cassettes made of an acceptor and a donor. In these cases, the emission band of the acceptor is far red-shifted with respect to the absorption band of the acceptor. The mechanism behind this is FRET phenomenon. This takes place when an energy transfer occurs, either through the electromagnetic field of the oscillating dipole of the donor [39] or through the covalent bonds by which the donor and acceptor are held together. The latter case happens even with minimal spectral overlap between the donor emission band and the acceptor absorption band [40].

Besides high brightness, a large Stokes shift is also a highly desired photophysical property as it allows high signal-to-noise ratios to be obtained (good limits of detection, sensitivities, and contrasts). Stokes shifts in aqueous solutions are usually larger if compared to organic solvents, but this comes at a cost. The fluorescence quantum yields (and consequently brightness) of fluorophores are in general smaller in aqueous solutions than those observed in non-polar solvents. The non-radiative decay rates of the electronically excited states are enhanced by the strong dipole–dipole interactions and hydrogen bonds between water molecules and fluorophores. Moreover, some fluorescent molecules also experience a large change on their dipole following both the photoexcitation and the photoemission (fluorescence). Basically, these events are followed by the phenomenon of re-alignment of the water molecules surrounding the fluorophore and a decrease of the total energy (electronic + vibrational + dipole–dipole interaction), as shown in Figure 2.15. The energy involved in the re-alignment following excitation is represented by E_1 and that with photoemission by E_2. This mechanism (among others) is explored in the design of fluorescent dyes with extra-large Stokes shift, for example, in coumarin, naphthalene, and isoquinoline derivatives.

2.7.6 Fluorescence Quantum Yield

Soon after the light-induced electronic excitation of a fluorophore in a liquid solution (the so-called quantum jump, related to the excitation of an electron of the π-system), many relaxation mechanisms start to compete with each other to bring the excited electron back to the ground state. Fluorescence [41] is just one of these possible mechanisms, wherein the sum of all of the probabilities of all possible events equals one. Therefore, the **fluorescence quantum yield** (Φ_f), which is defined as the number of emitted photons divided by the number of absorbed photons, cannot be larger than one ($0 \leq \Phi_f \leq 1$). One important decaying mechanism that competes with fluorescence occurs when the fluorophores are bound to absorbers or biomolecules. In this case, FRET may occur as energy is transferred from the donor (fluorophore) to a nearby existing chromophore (acceptor), leading to the deactivation of the donor whose fluorescence is quenched. This is very common when the nearby molecule is an identical fluorophore, especially when the fluorescence quantum yield of this fluorophore is low as a severe quenching effect is observed. If the acceptor is a different fluorescent molecule and its excitation band overlaps the emission band of the donor, then the acceptor may emit light exactly as if it had been excited directly. This strategy is used in many detection schemes in DNA sequencing and real-time PCR, and is also used to measure distances within macromolecules.

There are many techniques used to measure the fluorescence quantum yield of a dye [42]. The most popular techniques measure relative quantum yields by means of a spectrofluorometer and use a reference standard with a known fluorescence quantum yield, as proposed by [43].

2.7.7 Brightness

Fluorescence **brightness** (B) is one of the most important properties of a fluorophore. It is defined as the product of its molar decadic absorption coefficient at wavelength λ and its fluorescence quantum yield, $B(\lambda) = \varepsilon(\lambda)\ \Phi_f$. Its maximum value is given by $B_{max} = \varepsilon_{max}\ \Phi_f$. The fluorescence quantum yield of a given fluorophore ranges from 0 to 1 and expresses the probability of a fluorescence photon emission for each excitation event. However, even with a strong molar absorption coefficient, if the fluorescence quantum yield is zero ($\Phi_f = 0$) and the phosphorescence quantum yield is zero ($\Phi_p = 0$), then the behavior of the molecule will be the same as that of a common chromophore; i.e., there will be no emission (no fluorescence or phosphorescence). If this absorption band is located in the visible spectrum, then the molecule is termed a dye if it is organic and a pigment if it is inorganic.

2.7.8 Effective Brightness

In the great majority of cases, laser lines do not exactly match the maximum of the excitation bands. This happens for many reasons. First, because it is difficult to find a laser that exactly matches the $\lambda_{x,max}$ of the chosen fluorescent label or stain. Second, changes in the composition of the solution in which the experiments are performed may shift the band maximum, especially when organic modifiers are added. Finally, the labeling reactions and the microenvironment where these fluorescent labels end up also affects the excitation band maximum. Therefore, it is important to define effective brightness. **Effective brightness** ($B(\lambda)$) at wavelength λ is given by $B(\lambda) = \varepsilon(\lambda)\,\Phi_f$, where $\varepsilon(\lambda)$ is the molar absorption coefficient at wavelength λ. Molar absorption coefficient and brightness are expressed in the same unit ($\text{mM}^{-1}\text{cm}^{-1}$), as quantum yield ($\Phi_f$) is a dimensionless photophysical quantity. It is more convenient to express both molar absorption coefficients and brightness in $\text{mM}^{-1}\text{cm}^{-1}$ and not in $\text{M}^{-1}\text{cm}^{-1}$, as discussed in Section 2.7.2. Figure 2.17 illustrates the effective brightness in two situations: one laser at one-quarter of the maximum height of the left side of the excitation band ($\lambda_{a,-0.25}$) and a second laser line located at half of the maximum of the right side of the absorption band ($\lambda_{a,+0.5}$). In these cases, the effective brightnesses are one-quarter and one-half of the maximum, respectively, i.e., $B(\lambda_{a,-0.25}) = B_{max}/4$ and $B(\lambda_{a,+0.50}) = B_{max}/2$, respectively. The maximum brightness in this case will be $B_{max} = B(490\ \text{nm})$. A simple rule to predict the approximate values of $\lambda_{a,-0.25}$, $\lambda_{a,-0.5}$, $\lambda_{a,+0.5}$, $\lambda_{e,-0.25}$, $\lambda_{e,-0.5}$, $\lambda_{e,+0.50}$, $\lambda_{e,+0.25}$, and $\lambda_{e,+0.10}$, based on the class and $\lambda_{a,max}$ and $\lambda_{e,max}$, is shown in Chapter 6.

For the measurement of Φ_f of a dye, then the integral of the whole emission spectrum of that dye and a standard (with a known Φ_f) must be measured, no matter if the emission band is narrow or wide. The fluorometer light harvesting efficiency (Sections 2.6.1 and 2.6.2) and spectral responsivity (Section 2.4) is supposed to be identical for both, in order to have a precise measurement.

In practical situations, high numerical apertures are important when weak fluorescence signals have to be measured. There is a second even more tricky detail: if the filters used to block the excitation radiation also block part of the emission radiation (fluorescence), then an even lower signal should be expected. For example, let us suppose that two dyes (1 and 2) with $\lambda_{a,max,1}$ and $\lambda_{a,max,2}$ have the same brightness ($B_{max,1} = B_{max,2}$), they are excited by two lasers operating with the same power, at exactly $\lambda_{a,max,1}$ and $\lambda_{a,max,2}$, and are simultaneously observed on the same instrument. Even in this case the effective brightness may be very different from each other when the filter set used blocks more emission range of one than the other. Therefore, in general, the wider the emission range detected, the better. In any situation, maximizing the signal-to-noise ratio should be the parameter to follow. Note that color filters and interference filters are discussed in Sections 2.3.1 and 2.3.2, respectively.

2.7.9 Fluorescence Mean-Lifetime

Fluorescence mean-lifetime (τ_f) can be defined as the time required for a population of excited molecules to decrease to $1/e$, where e is the Napier constant or the base of natural logarithms. The half-lifetime ($\tau_{1/2}$) is defined as the time required for a population of excited molecules (N_o) to decay by half ($N_o/2$). The photophysical parameter τ_f is related to decay rate (Γ), half-lifetime, N, and N_o as follows:

$$N = N_o e^{-\Gamma t}, \quad \tau_f = \frac{1}{\Gamma}, \quad \text{and} \quad \tau_f = \frac{\tau_{1/2}}{\ln(2)}. \tag{2.3}$$

It is interesting to note that the probability of the spontaneous decay of a "single molecule" during the small time interval dt is given by:

$$dp = \Gamma dt, \tag{2.4}$$

with the condition that $\Gamma dt \ll 1$. It does not matter how long this molecule has been at the first excited state "waiting to decay," as it has no memory (decay is a Markovian process). This is a fundamental principle of spontaneous decay in quantum theory. All of the equations in Equation 2.3 can be deduced from the simple principle stated in Equation 2.4.

Fluorescence mean-lifetime is measured using pulsed light sources with fast detection electronics and a fast-counting system called "time-correlated single photon counting." Figure 2.18 presents an example of

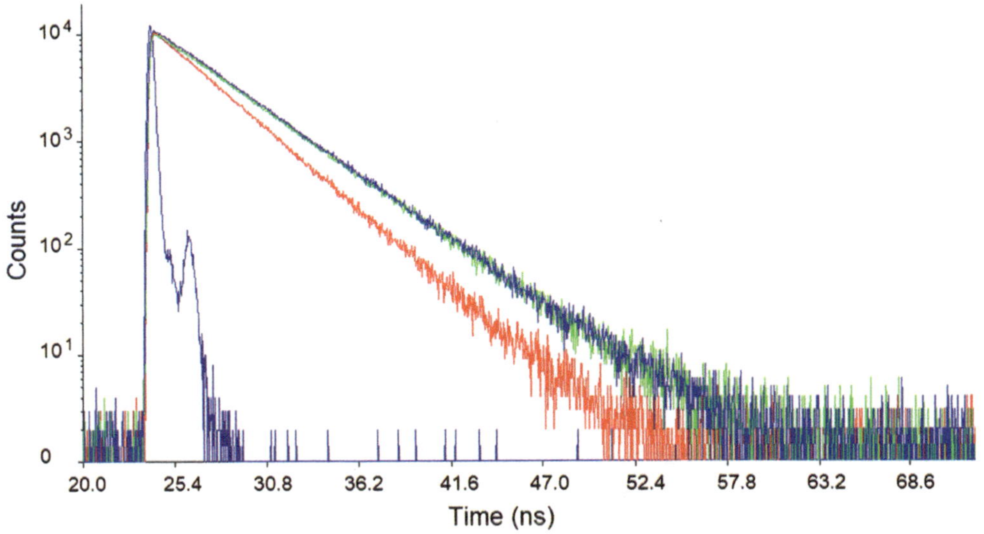

Figure 2.18 Measurement of the fluorescence mean-lifetime of three dyes using a time-correlated single photon counting instrument from PicoQuant (Berlin, Germany). The fluorescence mean-lifetimes are from Cascade yellow (red), Pacific blue (green), and 7-hydroxycoumarin (blue) in 20 mM tetraborate pH 9. (Unpublished results from the author.)

this and illustrates the noise observed over long lifetimes. There is more noise in long lifetimes because most of the events (counts) occur at shorter timescales. Note that the counts on the ordinate are expressed in a logarithmic scale.

2.7.10 Factors Affecting Fluorescence

The spectroscopic and photophysical variables ($\lambda_{a,max}$, $\lambda_{e,max}$, ε_{max}, Φ_f, and τ_f), as well as their derived quantities such as Stokes shift (Δ_S) and brightness (B_{max}), are affected by many factors. They will be briefly discussed in this section. Usually the environment (solvent used, viscosity, pH, and whether or not the dye is bound to larger structures) and temperature are the most important factors, followed by the constituents of the solvent (dissolved oxygen and quenchers). The number of solvents available are large and they present a great variety of properties (for water mixable solvents, see Section 3.2.2). Solvents are important because they are used to prepare the stock solutions of fluorescent labels and stains. Usually, they are found at ~1% (V/V) in the final stained or labeled sample. They are also used as components in the binary and ternary mobile phases of HPLC, together with aqueous buffers.

2.7.10.1 Effect of Microenvironment

There are a number of examples showing that holding a fluorophore within a constrained sub-nanometric cavity (or pore) maximizes its fluorescence quantum yield. In fluorescent proteins, for example, the fluorophore is held inside a protein (**steric impediment**) that restricts its stretching, bending, and twisting movements, thus minimizing the energy loss through non-radioactive decay channels after each excitation event. Another way to explain this phenomenon is to say that keeping the fluorophore confined in a small space minimizes the coupling constants that are responsible for the non-radioactive decays. In this way both the spontaneous and the stimulated decay mechanisms become the predominant phenomena, and therefore the fluorescence quantum yield is increased (Section 2.5.5). Remembering that the modes of the electromagnetic field of the vacuum are responsible for spontaneous decay, the occupancy of the modes within the cavity of a laser are responsible for stimulated emission [15]. This is why the fluorescence quantum yield is also affected when a fluorophore is positioned near a reflective surface, or near the interface between two dielectric media with different refractive indices. When this happens, the modes of the electromagnetic field of the vacuum are modified and, in addition, the field of the oscillating dipole of the fluorophore is reflected on the surface (or interface) and affects its own decay.

Another important example of fluorescence enhancement caused by both microenvironment change and steric impediments is the use of **inclusion complexes**. Cyclodextrins are an important class of such compounds and are well known for improving fluorescence quantum yield, and therefore the brightness of many dyes [44]. For a good review on fluorescent dyes and their **supramolecular host/guest complexes** with macrocycles in aqueous solution, see Dsouza et al. [45].

Finally, another example in which the fluorescence quantum yield of a small fluorophore is increased by a large factor due to the microenvironment is the use of **aptamers**. For example, the fluorescence quantum yield of DFHBI-IT ((*Z*)-4-(3,5-difluoro-4-hydroxybenzylidene)-2-methyl-1-(2,2,2-trifluoroethyl-1*H*-imidazol-5(4*H*)-one)) is only 0.00098 in plain water but increases to 0.94 (~960-fold) upon ligation to Spinash2 (an RNA aptamer) [46]. In this case, the RNA also restricts the stretching, bending, and twisting modes of the fluorescent dye. This is an extreme case to illustrate the effect of the fluorophore's local environment on its photophysical properties.

2.7.10.2 Influence of Liquid Viscosity on Fluorescence Quantum Yield and Fluorescence Mean-Lifetime

A large number of studies and measurements have shown that both fluorescence quantum yield (Φ_f) and fluorescence mean-lifetime (τ_f) increase with the dynamic viscosity of a solvent. The dynamic viscosity (η) of aqueous buffer solutions such as PBS, phosphate, carbonate, and Tris, among others, varies in a narrow range. This range depends on the temperature of the solution (from 0–100°C), as well as the concentration of the salts and the weak acids and/or bases used to prepare the buffer solution. The following equation have been proposed [47] as relationships between the cited photophysical parameters (τ_f and Φ_f) and **dynamic viscosity**:

$$\tau_f = A\eta^{\alpha} \text{ and } \Phi_f = B\eta^{\alpha} \tag{2.5}$$

with $0.2 < \alpha < 1$. Both A and B are positive constants and are related to both the fluorescent dye and solution used. Most measurements show that using $\alpha = 2/3$ provides a good approximation, as proposed by Förster and Hoffmann [47]. The dynamic viscosity of Equation 2.5 is related to free solutions and not, for instance, to diluted polymer solutions which may exhibit a very high viscosity due to polymer entanglement. In polymer solutions, the "frictional coefficients involved in diffusional transport are different to the frictional coefficients involved in hydrodynamic transport." In other words, the effective microenvironment viscosity experienced by the fluorescent dyes in a polymer solution is much lower than the dynamic viscosity of the solution that is measured, for instance, by a Ubbelohde viscometer. Therefore, caution should be taken as some confusion in the literature has been observed.

There is at least one important molecular mechanism that contributes to the positive correlation between dynamic viscosity and fluorescence quantum yield of long fluorescent dyes such as cyanines, squaraines, and merocyanines. Water molecules are much smaller than these dyes. Therefore, the high collisional rates induce strong bending on these structures and this in turn strongly contributes to the downward non-radiative transitions. This is corroborated by the negative correlation of temperature and fluorescence quantum yield (Section 2.7.10.4). High viscosity microenvironments, on the other hand, are made of larger molecules (sometimes longer and stiffer molecules that slip side-by-side) and these produce a much lower rate of molecular motions (bending and twisting events) on these long dyes.

2.7.10.3 Influence of Electric Permittivity and Hydrogen Bonding

Electric permittivity and the hydrogen bonding capability of the solvent are known to affect absorption and emission maxima (**solvatochromism effect**), especially for dyes which exhibit a significant difference between the electric dipole moments of their ground and excited states. Moreover, the fluorescence quantum yield of the great majority of these dyes drops significantly in polar solvents, especially in aqueous solutions. The large electric dipole moment of water and the hydrogen bonding-rich environment within water strengthens the coupling constants responsible for the non-radiative decays. This emphasizes the relevance of the present database, as only dyes with medium to high brightness in aqueous solutions are considered.

Solvatochromism can be used to sense local physical and chemical parameters. For instance, low water content in organic solvents can be detected with high sensitivity [48]. However, in most cases of label and stain applications, solvatochromism is an undesired effect. Aqueous solutions impose a significant Stokes shift increase in some dyes (e.g., coumarins), while a smaller increase is observed with others (cyanine and rhodamines). Lippert (1957) has proposed an elaborate equation that relates the Stokes shift to many dye and solvent parameters [49]. For a detailed review showing the quantitative relationship between solvatochromism and molecular mechanisms, see Suppan (1990) [50].

2.7.10.4 Effects of Temperature

Usually, temperature has a small effect on the wavelengths of excitation maximum and emission maximum. However, the effect of temperature on fluorescence quantum yield and fluorescence mean-lifetime is normally severe. Figure 2.19 illustrates the effect of temperature on these photophysical parameters for the pyronin Y (structure shown in inset).

At 20°C, fluorescence quantum yield is 0.84, while at 50°C, it is 0.44. Therefore, a temperature change of 30°C causes fluorescence quantum yield to decrease by almost 50%. Note that the fluorescence mean-lifetime also decreases with temperature, in a similar pattern to fluorescence quantum yield. This is a sign that higher temperatures decrease the probability of long-living fluorescence events, as they are lost to non-radioactive decay events. An increase of molecular motions in the excited state is the main mechanism behind fluorescence depletion by temperature.

2.7.10.5 Quenching

Quenching is a mechanism that increases the nonradiative decay of a given fluorophore when it is located close to another fluorophore or chromophore. The mechanisms of energy transfer to a nearby acceptor may be based on the direct action of the electromagnetic field of the oscillating dipole (in this case it may be an intermolecular

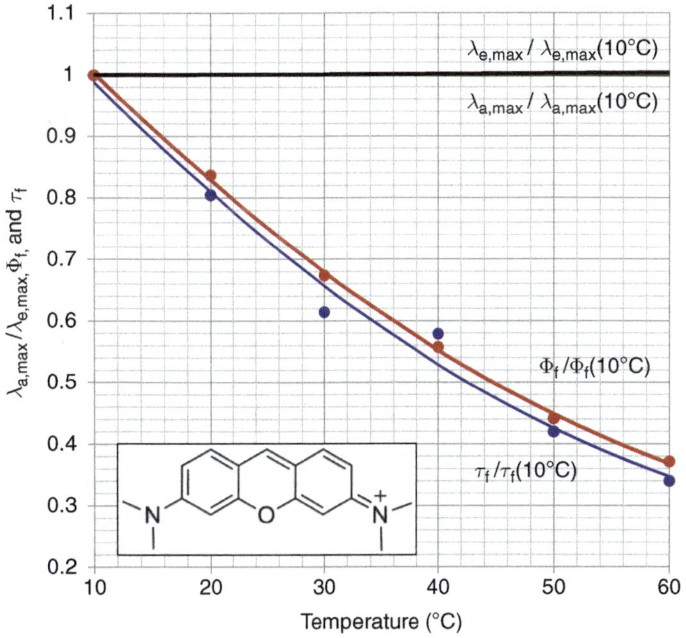

Figure 2.19 Effect of temperature on the $\lambda_{a,max}$, $\lambda_{e,max}$, Φ_f, and τ_f of pyronin (inset) in water. As temperature increases there is a small increase in the Stokes shift, but fluorescence quantum yield and mean-lifetime decrease. Data from Beşer et al. (2019) [51].

or intramolecular transfer) [39], or through the chain of atoms that connect the fluorophore to the acceptor (intramolecular). The latter case happens even with minimal spectral overlap between the donor emission band and the acceptor absorption band [40].

2.7.10.6 Self-Quenching

Self-quenching is related to a quenching effect caused by identical fluorophores that are in close proximity. This may happen when the labeling sites are closely located within a target molecule. This happens when the amino group is labeled in proteins, peptides, amino acids (e.g., lysine, arginine, and cystine, an oxidized form of cysteine made of two cysteines connected by a sulfur bond), and metabolites possessing two or more targetable amino groups in close proximity. This self-quenching is very pronounced in fluorophores with small Stokes shifts, a high overlap between their absorption and emission bands, and when two or more fluorophores are within a Förster radius of each other. This is also noticeable in a free solution when the free diffusing/moving dyes are at high concentrations. In this case, the average distance between identical dyes decreases with concentration and starts to produce a perceivable self-quenching effect. A method was recently introduced to precisely measure this effect without the inconvenience of auto-absorption [52], which normally occurs at high concentrations when a common cuvette with a path length in the millimeter range (1–10 mm) is used.

2.7.10.7 Singlet Oxygen Production by Sensitizer Dyes

Upon excitation by a photon, many dyes have a certain probability to be converted into a relatively long-lived state called the triplet state. These excited, long-lived molecules are capable of transferring part of their electronic energy to nearby dissolved ground state oxygen molecules (O_2). This excites the oxygen molecules to an excited state called **singlet oxygen** (lying 22.4 kcal mol^{-1} above the ground state). In pure water the mean-lifetime of the singlet oxygen state can be as long as a few seconds, but in C–H bond-rich environments it can be as short as a few microseconds. The physical properties of both oxygen molecules are similar to each other, but the singlet state is far more reactive toward organic compounds. Ultimately, the dyes themselves are also oxidized by the singlet oxygen. Examples of dyes that are likely to enter the triplet state include acridines (including methylene blue and its derivatives), thiorhodamines, and dyes containing heavy atoms, e.g., brominated and iodinated derivatives of fluorescein (rose bengal) and borondipyrromethenes [51, 53].

Therefore, in addition to the fluorescence quantum yield (Φ_f), a second very important parameter of these dyes is the triplet quantum yield (Φ_t), which represents the probability of producing a triplet state at each excitation event. In solids this is responsible for the phosphorescence phenomena. Note that the sum of fluorescence quantum yield and triplet quantum yield cannot exceed one (this is valid in both solids and liquids). In summary, a good fluorophore (with high brightness) has a high fluorescence quantum yield (close to one) and a low triplet quantum yield. On the other hand, a good sensitizer dye has a low fluorescence quantum yield and a high triplet quantum yield.

2.8 Photostability

Photostability (or **photorobustness)** is an expression used to denote the chemical stability of chromophores and fluorophores (in general) when subjected to light exposure. **Bleaching** is a general term that refers to the loss of absorption and/or emission due to a physical factor, chemical factor, or a combination of factors. When electromagnetic radiation (in general UV, but also light, which is the visible part of the spectrum) participates in the bleaching process, then it is termed **photobleaching**. Monitoring single molecules provides access to the **bleaching number**, which is defined as the average number of photons a fluorophore emits until photobleaching occurs (also called photoburst). On the other hand, **fluorescence fading** is observed in an ensemble of fluorescent dyes. It is usually characterized by a decreasing exponential of the fluorescence radiance (or intensity) of a small volume of a sample that is homogenously illuminated by a constant radiance (in space and time). **Photobleaching mean-lifetime** (or mean-lifetime before photobleaching) (τ_{pb}) is the time required for the fluorescence fading decreasing exponential to fall to 1/e, where *e* is Napier's number or the base of natural logarithms. The **quantum yield of photobleaching** (Φ_{pb}) expresses the number of molecules bleached per absorbed photon (a dimensionless quantity within the 10^{-5}–10^{-7} range for most dyes in degassed plain water).

Uncontrolled **blinking** is an undesired effect in single molecule tracking schemes and many other systems. In this phenomenon the fluorescent dye ceases to fluoresce for a while before returning to its fluorescent form. This is an uncontrolled intermittence in the fluorescence phenomenon. Conversely, in **photoswitching** the active and inactive forms can be controlled by many means, one of which involves the use of caged fluorescent dyes; for example, in STORM nanoscopy [26]. **Caged compounds** are structures that can be rapidly photoconverted from their inactive form into their active form by light. Typically, **photocaged compounds** have a covalently attached group that can be **photocleaved** by electromagnetic radiation of a specific wavelength. They are used in fluorescence nanoscopy to convert a non-fluorescent dye into a fluorescent dye (or the opposite), sometimes in a reversible manner.

Photochemical degradation refers to the transformation of a molecule, for example a fluorophore, into species with a lower molecular weight. This usually occurs through an oxidation process involving free radicals, oxygen reactive species, and singlet oxygen (Section 2.7.10.7) [52, 53]. The effect of photochemical degradation is usually more severe in solutions containing dissolved oxygen [54, 55, 56, 57]. The term gained popularity when it was discovered to destroy (oxidize) pollutants in water with the aid of UV irradiation. The term is also well known in polymer sciences, where the changes of the chemical and physical characteristics of polymers exposed to UV and visible radiation are noticeable over time. A detailed and remarkable list of photochemical definitions is given in the "Glossary of Terms used in Photochemistry" [1].

References

1 Braslavsky, S.E. (2007). Glossary of terms used in photochemistry, 3e. *Pure and Applied Chemistry* 79: 293–465.
2 IUPAC (2014). *Compendium of Chemical Terminology (The "Gold Book"),* Version 2.3.3 (compiled by A.D. McNaught and A. Wilkinson). Oxford, UK: Blackwell Scientific Publications. doi: 10.1351/goldbook.
3 Kist, T.B.L. (1996). Stochastic Schrödinger equations applied to dissipative quantum systems in optics. PhD Thesis, Rio de Janeiro: PUC-RJ.
4 Kist, T.B.L., Orszag, M., Brun, T.A., and Davidovich, L. (1999). Stochastic Schrödinger equations in cavity QED: physical interpretation and localization. *Optical B: Quantum Semiclass Optical* 1: 251–263.
5 Abbe, E. (1873). Beiträge zur theorie des mikroskops und der mikroskopischen Wahrnehmung. *Archiv fur Mikroskopische Anatomie* 9: 413–418.
6 Liu, C., Zeng, J., Liu, X. et al. (2021). Vortex-beam photoacoustic microscopy for enhancing the depth of field. *Proc. SPIE 11642, Photons Plus Ultrasound: Imaging and Sensing,* 116423R. doi: 10.1117/12.2589253.
7 Padgett, M. and Allen, L. (2000). Light with a twist in its tail. *Contemporary Physics* 41 (5): 275–285.
8 Bliokh, K.Y. and Nori, F. (2015). Transverse and longitudinal angular momenta of light. *Physics Reports* 592: 1–38.
9 Tanaka, H., Inoue, Y., and Mori, T. (2018). Circularly polarized luminescence and circular dichroisms in small organic molecules: correlation between excitation and emission dissymmetry factors. *ChemPhotoChem* 2: 386–402. doi: 10.1002/cptc.201800015.

10 Chen, Y. (2022). Circularly polarized luminescence based on small organic fluorophores. *Materials Today Chemistry* 23 (2022): doi: 10.1016/j.mtchem.2021.100651.

11 García-Parajó, M.F., Veerman, J.A., Bouwhuis, R. et al. (2001). Optical probing of single fluorescent molecules and proteins. *ChemPhysChem* 2: 647–360.

12 Galievsky, V.A., Stasheuski, A.S., and Krylov, S.N. (2016). Getting the best sensitivity from on-capillary fluorescence detectionin capillary electrophoresis: a tutorial. *Analytica Chimica Acta* 935: 58–81.

13 van de Linde, S., Heilemann, M., and Sauer, M. (2012). Live-cell super-resolution imaging with synthetic fluorophores. *Annual Review of Physical Chemistry* 63: 519–540. doi: 10.1146/annurev-physchem-032811-112012.

14 Liu, Z., Liu, J., Wang, X. et al. (2020). Fluorescent bioconjugates for super-resolution optical nanoscopy. *Bioconjugate Chemistry* 31: 1857–1872.

15 Vogel, W. and Welsch, D. (2006). *Quantum Optics*. WILEY-VCH Verlag GmbH & Co.

16 Valeur, B. and Berberan-Santos, M.N. (2012). *Molecular Fluorescence: Principles and Applications*, 2e. Hoboken: Wiley.

17 Lacowicz, J.R. (2006). *Introduction to Fluorescence – Principles of Fluorescence Spectroscopy*, 3e. Springer.

18 França, L.T.C., Carrilho, E., and Kist, T.B.L. (2002). A review of DNA sequencing techniques. *Quarterly Reviews of Biophysics* 35: 169–200.

19 Epstein, J.R., Biran, I., and Walt, D.R. (2002). Fluorescence-based nucleic acid detection and microarrays. *Analytica Chimica Acta* 469: 3–36.

20 Liu, W., Huang, S., Liu, N. et al. (2017). Establishment of an accurate and fast detection method using molecular beacons in loop-mediated isothermal amplification assay. *Scientific Reports* 7: 40125. doi: 10.1038/srep40125.

21 Yi, M., Zhao, Y., Jia, L.et al. (2014). Performance comparison of SNP detection tools with illumina exome sequencing data: an assessment using both family pedigree information and sample-matched SNP array data. *Nucleic Acids Research* 42 (12): e101. doi: 10.1093/nar/gku392.

22 Lea, W.A. and Simeonov, A. (2012). Fluorescence polarization assay. *Expert Opinion on Drug Discovery* 6: 17–32.

23 Lee, L. (1997). New energy transfer dyes for DNA sequencing. *Nucleic Acids Research* 25: 2816–2822.

24 Wuethrich, A. and Quirino, J.P. (2016). Derivatisation for separation and detection in capillary electrophoresis (2012–2015). *Electrophoresis* 37: 45–55.

25 Sauer, M., Hofkens, J., and Enderlein, J. (2011). *Handbook of Fluorescence Spectroscopy and Imaging*. WILEY-VCH Verlag GmbH & Co.

26 van de Linde, S. and Sauer, M. (2014). How to switch a fluorophore: from undesired blinking to controlled photoswitching. *Chemical Society Reviews* 43: 1076–1087.

27 Bingen, P., Reuss, M., Engelhardt, J., and Hell, S. (2011). Parallelized STED fluorescence nanoscopy. *Optical Express* 19: 23716.

28 Turkowyd, B., Virant, D., and Endesfelder, U. (2016). From single molecules to life: microscopy at the nanoscale. *Analytical and Bioanalytical Chemistry* 408: 6885–6911.

29 Rust, M.J., Bates, M., and Zhuang, X.W. (2006). Sub-diffraction-limit imaging by stochastic optical reconstruction microscopy (STORM). *Nature Methods* 3: 793–795.

30 Tam, J. and Merino, D. (2015). Stochastic optical reconstruction microscopy (STORM) in comparison with stimulated emission depletion (STED) and other imaging methods. *Journal of Neurochemistry* 135 (4): 643–658. doi: 10.1111/jnc.13257.

31 Weber, M., Leutenegger, M., Stoldt, S. et al. (2021). MINSTED fluorescence localization and nanoscopy. *Nature Photonics* 15: 361–366.

32 Marx, V. (2015). Probes: paths to photostability. *Nature Methods* 12: 187–190.

33 van der Velde, J.H.M., Oelerich, J., Huang, J.et al. (2016). A simple and versatile design concept for fluorophore derivatives with intramolecular photostabilization. *Nature Communications* 7: 10144. doi: 10.1038/ncomms10144.

34 Zheng, Q., Juette, M.F., Jockusch, S.et al. (2014). Ultra-stable organic fluorophores for single-molecule research. *Chemical Society Reviews* 43: 1044–1056.

35 Meagher, R.J., Won, J.-I., McCormick, L.C. et al. (2005). End-labeled free-solution electrophoresis of DNA. *Electrophoresis* 26: 331–350. doi: 10.1002/elps.200410219.

36 Chalmers, J.M. and Griffiths, P.R. (2002). *Handbook of Vibrational Spectroscopy*, 1e. Hoboken: Wiley.

37 Ren, T.-B., Xu, W., Zhang, W.et al. (2018). A general method to increase Stokes shift by introducing alternating vibronic structures. *Journal of the American Chemical Society* 140: 7716–7722. doi: 10.1021/jacs.8b04404.

38 Sednev, M. (2015). Fluorescent dyes with large Stokes shifts of 80—200 nm for optical microscopy and nanoscopy. PhD Thesis. Georg-August-Universität Göttingen.

39 Wu, L., Huang, C., Emery, B.P.et al. (2020). Förster resonance energy transfer (FRET)-based small-molecule sensors and imaging agents. *Chemical Society Reviews* 49: 5110–5139.

40 Lin, W., Yuan, L., Cao, Z. et al. (2010). Through-bond energy transfer cassettes with minimal spectral overlap between the donor emission and acceptor absorption: Coumarin–Rhodamine dyads with large pseudo-Stokes shifts and emission shifts. *Angewandte Chemie* 49: 375–379.

41 Würth, C., Grabolle, M., Pauli, J. et al. (2013). Relative and absolute determination of fluorescence quantum yields of transparent samples. *Nature Protocols* 8: 1535–1550.

42 Rurack, K. (2008). Fluorescence quantum yields: methods of determination and standards. *Springer Series on Fluorescence* 5: 101–145.

43 Parker, C.A. and Rees, W.T. (1960). Correction of fluorescence spectra and measurement of fluorescence quantum efficiency. *Analyst* 85: 587–600.

44 Elbashir, A.A., Dsugi, N.F.A., Mohmed, T.O.M., and Aboul-Enein, H.Y. (2014). Spectrofluorometric analytical applications of cyclodextrins. *Luminescence* 29: 1–7. doi: 10.1002/bio.2504.

45 Dsouza, R.N., Pischel, U., and Nau, W.M. (2011). Fluorescent dyes and their supramolecular host/guest complexes with macrocycles in aqueous solution. *Chemical Reviews* 111: 7941–7980.

46 Song, W., Strack, R.L., Svensen, N., and Jaffrey, S.R. (2014). Plug-and-play fluorophores extend the spectral properties of spinach. *Journal of the American Chemical Society* 136: 1198–1201. doi: 10.1021/ja410819x.

47 Förster, T. and Hoffmann, G. (1971). Die Viskositätsabhängigkeit der Fluoreszenzquantenausbeuten einiger Farbstoffsysteme. *Zeitschrift Für Physikalische Chemie* 75: 63–76. doi: 10.1524/zpch.1971.75.1_2.063.

48 Chen, W., Zhang, Z., Li, X. et al. (2015). Highly sensitive detection of low-level water content in organic solvents and cyanide in aqueous media using novel solvatochromic AIEE fluorophores. *RSC Advances* 5: 12191–12201.

49 Lippert, E. von (1957). Spektroskopische Bestimmung des Dipolmomentes aromatischer Verbindungen im ersten angeregten Singulettzustand. Zeitschrift für Elektrochemie *Berichte der Bunsengesellschaft für physikalische Chemie* 61(8): 962–975.

50 Suppan, P. (1990). Solvatochromic shifts: the influence of the medium on the energy of electronic states. *J. Photochem. Photobiol.* A50: 293–330.

51 Beşer, B.M., Arik, M., and Onganer, Y. (2019). Photophysical and photodynamic properties of Pyronin Y in micellar media at different temperatures. *Luminescence* 34: 415–425. doi: 10.1002/bio.3624.

52 Bae, W., Yoon, T.-Y., and Jeong, C. (2021). Direct evaluation of self-quenching behavior of fluorophores at high concentrations using an evanescent field. *PLoS ONE* 16 (2): e0247326.

53 Leen, V., Braeken, E., Luckermans, K. et al. (2009). A versatile, modular synthesis of monofunctionalized BODIPY dyes. *Chemical Communications* 30: 4515–4517. doi: 10.1021/acs.bioconjchem.0c00320.

54 Dobson, D.E., Mahoney, E.R., Mach, T.P. et al. (2020). Pentamethine sulfobenzoindocyanine dyes with low net charge states and high photostability. *Photochem. Photobiol. Sci.* 19: 56–65.

55 Ziessel, R., Ulrich, G., and Harriman, A. (2007). The chemistry of bodipy: a new El Dorado for fluorescence tools. *New Journal of Chemistry* 31: 496–501.

56 Kuznetsova, N.A., Kaliya, O.L., and Luk'yanets, E.A. (1995). Photochemistry of laser dyes for the visible spectrum. *Proc. SPIE (International Conference on Atomic and Molecular Pulsed Laser)* 2619: 161–165 doi: 10.1117/12.216902.

57 Nani, R.R., Kelley, J.A., Ivanic, J., and Schnermann, M.J. (2015). Reactive species involved in the regioselective photooxidation of heptamethine cyanines. *Chemical Sciences* 6: 6556–6563. doi: 10.1039/c5sc02396c.

3

Target-Fluorophore Binding

3.1 Introduction

This chapter begins by presenting the specifications of laboratory water with different purity grades and the properties of the water miscible solvents that are used to prepare stock solutions of fluorescent labels and stains. Some water miscible solvents that are less well known but have great potential are presented in Section 3.2. The chemical structures of the main "fluorogenic derivatization reagents" and their respective reactions are given in Section 3.3. All entities, neutral and ionic, that participate in the reactions, are shown. This provides a good view of the reactions in terms of the enthalpy change and the entropy change, which are related to the reaction solvent used and both the number and nature of the entities in the initial and final products of the reaction. However, the intermediate steps of the reactions, intermediate products, and detailed mechanisms are not shown.

The large group of labeling reactions are tabulated in Section 3.4. Forty-four labeling reactions (reactive moiety plus target functional group) are tabulated but not explicitly shown as it would take too much space to detail them all. In this case, the target molecules (biomolecules) are labeled (bioconjugated) with a great variety of fluorescent dyes, many of them exhibiting very high brightness in aqueous solutions (over 700 such dyes are presented in Chapters 4, 5, and the Appendices). The present chapter ends with a brief discussion of immunostaining.

As in all chapters of this book, the functional groups of the molecular structures (reagents and products) are shown in the ionization states that are expected in an aqueous solution at the respective reaction pH, which is usually neutral or mild basic.

3.2 Choosing the Right Solvent

3.2.1 Water and PBS

Pure water, PBS (phosphate-buffered saline), aqueous buffer solutions, and many water-organic mixtures are widely used in a great variety of protocols. Therefore, it is important to know the different grades of available laboratory water to prepare these solutions.

Laboratory water has many quality grades, and each grade is suitable for certain applications. In the scientific literature authors usually only mention the source of the water used in their experiments. Distilled water (that has or has not been filtered) is the most cited, followed by de-ionized water (produced by ion exchangers and then filtered), water produced by reverse osmosis, or water purchased from a supplier that guarantees a given quality level. It is worth mentioning that the quality of the water depends not only on the method used to produce it, but also on how it is handled after production and/or how the flask from the supplier is handled after being opened.

Fluorescent Dye Labels and Stains: A Database of Photophysical Properties, First Edition. Tarso B. Ledur Kist.

Producing a liter of pure water has a high built-in cost, and the cost of checking its quality may be higher still. Nevertheless, resistivity (or its inverse, conductivity) measurement is a fast and cheap way to estimate the level of ions present in water. However, this does not give any information about the levels of the organic compounds, colloids, and even the biological contaminants (i.e., bacteria, fungi, and other microorganisms) present as they little contribute to conductivity.

Many organizations have established laboratory water specifications or technical standards on water quality to facilitate good laboratory practices and increase quality standards for chemical, biochemical, and clinical analyses. Examples of such organizations include the International Organization for Standardization (ISO), American Society for Testing and Materials (ASTM), United States Pharmacopoeia (USP), and the Clinical and Laboratory Standards Institute (CLSI), among others. The ISO released the norm ISO 3696: 1987 entitled "Water for analytical laboratory use: Specifications and test methods," which was reviewed and confirmed in 2018. This norm provides the specifications for three grades of water purity that can be used for analytical purposes in laboratories (Table 3.1).

Water grade 3 can be produced by single distillation, reverse osmosis, or de-ionization. It is commonly used for ordinary analytical applications. Water grade 2 can be produced by multiple distillation, de-ionization, or reverse osmosis followed by distillation. It has a very low inorganic, organic, or colloidal material content and is suitable for sensitive analytical purposes. Water grade 1 is produced by further treating water grade 2. This can be done, for example, by redistilling the water grade 2 through a fused silica distillatory device, then filtering it through a 0.2 μm pore size membrane filter.

Table 3.1 The specifications of laboratory water.

Parameters	Grade 1	Grade 2	Grade 3
pH value at 25°C Inclusive range	Not app.[a]	Not app.[a]	5.0–7.5
Electrical conductivity μS/m at 25°C, max.	0.1[b]	1.0[b]	5.0
Oxidizable matter mg/L, mg of oxygen per liter of water, max.	Not app.[c]	0.08	0.4
Absorbance at 254 nm and 1 cm optical path length, absorbance units, max.	0.001	0.01	Not specified
Residue after evaporation on heating at 110°C mg/kg, max.	Not app.[c]	1	2
Silica (SiO_2) content mg/L, max.	0.01	0.02	Not specified

[a] Not applicable (Not app.) due to the difficulties associated with measurement of the pH value of high-purity water, and the doubtful significance of the value obtained.
[b] Values of conductivity for grade 1 and grade 2 water apply to freshly prepared water because there are many sources of contamination during storage.
[c] A limit for "oxidizable matter" and "residue after evaporation" for grade 1 water is not specified because of the difficulty of testing for compliance at this level of purity.

Adapted from BS EN ISO 3696: 1995. Water for Analytical Laboratory Use Specification and test methods. Last accessed 5 August 2022.

The ASTM has released the norm "D1193-06 (2011) – Standard Specification for Reagent Water," which establishes four types of water (I–IV) and three sub-grades (A–C) (Table 3.2). These sub-grades are important when the levels of viable bacteria and fungi, as well as the level of endotoxins, need to be known. When viable bacterial or fungi are not important then, according to the ASTM norm D1193-06 (2011), resistivity and levels of TOC, sodium, chlorine, and silicon will define the water as type I, II, III, or IV (Table 3.2). Ranking laboratory water purities in such a way is expected to increase the reproducibility and repeatability of analyses, assays, and imaging techniques.

The amount of dissolved oxygen (O_2) is not mentioned in the above cited norms, and the measured parameters do not indicate the concentration of molecular oxygen dissolved in the water. Upon excitation, many dyes end up in a long-lived state called the triplet state. In this state, they are good at promoting the excitation of ground state molecular oxygen which is dissolved in the solution to its highly reactive form called singlet oxygen (O_2*). This excited form, which may even have a lifetime of a few seconds in pure water, is very efficient at oxidizing the dyes themselves (see Section 2.7.10.7 for more details). Therefore, it is important to de-air any water that has been standing still for a long time before using it, as it may contain dissolved molecular oxygen (originating from air absorption or other processes).

PBS solutions are prepared to fix the pH level (acting as a buffer solution) and to imitate the average osmotic pressure of human cells. They are buffer solutions with a pH, ion concentration, and osmolarity that matches that of human cells. There are a few distinct protocols used to produce PBS, but the most widely used is known as the Dulbecco and Vogt protocol [1]. Each liter of this solution contains 8.01 g of NaCl, 0.20 g of KCl, 1.42 g of Na_2HPO_4, and 0.24 g of KH_2PO_4. Consequently, the following molar concentrations are obtained: 136.0 mM of NaCl; 2.7 mM of KCl; 10.0 mM of Na_2HPO_4; and 1.8 mM of KH_2PO_4. Solutions containing polyvalent cations should be added with care as most of them will precipitate with the phosphate ion. There are a few rhodamines with some extra positive charges in addition to quaternary amines that are especially sensitive to aggregation and precipitation.

3.2.2 Water Miscible Organic Solvents

Many dyes are less stable in aqueous solutions than in organic hydrophilic solvents. Moreover, many are not soluble enough in pure water or aqueous solutions to prepare stock solutions at the concentrations required. Therefore, hydrophilic organic solvents are also very important. Choosing the best solvent to prepare the stock solution of a given label or stain requires some attention. When selecting a solvent, it is important to consider its effects on the samples, effects on the reactions (when fluorogenic or labeling reactions are used), cost, green credentials, shelf lifetime, smell, toxicity, and the solubility characteristics of the reagents and products. For security reasons it is also important to know the vapor phase characteristics of the solvent. The two most important characteristics are vapor pressure at any given temperature and flash point. Flash point is the temperature at which a vapor can, in principle, burn when ignited by a spark. These properties are of significance to method developers because the safety restrictions of a flammable liquid (flash point <37.8°C) are much more stringent than those of a combustible liquid (37.8° < flash point <~ 100°C), which are, in turn, more stringent than those of a solvent with a flash point greater than ~100°C.

Most of the solvents shown in Table 3.3 are miscible in water at room temperature. 2-Butanol, 2-butanone, and 3-pentanone are exceptions as they are only slightly miscible in water at room temperature. Certain solvents are less commonly used to prepare stock solutions of labels and stains for many reasons. For instance, under some conditions, dimethoxyethane (Glyme) and bis(2-methoxyethyl)ether (Diglyme) act as bidentate ligands in the presence of some metal cations. The same behavior is expected from 2-methoxyethanol (2-ME), 2-ethoxyethanol (2-EE), 2-propoxyethanol (2-PE), 2-butoxyethanol (2-BE), and 2-Methoxyethanol (2-ME), although it is less pronounced. Bis(2-methoxyethyl)ether (Diglyme) has been reported to harm the human reproductive system [2]. Some polyalcohols, although water soluble, are not included in Table 3.3 because they are too viscous for fast pipetting and handling. Examples of these include ethylene glycol (ethane-1,2-diol), propylene glycol (propane-1,2-diol),

Table 3.2 Laboratory water purity grades according to the ASTM norm D1193-06(2011) – "Standard Specification for Reagent Water."

	Type	Grade	Conductivity (μS/cm), max.	Resistance (MΩ cm), min	pH	TOC (μg/L), max.	Sodium (μg/L), max.	Chloride (μg/L), max.	Silicon (μg/L), max.	Bacteria (CFU/mL), max.	Endotoxins (EU/L), max.
Ultrapure Water	I		0.056	18.0	–	50	1	1	3	–	–
	I	A	0.056	18.0	–	50	1	1	3	10/1000	0.03
	I	B	0.056	18.0	–	50	1	1	3	10/100	0.25
	I	C	0.056	18.0	–	50	1	1	3	100/10	–
Pure Water	II		1.0	1.0	–	50	5	5	3	–	–
	II	A	1.0	1.0	–	50	5	5	3	10/1000	0.03
	II	B	1.0	1.0	–	50	5	5	3	10/100	0.25
	II	C	1.0	1.0	–	50	5	5	3	100/10	–
	III		0.25	0.4	–	250	10	10	500	–	–
	III	A	0.25	0.4	–	250	10	10	500	10/1000	0.03
	III	B	0.25	0.4	–	250	10	10	500	10/100	0.25
	III	C	0.25	0.4	–	250	10	10	500	100/10	–
	IV		5.0	0.2	5.0–8.0	–	50	50	–	–	–
	IV	A	5.0	0.2	5.0–8.0	–	50	50	–	10/1000	–
	IV	B	5.0	0.2	5.0–8.0	–	50	50	–	10/100	–
	IV	C	5.0	0.2	5.0–8.0	–	50	50	–	100/10	–

Note: TOC is the acronym for total organic carbon content, CFU stands for colony-forming unit, and CFU/mL is a measure of the viable bacteria and fungi per unit volume of water.

1,3-propanediol (propane-1,3-diol), glycerin (propane-1,2,3-triol), 1,2-butanediol (butane-1,2-diol), 1,3-butanediol (butane-1,3-diol), 1,4-butanethiol (butan-1,4-diol), and methyl diethanolamine (bis(2-hydroxyethyl)(methyl) amine). 1-Butanol was not included in the table as it is much less soluble in water than 2-butanol. The values of the four physical parameters (parameters listed in Table 3.3) of butanone (Butan-2-one) lie between those of acetone and 3-pentanone; however, butanone is less soluble in water. Some solvents have a boiling point that is too low for it to be included in Table 3.3; e.g., acetaldehyde (20.2°C), methyl isocyanate (40°C, this should not be confused with methyl isocyanide which exhibits a boiling point of 60°C), and dimethoxymethane (42°C). Solvents containing a primary or secondary amine group were not added because, when mixed with water, they have an ionizable group close to neutral pH. In addition, many of them form adducts and even react with many derivatizing reagents. Some solvents were not included because they are unstable, e.g., some aldehydes. Other solvents are both toxic and unstable, e.g., methyl isocyanide (or isocyanomethane) and methyl isothiocyanate, so were also not included.

As seen in Table 3.3, most of the solvents and the stock solutions of labels and stains made from these solvents can be stored in freezers at −8°C without freezing. This happens because their freezing point is lower than −8°C. The exceptions are water, 1,4-dioxane, dimethylsulfoxide, formamide, and *N*-methylformamide. This is advantageous because they can quickly be warmed out and used straightaway, without the need to wait for a long time for the solid to fuse and reach working temperature (usually between 20 and 24°C).

The evaporation rates of solvents are given in volume/time or mass/time, and express the volume or mass lost per time under a given condition (flask shape, air temperature, and air velocity, among other factors). Note that some solvents are also hygroscopic. Therefore, in their anhydrous form they may gain volume and/or mass over time if the rate of water absorption is higher than the rate of solvent loss per unit time (this is the case for anhydrous DMSO, PC, and some formamides and acetamides).

If the evaporation rates of the solvents used to prepare stock solutions of labels and stains are too high, then some problems can be expected in some practical situations. First, it may be difficult to take precise quantitative measurements of some photophysical parameters, e.g., molar absorption coefficients, as the concentration of the label or stain may change during the handling time. Second, fast evaporation may compromise the robustness of derivatization reaction methods if the solvent used to prepare the standard stock solution is too volatile, as the concentration of the stock solution may change over time. Third, the water content of hygroscopic solvents increases over time, which may decrease the shelf life of some dyes and affect some reactions (e.g., fluorogenic and labeling reactions). Finally, in some experiments, the solvent needs to be removed and replaced by another solvent. In this case, it is very advantageous to be able to remove a solvent using a simple procedure, e.g., by gently flowing nitrogen gas over it. In other words, solvents that are easily removed, and subsequently easily replaced, are sometimes preferred. Therefore, it is important to know the evaporation rate of the solvents under normal working conditions. In a broader sense, it is important to know the change of volume, mass, and composition of solvents over time under normal working conditions. Note that the absorption of water by hygroscopic solvents may occur without a change in the net mass and/or volume of the stock solution when under some special conditions (temperature, pressure, and relative humidity).

The evaporation rate of a solvent is related to its vapor pressure. Nonetheless, it must be measured because the relationship between the evaporation rate of a solvent and its vapor pressure is not simple. Evaporation rates of solvents depend on many factors. Examples of such factors include: the shape of the flask used to manipulate the stock solution (e.g., a beaker); the area of liquid surface in contact with the air (internal diameter of the beaker); the distance between the liquid level and the top edge of the flask (beaker); the temperature of the flowing air; and the air velocity (e.g., of a fume hood). The relative humidity (expressed in %) of the air is also important for solvents such as water, PBS, and aqueous solutions in general. The relative humidity also affects the rate of mass change of hygroscopic solvents, as the total mass gained per unit time is expected to increase with increasing relative humidity.

Figure 3.2 shows the mass change of some water miscible solvents when 20 mL of a solvent was placed in 50 mL beakers located at the entrance of a fume hood with a grid (made of poly(methyl methacrylate) that assures

Table 3.3 General properties of water miscible solvents that can be used as reaction media or to prepare stock solutions of water-soluble fluorescent dyes.

	Name	IUPAC Name	Melting Point (°C)	Boiling Point (°C)	Vapor Pressure at 20°C (kPa)	Viscosity at 25°C (mPa.s or cP)
1	Acetone	Propan-2-one	−94.7	56.1	24.6	0.295
2	Dimethylacetal	1,1-Dimethoxyethane	−113	64.5	8.13	a
3	Acetonitrile (ACN)	Ethanenitrile	−45	82	9.71	0.334
4	Dimethoxyethane (Glyme)	1,2-Dimethoxyethane	−58	85	6.4	0.371
5	2-Butanone (MEK)	Butan-2-one	−86	79.6	10.4	0.43
6	3-Pentanone	Pentan-3-one	−39	102	4.66	0.444
7	Tetrahydrofuran (THF)	1,4-Epoxybutane	−108.4	66	17.60	0.48
8	Methanol	Methanol	−97.6	64.7	13.02	0.545
9	Water (H_2O)	Water	0	100	2.30	0.891
10	Dimethylacetamide (DMA)	*N,N*-Dimethylacetamide	−20	165	0.33	0.92
11	Bis(2-methoxyethyl) ether (Diglyme)	1-Methoxy-2-(2-methoxyethoxy)ethane	−68	162	0.33	1.09
12	Ethanol	Ethanol	−114.1	78.2	5.95	1.2
13	Dioxane	1,4-Dioxane	11.8	101.1	3.86	1.2
14	N-Methyl-2-pyrrolidinone (NMP)	1-Methylpyrrolidin-2-one	−24	203	<0.1	1.67
15	Dimethylformamide (DMF)	*N*,**N**-Dimethylmethanamide	−61	153	0.45	1.68
16	Methylformamide (MFA)	*N*-Methylmethanamide	−4	182.6	<0.1	1.68
17	2-Methoxyethanol (2-ME)	2-Methoxyethan-1-ol	−85	124	0.80	1.70
18	1-methoxy-2-propanol	1-Methoxypropan-2-ol	−97	120	1.54	1.70
19	1-Propanol	Propan-1-ol	−126	97.5	1.99	1.96
20	2-Ethoxyethanol (2-EE)	2-Ethoxyethan-1-ol	−90	135	0.40	1.96[b]
21	Dimethylsulfoxide (DMSO)	Dimethyl(oxido)sulfur	19	189	<0.1	1.97
22	2-Propoxyethanol (2-PE)	2-Propoxyethan-1-ol	−90	151	<0.1	2.3[b]
23	2-Propanol (Isopropanol)	Propan-2-ol	−89	82.6	4.40	2.41
24	Propylene carbonate (PC)	4-Methyl-1,3-dioxolan-2-one	−48.8	242	<0.1	2.5
25	2-Butanol	Butan-2-ol	−115	99	1.67	2.57
26	2-Butoxyethanol (2-BE)	2-Butoxyethan-1-ol	−77	171	<0.1	2.9
27	Formamide	Methanamide	2	210	1.07	3.23
28	Tetrahydrofuran-2-methanol (THF-2-methanol)	(Oxolan-2-yl)methanol	−80	178	0.11	4.62
29	2-Hydroxyethyl acetate (2-HEA)	2-Hydroxyethyl acetate	−61.7	183	<0.1	a

[a] Value not found, but the ranking was determined by an estimated value.
[b] Measured at 20°C. The molecular structures of some of these solvents are shown in Figure 3.1. Most of these solvents are miscible in water, with the only exceptions being 2-butanol (290 g/L at 25°C), 2-butanone (275 g/L at 25°C), and 3-pentanone (35 g/L at 20°C). The solvents are ranked by increasing viscosity values. Data was taken from https://pubchem.ncbi.nlm.nih.gov, http://www.chemspider.com, https://www.chemsrc.com, and http://www.ddbst.com (all pages were accessed in January 2021), and from printed sources [3].

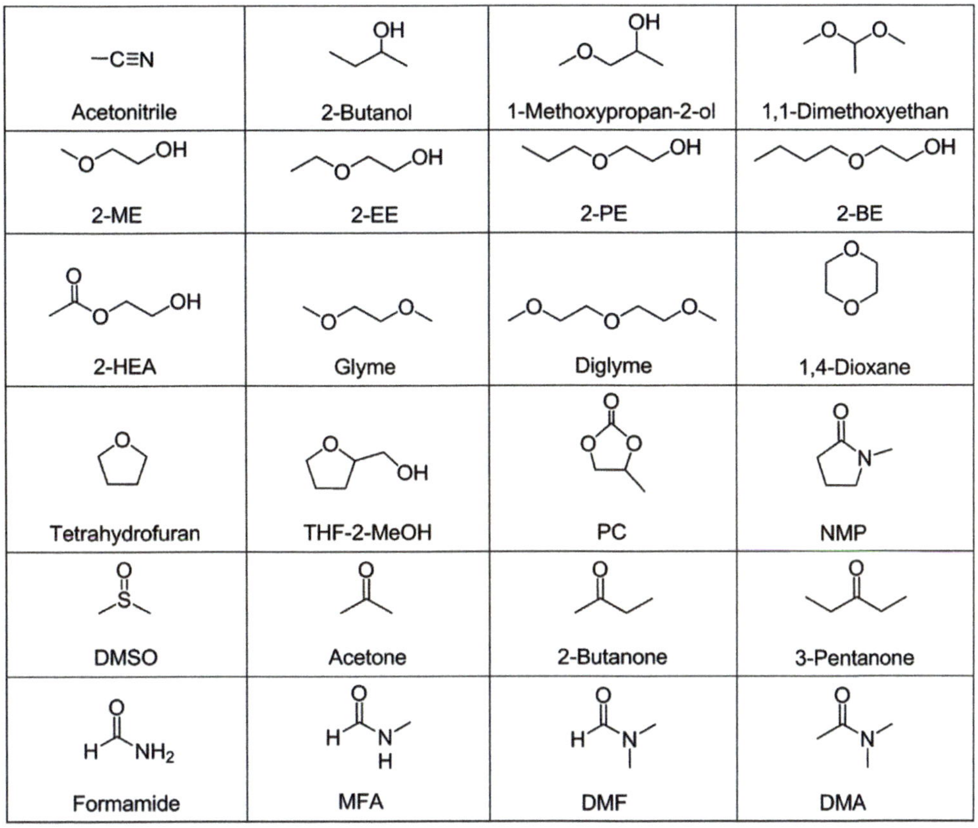

Figure 3.1 The molecular structure of some common water miscible solvents used to prepare stock solutions of water-soluble dyes. Some of the solvents can also be used as organic modifiers in the reaction media for derivatization and labeling reactions. Most of the listed solvents are miscible in water, with the exception of 2-butanol (290 g/L at 25 °C), 2-butanone (275 g/L at 25°C), and 3-pentanone (35 g/L at 20°C).

identical air velocity profiles over all samples. The fume hood was set to medium velocity, room temperature was fixed at 22°C (using air conditioning), and a dehumidifier kept the relative humidity constant at 60%. Although small vials are usually used to handle stock solutions, it is easier to verify evaporation rates when a beaker is used. The water miscible solvents used are presented in Table 3.3, Figure 3.1, and in footnote.[1] Note that the mass of DMSO, formamide, dimethylformamide, and propylene carbonate increases over time due to the absorption of water from the air. *N*-methyl-2-pyrrolidinino (NMP) (this is not plotted in Figure 3.2) exhibited a behavior like that of PC. Methanol produced a curve similar to that of acetonitrile, and tetrahydrofuran produced a curve similar to that of acetone. These two solvents were not plotted as this would produce a very crowded graph. For details, see the relative evaporation rates in Table 3.4.

1 **2-Propanol** (Fluka, 34965); **acetone PA** (Vetec, V000187); **acetonitrile** (J.T. Baker, 9012-03); **DMSO, dimethyl sulfoxide** (Sigma, D4540); **ethanol PA** (Merck, 1.00983); **formamide** (Sigma-Aldrich, 295876); **methanol** (Merck, 1.06007); ***N,N*-dimethylacetamide** (Sigma-Aldrich D137510); ***N,N*-dimethylformamide** (Sigma-Aldrich D137510); **NMP, 1-methyl-2-pyrrolidinone** (Sigma-Aldrich 494496); **PBS pH 7.4** (Sigma, P3813-1PAK); **propylene carbonate** (Sigma-Aldrich, 310328); **tetrahydrofurfuryl alcohol** (Sigma-Aldrich T12653), and **water** (Sigma-Aldrich 270733).

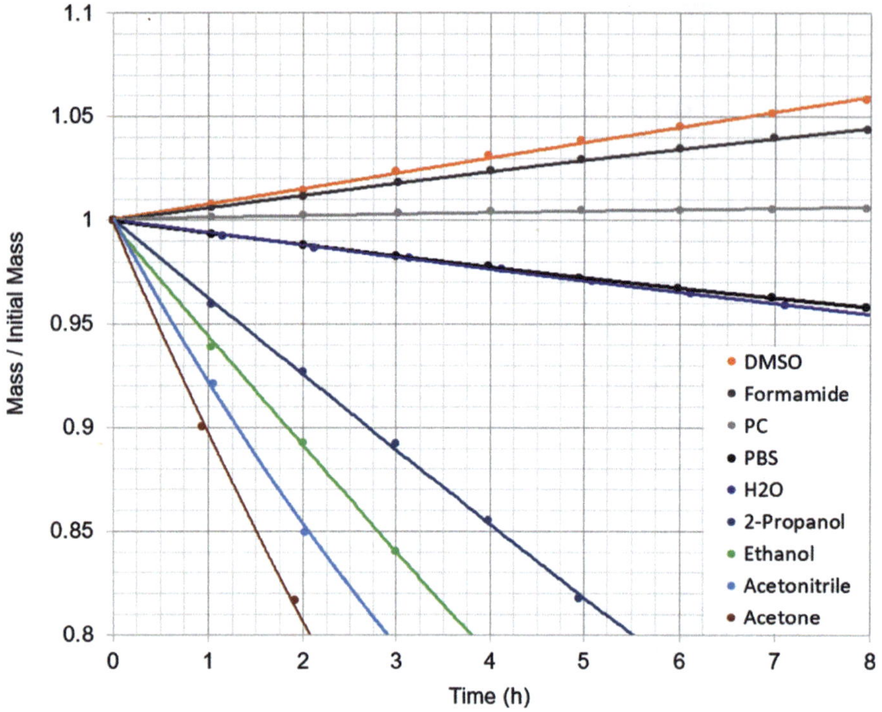

Figure 3.2 Evaporation rate of water-miscible solvents. The conditions of the experiment were as follows: 20 mL of each solvent was placed in identical 50 mL beakers. These beakers were placed at the entrance of a fume hood with a grid made of PMMA to assure identical air velocity profiles over all samples. The fume hood was set at medium velocity, room temperature was fixed at 22°C (using air conditioning), and the relative humidity was kept at 60% with a dehumidifier. (Unpublished results from the author.)

Table 3.4 Table comparing the absolute and relative mass or volume change of the solvents presented in Table 3.3, Figure 3.1, and Figure 3.2 (and details are listed in footnote[1]). (Unpublished results from the author.)

Water Miscible Solvent	Absolute Mass or Volume Change (%/h)	Evaporation Rate Relative to Water
Methanol	−10.00	17.54
Acetone	−9.55	16.75
Tetrahydrofuran (THF)	−7.79	13.67
Acetonitrile	−7.47	13.11
Ethanol	−5.27	9.25
2-Propanol	−3.64	6.39
Water	−0.57	1.00
PBS	−0.53	0.93
Propylene carbonate (PC)	0.08	−0.14
N,N-Dimethylformamide	0.10	−0.18
1-Methyl-2-pyrrolidinone	0.56	−0.98
Formamide	0.57	−1.00
Dimethylsulfoxide (DMSO)	0.75	−1.32

3.3 Fluorogenic Reactions

Fluorogenic reactions, which belong to a wide class of derivatization reactions [4], are characterized by nonfluorescent reagents that react with a target functional group to produce fluorescent derivatives. Fluorescent labeling reactions (Section 3.5), on the other hand, use a pre-existing fluorescent dye (see Chapters 4, 5, and the Appendices) to bind to a target functional group. This book is dedicated to organic small-molecule fluorescent dyes with medium to high brightness in aqueous solutions, that are used for labeling and staining. However, fluorogenic reactions and fluorescent labeling reactions should not be confused with fluorescent probes. Fluorescent probes are engineered dyes that, in general, only fluoresce when they interact with a specific target molecule. They are known as chemical sensors. Another class of probes change their fluorescence characteristics in response to a local physical parameter and, therefore, are called physical sensors. There are books dedicated to these chemical and physical sensing probes, [5] where more information can be found.

The reagents used in fluorogenic derivatization reactions exhibit a very low fluorescence or even no fluorescence at all. A product with a red-shifted absorption band is created in the course of a reaction with a target functional group, producing a new fluorophore which usually exhibits a high molar absorption coefficient and a high fluorescence quantum yield, i.e., high brightness. Such derivatization techniques are very interesting in many fields of investigation and application, as only the target functional groups can be detected after the reaction without the disturbance of any excess reagent. This is very convenient as there is no need to remove any excess reagent from the product of the reactions. As a result, good contrasts are produced in fluorescence microscopy and scanning confocal microscopy. These reactions also produce clean chromatograms from HPLC, clean gel images from gel electrophoresis, and clean electropherograms from capillary electrophoresis and microchip electrophoresis. However, the number of such fluorogenic derivatization reagents is smaller than the number of labeling reagents described in Section 3.5. Nevertheless, the most popular fluorogenic derivatization reagents that result in products with a high brightness in aqueous solutions are described in the next sections.

The chemical structures of the main fluorogenic derivatization reagents are given in this section, as well as the schemes of the reactions. All entities (neutral and ionic species) participating in the derivatization reactions are shown. This gives a better view of the reactions in terms of enthalpy change and entropy change, which can ultimately be used to maximize the speed and completeness of the reactions. However, the intermediate steps, intermediate products, and detailed reaction mechanisms will not be presented as this would extend the text too much and is outside the scope of the present book.

3.3.1 Primary Amines

Primary amines have the largest number of fluorogenic derivatization reagents. Note that primary amines are found in many classes of compounds, such as amino acids, aminated carbohydrates, pharmaceuticals, metabolites, peptides, and proteins. Fluorogenic reactions with primary amines are used for sensitive and quantitative fluorescence detections in microplate readers, flow cytometers, HPLC, capillary electrophoresis, microchip electrophoresis, and imaging techniques which use many types of optical microscopes and nanoscopes.

3.3.1.1 Fluorogenic Reactions of Primary Amines With Homocyclic *o*-Phthaldihaldehydes

The reaction of primary amines with *o*-phthaldihaldehyde (OPA), in the presence of a nucleophile (e.g., sulfhydryl), produces isoindole derivatives within a few seconds. The short reaction time of OPA, combined with its very low cost, makes it a good choice for fluorescence detection in assays, high performance liquid chromatography, microchip electrophoresis, and capillary electrophoresis (pre-capillary, in-capillary, and post-capillary). An excess of a thiol is required for this reaction to be completed in the shortest time, as shown in Figure 3.3. The thiol-containing molecule used can be chosen in order to fine-tune some desired properties of the derivative, such as the stability of the fluorescent derivative and its size (diffusion coefficient), hydrophilicity, charge (mobility, separation efficiency, and time efficiency in electrodriven separations), and complexing ability, to mention but a few. 2-Mercaptoethanol was used as the default thiol co-reagent. More recently, *N*-acetyl-L-cysteine, 2,3,4,6-tetra-O-acetyl-l-thio-[3-*D*-glucopyranose] [6], and *N*-isobutyryl-L-cysteine [7] have started to be used as they produce more stable and stereoselective derivatives. It has been shown that the use of other nucleophiles, such as NCS^-, CN^-, and HSO_3^-, produces derivatives with a lower brightness [8]. When α-ketothiols are used, such as *N*-(2-mercaptopropionyl)glycine, mercaptosuccinic acid, 2-mercaptopropionic acid, or mercaptoacetic acid, then the observed fluorescence is very low. This is a result of the low fluorescence quantum yield of the derivatives and is not due to incomplete reactions or other reasons [9]. Nevertheless, the brightness of some thiol (2-mercaptoethanol, *N*-acetyl-L-cysteine, and *N*-isobutyryl-L-cysteine) derivatives is higher in aqueous solutions than the brightness of other dyes whose longest wavelength excitation band is located in this region ($\lambda_{a,max}$ between 330 and 340 nm).

Figure 3.3 shows the fluorogenic derivatization reaction of OPA with amines and Table 3.5 gives an optimized protocol for this reaction. A similar reaction using 1,2-diacethylbenzene instead of OPA has been observed; however, out of the 18 amino acids tested only the reactions with taurine, histamine, lysine, ornithine, and 2,2-diamino propionic acid produced bright derivatives [10]. These products have an absorption band of $\lambda_{a,max}$ ~355 nm and an emission band of $\lambda_{e,max}$ ~445 nm in aqueous solutions containing some organic modifiers.

The OPA derivatives are considered moderately stable because the indole group slowly changes at room temperature [11, 12]. This can be mitigated using cyclodextrin inclusion complexes [13, 14] and/or a thiol-containing compound that confers superior stability to the fluorescent derivative [15].

The main disadvantages of OPA are its short wavelength of excitation maximum (in the UVA), which makes auto-fluorescence an issue in some applications, and the low number of excitation sources (LEDs and lasers) available in this spectral range. Finally, the derivative is only moderately stable if the correct temperature, pH, and thiol containing co-reagent are used (Table 3.5).

Naphthalene-2,3-dicarboxaldehyde is the second homolog of the *o*-dicarboxaldehyde homocyclic series. This reagent reacts faster with primary amines when the cyanide nucleophile is used instead of the commonly used thiol-containing compounds. The reaction of OPA with primary amines produces an indole with the elimination of two water molecules, while the reaction of NDA with primary amines produces *N*-substituted 1-cyanobenz[f]isoindoles with the elimination of one water molecule and one hydroxyl anion. The latter is still enthalpically very favorable, but the reaction is only complete at room temperature after ~20 min in the presence of at least a 5-fold excess of NDA and cyanide. The resulting *N*-substituted 1-cyanobenz[f]isoindoles are more stable, exhibit larger molar absorption coefficients, and larger fluorescence quantum yields than indoles. The brightness of the derivatives is slightly dependent on the target amine, but they all exhibit a very broad absorption band (with one peak at 420 nm and a second at 440 nm) and a wide emission band (>490 nm), as shown in Figure 2.16 in Chapter 2. Broad absorption bands offer more flexibility as many lasers and LEDs can be used as excitation sources in this spectral range (420–440 nm), and as an added benefit the detectors (photomultiplier tubes, silicon photomultipliers, avalanche photodiodes, and CCDs) commonly used are already sensitive in this fluorescence emission range (above 490 nm). It is important to note that this broad absorption band and its relatively high molar absorption coefficient makes these derivatives eligible for absorption detection applications in chromogenic-like applications; i.e., reactions that develop a color that has an absorbance that is proportional to the concentration of the target molecule or ion.

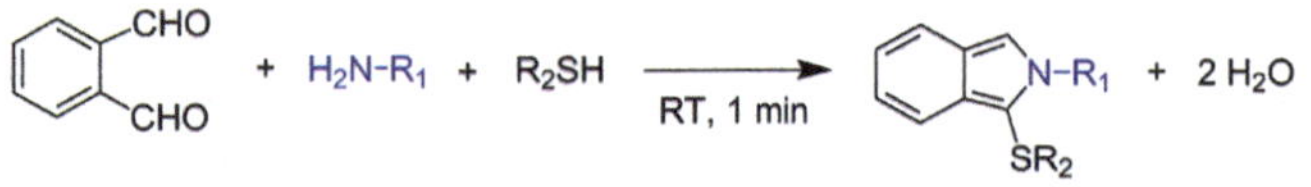

Figure 3.3 Fluorogenic derivatization reaction of primary amines with *o*-phthaldialdehyde (OPA). The product has an absorption band of $\lambda_{a,max}$ ~337 nm and an emission band of $\lambda_{e,max}$ ~450 nm in aqueous solutions containing some organic modifiers.

Table 3.5 A protocol for the fluorogenic derivatization reaction of primary amines with *o*-phthaldialdehyde (OPA). (Unpublished results from the author.)

Add the following reagents to a 1.0 or 1.5 mL microtube with a conical bottom in this order:

- 10 μL of the sample.
- 60 μL of 50 mM borate buffer pH 9 (prepared in de-ionized water).
- 10 μL of 10 mM 2-mercaptoethanol (prepared in de-ionized water with the pH adjusted to 9). Alternatively, a solution of *N*-acetyl-L-cysteine or another thiol containing compound can be used. Mix very well and bring the content (a single drop) to the bottom. It is worth checking that the pH of the mixture is at least ≥8.8.
- 10 μL of 10 mM *o*-phthaldialdehyde (OPA) in acetonitrile, mix well, and bring the content to the bottom.

Keep at room temperature (20–25°C) for 2 min.

Before injection for analysis add 10 μL of de-ionized water or, alternatively, 10 μL of a dye solution prepared with de-ionized water (e.g., fluoresceine) to work as an internal standard.

Total volume ≈ 100 μL.

Note 1: It produces a colorless solution. The presence of turbidity is a signature of insoluble derivatives.

Similarly to OPA, a disadvantage of NDA is that it does not react with secondary amines. Additionally, the resulting fluorophores from NDA reactions are moderately stable and less water-soluble than their OPA derivative counterparts (Figure 3.4). For the reaction of NDA with amino acids, for instance, the reaction medium must always contain an organic modifier (see Figure 3.1 for a list of options) in order to keep the fluorescent derivative soluble. Acetonitrile is usually used for this purpose and Table 3.6 gives a protocol that works well for most of the proteinogenic amino acids that have concentrations up to the micro-molar range in the final reaction solution. Another disadvantage of NDA is that lysine and other diamine targets are strongly subjected to fluorescence quenching (Section 2.7.10.6), leading to NDA derivatives with poor detection limits. Moreover, thiol containing amines may oxidize and produce dimmers held together by disulfide bonds. This happens with the amino acid cysteine, which produces cystine. The resulting two intramolecular nearby fluorophores are also subjected to a strong quenching effect. Worse than this, the nucleophilic thiol group of the thiol containing amines competes with cyanide during the reaction, producing weak or non-fluorescent adducts. Therefore, attention must be paid when using thiol-rich samples, especially when the added excess amount of cyanide is not high enough to compete with the thiols.

Figure 3.5 shows the electropherogram of a mixture of NDA-derivatized amino acids. The derivatization was performed according to the protocol in Table 3.6. The mixture was separated in a capillary electrophoresis instrument using the micellar electrokinetic chromatography mode of separation. 25 mM sodium borate with 50 mM SDS at pH 9 was used as the running buffer. Note that for the derivatization reaction (Table 3.6) a buffer without SDS must be used. The complete electropherogram was plotted to show that fluorogenic derivatization reactions produce clean electropherograms; i.e., the excess reagent does not disturb the baseline, does not occupy space in chromatograms and electropherograms, and does not saturate the detectors. A capillary with a 50 μm internal diameter and a 360 μm outer diameter was used. The small peaks at around 30 min are produced by cystine, lysine (which are subjected to quenching), and other amine containing compounds. An improved separation of this mixture of NDA-derivatized amino acids was recently achieved by Celá et al. [17].

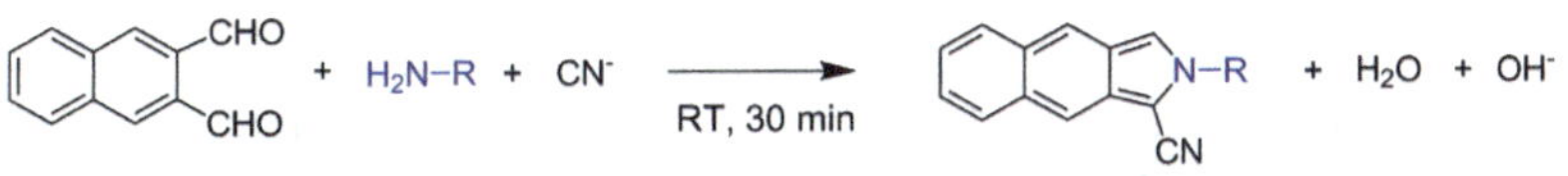

Figure 3.4 Fluorogenic reactions of primary amines with naphthalene-2,3-dicarboxaldehyde (NDA). This derivative has a wide absorption band in the 420–440 nm range and emits a band with $\lambda_{e,max}$ ~500 nm in aqueous solutions containing some organic modifiers (e.g., acetonitrile). The organic modifier must be used because most of the derivatives exhibit a low solubility in pure water.

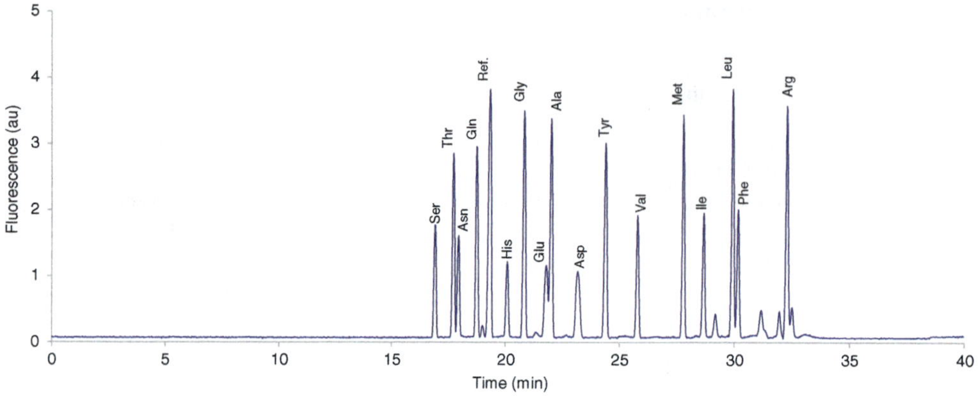

Figure 3.5 Capillary electrophoresis separation of NDA derivatized amino acids. The derivatization was performed according to the protocol in Table 3.6. The total length of the capillary was 67 cm (65 cm to the detector) and the sample was electrokinetically injected by applying 5 kV for 15 s. 25 mM sodium borate with 50 mM SDS at pH 9 was used as the running buffer. Sample components were separated by applying an electrostatic potential difference of 15 kV between the outlet and the inlet (positive voltage was applied at the inlet). The temperature of the capillary was kept at 15°C with a jacket (10 mm silicone tube) containing recirculating mineral oil from a thermal bath. The observed ionic current was 11.2 μA along the 67 cm capillary when 15 kV was applied. Fluorescence was induced by an LED and detection in the 490–550 nm range was performed by a photomultiplier tube. (Unpublished results from the author.)

Table 3.6 A protocol for the fluorogenic derivatization reaction of primary amines with naphthalene-2,3-dicarboxaldehyde (NDA). Clean glassware and ultra-pure de-ionized water should be used for trace analyses. (Unpublished results from the author; Figure 3.5.)

Add the following reagents to a 1.0 or 1.5 mL microtube with a conical bottom in this order:

- 10 μL of the sample.
- 60 μL of 50 mM borate buffer pH 9.2 (in de-ionized water). For more information about the effect of pH, see figure 1 in Montigny et al. [8].
- 10 μL of 20 mM KCN with pH adjusted to 9.2 [in de-ionized water]. A fresh solution should be prepared every week. Add this to the mixture prior to the addition of NDA (see Montigny et al.) [8]). Mix very well and bring the content (a single drop) to the bottom. It is worth checking that the pH of the mixture is at least ≥9.0.
- 10 μL of 10 mM NDA in acetonitrile. Mix and bring the content to the bottom.

Keep at 25°C in a dry bath and in the dark for 30 min.

Before injection for analysis add 10 μL of de-ionized water or, alternatively, 10 μL of a dye solution prepared with de-ionized water (e.g., fluoresceine) to work as an internal standard.

Total volume ≈ 100 μL

Note 1: It produces a greenish-blue bright and translucid solution. A pale yellow or orange opaque solution indicates that one or more of the derivatives (e.g., ammonia/ammonium) is not completely soluble.

Note 2: Cyanide should not be premixed with NDA before the addition of the amine containing sample. If premixed, a benzoine-type condensation reaction can take place, in which NDA molecules couple in the presence of cyanide and methanol or others [16].

Target molecules such as glutathione and γ-glutamylcysteine, which bear both a primary amine and a reduced thiol group, also react with NDA. In this case, the sulfhydryl group acts as the nucleophilic entity; i.e., there is no need to add the cyanide ion. This is shown in Figure 3.6. A similar reaction with glutathione, that used OPA instead of NDA, has been demonstrated by Cohn and Lyle [18]. The fluorescent derivatives exhibited $\lambda_{a,max}$ ~350 nm and an emission band with $\lambda_{e,max}$ ~420 nm in aqueous solutions containing some organic modifiers.

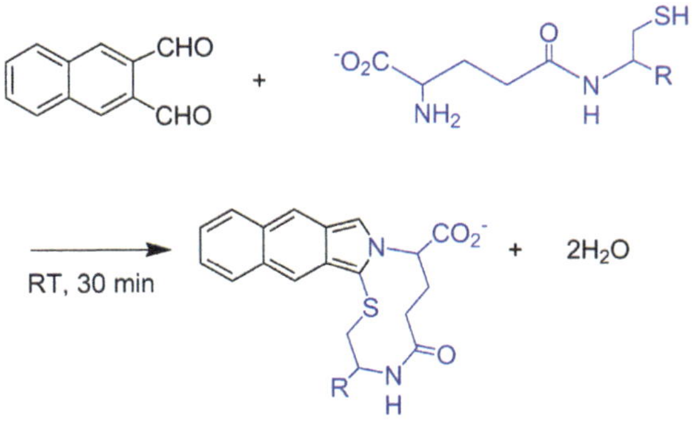

Figure 3.6 Fluorogenic reaction of NDA with glutathione and γ-glutamylcysteine (the only difference between them is the radical R), both containing a primary amine and a thiol group separated from each other by a flexible chain with a length of eight covalent bonds. As in Figure 3.4, this derivative has a large absorption band in the 420–440 nm range and an emission band with $\lambda_{e,max} \approx 500$ nm.

Antracene-2,3-dicarboxaldehyde (ADC) is the next homolog in the homocyclic *o*-dicarboxaldehydes series. It reacts with primary amines in the presence of the cyanide anion, just like NDA (Figure 3.7). However, the products of this reactions are much less soluble in aqueous solutions than the NDA derivatives. Only derivatives of highly hydrophilic primary amines (H_2N-R) are soluble in aqueous solutions. Therefore, the reactions should be conducted in water-miscible organic solvents containing a high content of the organic modifier (~50% V/V or more). For a recent review of OPA, NDA, and ADC, see Celá and Glatz [19].

3.3.1.2 Fluorogenic Reactions of Primary Amines With Heterocyclic *o*-Dicarboxaldehydes

Derivatization reagents based on *o*-dicarboxaldehydes that are positioned over the quinoline scaffold have been developed over the years. One such reagent is 3-benzoyl-2-quinoline-carboxaldehyde (BQCA), which was developed by Prof Novotny's group [20, 21] and applied to the analysis of amino acids (Figure 3.8). 3–2-(furoylquinoline-2-carboxaldehyde), also called 3-(2-furoyl)quinoline-2-carboxaldehyde (FQCA) [22], is a similar structure which contains a furan in place of the pendant phenyl group (Figure 3.8). This is sold under the name ATTO TAG FQ.[2] A third fluorogenic derivatization reagent of this class is 3-(4-carboxybenzoyl)-2-quinoline-carboxaldehyde (CBQCA) [23]. This reagent is similar to the previous two, but it has a benzoic acid in place of the furan (FQCA) or benzene (BQCA) groups. CBQCA-derivatives have the advantage of being more hydrophilic than the derivatives of the two previous reagents; however, lower fluorescence quantum yields and poorer limits of detection have been reported [24].

3.3.1.3 Fluorogenic Reactions of Primary Amines With Other Reagents

Primary amines also react with dicarboxaldehydes that are not *ortho*-dicarboxaldehydes, as long as the spatial geometry still allows for the formation of a nitrogen containing heterocycle. 1-Methoxycarbonylindolizine-3,5-dicarbaldehyde (IDA) (163556-04-5) is one such compound and was synthesized from 2-(1,3-dioxolan-2-yl)pyridine by Oguri et al. [25]. It reacts with primary amines at room temperature within 15 min (Figure 3.9) and produces derivatives that are three times brighter than their OPA derivative counterparts. The derivatives of 15 amino acids,

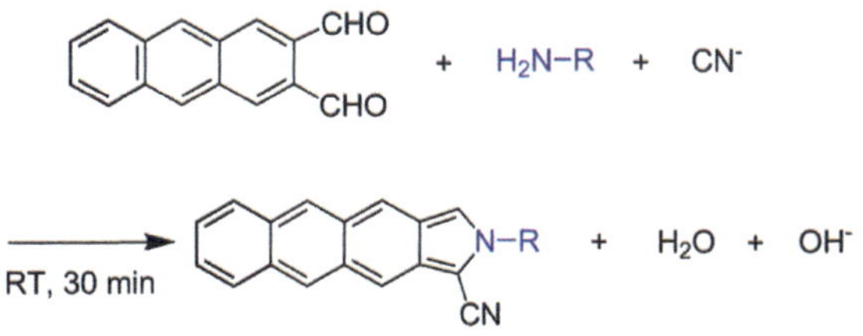

Figure 3.7 Fluorogenic reaction of primary amines with antracene-2,3-dicarboxaldehyde (ADC).

2 A trade mark of Thermo Fisher Scientific (Waltham, MA, EUA).

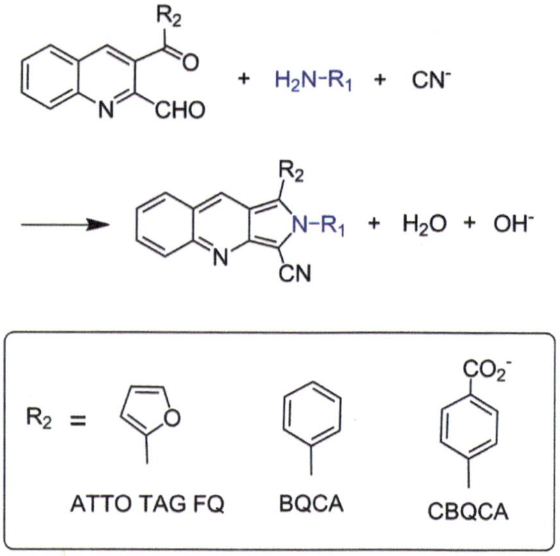

Figure 3.8 Fluorogenic derivatization reaction of heterocyclic *o*-dicarboxaldehydes with primary amines. CBQCA is by far the most used reagent, especially in electrodriven separation techniques. This is due to the extra charge introduced by the derivatization reaction of primary amines.

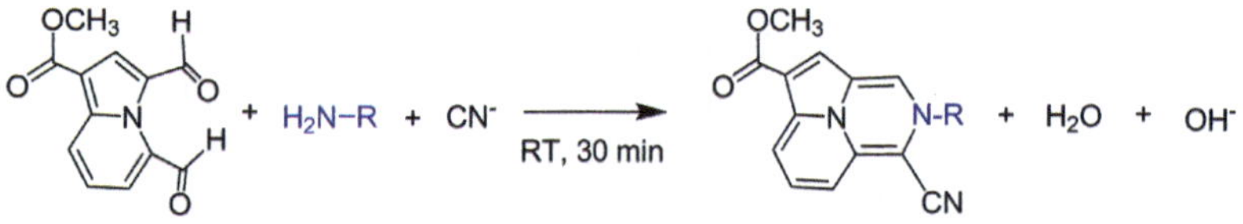

Figure 3.9 Fluorogenic reaction of primary amines with 1-methoxycarbonylindolizine-3,5-dicarbaldehyde (IDA). The IDA-alanine derivative; for instance, shows $\lambda_{a,max}$ = 409 nm (ε_{max} = 2.18 × 10^4 $M^{-1}cm^{-1}$) and an emission band with $\lambda_{e,max}$ ~482 nm in aqueous solutions containing an organic modifier (25–50% v/v ethanol).

as well as that of ammonia, showed good stability over a period of 2 hours. On average, the fluorescence intensity was higher at pH 4 than at pH 6, and higher at pH 6 than at pH 8 [26]. IDA is also suitable for an automated in-capillary derivatization reaction using an injected plug of sample sandwiched between the reagents [27].

Fluorescamine (38183–12-9) is a special case and was introduced to the fluorometric quantitation of primary amines in the 1970s [28]. The same authors proposed a method to convert secondary amines (proline and hydroxyproline) into primary amines, which can then be assayed with the same reagent [29]. Fluorescamine can be considered to be a fluorogenic reagent, not because the excess reagent does not fluoresce, but because it hydrolyzes rapidly to non-fluorescent products [30] (Figure 3.10). Fluorescamine reacts with a wide range of nucleophilic

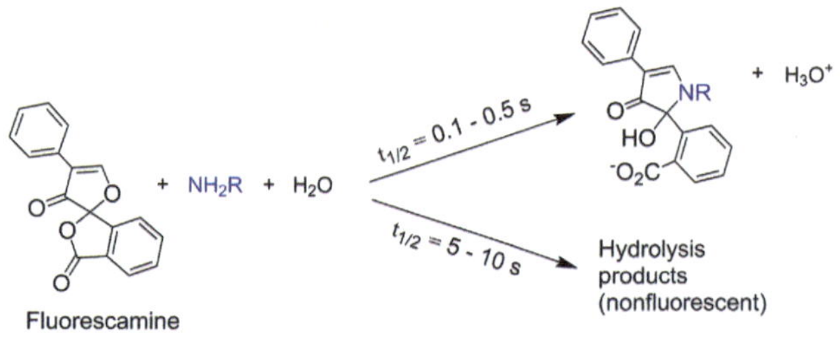

Figure 3.10 Derivatization reaction of primary amines with fluorescamine (4′-phenylspiro[2-benzofuran-3,2′-furan]-1,3′-dione). The excess fluorescamine is hydrolized to non-fluorescent products. In aqueous solution, the amino acid derivatives exhibit $\lambda_{a,max}$ = 385 nm and an emission band with $\lambda_{e,max}$ ~500 nm.

compounds, mainly primary amines and, to a lesser extent, with alcohols and thiols. When present in high concentrations, secondary amines and thiols have a negative effect on the fluorescence intensity of the reaction products of primary amines. Although fluorescamine reactions occur in the pH range of 3 to ~11, neutral pH is the most used. However, the pH should always be higher than the pK_a of the conjugated acid of the primary amine that has the lowest pK_a of the mixture. For a recent review, see Derayea and Samir [31]. Figure 3.11 shows the absorption spectra of some fluorescamine-derivatized amines.

The reaction of fluorescamine with primary amines is fast and the products are more hydrophilic (Figure 3.10) than the homocyclic *o*-dicarboxaldehydes derivatives. Five derivatization protocols were applied to 19 amino acids and were compared, in terms of relative fluorescence intensity, by Bantan-Polak et al. [32]. Table 3.7 shows a similar protocol, but it is adapted for larger sample volumes and a smaller final total volume of 100 μL.

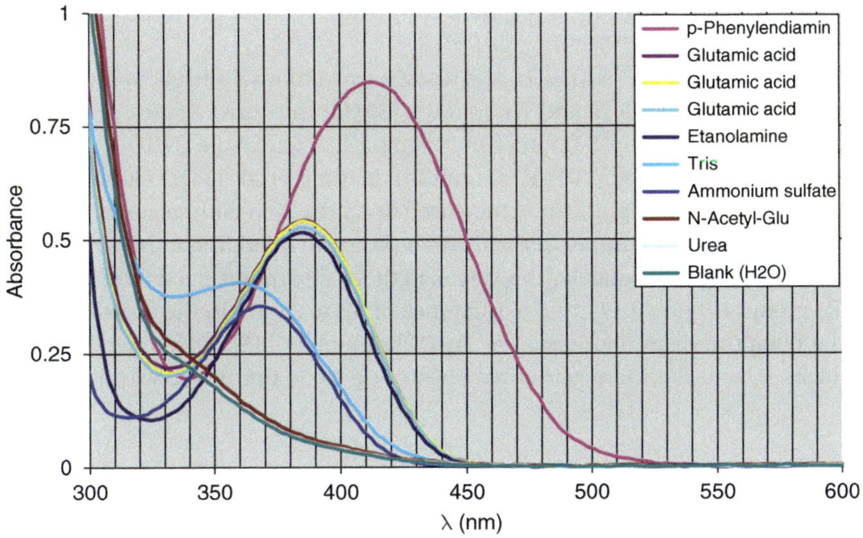

Figure 3.11 Absorption spectra of the products of the derivatization reaction of fluorescamine (according to Table 3.7) with p-phenylendiamin, glutamic acid (triplicate), ethanolamine, Tris, ammonium ion, *N*-acetyl-glutamic acid (which is expected to not react because the amine is blocked), urea (which is an amide and not an amine), and a blank (water). (Unpublished results from the author.)

Table 3.7 Protocol for the fluorogenic derivatization reaction of fluorescamine with primary amines. (Unpublished results from the author.)

Add the following reagents to a 1.0 or 1.5 mL microtube with a conical bottom in this order:

- 10 μL of the sample.
- 60 μL of 0.1 M borate buffer pH 8.5 (prepared in de-ionized water). Mix very well and bring the content (a single drop) to the bottom. It is worth checking that the pH of the mixture is at least ≥8.3.
- 30 μL of 0.1% fluorescamine in acetonitrile (or equivalently 36 mM fluorescamine in acetonitrile). Mix very well and bring the content to the bottom.

Keep in the dark at room temperature (25°C) for 5 min.

The final pH should not be less than 7.5, otherwise a smaller sample volume or a more concentrated borate buffer is needed.

Total volume ≈ 100 μL

Note 1: It produces a bright yellowish translucid solution. A pale yellow or orange opaque solution indicates that one or more of the derivatives is not completely soluble.

Note 2: If necessary, the original sample should be diluted to assure a molar excess of 10 times more fluorescamine than the total molar content of primary amines.

3.3.2 Secondary Amines

The reactions shown in Section 3.3.1 can be summarized as follows: the two hydrogens of the primary amines and the two pendant aldehydes of the derivatizing reagent are used to produce a new aromatic heterocyclic ring that is fused to the original aromatic mono- or polycyclic ring of the reagents. This significantly extends the length of the original π-electron system, creating a very large red-shift in the absorption band of the longest wavelength of absorption maximum. Moreover, the presence of the added heteroatom in the newly created aromatic ring increases the molar absorption coefficient (the nitrogen increases the electric dipole moment) while keeping the rigidity of the final structure. This is important to assure high fluorescence quantum yields. Such desired results are more difficult to obtain when the target molecules are secondary amines. This is based on the fact that there is only one hydrogen atom present and available for reactions in the secondary amine functional group. It is much more difficult to produce large red-shifts, to increase the molar absorption coefficient, and to keep high fluorescence quantum yields in a single atom or radical substitution reaction. Nevertheless, there are some exceptions which are listed in the following paragraphs.

The reagent 7-chloro-4-nitrobenzo-2-oxa-1,3-diazole (NBD-Cl), also known as 7-chloro-4-nitrobenzoxadiazole or 7-choloro-4-nitrobenzofurazan, was introduced by Gosh and Whitehouse [33] for the derivatization of primary and secondary amines. Later, the more reactive version 7-fluoro-4-nitrobenz-2,1,3-oxadiazole (NBD-F) became even more popular. (For a comparison of NBD-Cl and NBD-F, see table 1 in Jung et al. [34]). These NBD-X reagents have a very small native fluorescence that is significantly increased upon reaction with secondary or primary amines. See Figure 3.12 and Table 3.8 for an optimized derivatization protocol. In addition, the absorption maximum is red-shifted a few dozens of nm. Unfortunately, the wavelength of absorption maximum of these derivatives depends on the R_1 and R_2 groups (Figure 3.11), and a difference of up to 23 nm in the wavelength of absorption maximum was observed among the six amines tested by Al-Dirbashi et al. [35]. This difference was found to be entirely due to the radicals R_1 and R_2. Moreover, a red-shift of up to 56 nm in the wavelength of

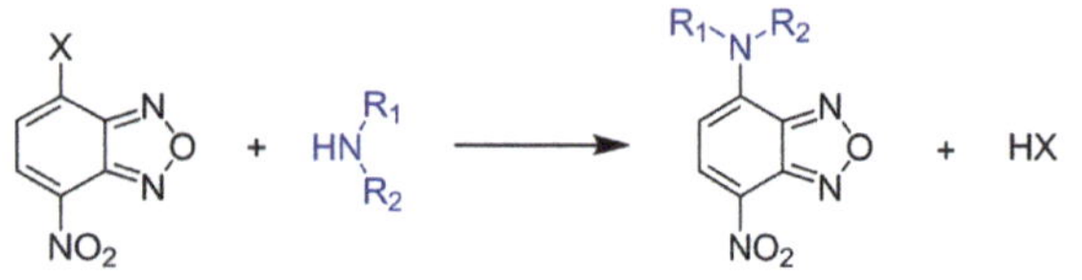

1 X = Cl (NBD-Cl)
2 X = F (NBD-F)

Figure 3.12 Derivatization reaction of primary and secondary amines with NBD-Cl and NBD-F. The wavelengths of absorption and emission maxima depend on the R_1 and R_2 groups, and also on the polarity of the solvent. Fluorescence quantum yield is highly dependent on the polarity of the solvent (it decreases with polarity).

Table 3.8 Protocol for the fluorogenic derivatization reaction of NBD-F with primary and secondary amines.

Add the following to a 1.0 or 1.5 mL screw-capped microtube with a conical bottom in this order:

- 10 μL of the sample.
- 40 μL of 50 mM phosphate buffer pH 7.0 (prepared in de-ionized water). A buffer with a slightly higher pH may be used as long as its pH is higher than the pK_a of the target amines. This is because the regent only reacts with the unprotonated fraction of the amines. Mix very well and bring the content (a single drop) to the bottom. It is worth checking that the pH of the mixture is at least ≥6.8.
- 50 μL of 20 mM NBD-F in ethanol. Mix very well and bring the content to the bottom.

Keep in the dark at 80°C for 10 min.

The final pH should not be less than 7.5, otherwise a smaller sample volume or a more concentrated buffer is needed.

Total volume ≈ 100 μL

absorption maximum was observed when going from a low electric permittivity solvent (n-hexane) to a high electric permittivity solvent (water).

Finally, the fluorescence quantum yields of the derivatives are very sensitive to the polarity of the medium. For instance, the hydroxyproline derivative of NBD-F shows $\lambda_{a,max}$ = 467 nm, $\lambda_{e,max}$ = 519 nm, ε_{max} = 20,000 $M^{-1}cm^{-1}$, and Φ_f = 0.80 in isobutyl methyl ketone. In methanol with 1 mM HCl, the values observed were 470 nm, 526 nm, 22,000 $M^{-1}cm^{-1}$, and 0.24, respectively. Finally, in water with pH 9, the values observed were 500 nm, 536 nm, 31,000 $M^{-1}cm^{-1}$, and 0.010, respectively [36]. From this, the observed brightness ($B_{max} = \varepsilon_{ma} \Phi_f$) in isobutyl methyl ketone is calculated as 16 $mM^{-1}cm^{-1}$ and 0.31 $mM^{-1}cm^{-1}$ in water; an almost 52-fold difference. There is an elegant article with a detailed study of the effect of solvent polarity in modulating the non-radiative decay channels that compete with fluorescence in NBD derivatives and other nitroaromatics in [37]. There is an extensive list of the applications of NBD-Cl as chromogenic and fluorogenic reagents for the determination of amine-containing pharmaceuticals using spectrofluorimetric and spectrophotometric methods in [38].

3.3.3 Thiols

There are well-known and well-stablished fluorescent labeling reactions for thiols (Section 3.4.2). Usually, a pendant carboxylic group of a fluorescent structure is modified to maleimide or iodoacetamide, which is then used to react with thiols under mild conditions. However, there is a lack of non-fluorescent reagents that react with all thiols to produce fluorescent derivatives. There are a few reagents that react with thiols; however, only some thiols produce a fluorescent product upon reaction with these reagents. One example of such a reagent is OPA, as shown in Figure 3.3. An excess of a primary amine must be added to the reaction, and the thiol containing compounds of the samples are expected to react within a few seconds (as shown in Figure 3.3). However, some thiols do not produce a fluorescent derivative with a high brightness. This is related to the R radical to which the thiol group is bound. For instance, radicals with a strong electron donating characteristic, such as $HSCO_2^-$ and $HSCH_2CO_2^-$, do not produce bright derivatives with OPA. However, $HSCH_2CH_2CO_2^-$ and $HSCH_2CH_2OH$ do react with OPA and amines to produce bright fluorescent derivatives [8].

3.3.4 Cyanide

The case of the cyanide ion (CN^-) analyte is similar to the case of thiols discussed in Section 3.3.3. The nucleophile anion of CN^- reacts with NDA in the presence of an excess of primary amines to produce a fluorescent *N*-substituted 1-cyanobenz[f]isoindole, as depicted in Figure 3.4. Alternatively, ADC or IDA can also be used in the presence of an excess of primary amines to react with the cyanide anion to produce fluorescent derivatives, as shown in Figures 3.6 and 3.8, respectively. Finally, ATTO TAG FQ, BQCA, or CBQCA are additional options that can react with cyanide in the presence of an excess of primary amine; however, they are more expensive. It has been shown that other nucleophile reagents (such as thiols, N_3^-, NCS^-, HSO_3^-) may also react, to some extent, with the above-mentioned reagents (NDA, ADC, IDA, TAG FQ, BQCA, or CBQCA) in the presence of an excess of primary amines [8]. However, the vast majority produce unstable derivatives or derivatives with low fluorescence quantum yields. Nonetheless, they are interferents that deserve attention because they compete for the reagents.

3.3.5 α-Dicarbonylic Compounds

Glyoxylic acid is an α-dicarbonyl compound and an important mammalian metabolite. It reacts with 2-aminobenzenthiol (2-ABT) to produce 4H-benzo[b][1,4]thiazin-3-ol (Figure 3.13), a fluorescent bicyclic thiazine (a hidroxythiazine ring fused with a benzene) that is moderately soluble in water. Despite this negative property, it exhibits an absorption band with $\lambda_{a,max}$ = 290 nm and an emission band with $\lambda_{a,max}$ = 390 nm [39]. The reaction occurs in an acidic medium at 90°C, takes no more than 30 min to complete, and the product of the reaction exhibits a much higher brightness than the reagents. By definition, this is a truly fluorogenic derivatization reaction.

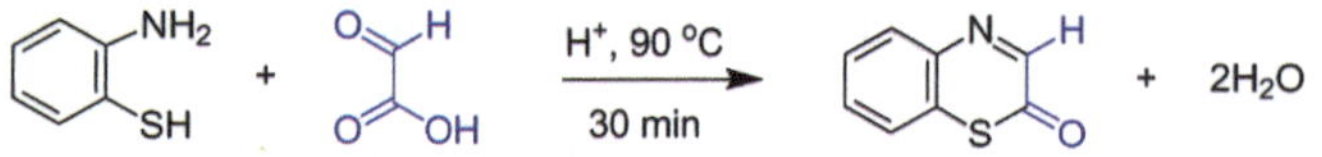

Figure 3.13 Reaction of glyoxylic acid (a α-dicarbonylic compound) with 2-aminobenzenthiol (2-ABT) and the proposed product of the reaction (proposed in a fruitful discussion between the present author and the corresponding author of [39]). The product 4H-benzo[b][1,4]thiazin-3-ol has $\lambda_{a,max}$ = 290 nm and $\lambda_{e,max}$ = 390 nm in a methanol–water mixture (70:30, v/v %).

Glyoxylic acid has been known for its reaction with other amines for a long time. Its reaction with biogenic amines, both in tissues and *in vitro*, allows very sensitive histochemical demonstrations of these amines to be performed [40, 41]. Note that glyoxylic acid also reacts with other reagents to produce fluorescent derivatives, but many of them are not soluble in aqueous solutions. For example, glyoxylic acid reacts with 1-pyreneboronic acid (non-fluorescent) in the presence of an amine (e.g., *N*-methylbutylamine) to produce a fluorescent α-amino acid [42]. However, this product exhibits poor solubility in water, PBS, and any aqueous solution that does not contain organic modifiers, a surfactant, or an inclusion complexing agent.

Chromogenic reactions is a field with many more reaction options. However, the detection limits and resolutions produced by chromogenic reactions in microscopy are poorer than those produced by fluorescence reactions. Additionally, the range of linear responses (linear working range) observed on concentration vs. absorbance graphs is much more limited than the range of linear responses on concentration vs. fluorescence graphs. This field of chromogenic reactions cannot be reviewed here because it is outside of the scope of this book. Nonetheless, it is worth mentioning that every fluorogenic reaction is almost always a chromogenic reaction. Therefore, the reactions presented in this section can be performed by those using absorbance measurement methods based on absorbed and/or transmitted colors.

3.4 Labeling Reactions

Labeling reactions in general are performed using a "label molecule" that has an arm linker with a reactive moiety at the end. This end moiety reacts specifically with a functional group of the target molecules. These reactions are also known as bioconjugation when the target molecules are biomolecules. However, these labeling reactions have a very broad range of applications, not just for fluorescent labeling, where the label is a fluorescent structure, but also for a great variety of other applications. Some examples are the following: i) radioisotope labeling in pharmaceutical sciences. In this case, radioactive isotope-containing structures are covalently bound to specific target molecules. This is used, for example, to produce radiopharmaceuticals for both treatment (radiotherapy, ^{225}Ac labeled antibodies for radioimmunotherapy) and diagnosis (scintigraphy, positron emission tomography); ii) for isotope labeling in mass spectrometry, including new isobaric labels [43]; iii) chemiluminescent and electrochemiluminescent labeling of molecules; and iv) to bind molecules used in chemotherapy onto antibodies for target directed chemotherapy, to mention a few.

Fluorescent labels are dotted with a short arm linker such that the spectral and photophysical characteristics of the fluorescent dye remain unchanged upon reaction. Consequently, aliphatic chains with two to six carbons are commonly used. As for fluorogenic reactions, labeling reactions are sometimes demanding and a source of errors in assays and analytical methods. An ideal labeling derivatization reaction is expected to react rapidly, under mild conditions, be specific for the desired target molecules or functional groups, and resist small changes in the reaction conditions (mixture of solvents used, pH, temperature, and interferents). Moreover, the reaction product (a single fluorescent derivative or labeled molecule) is expected to exhibit a molecular structure with long-term stability and stable spectral and photophysical characteristics. Such almost ideal conditions are only met by a few labeling reagents, as most reactions are still time-consuming and sensitive to the reaction conditions.

Table 3.9 shows a list of 44 common target functional groups of analyte molecules (column A), the respective reactive moiety (functional groups) of the fluorescent labels (column B), and the functional group produced by

Table 3.9 Common functional groups and the respective reactive moieties used to promote labeling. Nucleophilic functional groups are indicated in red and electrophilic functional groups in blue. Column A shows the main target functional groups and a second functional group, that may also react with the reactive moiety of column B under certain conditions, is given in parentheses where applicable.

	A	B	C
	Common functional groups found on the target molecules	**Reactive moiety (functional groups) of the fluorescent labels**	**Functional group produced by the resulting covalent linkage**
1	Alcohols	Carboxylic acids	Esters
		Phosphoramidites	Phosphite esters
		Silyl halides	Silyl ethers
		Sulfonate esters	Ethers
2	Alcohols (and phenols)	Acyl halides	Esters
		Acyl nitriles	Esters
		Alkyl halides	Ethers
		Alkyl sulfonates	Ethers
		Anhydrides	Esters
		Halotriazines	Triazinyl ethers
		Isocyanates	Urethanes
3	Aldehydes	Amines/anilines	Imines
4	Aldehydes or ketones	Hydrazines	Hydrazones
		Hydroxylamines	Oximes
5	Amines	Aryl halides	Aryl amines
6	Amines (and anilines)	Activated esters	Carboxamides
		Acyl azides	Carboxamides
		Acyl halides	Carboxamides
		Acyl nitriles	Carboxamides
		Alkyl halides	Alkyl amines
		Anhydrides	Carboxamides
		Halotriazines	Aminotriazines
		Imido esters	Amidines
		Isocyanates	Ureas
		Isothiocyanates	Thioureas
		Sulfonate esters	Alkyl amines
		Sulfonyl halides	Sulfonamides
7	Carboxylic acids	Alkyl halides	Esters
		Alkyl sulfonates	Esters
		Carbodiimides	N-acylureas or anhydrides
		Diazoalkanes	Esters
		Sulfonate esters	Esters
8	Carboxylic acids	Amines/anilines	Carboxamides
		Hydrazines	Hydrazides

(*Continued*)

Table 3.9 (Continued)

	A	B	C
9	Glycols	Boronates	Boronates esters
10	Phenol (and alcohols)	Sulfonyl halides	Sulfonate esters
11	Thiols	Alkyl halides	Thioethers
		Alkyl sulfonates	Thioethers
		Aryl halides	Thiophenols
		Aziridines	Thioethers
		Epoxides	Thioethers
		Haloacetamides	Thioethers
		Maleimides	Thioethers
		Sulfonate esters	Thioethers

the resulting covalent linkage (column C) from the reaction of A with B. The nucleophilic functional groups are indicated in red and the electrophilic functional groups in blue. Sometimes the analyte possesses a functional group shown in B, and therefore the corresponding reactive moiety of the fluorescent label is given under A. For example, glycols may be used to label boronates and amines may be used to label imido esters.

3.4.1 Covalent Labeling of Amines

A large number of reactive moieties are available to label amine containing biomolecules with fluorescent labels (Figure 3.14). Initially, fluorescein and rhodamine isothiocyanates were the most popular reagents for the

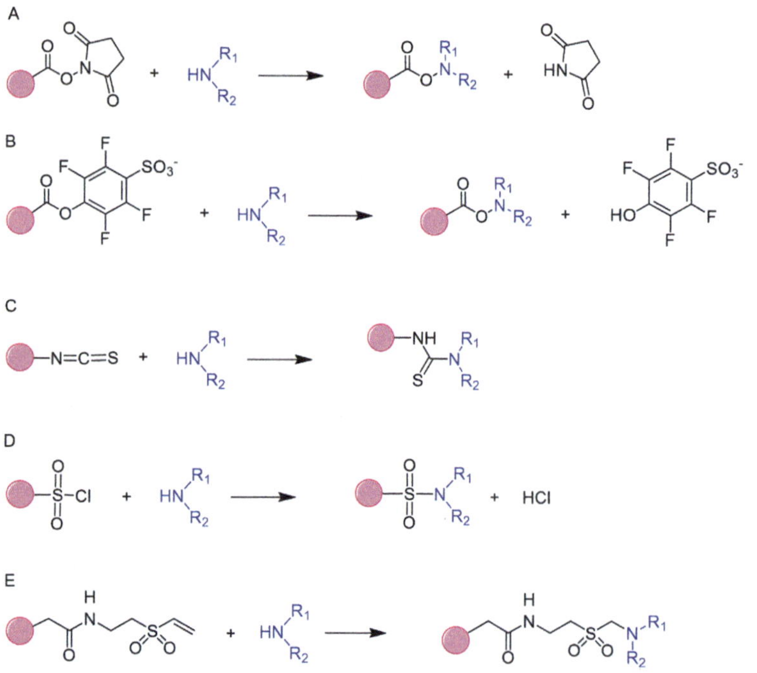

Figure 3.14 Some labeling derivatization reactions of primary and secondary amines with (A) succinimidyl ester, (B) 4-sulfotetrafluorophenyl (STP) ester, (C) isothiocyanate, (D) sulfonilchloride, and (E) vinylsulfone.

Table 3.10 Protocol for the labeling reaction of amines with succinimidyl ester.

Add the following to a 1.0 or 1.5 mL screw-capped microtube with a conical bottom in this order:

- 5–50 μL of the sample (depending on the total amine concentration of the sample) with pH adjusted to 9.
- 265–220 μL of 0.1 M carbonate buffer pH 8.3.
- 30 μL of 5 mM of the succinimidyl ester reagent prepared in anhydrous *N*,*N*-dimethylformamide. Mix it immediately and bring the content to the bottom.

Keep in the dark for 1 h at room temperature.

Total volume ≈ 300 μL

fluorescent labeling of proteins, peptides, amino acids, metabolites, amino sugars, and pharmaceutical compounds. However, in recent years, succinimidyl ester has overtaken isothiocyanate as it has the advantage of producing a more stable carboxamide derivative than the thiourea derivative (Table 3.10 gives an optimized derivatization protocol). It is worth mentioning that stock solutions of *N*-succinimidyl esters must be prepared in a water-free (but water miscible) solvent, as they are labile to water. Examples of such water miscible solvents include *N*,*N*-dimethylformamide and DMSO (but note that DMSO is very hygroscopic), among others (Figure 3.1). The 4-sulfotetrafluorophenyl (STP) ester is also an interesting alternative as the fluorescent labels carrying this reactive moiety are more soluble in aqueous solutions than the *N*-succinimidyl ester counterpart [44].

3.4.2 Covalent Labeling of Thiols

There are many reactive moieties used to bind fluorescent labels to thiols. The two most popular are the maleimide functional group and the iodoacetamide functional group. Maleimide is the most used, while both produce stable thioethers. Their reactions are depicted in Figures 3.15A and B, respectively. Table 3.11 gives an optimized labeling potocol for the labeling of thiols with maleimide. There are reviews covering the applications of thiol derivatization, both *in vitro* [45] and *in vivo* [46, 47]. The aziridine moiety is less used (Figure 3.15C). It was shown that dansylaziridine, for example, reacts selectively with thiols at pH 8.2 at 60°C in 1 h [48].

3.4.3 Covalent Labeling of Carboxylic Acids

The bromo-methylphenyl reactive moiety is widely used for the labeling (derivatization) of carboxylic acids in organic hydrophilic media and is still being improved [49]. (Table 3.12 gives an optimized protocol). One such good

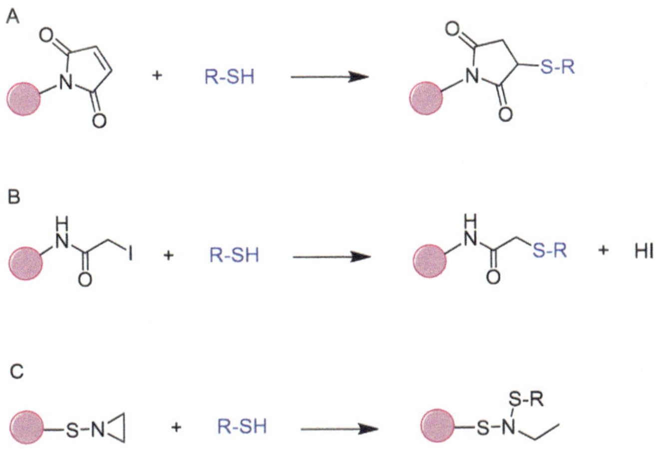

Figure 3.15 Three common labeling (or derivatization) reactions of thiols: (A) iodoacetamide, (B) maleimide, and (C) aziridine reactions.

Table 3.11 Protocol for the labeling reaction of thiols with maleimide.

Add the following to a 1.0 or 1.5 mL screw-capped microtube with a conical bottom in this order: • 50 μL of the sulfhydryl-containing sample (with an expected concentration of less than 10 mg/ml) with pH adjusted to ~7.2. • 40 μL of 0.1 M sodium phosphate buffer pH 7.2. • 10 μL of the 10 mM maleimide reagent prepared in anhydrous *N,N*-dimethylformamide (assure a 25-fold molar excess of maleimide in the reaction mixture). Mix it immediately. Keep in the dark for 2 h at room temperature. (Alternatively, it may be kept at 0–4°C, but in this case it needs to be left in the dark for 8 h.) Total volume ≈200 μL.

Table 3.12 Protocol for the labeling reaction of carboxylic acids with alkyl halides such as bromo-methylphenyl (adapted from Rodrigues et al. [49]).

Add the following to a 1.0 or 1.5 mL screw-capped microtube with a conical bottom in this order: • 5.0 mg (±0.2 mg) of anhydrous $KHCO_3$. • 171 μL of anhydrous propylene carbonate. • 15 μL of the anhydrous solution of the carboxylic acids (e.g. in methanol). • 27 μL of 10 mM 18-crown-6 in acetonitrile. • 27 μL of 5 mM of alkyl halide in acetonitrile (e.g., MPAC-Br). • A 6 mm × 2.8 mm magnetic bar. Close the tube and insert it into a 25 mL beaker containing a second magnetic stirrer and 20 mL of mineral oil kept at 75°C with a hot plate. Keep it at 75°C for 40 min, stirring continuously and vigorously. Remove the magnetic bar and centrifugate the mixture at 1000 g for 10 min. Two phases are produced: a fluorescent supernatant and a white pellet (excess $KHCO_3$). The labeled products will be found on the supernatant.

reagent is 3-[4-(bromomethyl)phenyl]-7-(diethylamino)coumarin (MPAC-Br) (Figure 3.16A). The derivatives are bright in aqueous solutions, stable for at least 24 h, and can be analyzed using HPLC and capillary electrophoresis, among other techniques. Another less used reaction is the condensation reaction with amines (Figure 3.16B).

3.4.4 Covalent Labeling of Alcohols

Most alcohols have a weak nucleophilic character and are therefore more difficult to label specifically in the presence of other electrophilic functional groups. One reactive moiety used is isocyanate, which should not be confused with isothiocyanate. However, isocyanate is readily hydrolyzed to amine with the evolution of CO_2 in the presence of water ($R\text{-}NCO + H_2O \rightarrow R\text{-}NH_2 + CO_2$). Therefore, a better strategy is to prepare a stock solution of the fluorophore with an acyl azide reactive moiety. This acyl azide can then be conveniently converted into isocyanate by just heating up the mixture of the target alcohols with isocyanate (Figure 3.17). Note that isocyanate also reacts with amines and to a less degree with phenols. Moreover, there should be no alcohol and amine functional groups present on the fluorescent structure to avoid polymerization. But if they are present in the structure of the chosen fluorescent label or stain, then they should be blocked first (alcohols to ethers and amines to alkyl amines) (see *The Molecular Probes Handbook* [50], 19).

3.4.5 Covalent Labeling of Reducing Saccharides

Monosaccharides with four, five, or six carbons have a great diversity of stereoisomers as they contain up to two, three, and four asymmetric carbons, respectively. Moreover, they form oligomers and polymers that have a great

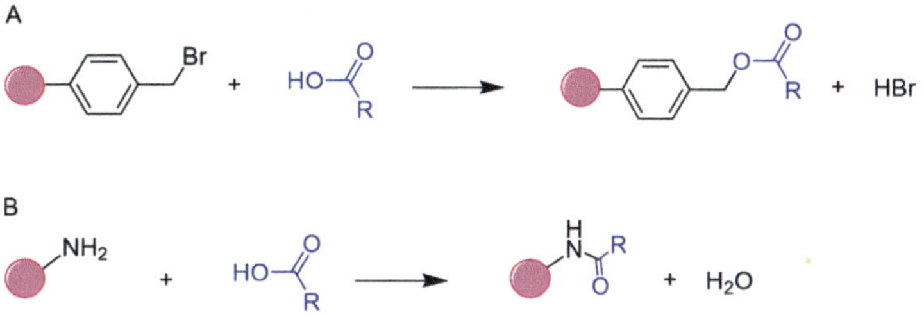

Figure 3.16 Fluorescent labeling reactions of carboxylic acids: (A) using a bromomethylphenyl moiety; and (B) a condensation reaction with amines.

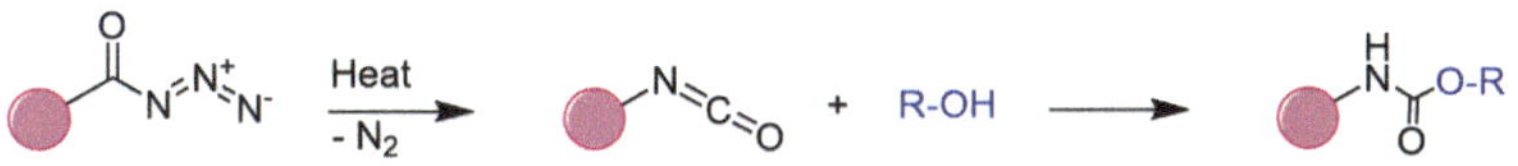

Figure 3.17 Fluorescent labeling reactions of alcohols with isocyanate which is produced at the instant of labeling through the conversion of an acyl azide to isocyanate by heating.

diversity of sequences (including different positional isomers) and ligation forms (alpha or beta). Such diversity produces one of the richest classes of compounds. Imaging or separation and analysis is made even more complex as most of these mono-, di-, and oligo-accharides only absorb radiation in the UVC range of the electromagnetic spectrum.

Labeling reducing saccharides with 8-aminonaphthalene-1,3,6-trisulfonic acid (ANTS) or 8-aminopyrene-1,3,6-trisulfonic acid (APTS) are widely used procedures [51] that confer a high fluorescence and charge to the resulting derivatives. For a detailed review on the separation and analysis of glycans by capillary electrophoresis, see Lu et al. [52]. Derivatives with a high "net charge-to-friction coefficient" ratio can be separated with high separation efficiencies and time efficiencies in electrodriven separation techniques such as capillary electrophoresis [53]. Another labeling reagent that is highly reactive to reducing saccharides and that confers an even higher electrophoretic mobility than APTS is found among the many acridine-triphosphates introduced by Fomin et al. [54]. The reagent is 3-((2-amino-7-(*N,N*-bis(2-((hydroxyoxidophosphoryl)oxy)ethyl)sulfamoyl)acridin-9-yl)amino) propyl hydrogen phosphate, hereby termed "acridine-triphosphate F-H 2" (this short name was created using the first letter of both first and last author names and the numbering system used in the original publication). The labeling reaction of glucose with the fluorescent reagent acridine-triphosphate F-H 2 (reductive amination) is shown in Figure 3.18 and the reaction protocol is given in Table 3.13.

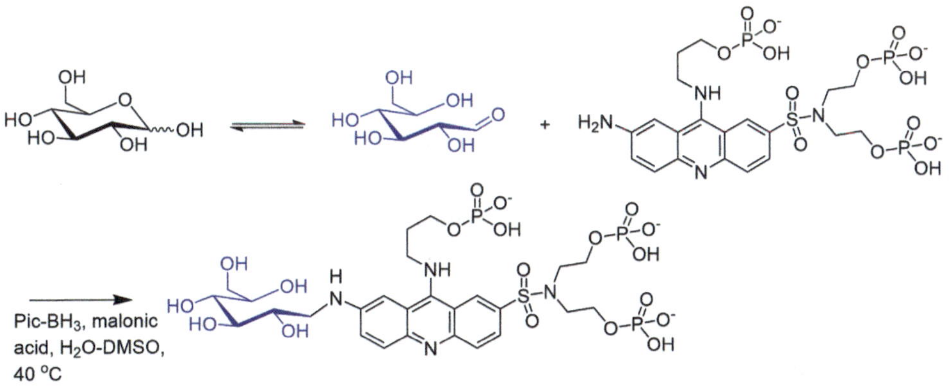

Figure 3.18 Fluorescent labeling reactions of reducing saccharides with acridine-triphosphate F-H 2. This derivative exhibits $\lambda_{a,max}$ = 466 nm, $\lambda_{e,max}$ = 626 nm, Δ_S = 160 nm, ε_{max} = 5.5 mM^{-1}cm^{-1}, Φ_f = 0.02, and B_{max} = 0.1 mM^{-1}cm^{-1} (in water).

The above-mentioned labeling reagents (ANTS, APTS, and acridine-triphosphate F-H 2) are all polyanionic labels. They are not the ideal label in at least two cases: first, when the analytes already have a positive net charge. In this case, the sum of the net charge of the labels plus the charge of the analyte may result in analytes with a low or null mobility in electrodriven techniques. Even worse, mixing polycationic analytes with polyanionic labels leads to severe molecular aggregation. Second, the hyphenated LC-MS and CE-MS techniques are largely used in laboratory protocols as the second step of analyte identification and quantification. However, anionic and polyanionic analytes are not suitable for these procedures, while cationic analytes are.

Two poly-cationic labels based on 8-aminopyrene were recently developed and are being utilized for the labeling, separation, and MS detection of reducing mono-, di-, and oligo-saccharides. They are termed APTMP (4,4′,4″-(8-aminopyrene-1,3,6-trisulfonyl)tris(1-methylpiperazine)) and APTDP (4,4′,4″-(8-aminopyrene-1,3,6-trisulfonyl)tris(1,1-dimethylpiperazinium) triiodide) (structures are shown in Figure 3.19). APTMP is a sulfonamide derivative of APTS containing three tertiary amines. APTDP was synthesized by the further methylation of APTMP to produce the three permanently charged quaternary ammonium moieties. The labeling protocol shown in Table 3.13 can be used for the labeling reaction of reductive mono-, di-, and oligo-saccharides with APTMP and APTDP; however, the long (probably overnight) incubation time necessary to achieve a satisfactory labeling degree should be noted.

3.4.6 Others

Dozens of other labeling and bioconjugation techniques already exist [55]. These include the biotin-streptavidin, phalloidin, and alkyne systems. Moreover, new techniques are continuously being developed [56].

Many label manufacturers provide labeling protocols and applications notes of their products. Moreover, there are well-written and well-illustrated manuals explaining the many applications of ligation, bioconjugation, and cross-linking technologies [57].

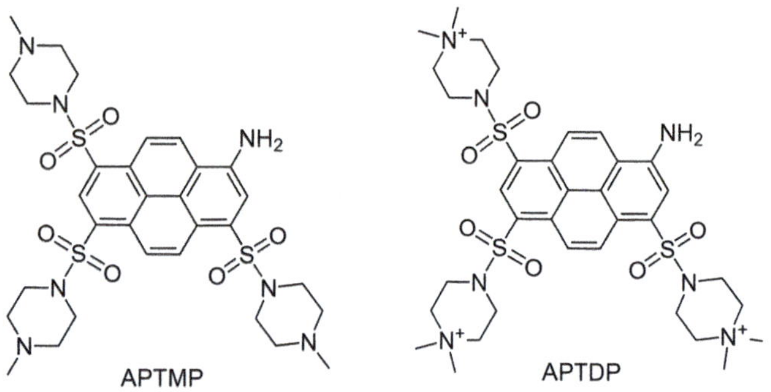

Figure 3.19 Molecular structures of the polycationic labels APTDP (4,4′,4″-(8-aminopyrene-1,3,6-trisulfonyl)tris(1,1-dimethylpiperazinium) triiodide and APTMP (4,4′,4″-(8-aminopyrene-1,3,6-trisulfonyl) tris(1-methylpiperazine)).

Table 3.13 Protocol for the labeling of reducing mono-, di-, and oligo-saccharides with acridine-triphosphate F-H 2 (adapted from Fomin et al. [54]).

Add the following to a 1.0 or 1.5 mL screw-capped microtube with a conical bottom in this order:

- acridine-triphosphate F-H 2 (2 μL of 0.1 M solution in water).
- saccharide (2 μL, less than 0.02 M).
- malonic acid (2 μL, 1 M solution in DMSO).

Remove the solvents with a freeze-dryer (pressure <0.2 mbar, temp. of the cooling coil −80°C).

- add 2-picoline-borane complex (4 μL, 0.5 M solution in DMSO).

Mix immediately by vortexing for 10 s and then incubate with shaking (400–600 rpm) at 40°C for 6 h.

3.5 Immunofluorescence

Immunofluorescence, including the variants immunohistofluorescence and immunocytofluorescence, is one of the most common and valuable techniques used to localize proteins within cells and tissues at a quantitative level. It is used to decorate biological targets with a high density of fluorophores by utilizing the aid of antibodies. The samples are then examined using fluorescence microscopy, confocal scanning microscopy, optical nanoscopy, cell sorters, flow cytometers, and many other techniques used in standard research and clinical diagnoses. Nonetheless, sample preparation is demanding and technically challenging due to the many critical steps involved, such as sample fixation, sample permeabilization, antibody incubation, fluid exchange, and stability of the specimens. A review outlining ten methods to improve sample processing at different immunofluorescence stages was recently published [58].

Immunofluorescence is largely used in fluorescence nanoscopy where resolutions below the 10 nm range are already common. Immunoglobulin G antibodies are medium-sized proteins with molecular weights of ~150 kDa and are widely used as stains. However, they have a large size of approximately ~7 nm. This results in a total distance of 10–15 nm or more between the target protein and the fluorophore [59, 60], leading to an increase in the localization uncertainty because the orientation of the primary/secondary antibody conjugation is impossible to control. Moreover, the steric hindrance produced by the combination of primary/secondary antibodies also limits the achievable labeling density. All optical nanoscopic techniques work with fluorescence detection, therefore small-molecule dyes with a high brightness and photorobustness in aqueous solutions are highly desired.

References

1 Dulbecco, R. and Vogt, M. (1954). Plaque formation and isolation of pure lines with poliomyelitis viruses. *J. Exp. Med.* 99: 167–182. doi: 10.1084/jem.99.2.167.

2 Pitt, M.J. (2010). Chemical safety: dangers of diglyme. cen.acs.org (accessed 1 September 2021).

3 Bellussi, G., Bohnet, M., Bus, J. et al. ed. (2011). *Ullmann's Encyclopedia of Industrial Chemistry*, 7e. Hoboken: Wiley-VCH.

4 Yamaguchi, M. and Ishida, J. (1999). *Modern Derivatization Methods for Separation Sciences,* chapter 3) (ed. T. Toyo'oka). Chichester, UK: Wiley.

5 Valeur, B. and Berberan-Santos, M.N. (2012). *Molecular Fluorescence: Principles and Applications*, 2e. Hoboken: Wiley.

6 Tivesten, A., Lundqvist, A., and Folestad, S. (1997). Selective chiral determination of aspartic and glutamic acid in biological samples by capillary electrophoresis. *Chromatographia* 44: 623–633.

7 Kühnreich, R. and Holzgrabe, U. (2016). Indirect enantioseparation of amino acids by CE using automated in-capillary derivatization with ortho-phthalaldehyde and *N*-acetyl-l-cysteine. *Chromatographia* 79: 1013–1022.

8 Montigny, P., Stobaugh, J.F., Givens, R.S. et al. (1987). Naphthalene-2,3-dicarboxaldehyde/cyanide ion: a rationally designed fluorogenic reagent for primary amines. *Anal. Chem.* 59: 1096–1101.

9 Sano, A. and Nakamura, H. (2001). Fluorescence quenching properties and chemiluminescence responses of α-ketothiols derivatized with *o*-phthalaldehyde and primary amino compounds. *Luminescence* 16: 25–28.

10 Roth, M. (1971). Fluorescence reaction for amino acid. *Anal. Chem.* 43: 880–882.

11 Jacobs, W.A., Leburg, M.W., and Madaj, E.J. (1986). Stability of o-phthalaldehyde-derived isoindoles. *Anal. Biochem.* 156: 334–340.

12 van Eijk, H.M.H., Rooyakkers, D.R., and Deutz, N.E.P. (1996). Automated determination of polyamines by high-performance liquid chromatography with simple sample preparation. *J. Chromatogr. A* 730: 115–120.

13 Male, K.B. and Luong, J.H.T. (2001). Derivatization, stabilization and detection of biogenic amines by cyclodextrin-modified capillary electrophoresis-laser-induced fluorescence detection. *J. Chromatogr. A* 926: 309–317.

14 Wagner, B.D. and McManus, G.J. (2003). Enhancement of the fluorescence and stability of o-phthalaldehyde-derived isoindoles of amino acids using hydroxypropyl-β-cyclodextrin. *Anal. Biochem.* 317: 233–239.

15 Allison, L.A., Mayer, G.S., and Shoup, R.E. (1984). *o*-phthalaldehyde derivatives of amines for high-speed liquid chromatography/electrochemistry. *Anal. Chem.* 56: 1089–1096.

16 McGill, C.M., Swearingen, K.E., Drew, K.L.et al. (2005). Reaction of naphthalene-2,3-dicarbaldehyde with cyanide: a unique oxidative condensation product. *J. Heterocycl. Chem.* 42: 475–481.

17 Celá, A., Mádr, A., Ješeta, M. et al. (2018). Study of metabolic activity of human embryos focused on amino acids by capillary electrophoresis with light-emitting diode-induced fluorescence detection. *Electrophoresis* 39 (23): 3040–3048. doi: 10.1002/elps.201800265.

18 Cohn, V.H. and Lyli, J. (1966). A fluorometric assay for glutathione. *Anal. Biochem.* 14: 434–440.

19 Celá, A. and Glatz, Z. (2020). Homocyclic *o*-dicarboxaldehydes: derivatization reagents for sensitive analysis of amino acids and related compounds by capillary and microchip electrophoresis. *Electrophoresis* 41: 1851–1869.

20 Beale, S.C., Hsieh, Y.Z., Savage, J.C. et al. (1989). 3-benzoyl-2-quinolinecarboxaldehyde: a novel fluorogenic reagent for the high-sensitivity chromatographic analysis of primary amines. *Talanta* 36: 321–325.

21 Beale, S.C., Savage, J.C., Wiesler, D. et al. (1988). Fluorescence reagents for high-sensitivity chromatographic measurements of primary amines. *Anal. Chem.* 60: 1765–1769.

22 Beale, S.C., Hsieh, Y.Z., Wiesler, D., and Novotny, M. (1990). Application of 3-(2-furoyl)quinoline-2-carbaldehyde as a fluorogenic reagent for the analysis of primary amines by liquid chromatography with laser-induced fluorescence detection. *J. Chromatogr. A* 499: 579–587.

23 Liu, J.P., Hsieh, Y.Z., Wiesler, D., and Novotny, M. (1991). Design of 3-(4-Carboxybenzoyl)-2-quinolinecarboxaldehyde as a reagent for ultrasensitive determination of primary amines by capillary electrophoresis using laser fluorescence detection. *Anal. Chem.* 63: 408–412.

24 Ta, H.Y., Perquis, L., Sarazin, C. et al. (2021). 3-(4-Carboxybenzoyl)quinoline-2-carboxaldehyde labeling for direct analysis of amino acids in plasma is not suitable for simultaneous quantification of tryptophan, tyrosine, valine, and isoleucine by CE/fluorescence. *Electrophoresis* 42: 1108–1114.

25 Oguri, S., Uchida, C., Miyake, Y. et al. (1995). 1-Methoxycarbonylindolizine-3,5-dicarbaldehyde as a derivatization reagent for amino compounds in high-performance capillary electrophoresis. *Analyst* 120: 63–68.

26 Oguri, S., Uchida, C., Mishina, M. et al. (1996a). Determination of amino acids by pre-column fluorescence derivatization with 1-methoxycarbonylindolizine-3,5-dicarbaldehyde. *J. Chromatogr. A* 724: 169–177.

27 Oguri, S., Fujiyoshi, T., and Miki, Y. (1996b). In-capillary derivatization with 1-methoxycarbonylindolizine-3,5-dicarbaldehyde for high-performance capillary electrophoresis. *Analyst* 121: 1683–1688.

28 Weigele, M., DeBernardo, S., Tengi, J.P., and Leimgruber, W. (1972). A novel reagent for the fluorimetric assay of primary amine. *J. Am. Chem. Soc.* 94: 5927–5928.

29 Weigele, M., DeBernardo, S., and Leimgruber, W. (1973). Fluorometric assay of secondary amines. *Biochem. Biophys. Res. Commun.* 50: 352–356.

30 Udenfriend, S., Stein, S., Böhlen, P. et al. (1972). Fluorescamine: a reagent for assay of amino acids, peptides, proteins, and primary amines in the picomole range. *Science* 178: 871–872.

31 Derayea, S.M. and Samir, E. (2020). A review on the use of fluorescamine as versatile and convenient analytical probe. *Microchem. J.* 156: 104835. doi: 10.1016/j.microc.2020.104835.

32 Bantan-Polak, T., Kassai, M., and Grant, K.B. (2001). A comparison of fluorescamine and naphthalene-2,3-dicarboxaldehyde fluorogenic reagents for microplate-based detection of amino acids. *Anal. Biochem.* 297 (2): 128–136.

33 Ghosh, P.B. and Whitehouse, M.W. (1968). 7-Chloro-4-nitrobenzo-2-oxa-1,3-diazole: a new fluorigenic reagent for amino acids and other amines. *Biochem. J.* 108 (1): 155–156. doi: 10.1042/bj1080155.

34 Jung, E., Dong, T.A., and Cai, X. (2011). Improved synthesis of 4-amino-7-nitrobenz-2,1,3-oxadiazoles using NBD fluoride (NBD-F). *Tetrahedron Lett.* 52: 2533–2535. doi: 10.1016/j.tetlet.2011.02.111.

35 Al-Dirbashi, O., Kuroda, N., and Nakashima, K. (1998). Characterization of the fluorescence properties of 4-Cuoro-7-nitrobenzo-2-oxa-1,3-diazole derivatives of some primary and secondary sympathomimetic amines. *Anal. Chim. Acta* 365: 169–176.

36 Ahnoff, M., Grundevlk, I., Arfwidsson, A. et al. (1981). Derivatization with 4-Chloro-7-nitrobenzofurazan for liquid chromatographic determination of hydroxyproline in collagen hydrolysate. *Anal. Chem.* 53: 485–489.

37 Poronik, Y.M., Sadowski, B., Szychta, K., et al. (2022). Revisiting the non-fluorescence of nitroaromatics: presumption versus reality. *J. Mater. Chem. C* 10 (8): 2870–2904.

38 Elbashir, A.A., Suliman, F.E.O., and Aboul-Enein, H.Y. (2011). The application of 7-Chloro-4-nitrobenzoxadiazole (NBD-Cl) for the analysis of pharmaceutical-bearing amine group using spectrophotometry and spectrofluorimetry techniques. *Appl. Spectrosc. Rev.* 46: 222–241. doi: 10.1080/05704928.2011.557121.

39 Alia, M.F.B., Kishikawa, N., and Naotaka Kuroda, N. (2020). Development of HPLC method for estimation of glyoxylic acid after pre-column fluorescence derivatization approach based on thiazine derivative formation: a new application in healthy and cardiovascular patients' sera. *J. Chromatogr. B* 1143: 122054. doi: 10.1016/j.jchromb.2020.122054.

40 Guidry, G. (1999). A method for counterstaining tissues in conjunction with the glyoxylic acid condensation reaction for detection of biogenic amines. *J. Histochem. Cytochem.* 47: 261–264.

41 Svensson, L.-Å., Björklund, and Lindvall, O. (1975). Studies of the fluorophore forming reactions of various catecholamines and tetrahydroisoquinolines with glyoxylic acid. *Acta Chem. Scand. B* 29: 341–348.

42 Chihara, K., Kishikawa, N., Ohyama, K. et al. (2012). Determination of glyoxylic acid in urine by liquid chromatography with fluorescence detection, using a novel derivatization procedure based on the Petasis reaction. *Anal. Bioanal. Chem.* 403: 2765–2770. doi: 10.1007/s00216-012-6036-1.

43 Bachor, R., Waliczek, M., Stefanowicz, P., and Szewczuk, Z. (2019). Trends in the design of new isobaric labeling reagents for quantitative proteomics. *Molecules* 24: 701. doi: 10.3390/molecules24040701.

44 Gee, K.R., Eric, A., Archer, E.A., and Kang, H.C. (1999). 4-Sulfotetrafluorophenyl (STP) Esters: new water-soluble amine-reactive reagents for labeling biomolecules. *Tetrahedron Lett.* 40: 1471–1474.

45 Shimada, K. and Mitamura, K. (1994). Derivatization of thiol-containing compounds. *J. Chromatogr. B* 659: 227–241.

46 Álvarez, A.I. and Ruiz, A.M. (2011). Thiol redox proteomics seen with fluorescent eyes: the detection of cysteine oxidative modifications by fluorescence derivatization and 2-DE. *J. Proteom.* 75: 329–338.

47 Ravasco, J.M.J.M., Faustino, H., Trindade, A., and Gois, P.M.P. (2018). Bioconjugation with maleimides: a useful tool for chemical biology. *Chem. Eur. J.* 25: 43–59. doi.org/10.1002/chem.201803174.

48 Lankmayr, E.P., Budna, K.W., Mfiller, K., and Nachtmann, F. (1979). Bestimmung von Thiolverbindungen durch selektive Fluorescenzderivatisierung und HPLC. *Fresenius Z. Anal. Chem.* 295: 371–374.

49 Rodrigues, C.E., Tonial, M., Schäfer, L. et al. (2018). Performance of 3-[4-(bromomethyl)phenyl]-7-(diethylamino) coumarin as a derivatization reagent for the analysis of medium and long chain fatty acids using HPLC with LIF detection. *J. Chromatogr. B* 1100–1101: 50–57. doi: 10.1016/j.jchromb.2018.09.043.

50 Molecular Probes (2010). *The Molecular Probes Handbook – A Guide to Fluorescent Probes and Fluorescent Labeling Technologies*, 11e. Oregon: Molecular Probes – Invitrogen.

51 Szabo, Z., Guttman, A., Rejtar, T., and Karger, B.L. (2010). Improved sample preparation method for glycan analysis of glycoproteins by CE-LIF and CE-MS. *Electrophoresis* 31 (8): 1389–1395. doi: 10.1002/elps.201000037.

52 Lu, G., Crihfield, C.L., Gattu, S. et al. (2018). Capillary electrophoresis separations of glycans. *Chem. Rev.* 118: 7867–7885. doi: 10.1021/acs.chemrev.7b00669.

53 Kist, T.B.L. (2020). *Open and Toroidal Electrophoresis: Ultra-High Separation Efficiencies in Capillaries, Microchips, and Slabs*. Chichester, UK: Wiley.

54 Fomin, M.A., Seikowski, J., Belov, V.N., and Hell, S.W. (2020). Negatively charged red-emitting acridine dyes for facile reductive amination, separation, and fluorescent detection of glycans. *Anal. Chem.* 92: 5329–5336. doi.org/10.1021/acs.analchem.9b05863.

55 Algar, R., Dawson, P.E., and Medintz, I.L. (2017). *Chemoselective and Bioorthogonal Ligation Reactions: Concepts and Applications*, 1e. Chichester, UK: Wiley-VCH.

56 Yi, S., Wei, S., Wu, Q. et al. (2022). Azaphilones as activation-free primary-amine-specific bioconjugation reagents for peptides, proteins and lipids. *Angew. Chem.* 61: e202111783. doi: 10.1002/anie.202111783.

57 Thermo Fisher Scientific (2012). *Crosslinking Technical Handbook* (www.thermoscientific.com/pierce) and Bioconjugation technical handbook. http://assets.thermofisher.com/TFS-Assets/BID/Handbooks/bioconjugation-technical-handbook.pdf (accessed 22 March 2022).

58 Piña, R., Santos-Díaz, A.I., Orta-Salazar, E. et al. (2022). Ten approaches that improve immunostaining: a review of the latest advances for the optimization of immunofluorescence. *Int. J. Mol. Sci.* 23: 1426. doi: 10.3390/ijms23031426.

59 Früh, S.M., Matti, U., Spycher, P.R. et al. (2021). Site-specifically-labeled antibodies for super-resolution microscopy reveal *in situ* linkage errors. *ACS Nano* 15: 12161–12170. doi.org/10.1021/acsnano.1c03677.

60 Ganji, M., Schlichthaerle, T., Eklund, A.S. et al. (2021). Quantitative assessment of labeling probes for super-resolution microscopy using designer DNA nanostructures. *Chem. Phys. Chem.* 22: 911–914. doi.org/10.1002/cphc.202100185.

4

Classes and Molecular Structures

4.1 Introduction

This chapter presents the molecular structures and some photophysical parameters of over 700 fluorescent dye labels and stains that have high brightness in aqueous solutions (plain water, phosphate buffer, phosphate-buffered saline (PBS), and HEPES buffer, to name but a few). This chapter also presents the references from which this data was taken. The following three most important photophysical parameters, from a practical point-of-view, are presented in this chapter: excitation or absorption maximum ($\lambda_{x,max}$ or $\lambda_{a,max}$), Stokes shift (Δ_S), and molecular brightness (B_{max}). Molecular brightness is defined as the product of molar absorption coefficient and fluorescence quantum yield ($B_{max} = \varepsilon_{max}\,\Phi_f$). In the present book, molar absorption coefficient and brightness are both expressed in $mM^{-1}cm^{-1}$ instead of $M^{-1}cm^{-1}$. Here, the dyes are separated into classes to facilitate the use of this book and to produce the information-rich scattergrams in Chapter 5.

Each of the 777 fluorescent dyes in the present database (book) is presented with at least the following photophysical properties: $\lambda_{x,max}$ (or $\lambda_{a,max}$), $\lambda_{e,max}$, Δ_S, Φ_f, ε_{max}, and B_{max}. The fluorescence mean lifetime (τ_f) of almost half of the dyes is also included. These photophysical parameters, in addition to the molecular sizes (number of atoms or N_a), number of negatively charged groups (N_n), number of positively charged groups (N_p), net charge (q), and the class to which the dyes belong, are presented in scattergrams in Chapter 5 and in the Appendices.

A class of compounds is defined as a "set of entities that share some common structural features to which variable parts are attached" (e.g., hydrocarbyl or other groups) [1]. The fluorescent entities of this book have been grouped into classes and the first criterion used is the scaffold structure from which they are derived; e.g., coumarins, pyrenes, and borondipyrromethenes, among others. Figure 4.1 gives the core structure and atom numbering of some classes included in the present database. The super-class of xanthenes required further splitting to produce better scattergrams in Chapter 5. If the xanthenes class had not been split, the scattergrams would be too crowded and this superclass would be represented by only a single mark. In addition, fluoresceins, rhodamines, rhodols, and pyronines exhibit very different spectral characteristics. When the *ortho*-carboxylic group on the pendant aryl is absent, the fluoresceins, rhodamines, and rhodols become separate classes called fluorones, rosamines, and rosols, respectively. From the chemical point-of-view, the absence of an *ortho*-carboxylic or *ortho*-sulfonic group means that the fluorophores no longer assume the lactone (closed) form (Figure 4.2). Fluoresceins, rhodamines, and rhodols are also reclassified as sulfofluoresceins, sulforhodamines, and sulforhodols, respectively, when they have a sulfonic group instead of a carboxylic group (Figure 4.3). Finally, when the oxygen of the xanthene core structure in all the above classes is substituted by another atom, new classes are formed. These classes are called heteroatom-substituted (HAS) xanthenes. They are HAS-fluoresceins, HAS-rhodamines, HAS-rhodols, HAS-pyronines, and HAS-fluorones (see Figure 4.3 and Table 4.1 for details).

Fluorescent Dye Labels and Stains: A Database of Photophysical Properties, First Edition. Tarso B. Ledur Kist.

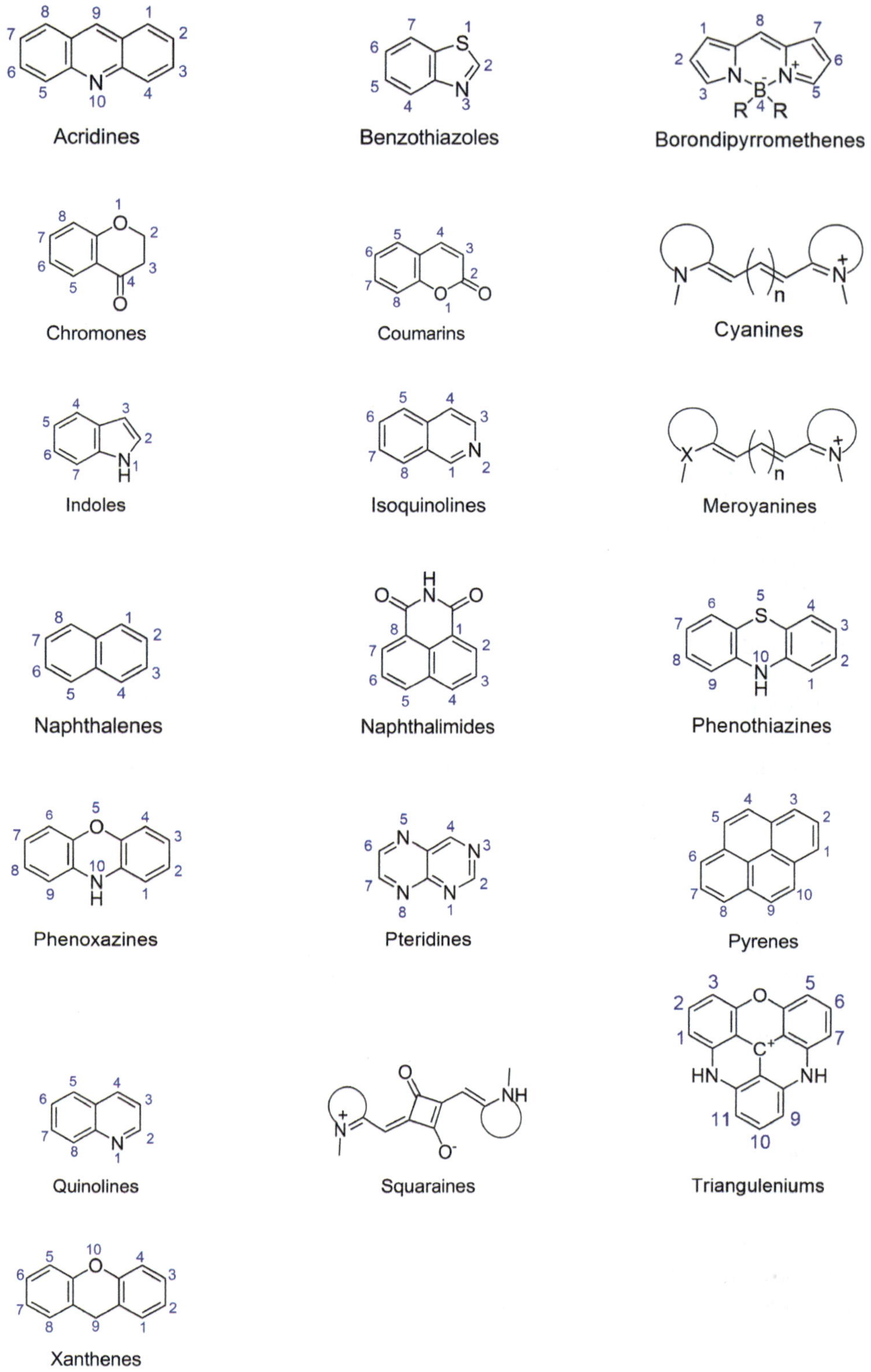

Figure 4.1 Core structures of the classes of fluorescent dyes included in this database (with at least some brightness in water, PBS or any bare aqueous buffer, i.e., without organic modifiers, inclusion complexing agents, or surfactants).

Figure 4.2 Atom numbering commonly used for sulfofluoresceins, fluoresceins, fluorones, sulforhodamines, rhodamines, rosamines, sulforhodols, rhodols, and rosols.

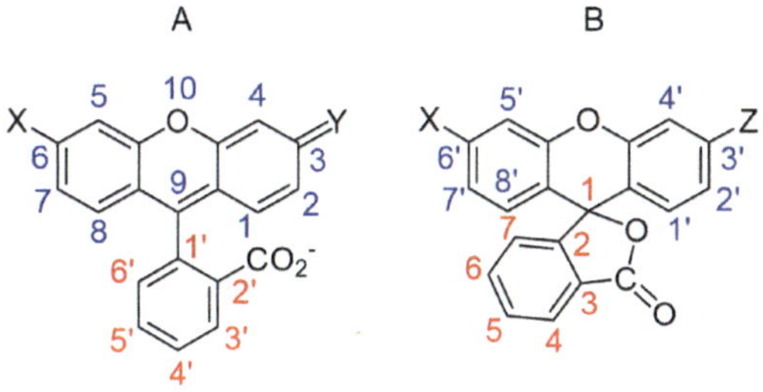

X	Y	Z	Class
OH	O	O^-	Fluoresceins
NH_2	NH_2^+	NH2	Rhodamines
NH_2	O	O^-	Rhodols

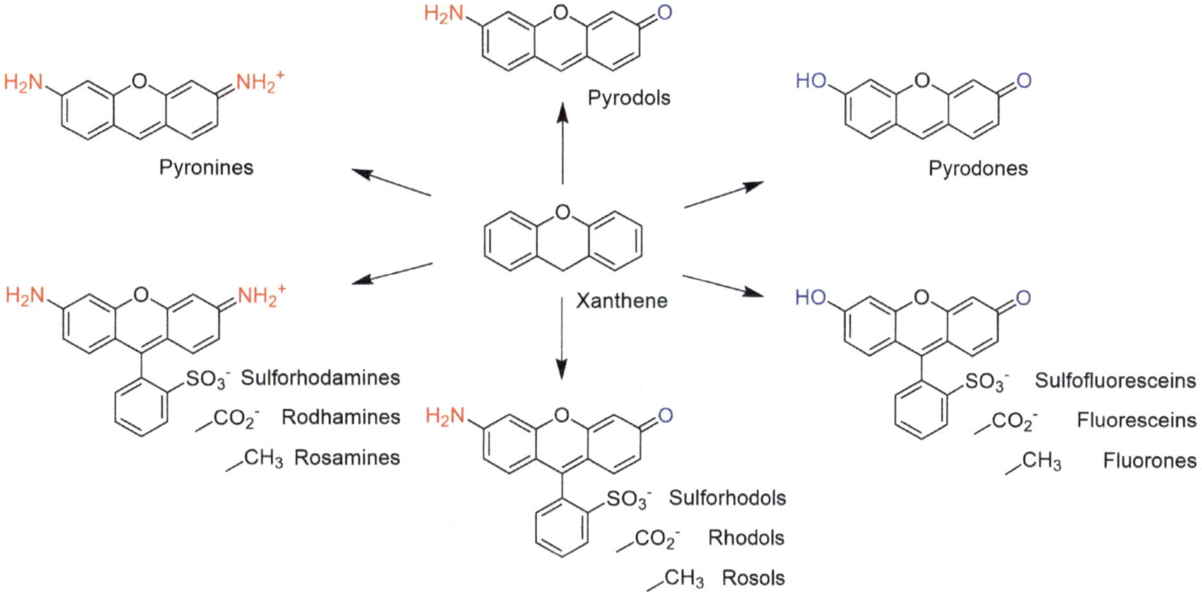

Figure 4.3 The xanthene structure and the current main classes of xanthene derivatives. Rosamines, rosols, and fluorones have a group (possibly a $-CH_3$ group or other) bound to carbons 2′ and/or 6′ that prevents the occurrence of the cyclic form (which occurs when a sulfonic or carboxylic group occupies one of these positions, see Figure 4.2).

Atom numbering is an important issue for fluoresceins (sulfofluoresceins), rhodamines (sulforhodamines), and rhodols (sulforhodols), which have the *ortho*-carboxyl (or *ortho*-sulfonic) group and therefore exist in an equilibrium between the open form (Figure 4.2A) and the lactone form (Figure 4.2B). For rhodamines they are called the zwitterionic and lactone form, respectively. The open form is colored and fluorescent, whereas the lactone (closed) form only absorbs in the ultra-violet region and is essentially non-fluorescent. Unfortunately, there are two numbering systems used in the literature; the first (Figure 4.2A) is commonly used when the fluorophore is applied as a label (where the open form is preferred), and the second (Figure 4.2B) is more appropriate when the fluorophore is initially used in its closed form (lactone) as a cell penetrant probe or a chemosensor. Both numbering systems will be used throughout the book, reflecting the choice of the original author(s).

Table 4.1 Structures of the HAS-rhodamines. Rhodamine is used as an example, but in principle it can be a HAS form of any of the classes of Figure 4.3 or others (mercaptanes instead of phenols, seminaphthoxanthene, naphthoxanthene, and others).

Rhodamines	X = O
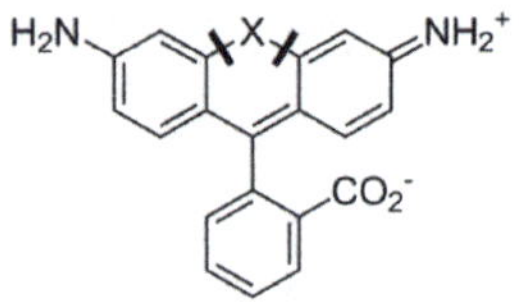	
HAS-rhodamines	
B-rhodamines **Bora-rhodamines**	
$B(OH)_2$-rhodamine Dihydroxyborate-rhodamine	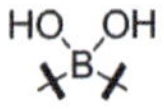
C-rhodamines **Carbo-rhodamines**	
$C(CH_3)_2$-rhodamine Dimethylcarbo-rhodamine	
Ge-rhodamines **Germanium-rhodamines**	
$Ge(CH_3)_2$-rhodamine Dimethylgermanium-rhodamine	
P-rhodamines **Phospha-rhodamines**	
PO_2^--rhodamine Phosphinate-rhodamine	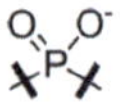
PO_2R-rhodamine Phosphine oxide-rhodamine	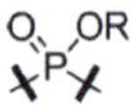
PO_2Ph-rhodamine Phenilphosphine oxide-rhodamine	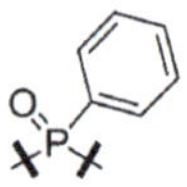
S- rhodamines **Sulfur-rhodamines**	
S-rhodamine Thio-rhodamine	

Table 4.1 (Continued)

SO_2-rhodamine Sulfone-rhodamine	
Se-rhodamines **Selenium-rhodamines**	
Se-rhodamine Seleno-rhodamine	
Si-rhodamines **Silico-rhodamines**	
$Si(CH_3)_2$-rhodamine Dimethylsilane-rhodamine	
$Si(OH)_2$-rhodamine Disilanol-rhodamine	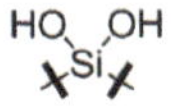
Te-rhodamines **Tellurium-rhodamines**	
TeO-rhodamine Telluroxide-rhodamine	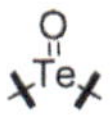

This book is primarily intended for users of fluorescent molecules and new technology developers. Clear, well defined, and informative short names are very important for every laboratory routine, facilitating communication and the readability of manuals and protocols. However, the names of some dyes are still too long. Here, they have been limited to a maximum of 30 characters for two reasons: i) to facilitate practical, everyday use; and ii) to maximize space in the tables and figures presented throughout the book, producing information-rich content on every page. Short names that are as informative as possible have been used. For example, the name "4,4-difluoro-2,6-disulfonate-1,3,5,7-tetramethyl boron-dipyrromethene" must be shortened to the limit of 30 characters, therefore the short name "2,6-sulfon.-1,3,5,7-Me-BODIPY" is used instead. However, the 29-character name "5(6)-carboxyfluorescein (FAM)" is used instead of just 5(6)-carboxyfluorescein or just FAM.[1] This was made to keep it as informative as possible and in recognition to the tradename. FAM and BODPY are tradenames of Thermo Fisher Scientific (Waltham, MA). Other tradenames are listed in the footnote.[1] The short names used throughout this book are given in alphabetical order in Appendix A, together with their respective synonymous, some IUPAC names, some CAS numbers, and molecular formulas.

Molecular structures contain a lot of information and are very useful to users. They are of particular use to dye developers (Organic Chemists), who continue to synthesize new, bright, photorobust dyes. In this chapter the molecular structures of fluorophores are depicted in their predominant ionic form, in the aqueous solution, and at the pH (highlighted in blue in each figure) at which their photophysical parameters were measured. Almost all parameters were measured when dyes were in neutral pH solutions, and only a few were measured in basic and/or acidic aqueous solutions. For simplicity, the carboxylic group has been represented by $-CO_2H$ or $-CO_2^-$, to save space and avoid potential confusion with the peroxide group (–COOH). Similarly, the sulfonic group has been represented by $-SO_3^-$, and phosphate esters by $-O\text{-}PO_3^{2-}$, $-O\text{-}PO_3H^-$, or $-O\text{-}PO_3H_2$ (depending on the pH).

1 Abberior Star is a trademark of Abberior GmbH (Göttingen, Germany). AlexaFluor, BODIPY, Cascade Blue, FAM, JOE, TET, HEX, Marina Blue, Oregon Green Pacific Blue, ROX, TAMRA, and Texas Red are trademarks of Thermo Fisher Scientific (Waltham, MA, USA). ATTO is a trademark of ATTO-Tec GmbH (Siegen, Germany). CF Dye is a registered trademark of Biotium, Inc. (Fremont, CA, USA). Cy3, Cy3.5, Cy5, and Cy5.5 are trademarks of GE Healthcare (Chicago, IL, USA). Dy is a trademark of Dyomics GmbH (Jena, Germany). Seta, SeTau, and Square dyes are trademarks of SETA BioMedicals (Urbana, IL, USA).

These sulphonic and phosphate groups are widely used to increase the water solubility of the dyes. The synthesis routes are not shown in this book; however, references which are dedicated to their development and synthesis are cited.

In this chapter the structures of the fluorescent dyes, some of their photophysical data, and respective references are presented. Each class has its own section, starting with the class that contains the entities with the highest brightness in aqueous solution (rhodamines), followed by the class that contains the second brightest entities (pyronines), and so on. The chapter ends with the classes containing the entities with the lowest brightness (e.g., pyridine and benzene derivatives). Some sections must be split into smaller sub-sections to facilitate the location of the references of each fluorescent dye. These sub-sections are cited in Chapters 5, 6, 7, and in the Appendices.

Only the spectral and photophysical properties measured in aqueous solutions (water or aqueous buffers), with a maximum of ~1% (v/v) organic modifiers or co-solvents, have been considered in this book. In some cases, these organic solvents are present in the stock solutions used to prepare the dye solutions. The number of synthetic and natural compounds that exhibit at least a weak fluorescence in such strict conditions (aqueous solutions) is still very large. The lower the brightness range considered, the larger the number of molecules that fit into the database. Therefore, the following cut-off criteria has been adopted in order to limit the number of fluorophores in this database:

1) Solubility of at least ~1 μM in aqueous solutions (without using inclusion complexing agents or surfactants, and with a maximum of 1% V/V organic modifiers or co-solvent present);
2) "Wavelength of absorption maximum of the longest wavelength band" located between 300 and 900 nm;
3) A brightness of at least 5% of the maximum brightness observed in the respective range. Exceptions to this rule were made for compounds with historical importance, dyes within a library of dyes, and those that are used as reference materials or standards for fluorescence quantum yield (Φ_F) or fluorescence mean-lifetime (τ_f);
4) Dyes are commercially available or with a published synthesis route (in articles, books, or patents).

All derivatives (modifications of the groups attached to core structures) of the above-mentioned classes that were made for the following purposes are included in the database:

1) To confer water solubility. For example, the following groups: sulfonic, phosphonic [2], sulfobetaine [3], carboxylic groups, or (ethylene glycol)methyl ether oligomers [4];
2) Added groups that fine tune the spectral properties or maximize brightness;
3) Arm linkers with a carboxylic group at the end, usually ethanoic acid, propanoic acid, butanoic acid, or hexanoic acid. They are used to attach reactive moieties at the carboxylic functional group, allowing a great variety of target structures to be fluorescently labeled. However, these second modifications on the arm linkers have not been considered in the present database, as this would excessively increase the number of entries. Most of these latter modifications (attached reactive moieties) do not significantly change the spectral and photophysical properties, especially if long arm linkers (e.g., hexanoic acid) are used to hold these reactive moieties. However, it is known that the photophysical properties of some dyes may change after conjugation to nucleic acids, proteins, and other biomolecules. Currently there are many reactive moieties that are used to bind fluorescent labels onto biomolecules (Chapter 3). For covalent ligation to amines, the following are used: NHS esters [5, 6], STP esters [7, 6], vinylsulfone [8], and isothiocyanates [6]. For covalent bonding to thiols, maleimido and iodoacetamido are used [6]. To bind fluorescent dyes to actin filaments, phalloidin is used [5]. Alkyl glycerol and phosphides [5] are used to promote the insertion of fluorescent dyes into lipid membranes, and the biotin and streptavidin system [6] is used to bind one to the other.

There are some discrepancies regarding the photophysical data presented by authors within the literature. This predominantly occurs with the values of fluorescence quantum yields (Φ_f), molar absorption coefficients (ε_{max}), and, to a lesser extent, with fluorescence emission maximum ($\lambda_{e,max}$) and wavelengths of excitation and absorption maxima ($\lambda_{x,max}$ and $\lambda_{a,max}$). There are many possible reasons for this, such as measurement errors, measurements taken at different temperatures, and uncertainties related to the purity of the dyes used in the measurements. Some dyes can be made up of a mixture of position isomers, and some have a flexible structure made up of a mixture of two or more conformational isomers in liquid solutions [9]. Some have a pK_a around neutral pH, which makes them sensitive to the exact pH used. A few are prone to frequent changes between oxidized and reduced forms. Some exhibit a low water solubility and, in addition, some are subjected to aggregation in aqueous solutions (a condition in which they lose fluorescence owing to π-stacking interactions; a phenomenon called aggregation-caused quenching). Nevertheless, the discrepancies in the published photophysical parameters are usually small (<10%) and do not compromise both the scatterplot analysis performed in Chapter 5 and the resulting properties that are revealed. It is possible to solve for these discrepancies in the published data by re-measuring the photophysical parameters using modern instruments under controlled temperatures and sample conditions, but this would take many years to complete. If this was done, the resulting scatterplot analysis and conclusions would not change significantly. Therefore, most of the data found in the literature was taken as reported and only ~5% of the spectral and photophysical parameters were measured and checked by the author.

The superclass xanthene (9*H*-10-oxaanthracene) contains the largest number of fluorescent dyes. It also contains the brightest entities among all currently available organic small-molecule fluorophores, some of which are also the most photostable (they exhibit the largest photobleaching relative mean-lifetime, as discussed in Chapter 7). Historically, fluorescein was synthesized more than one and a half centuries ago by the 1905 Nobel Prize winner in Chemistry, J.F.W.A. von Baeyer [10]. Its absorption maximum is in the blue region of the spectrum. Rhodamine [11] was synthesized next, with absorption maximum in the green range. Many more have been synthesized in the last few decades. Synthesis of new dyes boomed due to the need for good fluorescent dyes in the fields of dye lasers (in the 1980s), fluorescence microscopy (wide-field and confocal scanning), DNA sequencing (end of the 1990s and 2000s), real-time PCR (2000s), and more recently with the development of optical nanoscopy techniques. The most important classes derived from the xanthene superclass received the following short names: pyronines (6-amino-3H-xanthen-3-iminium), pyrodols (6-amino-3H-xanthen-3-one), and pyrodones (6-hydroxy-3H-xanthen-3-one); sulfofluoresceins, fluoresceins, and fluorones; sulforhodols, rhodols, and rosols; and finally, sulforhodamines, rhodamines, and rosamines (Figure 4.3). For each of these 12 classes, there is another class (totaling 24 classes), formed when the xanthene oxygen heteroatom of each class is substituted. In most cases, this produces a large bathochromic shift. In fact, each substituted chemical element can have its own class. For simplicity, all HAS-rhodamine derivatives (regardless of which substituent atom is used) are combined into one single class and the same is done for all other 11 classes of Figure 4.3.

Aside from the heteroatom, there are other possible modifications that can be performed on these 12 xanthene derived classes. Mercaptofluoresceins (fluoresceins whose phenol group has been replaced by a thiol group) have been synthesized and mercapto-substitutions can be performed to all phenol-containing classes (sulfofluoresceins, fluoresceins, flurones, sulforhodols, rhodols, rosols, pyrodols, and pyrodones). These may be considered as new classes. However, this is still not the end of the story as there are endless possibilities. For instance, the synthesis of seminaphthoxanthene and naphthoxanthene derivatives further increases the possibilities of new fluorescent dyes that can be derived from xanthene. In the case of seminaphthoxanthene, one of the benzenes that flanks the central pyran of the xanthene structure is replaced by a naphthalene to produce seminaphthoxanthene. Note that the asymmetric structures of sulforhodol, rhodol, rosols, and pyrodols further increase the number of classes of seminaphtho derivatives, as different structures are produced depending on whether the naphtalene is added to the amino or phenol side of these structures. Another possible modification of xanthenes is the replacement of both benzenes (on the left and right side of the pyran) by naphthalenes, producing naphthoxanthenes.

Autofluorescence is an important issue in bioimaging and in sensitive detections in bioanalytics, especially when short wavelengths (violet, purple, blue, and even green) are used for excitation. In addition, there are many options of reliable excitation sources (diode lasers) available at longer wavelengths (red and even near-infrared ranges of the spectrum). Finally, there are good photosensors which have a high spectral sensitivity in the red and near-infrared ranges. However, non-HAS xanthene dyes exhibit their longest wavelength absorption band maximum in the blue, green, and yellow ranges of the spectrum. Therefore, there are many reasons to push the excitation maxima of xanthene derivatives to the red and near-infrared ranges.

Recently, heteroatom-substituted (HAS) xanthene dyes have been synthesized. In these dyes, the oxygen atom of the xanthene structure (9H-xanthene) is replaced by another atom, producing small fluorophores with remarkably large bathochromic shifts (resulting in excitation maximum in the orange, red, and even near-infrared ranges of the spectrum). Some of these HAS derivatives have a high photostability and many maintain a high fluorescence quantum yield (the maximum value is 1). The most common structures of these HAS-xanthenes are listed in Table 4.1 as HAS-rhodamines; however, the rhodamine class was selected purely as an example. It could be any of the previously discussed xanthene derivatives (some of them are depicted in Figure 4.3).

Fluorescent dyes will be identified by a short name of no more than 30 characters from now on, for reasons previously explained. The number of available dyes within some classes is so large that a short name was created using the first letter of both first and last author names and the numbering used in the original work by these authors. Uncertainties regarding the identity of any dye can be resolved by referring to Appendix A. Moreover, most of the names and molecular structures shown in the figures of this chapter correspond to the predominant ionic form present at the pH indicated on the legend of each figure; i.e., the pH at which the photophysical parameters were measured. In all molecular structure figures of this chapter, the numbers in blue represent the photophysical parameters $\lambda_{x,max}$ (nm), Δ_S (nm), and B_{max} (mM^{-1}cm^{-1}), respectively. These symbols were not included on the figures to avoid overcrowding and excessive repetition.

The following photophysical and molecular parameters, presented in alphabetical order, are the essence of this database:

- B_{max} Fluorescence brightness at wavelength $\lambda_{a,max}$, calculated as $B_{max} = \varepsilon_{max}\,\Phi_f$ and expressed in units of mM^{-1}cm^{-1} to avoid excessively large numbers (as done for ε_{max}). For example, 82.0 mM^{-1}cm^{-1} $\equiv$ 82,000 M^{-1}cm^{-1} (where the symbol $\equiv$ denotes that they are identical, except for the significant numbers). See Section 2.7.2 for details.
- N_a Number of atoms of the fluorescent structure (counter-ions of salts are not counted) in its prevalent ionic form.
- N_n Number of negative elementary charges of a fluorescent structure in its prevalent ionic form in an aqueous solution at a specified pH.
- N_p Number of positive elementary charges of a fluorescent structure in its prevalent ionic form in an aqueous solution at a specified pH.
- q Number of net charges of a fluorescent structure in its prevalent ionic form in an aqueous solution at a specific pH ($q = N_p - N_n$). Note that a dye with $N_n = 0$ and $N_p = 0$ (exhibits $q = 0$) has different physical-chemical and physiological properties to a dye with $N_n = N_p > 0$. This also results in $q = 0$.
- Δ_S Stokes shift. Calculated using the following formula: $\Delta_S = \lambda_{e,max} - \lambda_{x,max}$ or $\Delta_S = \lambda_{e,max} - \lambda_{a,max}$.
- ε_{max} Decadic molar absorption coefficient at $\lambda_{a,max}$ in units of mM^{-1}cm^{-1}.
- $\lambda_{e,max}$, Wavelength of emission maximum.
- $\lambda_{x,max}$ Wavelength of excitation maximum at the longest wavelength band. When available, the maximum excitation wavelength was always used, otherwise the maximum absorption wavelength ($\lambda_{a,max}$) was used instead. In general, their divergence is less than ~2 nm at the longest wavelength band.
- τ_f Fluorescence mean-lifetime, expressed in nanoseconds (ns). Note that $\tau_f = \tau_{1/2}/\ln(2)$, where $\tau_{1/2}$ is the fluorescence half-lifetime.
- $\tau_{pb,r}$ Relative photobleaching mean-lifetime. Relative to a dye exposed to an identical radiance and other conditions.
- Φ_f Fluorescence quantum yield ($0.00 \le \Phi_f \le 1.00$). The measurement of Φ_f is also very unprecise (±5% at least). Nevertheless, two decimal points will be used throughout this book.

In summary, classes, molecular structures, references, and $\lambda_{a,max}$, Δ_S, and B_{max} are covered in this chapter. Almost all are covered in Chapter 5, where they are used to produce panoramic views (scattergrams) of the properties of over 700 fluorescent dye labels and stains which have a medium to high brightness in aqueous solutions.

4.2 Rhodamines

Rhodamines is the largest class, containing over 50 bright and photorobust fluorescent dyes. They are xanthene derivatives with one iminium group bound to carbon 3, one amino group bound to carbon 6, and a 2-benzoic acid bound to carbon 9. Tetramethylrhodamine [12] was among the first to be synthesized as a green absorber over a century ago. It has remarkably strong absorption, a high fluorescence quantum yield, and high photostability. Replacing one or more hydrogen atoms of this structure allowed the synthesis of many rhodamine derivatives that have absorption maximum in end of the blue, whole green, yellow, and even orange ranges of the spectrum. So far this has proved to be the most versatile scaffold, facilitating the synthesis of a great diversity of fluorescent entities, as shown in this section.

4.2.1 Rhodamines With Absorption Maximum Below 500 nm

There are only a few bright rhodamines that are soluble in aqueous solutions and whose longest wavelength band absorption maximum is 500 nm and below (Figure 4.4). They are rhodamine 110 (13558-31-1) [13, 14], Atto 488 [5], and fluorinated rhodamines 1e and 1j [15]. The fluorescent dyes DY-490 carboxylate ($\lambda_{a,max}$ = 491 nm) and Alexa Fluor 488 ($\lambda_{a,max}$ = 494 nm) are shown in Section 4.2.5 (rhodamines with a high net charge).

4.2.2 Rhodamines With Absorption Maximum Between 500 and 550 nm

There are a large number of rhodamines that have a wavelength of absorption maximum between 500 and 550 nm (Figure 4.5): Abberior Star 512 ([15, 16], molecular structures were kindly provided by the supplier); rhodamine RBzET [17]; rhodamine 123 (62669–70-9) [14, 18, 19]; DY-505 and DY-530 [20]; Atto 514, Atto 532, and Atto Rho6G, [5]; rhodamine 6G ([21]; there was a mistyping of $\lambda_{e,max}$, therefore, this value was taken from [14]); 530RH [22]; tetramethylrhodamine (TMR) [23]; 5(6)-carboxytetramethylrhodamine (TAMRA) in water [6] and in a phosphate buffer at pH 8–9 [24]; rhodamine 6G +2 [25]; fluorinated rhodamines 1f-5′-COOH, 1f-6′-COOH, 1k, 1r, 3-H, and 5-SO_3H; JF549 [26]; and JF525, JF526, and JF559 [27]. A few additional dyes (Abberior Star 488, rhodamine 6G +3, and Alexa Fluor 546) with absorption maximum between 500 and 550 nm and a high net charge are shown in Section 4.2.5 (rhodamines with a high net charge).

4.2.3 Rhodamines With Absorption Maximum Between 550 and 600 nm

The number of rhodamines that have $\lambda_{a,max}$ values between 550 and 600 nm is also large. This group contains the following (Figure 4.6): Abberior Star 580 ([15, 16, 28], structure was kindly provided by the supplier); Atto 550, Atto

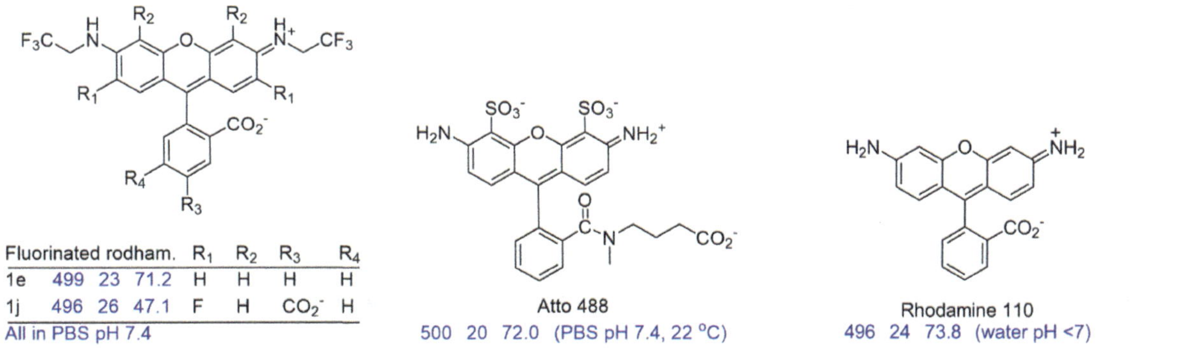

Figure 4.4 Structures of rhodamines with wavelength of absorption maximum of 500 nm and below. The text highlighted in blue in the figure insets give the values of $\lambda_{x,max}$ (nm), Δ_S (nm), and B_{max} ($mM^{-1}cm^{-1}$), respectively.

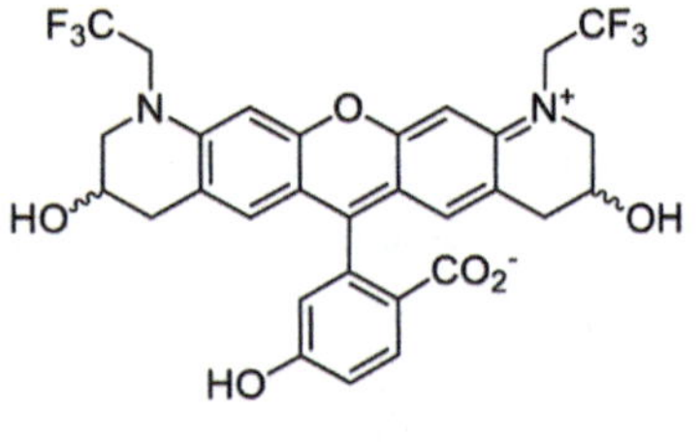

530RH
532 21 49.8 (PBS pH 7.4, RT)

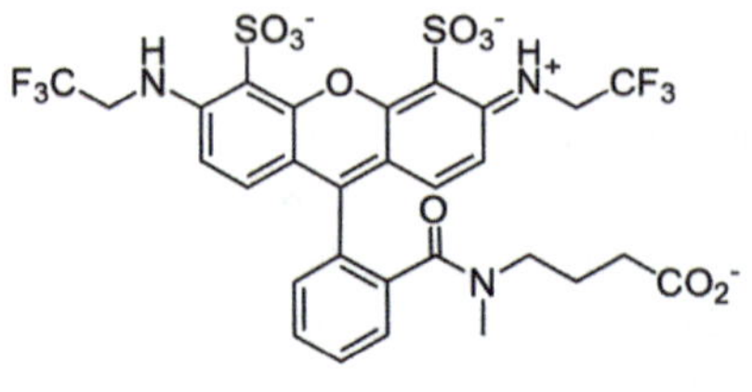

Abberior Star 512
511 19 68.8 (PBS pH 7.4)

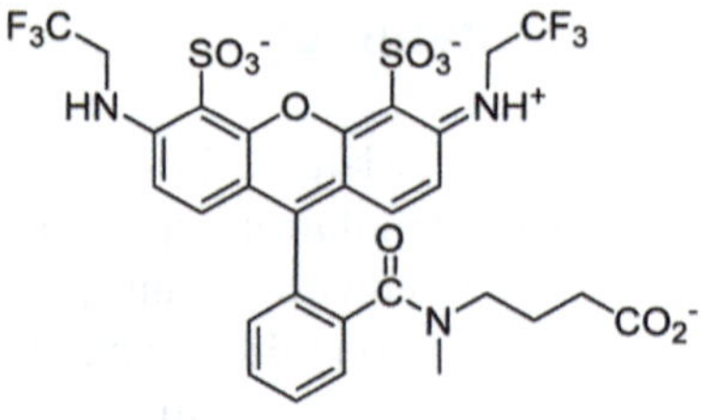

Atto 514
511 21 97.7 (PBS pH 7.4, 22 °C)

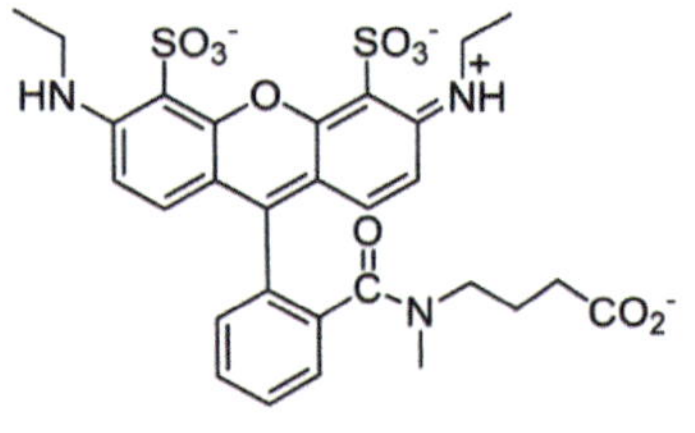

Atto 532
532 20 103.5 (PBS pH 7.4, 22 °C)

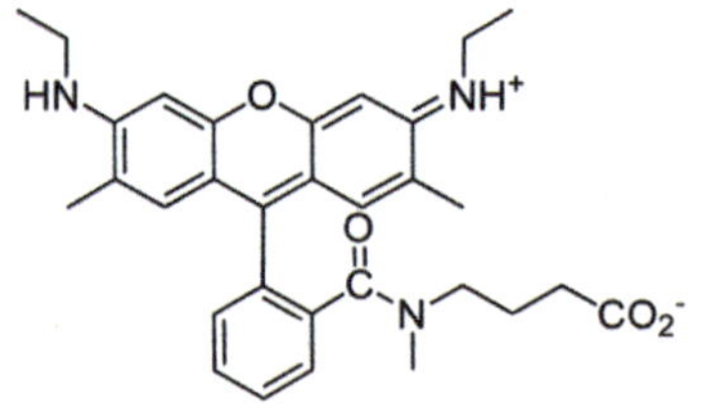

Atto Rho6G
533 24 108.1 (PBS pH 7.4, 22 °C)

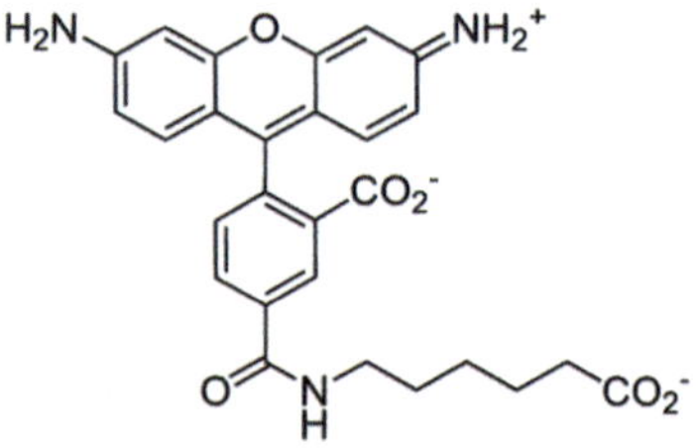

DY-505
505 25 63.2 (PBS pH 7.4)

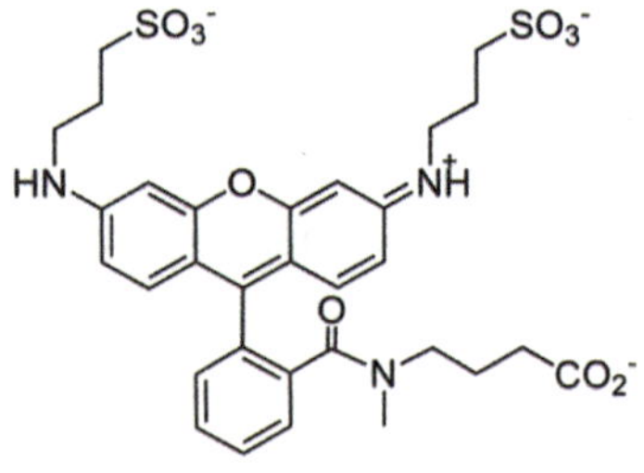

DY-530
533 21 80.6 (PBS pH 7.4)

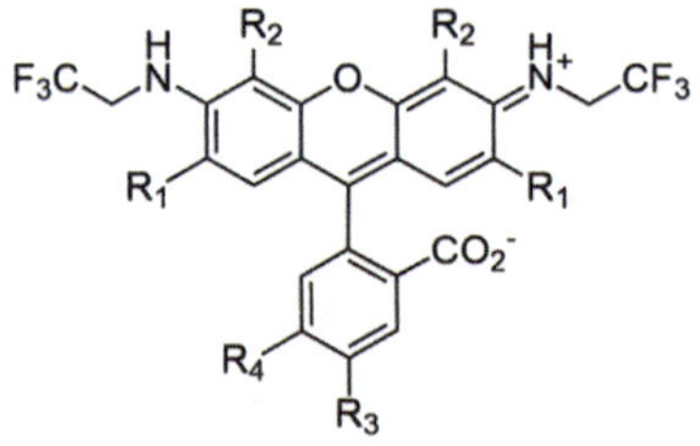

Fluorinated rodham.				R_1	R_2	R_3	R_4
1f5'	501	24	80.7	H	H	CO_2^-	H
1f6'	501	22	81.8	H	H	H	CO_2^-
1k	506	25	35.6	F	SO_2^-	H	H

All in PBS pH 7.4

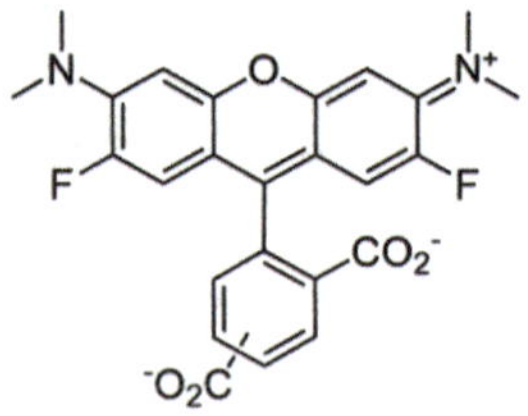

Fluorinated rhodamine 1r
525 29 8.6 (PBS pH 7.4)

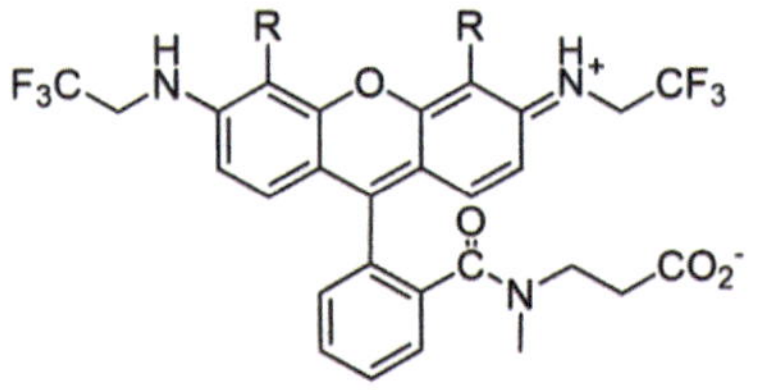

Fluorinated rodhamine				R
3H	510	21	49.0	H
5-SO_3H	512	18	44.7	SO_3^-

All in PBS pH 7.4

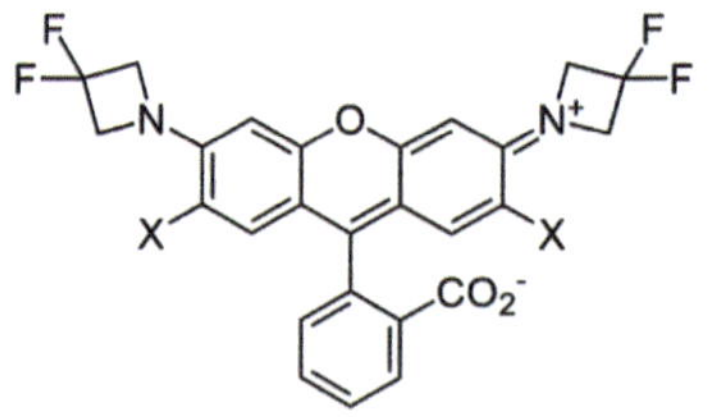

				X
JF525	525	24	85.5	H
JF526	526	24	16.5	F

All in 10 mM HEPES pH 7.3.

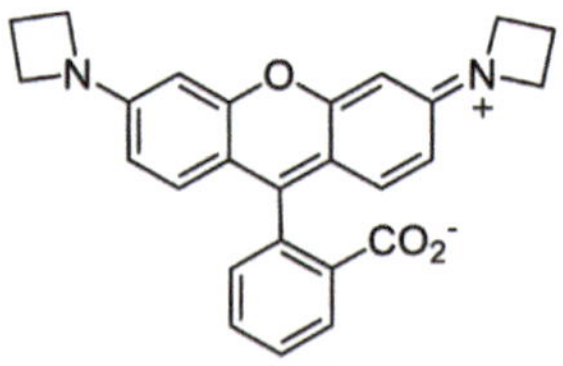

JF549
549 22 88.8 (HEPES pH 7.3)

Figure 4.5 Structures of the rhodamines that have a wavelength of absorption maximum between 500 and 550 nm. A few additional strtuctures are shown in Figure 4.8 (rhodamines with a high net charge). Legend as in Figure 4.4.

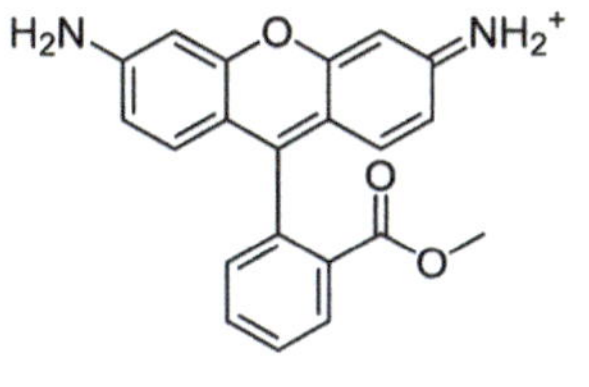

Rhodamine 123
502 26 67.5 (water pH < 7)

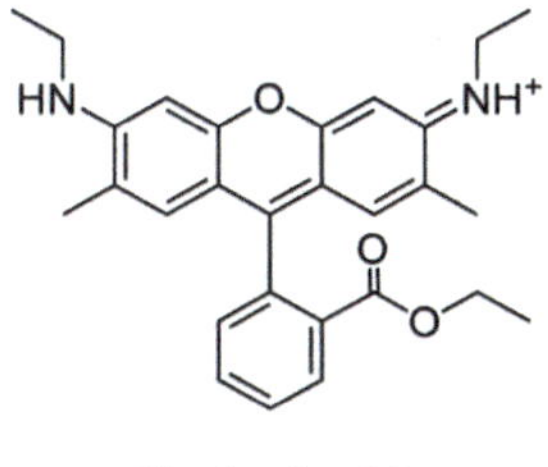

Rhodamine 6G
526.3 26 97.6 (water pH <7)
525.5 26 102.9 (D_2O pH <7)

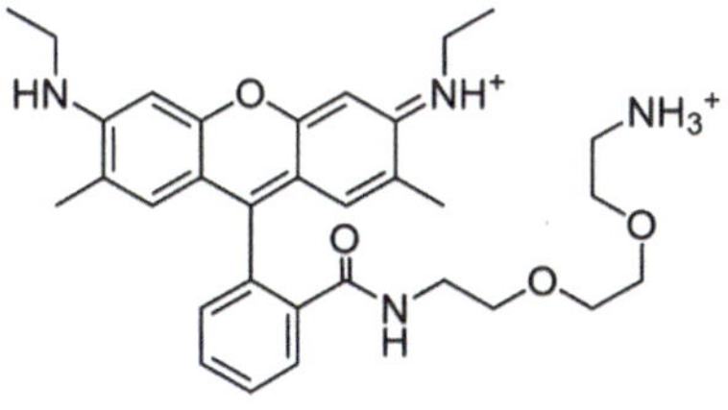

Rhodamine 6G +2
530 23 93.9 (water pH <7)

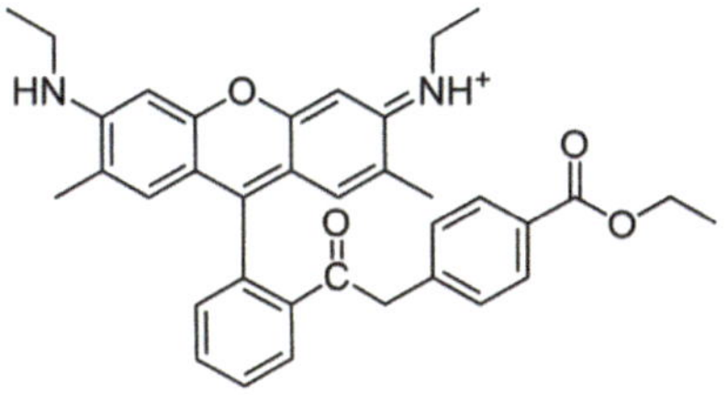

Rhodamine RBzEt
527 25 71.3 (water)

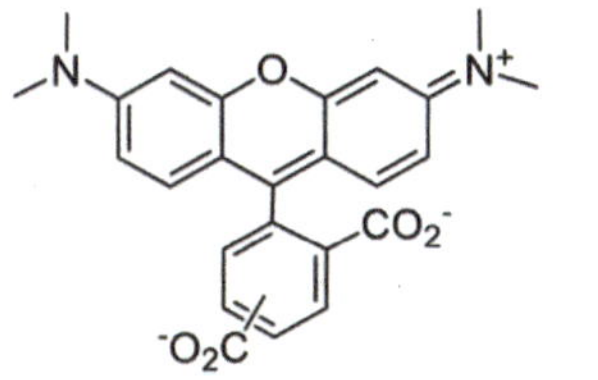

TAMRA
546 30 18.0 (water pH 7)
547 27 26.9 (50 mM P pH 8-9)

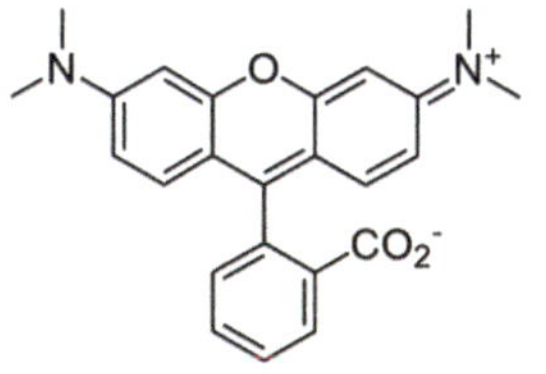

Tetramethylrhodamine
548 24 35.1 (10 mM HEPES pH 7.3)

Figure 4.5 (Continued)

565, Atto Rho3B, Atto Rho11, Atto Rho12, Atto Rho101, and Atto 590 [5]; DY-555 [20]; JF552, JF571, and JF574 [26]; tetraethylrhodamine (Rhodamine B) [6]; 575RH [22]; W-B 15 [29] (name given using the initials of the authors plus the numbering used in the original work. Note that this dye can also form a closed lactone form. In this sense it can also be considered a rhodamine); and 5(6)-carboxy-X-rhodamine (ROX) [6]. Note that the structures of Abberior Star 580 and Atto Rho101 are identical, but with slightly different brightness (probably due to some DMSO in the working solution that comes from the stock solution used). A few additional dyes (Alexa Fluor 568 and Alexa Fluor 594) with a high net charge are shown in Section 4.2.5. Wu and Burgess [29] examined the effect of the amine substituents on the photophysical properties of rosamines. The same trends can be expected to happen with rhodamines due to the similarity of the structures between rhodamines and rosamines (Section 4.17).

4.2.4 Rhodamines With Absorption Maximum Above 600 nm

All currently existing optical nanoscopy techniques rely on fluorescence and most of them require sample excitation with high radiance (w/m^2). Excitation photons with low energy are highly advantageous as less background light from autofluorescence is observed and sample integrity after long exposure is also maintained. There are only a few rhodamines that have an absorption maximum of above 600 nm, which is the end of the orange and the beginning of the red range of the spectrum (Figure 4.7). These are: Atto 594, Atto Rho13, and Atto Rho14 [5]; Abberior Star 600, Abberior Star 635, Abberior Star 635P, and Abberior Star Red ([15, 16, 28], structures were kindly provided by the supplier); KK9046 and KK1119 [2]; and KK114 [28]. The molecular structure of Abberior Star 580 carboxylic acid is identical to the structure of ATTO Rho101 carboxylic acid, but the supplied photophysical parameters of Abberior Star 580 ($\lambda_{a,max}$ = 587 nm, Δ_S = 20 nm, and B_{max} = 76.5 $mM^{-1}cm^{-1}$) are somehow different from those of Atto Rho101 ($\lambda_{a,max}$ = 587 nm, Δ_S = 22 nm, and B_{max} = 96.0 $mM^{-1}cm^{-1}$), despite the fact that the same buffer (PBS pH 7.4) was used for both measurements. A potential explanation is that some DMSO from the stock solution could have remained in the working solution in one of the cases.

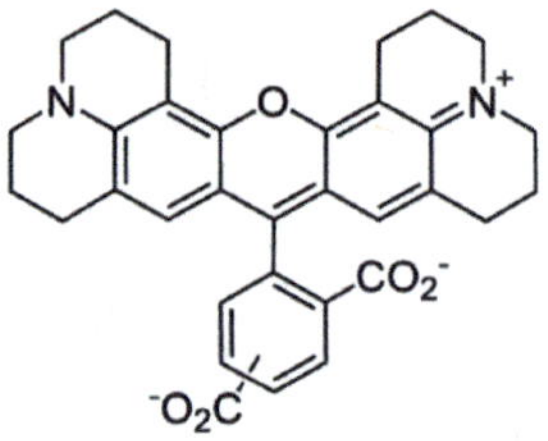

5(6)-Carboxy-X-rhodamine (ROX)
576 25 57.4 (water pH <7)

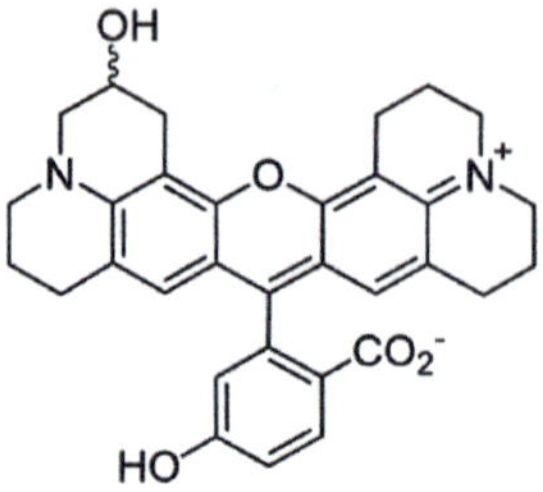

575RH
574 23 40.7 (PBS pH 7.4, RT)

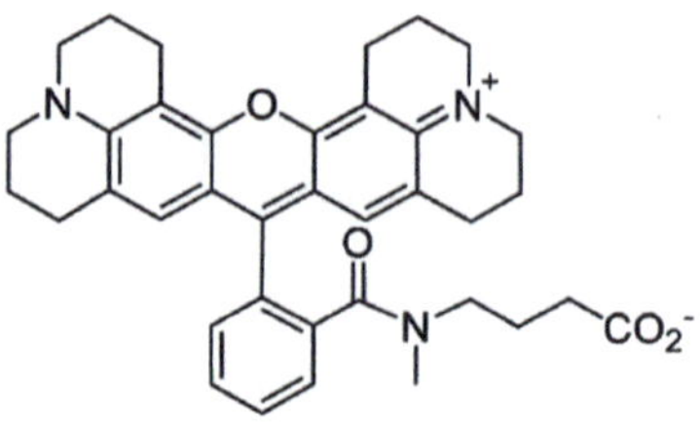

Abberior Star 580
587 20 76.5 (PBS pH 7.4)

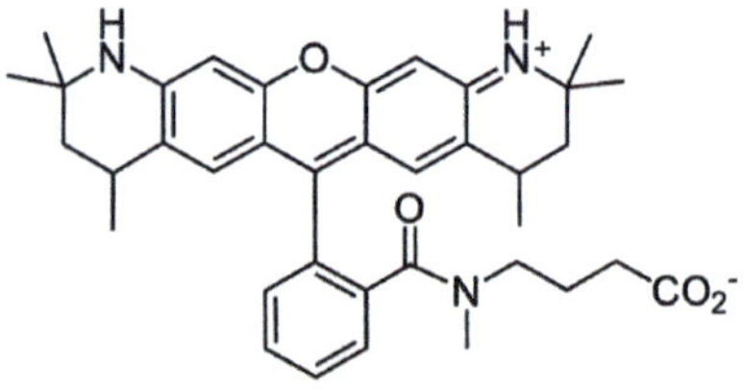

Atto 550
554 22 96.0 (PBS pH 7.4, 22 °C)

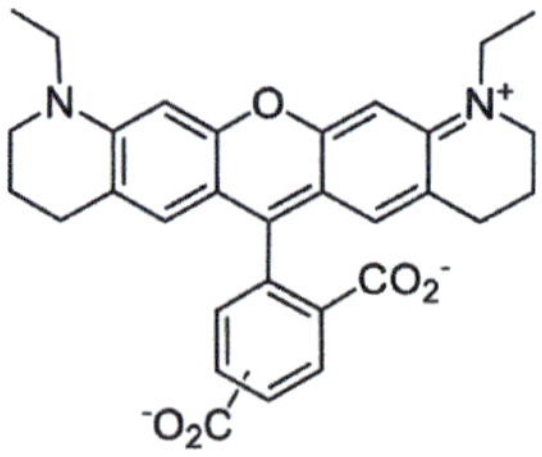

Atto 565
564 26 108.0 (PBS pH 7.4, 22 °C)

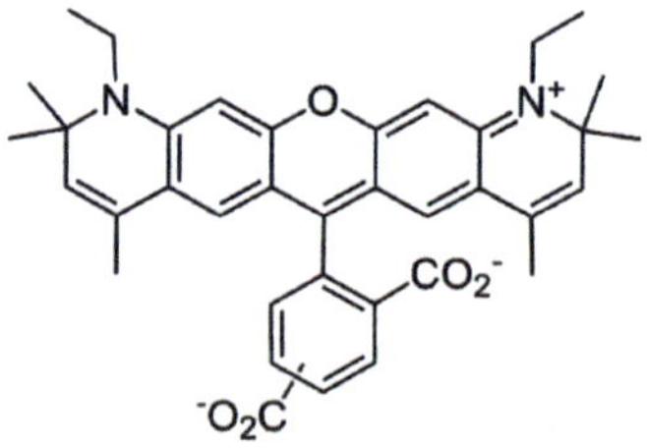

Atto 590
593 29 96.0 (PBS pH 7.4, 22 °C)

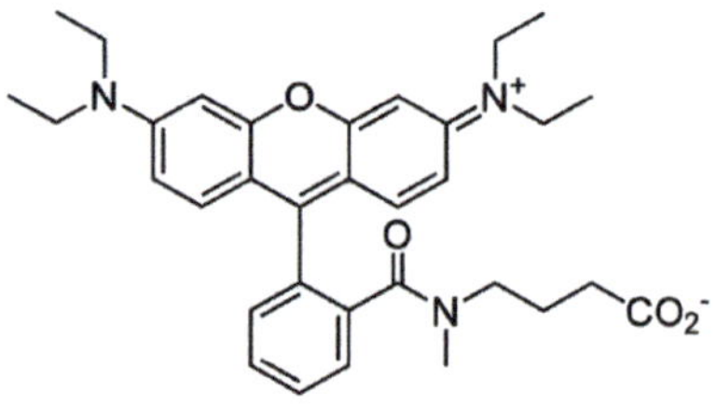

Atto Rho3B
566 23 60.0 (PBS pH 7.4)

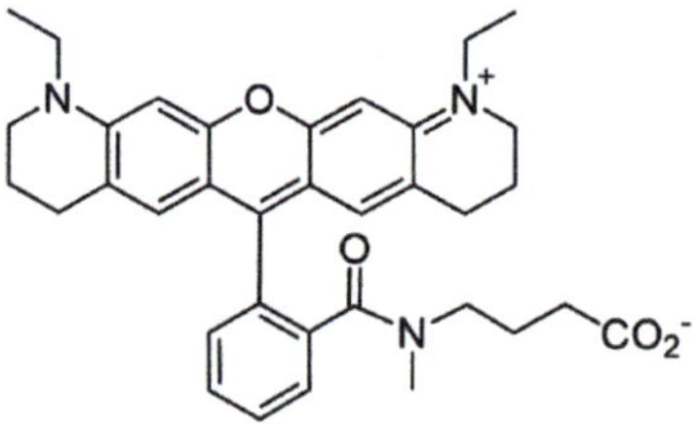

Atto Rho11
572 23 96.0 (PBS pH 7.4, 22 °C)

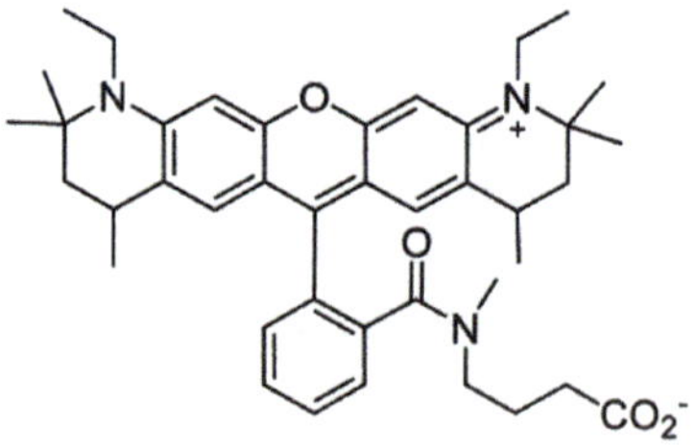

Atto Rho12
577 23 96.0 (PBS pH 7.4, 22 °C)

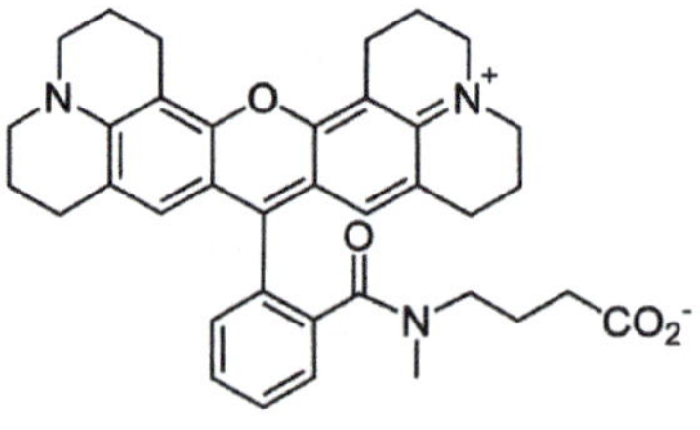

Atto Rho101
587 22 96.0 (PBS pH 7.4, 22 °C)

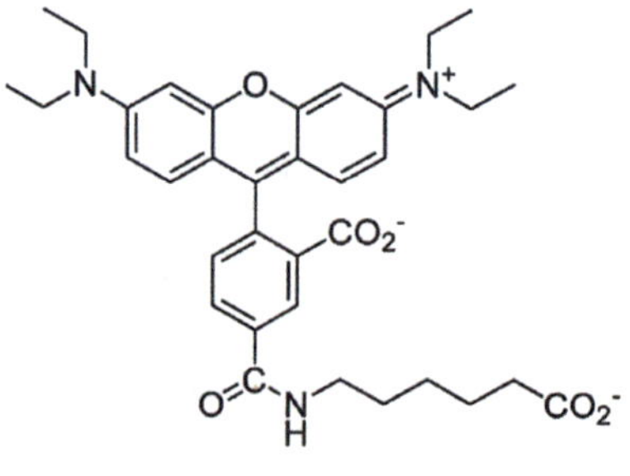

DY-555
555 25 31.7 (PBS pH 7.4)

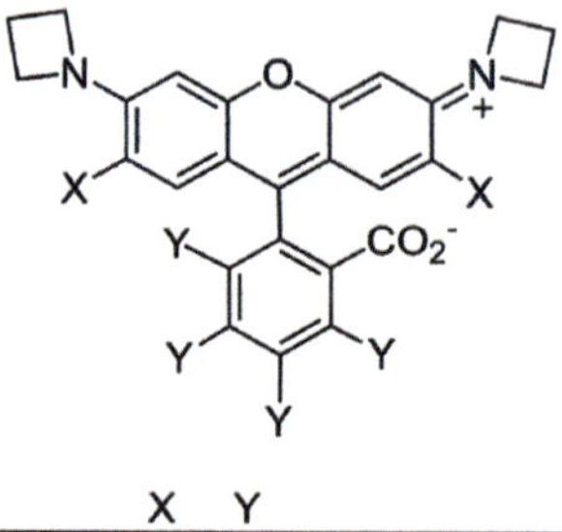

	X	Y			
JF552	F	H	552	23	79.0
JF571	H	F	571	19	78.7
JF574	F	F	574	20	80.3

(All in HEPES pH 7.3)

Figure 4.6 Structures of the rhodamines that have a wavelength of absorption maximum between 550 and 600 nm. Legend as in Figure 4.4.

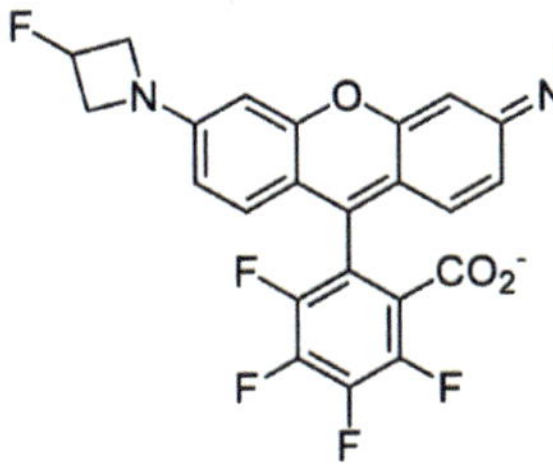

JF559
559 20 90.1 (10 mM HEPES pH 7.3)

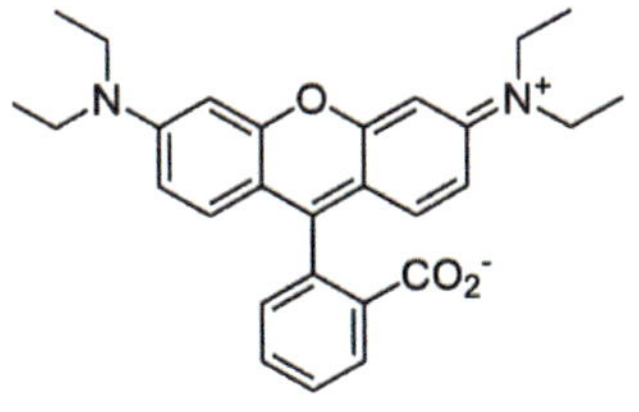

Tetraethylrhodamine
557 23 25.2 (Water pH 7)

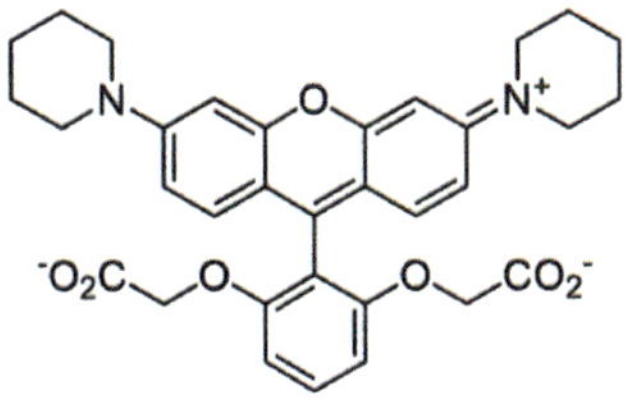

W-B 15
567 27 9.9 (100 mM P pH 7.4)

Figure 4.6 (Continued)

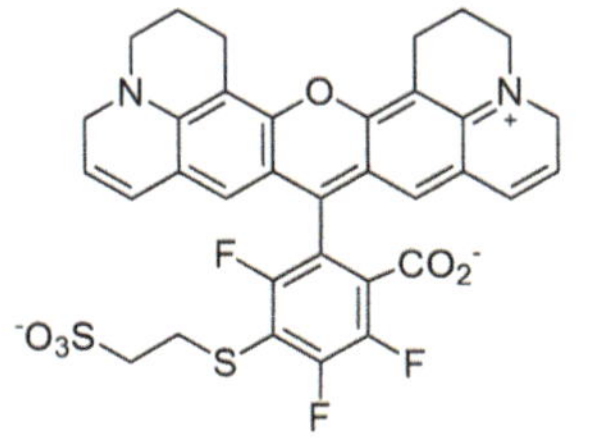

Abberior Star 600
604 19 74.4 (PBS pH 7.4)

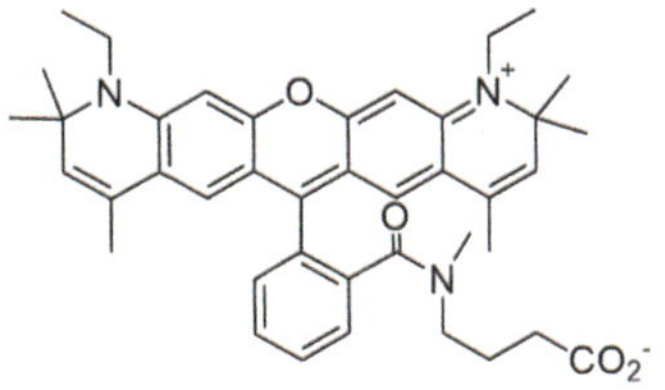

Atto Rho13
603 24 96.0 (PBS pH 7.4, 22 °C)

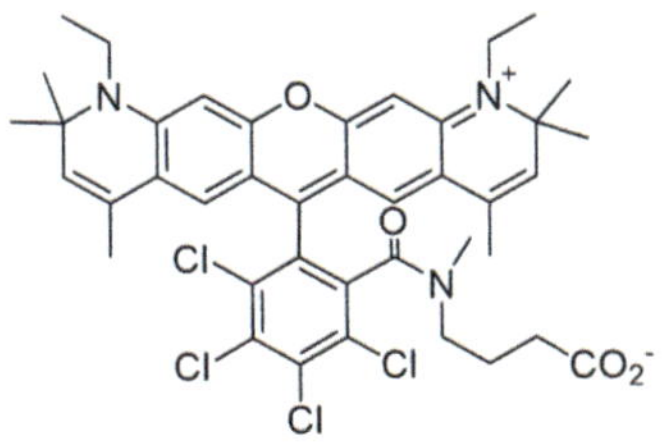

Atto Rho14
626 20 112.0 (PBS pH 7.4, 22 °C)

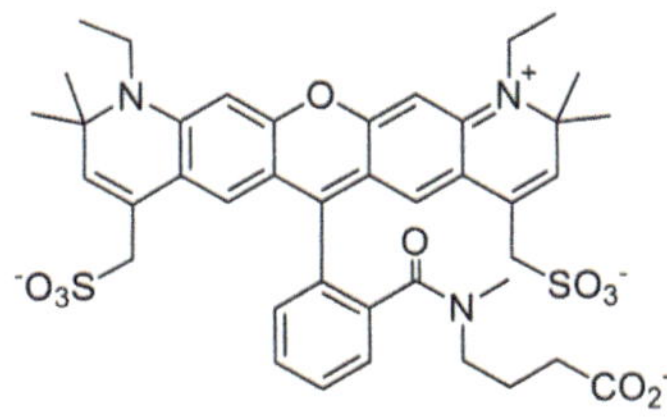

Atto 594
603 23 102.0 (PBS pH 7.4, 22 °C)

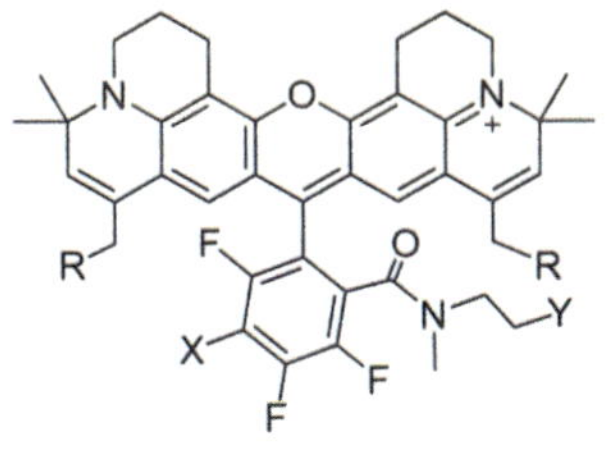

	X	Y	R			
KK1119	F	$CH_2CO_2^-$	SO_3^-	637	23	49.5
KK9046	F	$CH_2CO_2^-$	OH	632	22	40.5
KK114	F	CO_2^-	SO_3^-	637	23	50.6
Star 635	Z	$CH_2CO_2^-$	OH	634	20	32.1
Star 635P	F	$CH_2CO_2^-$	OPO_3H^-	635	20	44.0
Star Red	F	OCO_2^-	SO_3^-	638	17	66.0

(All in PBS pH 7.4). Z = $S(CH_2)_2SO_3^-$

Figure 4.7 Rhodamines that have an absorption maximum of above 600 nm. Legend as in Figure 4.4.

4.2.5 Rhodamines With a High Net Charge

Labels with a high net charge are highly desired in many applications. One field is the so-called electrodriven separations (separations promoted by the application of an electric field) which may be performed on slab of gels (gel electrophoresis), microchannels on chips (microchip electrophoresis), or using fused silica capillaries (capillary electrophoresis). The use of fluorescent labels increases the sensitivity (detection limits) and separation selectivity. Moreover, labels with a high net charge maximizes both the separation efficiency (number of theoretical plates) and time efficiency (number of theoretical plates divided by migration time squared) (see chapter 5 of [30]).

Within this database there are only a few classes that contain highly charged entities, the most outstanding being cyanines, merocyanines, and pyrenes. The rhodamines, which contain the brightest entities among all small organic dyes, have only a few dyes with a high net charge: Abberior Star 488 carboxylate ([15, 16] molecular structures were kindly provided by the supplier); Alexa Fluor 488, Alexa Fluor 546, Alexa Fluor 568, and Alexa Fluor 594 [6, 19]; DY-490 carboxylate [20]; and Rhodamine 6G +3 [25] (shown in Figure 4.8).

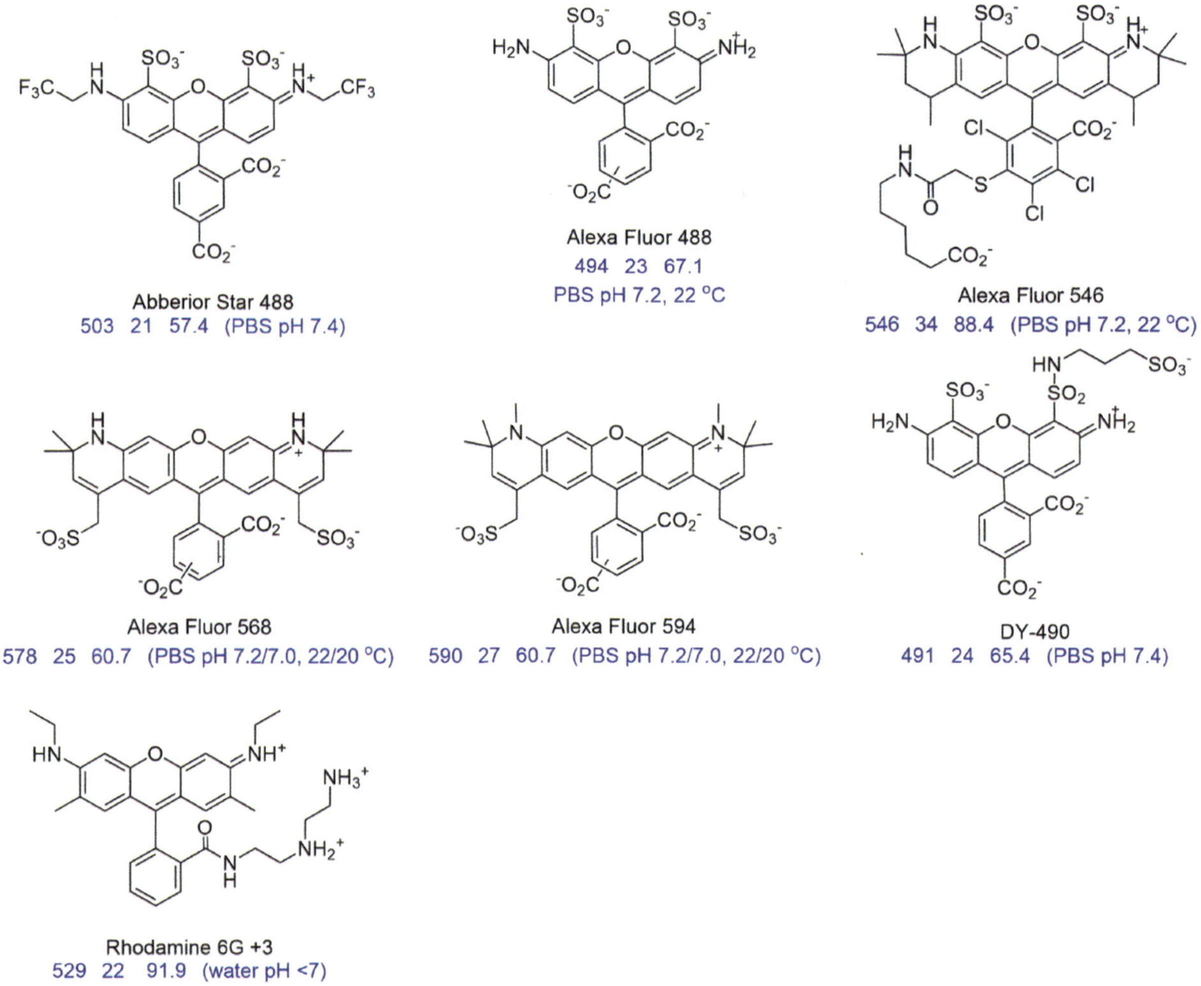

Figure 4.8 Rhodamines that have a high net charge. Legend as in Figure 4.4.

4.3 HAS-Rhodamines

In HAS-rhodamines, the oxygen atom of the xanthene core structure is replaced by another atom such as B, C, Ge, P, S, Si, and even Te (Figure 4.9). This is the main structural difference between rhodamines and HAS-rhodamines, and it produces an average red-shift of about 60 nm (depending on the new heteroatom) in the wavelength of excitation maximum of the longest wavelength band. This heteroatom substitution is an interesting way to increase the number of rhodamine derivatives that have absorption maxima deep in the red and infrared regions of the spectrum. Over 40 HAS-rhodamines have already been synthesized and some of them are already commercially available from many suppliers.

4.3.1 Carbo-Rhodamines

The largest group of HAS-rhodamines is made of dimethylcarbo-rhodamines ($C(CH_3)_2$-rhodamines). This group includes: carborhodamine [22, 23]; tetramethyl-carborhodamine or CTMR and JF608 [26, 27]; pyrrolidinyl-carborhodamine [23]; and JF585 [27]. The R in the molecular structure of 560CP [22] indicates a ligand or linker-ligand in dye-ligand conjugates. Kolmakov et al. [31] synthesized a series of carborhodamines with many degrees of hydrophilicity, which form dicationic, monocationic, zwitterionic, and up to dianionic species in the neutral pH range (see carborhodamines 1b, 1c, 16a, 20a, 20b, 21a, 21b, and 22 in Figure 4.9). Finally, there are also some commercial carborhodamines, such as Atto 647N [2, 5], Atto 620, and Atto 633 [5].

4.3.2 Silico-Rhodamines

The second largest group of HAS-rhodamines is made of silico-rhodamines (Si-rhodamines). It contains only dimethylsilane-rhodamines ($Si(CH_3)_2$-rhodamines) and no one disilanol-rhodamine ($Si(OH)_2$-rhodamine) (Figure 4.10). Examples include: SiR and 640SiRH [22]; SiRF and SiR-COOH [28]; thienyl-JF648, JF669, JF669-6-SCH2CO2Et, and JF669-6-SCH2CO2H [32]; and JF646 [27]. JF646 should not be confused with S-JF646 (a HAS-pyronine) or with Sulfo-JS646 (a HAS-sulforhodamine).

4.3.3 Other HAS-Rhodamines

Ge-Rhodamines (germanium-rhodamines) contain a germanium atom as a heteroatom and $Ge(CH_3)_2$-rhodamines are such an example. Both 630GeRH and GeR [22] belong to this sub-class. The R in the molecular structure of GeR [22] indicates a ligand or linker-ligand in dye-ligand conjugates (Figure 4.11). S-rhodamines are made of thio-rhodamines such as JF570 and JF593 [27] and sulfone-rhodamines such as JF724 [27]. There are also important P-rhodamines made of phosphinate-rhodamines JF688 and JF690 (Grimm 2020) and phenilphosphine oxide-rhodamines such as JF711 and JF722 [27]. These latter two rhodamine derivatives with this heteroatom substitution (X = POPh) exhibit an impressive redshift into the NIR while keeping a relatively small structure.

4.4 Pyronines

Pyronines are characterized by the presence of the three-ring xanthene (9H-10-oxaanthracene) and one iminium group bound to carbon 3 and one amino group bound to 6. This is exactly like rhodamines, but without the pendant aryl group bound to carbon 9. These structures are therefore smaller than rhodamines (Figure 4.12) and have moderate to high brightness. Examples of dyes belonging to this class include: Atto 520 [5]; acridine red (2465-29-4, despite the name widely adopted, it is a pyronine and not an acridine), pyronine B and pyronine Y (OP) in water [14]; pyronine Y (OP) in PBS pH 7.4 [33, 34]; pyronine 630 [35]; and AR116 [36]. The molar absorption coefficient of pyronine B was taken from Nau and Mohanty [37] and for acridine red and pyronine 630 they were estimated

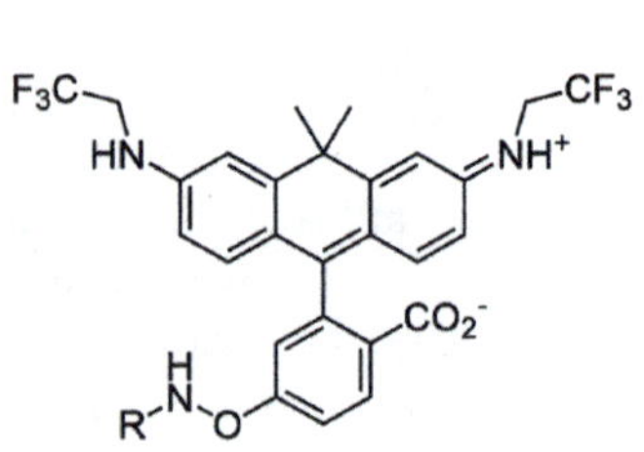

560CP
561 27 46.3 (PBS pH 7.4, RT)

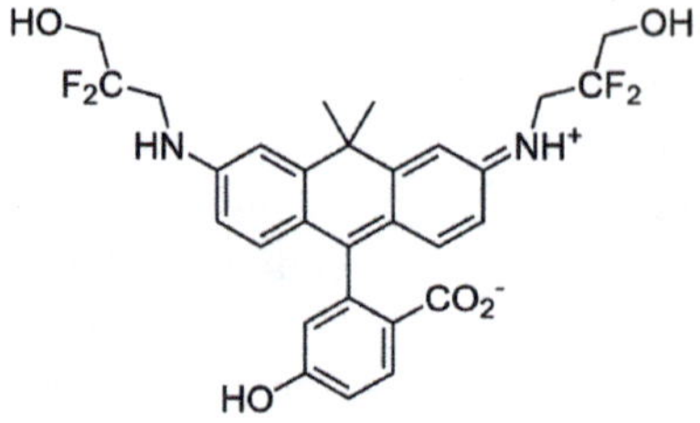

570CPH
571 29 56.0 (PBS pH 7.4, RT)

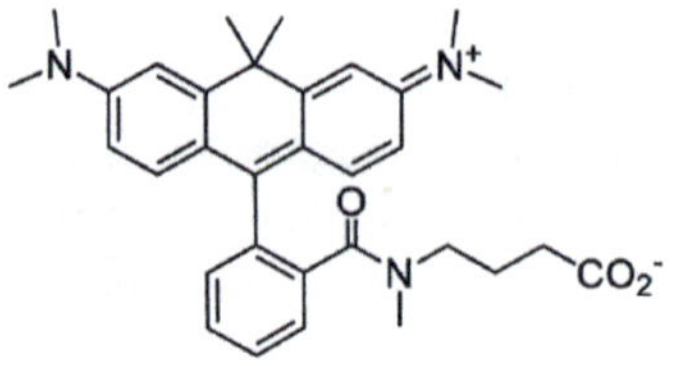

Atto 620
620 22 60.0 (PBS pH 7.4, 22 °C)

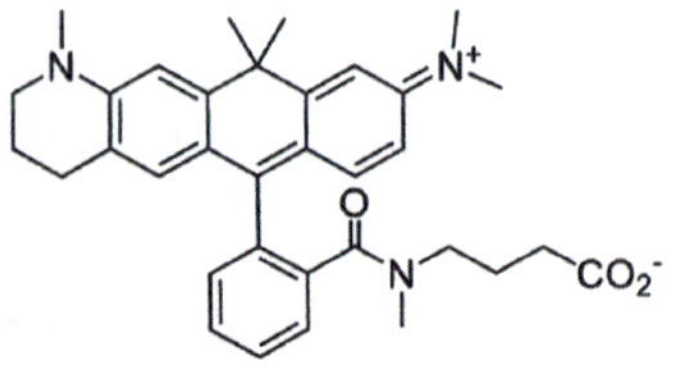

Atto 633
630 21 83.2 (PBS pH 7.4, 22 °C)

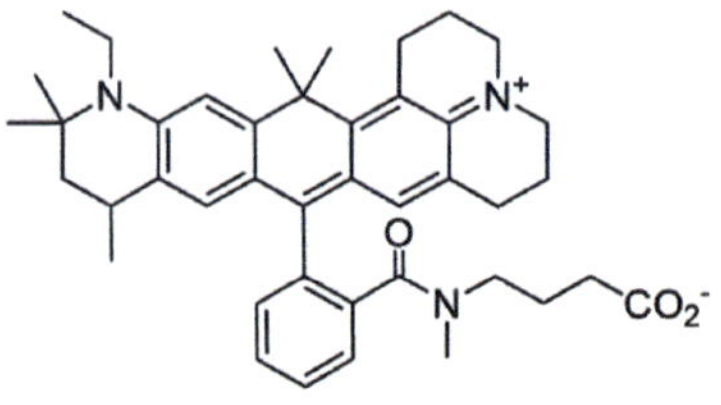

Atto 647N
646 18 97.5 (PBS pH 7.4, 22 °C)

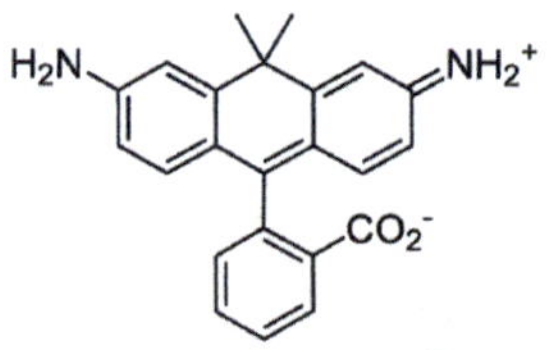

Carborhodamine
552 25 41.6 (10 mM HEPES pH 7.3)

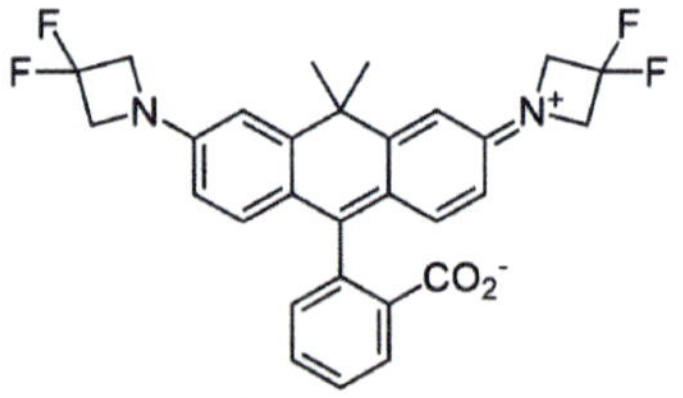

JF585 585 24 1.1
(10 mM HEPES pH 7.3)

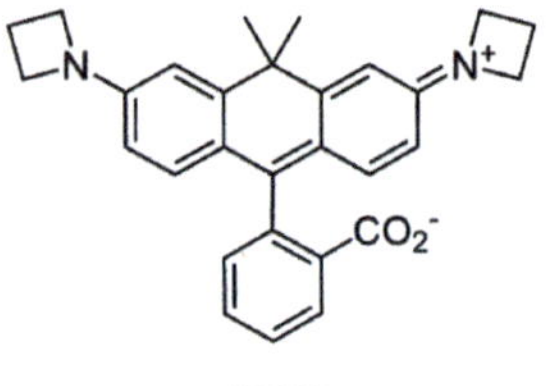

JF608
608 23 66.3 (10 mM HEPES pH 7.3)

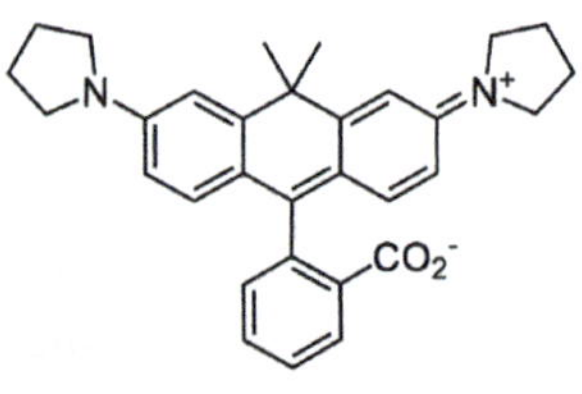

Pyrrolidinylcarborhodamine
613 20 46.9 (10 mM HEPES pH 7.3)

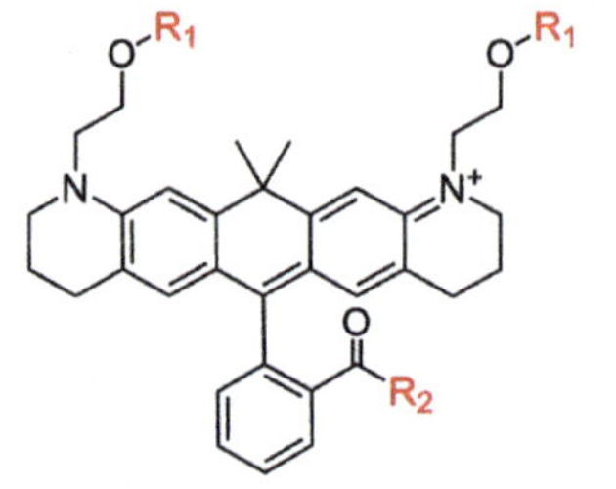

Carborhodamines			
1b	639	22	46.4
1c	640	22	47.1
16a	636	21	70.6
20a	640	21	37.1
20b	641	22	42.4
21a	640	23	39.2
21b	640	24	38.0
22	640	24	36.1

All in water

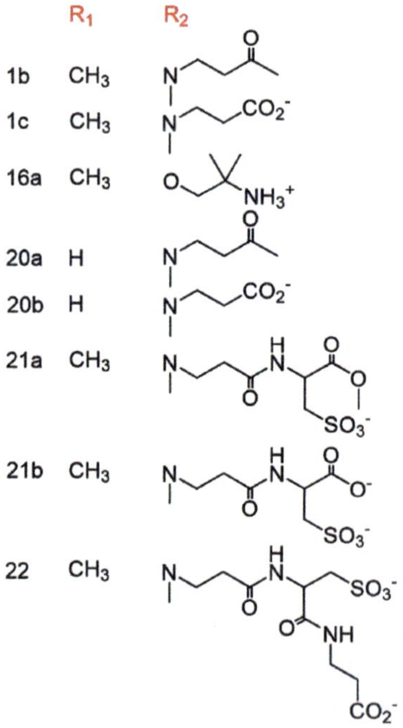

Figure 4.9 Structures and photophysical properties of carbo-rhodamines, more specifically dimethylcarbo-rhodamines ($C(CH_3)_2$-rhodamines). Numbers in blue represent $\lambda_{x,max}$ (nm), Δ_S (nm), and B_{max} ($mM^{-1}cm^{-1}$), respectively.

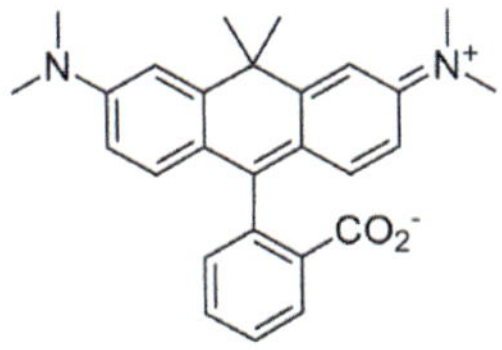

Tetramethylcarborhodamine
606 20 62.9 (10 mM HEPES pH 7.3)

Figure 4.9 (Continued)

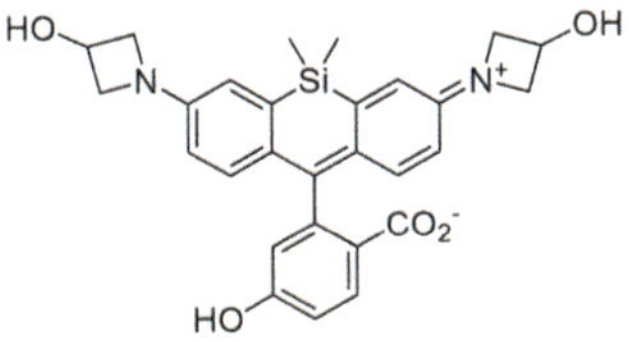

640SiRH
641 21 21.4 (PBS pH 7.4)

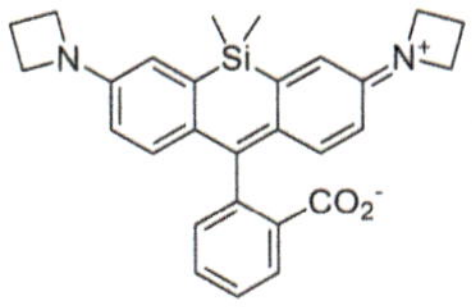

JF646 646 18 3.0
(10 mM HEPES pH 7.3)

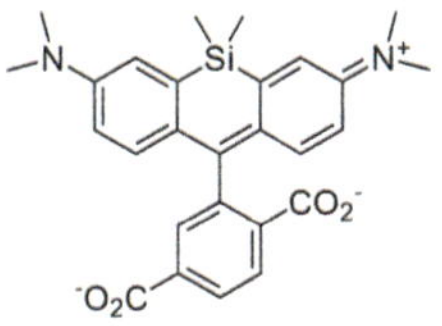

SiR-COOH
645 16 39.0 (water)

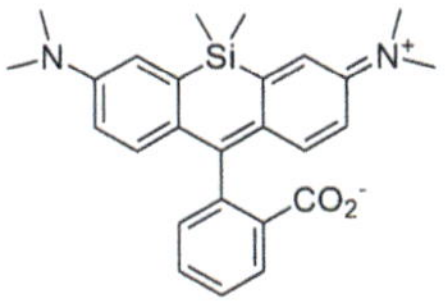

SiR
645 16 38.1 (PBS pH 7.4, RT)

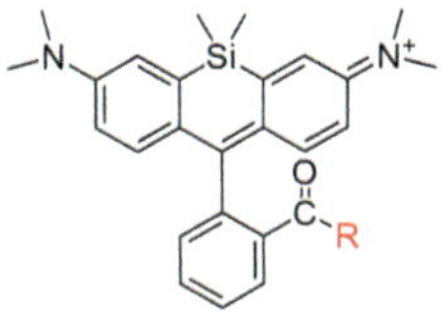

SiRF 1a-H: R = H

SiRH 1b-H: R =

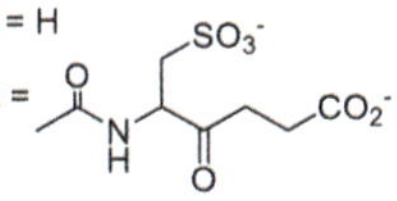

SiRF 1a-H	662	18	77.8	(water)
SiRF 1b-H	663	17	63.0	(water)

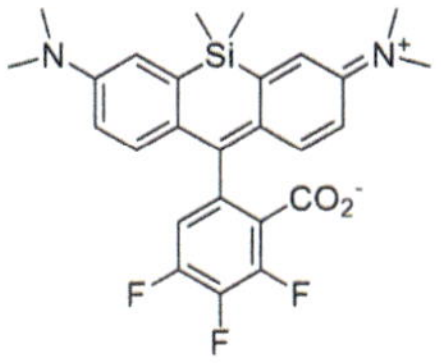

SiRF 13a
657 20 33.3 (water)

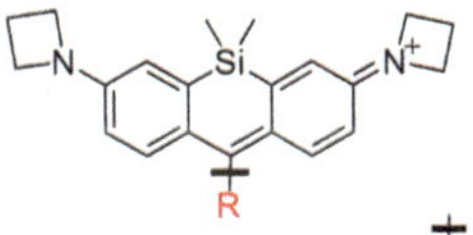

R =

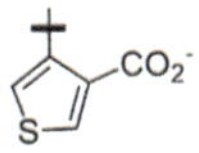

Thienyl-JF646

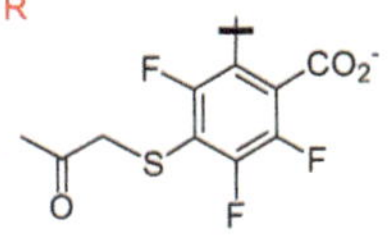

JF669-6-SCH$_2$CO$_2$Et

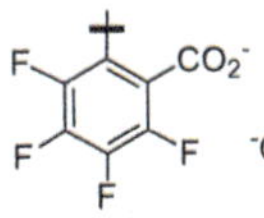

JF669

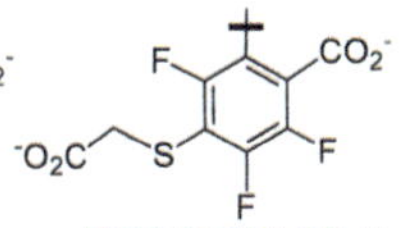

JF669-6-SCH$_2$CO$_2$H

Thienyl-JF646	651	16	45.2
JF669	669	13	41.4
JF699-6-SCH$_2$CO$_2$H	667	14	49.4
JF699-6-SCH$_2$CO$_2$Et	668	13	45.2

All in 10 mM HEPES pH 7.3.

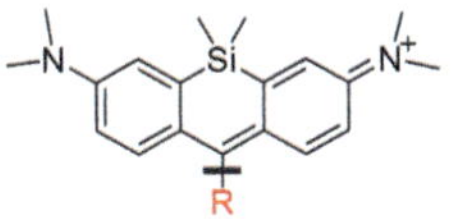

R =

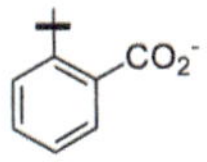

SiTMR

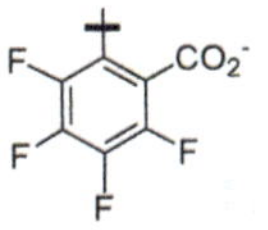

4,5,6,7-Tetrafluoro-SiTMR

SiTMR	643	19	11.5
4,5,6,7-Tetrafluoro-SiTMR	667	15	40.9

All in 10 mM HEPES pH 7.3.

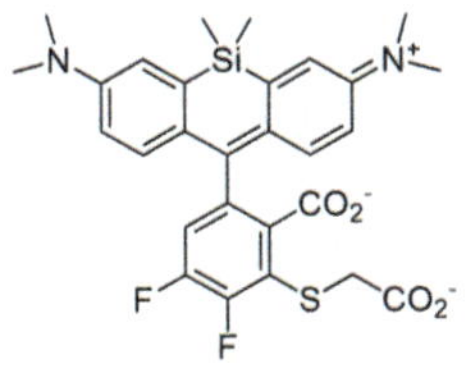

SiRF 15a
657 19 16.0 (water)

Figure 4.10 Structures and photophysical properties of silico-rhodamines. This set is made of many dimethylsilane-rhodamines (Si(CH$_3$)$_2$-rhodamines). Legend as in Figure 4.9.

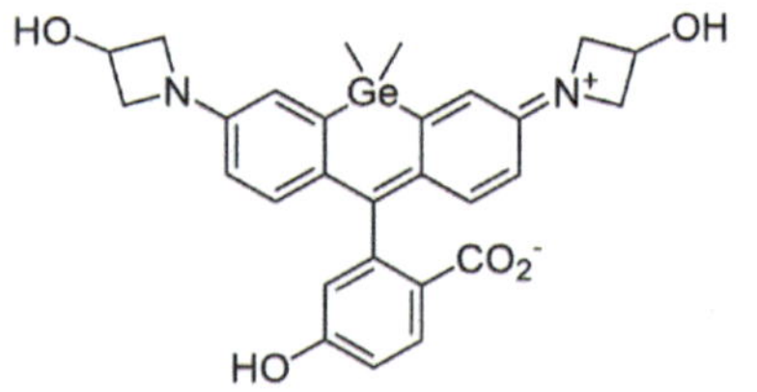

630GeRH
631 20 36.6 (PBS pH 7.4, RT)

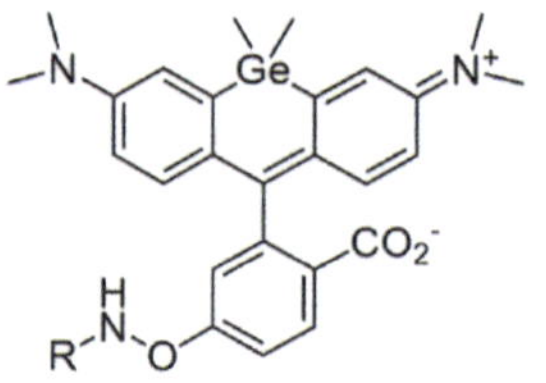

GeR
634 21 41.7 (PBS pH 7.4, RT)

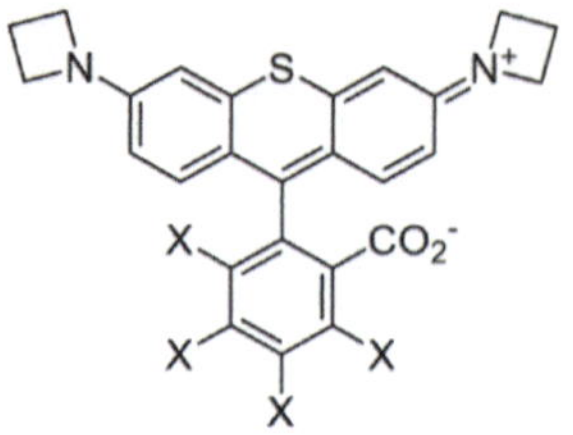

				X
JF570	570	23	52.6	H
JF593	593	19	49.6	F

All in 10 mM HEPES pH 7.3

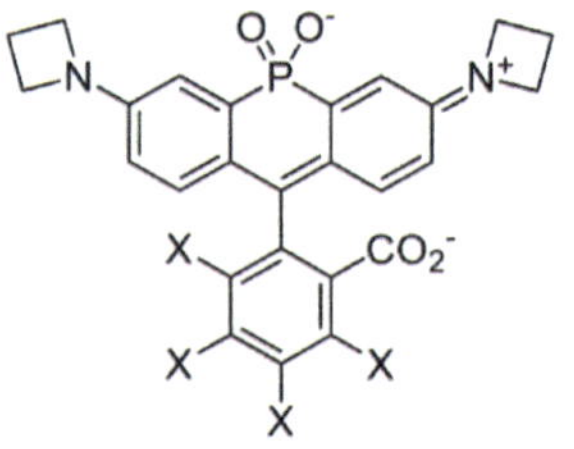

				X
JF668	668	19	9.0	H
JF690	690	17	36.0	F

All in 10 mM HEPES pH 7.3

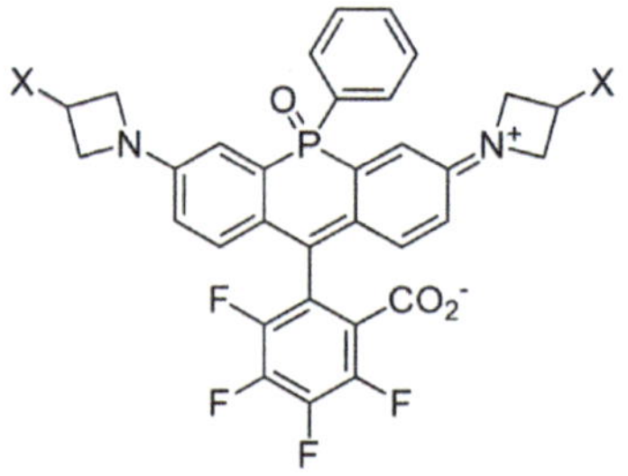

				X
JF711	711	21	2.1	F
JF722	722	21	9.5	H

All in 10 mM HEPES pH 7.3

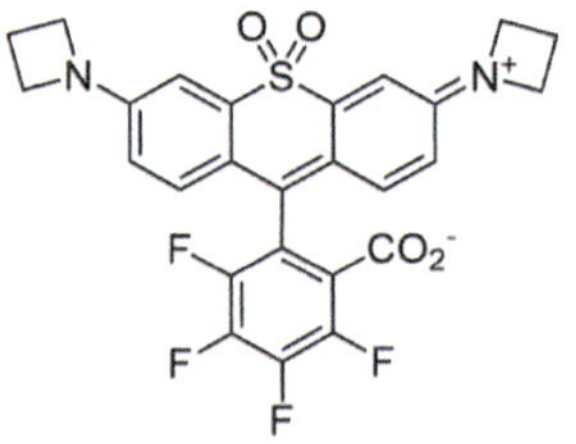

JF724
724 24 0.3 (10 mM HEPES pH 7.3)

Figure 4.11 Structures and photophysical properties of additional HAS-rhodamines. Legend as in Figure 4.9.

from similar structures by the author of the present book. The measurement of some photophysical parameters of pyronines, including the molar absorption coefficients, is challenging, as they tend to form aggregates in aqueous solutions. Pyronine B and pyronine Y form H- and J-aggregates. The changes in the photophysical properties of these aggregates in aqueous polystyrene sulfonate solutions are presented in Arik et al. [38].

Another series of new pyronines termed PONy (because they contain a phosphorus atom, an oxygen and a nitrogen in close proximity) are PONy P1, PONy P1-Halo, PONy P2, PONy P5, PONy P6, PONy P9, PONy P10, and PONy P13 [33]. They contain phosphinate, phosphonate, or phosphonamidate groups bound to the position 9 of pyronine and are compact dyes with redshifted absorption and emission bands and Stokes shifts up to 44 nm (8200 cm^{-1}).

4.5 HAS-Pyronines

Heteroatom-substituted pyronines (HAS-pyronines) are similar to HAS-rhodamines, the only difference being the absence of the pendant aryl group. They exhibit significant red-shifted wavelengths of absorption and emission maxima when compared to pyronines. This makes HAS-pyronines a good choice when a small, bright, and red-absorbing and emitting label is desired. C-pyronines or carbopyronines [31] was one of the first subclasses and continues to be important today. The HAS-pyronine class still receives the addition of new important entities, such as B-pyronines and Si-pyronines (Figure 4.13).

B-Pyronines (bora-pyronines) contain a boron atom as a heteroatom. In $B(OH)_2$-pyronines (dihydroxyborate-pyronines), the boron appears as a borinate, which is made of two B–C bonds and two B–O bonds. This bonding is the result of the coordination of a boronic acid molecule (made of two B–C bonds and one B–O bond) and an OH^- anion. The compound JS-R ($B(OH)_2$-pyronine) [39, 40] is one such example (Figure 4.13).

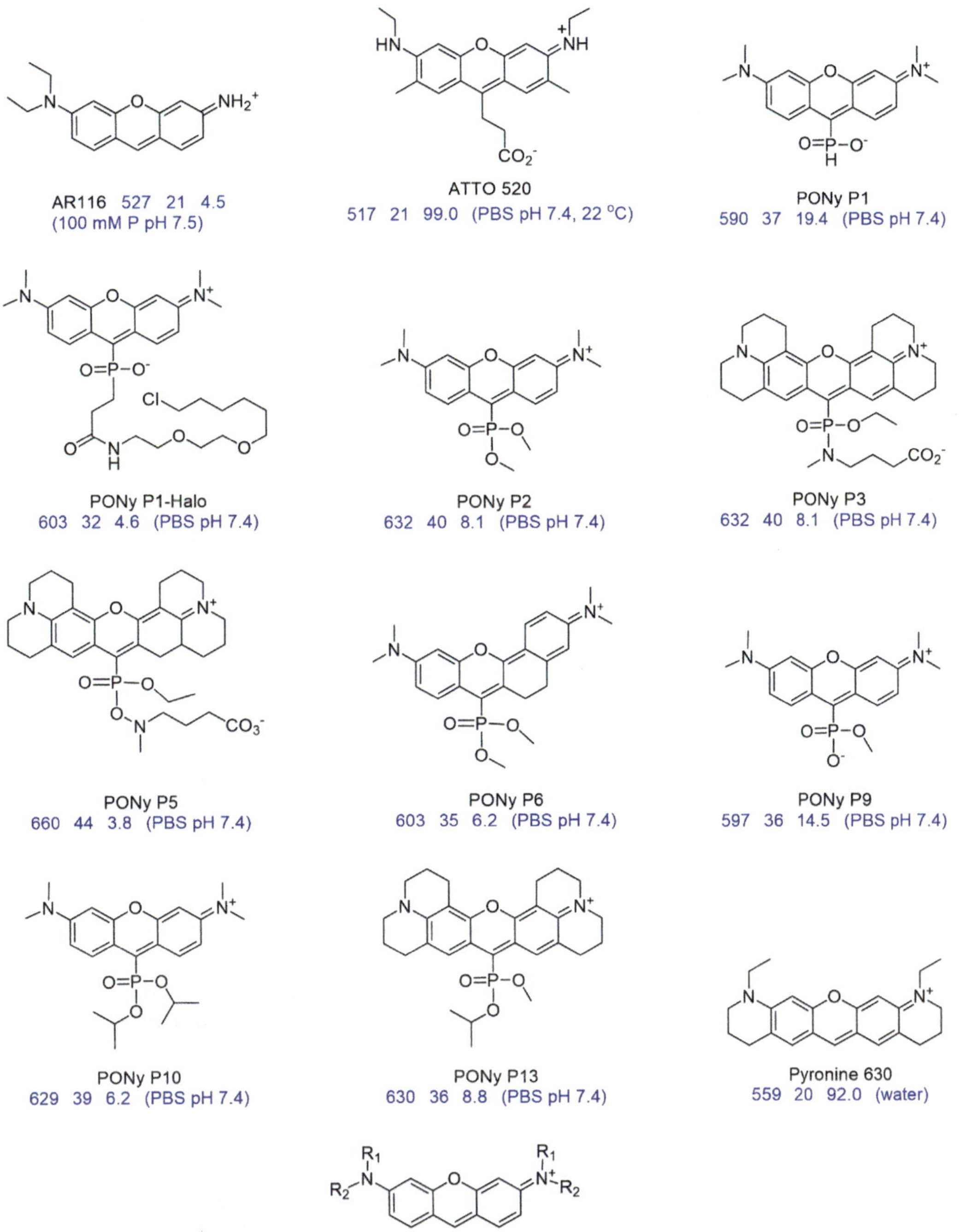

	R_1	R_2				
Pyronin B	CH_2CH_3	CH_2CH_3	552	20	16.9	(water)
Pyronin Y	CH_3	CH_3	546	20	16.9	(water)
Pyronin Y	CH_3	CH_3	546	23	27.7	(PBS pH 7.4)
Acridine red	H	CH_3	528	34	51.0	(water)

Figure 4.12 Molecular structure and some photophysical properties of pyronine derivatives in aqueous solutions. Numbers in blue represent $\lambda_{x,max}$ (nm), Δ_S (nm), and B_{max} (mM^{-1}cm^{-1}), respectively.

C-Pyronines (carbo-pyronines) such as $C(CH_3)_2$-pyronines (dimethylcarbo-pyronines) contain a few of the brightest structures currently available, rivaled only by rhodamines. Examples of such dyes include Atto 635 [3], Atto 610, and Atto 647 [5]. The structure of Atto 665 has not yet been disclosed by the supplier, but the photophysical data of this database suggests that it is probably a carbopyronine. Another dye of this class and with an innovative structure is PONy CP1 (Butkevich et al. 2018) [33].

Si-Pyronines (silico-pyronines) are mainly made of $Si(CH_3)_2$-pyronines (dimethylsilane-pyronines or simply sila-pyronines). Examples of entities in this class include 3,6-azetid.-Si-pyronine, 3,6-azetid.-9-CA-Si-pyronine, and S-JF646 (3,6-Bis(azetidinyl)-9-propionic acid-Si-pyronine) [32]. S-JF646 is an interesting structure despite its low brightness; it may be classified as an HAS-rhodamine if the pendant carboxy group takes the β-propiolactone form (3,7-di(azetidin-1-yl)-5,5-dimethyl-3′,4′-dihydro-5*H*,5′*H*-spiro[dibenzo[b,e]siline-10,2′-furan]-5′-one). This suggests that at a higher pH, S-JF646 may exhibit a higher brightness than at a lower pH, in which it is predominantly in its β-propiolactone form. The dimethylsilane-pyronines SiP-F, SiP-Pr, and SiP-Ph have good fluorescence quantum yield values (0.20, 0.21, and 0.23, respectively) [41]; however, their molar absorption coefficients are unknown. Therefore, the average value of similar HAS-pyronines is used in the database. The fluorescence quantum yield of TMDHS (or SiP-Me) was measured as 0.18 in water by Fu et al. [42] and as 0.41 in 50 mM PBS 7.4 by Nie et al. [41]. Throughout the database, fluorescence quantum yield values tend to be slightly higher when measured in PBS than when measured in plane water. Si-9-Iminopyronines were synthesized by Horváth et al. [43] with some of them exhibiting extra-large Stokes shifts (Δ_S = 205 nm for Si-9-iminopyronine 2f at pH 11 and ionic strength of 0.1 mol dm^{-3} by adding KCl. 1% of DMSO was added as a cosolvent).

4.6 Sulforhodamines

Sulforhodamines are rhodamines that contain an *ortho*-sulfonate in place of the *ortho*-carboxylate. Note that the sulfonic group exhibits a higher rotation energy barrier against rotation around the aryl bond, which is important to keep the fluorescence quantum yield high. Moreover, sulfonic acid exhibits a much lower pK_a and molecules with a sulfonic group are much more water soluble than their carboxylic acid counterparts. In addition to solvation, which is observed in both carboxylate ($-CO_2^-$) and sulfonate ($-O\text{-}SO_3^-$) groups, sulfonate forms more hydrogen bonds with water molecules than carboxylate does.

The main entities of this small class are sulforhodamine 101 (Texas red) [6]; sulforhodamine B [20, 44, 45, 46]; SR101-110, SR101-6G, SR101-Me_2, and SR101-B [47]; and Alex Fluor 633 [6] (Figure 4.14). The molecular structure of Alex Fluor 633 was only recently disclosed [48].

SR101-NaphH_2 is considered a sulforhodamine analog as the π-electron system is significantly elongated. These benzoxanthene derivatives are also known as unsymmetrical sulfoxanthenes. Their core structure includes a benzene ring fused to the xanthene scaffold [47] (Figure 4.14).

4.7 HAS-Sulforhodamines

HAS-sulforhodamines are heteroatom-substituted sulforhodamines. Sulfo-JF646 (Figure 4.15), synthesized by the group of L.D. Lavis [32], is one entity belonging to this class. It absorbs deep in the red range of the spectrum ($\lambda_{a,max}$ = 655), matches with reliable and high-quality diode lasers operating in this region, exhibits a high molar absorption coefficient (ε_{max} = 124 $mM^{-1}cm^{-1}$), and possesses a good fluorescence quantum yield (Φ_f = 0.51). Another example of HAS-sulforhodamine is SMSiR, with $\lambda_{a,max}$ = 652 nm and was synthesized by the group of Prof T. Nagano [34].

4.8 Fluoresceins

Fluorescein was synthesized back in the 1870s by A. Baeyer [10] and its structure is a xanthene with a keto group bound to carbon 3, a phenol group bound to carbon 6, and a 2-benzoic acid bound to carbon 9. Both the carboxylic

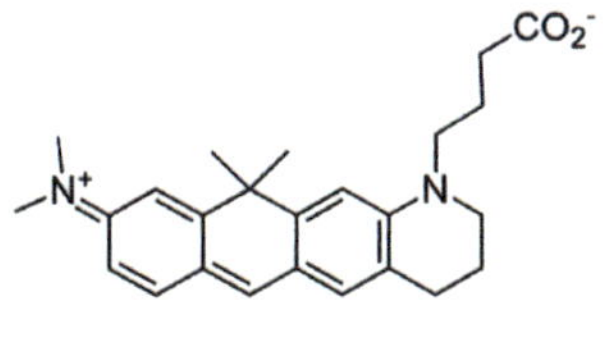

Atto 610
616 17 105.0 (PBS pH 7.4, 22 °C)

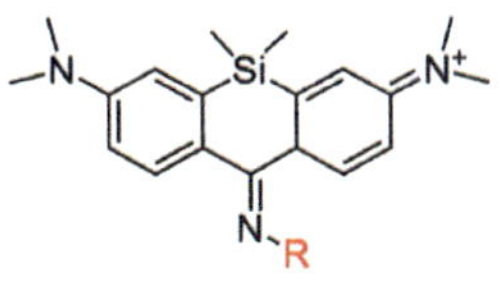

2c) R = 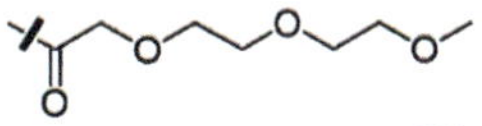

2f) R = 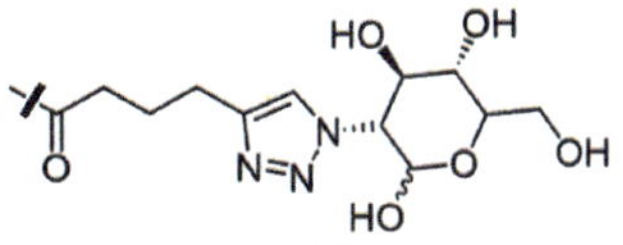

2g) R = 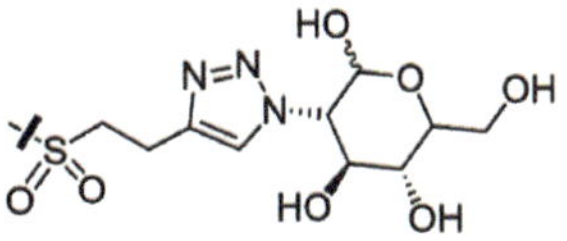

Si-9-Iminopyronines

2c)	666	15	23.3	(water pH 3 w/ 1% DMSO)
2c)	418	189	5.9	(water pH 11 w/ 1% DMSO)
2f)	664	18	31.9	(water pH 3 w/ 1% DMSO)
2f)	401	205	6.2	(water pH 11 w/ 1% DMSO)
2g)	469	162	4.7	(PBS pH 7.4)

Note: pH was adjusted with HCl or NaOH.

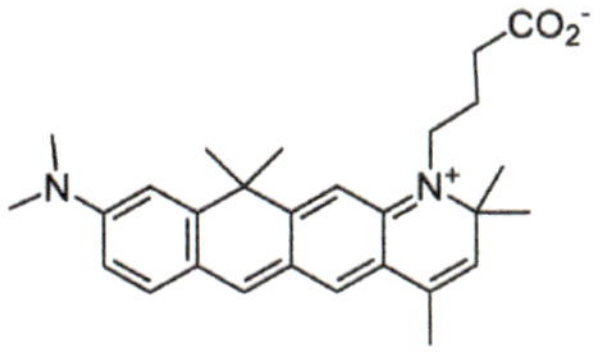

Atto 635
635 24 32.5 (PBS pH 7.4, 22 °C)

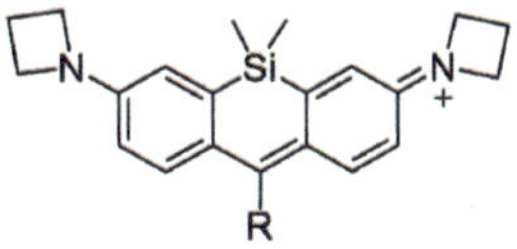

	R			
3,6-Azetid.-Si-pyronine	H	636	13	60.2
3,6-Azetid.-9-CA-pyronine	CO_2^-	641	16	31.2
S-JF646	$CH_2CH_2CO_2^-$	652	16	5.3

All in 10 mM HEPES pH 7.3

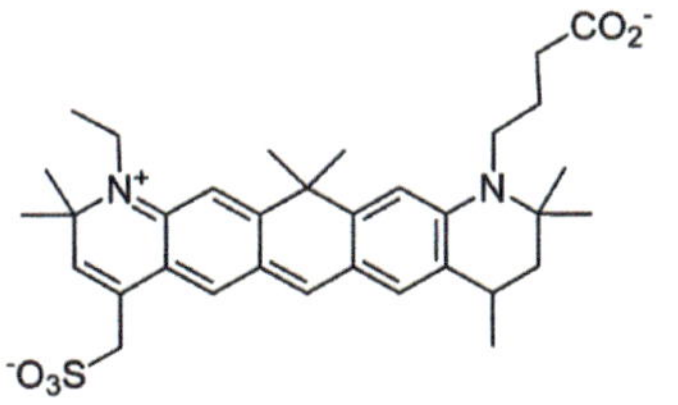

Atto 647
647 20 24.0 (PBS pH 7.4, 22 °C)

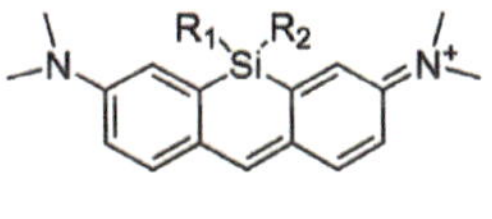

SiP-Me	633	21	26.3,	$R_1 = R_2 = CH_3$
SiP-F	637	14	12.8,	$R_1 = CH_3$, $R_2 = (CH_2)_2CF_3$
SiP-Ph	649	16	18.4,	$R_1 = R_2 = Ph$
SiP-Pr	632	20	13.4,	$R_1 = R_2 = CH(CH_3)_2$

All in 30 mM PBS pH 7.4

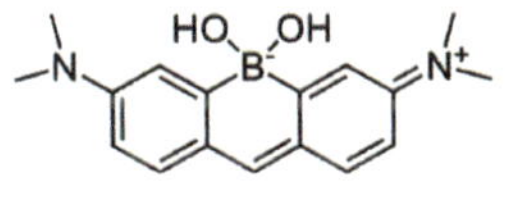

JS-R
611 20 76.7 (10 mM HEPES pH 7.3)

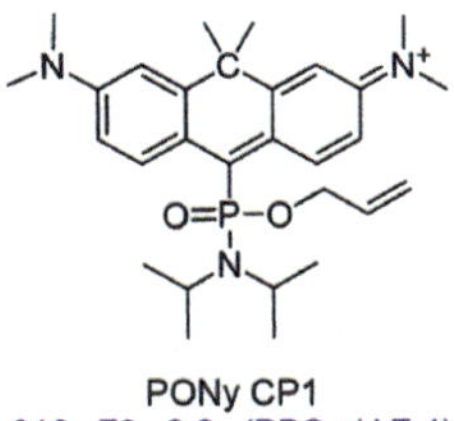

PONy CP1
610 73 3.2 (PBS pH 7.4)

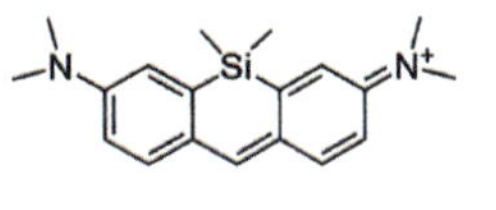

TMDHS
634 19 11.5 (water)

Figure 4.13 Structures and photophysical properties of some HAS-pyronines. Numbers in blue represent $\lambda_{x,max}$ (nm), Δ_S (nm), and B_{max} ($mM^{-1}cm^{-1}$), respectively.

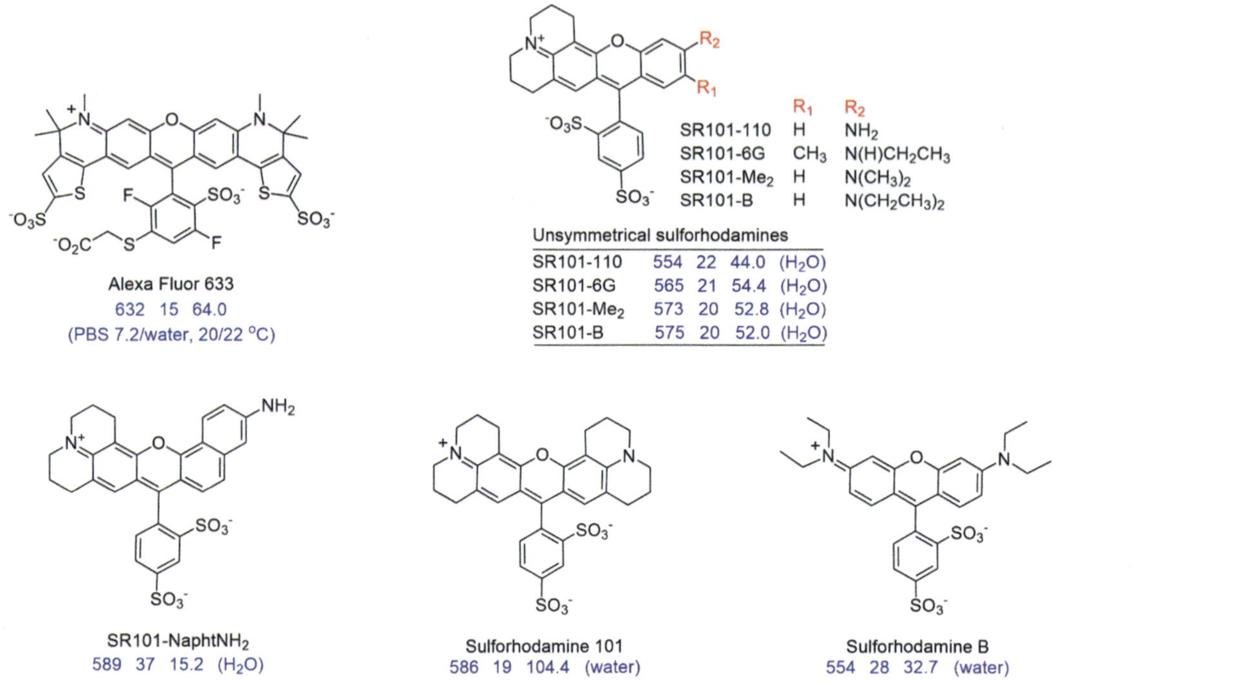

Figure 4.14 Structures and some photophysical properties of sulforhodamine 101, shulforhodamine B, Alexa Fluor 633 and some additional sulforhodamines and one sulforhodamine analog. These are highly hydrophilic, bright, and photorobust fluorophores. The numbers in blue show the photophysical parameters $\lambda_{x,max}$ (nm), Δ_S (nm), and B_{max} (mм$^{-1}$cm^{-1}).

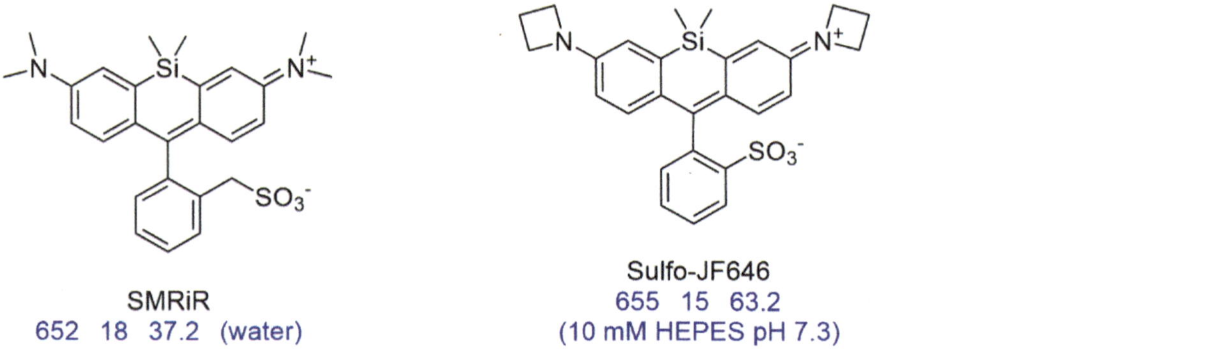

Figure 4.15 The structures of Sulfo-JF646 and SMRiR, which are HAS-sulforhodamine. The numbers in blue are $\lambda_{x,max}$ (nm), Δ_S (nm), and B_{max} (mм$^{-1}$cm^{-1}), respectively.

acid and phenol functional groups of fluorescein (Figure 4.16) are almost totally ionized above pH 9. This di-anion exhibits strong absorption in the blue region of the visible spectrum ($\varepsilon_{max} = \varepsilon_{490} = 76.9$ mм$^{-1}$m^{-1}) and strong fluorescence emission in the green ($\Phi_f = 0.93$). At neutral pH, the monoanionic and dianionic species coexist, although the former is more predominant. Acidification of the dianion first protonates the phenol group ($pK_3 = 6.43$) to yield the monoanion. This monoanion has lower molar absorption coefficient and fluorescence quantum yield values ($\varepsilon_{453} \approx \varepsilon_{472} = 29$ mм$^{-1}$m^{-1} and $\Phi_f = 0.38$, respectively). Further acidification then neutralizes the carboxylic acid group ($pK_2 = 4.31$) and produces a neutral structure with a lower molar absorption coefficient

($\varepsilon_{434} = 11\ \text{mM}^{-1}\text{m}^{-1}$) and fluorescence ($\Phi_f = 0.37$). Further acidification leads to the cationic form, which has an even lower fluorescence value ($pK_1 = 2.08$) [49]. Therefore, absorption and fluorescence are strongly pH dependent (Figure 4.16). These are useful properties for sensing tools in many applications; however, they may compromise method robustness in others. Moreover, fluoresceins are not photo-robust structures when compared to the great majority of fluorescent bright dyes. Despite these drawbacks, fluoresceins became very popular in the mid-twentieth century and are still used today in many areas, including as an intravenous marker in fluorescein angiography, which is a procedure used in ophthalmology to examine the retina (to show abnormal blood vessel growth and other conditions).

The use of halogen atoms [50], and other electron withdrawing groups as substituents to two or more hydrogen atoms of the parent xanthene structure, decreases the p*K*a of the phenol group, making them more pH insensitive around neutral pH. This knowledge, when combined with the use of other substituents, resulted in the synthesis of a large number of bright derivatives (fluoresceins) [51], which have played an important role in fluorescence microscopy, fluorescence confocal microscopy, fluorescence *in-situ* hybridization techniques, automated DNA sequencing of the first genome projects, real-time-PCR techniques, and many others.

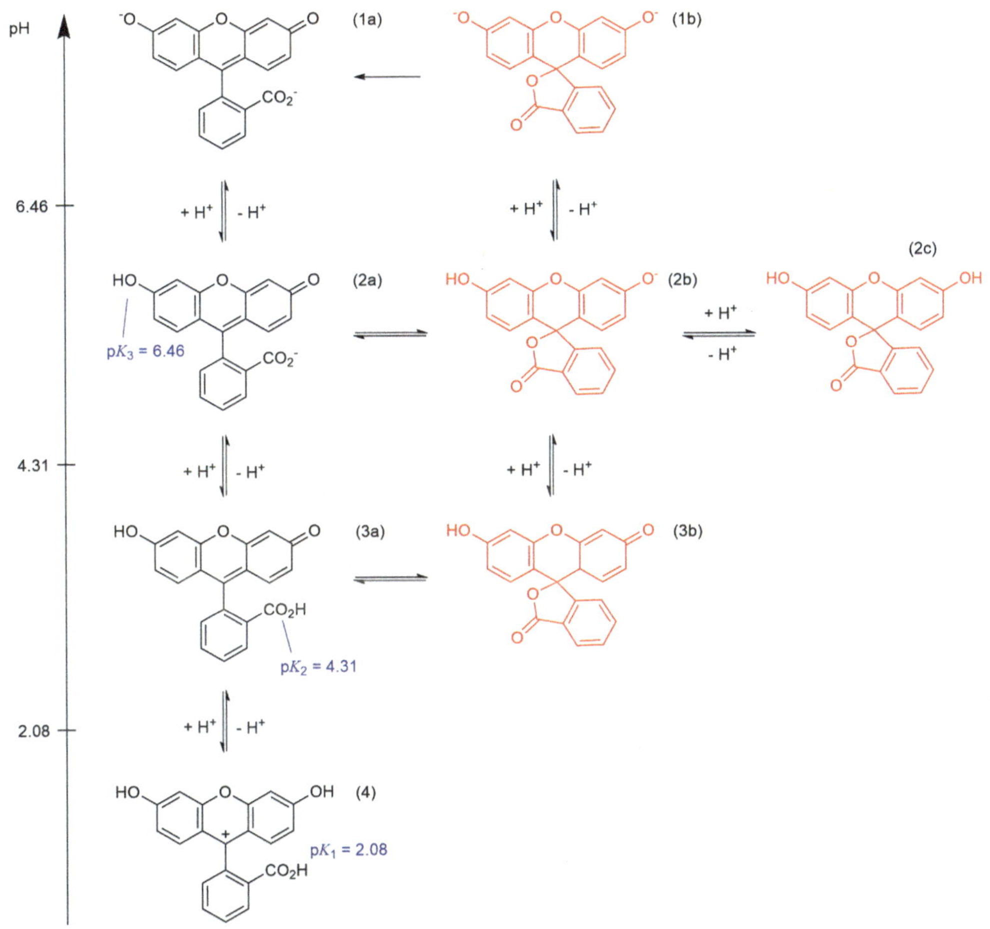

Figure 4.16 Some ionic forms of fluorescein and two lactonization reactions (2a and 3a). This reaction occurs in a wide range of pH and the products (lactones) exhibit a very low absorption and fluorescence in the visible.

Lactonization is also an important effect (red structures in Figure 4.16) as these lactones (closed forms) exhibit basically no absorption and fluorescence in the visible range of the spectrum. Figure 4.17 shows the electronic mechanism involved in the lactonization of one of the ionic species (monoanionic). This occurs not only with fluoresceins, mercaptofluoresceins, rhodols, and rhodamines, but also with sulfofluoresceins, mercaptosulfofluoresceins, sulforhodols, and sulforhodamines (when a sulfonic group is present instead of the carboxylic group). A potential example of a sulfur-rich class of dyes could be mercaptosulfothiofluoresceins; in this (probably stinky!) compound, the phenol has been replaced by a mercaptan, the *ortho*-carboxylic acid group by a sulfonic acid, and the heteroatom (O) of the xanthene core by a sulfur atom (thioxanthene).

4.8.1 Non-Halogenated Fluoresceins

The photophysical properties of fluorescein were measured and published by a few dozen authors. The following are a few examples: fluorescein in 0.1 M NaOH in water [21, 51, 52, 53]; fluorescein in 0.1 M NaOH in D_2O ([21] – there is a mistyping of $\lambda_{e,max}$ in this reference); fluorescein in phosphate pH 9 [50]; and fluorescein in aqueous buffer at 25°C and pH 9.18 [54] (Figure 4.18).

Non-halogenated fluorescein derivatives have a moderate to low photostability, exhibit pK_a values around neutral pH, and are more prone to lactonization. This makes these derivatives less suitable to be applied as fluorescent labels. Nevertheless, a few are interesting for some applications, such as intracellular pH indicators and singlet oxygen sensors, among others (for a review of fluorescent indicators for intracellular pH, see Han and Burgess [55]). Some examples of non-halogenated fluorescein derivatives are: DY-495 [20]; 5(6)-carboxyfluorescein (FAM) [6]; 5-carboxyfluorescein (S-H 2) [50]; NX, DMAX-EP, and 4-benzamidofluorescein [53]; 1,2,7,8-dibenzofluorescein [56]; BCPCF; and the highly charged fluorescein derivative BCECF [57, 58] (Figure 4.18). Small and highly charged fluorescent labels find interesting applications in electrodriven separations as they improve detection limits, selectivity, separation efficiencies, and time efficiencies all together (described in chapter 5 of [30]).

There are also many aminofluoresceins and nitrofluoresceins reported in the literature; however, their fluorescence quantum yield is low in aqueous solutions. Aminofluorescein is used in fluorescence-guided glioma surgery [59], for labeling and separation of saccharides and oligosaccharides [60, 61] and other analytes.

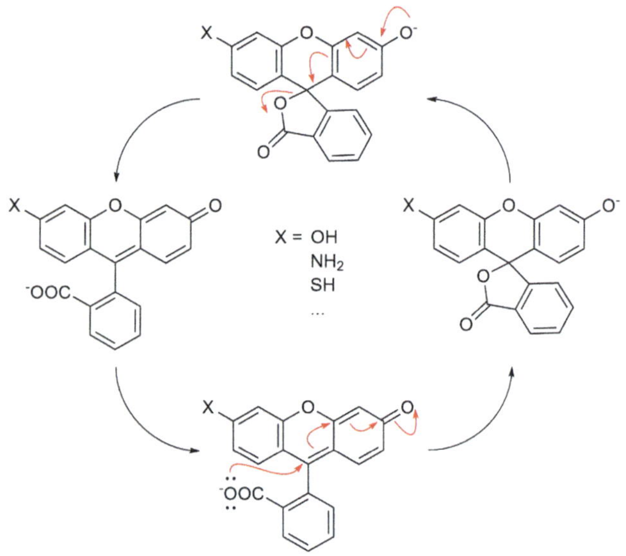

Figure 4.17 Detailed mechanism of lactonization and delactonization reactions of some monoanionic xanthene derivatives (fluorescein, rhodol, and mercaptofluorescein).

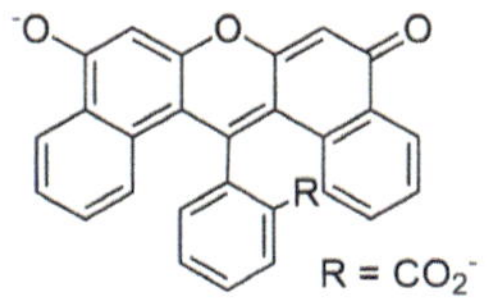

1,2,7,8-Dibenzofluorescein
534 31 19.4 (C pH 9.5)

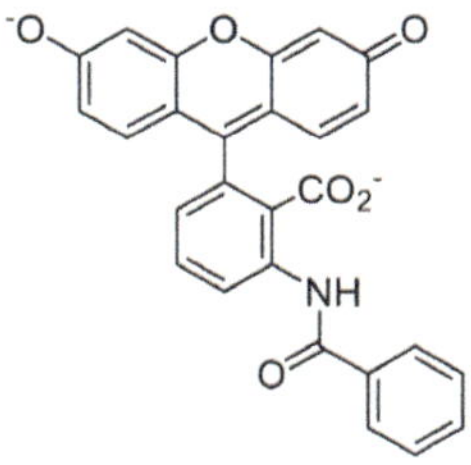

4-Benzamidofluorescein
490 27 63.2 (0.1 M NaOH)

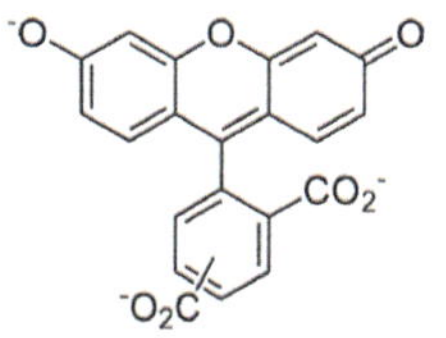

5(6)-Carboxyfluorescein (FAM)

495	25	74.7	(PBS pH 9)
492	24	77.9	(P pH 9)

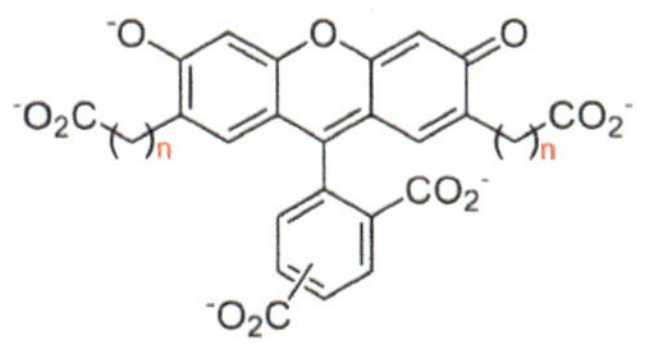

	n			
BCECF	2	505	22	56.2
BCPCF	3	505	22	55.6

All in NaOH 0.1 M

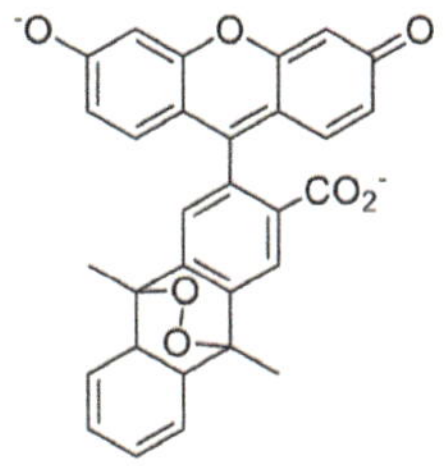

DMAX-EP
492 23 57.5
(100 mM P pH 7.5)

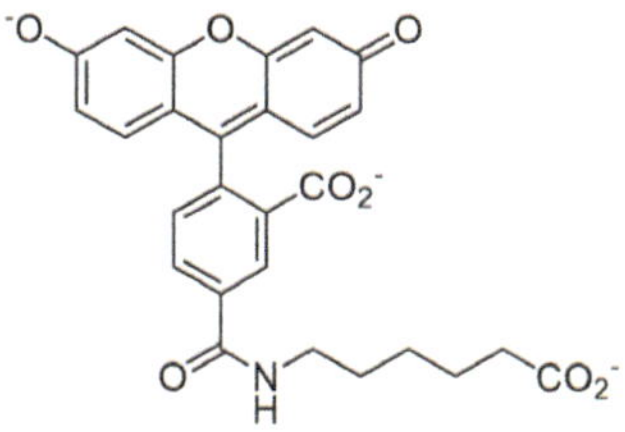

DY-495
494 27 53.2 (PBS pH 7.4)

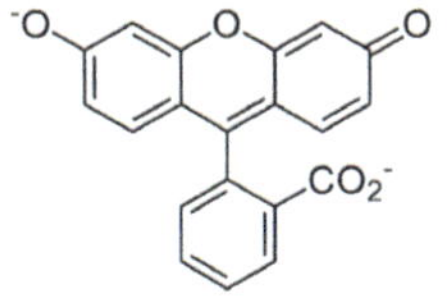

Fluorescein

490.2	26	61.6	(100 mM NaOH in water)
489.7	26	65.6	(100 mM NaOH in D_2O)
490	24	82.8	(Phosphate pH 9 in water)
492	22	70.5	(Aqueous buffer pH 9.18)

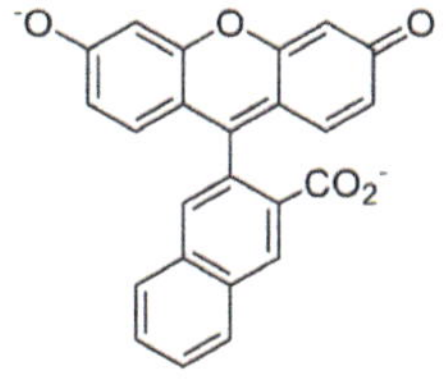

NX
492 20 57.2
(100 mM P pH 7.5)

Figure 4.18 Structures and some photophysical properties of "non-fluorinated fluoresceins." The numbers in blue show the photophysical parameters $\lambda_{x,max}$ (nm), Δ_S (nm), and B_{max} ($mM^{-1}cm^{-1}$).

4.8.2 Halogenated Fluoresceines

The photophysical properties of halogenated fluoresceins are less sensitive to pH around the neutral pH range. Examples of halogenated are: Oregon Green 488 [52, 62, 63]; Oregon Green 514 [6, 63]; 6-carboxy-2′,7′-dichlorofluorescein [63]; fluorinated fluoresceins S-H 17, 18, 19, and 20 [50]; fluorinated carboxyfluoresceins 14a, 14b, 15a, 15b, 16a, and 16b [50]; HEX, JOE, and TET [6, 64]; Yakima Yellow [65]; 2′,7′-Cl-4,5,6,7-F-fluorescein and 6-CA-MeThio-4,5,7-F-fluorescein [66]; 6-Azido-4,5,7-trifluorofluorescein [67]; and 1,2,7,8-Dibenzo-4,5-Br-fluorescein [56]. The structures of these halogenated fluoresceins are shown in Figure 4.19. The halogenated fluoresceins eosin Y (2-(2,4,5,7-Tetrabromo-6-oxido-3-oxo-3*H*-xanthen-9-yl)benzoate); eosin B (4′,5′-dibromo-3′,6′-dihydroxy-2′,7′-dinitro-1-spiro[isobenzofuran-3,9′-xanthene]one); erythrosine (2-(6-Hydroxy-2,4,5,7-tetraiodo-3-oxo-xanthen-9-yl)benzoic acid); and rose bengal (4,5,6,7-Tetrachloro-3′,6′-dihydroxy-2′,4′,5′,7′-tetraiodo-3*H*-spiro[[2]benzofuran-1,9′-xanthen]-3-one) are also well known, but their brightness is low in aqueous solutions and therefore not included in this database.

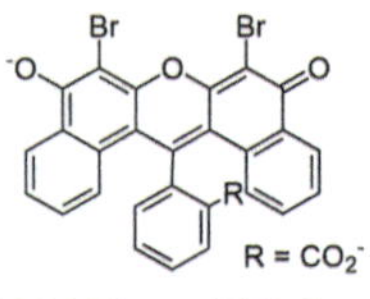

1,2,7,8-Dibenzo-4,5-Br-fluorescein
549 32 1.2 (C pH 9.5)

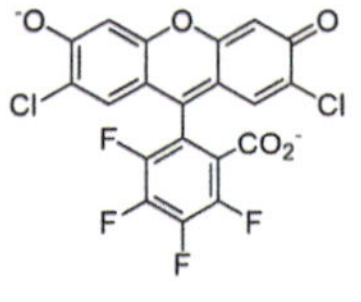

2',7'-Cl-4,5,6,7-F-fluorescein
520 18 70.8 pK_a = 4.2
(0.1 M NaOH)

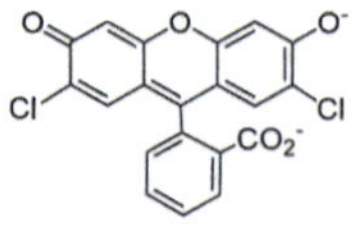

2',7'-Dichlorofluorescein
490 24 79.9 (PBS pH 9)

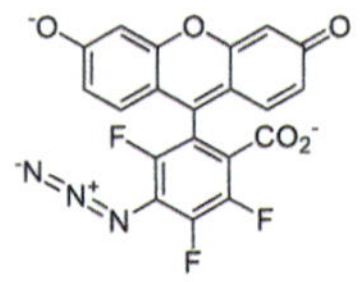

6-Azido-4,5,7-trifluorofluorescein
507 20 65.5 (0.1 M NaOH)

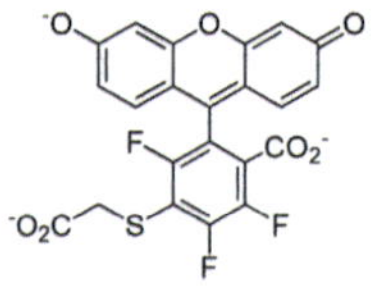

6-CA-MeThio-4,5,7-F-fluorescein
507 19 55.8 pK_a = 4.5
(0.1 M NaOH)

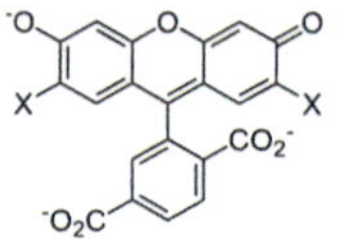

6-Carboxy-2',7'-dichlorofluorescein
X = Cl 492 24 79.0 (100 mM T pH 8.7)
6-Carboxy-2',7'-difluorofluorescein
X = F 492 25 79.0 (100 mM P pH 9)

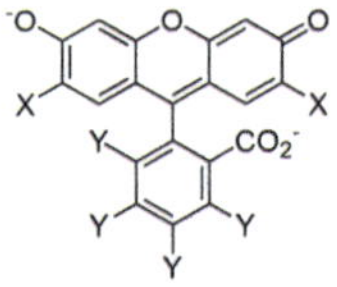

	X	Y				pK_a
G-C 1	Cl	H	503	19	88.5	4.7
G-C 5	H	Cl	511	16	76.9	6.0
G-C 6	Cl	Cl	522	15	54.6	4.3

All in aqueous buffer pH 9.18 and 25 °C.

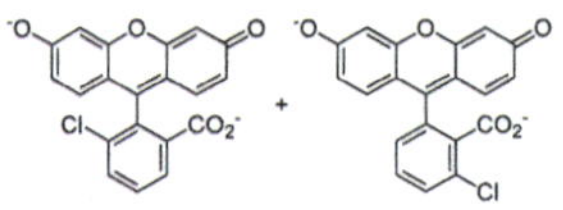

4(7)-Chlorofluorescein pK_a
G-C 2(3) 499 19 45.0 6.3
Aqueous buffer pH 9.18 and 25 °C.

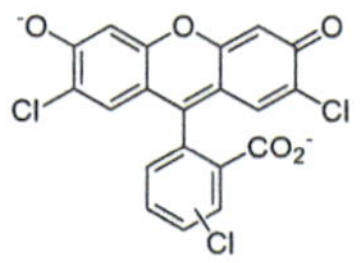

G-C 4 512 16 81.6 pK_a = 4.6
Aqueous buffer pH 9.18 and 25 °C.

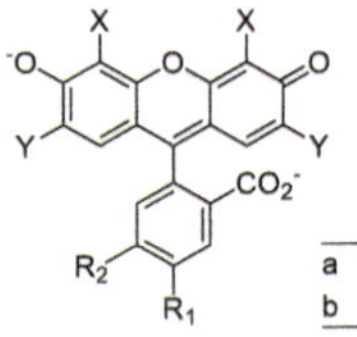

	R1	R2
a	CO_2^-	H
b	H	CO_2^-

	X	Y				pK_a
S-H 14a,b	H	F	492	24	79.9	4.8
S-H 15a,b	F	H	510	24	36.2	5.2
S-H 16a,b	F	F	510	24	46.0	3.7

All in phosphate pH 9.0.

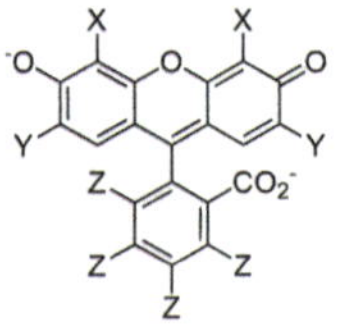

	X	Y	Z				pK_a
S-H 17	H	F	H	490	24	79.9	4.8
S-H 18	H	H	F	508	19	72.7	6.1
S-H 19	H	F	F	508	19	77.9	4.5
S-H 20	F	F	F	535	18	39.2	3.3

All in phosphate pH 9.0.

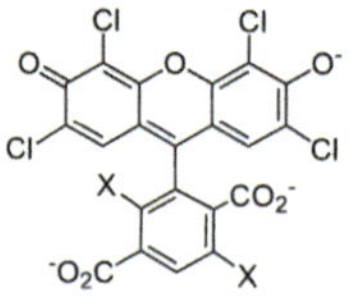

	X			
TET	H	521	15	51.1
HEX	Cl	535	24	68.6

Water pH 9.

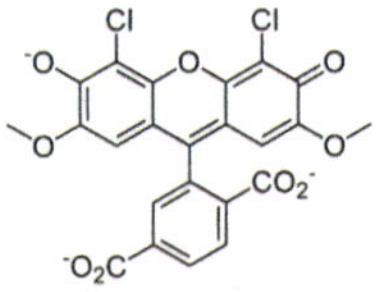

JOE
520 28 43.8 (water pH 9)

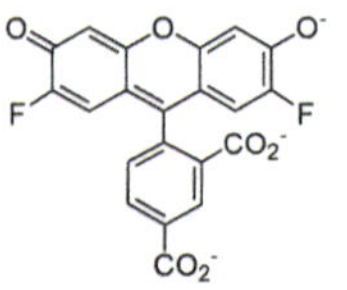

Oregon Green 488
490 24 73.7 (water pH 9)

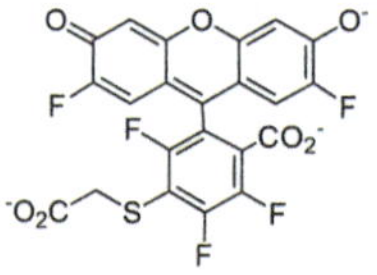

Oregon Green 514
511 19 84.4
PBS pH 7.2 22 °C.

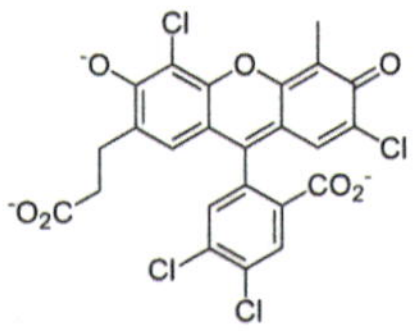

Yakima Yellow 530 19 80.6
100 mM Tris/P pH 7.0/7.4

Figure 4.19 Structures and some photophysical properties of "halogenated fluoresceins." Legends as in Figure 4.18.

4.8.3 Mercaptofluoresceins

Mercaptofluoesceins contain a thiol in place of the phenol group of fluorescein. The compounds termed mercaptofluoresceins MK-43, ML-69, MK-61, MK-67, and MK-75 are such examples [68]. They should not be confused with thiofluoresceins which are HAS-fluoresceins as they contain a sulfur in place of the oxygen at position 10 of the xanthene core structure (Figure 4.20). One remarkable property of these fluorescent dyes is their photostability. They are much more stable than fluorescein and 2′,7′-dichlorofluorescein (Chapter 7).

4.8.4 Fluorescein-Analogs

Naphthofluorescein and 4,10-dibromonaphthofluorescein are classified as fluorescein analogs [56, 69]. In these, the π-electron system is significantly extended and large bathochromic shifts and Stokes shifts are observed but with moderate values of molar absorption coefficient (44 and 32 $\text{mM}^{-1}\text{cm}^{-1}$, respectively) and quantum yield (0.14 and 0.05, respectively); see Figure 4.20.

4.9 HAS-Fluoresceins

In heteroatom-substituted fluoresceins, the xanthene's oxygen is replaced by another atom. C-fluoresceins (carbofluoresceins) are one important category in this class, an example of which are the well-known $C(CH_3)_2$-fluoresceins (dimethylcarbo-fluoresceins). The pK_a of the phenol group of carbofluorescein [23] is usually ~7.4 (close to the physiological pH), while that of fluorescein is ~6.4. This is important for practical applications, both in labeling and staining. Fluorescence mean-lifetime of carbofluorescein was taken from Whitaker et al. [24]. 2′,7′-Difluorocarbofluorescein (Virginia orange) [70] is another example of a $C(CH_3)_2$-fluorescein.

The largest set of HAS-fluoresceins consists of Si-fluoresceins (silico-fluoresceins); for example, $Si(CH_3)_2$-fluoresceins (dimethylsilane-fluoresceins) (Figure 4.21). Examples of these dyes include: Si-Fluorescein (compound 1); 2′,7′-Difluor-Si-fluorescein (compound 24); 2′,4′,5′,7′-tetrafluoro-Si-fluorescein (Maryland red); 3′,4′,5′,6′-tetrafluor-Si-fluorescein (compound 26); and 2,3′,4′,5′,6′,7-hexafluor-Si-fluorescein (compound 27) [32].

4.10 Sulfofluoresceins

Sulfofluoresceins contain a sulfonic group in place of the *ortho*-carboxylic group at the benzene moiety. These derivatives are expected to exhibit higher water solubility if compared to their carboxylic counter-parts. The

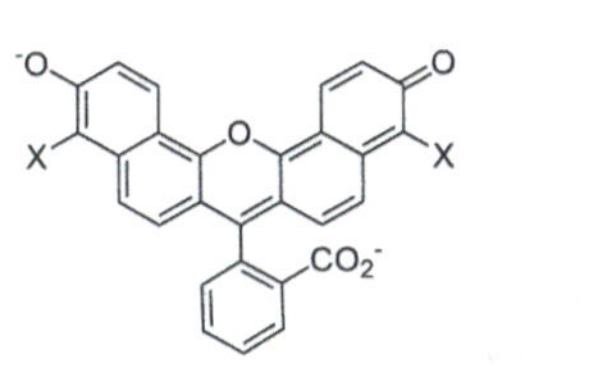

Naphthofluorescein (X = H)
595 65 6.1 (pK_a = 7.99)
4,10-Dibromonaphthofluorescein (X = Br)
605 77 1.6 (pK_a = 7.45)
All in carbonate pH 9.5.

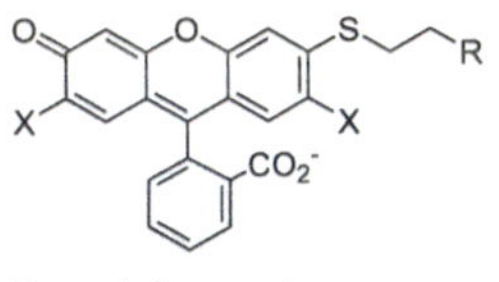

Mercaptofluorescein

	X	R			
MK-43	H	OH	495	31	7.4
MK-69	H	NH_2	495	31	5.0
MK-61	H	N_3	495	32	9.1
MK-67	Cl	OH	502	25	4.5
MK-75	Cl	N_3	502	31	1.9

(All in 1 mM NaOH)

Figure 4.20 Structures and some photophysical properties of fluorescein analogs. Legends as in Figure 4.18.

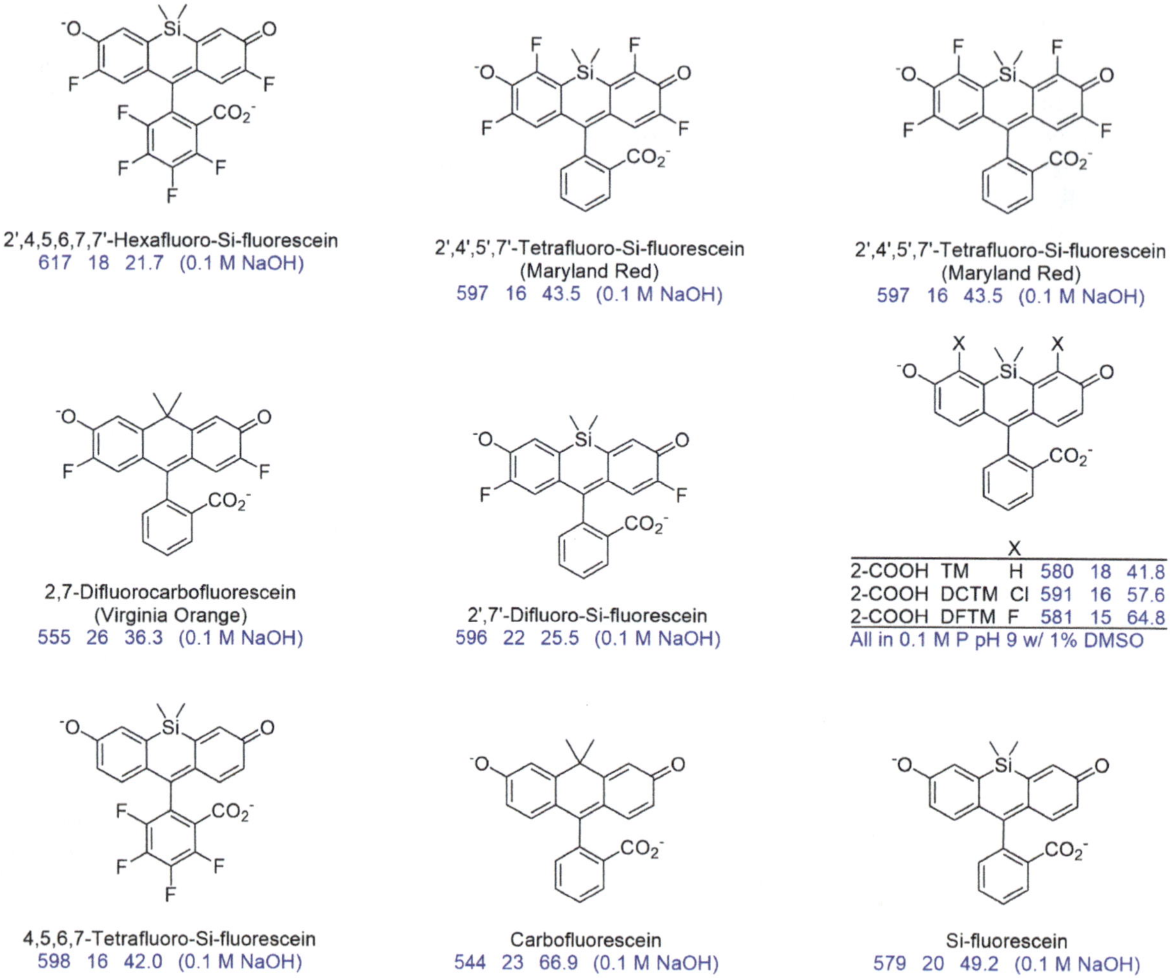

Figure 4.21 Structures and some photophysical properties of HAS-fluoresceins. The numbers in blue show the photophysical parameters $\lambda_{x,max}$ (nm), Δ_S (nm), and B_{max} (mM^{-1}cm^{-1}).

simplest entity of this class is sulfofluorescein [56] with a brightness of 71.1 mM^{-1}cm^{-1} (Figure 4.22). The halogenated sulfofluoresceins 2,4,5,7-tetra-F-sulfofluorescein (all photophysical data from [71] with exception of Φ_f, which was taken from a similar structure, 2,4,5,7-tetrafluoro-fluorescein) and 2,4,5,7-tetra-Br-sulfofluorescein [72], are also water-soluble and exhibit lower pK_a values for the phenol group than sulfofluorescein. The compound 1,2,7,8-dibenzosulfofluorescein, despite the two additional benzene rings fused to the xanthene core structure, it is still soluble in water and with a brightness that is ~2/3 that of sulfofluorescein [56]. Many more sulfofluoresceins were synthesized, but with incomplete photophysical data.

Vita blue is a sulfofluorescein analog with a large bathochromic shift caused by the elongated π-electron system (that runs from the two flanking oxygens – from the phenolate group to the ketone group) [56].

4.11 Fluorones

Fluoresceins, rhodamines, and rhodols exist in equilibrium between their bright fluorescent quinoid (in fluoresceins and rhodols) or zwitterionic (in rhodamines) forms and their colorless lactone form (closed form). Fluorones are similar to fluoresceins but characterized by the lack of a carboxylic group at carbon-2′ of the pendant aryl group. The

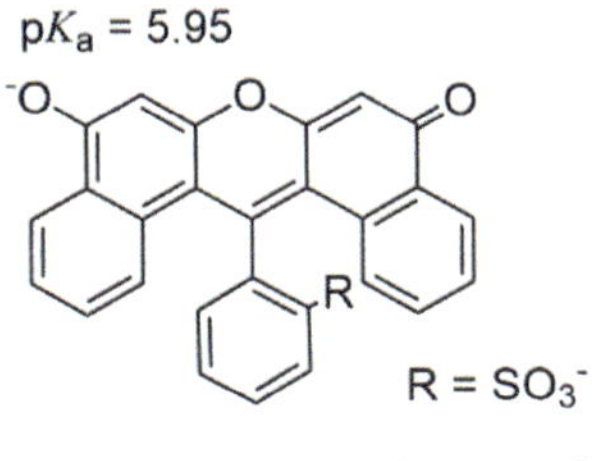

1,2,7,8-Dibenzosulfofluorescein
541 26 38.8 (Carbonate pH 9.5)

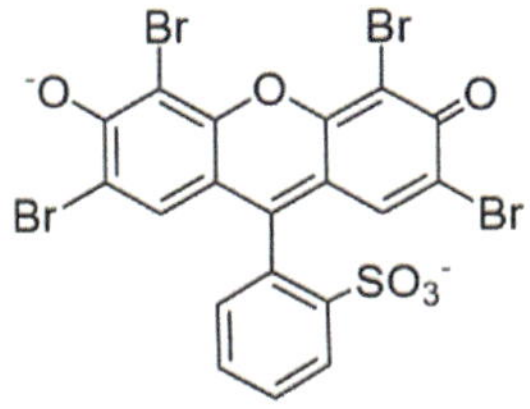

2,4,5,7-Tetra-Br-sulfofluorescein
524 22 30.4 (100 mM NaOH pH 12)

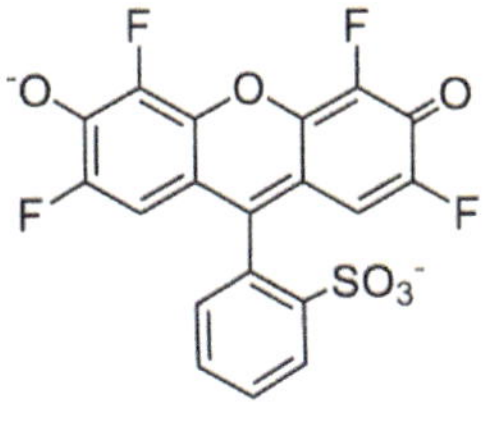

2,4,5,7-Tetra-F-sulfofluorescein
515 22 50.5 (P pH 9.18)

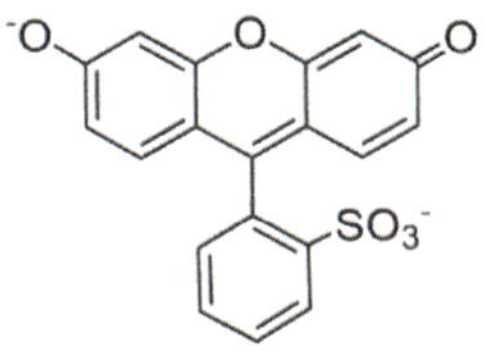

Sulfofluorescein
497 20 71.1 (C pH 9.5)

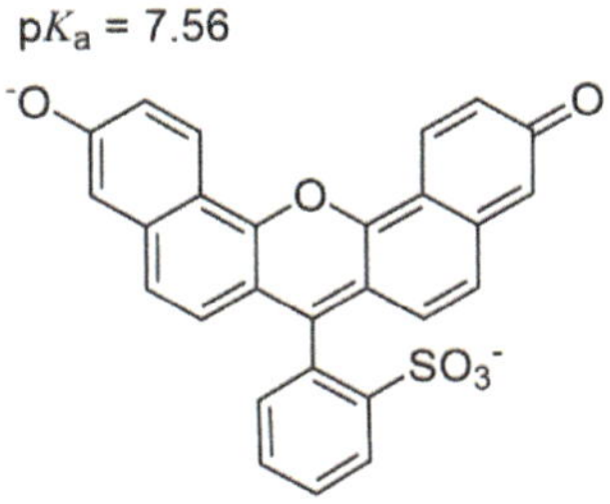

Vita Blue 609 56 5.4
(carbonate pH 9.5)

Figure 4.22 Structures and some photophysical properties of a sulfofluorescein. Numbers in blue show the photophysical parameters $\lambda_{x,max}$ (nm), Δ_S (nm), and B_{max} (mM^{-1}cm^{-1}), respectively.

substituent of this carbon, when it exists, may be a small aliphatic group like methyl, methoxy, or others. Therefore, the dyes of this class are not subjected to the non-fluorescent lactone form. Examples of fluorones include (Figure 4.23): 2-Me TokyoGreen, 2,4-DiMe TokyoGreen, 2,5-DiMe TokyoGreen, 2-Me TokyoGreen, 2-Me-4-OMe TokyoGreen, 2-OMe TokyoGreen, 2-OMe-5-Me TokyoGreen [51] (a molar absorption coefficient of ~85 mM^{-1}cm^{-1} was estimated for all TokyoGreens), and Pennsylvania green and 4-carboxy Pennsylvania green [62, 52].

4.12 HAS-Fluorones

HAS-fluorones are identical to fluorones (they do not form the lactone form), but the heteroatom oxygen of their xanthene structure is replaced by another atom. They exhibit a significant bathochromic shift which depends on the heteroatom substitution. Examples of HAS-fluorones include (Figure 4.24): NR 600 and NR 604 [73], POF [74, 75], 2-Me Tokyo Magenta (2-Me TM), and 2-Me DCTM [76, 77, 78, 79].

4.13 Cyanines

Cyanines are made of two nitrogen containing heterocycles (usually two pyrrole rings) connected by a polymethine bridge [80, 81]. The two pyrroles are in fact a 3*H*-pyrrole connected to a 2,3-dihydro-1*H*-pyrrole. The delocalized electrons of the conjugated bridge make the two ends of the bridge oscillate between these two pyrrole structures. The length of this bridge is the key to tuning the excitation/emission maxima of the dyes. The excitation and emission maxima are subjected to a bathochromic shift by approximately 100 nm with every additional vinylene unit (or two methines) in the polymethine chain. The Woodward–Fieser rules can also be used to calculate the fine-tuning effects that the substituents cause on the wavelength of absorption maximum.

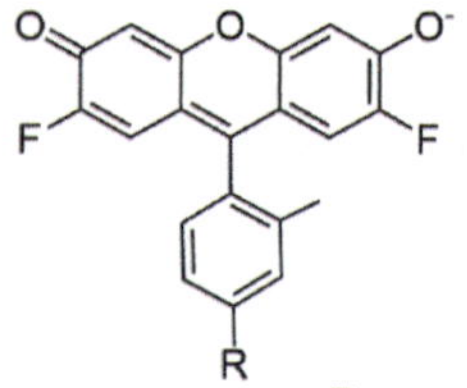

	R			
Pennsylvania Green	H	494	20	74.6
4-Carboxy Pennsylvania Green	CO_2^-	494	21	55.1

All in PBS pH 7.4.

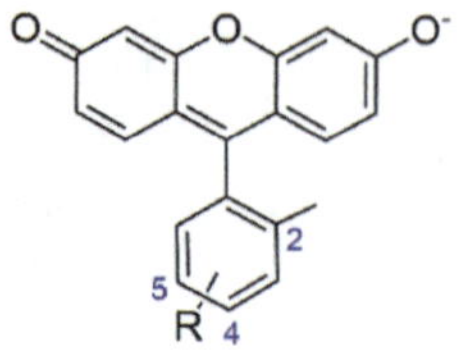

	R			
2-Me TokyoGreen	H	491	19	72.2
2,4-DiMe TokyoGreen	CH_3	491	19	73.5
2,5-DiMe TokyoGreen	CH_3	491	19	75.4
2-Me-4-OMe TokyoGreen	OCH_3	492	17	71.4

All in 100 mM NaOH.

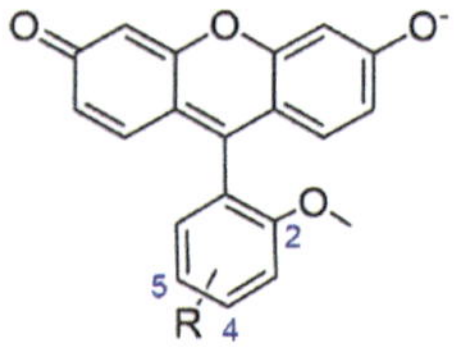

	R			
2-OMe TokyoGreen	H	494	21	73.1
2-OMe-5-Me TokyoGreen	CH_3	494	20	42.5
2,4-DiOMe TokyoGreen	OCH_3	494	19	17.0
2,5-DiOMe TokyoGreen	OCH_3	494	18	0.8

All in 100 mM NaOH.

Figure 4.23 Structures and some photophysical properties of HAS-fluorones. The numbers in blue show the photophysical parameters $\lambda_{x,max}$ (nm), Δ_S (nm), and B_{max} (mM^{-1}cm^{-1}).

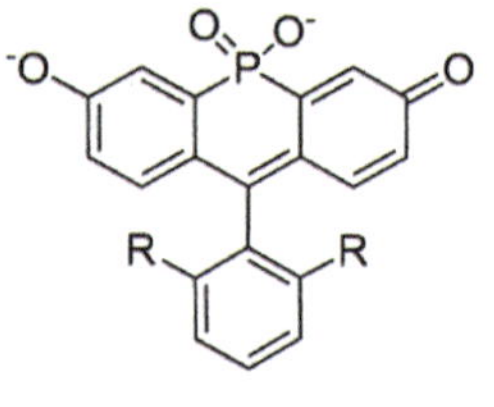

	R				pK_a
NR 600	-CH_3	600	19	45.9	6.75
NR 604	-OCH_3	604	23	21.9	6.71

All in 10 mM PBS pH 7.4 with 1% DMSO

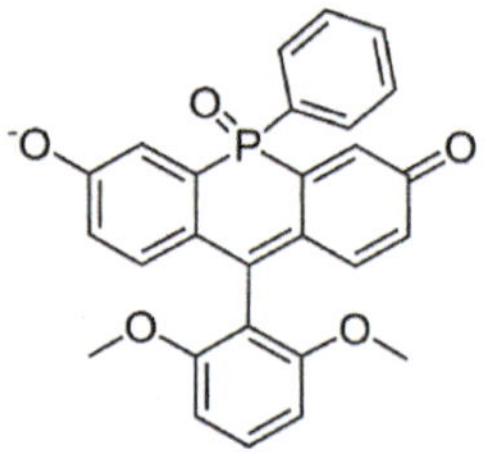

POF
632 29 13.0 (PBS w/ 1% DMSO)

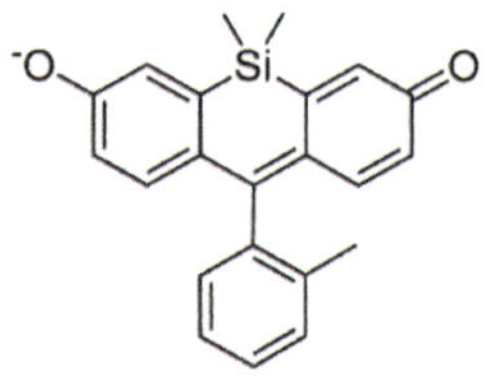

2-Me-Si-fluorescein
(2-Me Tokyo Magenta)
582 16 46,2 (0.1 M NaOH)

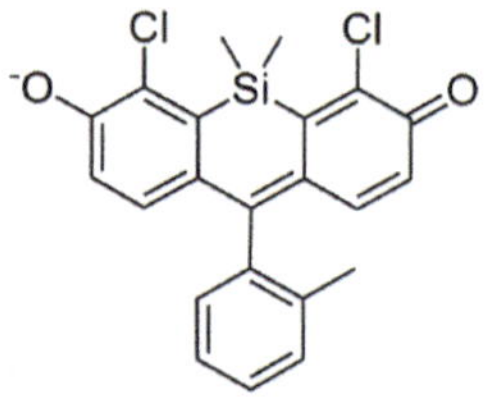

2-Me DCTM
597 10 62.4
(0.1 M P pH 9 w/ 1% DMSO)

Figure 4.24 Structures and some photophysical properties of HAS-fluorones. The numbers in blue show the photophysical parameters $\lambda_{x,max}$ (nm), Δ_S (nm), and B_{max} (mM^{-1}cm^{-1}).

Long and flexible polymethine chains are more prone to non-radiative decays due to the high rate of water molecule collisions at room temperature and the subsequent bending, twisting, and stretching effects caused by these very fast collision events powered by thermal energy. The Cy3 fluorophore, for instance, has a fluorescence quantum yield of only 0.04 in aqueous solutions. By synthesizing a rigidified trimethine bridge (Cy3B) of this dye, the fluorescence quantum yield increases up to 0.67. The fluorescence of an equimolar solution of Cy3B was observed to be eight times larger than the fluorescence of Cy3 [82] (Figure 4.25). However, it may be challenging to find a way to synthesize rigid structures for pentamethine and heptamethine chains.

Cyanines are known to exhibit a narrow Stokes shift in both organic and aqueous solutions. The average value in aqueous solutions (calculated from the present database) is typically $\Delta_S \sim 21$ nm. Small Stokes shifts, together with narrow excitation and emission bands, make it more difficult to filter out excitation radiation (especially when excitation at $\lambda_{x,max}$ is used) from fluorescent emission. A panoramic view of Stokes shift values is given in Section 5.2.5 and Appendix D.

4.13.1 Trimethine Cyanines

Examples of trimethine cyanines include Cy3.10-OH, Cy3.12-OH, and Cy3.13-OH synthesized by Southwick et al. [83], and Cy3.205 [84] (Figure 4.25). Sulfo-Cyanine 3 differs from Cy3 by the presence of one aminohexanoate, while Cy3.18.OH [85] has two aminohexanoate. There are dozens of Cy3 variants in the literature, their only differences being the length of the two aminoalkyl groups. All of these variations were not included in the present database to avoid excessive redundant entries. The photophysical parameters of Cy3, for instance, has been measured by many authors (e.g., Zheng et al. [86]). Its average quantum yield in water is 0.057 and it has an average mean fluorescence lifetime of 0.19 ns (both averages calculated by the present author). Neo-Cy3 [87] is significantly different to Cy3, but unfortunately exhibits a low fluorescence quantum yield ($\Phi_f = 0.02$). Cy3.5, with four sulfonic groups, is at least two times brighter than Cy3 with two sulfonic groups ([86] and references therein). Note that Cy3′ significantly differs from Cy3 [37] and was therefore included in the present database (structure shown in Figure 4.25). Cy2 is also a trimethine cyanine but with two benzoxazoles instead of two benzopyrroles ([86] and references therein). This Cy2 cyanine dye is much cited but no reference was found where the photophysical data of this dye was measured in aqueous solutions. Therefore, it was not included in the present printed database.

Alexa Fluor 555 and AF555 [6] are also trimethine cyanines. Several sources provide the wrong molecular structure for Alexa Fluor 555. The differences between AF555 and Alexa Fluor 555 are the positions of the alkylsulfonate and hexanoate sidechains. These structures have only recently been disclosed and validated [88]. It seems that AF555 is a subtle replacement of Alexa Fluor 555, which can be used in applications where similar spectroscopic parameters are desired. Cy3 [82] is the most known trimethine cyanine and Cy3B is the above-mentioned rigidized form of Cy3 synthesized by Cooper et al. [82].

Many suppliers omit the information of liquids in which the photophysical parameters were measured, even for water soluble dyes. Fluorescence quantum yield is usually much higher in DMSO and DMF than in water or PBS. However, in most cases, imaging in DMSO and DMF is not realistic. Therefore, a large quantity of photophysical data could not be included in this database due to the lack of precise data of the measurement conditions used.

4.13.2 Pentamethine Cyanines

A large number of water-soluble pentamethine cyanines have already been synthesized. Back in 1990, Southwick et al. [83] synthesized Cy5.10-OH, Cy5.12-OH, Cy5.11-OH, and Cy5.13-OH. In 1993, Mujamdur et al. [85] synthesized Cy5.29.OH and, in 1996 [84], Cy5.203, Cy5.204, and Cy5.205. NIR2 and NIR are also pentamethines [88]. They exhibit high molar absorption coefficients, but the quantum yields in table 1 of Lin et al. [88] may require further validation (confirmed by the corresponding author). Just like Cy3, there are many innocuous variations of

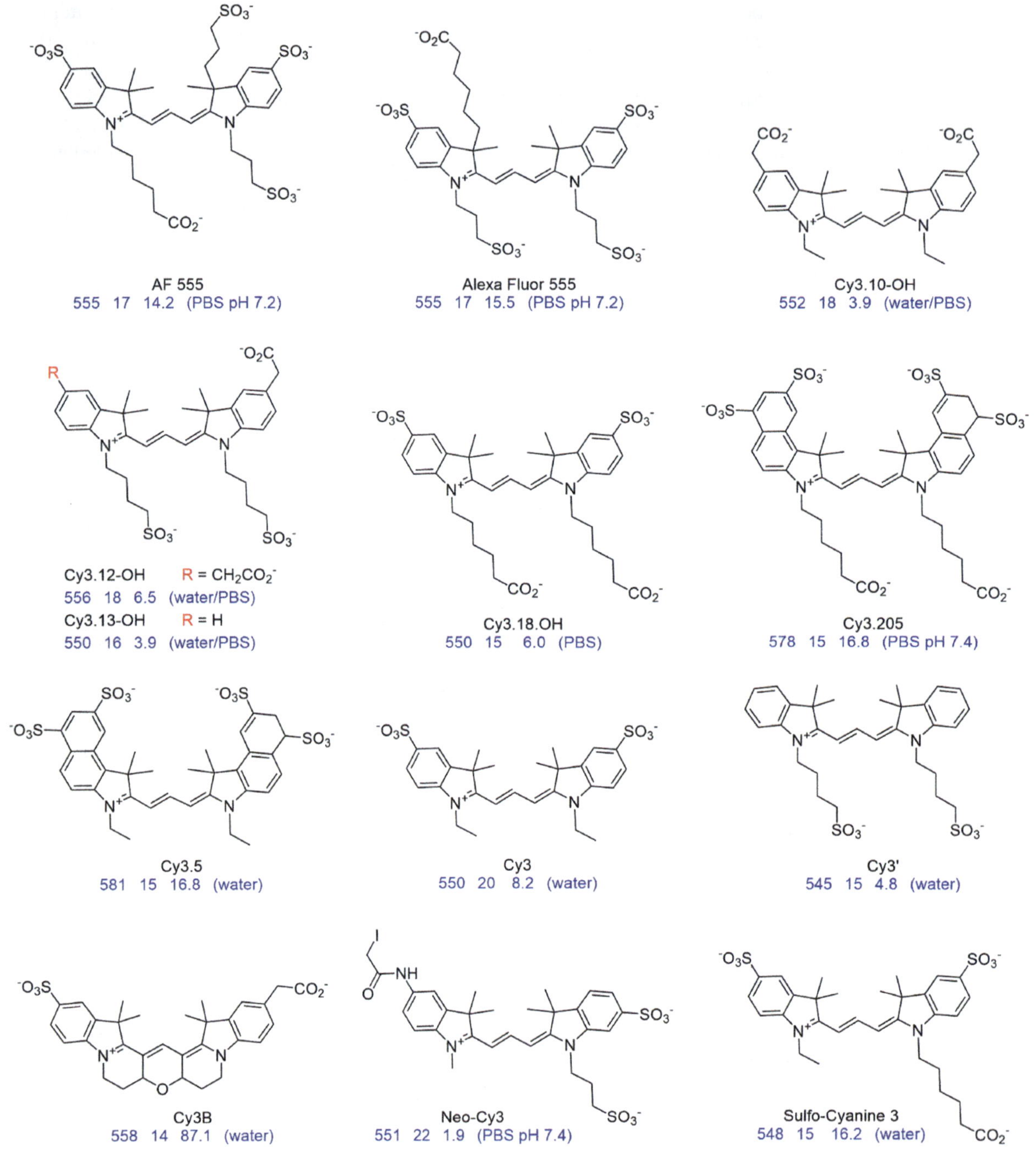

Figure 4.25 Molecular structures and some photophysical parameters of trimethine cyanines. The numbers in blue show the photophysical parameters $\lambda_{x,max}$ (nm), Δ_S (nm), and B_{max} ($mM^{-1}cm^{-1}$), respectively.

Cy5. Therefore, only the photophysical parameters of Cy5 shown in Figure 4.26 (from Nau and Mohanty [37]) were included in the present database. Cy5′ [37] and Neo-Cy5 [87] are significantly different from normal Cy5 and were included in the database. The same can be said about CypHer5 [89] and Alexa Fluor 680 ([6], where the structure was elucidated by McCann et al. [90]). The structures of Sulfo-Cyanine5 [91] and Cy5.18.OH [85] are identical. Zy5 [85, 92, 93] is similar to Cy5 as it has one hexanoic acid bound to a nitrogen, which is a convenient arm linker for reactive moieties. Zy5.5 [85, 92] is similar to Cy5.5 in the same way. DY-679P [20] is also a pentamethine. More recently, AF647 and AFD647 have been introduced as a replacement of Alexa Fluor 647 [6]. The molecular structures of these three dyes have only recently been disclosed and validated by Gebhardt et at. [88]. The only differences between these three dyes are the positions and lengths of the alkylsulfonate sidechains and the position of the hexanoate (similar to AF555 and Alexa Fluor 555) (Figure 4.25).

In order to increase the water solubility of a Cy5 dye, Wycisk et al. [94] equipped its two nitrogen atoms with the following modifications: i) dendrons (of generation 1) [G1.0] with a total of eight hydroxy functional groups; ii) linear polyethylene glycol chains; and iii) a common alkylsulfonate for comparison. These dyes, synthesized by Wycisk et al. [94], are hereby termed W-L 8a, W-L 8b, W-L 8c, W-L 8d, and W-L 8e (initials of the names of first and last author plus the numbering used in the original work) (Figure 4.26).

4.13.3 Heptamethine Cyanines

Examples of heptamethine cyanines include: NIR1 and NIR3 [89]; TG170 [95]; CW800, RS800, ICG, ZW800-1, and ZW800-3a [96]; HCDS, HCDO, and HCDN [97]; ICG [96, 98, 99]; DY-747P1 [20]; IRDye78-CA [100]; Zy7 carboxy [93]; Cy7 monosulfonic acid [101]; and LS479 and LS482 [102] (Figure 4.27).

Cypate has a low solubility in plain water or aqueous buffers. Ye et al. [103] increased the solubility of this dye by taking advantage of the two carboxylic groups and producing dendrons of generation-one, as well as carboxylates of generation-two. The disadvantage of these modifications is that they may work as chelating agents of polyvalent cations. In the same paper, the authors also produced neutral carbohydrate dendrons of generation-one and generation-two. These hydrophilic structures are hereby termed Y-A 1 (cypate), Y-A 2, Y-A 3, Y-A 4, Y-A 5, Y-A 6, Y-A 7, Y-A 8, Y-A 9, Y-A 10, and Y-A 11 (initials of the names of first and last author and numbers used in the original publication).

4.14 Borondipyrromethenes

Borondipyrromethenes are better known as BODIPYs™, which is a portmanteau of **bo**ron-**dipy**rromethene or 4,4-difluoro-4-bora-3a,4a-diaza-s-indacene, as shown in Figure 4.1. BODIPY is a registered trademark of Thermo-Fisher Scientific (Waltham, MA, EUA) that became a technical common name of a chemical class of compounds. The fluorescent dyes of this class exhibit medium molar absorption coefficients but have high fluorescence quantum yields due to their rigid structures. They can be seen as boron-stabilized or boron-rigidized monomethine chains. The two five-ringed pyrroles are held together by both a single methine group and a borinate complex. Borinates, which are made of two B–C bonds and two B–O bonds, are the result of the coordination of boronic acid (made of two B–C bonds and one B–O bond) and an OH^- anion. Consequently, borondipyrromethenes are compact structures with a relatively long π-electron system, as the unlocalized pairs of electrons are located around the outside of the structure from one nitrogen to the other nitrogen. This is different to cyanines, where the conjugated chain (π-electron system) runs along the shortest distance between the two nitrogen atoms. Size details (number of atoms or N_a) of the fluorophores are given in Section 5.2.1. For reviews and review-like articles on this class, see Loudet and Burgess [104], García-Moreno et al. [105], Umezawa et al. [106], and Bañuelos [107].

New synthesis routes of BODIPYs are also important as they increase the number of possible new derivatives with this core structure. An improved synthesis route for this parent structure and derivatives has been demonstrated by Tram et al. [108] and Arroyo et al. [109]. The latter paper provided the complete photophysical data of some derivatives in water, but does not provide molar absorption coefficients and fluorescence lifetimes. The photophysical data are not tabulated with the other solvents used (cyclohexane, toluene, chloroform, tetrahydrofuran, ethyl acetate, and ethanol), but can instead be found within the text.

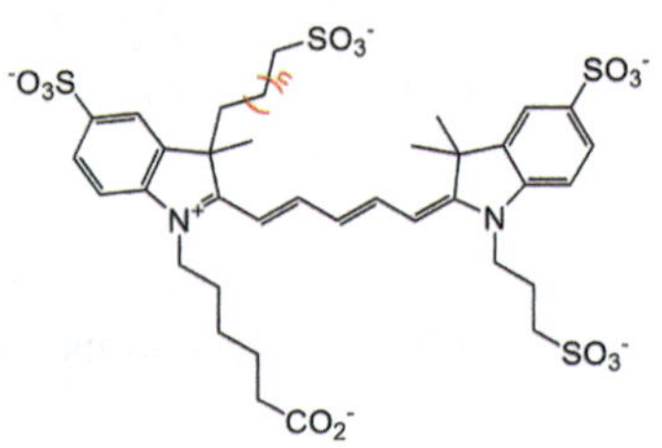

	n			
AFD647	1	650	22	89.1
AF647	2	650	22	89.1

(PBS pH 7.2)

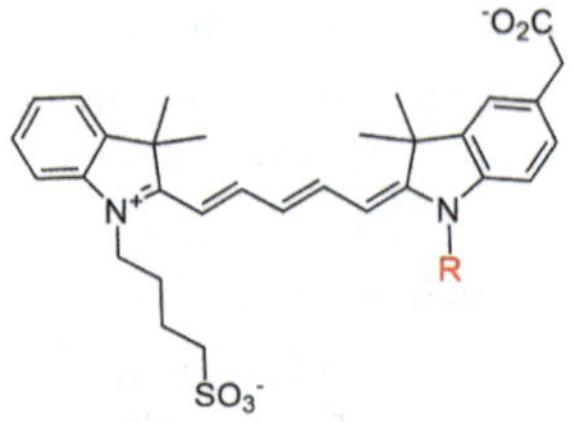

Cy5.11-OH R = CH_2CH_3
646 21 12.9 (water/PBS)
Cy5.13-OH R = $(CH_2)_4SO_3^-$
648 21 25.8 (water/PBS)

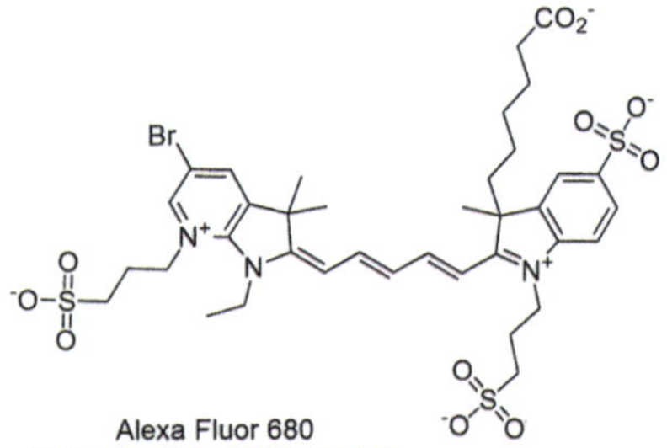

Alexa Fluor 680
679 23 65.8 (PBS pH 7.4)

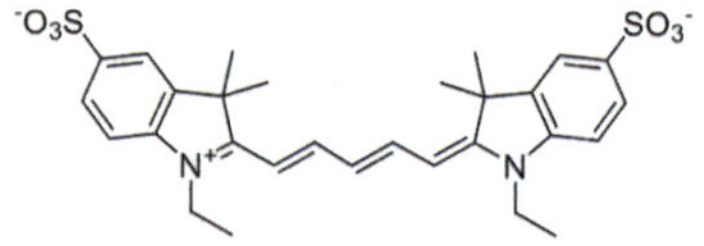

Cy5
647 16 67.5 (water)

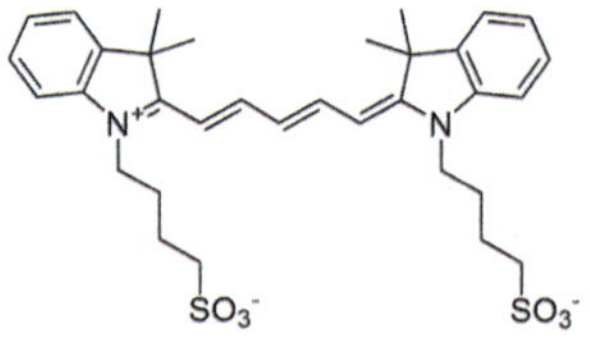

Cy5'
642 18 23.4 (water)

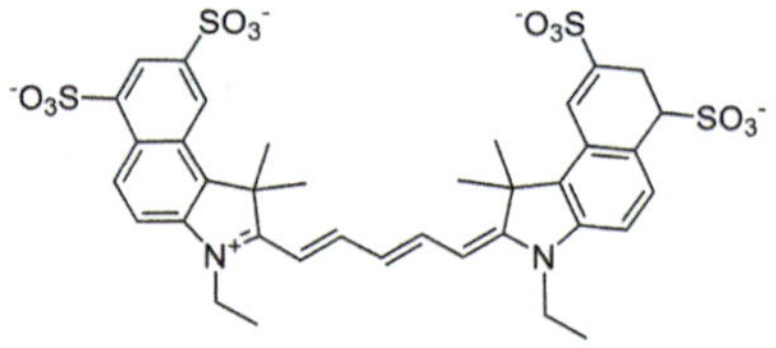

Cy5.5
674 20 43.7 (water)

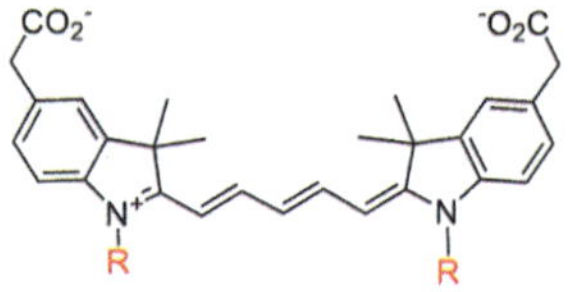

Cy5.10-OH R = CH_2CH_3
648 19 19.3 (water/PBS)
Cy5.12-OH R = $(CH_2)_4SO_3^-$
650 24 27.9 (water/PBS)

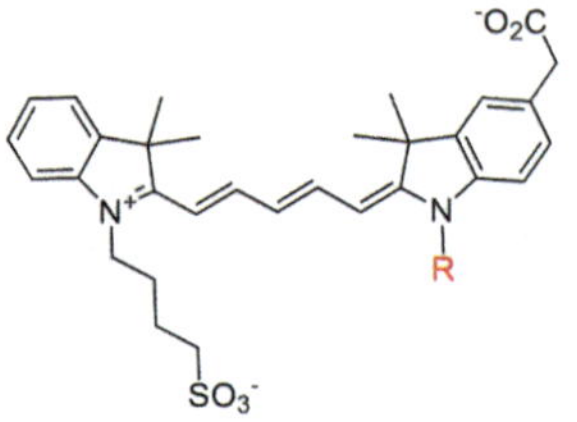

Cy5.11-OH R = CH_2CH_3
646 21 12.9 (water/PBS)
Cy5.13-OH R = $(CH_2)_4SO_3^-$
650 21 25.8 (water/PBS)

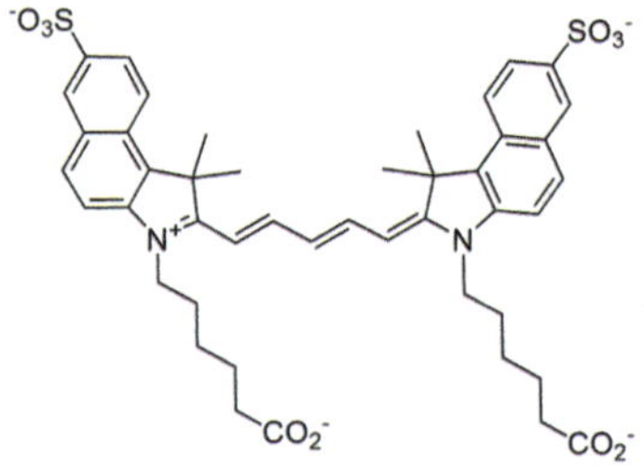

Cy5.203
680 22 19.0 (PBS pH 7.4)

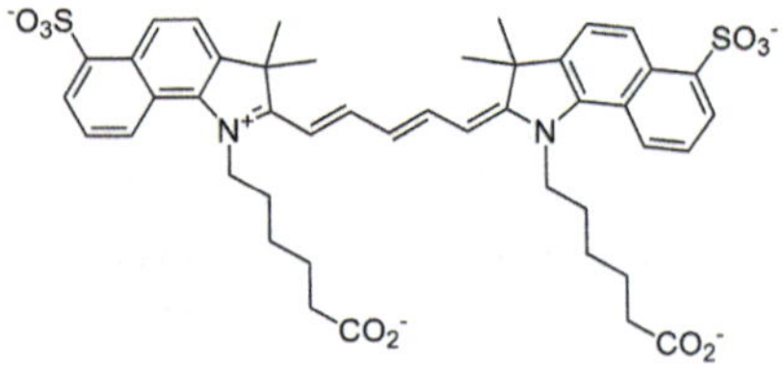

Cy5.204
682 25 19.7 (PBS pH 7.4)

Figure 4.26 Molecular structures and some photophysical parameters of pentamethine cyanines. Legends as in Figure 4.25.

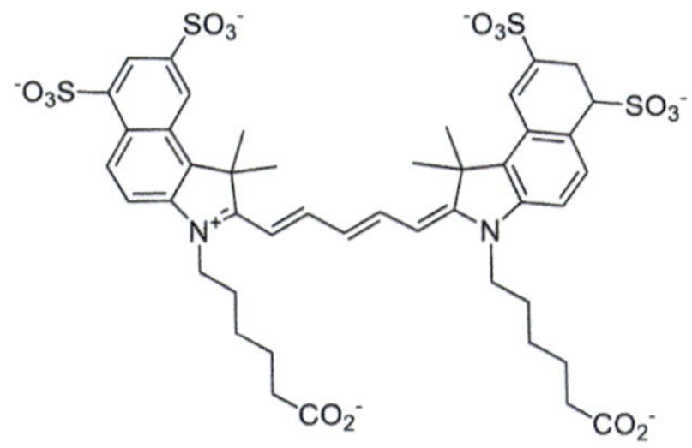

Cy5.205
674 20 43.7 (PBS pH 7.4)

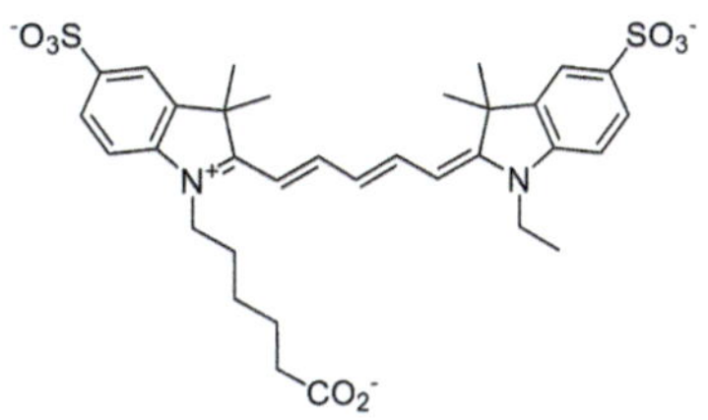

Cy5.29.OH
650 16 50.0 (PBS)

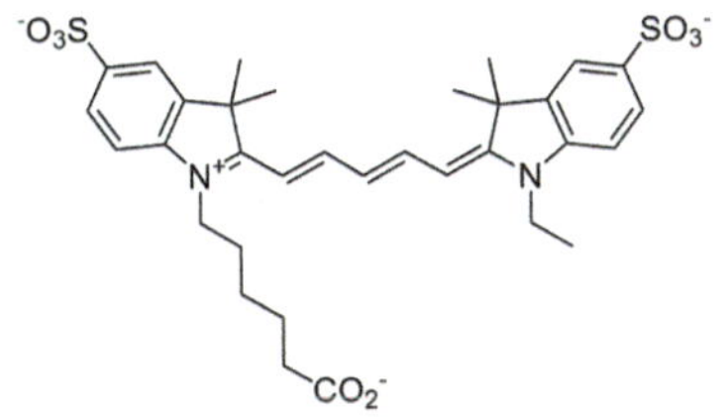

Cy5.29.OH
650 16 50.0 (PBS)

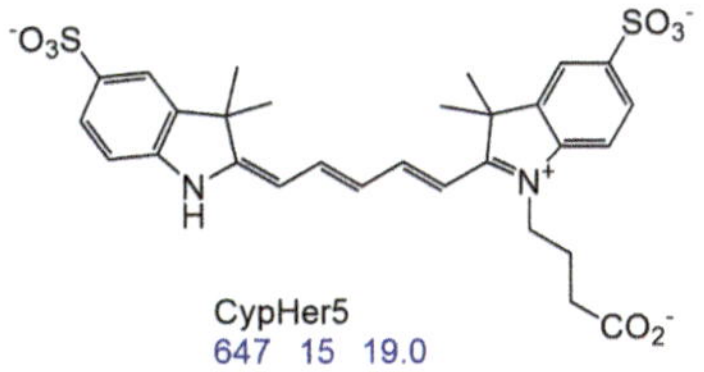

CypHer5
647 15 19.0
In acetate pH 4.6

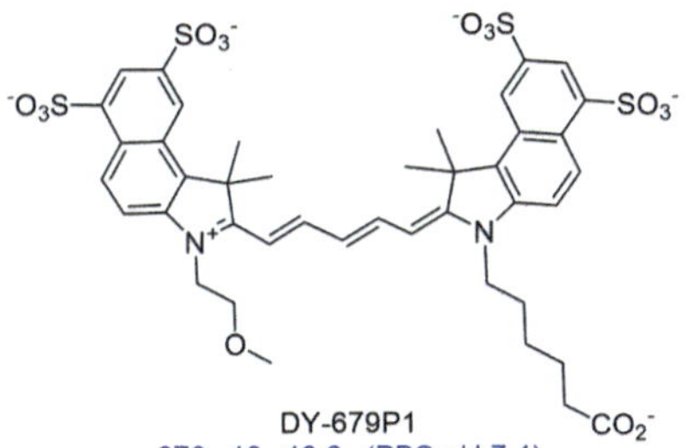

DY-679P1
679 18 46.6 (PBS pH 7.4)

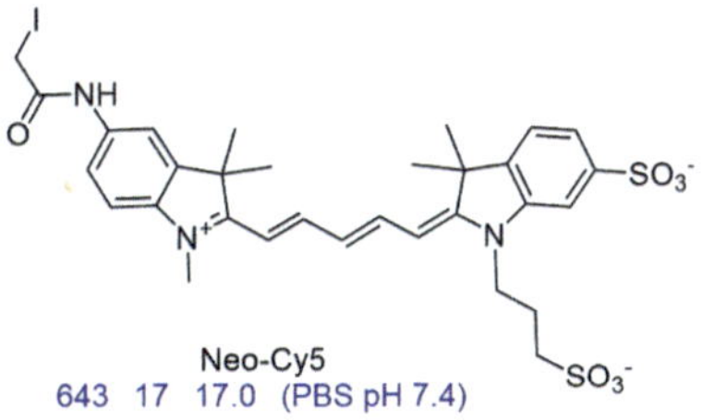

Neo-Cy5
643 17 17.0 (PBS pH 7.4)

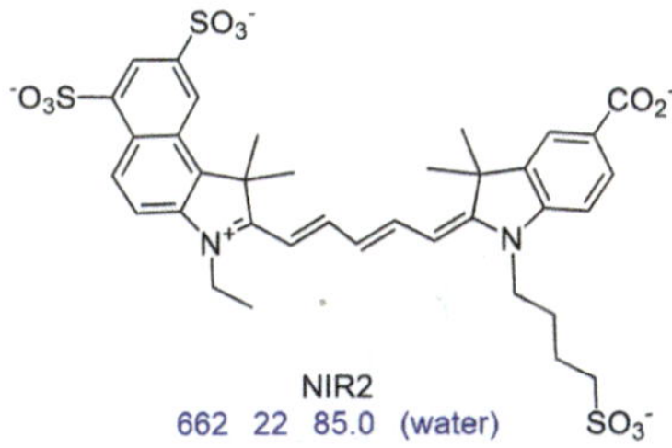

NIR2
662 22 85.0 (water)

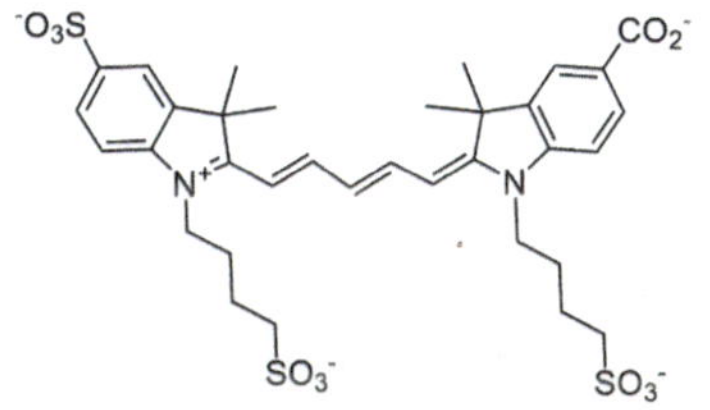

NIR4
650 21 111.8 (water)

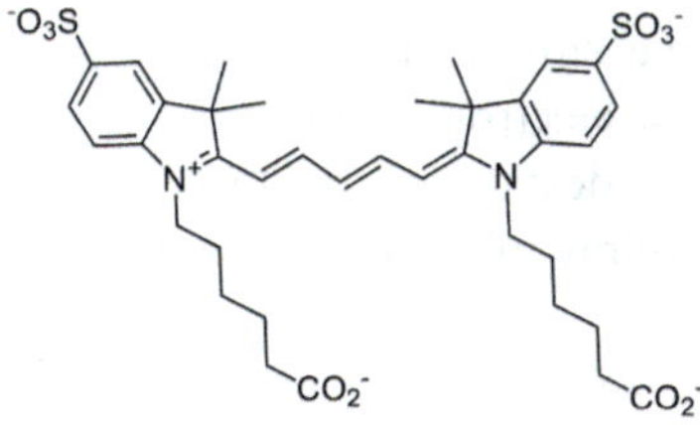

Sulfo-Cyanine5 646 16 54.2 (water)
Cy5.18.OH 650 17 67.5 (PBS)

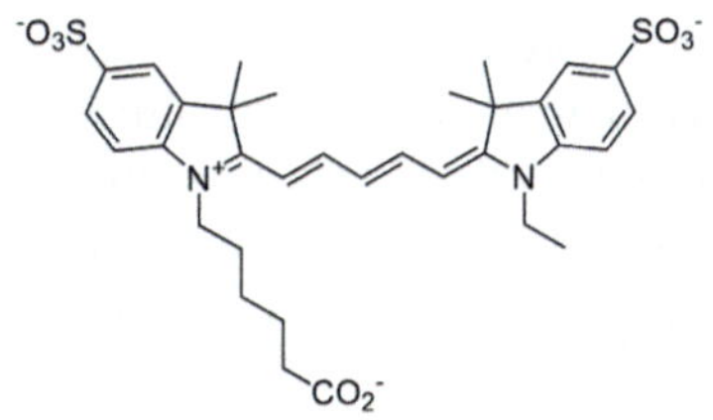

Zy5
647 18 67.5 (water)

Figure 4.26 (Continued)

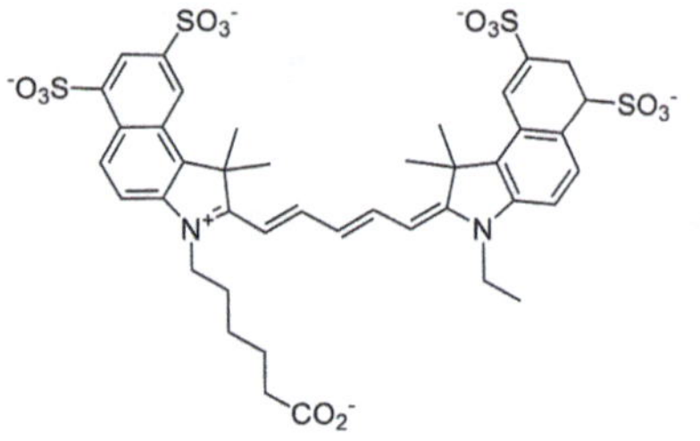

Zy5.5
676 18 43.7 (phosphate pH 7.4)

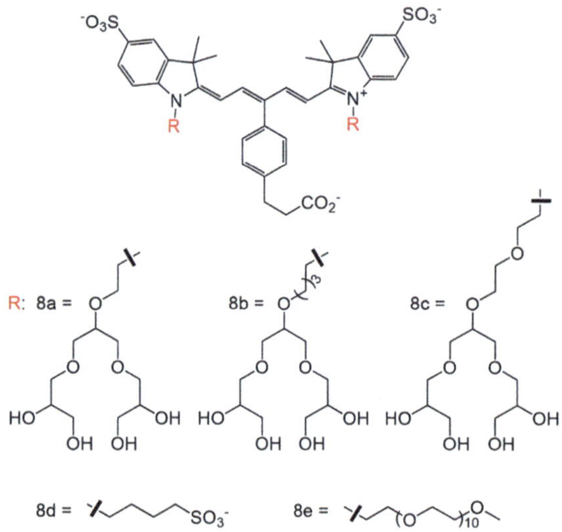

R			
W-L 8a	643	24	10.4
W-L 8b	644	24	28.3
W-L 8c	643	23	12.8
W-L 8d	643	23	21.9
W-L 8e	643	22	17.2

All in PBS pH 7.4

Figure 4.26 (Continued)

There are large errors in the literature associated with the measurement of ε_{max} and Φ_f values of lipophilic BODIPY dyes in water. The main reason for this is the formation of aggregates by these almost planar (with the exception of the two fluorine atoms) and lipophilic compounds in aqueous media. A decrease of absorbance is always associated with a broadening of the absorption bands, as evidenced by the increase of the full width at half maximum (FWHM) in water, which does not happen in organic solvents. This can be mitigated by adding hydrophilic substituents [3] to the core structure or raising the temperature during the measurements. Substituents are added to the parent structure to manage water solubility and the energy levels of the electronic transitions. In the case of BODIPY derivatives, the effect of electron donating substituents at position 8 destabilizes the LUMO level and causes a blue shift in the absorption maxima. On the other hand, electron withdrawing substituents stabilize the LUMO level and promote a red-shift in the absorption maxima [110].

4.14.1 Small Water-Soluble Borondipyrromethenes

The BODIPY scaffold is a very versatile structure used to produce dyes for a great diversity of applications. Their fluorescent derivatives are known to exhibit very narrow Stokes shifts, both in both organic solvents (usually

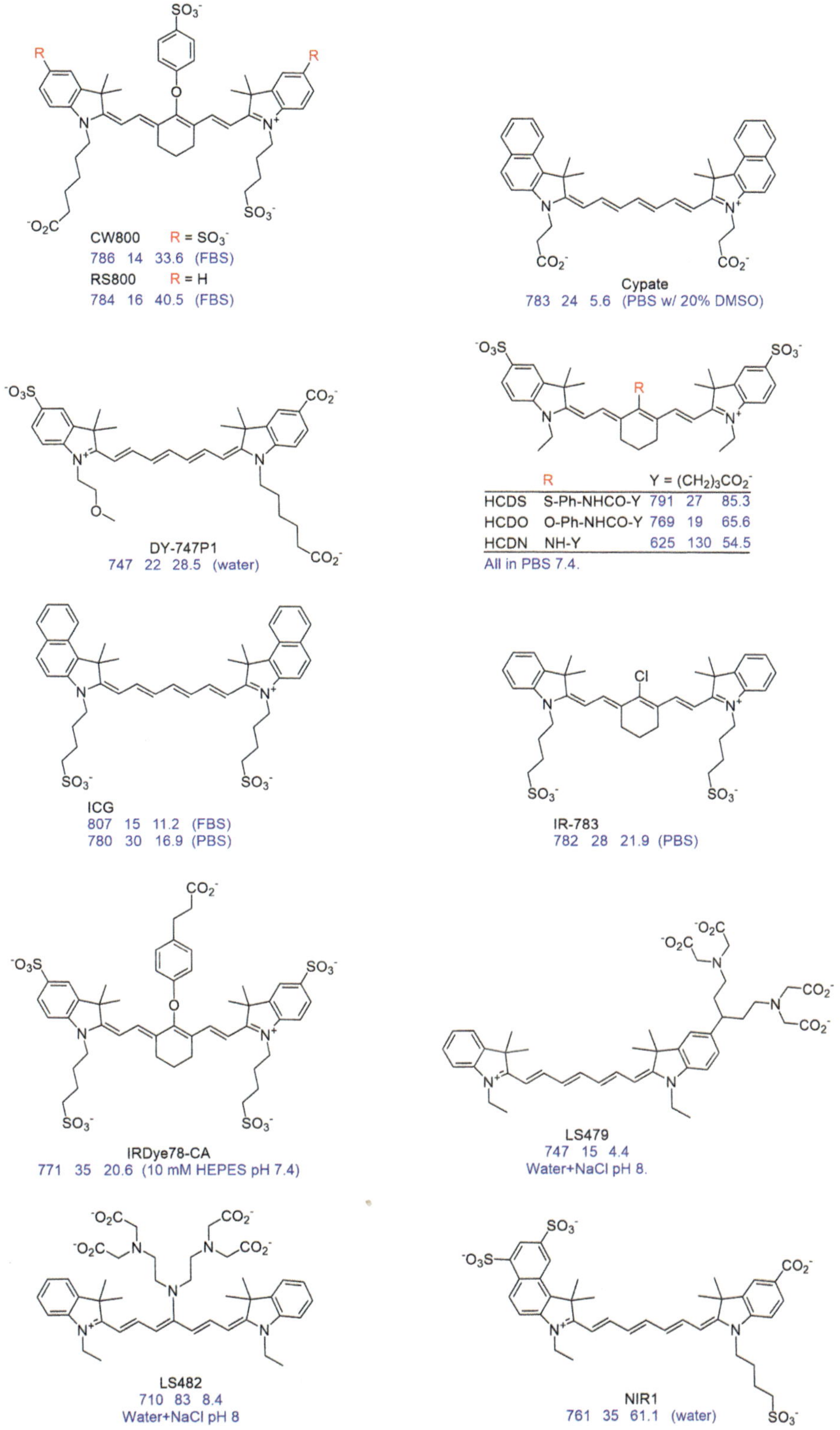

Figure 4.27 Molecular structures and some photophysical parameters of heptamethine cyanines. Legends as in Figure 4.25.

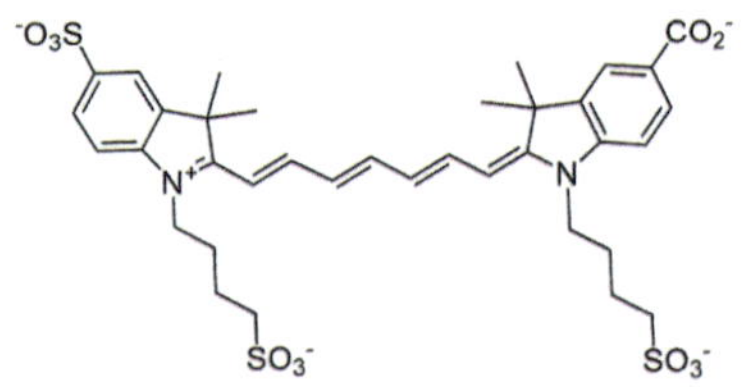

NIR3
750 27 77.0 (water)

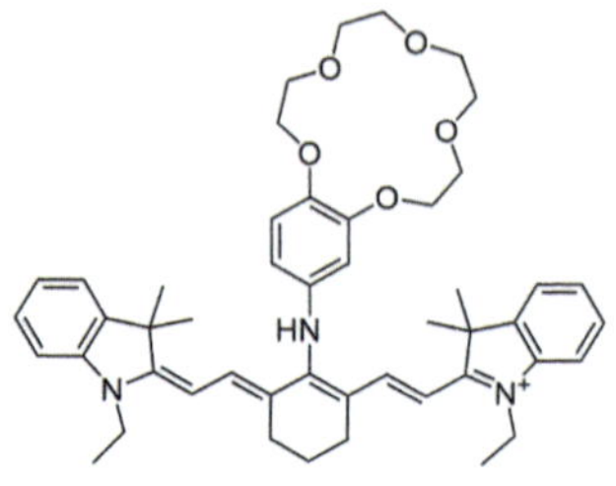

TG170 791 12 7.4 (water pH 5.9)

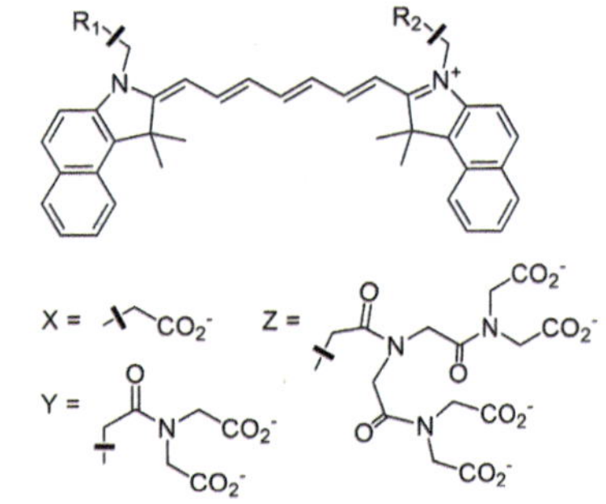

	R1	R2			
Y-A 1	X	X	778	27	6.7
Y-A 2	X	Y	778	28	5.7
Y-A 3	Y	Y	777	30	6.4
Y-A 4	Y	Z	778	27	6.0
Y-A 5	Z	Z	778	26	5.5

All in PBS pH 7.2.

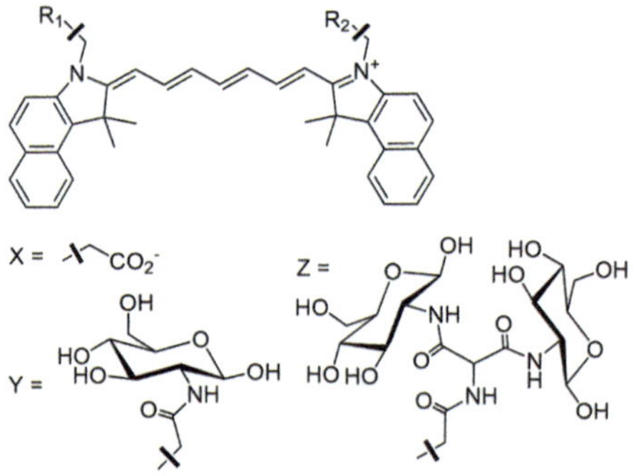

	R1	R2			
Y-A 6	X	Y	778	27	6.7
Y-A 7	Y	Y	779	26	6.7
Y-A 8	Y	Z	780	26	5.9
Y-A 9	Z	Z	779	30	5.6

All in PBS pH 7.2.

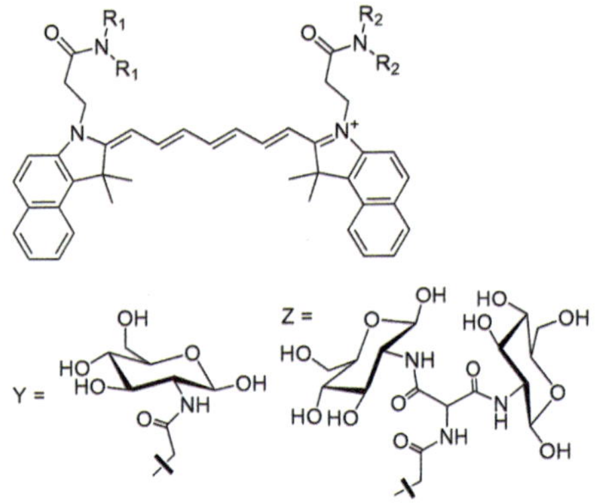

	R1	R2			
Y-A 10	Y	Z	780	28	6.5
Y-A 11	Z	Z	780	27	5.9

All in PBS pH 7.2.

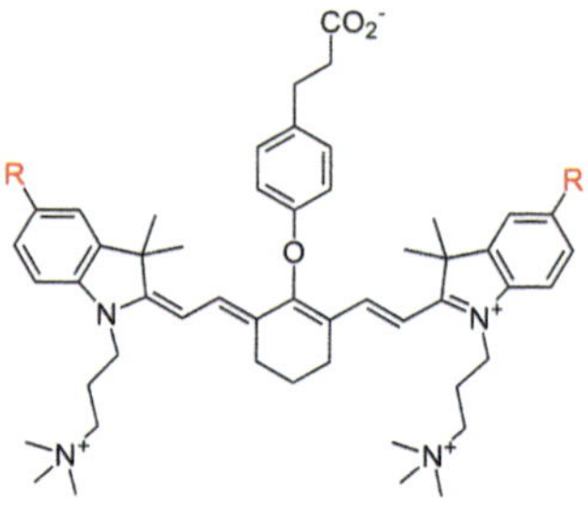

ZW800-1 R = SO_3^-
772 16 37.6 (FBS)
ZW800-3a R = H
774 16 49.7 (FBS)

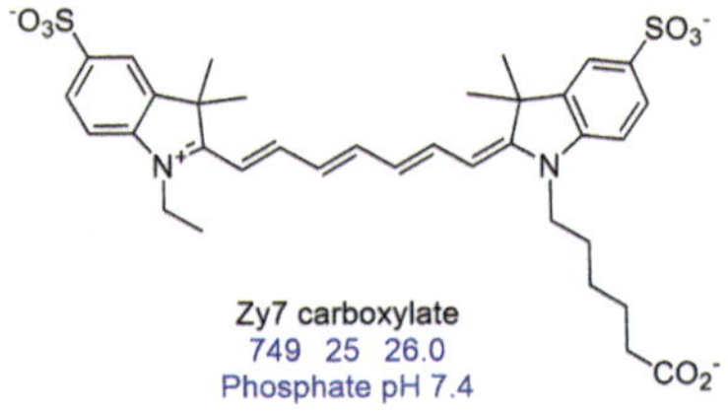

Zy7 carboxylate
749 25 26.0
Phosphate pH 7.4

Figure 4.27 (Continued)

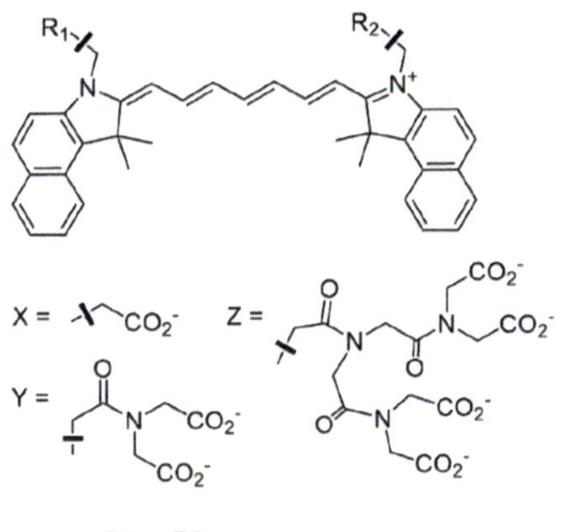

	R1	R2			
Y-A 1	X	X	778	27	6.7
Y-A 2	X	Y	778	28	5.7
Y-A 3	Y	Y	777	30	6.4
Y-A 4	Y	Z	778	27	6.0
Y-A 5	Z	Z	778	26	5.5

All in PBS pH 7.2.

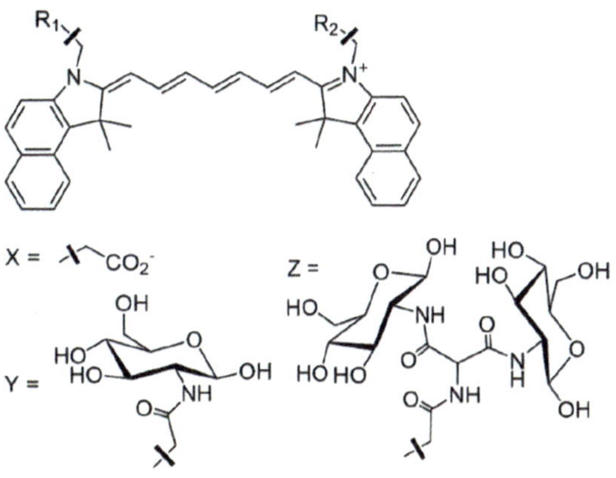

	R1	R2			
Y-A 6	X	Y	778	27	6.7
Y-A 7	Y	Y	779	26	6.7
Y-A 8	Y	Z	780	26	5.9
Y-A 9	Z	Z	779	30	5.6

All in PBS pH 7.2.

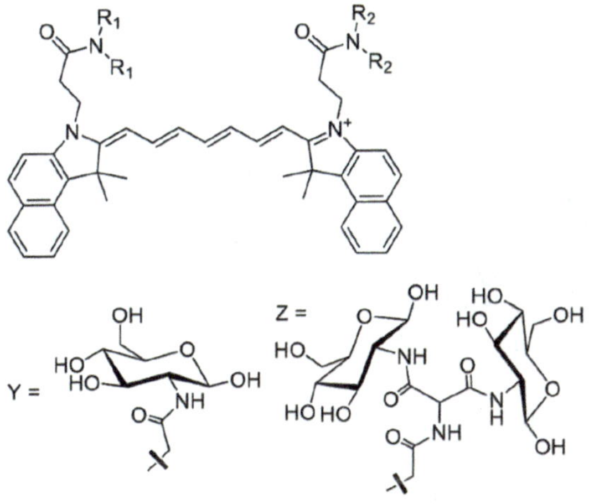

	R1	R2			
Y-A 10	Y	Z	780	28	6.5
Y-A 11	Z	Z	780	27	5.9

All in PBS pH 7.2.

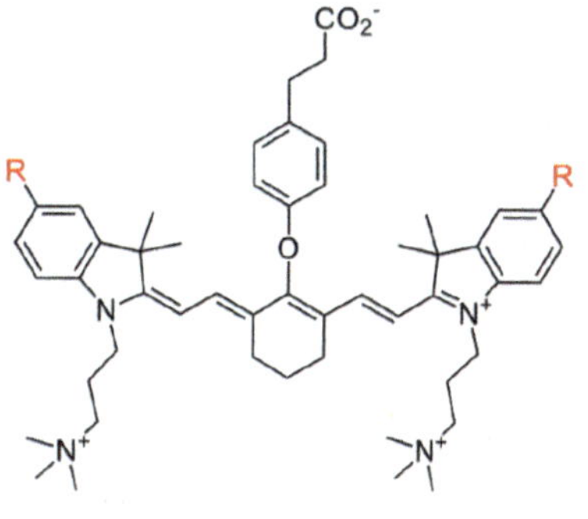

ZW800-1 R = SO_3^-
772 16 37.6 (FBS)
ZW800-3a R = H
774 16 49.7 (FBS)

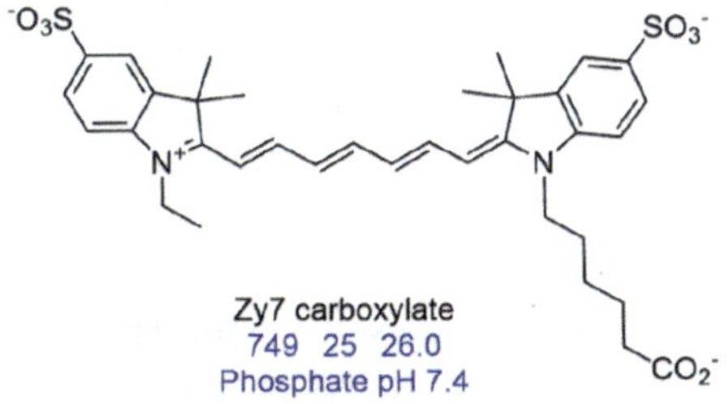

Zy7 carboxylate
749 25 26.0
Phosphate pH 7.4

Figure 4.27 (Continued)

~10 nm) and in aqueous solutions (usually ~17 nm) (Section 5.2.5). Plain BODIPY and methyl and ethyl substitute BODIPY dyes exhibit a very low solubility in water. Umezawa et al. [106] compiled and reviewed the structures and some photophysical parameters of 226 BODIPY dyes in total; however, almost all of the measurements were taken in organic solvents. Examples of plain borondipyrromethenes that are only slightly soluble in water include the following (Figure 4.28): BODIPY, which is the simplest (smallest) of all BODIPYs [109]; NBDY or BODIPY 507/545 [111]; BODIPY FL ([30], the Stokes shift in water was confirmed by the present author to be only 9 nm [6]); BODIPY 630/650 ($\lambda_{a,max}$, $\lambda_{e,max}$, and ε_{max} from Buschmann et al. [112], Φ_f was estimated to be 0.5 because the supplier does not specify whether measurements were taken in water or an organic solvent); and DAMBO-T and DAMBO-P^H-T [113]. The values of $\lambda_{a,max}$ and $\lambda_{e,max}$ in aqueous solutions of many BODIPY dyes are known; however, the values of ε_{max} and Φ_f remain uncertain because their low water-solubility and tendency to aggregate produces conflicting values.

An interesting strategy to increase the water solubility of BODIPY dyes is to covalently bind sulfonic groups to their core structure. This produces small, bright, and negatively charged fluorescent dyes. Wories et al. [114] synthesized the monosulfonate derivative, hereby termed 6-sulfonate-1,4,5,8-Me-BODIPY. Shah et al. [115] synthesized the disulfonate 2,6-disulfonate-1,3,5,7,8-pentamethyl-BODIPY in 1990 and prepared it as a salt of Na^+, K^+, Rb^+, Cs^+, and $NH_4{}^+$. The measured photophysical parameters of these salts in water produced almost the same values. Their average values are $\lambda_{a,max} = 496$ nm, $\lambda_{e,max} = 533$ nm, $\varepsilon_{max} = 83.13$ mM^{-1}cm^{-1}, and $\Phi_f = 0.83$. Arbeloa et al. [17] resynthesized this same dye and named it PM556. However, the results of the photophysical parameters measured by Arbeloa are slightly different (narrower Stokes shift). The data obtained by Arbeloa et al. was included in the present database (Figure 4.28). Li et al. [116] synthesized hydrophilic derivatives of 8-trifluoromethyl-BODIPY dyes, which are hereby termed 8-trifluoromethyl-BODIPY 10, 8-trifluoromethyl-BODIPY 11, and 8-trifluoromethyl-BODIPY 12. The research group of K. Burgess also synthesized a series of water-soluble BODIPYs (mono- and disulfonated tetramethyl-BODIPYs) bearing four different derivatization moieties: 4-iodobenzene (for further functionalization via organometallic cross-coupling reactions); 4-bromobenzene (reactive toward nucleophiles); 4-azidebenzene (for copper-mediated cycloadditions to alkynes); and an arm-linker with a carboxylic group (for coupling to amines). The dyes with a reactive moiety synthesized by Li et al. [116] are hereby termed L-B 1a, L-B 1b, L-B 2a, L-B 2b, L-B 4a, L-B 4b, and L-B 5 (using the first letter of both first and last author names and the numbering used in the original work). The same ε_{max} value of L-B 1b was attributed to L-B 1a. The structures are shown in Figure 4.28.

It is worth mentioning that the sulfonated BODIPY dyes have the potential to greatly increase separation efficiency, time efficiency, and detection selectivity, all at the same time, when used as fluorescent labels in electrodriven separation techniques such as electrophoresis along gel slabs, microchip channels, and fused silica capillaries (see table 5.5 and section 8.4.3 of [30]). This is due to their small size and high charge (−2). Labels with small sizes and high charges provide high electrophoretic mobility, which increases separation efficiency and time efficiency.

Another discovery are the *meso*-carboxy BODIYs. They exhibit some water solubility and high brightness in aqueous solution for all the free carboxylate and the amine derivatives obtained by the reaction of the pentafluorophenyl ester and succinimidyl ester with amines [117]. These fluorescent *meso*-carboxyBODIPYs are here termed J-K 1, J-K 3a, J-K 3b, and J-K 3c (Figure 4.28).

4.14.2 Medium-Sized, Water-Soluble Borondipyrromethenes

A second efficient strategy to increase the water solubility of BODIPY dyes is to attach sulfobetaine structures to boron (at position 4) and/or to some carbons (mainly 3, 5, and 8). Nie et al. [41] synthesized such dyes (Distyryl-BODIPY 11, 12, 13, 17, and 18; Figure 4.29). Two of them (17 and 18) exhibit high quantum yields for a red absorbing dye. One explanation for this may be the electrostatic attraction between the quaternary amines and the adjacent sulfonic groups of the four pendant sulfobetaines of these BODIPY dyes (17 and 18). This adds some rigidity to both the two pendant 2-butyne groups and the two pendant styryl groups, the latter of which is part of the extended π-electron system. Niu [3] also synthesized the BODIPY dyes, hereby termed Niu-18, Niu-25, Niu-27, and Niu-29 (Figure 4.29).

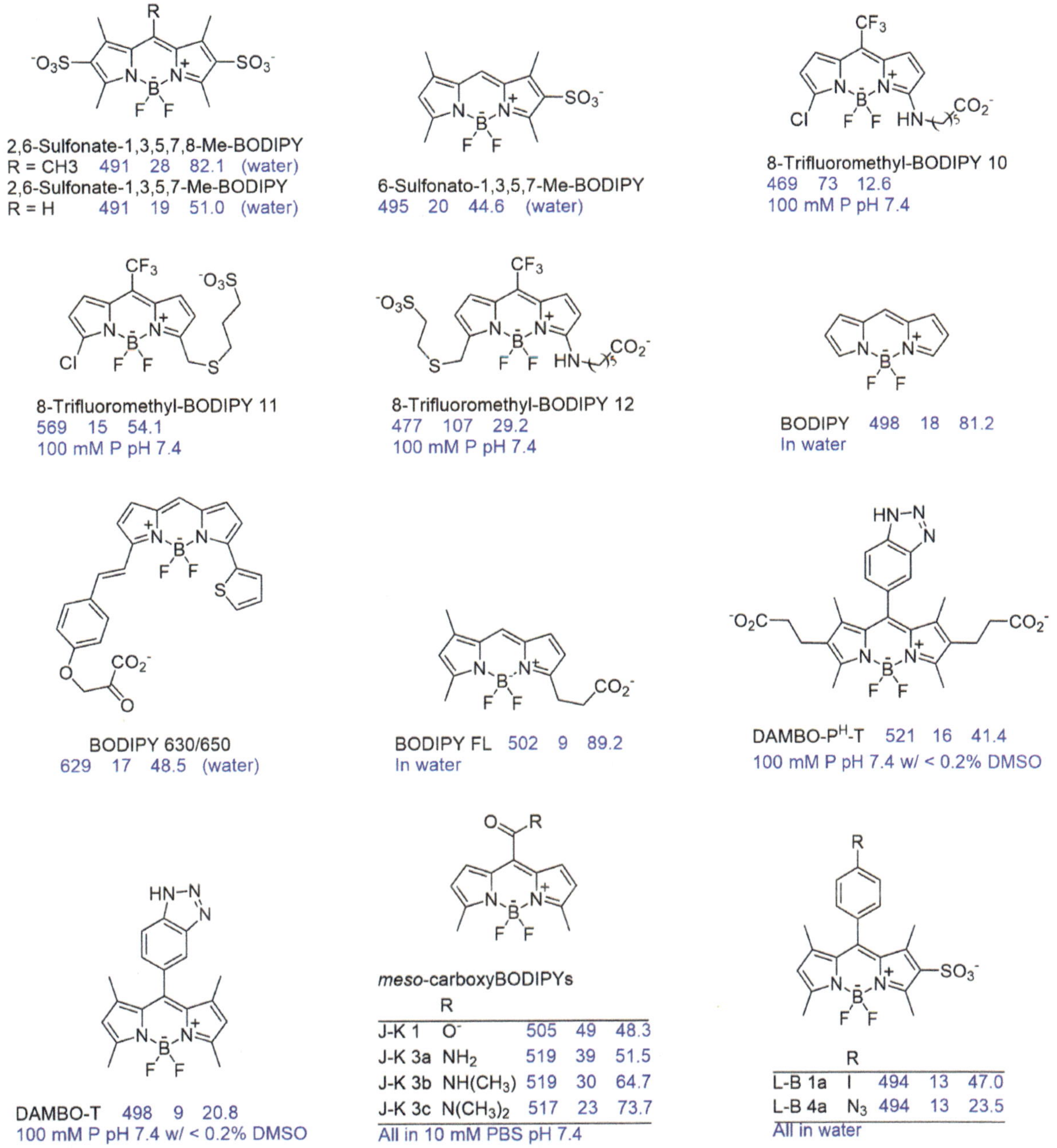

	R			
J-K 1	O^-	505	49	48.3
J-K 3a	NH_2	519	39	51.5
J-K 3b	$NH(CH_3)$	519	30	64.7
J-K 3c	$N(CH_3)_2$	517	23	73.7

All in 10 mM PBS pH 7.4

	R			
L-B 1a	I	494	13	47.0
L-B 4a	N_3	494	13	23.5

All in water

Figure 4.28 Molecular structures and photophysical properties of small borondipyrromethenes. Numbers in blue show the photophysical parameters $\lambda_{x,max}$ (nm), Δ_S (nm), and B_{max} ($mM^{-1}cm^{-1}$), respectively.

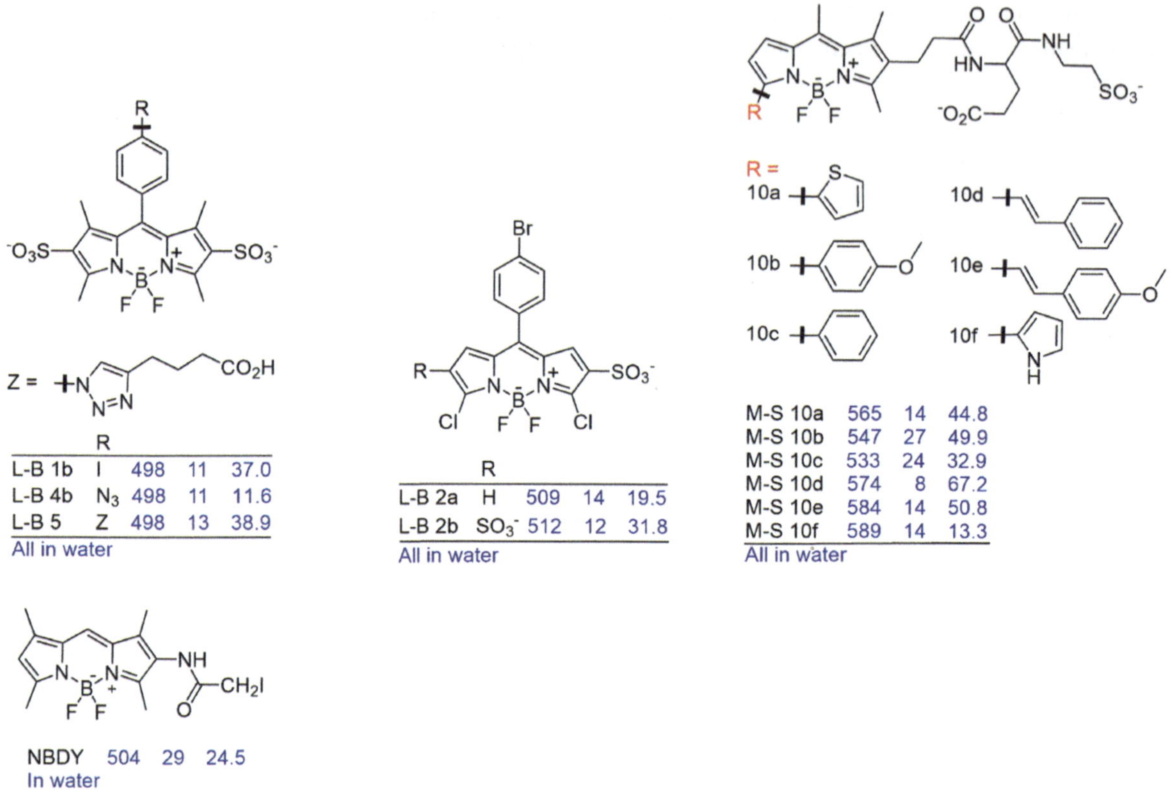

Figure 4.28 (Continued)

A third strategy to increase the water solubility of BODIPY dyes is to add carbohydrates or alcohols. For a review of the available methods for the synthesis of BODIPY-carbohydrate hybrids, see Gómez and López [118]. These structures, despite being larger than sulfonic derivatives, are neutral and exhibit a good biocompatibility and reduced citotoxicity. For examples of such dyes, which are G-L 13a, G-L 13b, and G-L 14 [118], see Figure 4.29. The initials of the names of the first and last authors, as well as the numbering used in the original article, were used to identify these BODIPY dyes.

4.14.3 Large Water-Soluble Borondipyrromethenes

A fourth strategy to increase water solubility is to attach branches of oligo(ethylene glycol)methyl ether groups at positions 2, 8, and 6, or at position 4 of the BODIPY core structure (Zhu et al. [4]). The dyes synthesized by Zhu et al. [4] are hereby termed Z-L (using the first letter of both first and last author names and the alphabetical characters used in the original work), i.e., Z-L A, Z-L B, Z-L C, Z-L D, Z-L E, Z-L F, Z-L G, Z-L H, Z-L I, and Z-L J (ε_{max} was estimated to be ~50 $mM^{-1}cm^{-1}$). Figure 4.30 shows these structures and some of their photophysical parameters. The Stokes shift of the dye M-S 10d was reported to be only 8 nm in the aqueous solution used (0.1% Triton X-100 in water). Bura and Ziessel [119] used a combination of very short (ethylene glycol) methyl ether oligomers and phosphonate fragments to produce bright water-soluble BODIPY dyes (hereby termed B-Z 7a, B-Z 7b, B-Z 14, and B-Z 19) (Figure 4.30). The Stokes shift of dye B-Z 7a was also reported to be small (only 9 nm) in the used solvent (plain water).

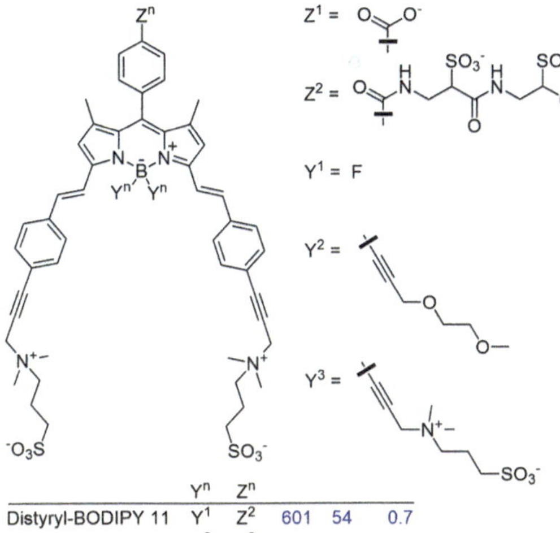

	Y^n	Z^n			
Distyryl-BODIPY 11	Y^1	Z^2	601	54	0.7
Distyryl-BODIPY 12	Y^2	Z^2	642	15	12.1
Distyryl-BODIPY 13	Y^3	Z^1	641	16	14.5

All in PBS pH 7.5

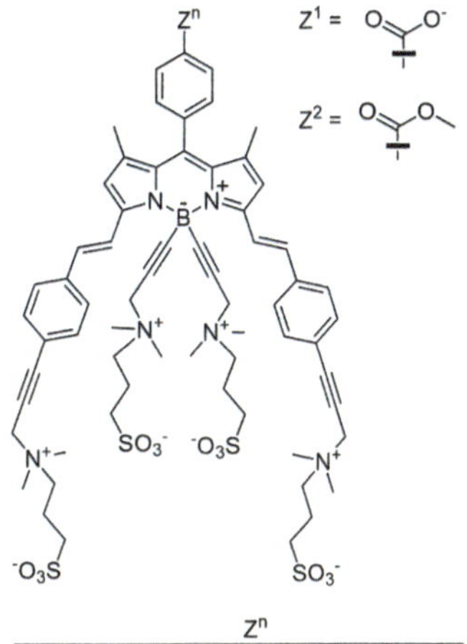

	Z^n			
Distyryl-BODIPY 17	Z^1	630	13	22.9
Distyryl-BODIPY 18	Z^2	628	13	26.9

All in water

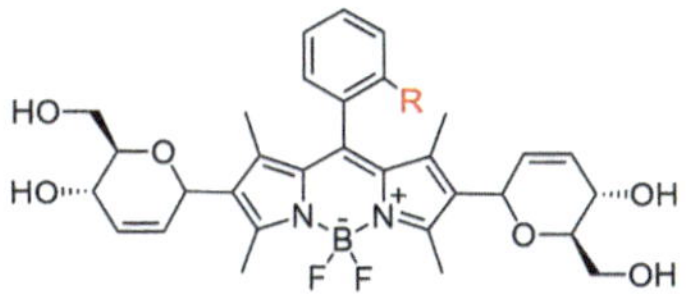

13a R = H

13b R = CH_2N_3 14 R =

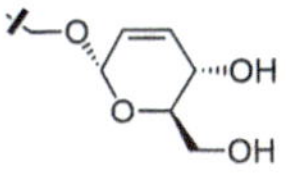

G-L 13a	505	12	16.4
G-L 13b	509	13	37.5
G-L 14	508	13	45.4

All in water

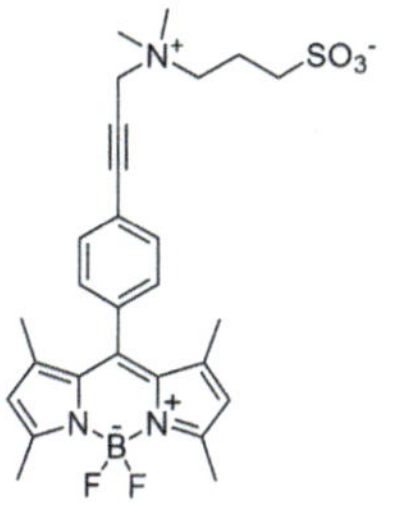

Niu 18 496 15 15.1
In water

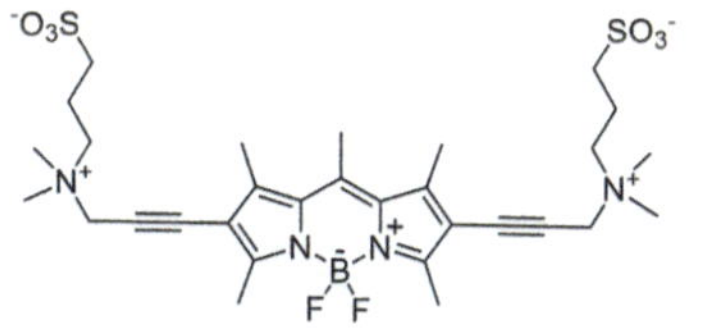

Niu 25 521 19 48.2
In PBS

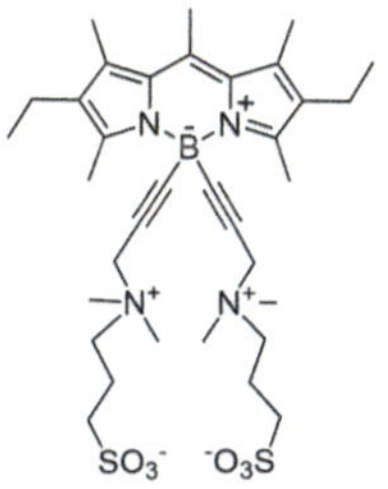

Niu 27 510 18 51.2
In water

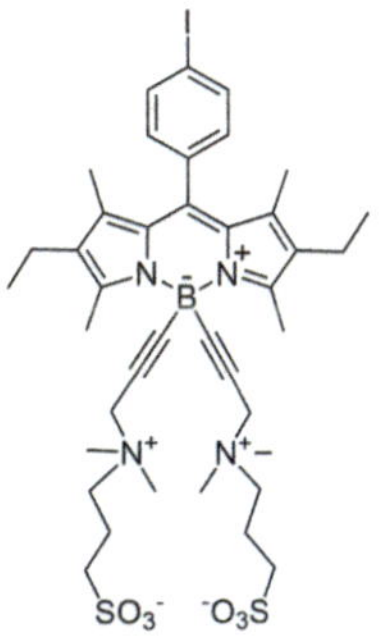

Niu 29 519 11 17.0
In water

Figure 4.29 Molecular structures and photophysical properties of medium size borondipyrromethene dyes. Legends as in Figure 4.28.

Blázquez-Moraleja et al. [120] presented a tethering reaction to produce tagged BODIPY dyes with even longer (ethylene glycol)methyl ether oligomers. Unfortunately, these dyes are too large to be included in the present database.

4.14.4 Other Classes Derived From Borondipyrromethene

This section contains the modified structures of the basic BODIPY scaffold, also known as BODIPY analogs. They should not be confused with the borondipyrromethene derivatives shown in the previous section, because they have undergone substantial changes in their π-electron system and, consequently, exhibit slightly different photophysical and/or chemical properties. Examples of such modified BODIPYs include 2,6-restricted BODIPYs, aza-BODIPYs, B,O-chelated BODIPYs, BOIMPYs, BOPAHYs, and BOPHYs, among others (depicted in Figure 4.31).

Figure 4.32 shows some water soluble and bright derivatives of BOPAHY, which is one of the modified borondipyrromethenes of Figure 4.31. Indeed, these are derivatives from 1,2,3-triazolium BOPAHY and are termed

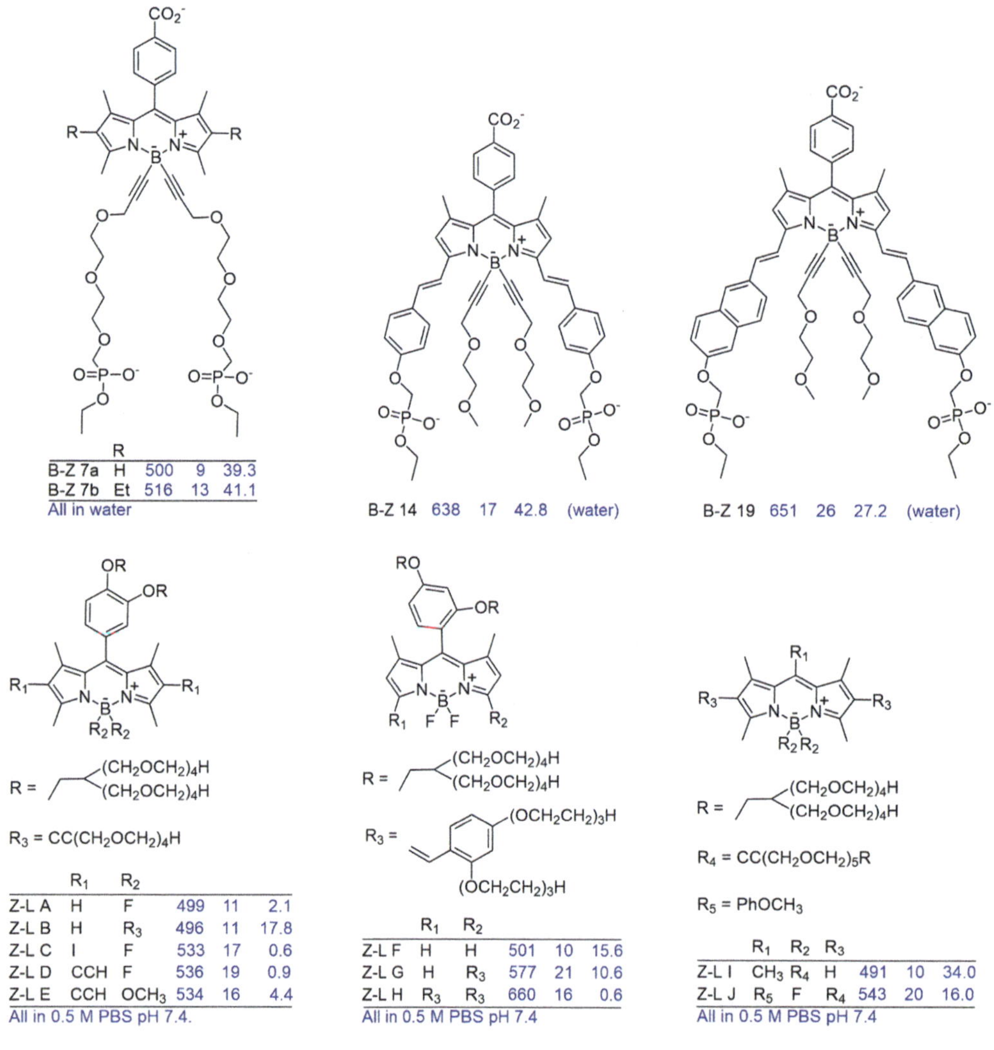

Figure 4.30 Molecular structures and photophysical properties of common borondipyrromethenes dyes functionalized with branched oligo(ethylene glycol)methyl ether groups to increase their water solubility. Legends are as in Figure 4.28.

BOPAHY H-D 18ab′, H-D 18bb′, H-D 19aa′, H-D 19ba′, H-D 19ba′, H-D 19bb′, and H-D 20 bb′ (Figure 4.32). The only difference between H-D 19bb′ and H-D 20 bb′ is the counter-ion (anion) of these fluorescent structures (cations), iodide and triflate, respectively.

4.15 Rhodols

Rhodols are xanthene derivatives with one keto group bound to carbon 3, one amino group bound to carbon 6, and a 2-benzoic acid bound to carbon 9 of the xanthene moiety. The structure of rhodols is similar to that of fluoresceins, with the exception of the phenol group that is replaced by an amino group. The pK_a of the conjugated acid of this amino group is much lower than that of the phenol group of the fluorescein counterparts, which results in photophysical parameters that are less sensitive to pH in the neutral pH range. Moreover, rhodols have been reported to have a higher photostability than fluorescein, while the other photophysical parameters (e.g., wavelength of absorption maxima and brightness) are similar to those of fluoresceins.

4.15.1 The First Rhodols Synthesized

6-Dimethylaminorhodol was synthesized in the pioneering work of Lee et al. [56]. Whitaker et al. [24] also re-synthesized 6-dimethylaminorhodol within a library of rhodols, some of which have an impressively high brightness. The dyes whose photophysical properties were measured by Whitaker et al. [24] are: 5(6)-carboxyrhodol, 2,7-diCl-6-methylaminorhodol, 2,7-diMe-6-EtAm-carboxyrhodol, 2-Cl-7-Me-6-EtAm-carboxyrhodol, 7-Me-6-EtAm-carboxyrhodol, 2-Et-7-Me-6-EtAm-carboxyrhodol, 4,7-diMe-6-EtAm-carboxyrhodol, 2-PrAc-7-Me-6-EtAm-rhodol, 6-piperidinocarboxyrhodol, 2-Br-7-Me-6-EtAm-carboxyrhodol, 2-Cl-6-piperidinocarboxyrhodol, 6-quinolinocarboxyrhodol, 2-cloro-6-quinolinocarboxyrhodol, 2-ethyl-6-quinolinocarboxyrhodol, and 2-PrAc-6-quinolinorhodol (Figure 4.33).

4.15.2 Rhodols Synthesized More Recently

Peng and Yang [122] have also produced a library of rhodols. Among other things, this library shows the effects of various 6-amino modifications on the photophysical properties of rhodols (Figure 4.34). The same effects can be expected for other anime-containing dyes in which the nitrogen is located at the end of the π-electron system (e.g., rhodamines, pyronines, aminocoumarins, pyrodols). The effects of nitrogen substituents on the photophysical properties of many dyes were studied by Singha et al. [121].

The brightest rhodols synthesized by Peng and Yang [122] are: rhodol (2-(6-Amino-3-oxo-3H-xanthen-9-yl) benzoate or simple rhodol) ($\Phi_f = 0.98$), 6-dimethylaminorhodol ($\Phi_f = 0.20$), 6-diethylaminorhodol ($\Phi_f = 0.15$), 6-dipropylaminorhodol ($\Phi_f = 0.10$), 6-dibutylaminorhodol ($\Phi_f = 0.08$), 6-propylaminorhodol ($\Phi_f = 0.94$), 6-butylaminorhodol ($\Phi_f = 0.93$), 6-allylaminorhodol ($\Phi_f = 0.92$), 6-pyrrolidinorhodol ($\Phi_f = 0.12$), 6-piperidinorhodol ($\Phi_f = 0.10$), 6-benzylaminorhodol ($\Phi_f = 0.95$), 6-acrylamidorhodol ($\Phi_f = 0.19$), 6-benzamidorhodol ($\Phi_f = 0.16$), 6-morpholinorhodol ($\Phi_f = 0.13$), 2,7-diCl-6-methylaminorhodol ($\Phi_f = 0.67$), 3-methoxymethyletherrhodol ($\Phi_f = 0.85$), and 3-MOMO-6-propylaminerhodol ($\Phi_f = 0.35$) (Figure 4.34). The 6-phenylaminorhodols, 2-(3-oxo-6-(phenylamino)-3*H*-xanthen-9-yl)benzoate, and 2-(6-(methyl(phenyl)amino)-3-oxo-3*H*-xanthen-9-yl)benzoate) synthesized by Peng and Yang [122], are not shown because their brightness was reported to be very low and the present database aims to only present fluorescent dyes with a medium to high brightness in aqueous solutions.

It is worth mentioning that 3-methoxymethylrhodol exhibits a wavelength of absorption maximum of 477 nm in 50 mM phosphate pH 8 with <0.1% acetonitrile as a co-solvent, which is shorter than that of fluorescein [122]. The brightest rhodols are 2-(6-Amino-3-oxo-3H-xanthen-9-yl)benzoate (rhodol) [122] and 5(6)-carboxyrhodol [67], with respective brightnesses of 68.6 $mM^{-1}cm^{-1}$ (in phosphate pH 8) and 61.9 $mM^{-1}cm^{-1}$ (in phosphate pH 9). Other bright rhodols of note are 6-azetidinylamino rhodol with B_{max} = 50.1 $mM^{-1}cm^{-1}$ in 10 mM HEPES pH 7.3 [26],

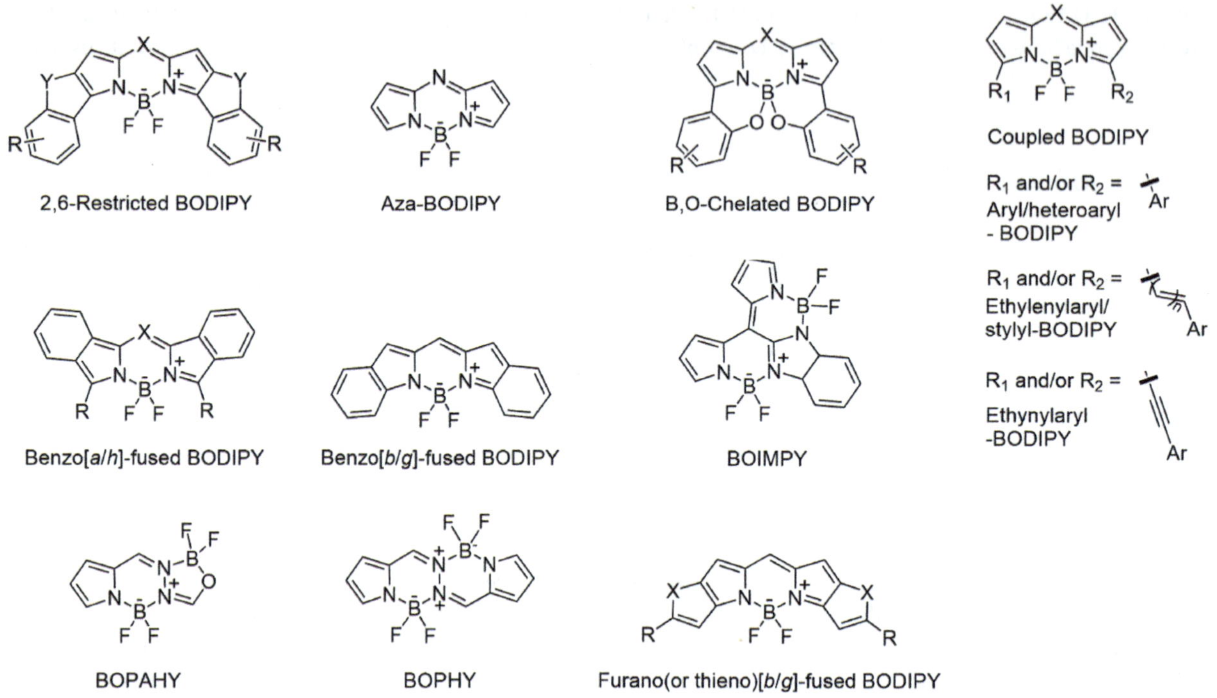

Figure 4.31 Core structure of borondipyrromethene (BODIPY) and a dozen modified borondipyrromethenes.

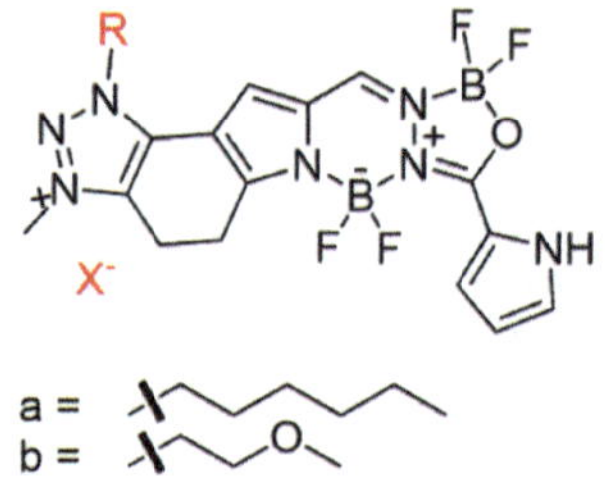

BOPAHY	R	X^-			
H-D 18ab'	a	I^-	429	24	28.0
H-D 18bb'	b	I^-	428	25	24.9

All in water

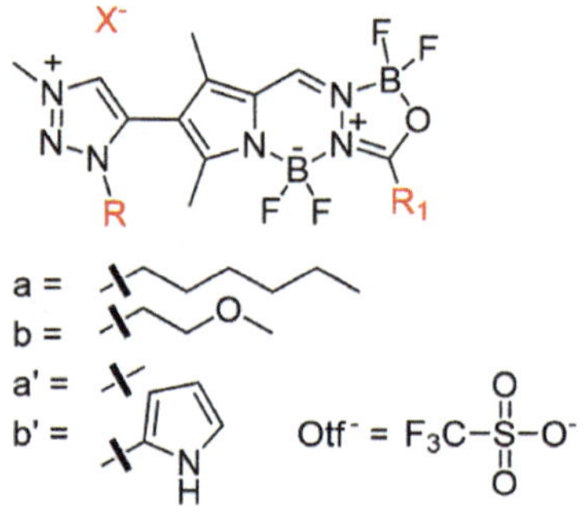

BOPAHY	R	R_1	X^-			
H-D 19aa'	a	a'	I^-	368	19	0.7
H-D 19ba'	b	a'	I^-	367	19	1.5
H-D 19ab'	a	b'	I^-	416	19	10.2
H-D 19bb'	b	b'	I^-	416	18	8.6
H-D 20bb'	b	b'	Otf^-	416	19	8.6

All in water

Figure 4.32 Molecular structures and photophysical properties of some BOPAHY derivatives. Legends as in Figure 4.28.

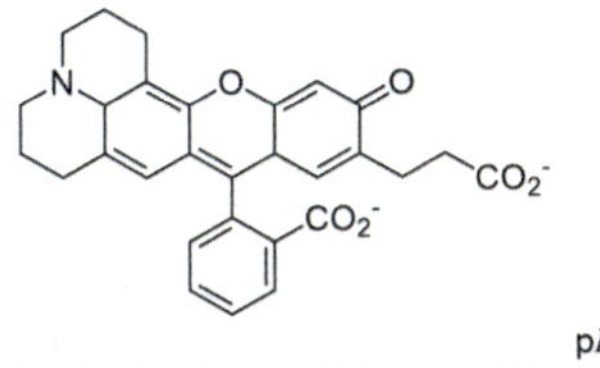

				pK_a
2-PrAc-6-Quinolinorhodol	541	31	28.5	6.63

50 mM phosphate pH 8-9

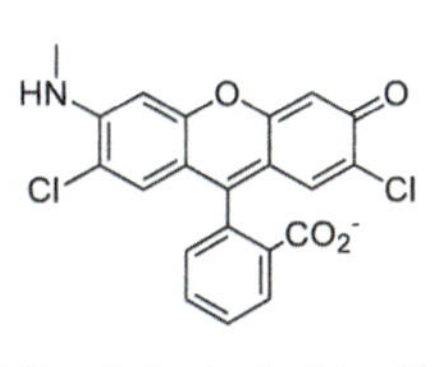

2,7-DiCl-6-methylaminorhodol 512 21 47.5

50 mM phosphate pH 8 with <0.1% acetonitrile.

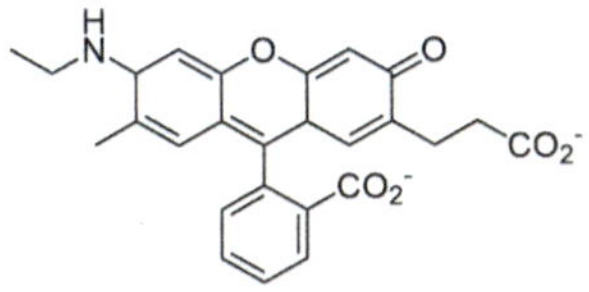

				pK_a
2-PrAc-7-Me-6-EtAm-rhodol	509	26	73.0	5.91

50 mM phosphate pH 8-9

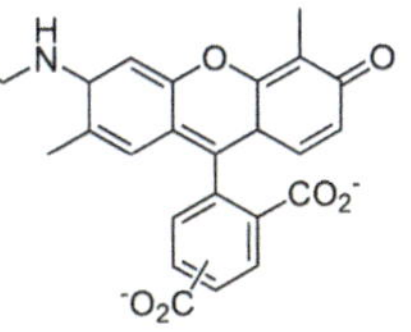

				pK_a
4,7-DiMe-6-EtAm-carboxyrhodol	516	28	24.4	5.97

50 mM phosphate pH 8-9

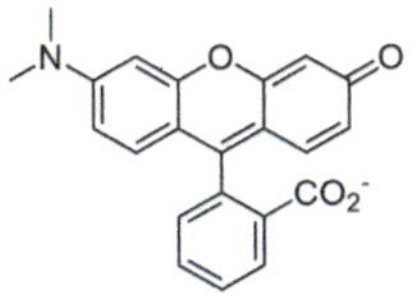

6-Dimethylaminorhodol

522 23 11.4

Carbonate pH 9.5

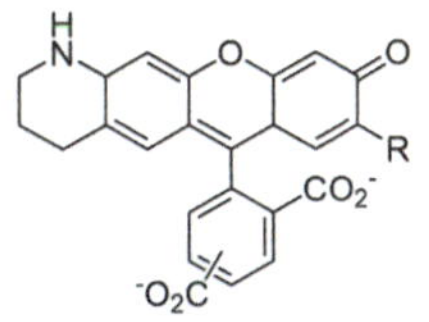

	R				pK_a
6-Piperidinocarboxyrhodol	H	516	25	63.9	6.12
2-Cl-6-piperidinocarboxyrhodol	Cl	524	24	55.7	4.56

All in 50 mM phosphate pH 8-9

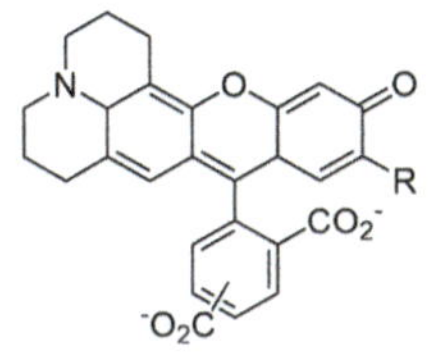

	R				pK_a
6-Quinolinocarboxyrhodol	H	540	30	28.7	6.41
2-Cloro-6-quinolinocarboxyrhodol	Cl	548	27	31.2	5.11
2-Ethyl-6-quinolinocarboxyrhodol	CH_2CH_3	543	31	38.9	6.24

All in 50 mM phosphate pH 8-9

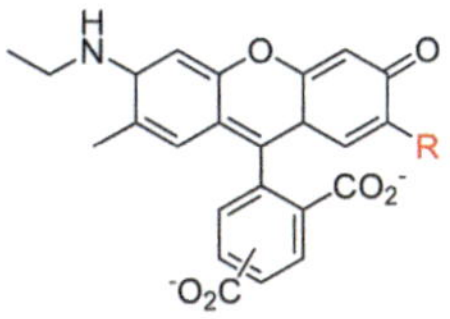

	R				pK_a
7-Me-6-EtAm-carboxyrhodol	H	508	26	64.0	5.90
2-Cl-7-Me-6-EtAm-carboxyrhodol	Cl	514	25	66.0	4.49
2-Br-7-Me-6-EtAm-carboxyrhodol	Br	516	23	49.3	4.30
2,7-Di-Me-6-EtAm-carboxyrhodol	Me	510	26	59.0	5.82
2-Et-7-Me-6-EtAm-carboxyrhodol	Et	511	27	61.3	5.67

All in 50 mM phosphate pH 8-9

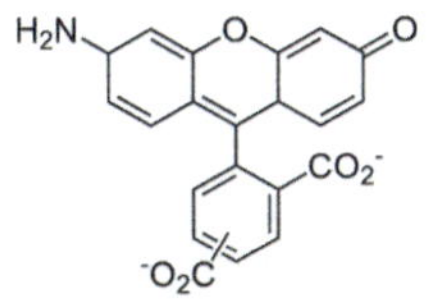

				pK_a
5(6)-Carboxyrhodol	494	26	53.3	5.59

50 mM phosphate pH 8-9

Figure 4.33 Molecular structures and the photophysical parameters of the first rhodols. The numbers in blue represent $\lambda_{x,max}$ (nm), Δ_S (nm), and B_{max} (mM^{-1}cm^{-1}).

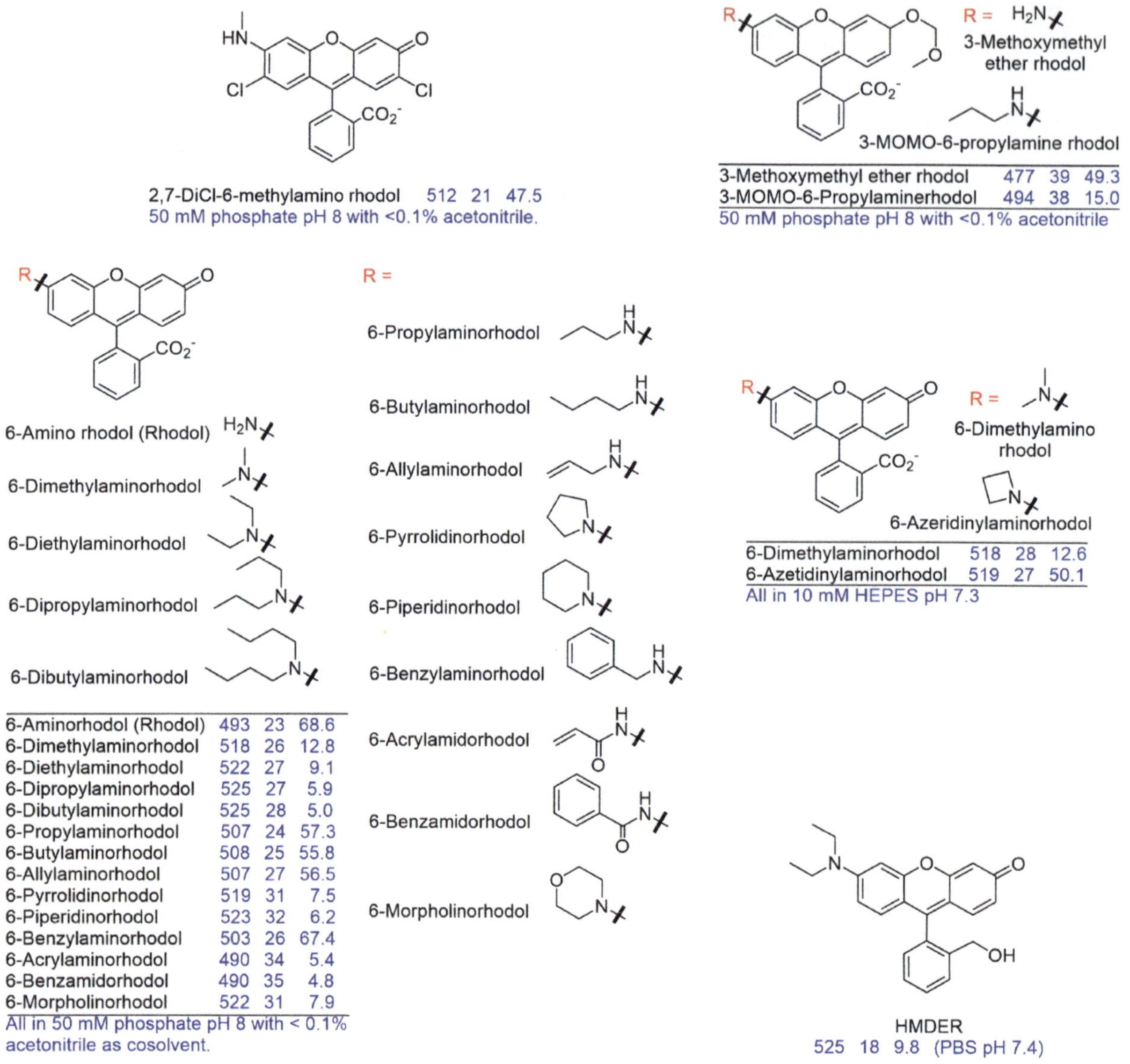

3-Methoxymethyl ether rhodol	477	39	49.3
3-MOMO-6-Propylaminerhodol	494	38	15.0

50 mM phosphate pH 8 with <0.1% acetonitrile

6-Aminorhodol (Rhodol)	493	23	68.6
6-Dimethylaminorhodol	518	26	12.8
6-Diethylaminorhodol	522	27	9.1
6-Dipropylaminorhodol	525	27	5.9
6-Dibutylaminorhodol	525	28	5.0
6-Propylaminorhodol	507	24	57.3
6-Butylaminorhodol	508	25	55.8
6-Allylaminorhodol	507	27	56.5
6-Pyrrolidinorhodol	519	31	7.5
6-Piperidinorhodol	523	32	6.2
6-Benzylaminorhodol	503	26	67.4
6-Acrylaminorhodol	490	34	5.4
6-Benzamidorhodol	490	35	4.8
6-Morpholinorhodol	522	31	7.9

All in 50 mM phosphate pH 8 with < 0.1% acetonitrile as cosolvent.

6-Dimethylaminorhodol	518	28	12.6
6-Azetidinylaminorhodol	519	27	50.1

All in 10 mM HEPES pH 7.3

Figure 4.34 Molecular structures and the photophysical parameters of more elaborated rhodols. Legend as in Figure 4.33.

6-benzylaminorhodol with B_{max} = 67.4 mM^{-1}cm^{-1} in phosphate pH 8, and 2,7-diCl-methylaminorhodol with B_{max} = 47.5 mM^{-1}cm^{-1} in phosphate pH 8 [122].

HMDER is a very special rhodol as it also exists in equilibrium between a closed (non-fluorescent) and open (fluorescent) form. However, here an ether occurs at the closed form due to the hydroxymethyl group bound to carbon 2′ instead of a carboxylic group [123]. For this reason, it is better classified as a rhodol instead of a rosol (Section 4.16).

4.15.3 Rhodol Analogs

Some dyes with a structure that resembles that of rhodols have been synthesized, but they have subtle differences involving the nature or spatial arrangement of the atoms belonging to their π-electron system. One such example is 2-OH-6-diMeAm-benzo[c]xanthene (Figure 4.35). It was synthesized by Jiao et al. [124], and despite its low brightness (probably due to the diethylamino bound at carbon 6), it shows the synthesis possibilities of a great variety of dyes. Another rhodol analog is SNARF-1 (Whitaker et al. [125], Figure 4.35).

4.16 HAS-Rhodols

There are only a few synthesized dyes belonging to the heteroatom-substituted rhodol class. The only member of this class whose photophysical parameters have been fully measured is termed carborhodol S-H 13-H [126] (Figure 4.36). This belongs to the type $C(CH_3)_2$-rhodol (dimethylcarbo-rhodol) of the C-rhodol (carbo-rhodol) subclass. The same structure, as well as many others, was synthesized by Poronik et al. [127]. More of this class of dyes can be expected to be synthesized during the coming years due to their high brightness, high photostability, and long wavelength of excitation maximum that is promoted by dimethylcarbone substitution.

There are also at least two dimethylsilane-rhodols termed 2′COOH-Si-rhodol and 2′COOH-4CH_2OH-Si-rhodol synthesized by Ito et al. [128]. They measured all photophysical parameters of these two dyes, with the exception of ε_{max} which was estimated to be ~70 ($mM^{-1}cm^{-1}$) from similar structures, so that these two dyes could be added to this database. More of this class of dyes can also be expected to be synthesized during the coming year due to potential high brightness, high photostability, and their long wavelength of excitation maximum that is promoted by dimethylsilane substitution.

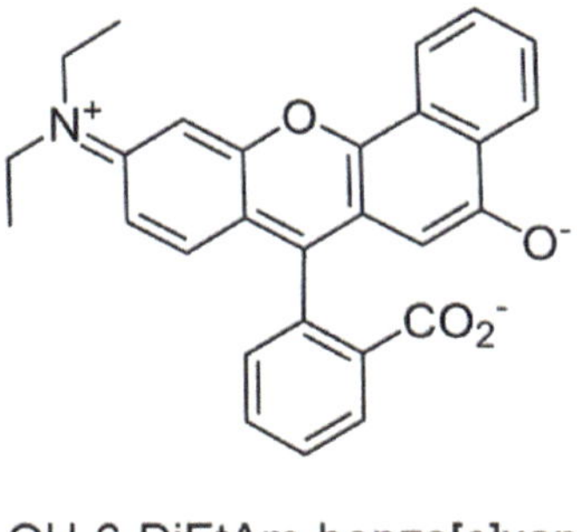

2-OH-6-DiEtAm-benzo[c]xanthen
578 42 6.8 (H_2O)

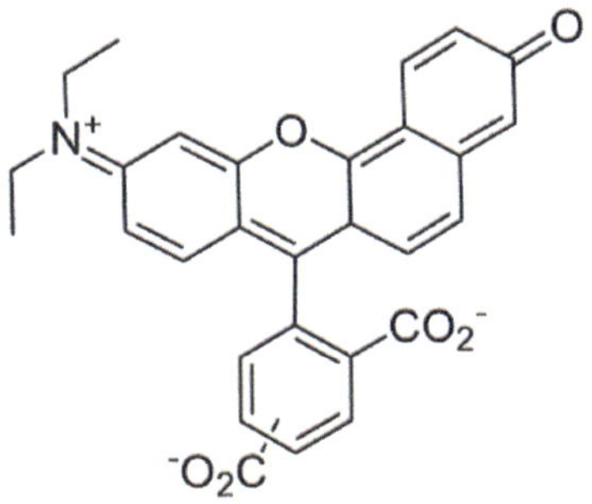

SNARF-1 573 58 4.0
Phosphate pH 10

Figure 4.35 Molecular structures and photophysical parameters of a rhodol-analog dye. Legend as in Figure 4.33.

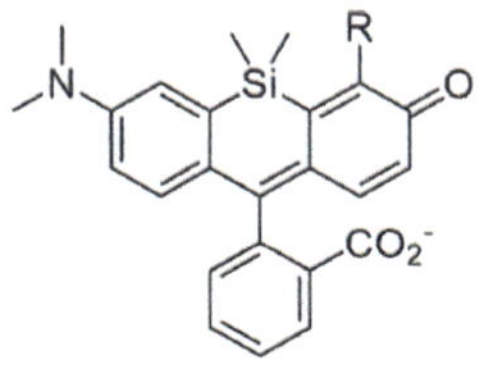

				R
2'COOH-Si-rhodol	609	21	15.4	H
2'COOH-4CH_2OH-Si-rhodol	610	20	14.7	CH_2OH

All in 100 mM P pH 7.4

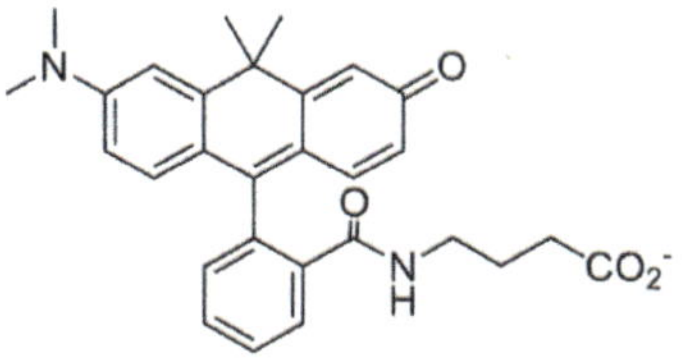

Carborhodol S-H 13-H
586 27 18.8 (PBS pH 7.4)

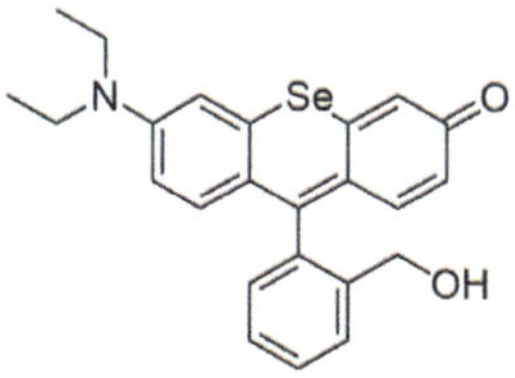

HMDESeR
558 32 0.5 (100 mM P pH 7.4)

Figure 4.36 Molecular structure and photophysical properties of the HAS-rhodols. The numbers in blue represent $\lambda_{x,max}$ (nm), Δ_S (nm), and B_{max} ($mM^{-1}cm^{-1}$).

HMDESeR is an HAS-rhodol (Figure 4.36) because its structure exists in equilibrium between the non-fluorescent ether (closed) form and the quinoid (open) form; i.e., the hydroxymethyl group is also subjected to cyclization in this case, similar to what happens with carboxylic groups [129]. Its fluorescence quantum yield is reported to be low (0.006) and this compound has been observed to be an effective photosensitizer.

4.17 Rosamines

Rosamines are very similar to rhodamines as they also have an aryl group bound to the carbon-9 of the xanthene scaffold. However, the substituent groups of this pendant aryl do not form closed lactone forms. Therefore, rosamines are advantageous for some applications; however, care must be taken to prevent the free rotation of the pendant aryl group (observed in some rosamines), as it has a detrimental effect on the fluorescence quantum yield. Therefore, the brightest rosamine derivatives produced so far have a methyl group grafted onto the C-2′ atom (Figure 4.3), as this significantly increases fluorescence by increasing the energy barrier for the rotation of the pendant aryl group. In this way some non-radioactive decay channels are mitigated.

In some rhodamine dyes the carboxylic group bound to the C-2′ atom of the pendant aryl group is modified to a carboxamide to prevent cyclization to the colorless and fluorescence inactive lactone form. Nevertheless, even with the presence of carboxamindes, these dyes must still be considered rhodamines and not rosamines. This is because the hydrolysis of this carboxamide makes cyclization possible again.

Rosamines are generally less soluble in aqueous solutions than their rhodamine (and sulforhodamines) counterparts; this happens due to the lack of the carboxylic (or sulfonic) group at the C-2′ atom of the pendant aryl group. The simplest rosamine was termed 2-Me RG by Piao et al. [130]; the tetramethylrosamine (TMRosamine) form, which should not be confused with tetramethylrhodamine (TMR), was reported to exhibit a lower brightness. This was measured by Zhou et al. [131] using a diluted solution of TMRosamine in 10 mM PBS pH 7.4 with 0.1% DMSO (Figure 4.37). A rosamine synthesized by Wu and Burgess [29], here termed W-B 5h, is much more soluble in aqueous solutions at neutral pH due to the partial protonation of the piperazine groups. Finally, Alexa Fluor 532 is a very bright and popular rosamine [6, 132] (Figure 4.37).

4.18 HAS-Rosamines

HAS-rosamines have undergone a heteroatom substitution at position 10 of the xanthene ring. The largest subclass is made up of silico-rosamines (Si-rosamines). This subclass contains one disilanol-rhodamine ($Si(OH)_2$-rhodamine), while all other members are dimethylsilane-rosamines ($Si(CH_3)_2$-rosamines) (Figure 4.38).

4.18.1 Silico-Rosamines

The detailed list and references for silico-rosamines is as follows: 2-Me SiR600, 2-Me SiR650, 2-Me SiR680, 2-Me SiR700, and 2-M2 SiR720 [79, 133, 134] and the dyes hereby termed 2-Me SiR620, 2-Me SiR640, 2-Me SiR654, 2-Me SiR666, 2-Me SiR670, 2-Me SiR671, and 2-Me SiR675 [135].

SiR-methyl [28] is not able to assume the closed lactone form as it has a methyl group bound to the C-2′ and only has a carboxylic group at the C-5′ position; therefore, by definition, this is an HAS-rosamine and not an HAS-rhodamine. $SiOH_2R$ is the only disilanol-rosamine (Figure 4.38) whose photophysical parameters were measured in aqueous solution (10 mM PBS pH 7.4 with 0.1% DMSO) [131].

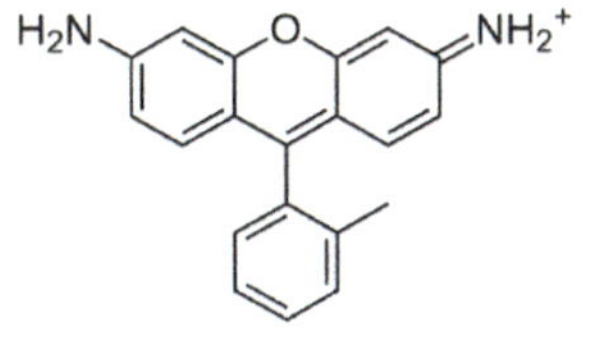

2-Me RG
498 22 68.8 (PBS pH 7.4)

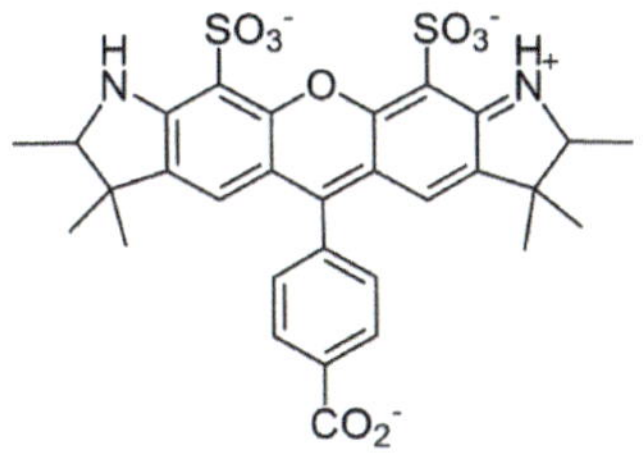

Alexa Fluor 532
531 23 49.4 (PBS pH 7.2 22 °C)

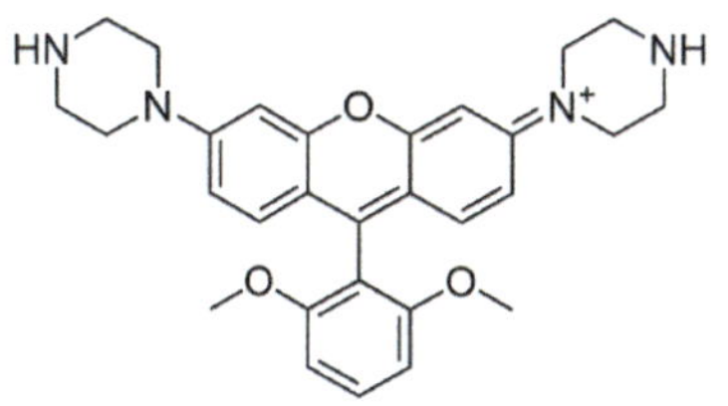

Rosamine W-B 5h
542 26 10.0 (100 mM P pH 7.4)

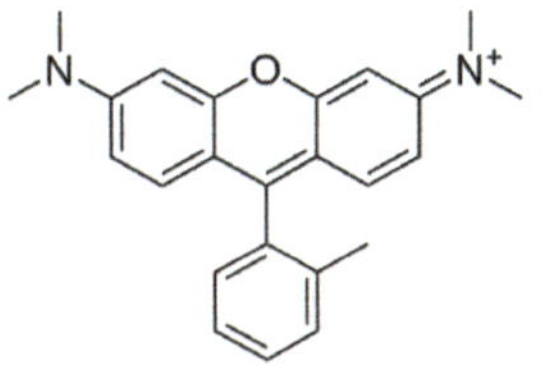

Tetramethylrosamine (TMRosamine)
550 20 23.1 (10 mM PBS pH 7.4)

Figure 4.37 Molecular structures and some photophysical parameters of rosamines. The photophysical parameters shown are $\lambda_{x,max}$ (nm), Δ_S (nm), and B_{max} (mM^{-1}cm^{-1}), presented in this order.

4.18.2 Phospha-Rosamines

P-rosamines (phospha-rosamines) are phosphorus-substituted rosamines and exhibit an extraordinarily long wavelength of absorption maximum due to the electron-withdrawing effect of the phosphorous moiety. PO_2^- rosamines (phosphinate-rosamines), POR-rosamines, and PO_2R-rosamines (phosphine oxide-rosamines) are such examples. The main entities of this subclass are: NR666, NR698, NR700, and NR744 [136]; PR, Me-PR, and tMe-PR [137]; and NR669 and NR675 [73]. These structures are shown in Figure 4.39.

4.18.3 Other HAS-Rosamines

B-rosamines (bora-rosamines) have a boron as the heteroatom. $B(OH)_2$-rosamines (dihydroxyborate-rosamines) are examples in which the boron appears as a borinate, which is made of two B–C bonds and two B–O bonds. This is the result of the coordination of boronic acid (made of two B–C bonds and one B–O bond) and an OH^- anion. An example of such a compound is RF620 [131], which can be seen in Figure 4.40.

Te-rosamines (tellurium-rosamines), such as TeO-rosamines (telluroxide-rosamines), contain the heavy atom tellurium at position 10 of the xanthene core structure. This interesting structure was synthesized by Koide et al. (2012) [138] and exhibits a wavelength of absorption maximum at the end of the visible and beginning of the infrared range (Figure 4.40).

S-rosamines (sulfur-rosamines) have a sulfur as the heteroatom. This subclass contains SO_2-rosamines (sulfone-rosamines) with, for example, the following dyes: SO_2R1, SO_2R2, SO_2R3, SO_2R4, and SO_2R5 [139]. Structures are all shown in Figure 4.40.

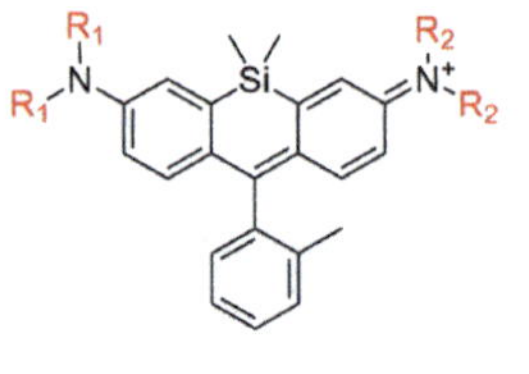

	R_1	R_2				
2-Me SiR600	H	H	593	20	34.5	(w)
2-Me SiR620	H	Me	620	18	20.1	(P)
2-Me SiR650	Me	Me	646	14	34.1	(w)

w = water; P = 100 mM phosphate pH 7.4 w/ 1% DMSO.

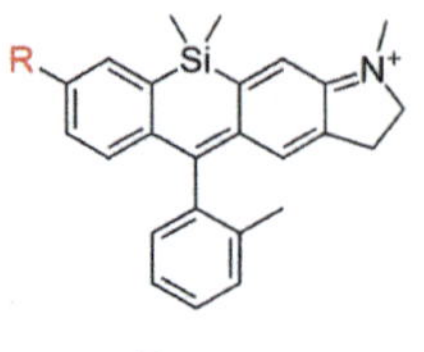

	R			
2-Me SiR640	NH_2	637	23	9.7
2-Me SiR654	NHEt	654	18	19.2
2-Me SiR666	$N(Me)_2$	666	18	20.0
2-Me SiR670	$N(Et)_2$	670	17	24.3

All in 100 mM phosphate pH 7.4 w/ 1% DMSO.

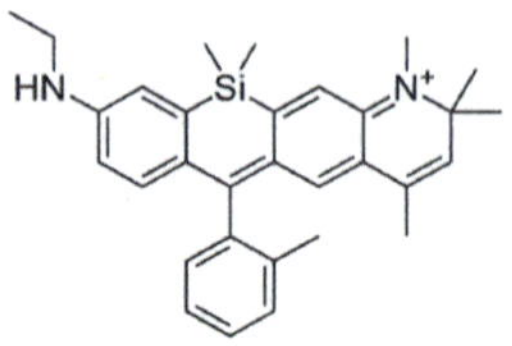

2-Me SiR675
675 23 6.1 (100 mM phosphate pH 7.4 w/ 1% DMSO)

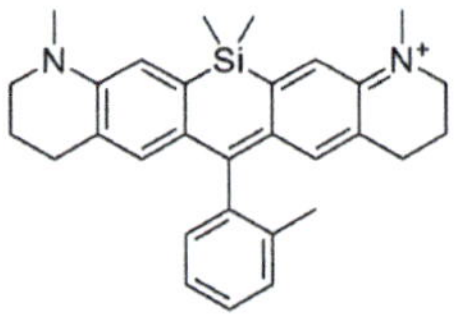

2-Me SiR680
674 15 45.5 (water)

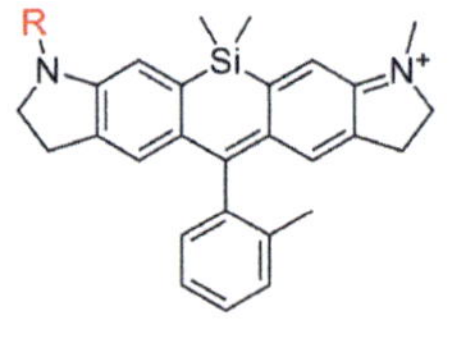

	R				
2-Me SiR700	CH_3	691	21	12.0	(w)
2-Me SiR671	H	671	16	28.7	(P)

w = water; P = 100 mM phosphate pH 7.4 w/ 1% DMSO.

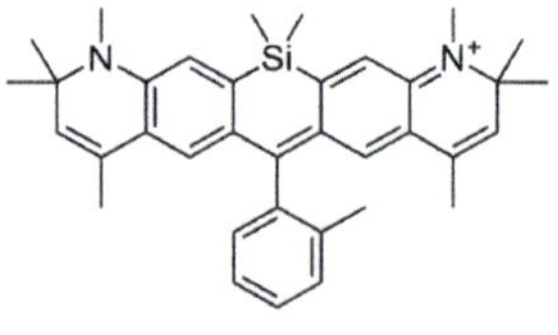

2-Me SiR720
721 19 8.0 (water)

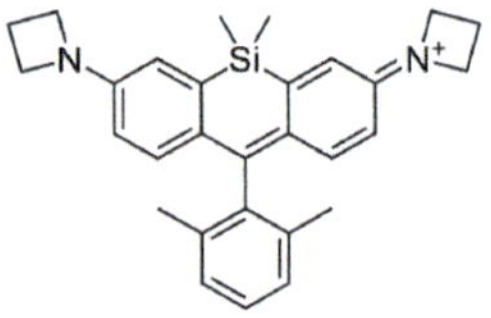

3,7-Dimethyl-JF646
651 13 45.7 (10 mM HEPES pH 7.3)

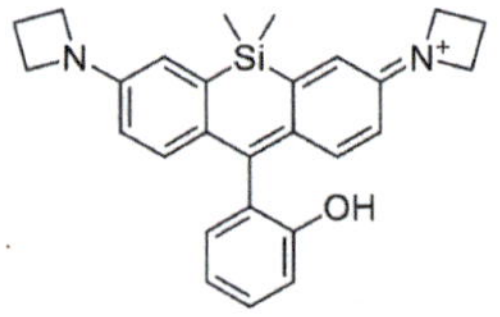

3-Hydroxy-JF646
651 15 65.2 (10 mM HEPES pH 7.3)

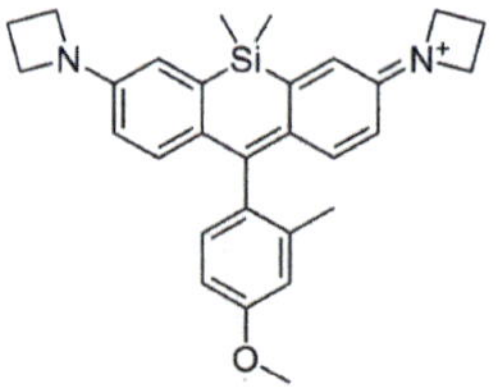

3-Methyl-5-Methoxy-JF646
649 15 53.9 (10 mM HEPES pH 7.3)

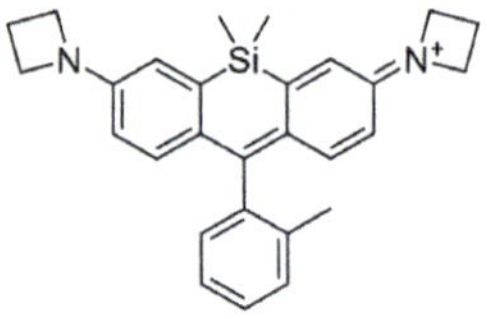

3-Methyl-JF646
649 17 55.4 (10 mM HEPES pH 7.3)

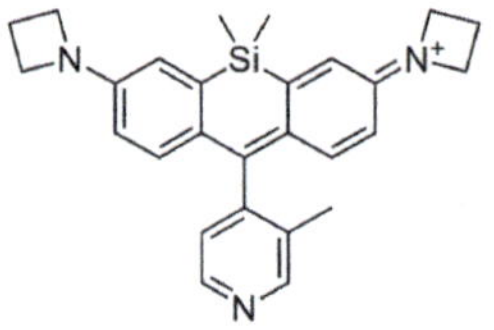

5-Aza-3-methyl-JF646
656 14 63.8 (10 mM HEPES pH 7.3)

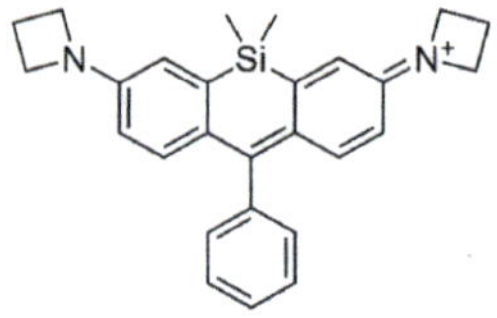

Des-carboxy-JF646
648 14 23.2 (10 mM HEPES pH 7.3)

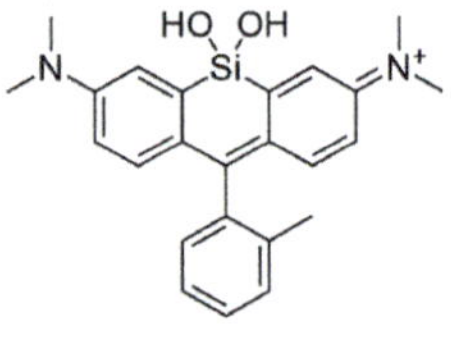

SiOH2R
663 18 45.1 (10 mM PBS pH 7.4 with 0.1% DMSO)

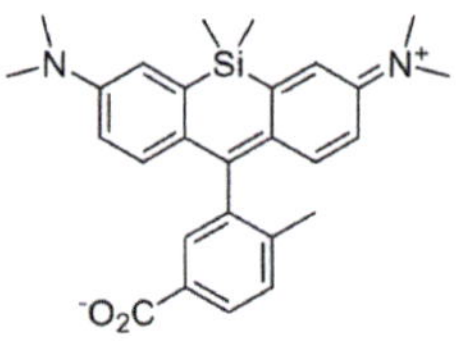

SiR-methyl
648 14 39.0 (water)

Figure 4.38 Structures and photophysical properties of Si-rosamines (one disilanol-rosamine and many dimethylsilane-rosamines). The photophysical parameters shown in blue are $\lambda_{x,max}$ (nm), Δ_S (nm), and B_{max} ($mM^{-1}cm^{-1}$), given in this order.

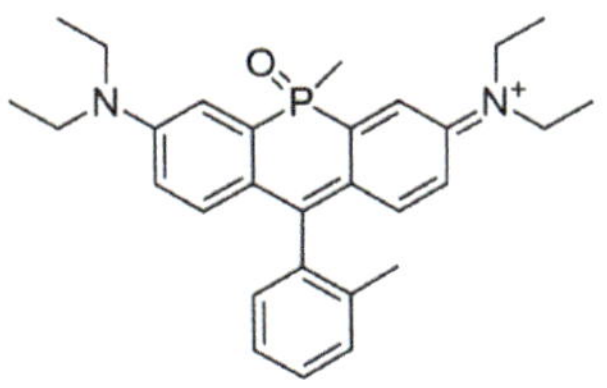

Me-PR
694 18 10.1 (PBS w/ 1% DMSO)

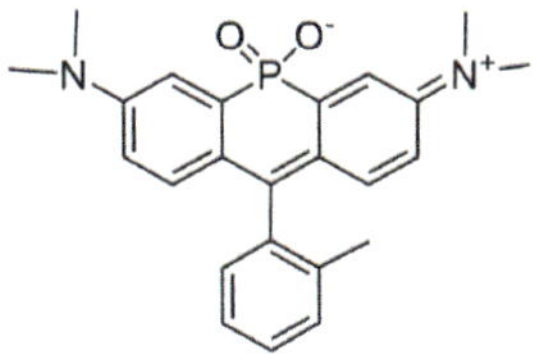

NR666
666 19 62.7 (PBS w/ 1% DMSO)

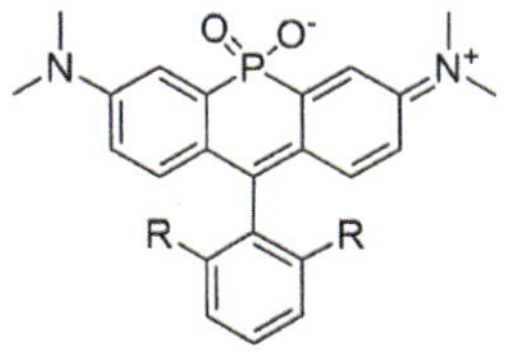

	R				pK_a
NR 669	$-CH_3$	669	20	23.1	<0.52
NR 675	$-OCH_3$	675	20	17.5	<0.59

All in 10 mM PBS pH 7.4 with 1% DMSO

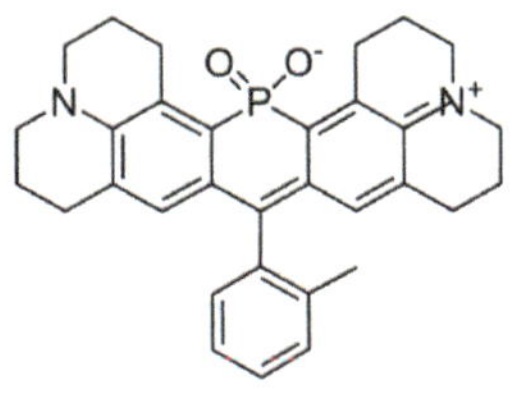

NR698
698 14 8.2 (PBS w/ 1% DMSO)

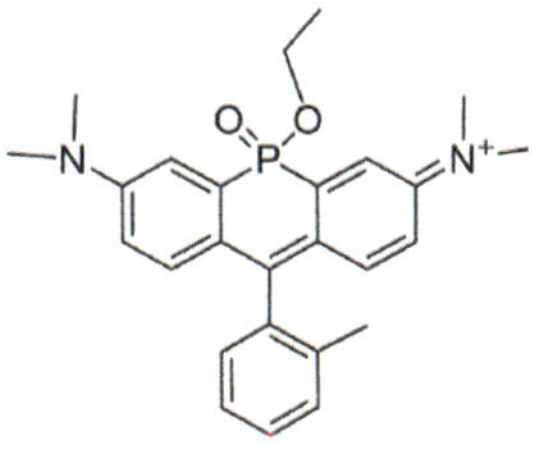

NR700
700 22 7.8 (PBS w/ 1% DMSO)

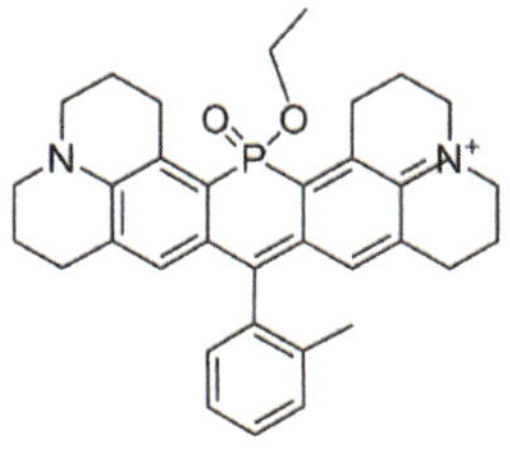

NR744
744 20 12.8 (PBS w/ 1% DMSO)

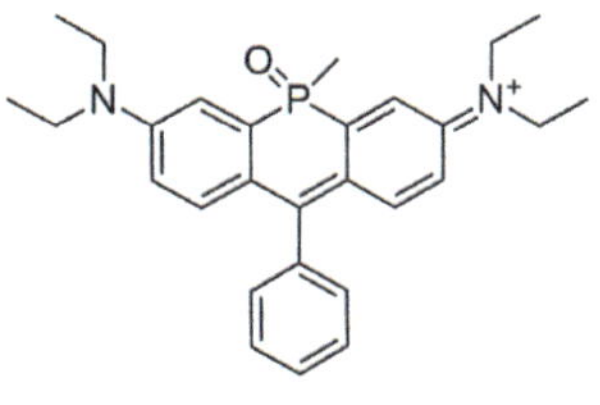

PR
694 18 3.9 (PBS w/ 1% DMSO)

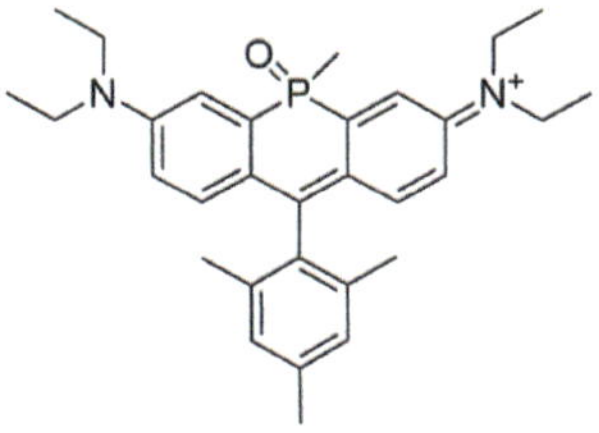

tMe-PR
696 17 10.2 (PBS w/ 1% DMSO)

Figure 4.39 Molecular structures of some HAS-rosamines with a phosphorous atom as the heteroatom. They are known as phospha-rosamines and have the following subclasses: PO_2^--rosamines (phosphinate-rosamines); POR-rosamines; and PO_2R-rosamines (phosphine oxide-rosamines). Legend as in Figure 4.38.

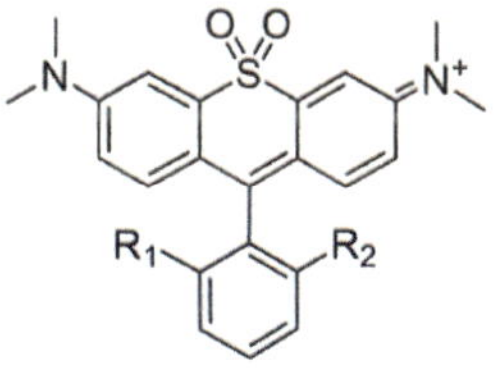

	R1	R2			
SO_2R1	H	H	703	33	2.6
SO_2R2	H	CH_3	703	39	6.9
SO_2R3	H	$O\text{-}CH_3$	707	40	3.4
SO_2R4	CH_3	CH_3	704	38	9.4
SO_2R5	$O\text{-}CH_3$	$O\text{-}CH_3$	710	42	8.7

10 mM PBS pH 7.4 w/ 0.25% CH_3CN.

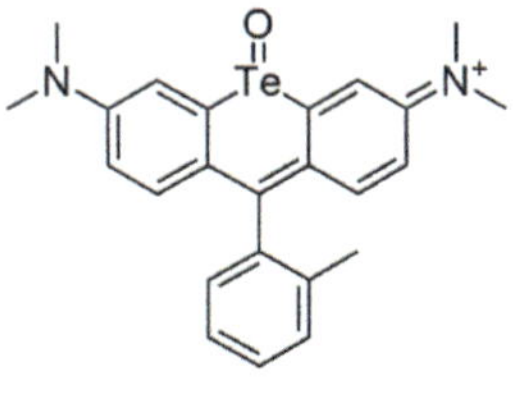

2-Me TeOR
669 17 21.6 (PBS pH 7.4)

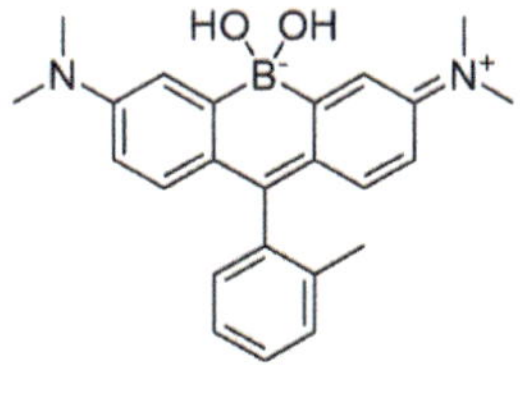

RF620
620 16 39.2 (10 mM PBS pH 7.4)

Figure 4.40 Molecular structures and photophysical properties of some HAS-rosamines in which the heteroatom is a boron (dihydroxyborate-rosamine), tellurium (telluroxide-rosamine), or a sulfur (sulfone-rosamines). Legend as in Figure 4.38.

4.19 Rosols

Rosols (6-amino-3H-xanthen-3-one) are structurally related to rhodols, just as rosamines are related to rhodamines and fluorones are related to fluoresceins, i.e., the absence of the carboxylic group at carbon 2′ (Figure 4.3). This prevents the formation of the lactone (cyclic) form and has many practical implications. Hettie et al. [140] synthesized a series of rosols they named Me-Rosol, Et-Rosol, Pip-Rosol, Jul-Rosol, and THQ-Rosol (Figure 4.41).

Vygranenko et al. [141] synthesized a series of rosols using a new method to directly transform coumarins into rosols (compounds 4, 5, 6, 7, 8, 9, 10, and 11). These dyes exhibit high Φ_f values in DMSO and CH_2Cl_2, as well as very intense orange-red fluorescence. However, their photophysical data was not measured in aqueous solutions, probably due to the low water solubility of these dyes.

4.20 HAS-Rosols

HAS-rosols are rosols whose heteroatom (oxygen) has been substituted by another atom. P-rosols (phospharosols) are a subclass of these compounds, containing the types POPh-rosol (phosphine oxide phenyl-rosols) and PO_2^--rosols (phosphinate-rosols). The fluorescent dye PORL, synthesized and named by Grzybowski et al. [75], is one example of phosphine-oxide-rosol (Figure 4.42). This dye has an absorption maximum at the beginning of the infrared ($\lambda_{a,max}$ = 670 nm). It was also synthesized by Poronik et al. [127], and both papers present the same photophysical properties. The previously mentioned are phosphine oxide phenyl-rosols; examples of PO_2^--rosols (phosphinate-rosols) are NR632 and NR638, synthesized by Fang et al. [73]. D.T. Gryko and his team have also synthesized many sulfone-rosols (which are HAS-rosols with a sulfur as the heteroatom). However, the synthesized entities are not soluble enough in aqueous solutions for their photophysical properties to be measured [127]. Among others, two interesting dimethylsilane-rosol, named 2′Me-Si-rosol and 2′Me-4CH_2OH-Si-rosol, have been synthesized by Ito et al. [128]. Their molar absorption coefficients were estimated to be ~70 $\text{mM}^{-1}\text{cm}^{-1}$ from similar structures so that they could be included in this database and $\lambda_{a,max}$, $\lambda_{e,max}$, ε_{max}, and Φ_f are provided in supplementary table 1 of Ito et al. [128].

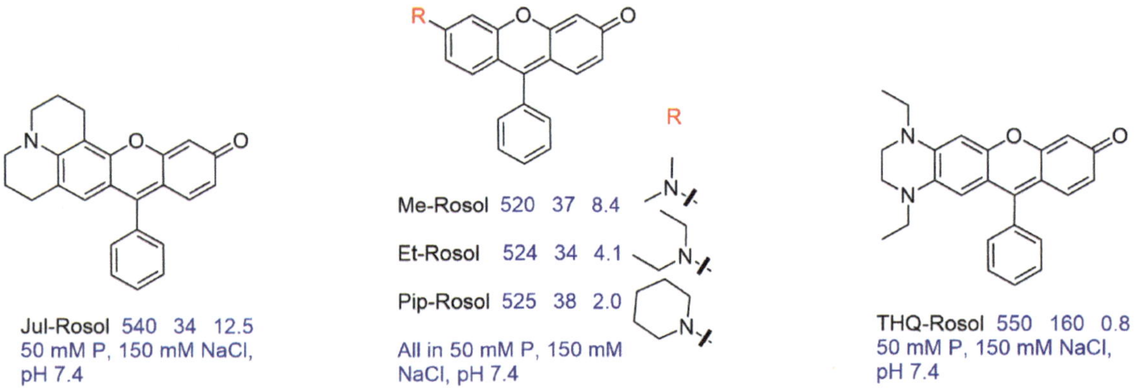

Figure 4.41 Molecular structure and photophysical properties of rosols. The photophysical parameters shown in blue are $\lambda_{x,max}$ (nm), Δ_S (nm), and B_{max} ($\text{mM}^{-1}\text{cm}^{-1}$), respectively.

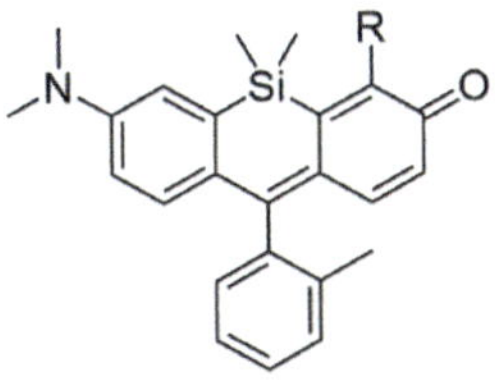

				R
2'Me-Si-rosol	612	18	19.6	H
2'Me-4CH$_2$OH-Si-rosol	613	18	21.0	CH$_2$OH

100 mM P pH 7.4

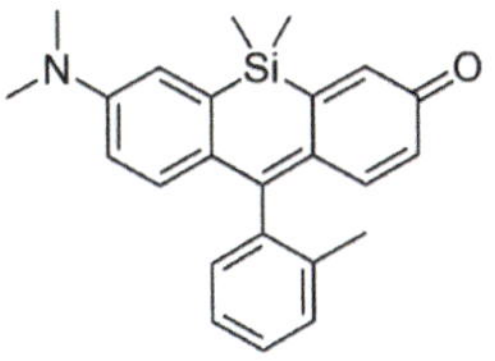

6-Dimethylamino Si(CH$_3$)$_2$-rosol 612 18 19.6
PBS pH 7.4

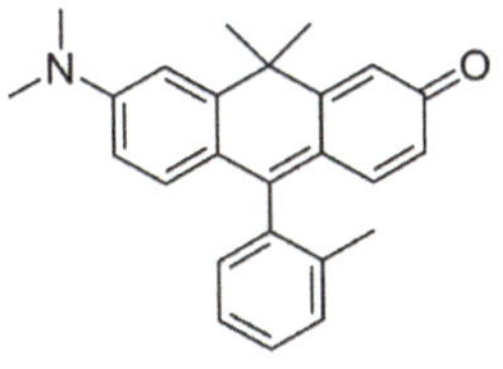

6-Dimethylamino-carborosol
606 20 62.9
10 mM HEPES pH 7.3

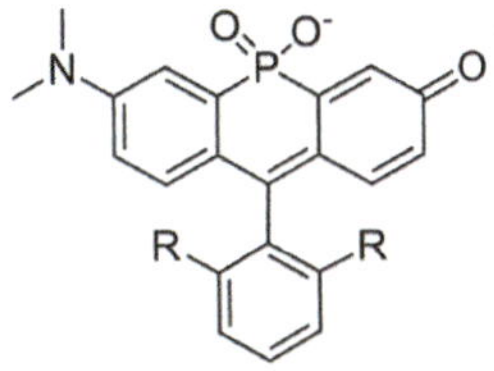

	R				pK_a
NR 632	-CH$_3$	632	23	16.6	5.20
NR 638	-OCH$_3$	638	24	11.6	5.16

All in 10 mM PBS pH 7.4 with 1% DMSO

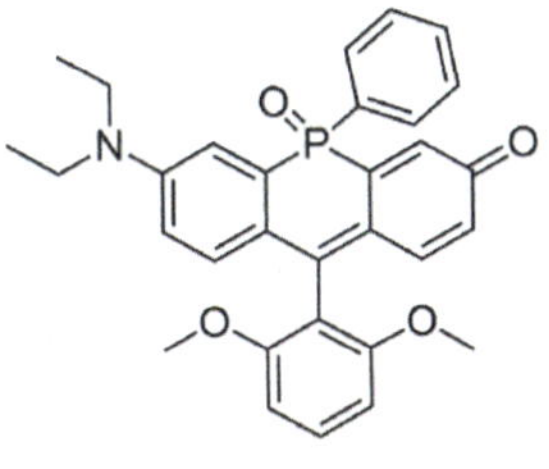

PORL
670 28 7.2
PBS w/ 1% DMSO

Figure 4.42 Molecular structure and photophysical properties of the HAS-rhosol. The photophysical parameters shown in blue are $\lambda_{x,max}$ (nm), Δ_S (nm), and B_{max} (mM^{-1}cm^{-1}), given in this order.

4.21 Pyrodols and Pyrodones

Pyrodol (6-amino-3*H*-xanthen-3-one) is a small, planar, and compact structure. The name "pyrodols" (here proposed for this class) comes from their resemble to both pyronines (lack of the aryl group) and rhodols which contain a xanthene with a ketone and an amino group bound to carbons 3 and 6, respectively. Examples of dyes belonging to this class are 6-quinoline-2-propanoic-pyrodol and 6-quinoline-2-pentanoic-pyrodol (these are short names) [24] (Figure 4.43).

Another class of compounds very similar to pyrodols are the (here termed) pyrodones (3-oxo-3*H*-xanthen-6-olate) [143]. They resemble pyrodols but with a ketone group in place of the amine group. They are also identical to fluorones but with the lack of the pandant aryl group. As far as the author knows, this is the only dye of this class with a published synthesis route in the literature and with some photophysical parameters measured (Figure 4.44).

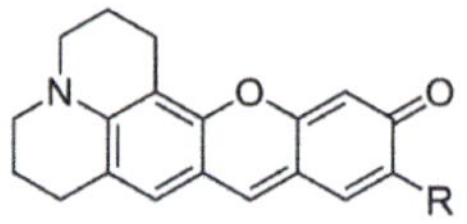

Pyrodols			
6-Quinoline-2-propanoic-pyrodol			
538	24	46.1	R = $(CH_2)_2CO_2^-$
6-Quinoline-2-pentanoic-pyrodol			
538	24	44.8	R = $(CH_2)_4CO_2^-$
All in 50 mM P 8-9			

Figure 4.43 Molecular structure and photophysical of two pyrodols. The photophysical parameters shown in blue are $\lambda_{x,max}$ (nm), Δ_S (nm), and B_{max} ($mM^{-1}cm^{-1}$), given in this order.

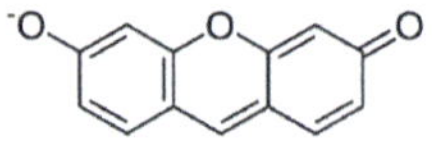

Pyrodone
472 28 22.2
100 mM P pH 7.4

Figure 4.44 Molecular structure and photophysical of a pyrodone. The photophysical parameters shown in blue are $\lambda_{x,max}$ (nm), Δ_S (nm), and B_{max} ($mM^{-1}cm^{-1}$), given in this order.

4.22 Trianguleniums

Trianguleniums is a new class of fluorescent dyes which exhibit a long fluorescence mean-lifetime (between 12 and 19 ns in water) and a high fluorescence quantum yield (~0.50 in water). The most promising sub-classes are the diazaoxatriangulenium (DAOTA) derivatives. The six DAOTA dyes with the highest brightness, so far, in aqueous solutions are DAOTA 1, DAOTA 2, DAOTA 3, DAOTA 4, DAOTA 5, and DAOTA 6 [144] (Figure 4.45). The brightnesses were calculated using a molar absorption coefficient estimated of 50 $mM^{-1}cm^{-1}$.

4.23 Acridines

Acridine is a planar and linear tricyclic structure like anthracene, but one of the central CH groups is replaced by nitrogen. It can also be seen as a dibenzopyridine, in which the six-membered heterocyclic pyridine is flanked by two benzene rings, or as a benzoquinoline, which is a quinoline structure fused to a benzene ring. This

	R_1	R_2			
DAOTA 1	H	H	558	29	15.5
DAOTA 2	SO_3^-	H	564	30	13.5
DAOTA 3	SO_3^-	SO_3^-	565	29	14.0
All in water					

	R_1	R_2			
DAOTA 4	H	CH_3	549	27	28.0
DAOTA 5	H	X	548	28	28.0
DAOTA 6	SO_3^-	CH_3	558	26	18.0
All in water					

Figure 4.45 Molecular structure and photophysical properties of triangulenium dyes DAOTA 1, 2, 3, 4, 5, and 6. The photophysical parameters shown in blue are $\lambda_{x,max}$ (nm), Δ_S (nm), and B_{max} ($mM^{-1}cm^{-1}$), given in this order.

three-ringed structure is the scaffold of a few fluorescent dyes (Figure 4.46), which have absorption maximum in the violet, purple, and blue ranges of the spectrum and relatively large Stokes shifts (the average value is $\Delta_S = 46$ nm).

4.23.1 Simple Acridines

The simplest acridines are 9-aminoacridine (134–50-9) [18, 145, 146] and 10-ethylacridine (948–43-6) [145, 146], which have brightness values of 9 and 3.4 $\text{mM}^{-1}\text{cm}^{-1}$ in water, respectively. A phosphorylated acridine termed PONy A1 was synthesized by Butkevich et al. [33]. Proflavine (3,6-diaminoacridine, 92–62-6) (13.2 $\text{mM}^{-1}\text{cm}^{-1}$ in water at pH 7) has some mutagenic activity [145, 147, 148] and is brighter than the previous dyes. The brightness of acridine orange (3,6-dimethylaminoacridine) in pure water was observed to be approximately 6 $\text{mM}^{-1}\text{cm}^{-1}$ [149]. The pK_a of acridine orange was found to be 9.8 [150].

The acridines reported with a brightness higher than 15 $\text{mM}^{-1}\text{cm}^{-1}$ are Atto 465 and Atto 495 [5]; 3,6-di(azetidin-1-yl)acridine [26]; and JF479 and JF502 (Grimm et al. 2020) (Figure 4.46).

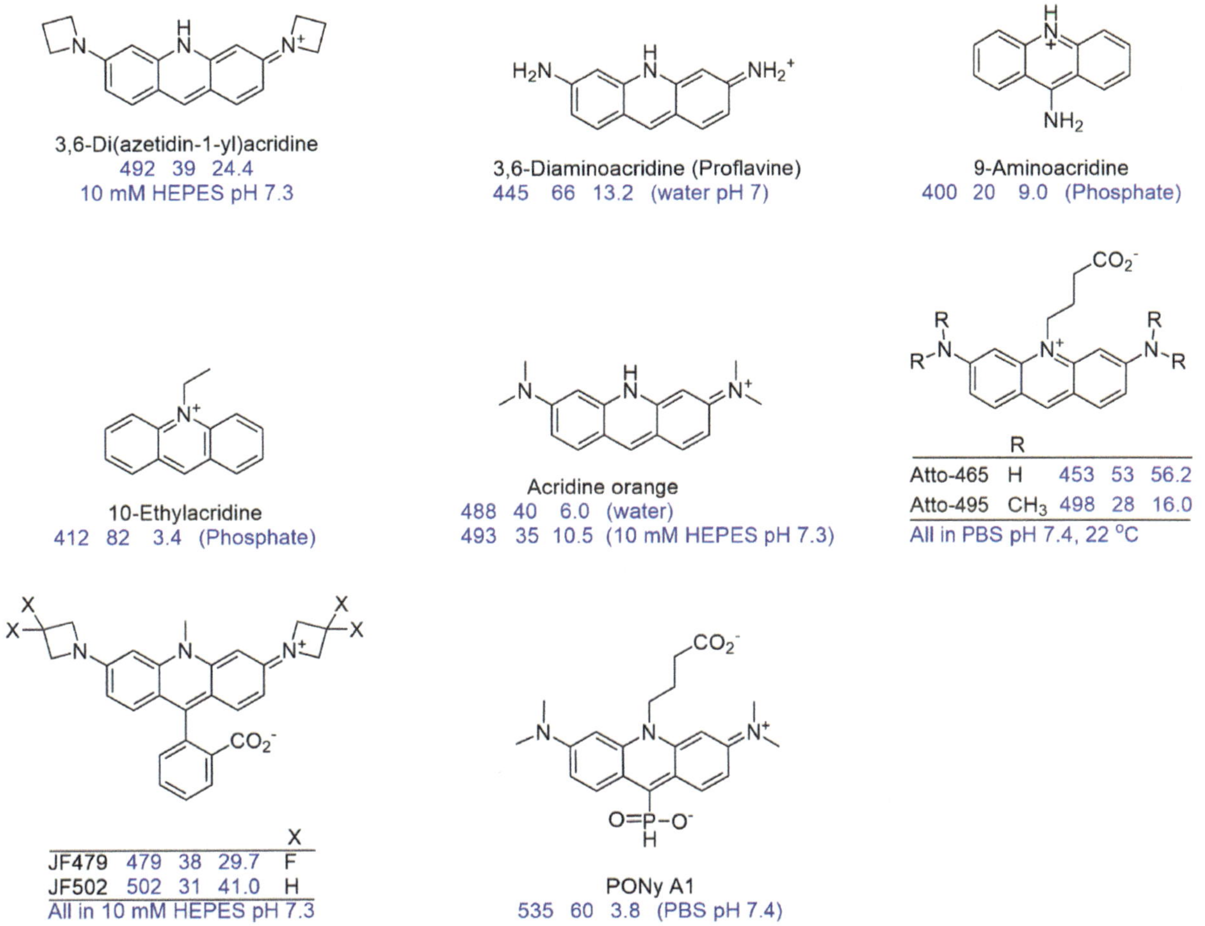

Figure 4.46 Structures of acridines which have high brightness in aqueous solutions. Parameters shown in blue are: $\lambda_{x,max}$ (nm), Δ_S (nm), and B_{max} ($\text{mM}^{-1}\text{cm}^{-1}$), respectively.

4.23.2 Acridones

Acridones (acridin-9[10*H*)-one) are acridines with a carbonyl group at position 9. Acridones are slightly different from acridines and deserve to be placed in a different class. On average, they exhibit smaller molar absorption coefficients and longer fluorescence mean-lifetimes. *N*-methyl acridone is an example of such a dye. The photophysical data of *N*-methylacridone was taken from table III of Hinze et al. [151]. Pereira et al. [152] synthesized a few fluorescent acridones which have higher water solubility: $ACSO_3H$, Met $ACSO_3H$, and Pet $ACSO_3H$ (Figure 4.47).

4.24 Merocyanines

The word merocyanine means half-cyanine and is appropriate as only one end of the π-electron system is a nitrogen-containing heterocycle (usually a pyrrole), while cyanines have a pyrrole at both ends (Section 4.12) (Figure 4.1). In merocyanines, the other end of the system consists of a benzopyrylium, a barbiturate ring, or another heterocycle (Figure 4.48). Like cyanines, these are push-pull fluorophores which contain a donor group belonging to one heterocycle and an acceptor group belonging to the other heterocycle. These groups are covalently connected by a polymethine chain (with an odd number of –CH= groups) [153]. The wavelength of maximum absorption of the longest wavelength band primarily depends on the length of the polymethine chain and, as for cyanines, can be fine-tuned by the strength of the electron withdrawing or electron donating character of the groups bound to the heterocycles. Moreover, the choice of donor/acceptor groups defines the ionic form of the structure, which can range from neutral, cationic, to zwitterionic.

The following merocyanines contain a benzopyrylium ion (as one of the heterocycles) connected by a polymethine chain to the second heterocycle (pyrrole): DY-635, DY-636, DY-651, DY-654, DY-679P1, DY-682, DY-681, DY-731, and DY-780 [20]. Many photophysical parameters of these dyes in aqueous solutions were kindly provided by the supplier (Dyomics) by personal communication.

4.25 Phenoxazines

Oxazines are six-membered heterocycles with one oxygen atom and one nitrogen atom replacing two of the carbons in this ring. Taking the oxygen as atom number 1, then the relative position of the nitrogen may be position 2, 3, or 4. There are also two double bonds within the oxazine structure. Consequently, there are eight possible isomers, depending on the relative position of the oxygen and nitrogen atoms and the two double bonds. For symmetry reasons and due to the high electronegativity of the oxygen atom, the brightest oxazine derivatives are

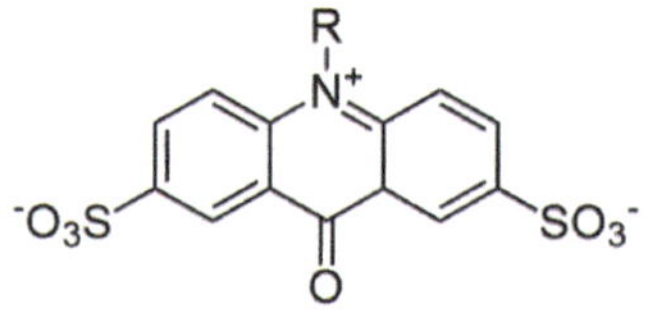

	R			
$ACSO_3H$	H	396	21	3.6
$MetACSO_3H$	CH_3	403	30	4.2
$PetACSO_3H$	nC_5H_{11}	406	25	5.6

All in water pH 5.5

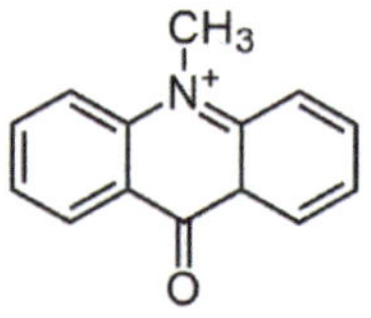

N-Methylacridone
411 20 8.7 (water)

Figure 4.47 Structures of acridones which have high brightness in aqueous solutions. Legend as in Figure 4.46.

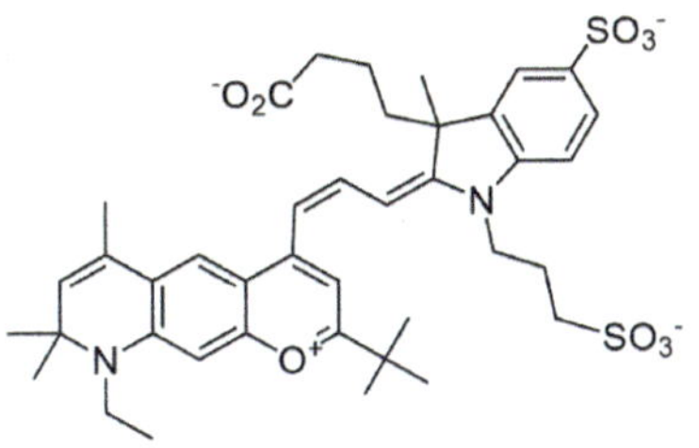

DY-651
653 25 24.1 (PBS pH 7.4)

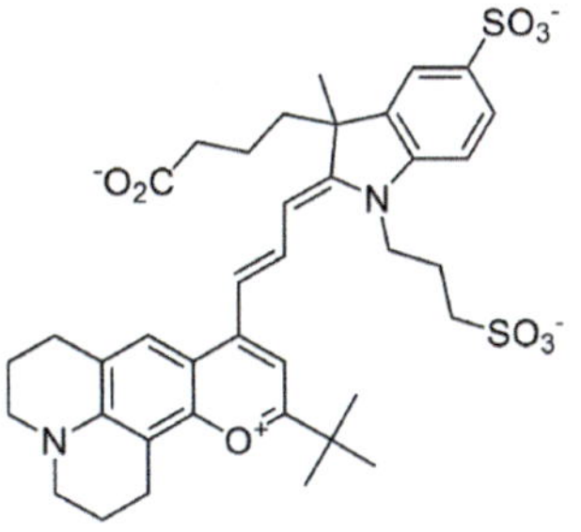

DY-636
645 26 16.3 (PBS pH 7.4)

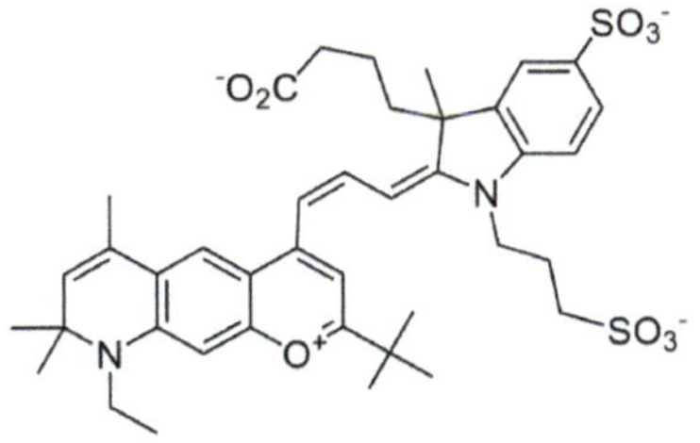

DY-651
653 23 24.1 (PBS pH 7.4)

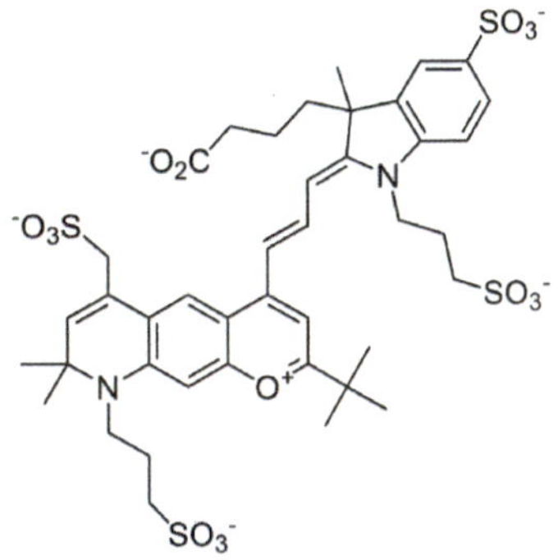

DY-654
653 24 35.2 (PBS pH 7.4)

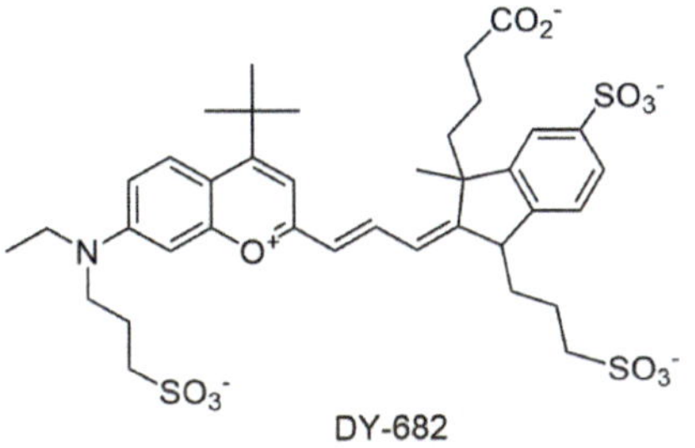

DY-682
690 20 48.6 (PBS pH 7.4)

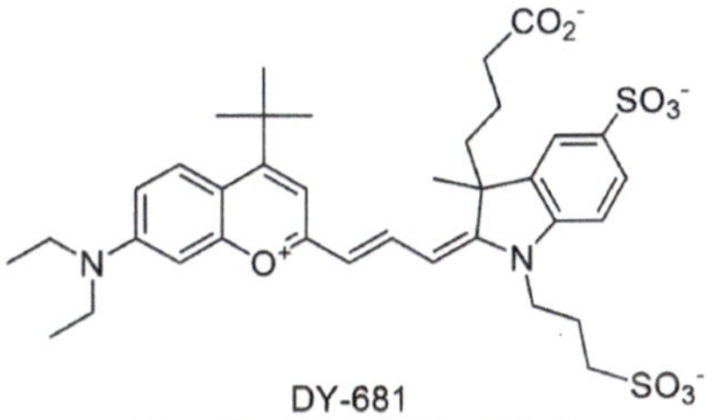

DY-681
691 17 11.2 (PBS pH 7.4)

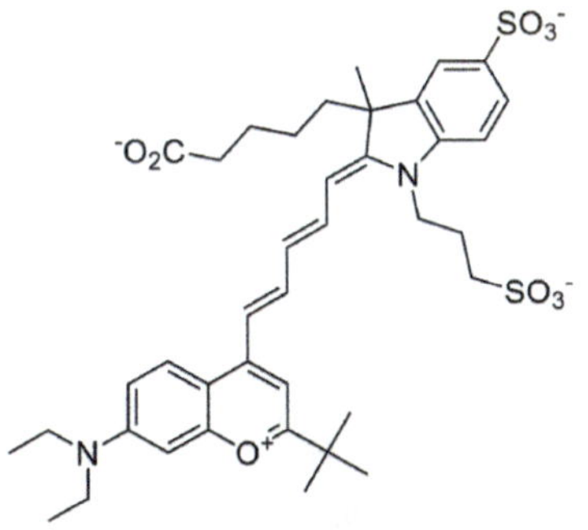

DY-731
736 23 23.6 (PBS pH 7.4)

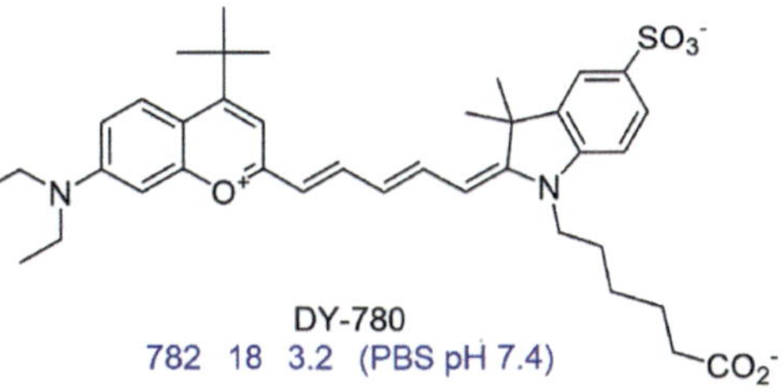

DY-780
782 18 3.2 (PBS pH 7.4)

Figure 4.48 Molecular structures and photophysical parameters of a few merocyanines. The photophysical parameters shown in blue are $\lambda_{x,max}$ (nm), Δ_S (nm), and B_{max} (mM^{-1}cm^{-1}), given in this order.

phenoxazines. Phenoxazines are made of a 2*H*-1,4-oxazine core flanked by two benzene rings (Eicher et al. [154], p. 442). The majority of the phenoxazines have absorption maximums of above 620 nm; i.e., they absorb in the red range of the visible spectrum and emit at the end of the visible and the beginning of the infrared. Figure 4.49 shows the molecular structures of some bright phenoxazines.

Examples of phenoxazines (Figure 4.48) include: Redmond red [65]; oxazine 170 [155]; EVOblue30 ([112], structure taken from figure 1 of Baker et al. [156], ε_{max} value found in a technical note from Innovative Optische Messtechnik GmbH, Berlin, Φ_f in water estimated to be lower than that of oxazine 170); 3,7-diethylaminophenoxazine and 3,7-di(azetidin)phenoxazine [26]; phosphorilated oxazine 2-H [28]; Bza-Oet, Bza-OH, Bzh-Cl, Bzh-H, Bpe-H, and Bzm-H [157]; cresyl violet [37]; Atto Oxa12 (the structure of this dye has not been disclosed, but is probably a phenoxazine because it is a lipophilic version of Atto 665); Atto 655; Atto 680; and Atto 700 [5]. The only structural difference between Atto 655 and Atto 680 is a double bond (Figure 4.49).

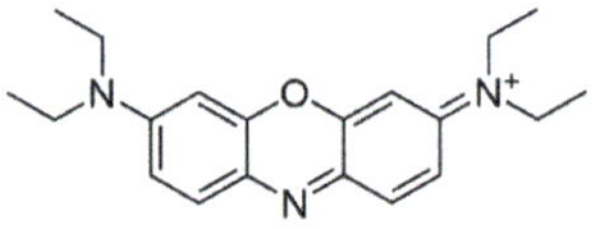

3,7-Diethylaminophenoxazine
665 14 7.7 (10 mM HEPES pH 7.3)

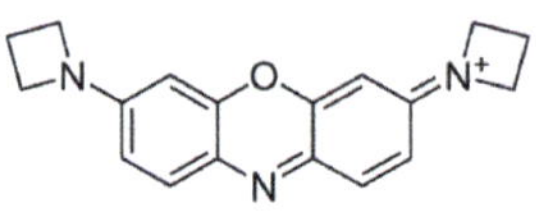

3,7-Di(azetidin)phenoxazine
647 14 23.7 (10 mM HEPES pH 7.3)

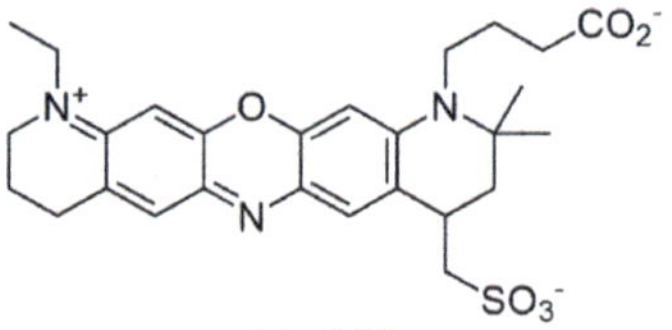

Atto 655
663 17 37.5 (PBS pH 7.4, 22 °C)

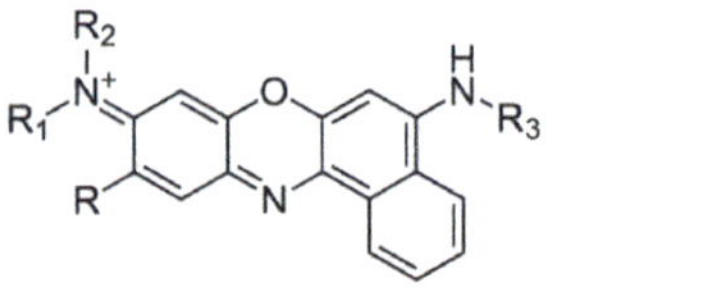

	R	R1	R2	R3
Bza-OEt	Me	H	Et	$(CH_2)_3CO_2CH_2CH_3$
Bza-OH	Me	H	Et	$(CH_2)_3CO_2H$
Bzh-H	Me	H	Et	$(CH_2)_3OH$
Bzh-Cl	Me	H	Et	$(CH_2)_3Cl$
Bpe-H	H	Et	Et	$(CH_2)_3OH$
Bzm-H	H	Me	Me	$(CH_2)_3OH$

Bza-OEt	625	28	8.7
Bza-OH	620	33	4.7
Bzh-H	620	31	4.9
Bzh-Cl	620	31	11.1
Bpe-H	640	41	11.0
Bzm-H	635	39	4.1

All in water pH 7.4

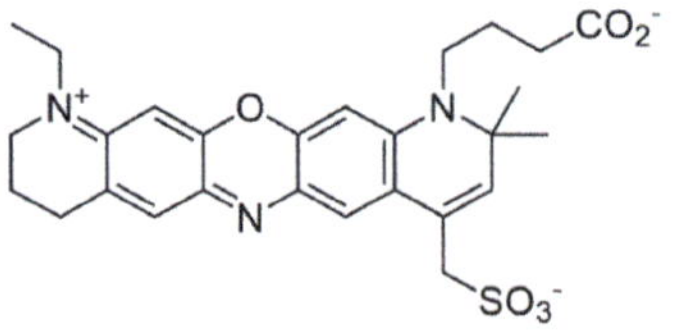

Atto 680
681 17 37.5 (PBS pH 7.4, 22 °C)

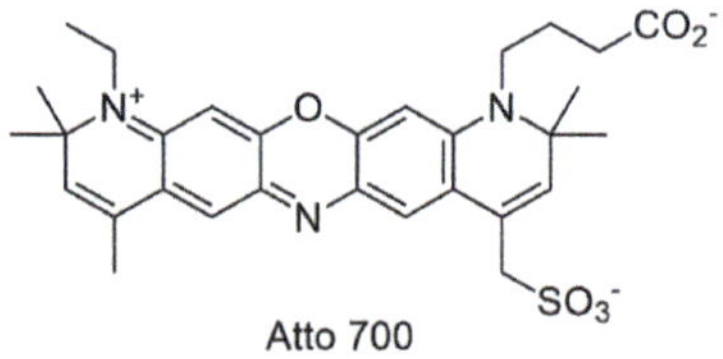

Atto 700
700 16 30.0 (PBS pH 7, 22 °C)

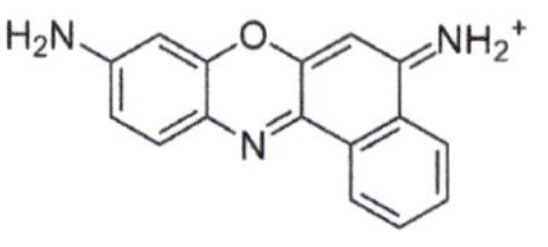

Cresyl violet
585 40 11.9 (water)

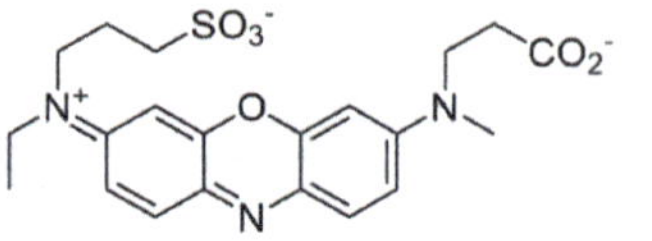

EVOblue30
650 17 10.0 (water)

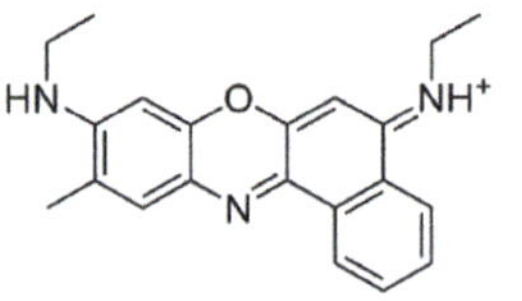

Oxazine 170
620 30 11.5 (water)

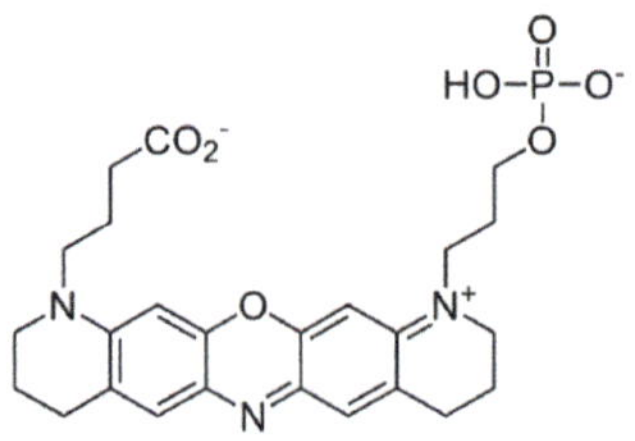

Phosphorilated oxazine 2-H
661 17 16.8 (water)

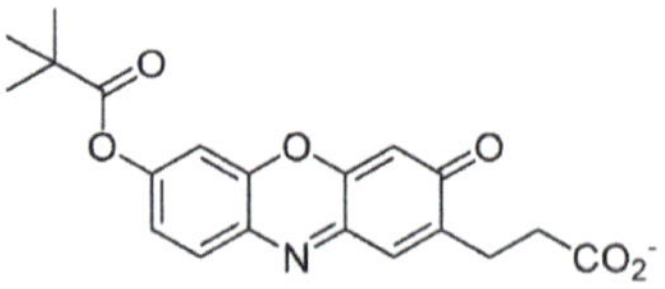

Redmond red
579 16 43.9 (water)

Figure 4.49 Structures and photophysical properties of some phenoxazines. The photophysical parameters shown in blue are $\lambda_{x,max}$ (nm), Δ_S (nm), and B_{max} (mM^{-1}cm^{-1}), given in this order.

4.26 Coumarins

Coumarins (2*H*-chromen-2-ones) owe their class name to "Coumarou," the vernacular name of the "tonka bean" from which coumarin itself was isolated in 1820. Coumarins are found in the seeds and tissues of many plants and are classified as a member of the benzopyrone family of compounds, all of which consist of a benzene ring joined to a pyrone ring or more exactly benzo-*a*-pyrones [158]. Fluorescent coumarins typically contain an electron-releasing group (commonly a phenolate or amino group) in the 7-position and an electron-accepting heterocyclic group on the opposite side (usually the ketone group at 4-position).

4.26.1 7-Hydroxy Coumarins

Many photophysical parameters of 7-hydroxy-4-methylcoumarin and 7-hydroxy-4-methylcoumarin-3-acetic acid were measured in water at both pH 5 and 10 ([159] PhotochemCAD database of [160] and references therein). The fluorescence mean-lifetime of 7-hydroxy-4-methylcoumarin was measured by Kist [161] (Figure 2.18) at pH 9. The pK_a of the phenol (7-hydroxy) group is lowered by the halogenation of the coumarin core structure. This resulted in the fluorescent dyes Pacific blue (6,8-difluoro-7-hydroxy-2-oxo-2H-chromene-3-carboxylic acid) ([6, 159] and fluorescence lifetime taken from Kist [161], see Figure 2.18 in Chapter 2); Marina blue (2-(6,8-difluoro-7-hydroxy-4-methyl-2-oxo-2*H*-chromen-3-yl)acetic acid) [159]; P-R 12, P-R 13, and P-R 14 [162] (Figure 4.50).

4.26.2 Small 7-Amino Coumarins

The following coumarins have an amino group bound to carbon-7 and exhibit photophysical properties that are even less sensitive to pH in the neutral range: coumarin 1, coumarin 2, coumarin 120, coumarin 152, and coumarin 307 (they are only slightly soluble in water). Nonetheless, Arbeloa et al. [17] managed to measure their photophysical parameters in water. Coumarin 1, coumarin 152, and coumarin 307 (shown in Figure 4.51) have a low brightness; nevertheless, they were kept in the database because important relationships between structure vs. photophysical properties can be understood from these dyes; however, coumarin 151 was excluded because

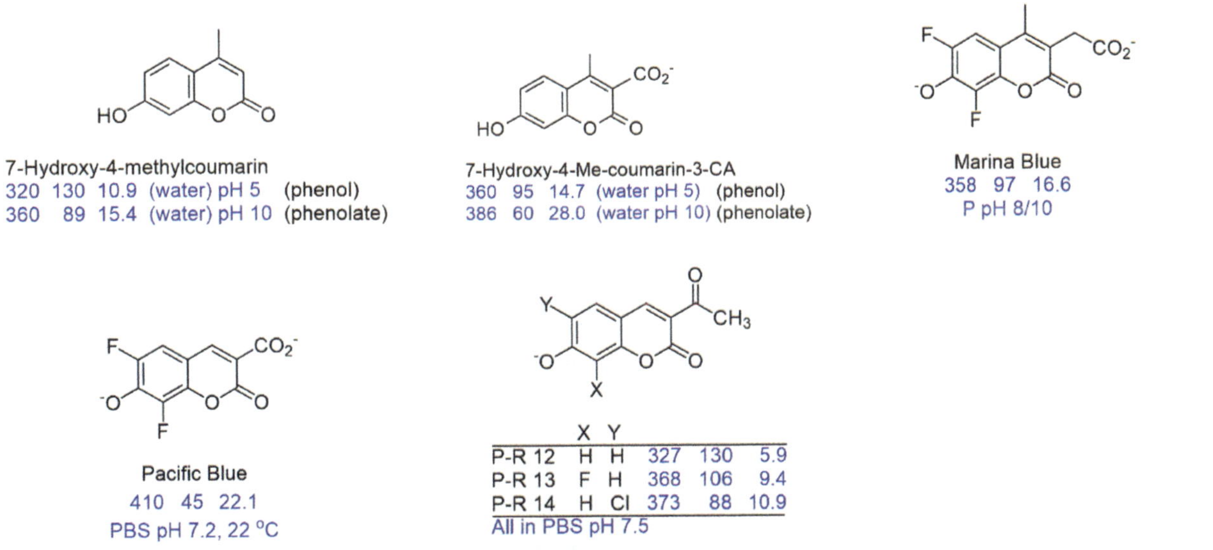

Figure 4.50 The molecular structure and photophysical properties of 7-hydroxy coumarins. The photophysical parameters shown in blue are $\lambda_{x,max}$ (nm), Δ_S (nm), and B_{max} (mM^{-1}cm^{-1}), given in this order.

the reported brightness is too low (only 0.1 mM^{-1}cm^{-1}). Coumarin 102 is also slightly soluble in water [37]. Φ_f was measured by Jones II et al. [163]. Singha et al. [121] also studied the effect of nitrogen alkyl substituents on the photophysical properties of many dyes, including the following coumarins: 7-diethylamino-coumarin; 7-ethylamino-coumarin; and 7-(4-hydroxy)cyclohexyl coumarin (Figure 4.51). Păunescu et al. [162] also synthesized halogenated 7-amino coumarins, hereby termed P-R 3, P-R 5, P-R 6, P-R 21, and P-R 22 (Figure 4.51).

4.26.3 More Elaborated 7-Amino Coumarins

Nizamov et al. [164] synthesized the coumarins, hereby termed sulfonated coumarin N-H 6-H, sulfonated coumarin N-H 7-H, phosphorylated coumarin N-H 21c, and phosphorylated coumarin N-H 22c (N and H are the

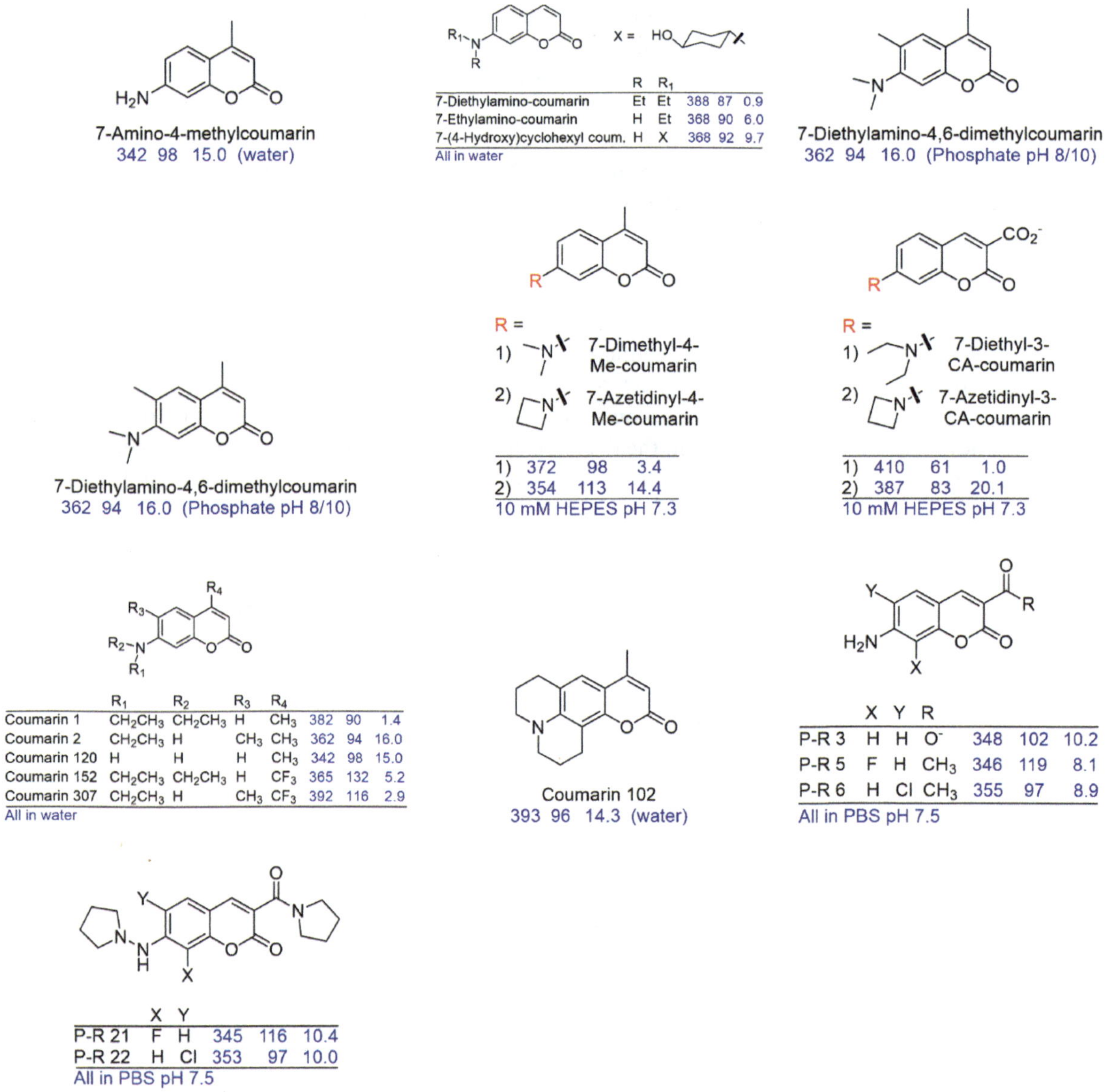

Figure 4.51 The molecular structure and photophysical properties of small 7-amino coumarins. Legend as in Figure 4.50.

initials of first and last author and 6-H, 7-H, 21c, and 22c is the numbering of the structures in the original article) [89, 164, 165].

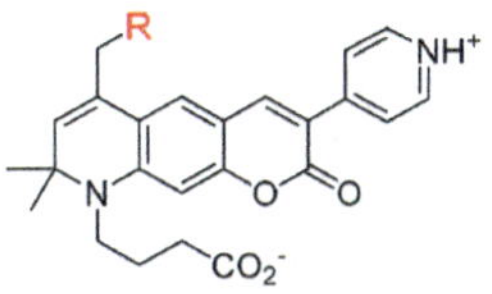

Abberior Star 440SXP
R = -O-PO_3H^- 436 79 15.4
Sulfonated coumarin N-H 6-H
R = -SO_3^- 436 79 17.6
All in PBS pH 7.4.

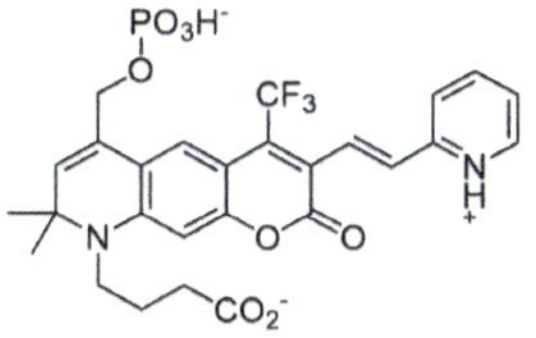

Abberior Star 470XP
472 152 3.4 (PBS pH 7.4)

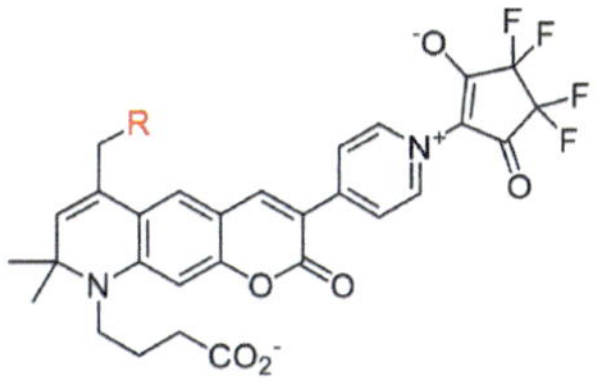

Abberior Star 520SXP
R = -O-PO_3H^- 522 110 6.3
Sulfonated coumarin N-H 7-H
R = -SO_3^- 520 119 2.4
All in PBS pH 7.4.

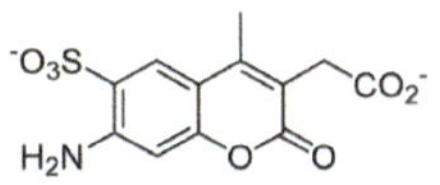

Alexa Fluor 350
346 99 16.9 (water)

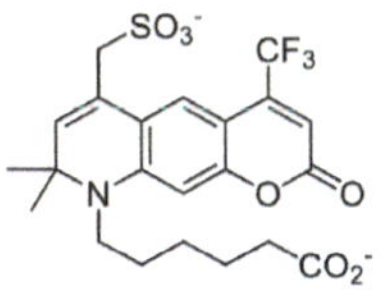

Alexa Fluor 430
431 110 8.8 (PBS pH 7.4)

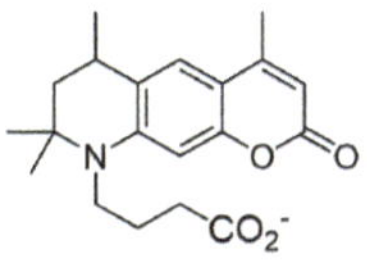

Atto 390
390 86 21.6 (PBS pH 7.4, 22 °C)

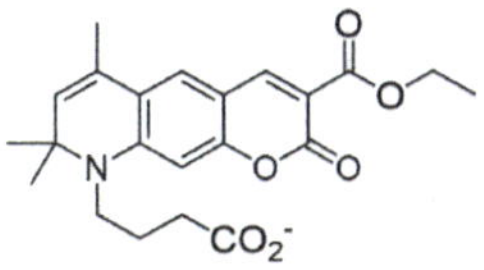

Atto 425
439 46 40.5 (PBS pH 7.4)

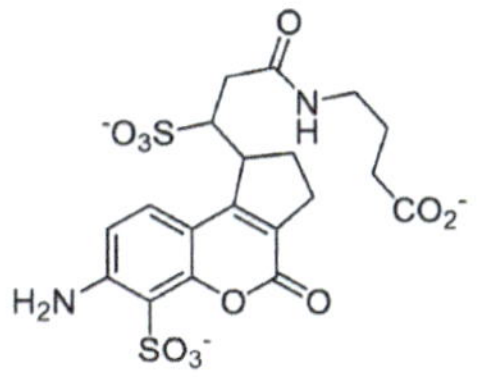

DY-350
353 79 15.9 (PBS pH 7.4)

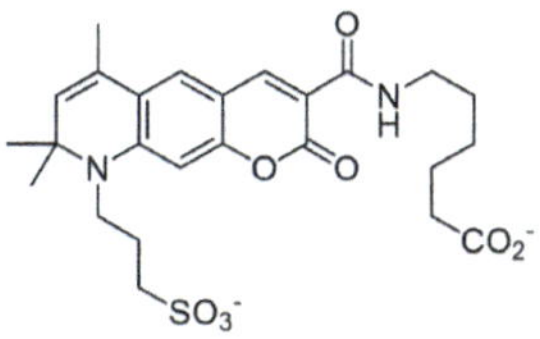

DY-430
444 43 26.6 (PBS pH 7.4)

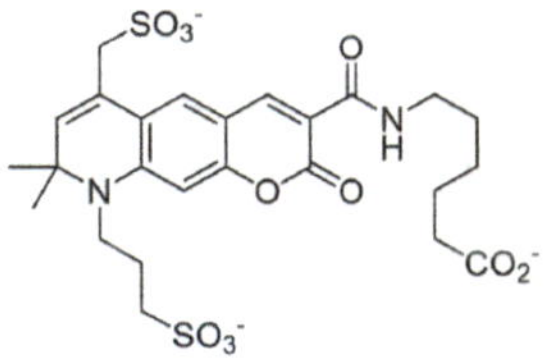

DY-431
442 54 26.2 (PBS pH 7.4)

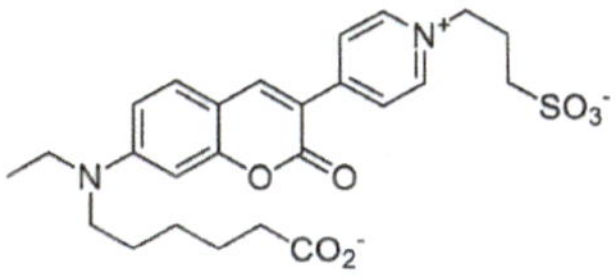

DY-485XL
482 78 - (water)

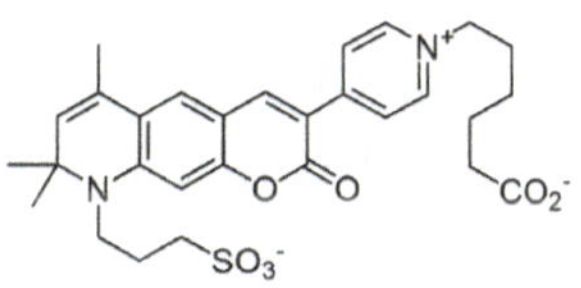

DY-510XL
493 92 - (water)

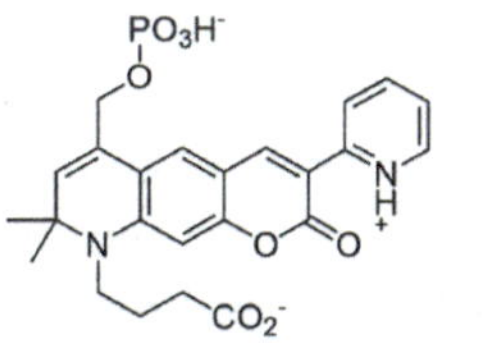

Phosphorilated coumarin N-H 21c
432 80 16.5 (PBS pH 7.4)

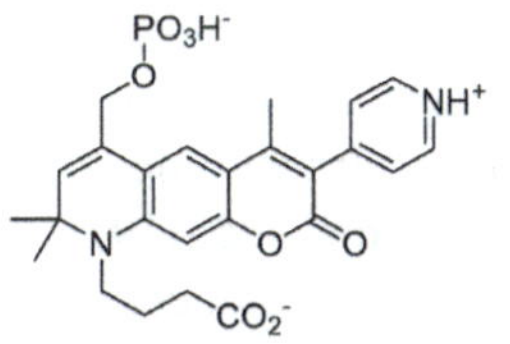

Phosphorilated coumarin N-H 22c
410 93 6.1 (PBS pH 7.4)

Figure 4.52 The molecular structure and photophysical properties of more elaborated 7-amino coumarins. Legend as in Figure 4.50.

Coumarins synthesized with an azetidinyl group instead of the traditional dimethylamino or diethylamino showed a much higher brightness in aqueous solutions: 7-dimethylamino-4-Me-coumarin, 7-azetidinyl-4-Me-coumarin, 7-diethylamino-3-CA coumarin, and 7-azetidinyl-3-CA coumarin [26].

Finally, four brands of well-known coumarins are presented (in alphabetic order): i) Abberior Star 440SXP, Abberior Star 470SXP, and Abberior Star 520SXP [16, 166, 167]; ii) Alexa Fluor 350 and Alexa Fluor 430 [6, 165]; iii) Atto 390 and Atto 425 [5]; and iv) DY-350, DY-431, DY-430, DY-485XL, and DY-510XL [20, 165]. Note that Abberior Star 470XP can be classified as a hybrid structure (a coumarin connected to a pyridine through a polymethine bridge), and few members of the above list consist of a coumarin connected to a pyridine with a single bond. The brightness of two dyes (DY-485-XL and DY-510XL) could not be calculated because their fluorescence quantum yield has not yet been disclosed. It is worth mentioning that some of the above coumarins have extra-large Stokes shifts in neutral pH aqueous solutions.

Coumarins are largely used as fluorescent probes for the fluorescent detection and quantification of a great variety of targets, mainly cations. There is a recent review covering the synthesis of these probes and their applications [168].

4.27 Sulforhodols

SR101-OH is a typical hydrophilic sulforhodol with $\lambda_{a,max} = 548$ nm, $\lambda_{e,max} = 572$ nm, and $\Phi_f = 0.35$, all measured in water [47]. The molar absorption coefficient is presumed to be approximately 80 $mM^{-1}cm^{-1}$ based on the absorption coefficients of similar rhodols. 2,7-DiCl-6-pyrrolidinosulforhodol is also a sulforhodol; however, it contains a polyene chain that makes it a voltage-sensitive probe (figure 16 of Poronik et al. [127]). Figure 4.53 also shows the structure and some photophysical properties of SR101-NaphtOH (synthesized by Chevalier et al. [47]). As pointed out by the authors, this is a π-extended rhodol-like dye. Therefore, strictly speaking, this dye belongs in another class, as its π-electron system was significantly extended (modified). Despite this, it will be kept in this section to avoid overcomplication.

4.28 Pyrenes

Pyrene is made of a four-ringed aromatic, planar, and rigid structure (Figure 4.54). Usually, three sulfonic groups are introduced to three of the following positions: 1, 3, 6, and 8. This confers water solubility and a large bathochromic shift. A phenol or amino group is introduced to the remaining position for further modification. Water solubility may also be conferred by using sulfobetaine bound to pyrene [3]. Unfortunately, pyrene is not as versatile as other dyes; for example, xanthene, borondipyrromethene, and coumarin. The four-ringed pyrine arrives at the synthesis lab for modification, and the only modifications that have been made is decorating it with different groups. In this sense, it is not a flexible scaffold like other classes. Nevertheless, a few bright pyrene derivatives have been synthesized.

Whitaker et al. [46] synthesized a library of 20 water-soluble fluorescent bright dyes by modifying the phenol group of 8-hydroxypyrene-1,3,6-trisulfonic acid, including Cascade blue and Alexa Fluor 405 [6] (Figure 4.54). The other well-known dyes of this class are: 1,3,6,8 pyrenetetrasulfonic acid ($\lambda_{a,max}$, $\lambda_{e,max}$, and ε_{max} taken from [19] and Φ_f and τ_f from [169]); 8-aminopyrene-1,3,6-trisulfonic acid ([19, 167, p. 11]. $\Phi_f \sim 0.90$, but may change after ligation to reducing sugars); DY-405 carboxylic acid [20]; sulfobetaine pyrene 21 [3]; and pyrene-1-sulfonic acid ([169] ε_{max} was estimated to be similar to sulfobetaine pyrene 21). The photophysical properties of 8-hydroxypyrene-1,3,6-trisulfonic acid (HPTS) are very sensitive to pH within the neutral range, as the pK_a of the phenol group is 7.3 at 22°C [19]. In this case, the absorption and emission spectra change significantly as the predominant ionic species goes from the tri-anionic form to the tetra-anionic form (at pH ~10). The photophysical data shown are in pH 2 and pH 10 ($\lambda_{a,max}$, $\lambda_{e,max}$, and ε_{max} taken from [19] and Φ_f and τ_f from [169]).

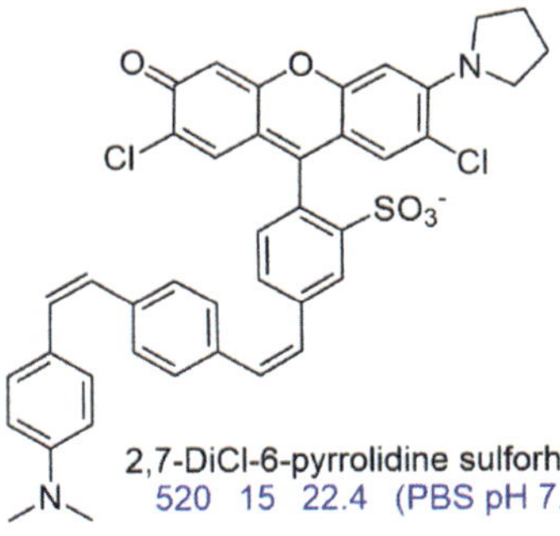

2,7-DiCl-6-pyrrolidine sulforhodol
520 15 22.4 (PBS pH 7.4)

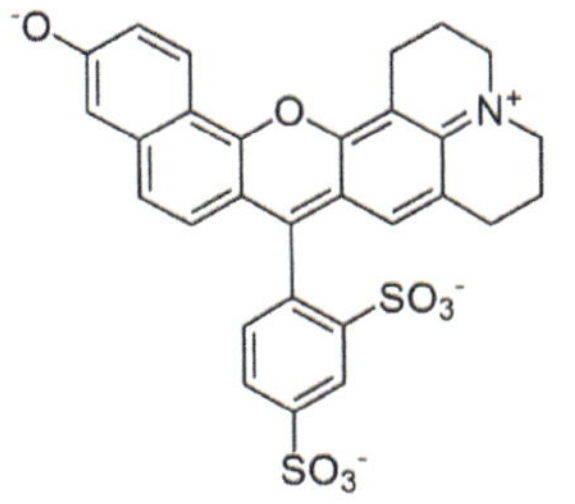

SR101-NaphtOH
580 39 12.0 (H_2O)

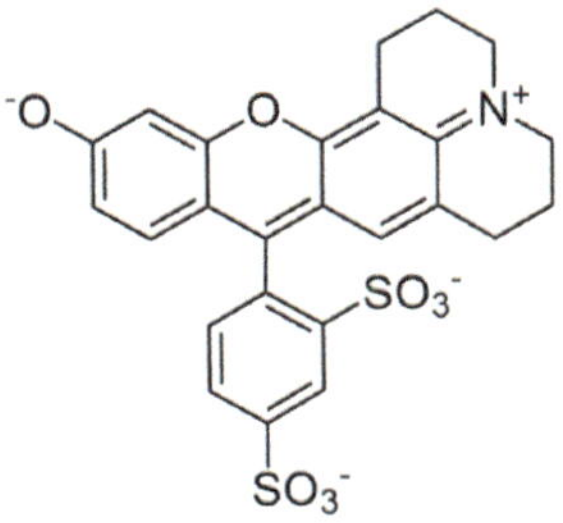

SR101-OH
548 24 28.0 (H_2O)

Figure 4.53 Molecular structure and some photophysical data of sulforhodols and one sulforhodol analog. The photophysical parameters shown in blue are $\lambda_{x,max}$ (nm), Δ_S (nm), and B_{max} ($mM^{-1}cm^{-1}$), given in this order.

The dyes of this class are widely used for fluorescent labeling, especially to promote electrodriven separations of uncharged analytes. For example, reducing sugars are labeled using 8-aminopyrene-1,3,6-trisulfonic acid and the highly negatively charged ($q = -3$) derivatized analytes are separated by electrophoresis. Toroidal capillary electrophoresis of such derivatives delivers very large numbers of theoretical plates (extremely high separation efficiencies) (Kist [30, 170]). Similarly, amino acids and peptides are labeled using succinimidyl esters produced from 8-hydroxypyrene-1,3,6-trisulfonic acid (HPTS) (such as Cascade blue and Alexa Fluor 405). The photophysical properties of HPTS derivatized amino acids and peptides are almost pH independent (in the pH >6 range). The only exception is 8-aminopyrene-1,3,6-trisulfonic acid, as the derivatized analyte is directly bonded to the nitrogen bound to carbon 8 of the pyrene core.

Another peculiar property of this class is their high photostability and photoacid behavior. For example, the pK_a of HPTS drops from 7.3 to 1.4 (pK_a^*) upon excitation [171]. Surprisingly, changing the three sulfonic acid groups to more electron-withdrawing substituents, such as 8-hydroxypyrene-*N*,*N*,*N*′,*N*′,*N*″,*N*″-hexamethyl-1,3,6-trisulfonic acid, lowers the pK_a to 5.6 and the pK_a^* even further to −0.8. Such dyes that have a pK_a^* of below zero are referred to as super-photoacids.

4.29 Quinolines

Quinolines (Figure 4.55) can be seen as benzopyridines [154, p. 386]. They are very common alkaloids found in plants and hundreds of biologically active quinoline derivatives were recently identified [172]. This class only has a few entities that have moderate to small brightness values. Nonetheless, they are important from a historical

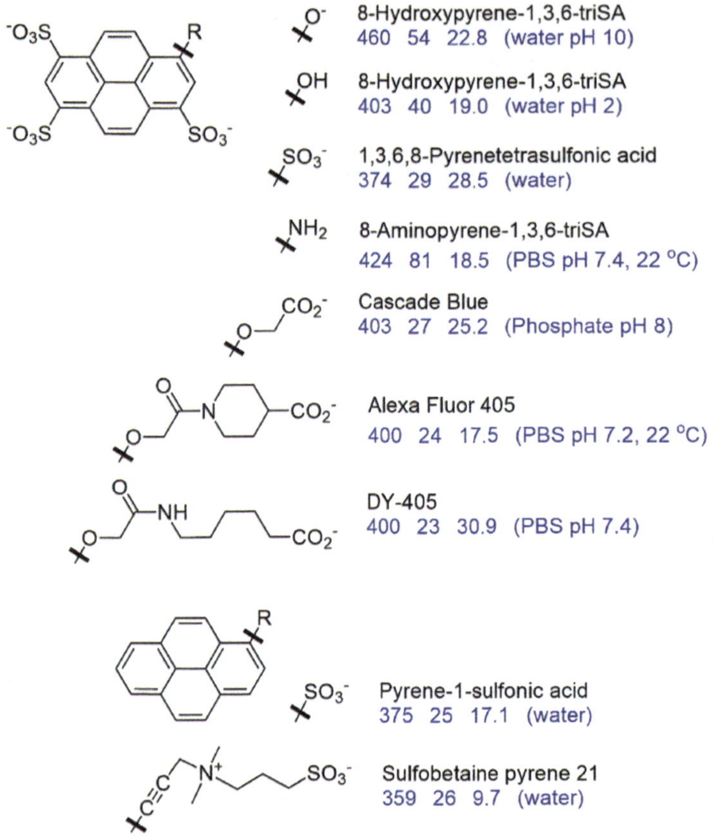

Figure 4.54 The most common fluorescent dyes that have a pyrene core structure. The photophysical parameters shown in blue are $\lambda_{x,max}$ (nm), Δ_S (nm), and B_{max} ($mM^{-1}cm^{-1}$), given in this order.

point-of-view because quinine sulfate was one the first fluorescent structures to be studied. The word fluorescence was introduced by G.G. Stokes in 1852 while he was studying the blue emission (at that time it was called *internal emission*) of quinine sulfate solutions [173]. He correctly identified fluorescence as an emission followed by an absorption process.

Quinine is still an important standard for the measurement of fluorescence quantum yields [18, 174]. In a 0.1 M perchlorate aqueous solution, quinine exhibits a fluorescence quantum yield of 0.60 ± 0.007, which is constant between 20 and 45°C. Therefore, this solution should be used as the fluorescence quantum yield standard (as proposed by Nawara and Waluk [175]) instead of sulfuric acid solutions. This is because the Φ_f of quinine in sulfuric acid solutions experiences large variations around room temperature (Φ_f changes by ~0.45% per degree Celsius). This probably explains the inconsistency of the Φ_f values reported for this solution in the literature. The temperature dependence of quinine sulfate may be related to the presence of bisulfate anions in the solution, which might be involved with collisional quenching. The effects of halide ions on the fluorescence of quinine sulfate was also studied by Guo et al. [176]. 7-Am-4-Me-2-hydroxyquinoline is also a known quinoline with medium to high brightness and with absorption maximum at 340 nm (PhotochemCAD database of [160] and references therein).

Carbostyril (quinolin-2(1*H*)-one) derivatives synthesized by Uray et al. (2000) showed higher pH stability and higher brightness values than other fluorescent dyes in their respective absorption range (UVA) [177]. Figure 4.55 shows the structures and some photophysical parameters of carbostyrils 1, 2, 4, 5, 6, 7, 9, carbostyril 124 (compound 10) from Uray et al, and carbostyril 23b from Strohmeier et al. [178]. The fluorescence quantum yield of

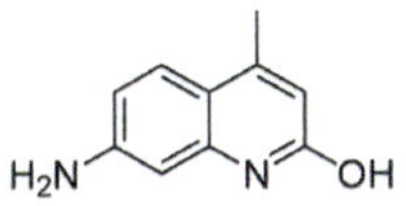

7-Am-4-Me-2-hydroxyquinoline
340 100 15.5 (water)

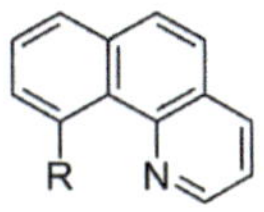

	R			
BQ	H	370	60	11.2
OBQ	OCH_3	385	85	15.0
HBQ	OH	390	90	5.2

All in water with 5 > pH > 2.

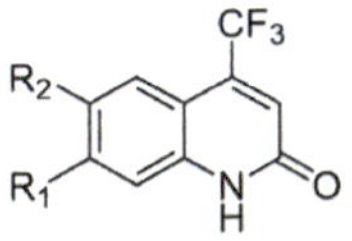

Carbostyril

	R1	R2			
1	OCH_3	OCH_3	358	74	2.7
4	OCH_3	NH_2	379	152	0.6
6	OCH_3	$NHCH_3$	389	155	0.6
7	OCH_3	$N(Et)_2$	366	214	0.6
10	NH_2	H	341	76	7.2

All in water.

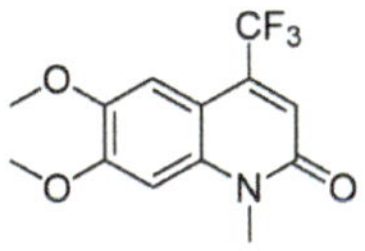

Carbostyril 2
360 70 0.9 (water)

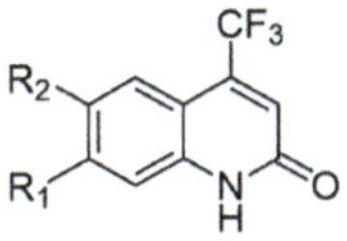

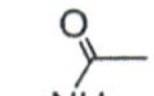

Carbostyril

X = —NH

	R1	R2			
5	OCH_3	X	353	87	1.0
9	X	OCH_3	340	42	0.9

All in water.

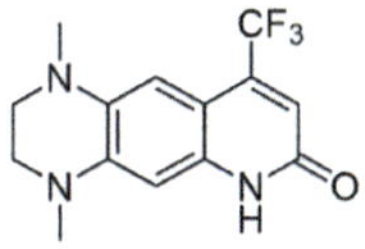

Carbostyril 13
420 136 4.5 (water pH 7)

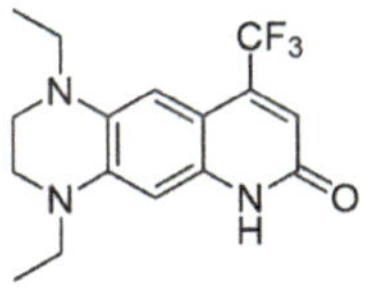

Carbostyril 23b
397 159 4.5 (water pH 12)
424 132 4.5 (water pH 7)
380 95 7.3 (water pH 2.8)

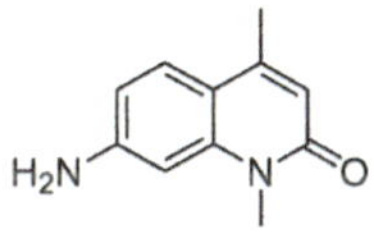

Carbostyril 124
341 76 7.2 (water)

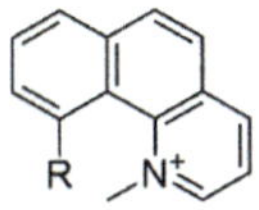

	R			
NBQ	OH	380	90	12.2
NOBQ	OCH_3	385	85	14.3

All in water with 5 > pH > 2.

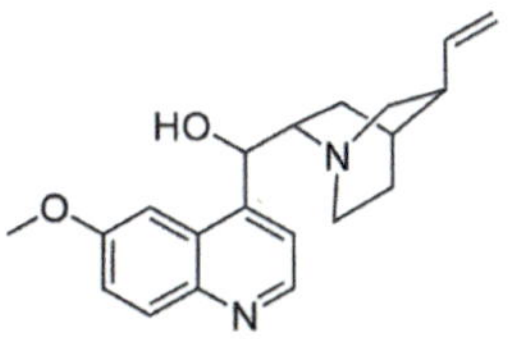

Quinine
331 51 2.8 (water pH 7)
347 104 3.1 (0.05 M H_2SO_4)
347 104 3.4 (0.1 M $HClO_4$)

Figure 4.55 Molecular structures and some photophysical properties of quinolines. The photophysical parameters shown in blue are $\lambda_{x,max}$ (nm), Δ_S (nm), and B_{max} ($mM^{-1}cm^{-1}$), given in this order.

carbostyril 124 in water was typed as 0.959 in Uray et al. (2000), when the correct value is probably 0.459 (based on the result obtained for this dye in DMSO, which was reported as $\Phi_f = 0.485$). Carbostyril 13 is very similar to carbostyril 23b and was synthesized by Kelterer et al. [179].

Many derivatives of 10-hydroxybenzo[*h*]quinoline were synthesized by Chou and Wei [180]. Some of them are slightly soluble in water and therefore some of their photophysical parameters were measured by Chou and Wei, with the exception of their molar absorption coefficient. The value of $\varepsilon_{max} = 35.0\ \text{mM}^{-1}\text{cm}^{-1}$ was attributed based on similar structures so that the dyes BQ, OBQ, HBQ, NBQ, and NOBQ could be included in the present database (Figure 4.55).

4.30 Benzothiazoles

Benzothiazoles consist of a five-membered 1,3-thiazole ring fused to a benzene ring. All nine atoms of the bicyclic structure and the attached substituents form a planar structure. This is the core structure of d-luciferin, the substrate for the bioluminescent phenomena found in fire-flies and hundreds of species of marine vertebrates, invertebrates, and microorganisms (such as bacteria). In bioluminescence, part of the energy of ATP is converted into luminous energy as luciferase, oxygen, and co-factors assist the conversion of d-luciferin into oxy-luciferin (the molecule that releases a bioluminescent photon).

The structure and some relevant photophysical properties of luciferin at pH 5 and pH 11, as well as dehydroluciferin at pH 4.5 and 11, are shown in Figure 4.56 (taken from Morton et al. [181]). Pacific orange is a commercial dye that is also a benzothiazole derivative [6]. The photophysical data of Pacific orange was taken from Abrams et al. [182], with the exception of fluorescence quantum yield which was estimated based on similar dyes ((-O)-luciferin with $\Phi_f = 0.62$). The dye Stafani HB-7 (3-(Benzo[*d*]thiazol-2-yl)-4-hydroxybenzenesulfonic acid) was recently synthesized and exhibits "excited state intramolecular proton transfer" (ESIPT) [183, 184]. This ESIPT phenomenon causes very large Stokes shifts, and increases photostability (as the absorption band changes when the proton has been transferred), among other advantages.

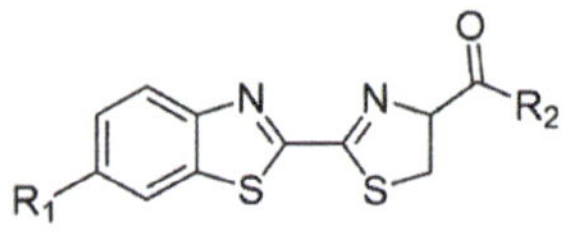

	R_1	R_2				pH
($^-$O)-Luciferin	O$^-$	O$^-$	384	153	11.4	11
(HO)-Luciferin	OH	O$^-$	328	209	4.5	5
(HO)-Luciferyl Adn	OH	Adn	335	213	1.6	5

All in water.
Adn = adenylate.

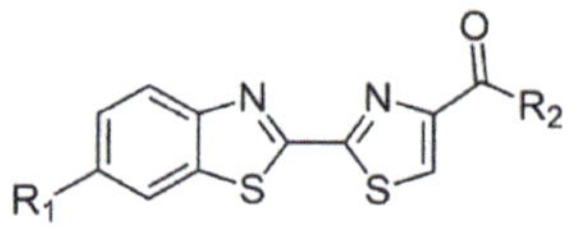

	R_1	R_2				pH
($^-$O)-Dehydroluciferin	O$^-$	O$^-$	396	159	11.2	11
(HO)-Dehydroluciferin	OH	O$^-$	348	207	4.7	4.5

All in water.

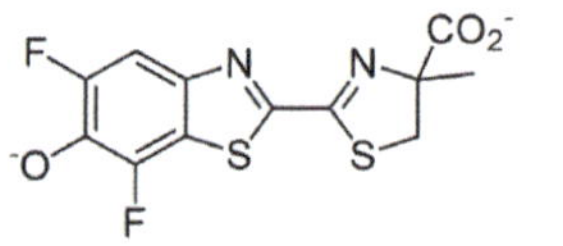

Pacific orange
404 149 15.5 (Carbonate pH 9.5)

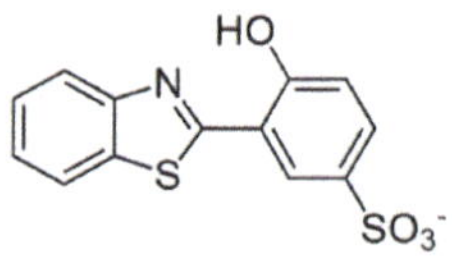

Stefani HB-7
331 118 3.6 (Water)

Figure 4.56 Molecular structure of some benzothiazole derivatives and their respective photophysical properties. The photophysical parameters shown in blue are $\lambda_{x,max}$ (nm), Δ_S (nm), and B_{max} ($\text{mM}^{-1}\text{cm}^{-1}$), given in this order.

4.31 Chromones

Chromone [154 p. 336] is an isomer of coumarin and, similar to coumarin, their derivatives are largely found in plants. However, it does not produce as many bright derivatives as coumarin. One reason for this is because the ketone group in chromone is too far away from the ether group, which makes it a less effective end of the push-pull scheme of the π-electron system.

Chromone derivatives are collectively known as chromones. Examples of fluorescent dyes of this class are: 3-(4-OH-Ph)-6-MeO-chromone (3-(4-hydroxyphenyl)-6-methoxy-4*H*-chromen-4-one); 3-buthyl-7-hydroxychromone (3-butyl-7-hydroxy-4*H*-chromen-4-one); 3-Bt-6-MeO-7-hydroxychromone (3-Buthyl-6-methoxy-7-hydroxychromone); and 3-AB-6-MeO-7-OH-chromone (3-(4-Azidobutyl)-6-methoxy-7-hydroxychromone) [185] (Figure 4.57). For a review on synthesis of chromones, see Gaspar et al. [186].

4.32 Naphthalimides

Naphthalimide is the imide of 1,8-naphthalic anhydride. This three-ringed structure has only a few fluorescent derivatives that are mainly used for staining. Examples of such fluorescent derivatives are lucifer yellow CH [165, 187, 188, p. 20], lucifer yellow VS [189], and brilliant sulfaflavine [190]. Note that brilliant sulfaflavine has only one methyl group in Aeby [190], while current suppliers sell it with two methyl groups bound to the benzene. The research group of L.D. Lavis also synthesized 2-(6-(Azetidin-1-yl)-1,3-dioxo-1*H*-benzo[*d*]isoquinolin-2(3H)-yl) acetic acid, which was given the short name of 6-azetidinylnaphthalimide [26] (Figure 4.58).

Singha et al. [121] synthesized three naphthalimides with different amino donors at position 4. Two (naphthalimides 10b and 10c) are shown in Figure 4.58. The third (containing a 4-dimethylamino) was not included in this database because its brightness in water was too low (<0.01 mM^{-1}cm^{-1}). Kelly et al. [191] also synthesized some naphthalimides, of which AI3 is the brightest (the concentration of AI3 in the experiments was estimated by the authors using published molar extinction coefficients of 15 mM^{-1}cm^{-1} for compounds with similar structures). Banerjee et al. [192] synthesized a naphthalimide with a pyridinium moiety. This water soluble naphthalimide derivative is hereby termed naphthalimide B-K 1 (Figure 4.58).

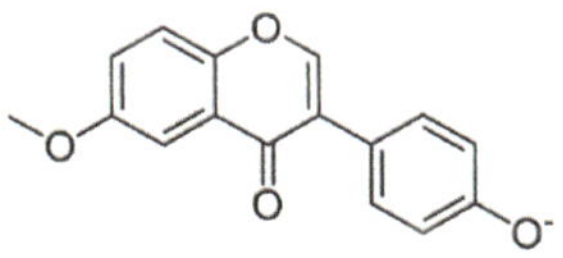

3-(4-OH-Ph)-6-MeO-chromone
311 157 0.6 (0.1 M Tris-HCl pH 8.0)

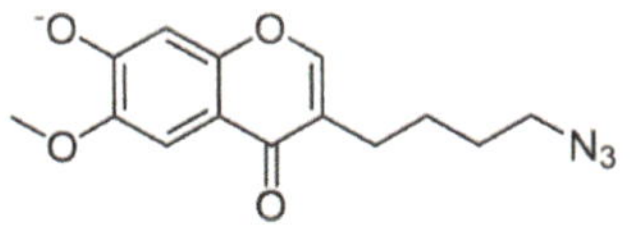

3-AB-6-MeO-7-OH-chromone
346 99 6.6 (0.1 M Tris-HCl pH 8.0)

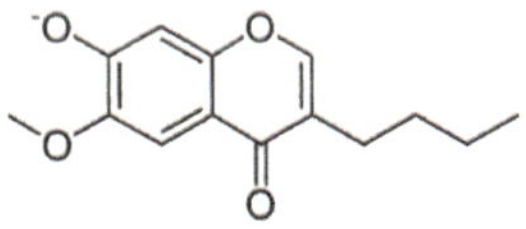

3-Bt-6-MeO-7-hydroxychromone
345 102 7.8 (0.1 M Tris-HCl pH 8.0)

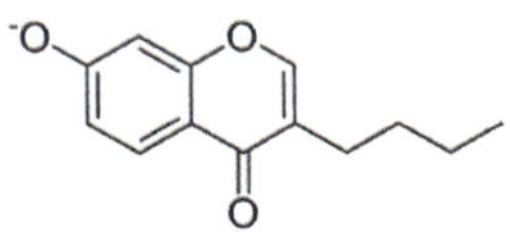

3-Buthyl-7-hydroxychromone
334 120 2.6 (0.1 M Tris-HCl pH 8.0)

Figure 4.57 Molecular structures of few a chromones and their respective photophysical properties shown in blue [$\lambda_{x,max}$ (nm), Δ_S (nm), and B_{max} (mM−1cm−1), given in this order].

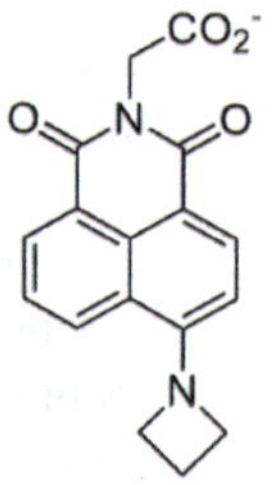

6-Azetidinyl naphthalimide
464 89 5.0 (HEPES pH 7.3)

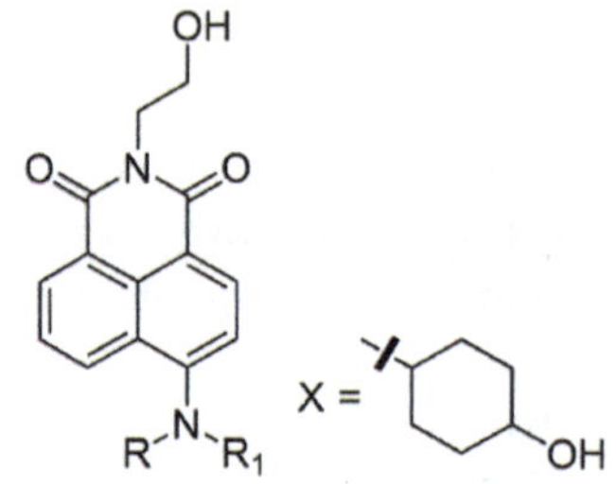

Naphthalimides

	R	R_1			
10b	H	CH_3	448	97	2.6
10c	H	X	453	90	5.1

All in water.

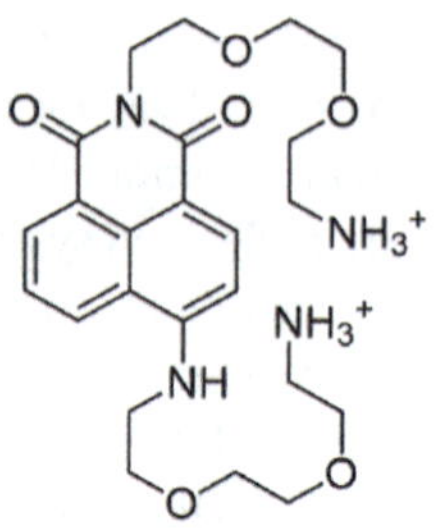

AI3
445 107 3.1 (water pH 7.4)

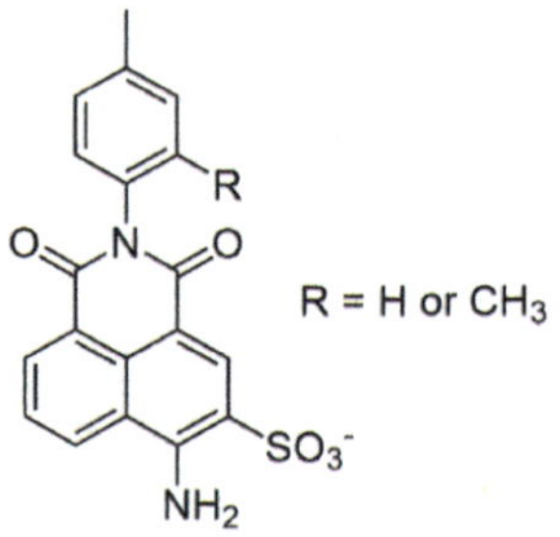

Briliant sulfaflavine
430 90 4.3 (water)

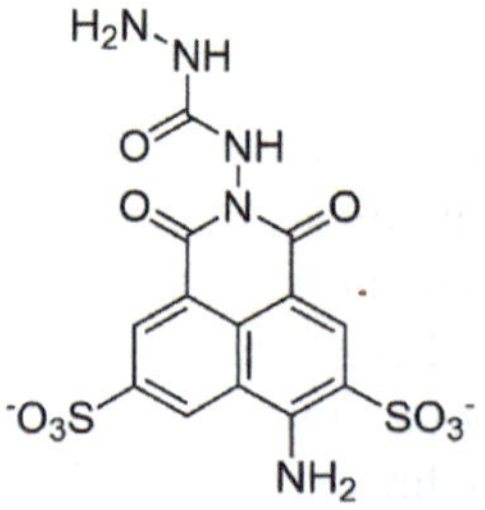

Lucifer yellow CH
429 111 2.3 (water)

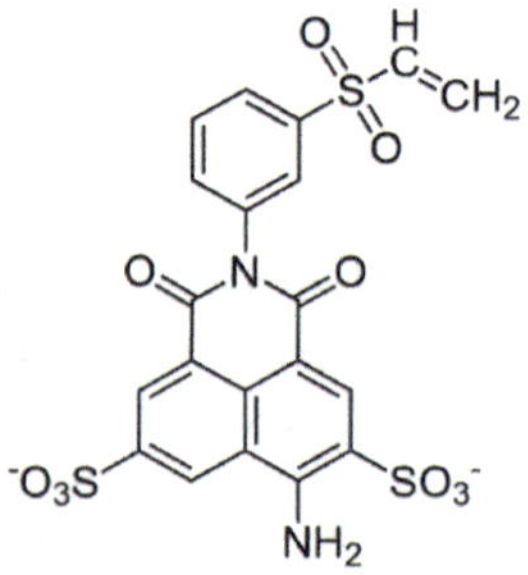

Lucifer yellow VS
430 105 2.7 (water)

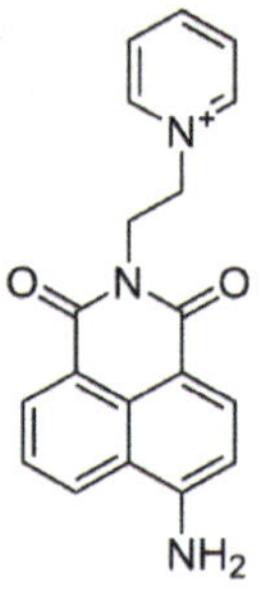

Naphthalimide B-K 1
436 116 1.0 (10 mM P pH 7.0)

Figure 4.58 Examples of fluorescent derivatives of naphthalimide. They are mostly used for staining in microscopy.

4.33 Indoles

1*H*-Indole (or 2,3-Benzopyrrole) is a bicyclic structure consisting of a benzene ring fused to a five-membered pyrrole ring (Figure 4.1; Eicher et al. [154], p. 125). Like coumarins and chromones, indoles also belong to the group of plant alkaloids and are commonly found in many plant species. The amino acid tryptophan is also an indole derivative; however, its wavelength of absorption maximum is below 300 nm and is therefore not included in this database.

One of the most detailed studies ever performed on the photophysical properties of harmine, 6-chloro-harmine, 8-chloro-harmine, and 6,8-dichloro-harmine was done by Rasse-Suriani et al. [193]. They measured many photophysical parameters in neutral water and in water at pH 4. Harmine, harmane, and harmaline were also studied by Rurack [18], Nafisi et al. [194], and Lakowicz [195]. Pardo et al. [196] also measured the fluorescence quantum yield and fluorescence lifetime of nor-harmane (or β-carboline), harmane, harmine, 2-methylharmine, and harmaline. All measurements were taken in 50 mM H_2SO_4 aqueous solutions, with the exception of harmaline which was measured in 5 mM H_2SO_4. The H_2SO_4 concentrations used in some references are expressed in normality, but note that for sulfuric acid the concentration in molarity (c(M)) is equal to half the concentration in normality (c(N)), i.e., c(M) = c(N)/2.

Figure 4.59 shows the structures and photophysical properties of some popular indoles. The absorption maxima, emission maxima, and molar absorption coefficients of harmine and harmaline in acidic conditions were taken from Savory and Turnbull [197]. The only difference between harmine and harmaline is a double bond. Note that under acidic conditions most of these indoles have two absorption bands, a shorter one with a larger molar absorption coefficient and a longer one with a smaller molar absorption coefficient.

4.34 Naphthalenes

Naphthalene derivatives are known for their moderate brightness and absorption maximum in the UVA, violet, and purple regions of the spectra. *Ullmann's Encyclopedia of Industrial Chemistry* gives an extensive list of naphthalene derivatives [198]. The fluorescent entities of this class usually contain a sulfonic group, which makes these derivatives water soluble and usually also participates in the push-and-pull effect on the π-electron system. The structures of this class exhibit moderate to low brightness.

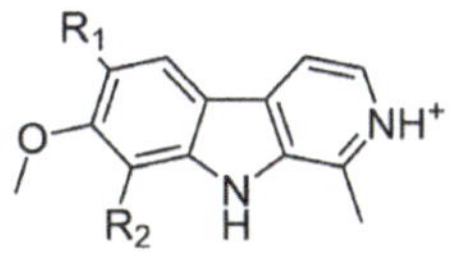

	R_1	R_2				pK_a
Harmine (Ha)	H	H	357	59	3.6	7.5
6-Cl-Ha	Cl	H	362	66	1.4	6.3
8-Cl-Ha	H	Cl	361	62	0.6	6.4
6,8-diCl-Ha	Cl	Cl	369	60	1.2	5.4

All in water pH 4.

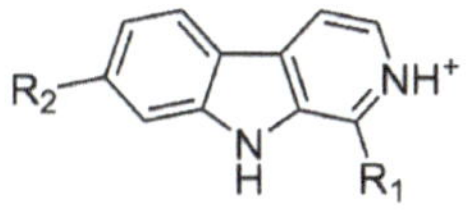

	R_1	R_2			
nor-Harmane	H	H	360	94	3.6
Harmane	CH_3	H	350	90	5.1
Harmine	CH_3	OCH_3	326	86	5.3

All in 50 mM H_2SO_4.

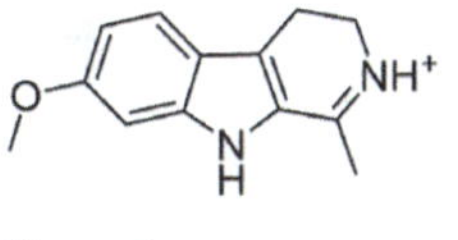

Harmaline

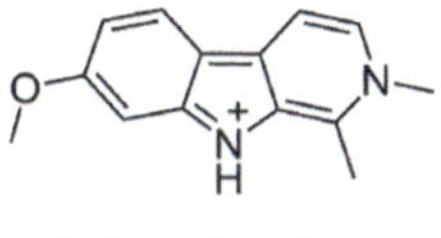

2-Methylharmine

2-Methylharmine	350	83	2.8	(5 mM H_2SO_4)
Harmaline	383	92	9.1	(50 mM H_2SO_4)

Figure 4.59 The structures of indoles in their predominant ionic form at the pH at which the photophysical parameters were measured. The photophysical parameters shown in blue are $\lambda_{x,max}$ (nm), Δ_S (nm), and B_{max} ($mM^{-1}cm^{-1}$), given in this order.

Some important examples of dyes belonging to this class are: acedan (2-acetyl-6-(dimethylamino)naphthalene) [121]; prodan (2-propionyl-6-(dimethylamino)naphthalene) [199]; acrylodan (6-acryloyl-2-dimethylaminonaphthalene [200, 201]; edans [202]; and acyman ((1*r*,4*r*)-4-(methyl(naphthalen-2-yl)amino)cyclohexan-1-ol)) [121, 203]. The molar absorption coefficient of acyman was kindly provided by one of the authors [203].

The dyes prodan, edans, acrylodan, acyman, and acedan (and the acedan amino-modifications 2 to 8; Figure 4.60), develop a very large dipole moment when they are excited to the first excitation state; i.e., the modulus of their dipole moment changes significantly following the upward photoexcitation. Therefore, in addition to the very large Stokes shifts observed, their photophysical properties are also very sensitive to the polarity of the microenvironment.

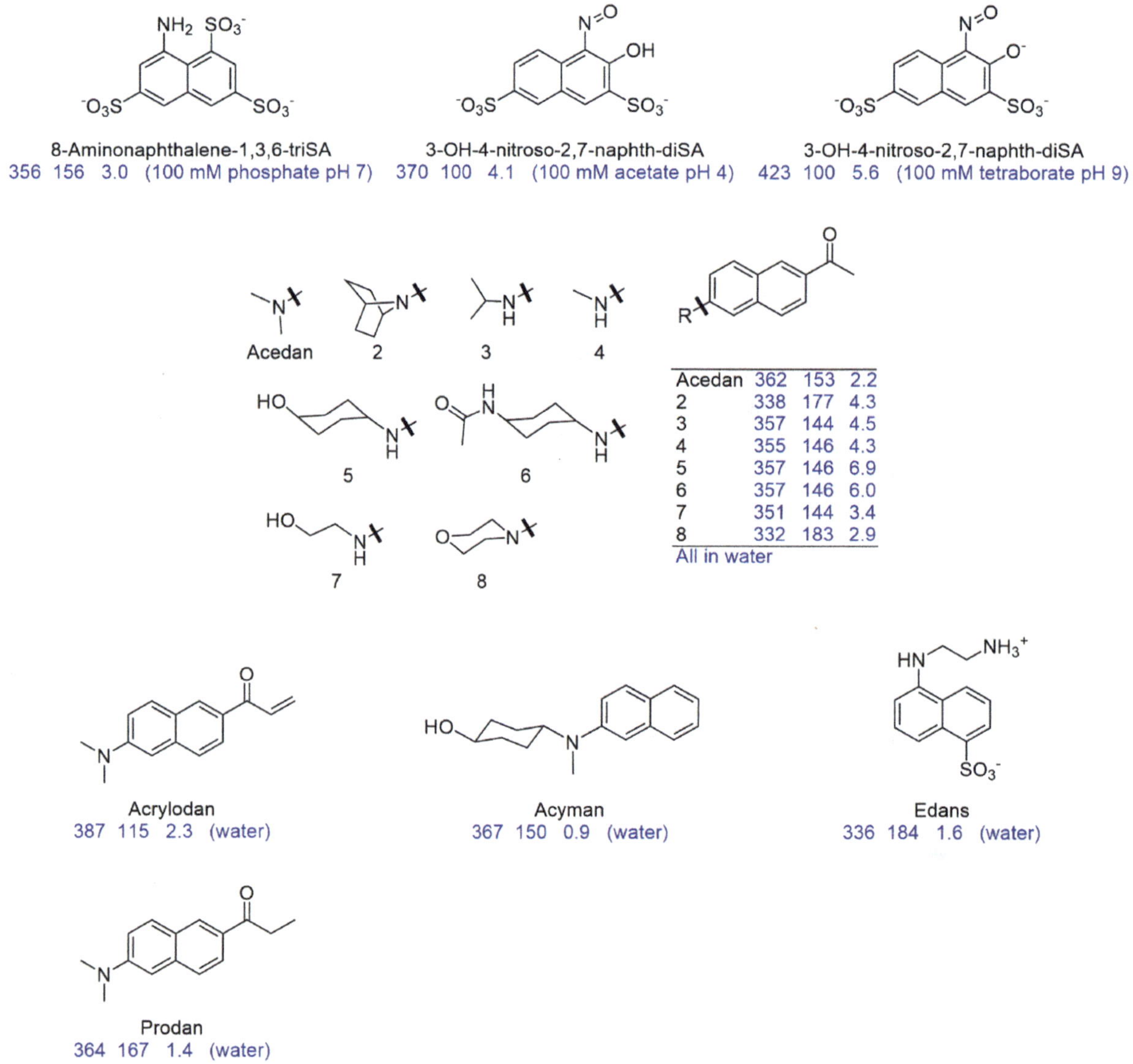

Acedan	362	153	2.2
2	338	177	4.3
3	357	144	4.5
4	355	146	4.3
5	357	146	6.9
6	357	146	6.0
7	351	144	3.4
8	332	183	2.9

All in water

Figure 4.60 Molecular structures of naphthalene fluorescent dyes. The photophysical parameters shown in blue are $\lambda_{x,max}$ (nm), Δ_S (nm), and B_{max} ($mM^{-1}cm^{-1}$), respectively.

Consequently, they are regarded as polarity sensors [201]. It is worth noting that prodan aggregates in water solutions, even when present at low concentrations [204].

A second group of dyes belonging to this class (including the already mentioned edans) have sulfonic groups and are therefore more soluble in aqueous solutions. A couple of examples are 8-aminonaphthalene-1,3,6-trisulfonic acid and 3-hydroxy-4-nitroso-2,7-naphthalenedisulfonic acid (unpublished data from the author).

4.35 Squaraines

Squaraines are similar to cyanines. The only difference is the presence of a squaric acid (3,4-dihydroxycyclobut-3-ene-1,2-dione) in the middle of the polymethine bridge that connects the two pyrrole rings. For detailed reviews on squaraines, see Ferreira et al. [205] (Figure 4.61).

Squaraines are important, not only for biological imaging, labeling, and analytics in general, but also in the field of photovoltaic energy. In a recent elegant article, the application of squaraines in both the photovoltaic field (dye-sensitized solar cells, organic solar cells, and perovskite solar cells) and the biomedical field (biosensors, biological imaging, and photodynamic/photothermal therapy reagents) was reviewed by He et al. [206].

Similar to cyanines and merocyanines, the conjugated system of squaraines is flexible and completely unprotected from the collisions of the water molecules of the aqueous media. Therefore, they too exhibit low fluorescence quantum yields. A rigidized cyanine called Cy3B was synthetized from Cy3 [82]. Its fluorescence quantum yield is eight times larger than that of Cy3. Nonetheless, it is difficult to synthesize rigidified connecting bridge structures for pentamethine, heptamethine, and squaraine. One solution is to protect the inner polymethine segments of the long squaraines, merocyanines, or cyanines using macrocycles, the so-called rotaxanes [207]. The macrocycle encircles and interlocks with the squaraine, merocyanine, or cyanine, mitigating the non-radiative relaxation mechanisms and leading to higher fluorescence quantum yields. Gassensmith et al. [208] used this strategy to improve the water solubility of squaraines with tetralactam macrocycles. This has the additional effect of preventing aggregation in aqueous solutions. Several additional functions of these macrocycles can be predicted. For example, it can be used as a labeling moiety, to facilitate cell penetration, to protect encircled dyes from photodegradation, and as a donor in FRET (when macrocycles are made up of many donors and the inner structure, fluorophore, works as an acceptor). In this case of FRET, the macrocycle works as a light harvesting accessory, giving the set a very high molar absorption coefficient. The inner structure functions as the acceptor with high fluorescence quantum yield and short fluorescence mean-lifetime. However, this topic is outside the scope of this book.

Examples of dyes belonging to this class include squaraines T-P 4a, T-P 4b, T-P 4c, T-P 4d, T-P 5a, and T-P 5b [209] and the highly negatively-charged squaraines nor-SqO1, nor-SqB1, nor-SqCN1, nor-SqO2, nor-SqB2, nor-SqCN2, nor-SqO3, nor-SqB3, nor-SqCN3, SqO1, SqB1, and SqCN1 [90]. The molecular structures of square-635-carboxylate, square-655, square-650-pH-carboxylate, square-660-carboxylate, and square-680-carboxylate were not disclosed by the manufacturer (Seta Biomedicals 2022).

It is worth noting that the brightness of some fluorophores increases by up to 20-fold when a protein (e.g., bovine serum albumin, BSA) is added to the solution [209]. This is mainly caused by an increase in the fluorescence quantum yield. One possible contribution to this comes from dye docking (non-covalent binding) between the α-helices segments of this protein. Consequently, the dyes are locked in and are also protected from being directly hit by water molecules. In this way the non-radioactive decay rates are decreased.

4.36 Pteridines

Pteridines are composed of a pyrimidine ring fused to a pyrazine ring (Eicher et al. [154], p. 487). This nitrogen-rich bi-heterocycle aromatic class of compounds have many derivatives with many biological roles. If compared to other fluorescent dyes with $\lambda_{x,max}$ in the same range, many dyes that belong to this class have a weak

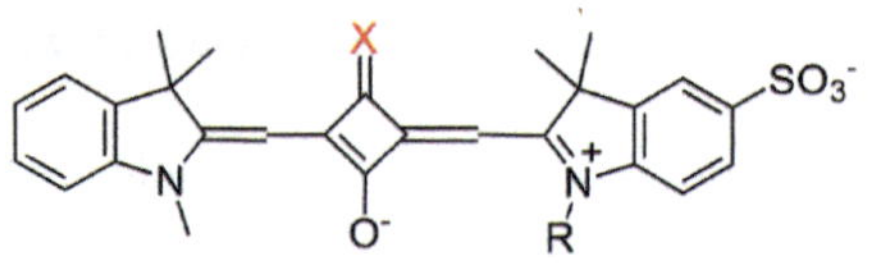

R = 

Squaraine				X
T-P 4a	628	14	4.8	S
T-P 4b	659	17	4.1	C(CN)$_2$
T-P 4c	624	11	4.9	O

All in 67 mM phosphate pH 7.4

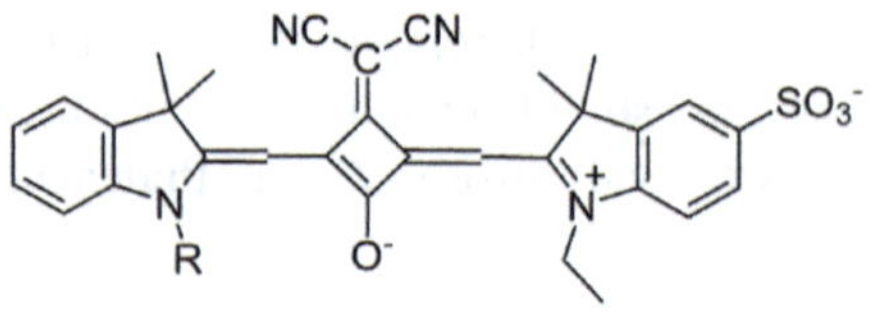

R = 

Squaraine
T-P 4d 658 19 5.4
67 mM phosphate pH 7.4

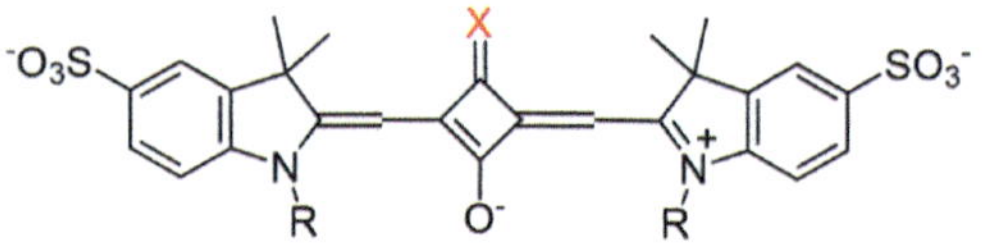

R = 

Squaraine				X
T-P 5a	636	12	12.9	S
T-P 5b	667	18	13.1	C(CN)$_2$
T-P 5c	632	10	15.9	O

All in 67 mM phosphate pH 7.4

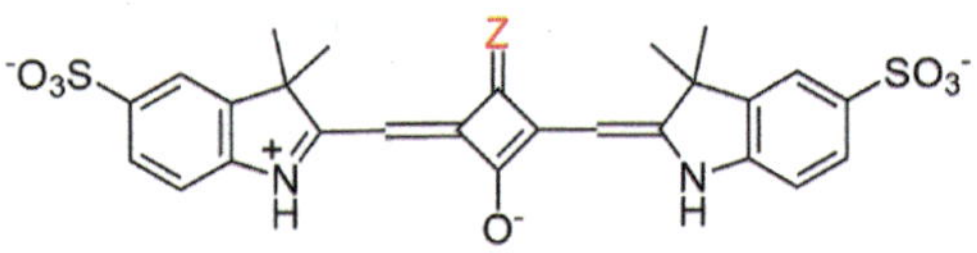

nor-SqO1; nor-SqB1; nor-SqCN1

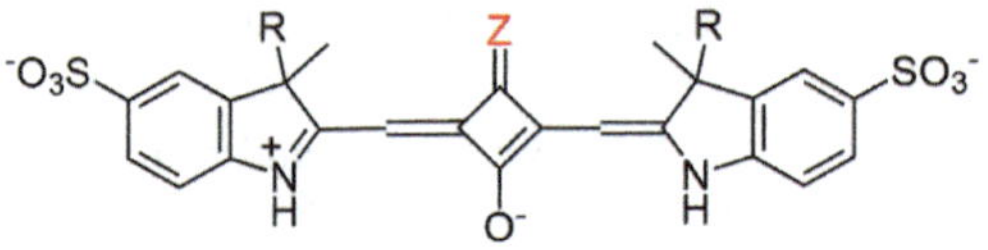

nor-SqO2; nor-SqB2; nor-SqCN2

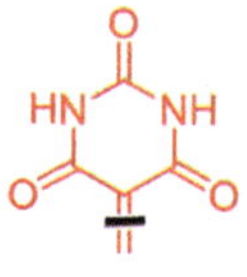

barbituric

R = 

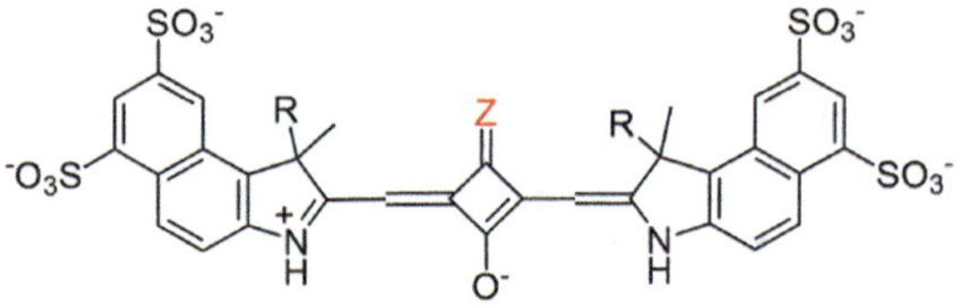

nor-SqO3; nor-SqB3; nor-SqCN3

				Z
nor-SqO1	636	17	47.6	O
nor-SqB1	638	28	17.6	barbituric
nor-SqCN1	670	23	23.3	C(CN)$_2$
nor-SqO2	639	18	60.3	O
nor-SqB2	638	28	25.0	barbituric
nor-SqCN2	670	23	31.7	C(CN)$_2$
nor-SqO3	662	17	26.0	O
nor-SqB3	662	22	18.3	barbituric
nor-SqCN3	693	21	16.9	C(CN)$_2$
SqO1	632	10	18.5	O
SqB1	627	23	7.5	barbituric
SqCN1	667	18	13.1	C(CN)$_2$

All in 67 mM phosphate pH 7.4

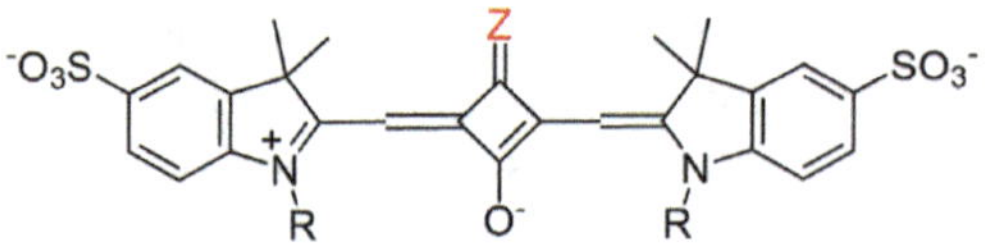

SqO1; SqB1; SqCN1

Figure 4.61 Molecular structures of squaraine fluorescent dyes. The photophysical parameters shown in blue are $\lambda_{x,max}$ (nm), Δ_S (nm), and B_{max} (mM^{-1}cm^{-1}), given in this order.

fluorescence brightness and only a few have a medium brightness (Figure 4.62). Flavin mononucleotide (FMN) [210, 211], DMPT synthesized by Saleh et al. [212], and tri-O-acetyl-P-luminoarosin and luminarine [213] all exhibit a large Stokes shift, but nonetheless have a moderate to low brightness. Isoalloxazines are derivatives of benzopteridine, and Visser and Müller synthesized a dozen of them in 1979. Figure 4.62 only shows the six brightest isoalloxazines synthesized by Visser and Müller [214].

4.37 Isoquinolines

Isoquinolines are isomers of quinolines. Both isoquinolines and benzoquinolines are benzopyridines (the product of the fusion of a benzene with a pyridine). Dye derivatives of this scaffold show large Stokes shifts but only

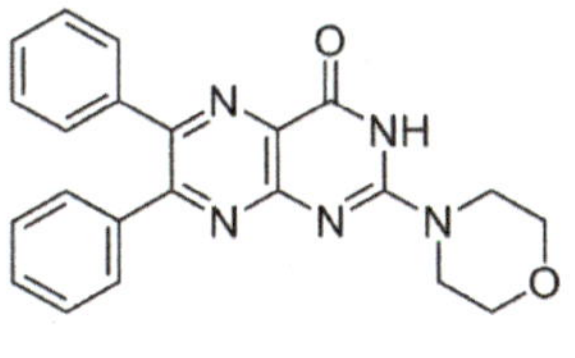

DMPT 382 93 0.7 (pK_a = 7.2)
(Water pH 7.0)

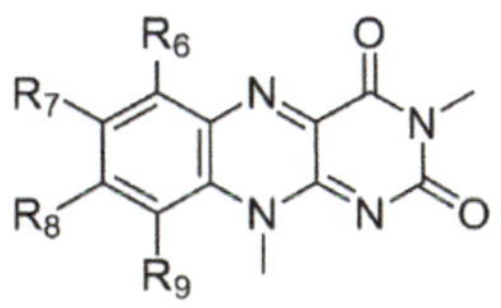

R_6	R_7	R_8	R_9	
H	CH_3	H	CH_3	3,7,9,10-TetraMe-isoalloxazine
CH_3	H	H	H	3,6,10-Trimethylisoalloxazine
H	CH_3	H	H	3,7,10-Trimethylisoalloxazine
H	CH_3	CH_3	H	3,7,8,10-TetraMe-isoalloxazine
H	H	CH_3	H	3,8,10-Trimethylisoalloxazine
H	H	H	H	3,10-Dimethylisoalloxazine
CH_3	CH_3	CH_3	H	3,6,7,8,10-PentaMe-isoalloxazine

3,7,9,10-TetraMe-isoalloxazine	452	119	0.6
3,6,10-Trimethylisoalloxazine	439	93	0.6
3,7,10-Trimethylisoalloxazine	446	86	3.1
3,7,8,10-TetraMe-isoalloxazine	444	85	3.1
3,8,10-Trimethylisoalloxazine	433	85	1.3
3,10-Dimethylisoalloxazine	435	85	1.4
3,6,7,8,10-PentaMe-isoalloxazine	448	123	0.8

All in water with < 1% acetonitrile.

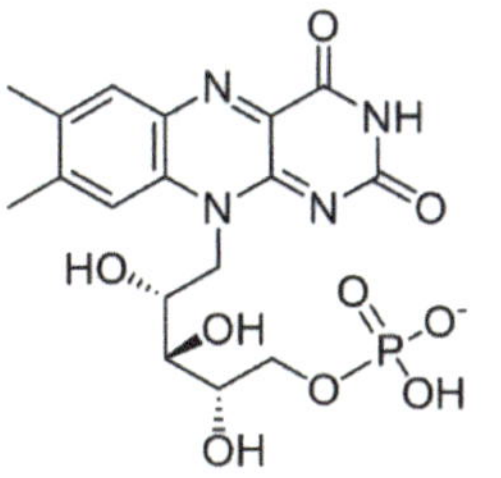

Flavin mononucleotide (FMN)
450 80 3.0 (Water)

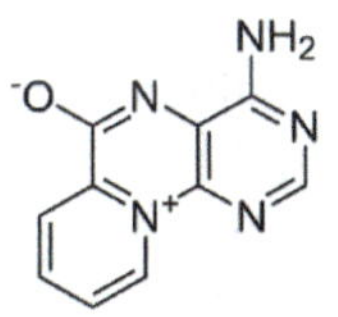

Luminarine 428 105 4.2
(Water pH 7.0)

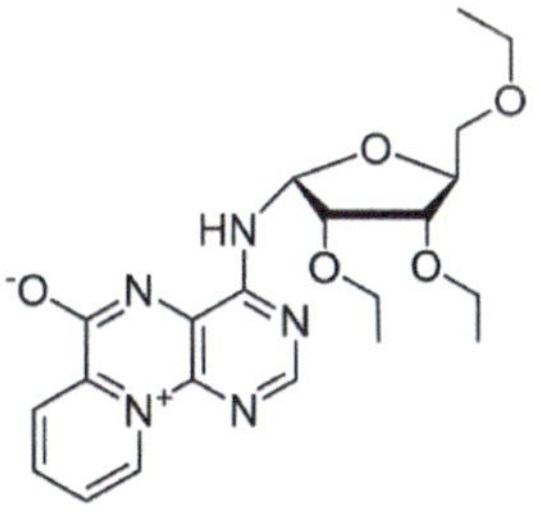

Tri-O-acetyl-P-luminoarosin
425 103 6.7
(Water pH 7.0)

Figure 4.62 Molecular structures and photophysical properties of pteridines. Numbers in blue are $\lambda_{a,max}$ (nm), Δ_S (nm), and B_{max} ($mM^{-1}cm^{-1}$), respectively.

medium to low brightness values if compared to other dyes with absorption maximum in this range of the spectrum. The chemistry of isoquinolines is explained in chapter 6.17 of Eicher et al. [154].

A series of isoquinolines were synthesized by Belyy et al. [215] and termed isoquinolines 3a-3o, 4b-c, 7a-f, 8, and 10b. This library of isoquinolines shows the effect of methyl substituents on the photophysical properties of this class of dyes, as demonstrated in Figure 4.63. Note that these authors expressed the absorption and fluorescence emission maxima in wavenumber ($\nu_{a,max}$ and $\nu_{e,max}$ in units of cm^{-1}) instead of wavelength ($\lambda_{a,max}$ and $\lambda_{e,max}$ in nm). Therefore, these wavenumbers (ν) were first converted to λ(cm) using $\lambda(cm) = 1/\nu(cm^{-1})$, and then to λ(nm) using $\lambda(nm) = \lambda(cm)\ 10^7$ (nm/cm).

4.38 Benzene Derivatives

Benzene derivatives are small molecules with low molar absorption coefficients (small absorption cross-sections) and modest fluorescence quantum yields. Therefore, they have low brightness values. Nevertheless, they are still among the brightest fluorophores whose absorption maximum of their longest wavelength band is located between 300 and 325 nm.

Figure 4.64 shows the molecular structures of some fluorescent dyes belonging to the class of benzenes: salicylic acid, 4-methylsalicylic acid2 (4-MeSA), 5-methylsalicylic acid2 (5-MeSA), and 5-methoxysalicylic acid2 (5-MeOSA) [216]; 2,5-bis(methylsulfonyl)-1,4-diaminobenzene (BMeS-p-A) [217]; and BMeS-p-A-COOH and BMeS-p-A-2COOH [218].

4.39 Other Single Structures

All classes which only have one or two bright entities are presented in this section. In general, the entities of these classes are not very bright and this is why they generally only have one or two representatives per class. Here they are presented in size order, starting with the smallest, and their structures are shown in alphabetical order.

4.39.1 Small Structures

2-Aminopyridine and 3-aminopyridine [18, 174, 219] only have 13 atoms and belong to the class of pyridines. They are among the smallest fluorescent structures whose absorption maximum of the longest wavelength band is located at around 300 nm. They have broad absorption bands and their exact maxima are a few nm below 300 nm. However, to prevent their exclusion from this database, they were rounded up to 300 nm. 4-Aminopyridine was not included in the present database because its brightness was reported to be only 0.01 $mM^{-1}cm^{-1}$.

The phenazine class is not shown in Figure 4.1 because it only has one fluorescent derivative (Safranin T) that is only slightly soluble in water (Figure 4.65) and has a modest brightness value. Nevertheless, Aygün et al. [149] succeeded in measuring the photophysical properties of this dye in water.

The acridizinium cation (or benzoquinolizinium, Eicher et al. [154], p. 420) significantly differs from acridine and some fluorescent derivatives of adridizinium have been synthesized and reported [220]. The reported brightness of acridiziniums I-L 5a and I-L 5b in aqueous solutions (Figure 4.65) is lower than that of the brightest acridine derivatives (Section 4.23), the main reason being the low fluorescence quantum yield of acridiziniums. The halogenated acridiziniums J-N 1a to J-N 1e showed even higher fluorescence quantum yields and brightness values [221]. Moreover, Jagt et al. [221] reported that the counter ion has some effect on fluorescence quantum yield. These dyes exhibit three overlapping absorption bands. The values of $\lambda_{x,max}$ and B_{max} of the band with the highest molar absorption coefficient was taken for the present database (Figure 4.65). It is important to note that some aminoacridizinium dyes intercalate into DNA and cause strand breaks and other damage upon exposure to UVA radiation [222].

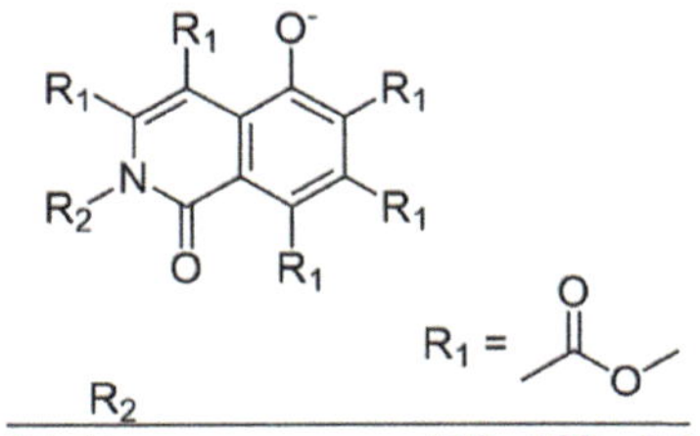

	R_2			
3d	H	418	102	6.2
3e	CH_3	408	116	6.4
3f	i-Propyl	405	114	6.0
3g	*cyclo*-C_4H_7	414	108	7.1
3h	*cyclo*-C_5H_7	406	119	5.2
3i	*cyclo*-C_6H_{11}	405	116	6.1
3j	Phenyl	409	116	8.4
3k	4-HOC_6H_4	411	112	2.0
3l	4-MeC_6H_4	410	112	7.0
3m	4-BrC_6H_4	408	118	6.5
3n	4-$MeOC_6H_4$	411	110	6.8
3o	2,4-$(MeO)_2C_6H_3$	410	111	1.7

All in PBS pH 7.4

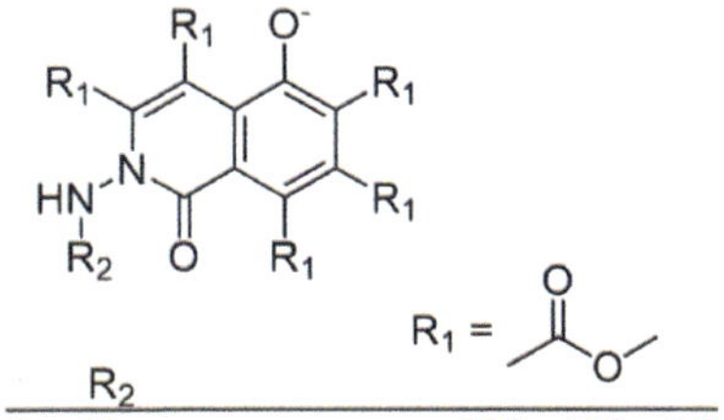

	R_2			
8	H	406	128	6.2
7a	Ac	407	121	4.7
7b	Boc	406	120	6.6
7c	$HOCH_2CO$	408	116	4.1
7d	COPh	408	118	2.1
7f	3,4-$(MeO)_2C_6H_3CO$	406	122	2.6

All in PBS pH 7.4

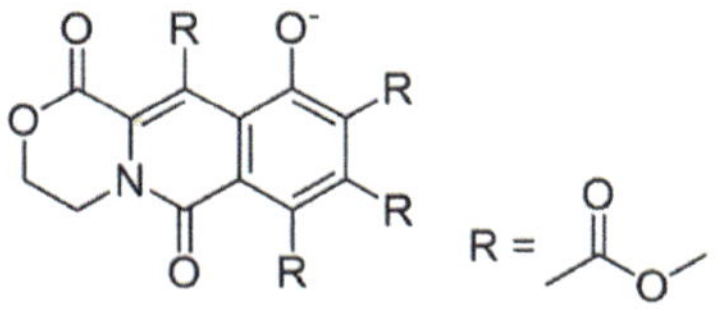

4b 436 114 5.5 (PBS pH 7.4)

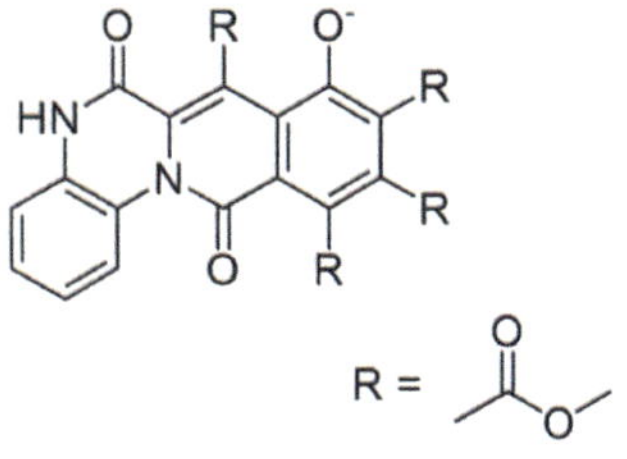

4c 453 76 5.3 (PBS pH 7.4)

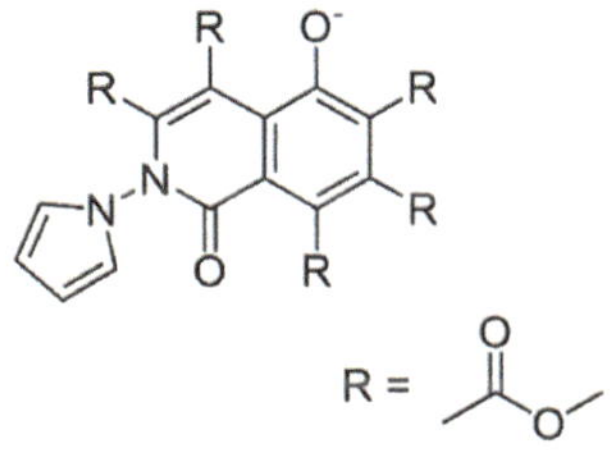

10b 410 116 2.6 (PBS pH 7.4)

Figure 4.63 Molecular structures and some photophysical parameters of isoquinoline derivatives. Parameters shown in blue are $\lambda_{x,max}$ (nm), Δ_S (nm), and B_{max} (mM^{-1}cm^{-1}), respectively.

Six hydrophilic diketopyrrolopyrroles (DPP) were synthesized by Heyer et al. [223] and all six dyes exhibited both large Stokes shifts and good brightness values for this region of the spectrum. Hereby they are termed DPP 5, DPP 6, DPP 7, DPP 8, DPP 9, and DPP 10 (Figure 4.65). An insufficient quantity of the dye DPP 10 was obtained by Heyer et al. and measurement of ε_{max} was not possible. In order to include this dye in the present database, the value of ε_{max}(DPP 10) was estimated to be 17.1 mM^{-1}cm^{-1}, assuming that ε_{max}(DPP 10)/ε_{max}(DPP 9) is approximately equal to ε_{max}(DPP 8)/ε_{max}(DPP 7).

Carbazoles is a class of dyes made of an indole flanked by two fused benzene rings, resulting in a rigid, planar heterotricyclic aromatic structure. They exhibit the longest wavelength absorption band in the UVA, with a

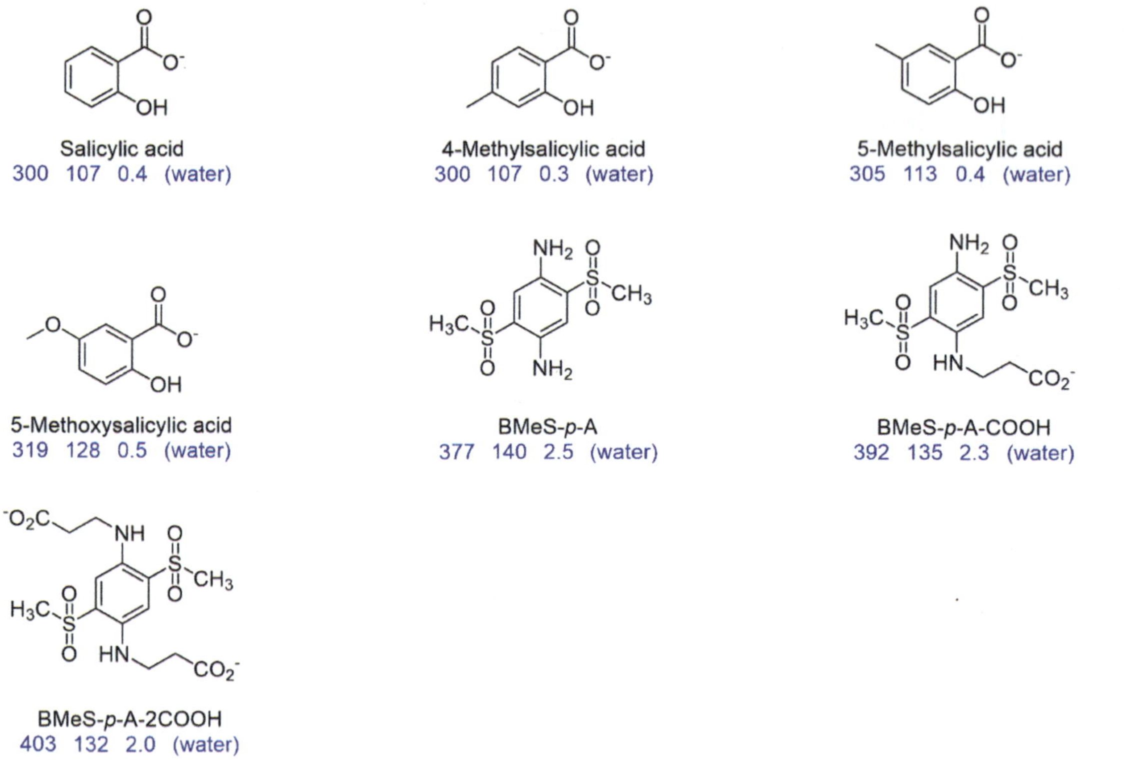

Figure 4.64 Molecular structures of benzene derivatives. Note that the absorption maximum of the longest wavelength band of these structures is located between 300 and 325 nm.

modest molar absorption coefficient and high fluorescence quantum yield. 3-Amino-9-ethylcarbazole 1 and its derivative "3-amino-9-ethylcarbazole 2" are such examples (Figure 4.65) [224].

4.39.2 Medium-Sized Structures

Epicocconone and epicocconone derivatives were synthesized and had their photophysical parameters measured by Peixoto et al. [225]. Only the brightest were included in the present database: epicocconone P-F 4b, P-F 4c, P-F 4d, P-F 17a, and P-F 17d (Figure 4.66).

Shaya et al. [226] synthesized a series of aminofluorene derivatives termed I6, PP1, PP2, PP3, PP4, PP5, PP6, PP6 PP8, PP9, PP10, PP11, PP12, PP3, PP14, and others. Only the four brightest in water were included in the present database (I-6, PP6, PP11, and PP12) and their structures are shown in Figure 4.66. All others exhibited $\Phi_f < 0.05$ and were therefore not included in the database. Aminofluorene derivatives exhibit a very large Stokes shift (almost all have Δ_S larger than 200 nm). This is the result of a large electric dipole moment change upon photoexcitation from the ground (S_0) to the first excited state (S_1).

4.39.3 Large Structures (N_a > 80)

Phospholes are a heavier analog of pyrroles and a few fluorescent phosphole dyes were recently synthesized [227]. As shown in Figure 4.67, the dyes *trans*-PO-PB430, *cys*-PO-PB430, *trans*-PO-PB460, and PO-PB430 exhibit large Stokes shifts, good photostability, and moderate brightness values in PBS pH 7.4.

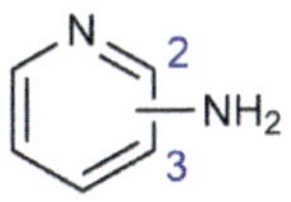

2-Aminopyridine	~300	63	2.9
3-Aminopyridine	~300	90	0.9

All in water

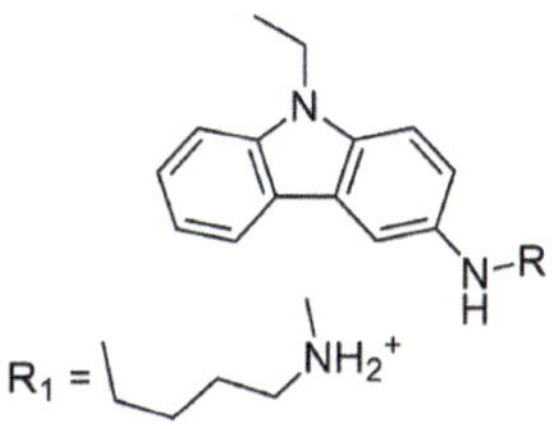

	R
3-Amino-9-ethylcarbazole 1	H
3-Amino-9-ethylcarbazole 2	R_1

Carbazoles

R				
H	364	88	1.1	(water)
R_1	364	94	1.3	(water)

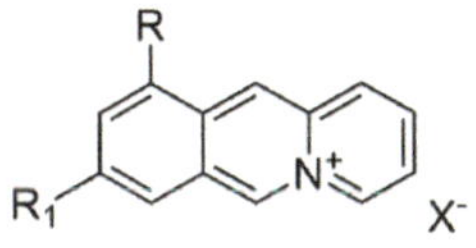

Acridiziniums

	R	R_1	X			
J-N 1a	H	H	Br	377	27	2.9
J-N 1b	H	F	Br	381	27	2.2
J-N 1c	F	F	Br	354	66	2.5
J-N 1d	H	Cl	Br	356	52	1.0
J-N 1e	H	OCH_3	Cl	355	100	1.3

All in water

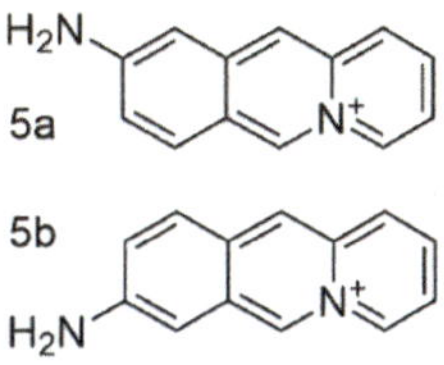

Acridiziniums

I-L 5a	387	120	2.0
I-L 5b	369	182	0.2

All in water

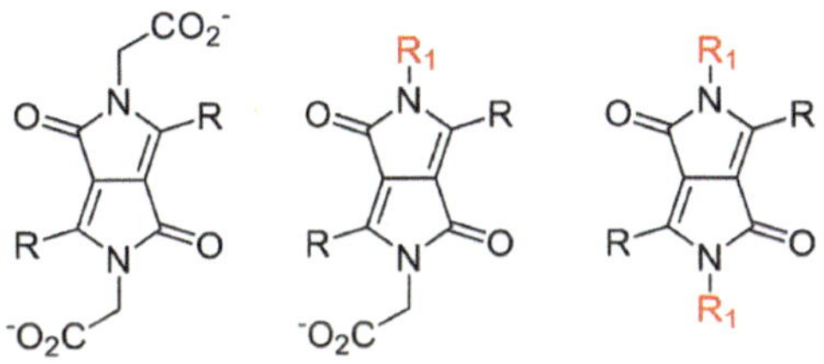

DPP 5
R = phenyl
DPP 6
R = 2-thienyl

DPP 7
R = phenyl
DPP 8
R = 2-thienyl

DPP 9
R = phenyl
DPP 10
R = 2-thienyl

Diketopyrrolopyrroles (DPP)

DPP 5	455	60	12.9
DPP 6	530	56	15.0
DPP 7	456	92	11.1
DPP 8	527	67	10.1
DPP 9	451	90	12.5
DPP 10	523	71	10.4

All in PBS

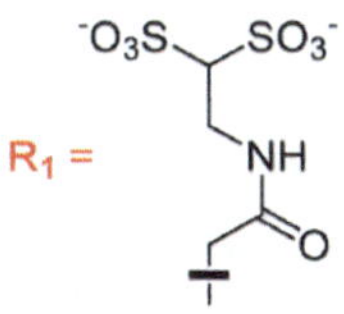

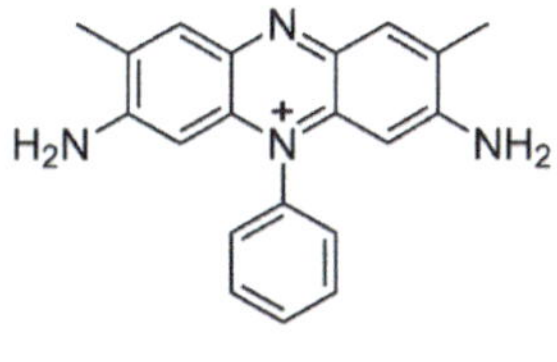

Safranin T
517 69 2.9 (water)

Figure 4.65 Molecular structures of "other" small-molecule fluorescent dyes. Parameters shown in blue are $\lambda_{x,max}$ (nm), Δ_S (nm), and B_{max} (mM^{-1}cm^{-1}), respectively.

ECXi is a pentacene derivative, which is a polycyclic aromatic hydrocarbon that consists of five linearly-fused benzene rings (Figure 4.65). This is a very large structure and was synthesized by Lei et al. [228]. It exhibits an absorption maximum at 887 nm (in the near-infrared) and a Stokes shift of 51 nm in aqueous buffer solutions.

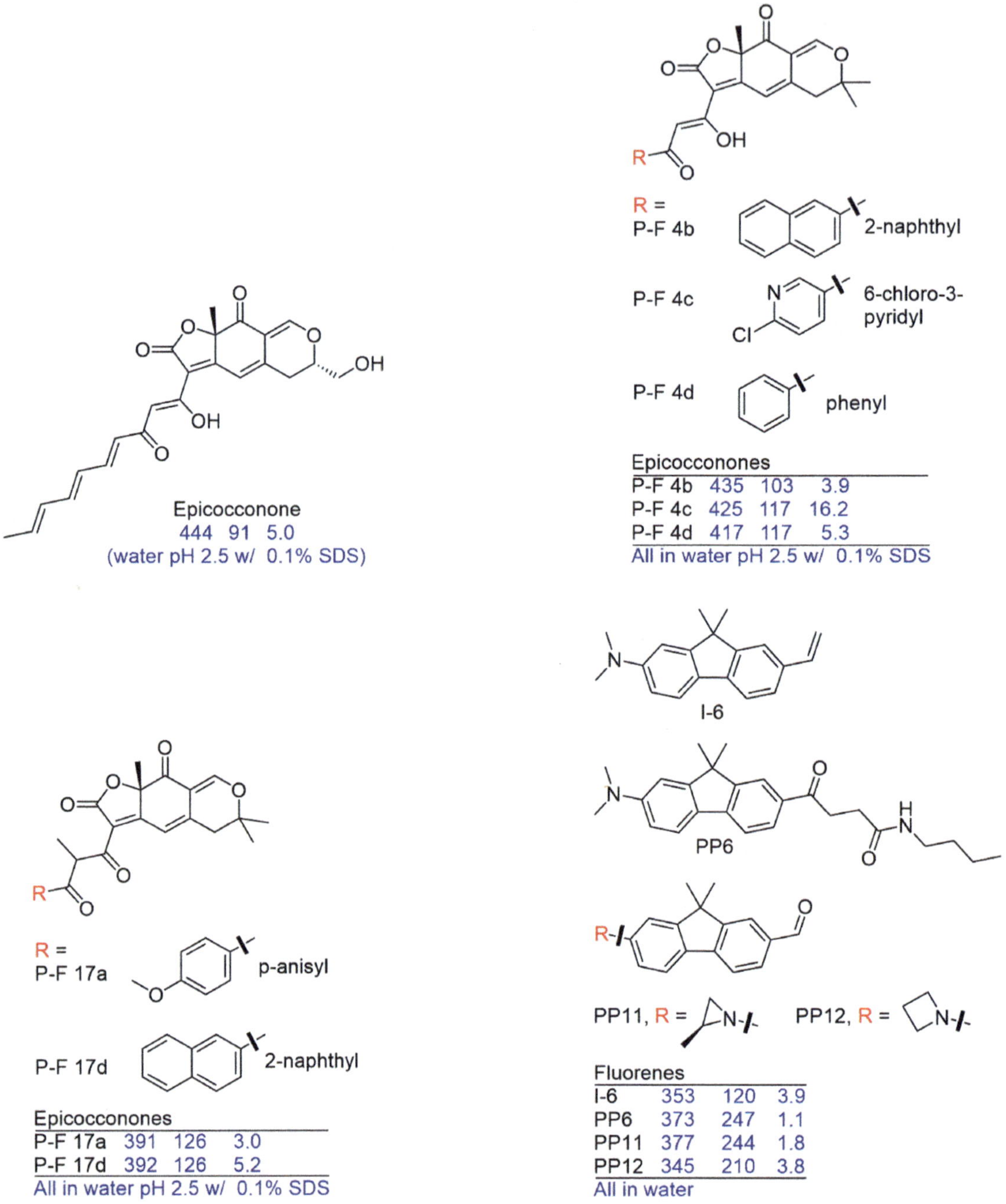

Figure 4.66 Molecular structures of "other" medium-sized fluorescent dyes. Legend as in Figure 4.65.

Spirobifluorene 1 [229] belongs to the fluorene class and is a large structure made of 169 atoms. This is on the edge of the inclusion limit of the present database. Its wavelength of maximum absorption of the longest wavelength band is only 378 nm and it exhibits a large Stokes shift (Δ_S = 102 nm) and a high brightness value (B_{max} = 25.6 mM^{-1}cm^{-1}, which is particularly high for this range of the spectrum (UVA)).

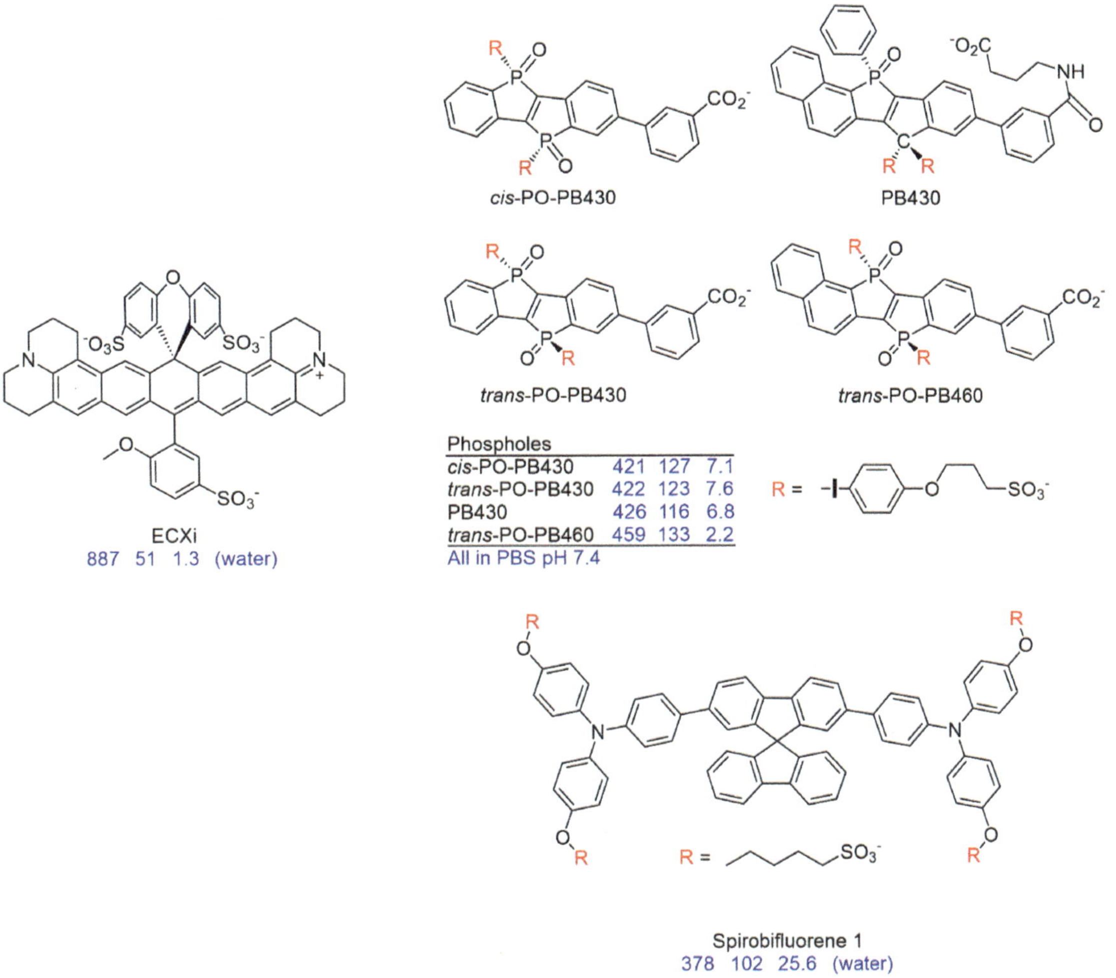

Phospholes			
cis-PO-PB430	421	127	7.1
trans-PO-PB430	422	123	7.6
PB430	426	116	6.8
trans-PO-PB460	459	133	2.2
All in PBS pH 7.4			

Figure 4.67 Molecular structures of "other" large fluorescent dyes. Legend as in Figure 4.65.

4.40 Hybrid Structures

Figure 4.1 shows the core structures of the most important organic small-molecule fluorescent dyes. Current bright derivatives are presented in Sections 4.2 to 4.39. However, there are three additional ways of creating new fluorescent dyes from existing ones: i) a new dye can be produced by fusing the rings of two existing dyes; i.e., fusing two of the above-mentioned structures (Sections 4.2 to 4.39); ii) connecting two of the above-mentioned structures by a single bond which, in general, belongs to the π-electron systems; and iii) finally, two of the above-mentioned structures (Sections 4.2 to 4.39) can be connected by a polymethine bridge.

The fusion or connection of two existing dyes produces a new and larger structure that may be fluorescent or not. They are known as π-extended systems and most of them are less soluble in aqueous solutions due to their larger sizes. However, there are some exceptions; i.e., dyes produced by i), ii), or iii) that are soluble in aqueous solutions and with a good brightness.

4.40.1 Hybrid Structures: Fusion of Two Existing Dyes

The dyes NCR and NCRMe are the result of the fusion of a coumarin (highlighted with red in Figure 4.68) with a rhodamine [230]. These dyes do not exhibit a high fluorescence quantum yield in any of the solvents tested by Huang et al. [230], including water. There are many potential explanations for the lack of fluorescence when two bright dyes are fused together. Nonetheless, NCR and NCRMe may produce brighter entities if, for example, the two ethyl groups that are bound to the nitrogen are replaced by other alkyl groups. This replacement could be performed, for instance, by either using *N*-propylamine or by the cyclization of the two propyl ends of dipropylamine with the neighboring carbons.

Different fusion products from the combination of coumarin and sulforhodamine were produced by Chevalier et al. [47] and named DEMAC-SR101, DEAC-SR101, and HC6H-SR101 (Figure 4.68). The fusion products produced by the authors were not sufficient for highly accurate measurements of absorption coefficients, therefore a value of ~80 $mM^{-1}cm^{-1}$ was attributed from similar structures to allow for comparisons (Figure 4.68).

Another example of fused structures is the fusion of chromene with quinoline, which results in the chromenoquinolines [231] and hereby termed chromenoquinolines L-S 4a to 4f (Figure 4.68).

4.40.2 Hybrid Structures: Single Bond Connected Dyes

Benzothiazol-coumarins are an example of a hybrid structure formed by a coumarin and a benzothiazole that are held together by a single covalent bond. The derivatives BC-OH, BC-H, and BC-BA depicted in Figure 4.69 were shown to be bright and photorubust [232]. These dyes have been used as probes for the detection of peroxynitrite and peroxide.

4-Aminophenol and 4-hydroxybenzenesulfonic acid (Figure 4.1) are small structures with small brightness values. Binding one of these two structures to a benzoxazole brings the absorption maximum of the longest wavelength band of the set to around 325 nm, and produces a higher brightness value than that of the bare parental structures. The dyes Stefani HB-6, Stefani HB-9, and Stefani HB-11 are such examples and were synthesized by Barros and Stefani [184] and Barros et al. [183]. These compounds are also characterized by an excited state intramolecular proton transfer (ESIPT), in which a proton is transferred from the proton donor group (–OH) to the proton receptor group (–N=). This ESIPT phenomenon produces the additional advantages of large Stokes shifts and higher photostability values (see Stefani HB-6, Stefani HB-9, and Stefani HB-11 in Figure 4.69).

Cascade yellow is historically important as its large Stokes shift was very impressive when it first entered the market [6]. It consists of an oxazole bound to a pyridine by a single bond, flanked by two benzenes. Its π-electron system runs between the oxygen of the 2-methoxy group (of the 2-methoxy benzenesulfonate), through the oxazole and up to the nitrogen atom of the pyridine. However, this "pyridyloxazole" flexible structure exhibits a modest molar absorption coefficient and a low fluorescence quantum yield. Fluorescence mean-lifetime was measure by the author (Figure 2.18 [161, 233]). Note that the dye PyMPO has a similar structure to Cascade yellow, but it does not have the sulfonic group [234].

Lucigenin is made of two acridines (10-methylacridinium) held together by a single covalent bond (carbon 9 to carbon 9). This is one of the brightest dyes that has an absorption maximum of the longest wavelength band in this range ($\lambda_{a,max}$ = 380 nm). The photophysical data of lucigenin were taken from table II of Hinze et al. [151]. Note that lucigenin and many other acridines are widely used as chemiluminescent and electrochemiluminescent probes for a great variety of targets [235].

A coumarin (4,7-dihydroxycoumarin) grafted onto carbon-9 of pyronine B produced an interesting bright dye synthesized by Renault et al. [36] and hereby termed "coumarin-pyronine B 9." A phosphorylated coumarin bound with a single bond to a benzothiazole also produced a hybrid named PONy C7 by Butkevich et al. [33]. All dyes mentioned in this section are shown in Figure 4.69. A few additional dyes of this class were classified as coumarins (Figure 4.52).

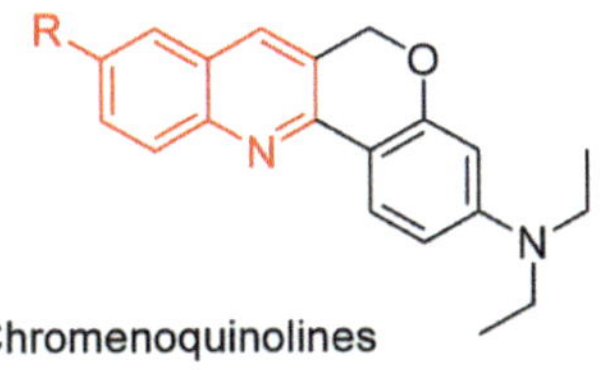

Chromenoquinolines

	R			
L-S 4a	CH_3	473	97	5.4
L-S 4b	OCH_3	477	102	5.2
L-S 4c	CN	500	91	1.1

All in water

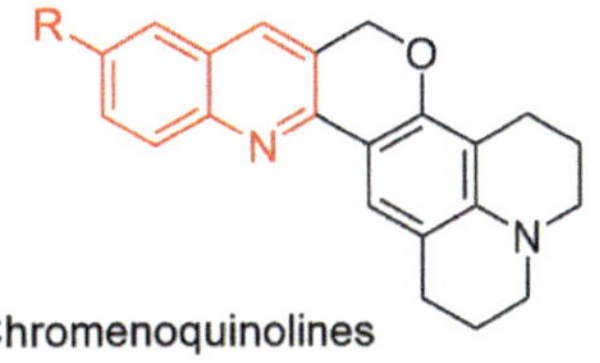

Chromenoquinolines

	R			
L-S 4d	CH_3	500	108	2.8
L-S 4e	OCH_3	502	111	2.2
L-S 4f	CN	529	93	1.9

All in water

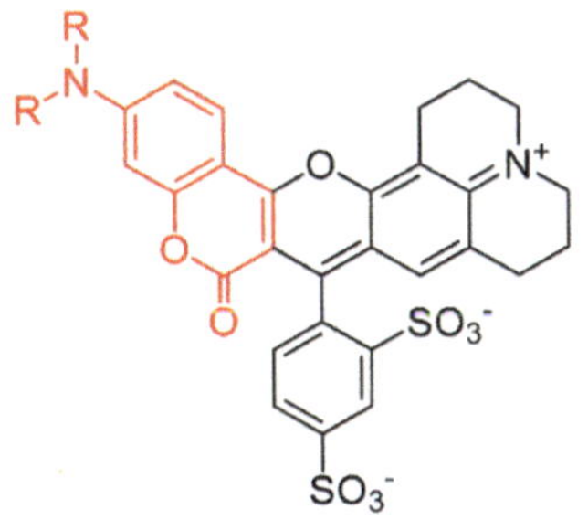

	R			
DMAC-SR101	CH_3	605	42	7.2
DEAC-SR101	CH_2CH_3	610	32	8.0

All in PBS

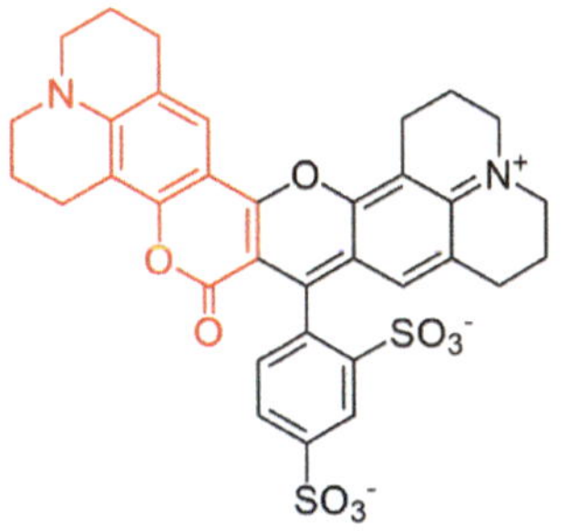

HC6R-SR101 624 30 12.0
(PBS+5% BSA)

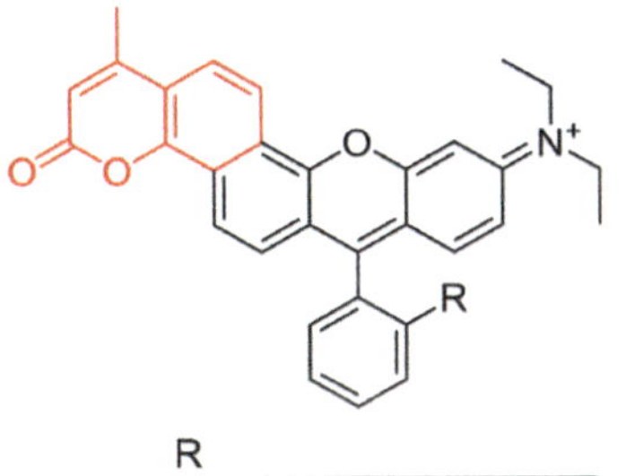

	R			
NCR	CO_2^-	565	53	-
NCRMe	CO_2CH_3	568	39	0.1

All in water

Figure 4.68 Molecular structures of fluorescent dyes produced by the fusion of two existing dyes. The photophysical parameters shown in blue are $\lambda_{x,max}$ (nm), Δ_S (nm), and B_{max} (mM^{-1}cm^{-1}), respectively.

4.40.3 Hybrid Structures: Polymethine Bridged Dyes

Connecting two fluorescent structures with a polymethine bridge is an interesting way to produce a new fluorescent dye with a longer wavelength of absorption maximum. Merocyanines do not fall into this category because their end hanging structures are not usually fluorescent as single entities in free solution.

The COUPY dyes (COUPY is a portmanteau for **Cou**marin-**py**ridin) are examples of such structures and consist of a coumarin that is connected to a pyridine ring by a polymethine bridge [236] (Figure 4.70).

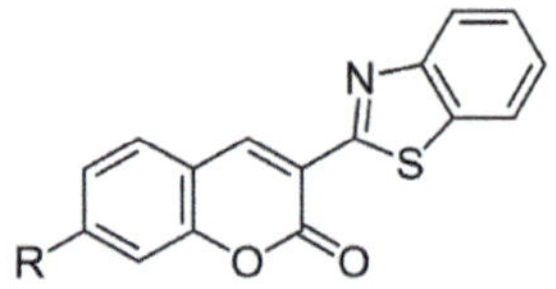

	R			
BC-OH	-O^-	442	46	32.3
BC-H	-H	365	105	2.4
BC-BA	-$B(OH)_2$	371	102	4.3

All in 100 mM P pH 7.4:ethanol, 4:1

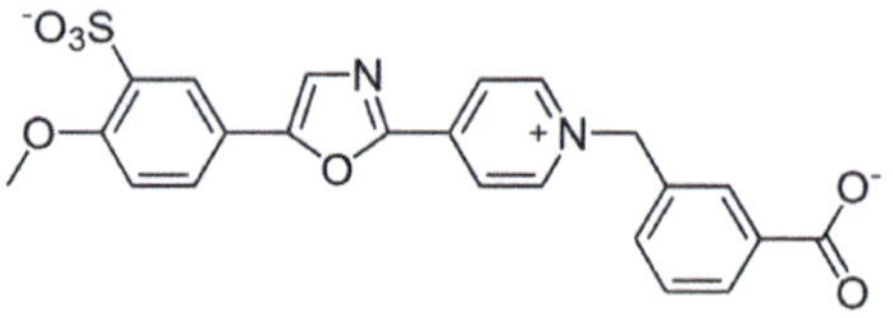

Cascade yellow
404 145 1.2 (water)

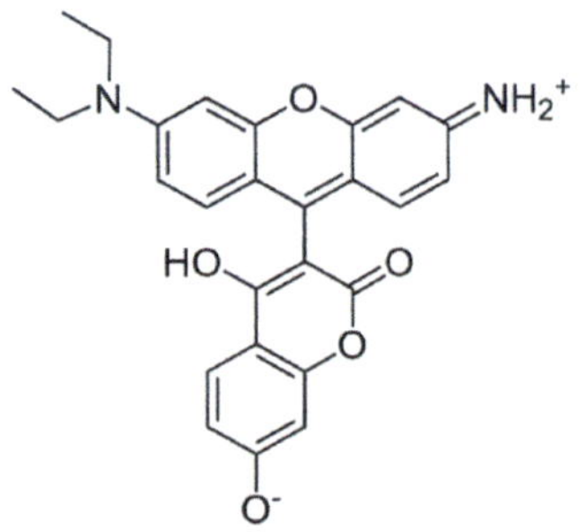

Coumarin-pyronine 9 560 19 17.2
(100 mM P pH 7.5)

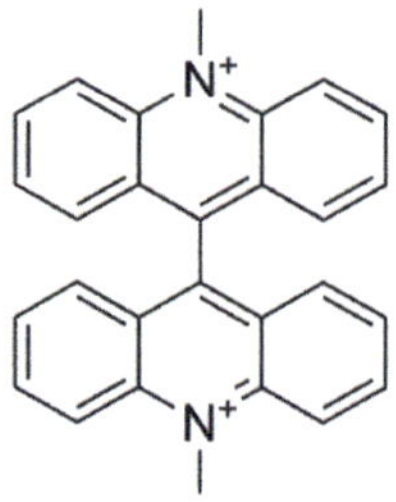

Lucigenin
368 117 18.8 (water)

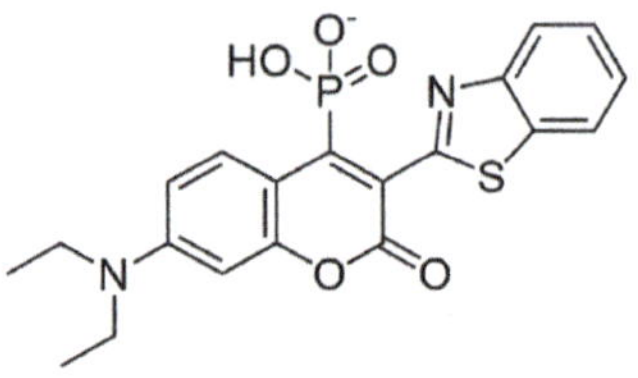

PONy C7
419 194 0.8 (water)

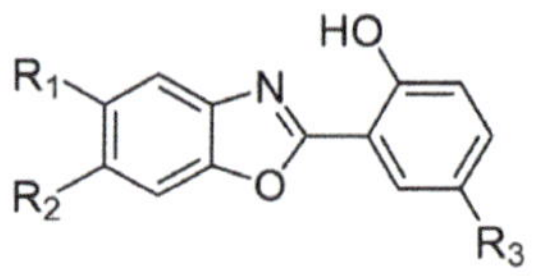

				R1	R2	R3
Stefani HB-6	317	48	0.9	H	H	SO_3^-
Stefani HB-9	320	103	4.2	H	CH_3	SO_3^-
Stefani HB-11	329	31	4.0	SO_3^-	H	NH_2

All in water

Figure 4.69 Molecular structures of fluorescent dyes produced from a single bond connection between two existing dyes. The photophysical parameters shown in blue are $\lambda_{x,max}$ (nm), Δ_S (nm), and B_{max} (mM^{-1}cm^{-1}), respectively, and "P" stands for phosphate buffer.

4.41 Non-Disclosed Structures

The molecular structures of 43 dyes out of a total of approximately 777 has not been disclosed by the suppliers (this is ~5.5% of the total). Nevertheless, they were kept in the database because the most important photophysical parameters ($\lambda_{a,max}$, $\lambda_{e,max}$, ε_{max}, and Φ_f) were disclosed by the suppliers or measured. It is not possible to accurately classify these dyes among the above 39 classes (Sections 4.2 to 4.40) without knowing their molecular structures.

The total list of these dyes and the references where the photophysical parameters were taken from are as follows: Atto 430LS carboxylate, Atto 490LS carboxylate, Atto 542 carboxylate, Atto 665 carboxylate, Atto 725 carboxylate, Atto 740 carboxylate, and Cyan 500 carboxylate [5]; Alexa Fluor 660, Alexa Fluor 700, and Alexa Fluor 750 [6];

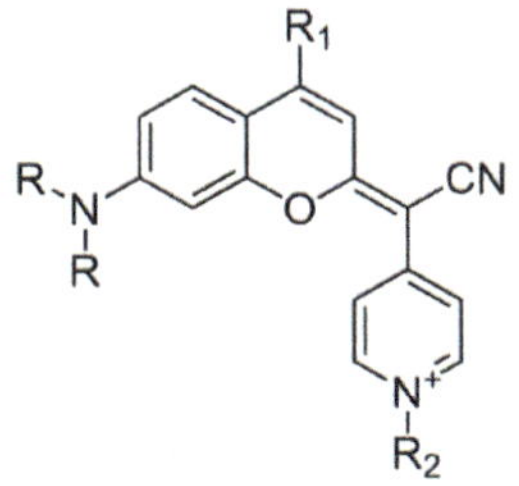

COUPY	R	R_1	R_2			
9	CH_2CH_3	CH_3	CH_3	543	62	4.6
10	CH_2CH_3	CH_3	CH_2CF_3	562	69	0.8
11	CH_3	CF_3	CH_3	567	95	0.2
12	CH_3	CF_3	CH_2CF_3	595	88	0.3

All in water

Figure 4.70 Molecular structure of a fluorescent dye produced by the polymethine bridge connection of two existing dyes. The photophysical parameters shown in blue are $\lambda_{x,max}$ (nm), Δ_S (nm), and B_{max} (mM^{-1}cm^{-1}), given in this order.

CF405S (pyrene), CF405M (coumarin), CF430 (cyanine), CF450 (cyanine), CF514 (rhodamine), CF532 (rhodamine), CF594 (rhodamine), CF620R (rhodamine), CF647 (cyanine), CF660C (cyanine), CF660R (rhodamine), CF680 (cyanine), and CF680R (rhodamine) [237] Φ_f measured by [25]; and Seta-375-carboxylate, SeTau-380-NHS, Seta-385-carboxylate, Seta-400-carboxylate, Seta-405-carboxylate, SeTau-405-NHS, Seta-470-carboxylate, Seta-555-carboxylate, Seta-580-carboxylate, Seta-632-NHS, Seta-633-NHS, Seta-635-pH-di-Carboxy, Seta-646-NHS, Seta-650-NHS, Seta-660-di-NHS, Seta-670-NHS, Seta-670-pH-di-Carboxy, Seta-680-NHS, Seta-700-NHS, Seta-700-di-NHS, Seta-750-NHS, and Seta-780-NHS (Seta Biomedicals 2022) [93].

4.42 Fluorescent Structures Other Than Small-Molecule Organic Dyes

The following groups of fluorescent structures were not considered in this database because they do not meet the selection criterion (single, organic, and small-molecule fluorescent dyes which have medium to high brightness values in aqueous solutions):

i) Structures based on large coordination compounds (e.g. porphyrins and phthalocyanines).
ii) Fluorescent DNA intercalating dyes [238]. These are not included in the database because their fluorescence brightness is highly dependent on their non-covalent ligation to the minor or major grooves of dsDNA.
iii) Macro- and micro-environment fluorescent chemosensors of physical (viscosity, polarity, and temperature) [239] and chemical (pH, ions, and molecules) variables [240]. A good fluorescent label should not be sensitive to micro-environmental conditions. On the other hand, a good chemosensor is highly selective and very sensitive to a given physical and chemical variable. Such dyes are called fluorescent probes and are largely used in cell biology [241] and analytical chemistry [242].
iv) Fluorescent chiral sensors [243]. This is a very specialized (chiral selective) group of chemosensors (also known as molecular probes).
v) Fluorescent ion indicators [6]. The fluorescence quantum yield of these dyes changes upon interaction with a specific ion. There is a book [244] dedicated to metal–fluorescent dye interactions and how this can be applied to metal detection and quantification.

vi) Fluorescent pH indicators. These are special fluorescence indicators of ion H_3O^+ concentrations, as the degree of ionization of a group (in general a phenol bound to the dye) changes almost all of its photophysical properties. This especially impacts the wavelength of absorption and emission maxima, but also changes the fluorescence quantum yields and molar absorption coefficients.

vii) Fluorescent proteins [245]. Fluorescent proteins were initially found in nature. In general, they consist of a fluorophore that is locked inside a protein. Despite these locked fluorophores being organic and small, they normally have very low fluorescence brightness values when in free solution.

viii) Membrane potential fluorescent sensors [246]. These fluorescent dyes are sensitive to transmembrane electric field strengths (observed across the cell membranes of neurons). They penetrate the lipid bilayer and their photophysical properties change, depending on whether they are at the resting membrane potential (or resting voltage) or the action potential.

ix) Pyrylium dyes [247, 248].

x) Quantum dots and nanoparticles [249].

xi) Dendrimers or oligomers of dyes [250].

xii) Donor-acceptor cassettes. Cassettes are made of donor structures that are covalently bound to acceptor-fluorophores. Energy transfer from donors to acceptors occurs via covalent bonds [251] or directly through the electromagnetic field of the oscillating dipole [252]. In both cases, the π-electron system does not run from one of the structures to the other structure, as happens with hybrid structures (Section 4.39.3).

References

1 Moss, G.P., Smith, P.A.S., and Tavernier, D. (1995). Glossary of class names of organic compounds and reactive intermediates based on structure. *Pure App. Chem.* 67 (8/9): 1307–1375.

2 Wurm, C.A., Kolmakov, K., Göttfert, F. et al.(2012). Novel red fluorophores with superior performance in STED microscopy. *Opt. Nanoscopy* 1: 1–7. doi: 10.1186/2192-2853-1-7.

3 Niu, S. (2011). Advanced water soluble BODIPY dyes: synthesis and application. PhD thesis. University of Strasbourg.

4 Zhu, S., Zhang, J., Vegesna, G. et al. (2011). Highly water-soluble neutral BODIPY dyes with controllable fluorescence quantum yields. *Org. Lett.* 13: 438–441.

5 Atto-Tec (2020). *Fluorescent Labels and Dyes. Catalogue 2019–2020*. Siegen, Germany: ATTO-TEC GmbH.

6 Molecular Probes (2010). *The Molecular Probes Handbook: A Guide to Fluorescent Probes and Fluorescent Labeling Technologies*, 11e. Oregon: Molecular Probes – Invitrogen.

7 Gee, K.R., Archer, E.A., and Kang, H.C. (1999). 4-Sulfotetrafluorophenyl (STP) esters: new water-soluble amine-reactive reagents for labeling biomolecules. *Tetrahedron Lett.* 40: 14711474.

8 Bioacts (2022). Namdong-gu, Incheon, South Korea. Datasheets accessed March 2022.

9 Bendig J., Schedler U., Harder T. et al. (1995). Photophysical and photochemical deactivation behaviour of streptopolymethines. *J. Photochem. Photobiol. A: Chemistry* 91: 53–57.

10 Baeyer, A. (1871). Uber eine neue Klasse von Farbstoffen (On a new class of dyes). *Berichte der Dtsch. Chem. Gesellschaft* 4: 555–558. doi: 10.1002/cber.18710040209.

11 Ceresole, M. (1887). Verfahren zur Darstellung von Farbstoffen aus der Gruppe des Meta-amidophenolphtaleins. German Patent No. 44002.

12 Noelting, E. and Dziewonski, K. (1905). Zur Kenntnis der Rhodamine. *Berichte der Deutsch. Chem. Gesellschaft* 38: 3516–3527.

13 Johnson, I. (1998). Fluorescent probes for living cells. *Molecul Probes, Eugene*. doi: 10.1023/a:1003287101868.

14 Zhang, X.F., Zhang, Y., and Liu, L. (2014a). Fluorescence lifetimes and quantum yields of ten rhodamine derivatives: structural effect on emission mechanism in different solvents. *J. Luminesc.* 145: 448–453. doi: 10.1016/j.jlumin.2013.07.066.

15 Mitronova, G.Y., Belov, V.N., Bossi, M.L. et al. (2010). New fluorinated rhodamines for optical microscopy and nanoscopy. *Chem. A Eur. J.* 16 (15): 4477–4488. doi: 10.1002/chem.200903272.

16 Abberior Dyes (2022). *Abberior Dyes for Superior Super-Resolution Microscopy Application*. www.abberior.com (accessed 12 February 2022).

17 Arbeloa, F.L., Arbeloa, T.L., and Arbeloa, I.L. (2001). Photophysical properties of laser dyes: correlations with the lasing characteristics. In: *Handbook of Advanced Electronic and Photonic Materials and Devices* (ed. H.S. Nalwa), Chapter 5, 209–245. Academic Press.

18 Rurack, K. (2008). Fluorescence quantum yields: methods of determination and standards. In: *Standardization and Quality Assurance in Fluorescence Measurements,* e1 (ed. U. Resch-Genger). Vol. 5, 101–145. Berlin, Heidelberg: Springer. doi: 10.1007/4243_2008_019.

19 Sabnis, W.W. (2015). *Handbook of Fluorescent Dyes and Probes.* Hoboken. Wiley.

20 Dyomics (2017). *Fluorescent Dyes for Bioanalytical and Hightech Applications*, 8e. Jena, Germany: Dyomics GmbH.

21 Magde, D., Wong, R., and Seybold, P.G. (2002). Fluorescence quantum yields and their relation to lifetimes of rhodamine 6G and fluorescein in nine solvents: improved absolute standards for quantum yields. *Photochem. Photobiol.* 75 (4): 327–334. doi: 10.1562/0031-8655(2002)0750327FQYATR2.0.CO2.

22 Butkevich, A.N., Belov, V.N., Kolmakov, K. et al. (2017). Hydroxylated fluorescent dyes for live-cell labeling: synthesis, spectra and super-resolution STED. *Chem. A Eur. J.* 23 (50): 12114–12119. doi: 10.1002/chem.201701216.

23 Grimm, J.B., Sung, A.J., Legant, W.R. et al. (2013). Carbofluoresceins and carborhodamines as scaffolds for high-contrast fluorogenic probes. *ACS Chem. Biol.* 8 (6): 1303–1310. doi: 10.1021/cb4000822.

24 Whitaker, J.E., Haugland, R.P., Ryan, D. et al. (1992). Fluorescent rhodol derivatives: versatile, photostable labels and tracers. *Anal. Biochem.* 207 (1): 267–279.

25 Kist, T.B.L. (2022). Absorption measured on an Agilent/Varian spectrophotometer, model Cary 5000, instrument version 2.24. Fluorescence emission spectra measured on a Shimatzu spectrofluorometer model RF-5301PC.

26 Grimm, J.B., English, B.P., Chen, J. et al. (2015). A general method to improve fluorophores for live-cell and single-molecule microscopy. *Nat. Methods* 12 (3): 244–250. doi: 10.1038/nmeth.3256.

27 Grimm, J.B., Tkachuk, A.N., Xie, L. et al. (2020). A general method to optimize and functionalize red-shifted rhodamine dyes. *Nat. Methods* 17: 815–821. doi: 10.1038/s41592-020-0909-6.

28 Kolmakov, K., Hebisch, E., Wolfram, T. et al. (2015). Far-red emitting fluorescent dyes for optical nanoscopy: fluorinated silicon-rhodamines (SiRF Dyes) and phosphorylated oxazines. *Chem. – A Eur. J.* 21 (38): 13344–13356. doi: 10.1002/chem.201501394.

29 Wu, L. and Burgess, K. (2008). Synthesis and spectroscopic properties of rosamines with cyclic amine substituents. *J. Org. Chem.* 73: 8711–8718.

30 Kist, T.B.L. (2020). *Open and Toroidal Electrophoresis: Ultra-High Separation Efficiencies in Capillaries, Microchips, and Slabs*. Chichester, UK: Wiley.

31 Kolmakov, K., Belov, V.N., Wurm, C.A. et al. (2010). A versatile route to red-emitting carbopyronine dyes for optical microscopy and nanoscopy. *Eur. J. Org. Chem.* 3593–3610. doi: 10.1002/ejoc.201000343.

32 Grimm, J.B., Brown, T.A., Tkachuk, A.N., and Lavis, L.D. (2017). General synthetic method for Si-fluoresceins and Si-rhodamines. *ACS Cent. Sci.* 3 (9): 975–985. doi: 10.1021/acscentsci.7b00247.

33 Butkevich, A.N., Sednev, M.V., Shojaei, H. et al. (2018). PONy dyes: direct addition of P(III) nucleophiles to organic fluorophores. *Org. Lett.* 20: 1261–1264.

34 Koide, Y., Urano, Y., Hanaoka, K. et al. (2011). Development of an Si-rhodamine-based far-red to near-infrared fluorescence probe selective for hypochlorous acid and its applications for biological imaging. *J. Am. Chem. Soc.* 133: 5680–5682.

35 Sauer, M., Han, K.-T., Miiller, R. et al. (1995). New fluorescent dyes in the red region for biodiagnostics. *J. Fluoresc.* 5: 247–261.

36 Renault, K., Chevalier, A., Bignon, J. et al. (2021). Coumarin-pyronin hybrid dyes: synthesis, fluorescence properties and theoretical calculations. *ChemPhotoChem* 5: 822–838.

37 Nau, W.M. and Mohanty, J. (2005). Taming fluorescent dyes with cucurbituril. *Int. J. Photoenergy* 7: 133–141.

38 Arik, M., Kassa, S.B., and Onganer, Y. (2020). Molecular aggregates of pyronin dyes with polyelectrolyte polystyrene sulfonate (PSS) in aqueous solution. *J. Photochem. Photobiol A: Chem.* 391: 112309.

39 Deng, F. and Xu, Z. (2019). Heteroatom-substituted rhodamine dyes: structure and spectroscopic properties. *Chin. Chem. Lett.* 30: 1667–1681. doi: 10.1016/j.cclet.2018.12.012.

40 Shimomura, N., Egawa, Y., Miki, R. et al. (2016). A red fluorophore comprising a borinate-containing xanthene analogue as a polyol sensor. *Org. Biomol. Chem.* 14 (42): 10031–10036. doi: 10.1039/c6ob01695b.

41 Nie, H., Qiao, L., Yang, W. et al. (2016). UV-assisted synthesis of long-wavelength Si-pyronine fluorescent dyes for real-time and dynamic imaging of glutathione fluctuation in living cells. *J. Mater. Chem. B* 4: 4826–4831. doi: 10.1039/c6tb00938g.

42 Fu, M., Xiao, Y., Qian, X. et al. (2008). A design concept of long-wavelength fluorescent analogs of rhodamine dyes: replacement of oxygen by silicon atom. *Chem. Commun.* 15: 1780–1782. doi: 10.1039/b718544h.

43 Horváth, P., Šebej, P., Šolomek, T., and Klán, P. (2015). Small-molecule fluorophores with large Stokes shifts: 9-iminopyronin analogues as clickable tags. *J. Org. Chem.* 80: 1299–1311.

44 Bedrick, A.E., Painter, J.L., and Bonitz, D.L. (1979). Spectral properties and relative quantum yields of new fluorescent stains. *Immunopharmacology* 2 (1): 33–37. doi: 10.1016/0162-3109(79)90018-3.

45 Kitamura, M., Murakami, K., Yamada, K. et al. (2013). Binding of sulforhodamine B to human serum albumin: a spectroscopic study. *Dyes Pigm.* 99 (3): 588–593. doi: 10.1016/j.dyepig.2013.06.011.

46 Whitaker, J.E., Haugland, R.P., Moore, P.L. et al. (1991). Cascade blue derivatives: water soluble, reactive, blue emission dyes evaluated as fluorescent labels and tracers. *Anal. Biochem.* 198 (1): 119–130. doi: 10.1016/0003-2697(91)90515-U.

47 Chevalier, A., Renard, P.-Y., and Romieu, A. (2014). Straightforward access to water-soluble unsymmetrical sulfoxanthene dyes: application to the preparation of far-red fluorescent dyes with large Stokes shifts. *Chem. – A Eur. J.* 20: 8330–8337.

48 Lavis, D.L. (2020). Personal communication to the social media. https://twitter.com/rhodamine110/status/1245924350173171716/photo/1 (accessed 12 February 2022).

49 Sjöback, R., Nygren, J., and Kubista, M. (1995). Absorption and fluorescence properties of fluorescein. *Spectrochim. Acta A* 51: L7–L21. doi: 10.1016/0584-8539(95)01421-P.

50 Sun, W.-C., Gee, K.R., Klaubert, D.H., and Haugland, R.P. (1997). Synthesis of fluorinated fluoresceins. *J. Org. Chem.* 62: 6469–6475.

51 Urano, Y., Kamiya, M., Kanda, K. et al. (2005). Evolution of fluorescein as a platform for finely tunable fluorescence probes. *J. Am. Chem. Soc.* 127: 4888–4894. doi: 10.1021/ja043919h.

52 Mottram, L.F., Maddox, E., Schwab, M. et al. (2007). A concise synthesis of the Pennsylvania Green fluorophore and labeling of intracellular targets with O^6-benzylguanine derivatives. *Org. Lett.* 9: 3741–3744.

53 Tanaka, K., Miura, T., Umezawa, N. et al. (2001). Rational design of fluorescein-based fluorescence probes. Mechanism-based design of a maximum fluorescence probe for singlet oxygen. *J. Am. Chem. Soc.* 123: 2530–2536.

54 Ge, F.Y. and Chen, L.-G. (2008). pH fluorescent probes: chlorinated fluoresceins. *J. Fluoresc.* 18: 741–747. doi: 10.1007/s10895-007-0305-y.

55 Han, J. and Burgess, K. (2010). Fluorescent indicators for intracellular pH. *Chem. Rev.* 110 (5): 2709–2728. doi: 10.1021/cr900249z.

56 Lee, L.G., Berry, G.M., and Chen, C.-H. (1989). Vita blue: a new 633-nm excitable fluorescent dye for cell analysis. *Cytometry* 10: 151–164.

57 Boens, N., Qin, W., Nikola Basaric, N. et al. (2006). Photophysics of the fluorescent pH indicator BCECF. *J. Phys. Chem. A* 110: 9334–9343.

58 Liu, J., Diwu, Z., and Klaubert, D.H. (1997). Fluorescent molecular probes III. 2′,7′-Bis-(3-carboxypropyl)-5-(and-6)-carboxyfluorescein (BCPCF): a new polar dual-excitation and dual-emission pH indicator with a pK_a of 7.0. *Bioorganic Med. Chem. Lett.* 7 (23): 3069–3072.

59 Senders, J.T., Muskens, I.S., Schnoor, R. et al. (2017). Agents for fluorescence-guided glioma surgery: a systematic review of preclinical and clinical results. *Acta Neurochir.* 159: 151–167. doi: DOI 10.1007/s00701-016-3028-5.

60 Kazarian, A.A., Smith, J.A., Hilder, E.F. et al. (2010). Development of a novel fluorescent tag O-2-[aminoethyl] fluorescein for the electrophoretic separation of oligosaccharides. *Anal. Chim. Acta* 662 (2): 206–213.

61 Ogamo, A., Matsuzaki, K., Uchiyama, H., and Nagasawa, K. (1982). Preparation and properties of fluorescent glycosaminoglycuronans labeled with 5-aminofluorescein. *Carbohydr. Res.* 105: 69–85.

62 Mottram, L.F., Boonyarattanakalin, S., Kovel, R.E., and Peterson, B.R. (2006). The Pennsylvania Green fluorophore: a hybrid of Oregon Green and Tokyo Green for the construction of hydrophobic and pH-insensitive molecular probes. *Org. Lett.* 8 (4): 581–584. doi: 10.1021/ol052655g.

63 Nunnally, B.K., He, H., Li, L.-C. et al. (1997). Characterization of visible dyes for four-decay fluorescence detection in DNA sequencing. *Anal. Chem.* 69 (13): 2392–2397. doi: 10.1021/ac961281p.

64 Chuvilin, A.N., Smirnov, I.P., Mosina, A.G. et al. (2016). A solution to the common problem of the synthesis and applications of hexachlorofluorescein labeled oligonucleotides. *PLOS One* 1–18. doi: 10.1371/journal.pone.0166911.

65 Glen Research (2017). Glen Research, Sterling, VA, USA. Datasheet accessed August 2017.

66 Gee, K.R., Poot, M., Klaubert, D.U. et al. (1997). Fluorinated Xanthene Derivatives. European Patent EP 0 853 647 B2, international application number PCT/US1997/006090, filed 11 April 1997.

67 Klaubert, D.H. and Gee, K.R. (2000). *Synthesis of Fluorinated Xanthene Derivatives.* US Patent 6,229,055, filed 3 August 2000 and issued 8 May 2001.

68 Kotásková, M. (2012). Synthesis of New Xanthene Derivatives. PhD Thesis. Johannes Gutenberg University of Mainz.

69 Lavis, D.L. and Raines, R.T. (2008). Bright ideas for chemical biology. *ACS Chem. Biol.* 3 (3): 142–155.

70 Grimm, J.B., Gruber, T.D., Ortiz, G. et al. (2016). Virginia orange: a versatile, red-shifted fluorescein scaffold for single- and dual-input fluorogenic probes. *Bioconjugate Chem.* 27 (2): 474–480. doi: 10.1021/acs.bioconjchem.5b00566.

71 Funabiki, K., Sugiyama, N., Iida, H. et al. (2006). Ring-fluorinated fluoresceins as an organic photosensitizer for dye-sensitized solar cells using nanocrystalline zinc oxide. *J. Fluor. Chem.* 127(2): 257–262. doi: 10.1016/j.jfluchem.2005.12.009.

72 Ponyaev, A.I., Martynova, V.P., and El'tsov, A.V. (2001). Photophysical and photocatalytic parameters of sulfo and tetrabromosulfo derivatives of fluorescein. *Russ. J. Gen. Chem.* 71 (11): 1744–1750.

73 Fang, Y., Good, G.N., Zhoua, X., and Stains, C.I. (2019). Phosphinate-containing rhodol and fluorescein scaffolds for the development of bioprobes. *Chem. Commun.* 55 (42): 5962–5965. doi: 10.1039/c9cc02492a.

74 Fukazawa, A., Suda, S., Taki, M. et al. (2016). Phospha-fluorescein: a red-emissive fluorescein analogue with high photobleaching resistance. *Chem. Commun.* 52 (6): 1120–1123. doi: 10.1039/c5cc09345g.

75 Grzybowski, M., Taki, M., and Yamaguchi, S. (2017). Selective conversion of P=O-bridged rhodamines into P=O rhodols: solvatochromic near-infrared fluorophores. *Chem. Eur. J.* 23: 13028–13032.

76 Egawa, T., Hirabayashi, K., Koide, Y. et al. (2013). Red fluorescent probe for monitoring the dynamics of cytoplasmic calcium ions. *Angew Chem. Int. Ed. Engl.* 52(14): 3874–3877. doi: 10.1002/anie.201210279. Epub 25 February 2013.

77 Egawa, T., Koide, Y., Hanaoka, K. et al. (2011). Development of a fluorescein analogue, TokyoMagenta, as a novel scaffold for fluorescence probes in red region. *Chem. Commun.* 47: 4162–4164.

78 Hirabayashi, K., Hanaoka, K., Takayanagi, T. et al. (2015). Analysis of chemical equilibrium of silicon-substituted fluorescein and its application to develop a scaffold for red fluorescent probes. *Anal. Chem.* 87 (17): 9061–9069. doi: 10.1021/acs.analchem.5b02331.

79 Kushida, Y., Nagano, T., and Hanaoka, K. (2015). Silicon-substituted xanthene dyes and their applications in bioimaging. *Analyst* 140 (3): 685–695. doi: 10.1039/c4an01172d.

80 Altman, R.B., Terry, D.S., Zhou, Z. et al. (2012). Cyanine fluorophore derivatives with enhanced photostability. *Nat. Methods* 9 (1): 68–71.

81 Schreiber, M., Fulscher, M.P., and Bub, V. (2001). The electronic spectra of symmetric cyanine dyes: a CASPT2 study. *Phys. Chem. Chem. Phys.* 3: 3906–3912.

82 Cooper, M., Ebner, A., Briggs, M. et al. (2004). Cy3B^{TM}: improving the performance of cyanine dyes. *J. Fluoresc.* 14 (2): 145–150.

83 Southwick, P.L., Ernst, L.A., Tauriello, E.W. et al. (1990). Cyanine dye labeling reagents: carboxymethylindocyanine succinimidyl esters. *Cytometry* 11: 418–430.

84 Mujumdar, S.R., Mujumdar, R.B., Grant, C.M., and Waggoner, A.S. (1996). Cyanine-labeling reagents: sulfobenzindocyanine succinimidyl esters. *Bioconjugate Chem.* 7 (3): 356–362. doi: 10.1021/bc960021b.

85 Mujumdar, R.B., Ernst, L.A., Mujumdar, S.R. et al. (1993). Cyanine dye labeling reagents: sulfoindocyanine succinimidyl esters. *Bioconjugate Chem.* 4 (2): 105–111.

86 Zheng, Q., Juette, M.F., Jockusch, S. et al. (2014). Ultra-stable organic fluorophores for single-molecule research. *Chem. Soc. Rev.* 43 (4): 1044–1056. doi: 10.1039/c3cs60237k.

87 Toutchkine, A., Nalbant, P., and Hahn, K.M. (2002). Facile synthesis of thiol-reactive Cy3 and Cy5 derivatives with enhanced water solubility. *Bioconjugate Chem.* 13 (3): 387–391. doi: 10.1021/bc015558q.

88 Gebhardt, C., Lehmann, M., Reif, M.M. et al. (2021). Molecular and spectroscopic characterization of green and red cyanine fluorophores from the Alexa Fluor and AF series. *ChemPhysChem* 22: 1566–1583. doi:10.1002/cphc.202000935.

89 Lin, Y., Weissleder, R., and Tung, C.-H. (2002). Novel near-infrared cyanine fluorochromes: synthesis, properties, and bioconjugation. *Bioconjugate Chem.* 13: 605–610. doi: 10.1021/bc0155723.

90 Hovor, I.V., Kolosova, O.S., Sanin, E.V. et al. (2019). Water-soluble norsquaraine dyes for protein labeling and pH-sensing applications. *Dyes Pigm.* 170: 107567.

91 McCann, T.E., Kosaka, N., Koide, Y. et al. (2011). Activatable optical imaging with a silica-rhodamine based near infrared (SiR700) fluorophore: a comparison with cyanine based dyes. *Bioconjugate Chem.* 22 (12): 2531–2538. doi: 10.1021/bc2003617.

92 Lumiprobe (2017). Lumiprobe GmbH, Hannover, Germany. Web datasheet accessed 8 October 2017.

93 Seta Biomedicals (2022). Specification sheet. www.setabiomedicals.com (accessed April 2020).

94 Wycisk, V., Pauli, J., Welker, P. et al. (2015). Glycerol-based contrast agents: a novel series of dendronized pentamethine dyes. *Bioconjugate Chem.* 26: 773–781. doi: 10.1021/acs.bioconjchem.5b00097.

95 Swamy, A.R., Mason, J.C., Lee, H. et al. (2006). Near-infrared absorption/luminescence measurements. In: *Encyclopedia of Analytical Chemistry* (ed. R.A. Meyers). 1–35. Hoboken: Wiley.

96 Choi, H.S., Nasr, K., Alyabyev, S. et al. (2011). Synthesis and *in vivo* fate of Zwitterionic near-infrared fluorophores. *J. Am. Chem. Soc.* 123: 2530–2536.

97 Bertolino, C.A., Caputo, G., Barolo, C. et al. (2006). Novel heptamethine cyanine dyes with large Stokes shift for biological applications in the near infrared. *J. Fluoresc.* 16: 221–225. doi: 10.1007/s10895-006-0094-8.

98 Pauli, J., Brehm, R., Spieles, M. et al. (2010). Novel fluorophores as building blocks for optical probes for *in vivo* near infrared fluorescence (NIRF) imaging. *J. Fluoresc.* 20 (3): 681–693. doi: 10.1007/s10895-010-0603-7.

99 Soper, S.A. and Mattingly, Q.L. (1994). Steady-state and picosecond laser fluorescence studies of nonradiative pathways in tricarbocyanine dyes: implications to the design of near-IR fluorochromes with high fluorescence efficiencies. *J. Am. Chem. Soc.* 116: 3744–3752.

100 Nakayama, A., Bianco, A.C., Zhang, C.-Y. et al. (2003). Quantitation of brown adipose tissue perfusion in transgenic mice using near-infrared fluorescence imaging. *Mol. Imaging* 2 (1): 37–49.

101 Interchim (2019). Moutinçon, France. FluoProbes catalog accessed on August 2019.

102 Berezin, M.Y., Guo, K., Akers, W. et al. (2011). Near-infrared fluorescence lifetime pH-sensitive probes. *Biophys. J.* 100: 2063–2072.

103 Ye, Y., Bloch, S., Kao, J., and Achilefu, S. (2005). Multivalent carbocyanine molecular probes: synthesis and applications. *Bioconjugate Chem.* 16 (1): 51–61. doi: 10.1021/bc049790i.

104 Loudet, A. and Burgess, K. (2007). BODIPY dyes and their derivatives: syntheses and spectroscopic properties. *Chem. Rev.* 107: 4891–4932.

105 García-Moreno, I., Wang, L., Costela, A. et al. (2012). Synthesis and optical and redox properties of symmetric and asymmetric BODIPYs. *ChemPhysChem* 13: 3923–3931. doi: dx.doi.org/10.1002/cphc.201200587.

106 Umezawa, K., Citterio, D., and Suzuki, K. (2014). New trends in near-infrared fluorophores for imaging. *Anal. Sci.* 30: 327–349.

107 Bañuelos, J. (2016). BODIPY dye, the most versatile fluorophore ever? *Chem. Rec.* 16: 335–348. doi: 10.1002/tcr.201500238.

108 Tram, K., Yan, H., Jenkins, H.A. et al. (2009). The synthesis and crystal structure of unsubstituted 4,4-difluoro-4-bora-3a,4a-diaza-s-indacene (BODIPY). *Dyes Pigm.* 82: 392–395.

109 Arroyo, I.J., Hu, R., Merino, G. et al. (2009). The smallest and one of the brightest. Efficient preparation and optical description of the parent borondipyrromethene system E. *J. Org. Chem.* 74: 5719–5722.

110 Petrushenko, I.K. and Petrushenko, K.B. (2015). Effect of meso-substituents on the electronic transitions of BODIPY dyes: DFT and RI-CC2 study. *Spectrochim. Acta. A: Mol. Biomol. Spectrosc.* 138: 623–627.

111 Isaksson, M., Kalinin, S., Lobov, S. et al. (2003). An environmental-sensitive BODIPY®-derivative with bioapplication: spectral and photophysical properties. *J. Fluoresc.* 13: 379–384.

112 Buschmann, V., Weston, K.D., and Sauer, M. (2003). Spectroscopic study and evaluation of red-absorbing fluorescent dyes. *Bioconjugate Chem.* 14 (1): 195–204. doi: 10.1021/bc025600x.

113 Gabe, Y., Urano, Y., Kikuchi, K. et al. (2004). Highly sensitive fluorescence probes for nitric oxide based on boron dipyrromethene chromophore: rational design of potentially useful bioimaging fluorescence probe. *J. Am. Chem. Soc.* 126 (10): 3357–3367. doi: 10.1021/ja037944j.

114 Wories, H.J., Koek, J.H., Lodder, G. et al. (1985). A novel water-soluble fluorescent probe: synthesis, luminescence and biological properties of the sodium salt of the 4-sulfonato-3,3′,5,5′-tetramethyl-2,2′-pyrromethen-1,1′-BF_2 complex. *Recl. Trav. Chim. Pays-Bas* 104: 288–291.

115 Shah, M., Thangaraj, K., Soong, M.-L. et al. (1990). Pyrromethene–BF2 complexes as laser dyes: 1. *Heteroat. Chem.* 1 (5): 389–399.

116 Li, L., Han, J., Nguyen, B., and Burgess, K. (2008). Functionalization of the 4,4-difluoro-4-bora-3a,4a-diaza-s-indacene (BODIPY) core. *Bioorganic Med. Chem. Lett.* 18: 3112–3116.

117 Jeon, S., Kim, T.-I., Jin, H. et al. (2020). Amine-reactive activated esters of meso-CarboxyBODIPY: fluorogenic assays and labeling of amines, amino acids, and proteins. *J. Am. Chem. Soc.* 142: 9231–9239.

118 Goméz, A.M. and Lopéz, J.C. (2019). Carbohydrates and BODIPYs: access to bioconjugatable and water-soluble BODIPYs. *Pure Appl. Chem.* 91(7): 1073–1083. doi: 10.1515/pac-2019-0204.

119 Bura, T. and Ziessel, R. (2011). Water-soluble phosphonate-substituted BODIPY derivatives with tunable emission channels. *Org. Lett.* 13 (12): 3072–3075. doi: 10.1021/ol200969r.

120 Blázquez-Moraleja, A., Álvarez-Fernández, D., Montero, R.P. et al. (2019). A general modular approach for the solubility tagging of BODIPY dyes. *Dyes Pigm.* 170: 107545.

121 Singha, S., Kim, D., Roy, B. et al. (2015). A structural remedy toward bright dipolar fluorophores in aqueous media. *Chem. Sci.* 6: 4335–4342. doi: 10.1039/c5sc01076d.

122 Peng, T. and Yang, D. (2010). Construction of a library of rhodol fluorophores for developing new fluorescent probes. *Org. Lett.* 12 (3): 496–499. doi: 10.1021/ol902706b.

123 Kamiya, M., Asanuma, D., Kuranaga, E. et al. (2011). β-galactosidase fluorescence probe with improved cellular accumulation based on a spirocyclized rhodol scaffold. *J. Am. Chem. Soc.* 133: 12960–12963. doi: 10.1021/ja204781t.

124 Jiao, X., Liu, C., Huang, K., Zhang, S. et al. (2015). Molecular design and synthesis of a pH independent and cell permeant fluorescent dye and its applications. *Org. Biomol. Chem.* 13: 6647–6653. doi: 10.1039/c5ob00448a.

125 Whitaker, J.E., Haugland, R.P., and Prendergasty, F.G. (1991). Spectral and photophysical studies of benzo[c] xanthene dyes: dual emission pH sensors. *Anal. Biochem.* 194 (2): 330–344.

126 Sednev, M.V., Wurm, C.A., Belov, V.N., and Hell, S.W. (2013). Carborhodol: a new hybrid fluorophore obtained by combination of fluorescein and carbopyronine dye cores. *Bioconjugate Chem.* 24 (4): 690–700. doi: 10.1021/bc3006732.

127 Poronik, Y.M., Vygranenko, K.V., Gryko, D., and Gryko, D.T. (2019). Rhodols: synthesis, photophysical properties and applications as fluorescent probes. *Chem. Soc. Rev.* 48: 5242–5265.

128 Ito, H., Kawamata, Y., Kamiya, M. et al. (2018). Red-shifted fluorogenic substrate for detection of lacZ-positive cells in living tissue with single-cell resolution. *Angew. Chem. Int. Ed.* 57: 15702–15706. doi: 10.1002/anie.201808670.

129 Ichikawa, Y., Kamiya, M., Obata, F. et al. (2014). Selective ablation of b-galactosidase-expressing cells with a rationally designed activatable photosensitizer. *Angew. Chem. Int. Ed.* 53: 6772–6775. doi: 10.1002/anie.201403221.

130 Piao, W., Tsuda, S., Tanaka, Y. et al. (2013). Development of azo-based fluorescent probes to detect different levels of hypoxia. *Angew. Chem. Int. Ed.* 52: 13028–13032. doi: 10.1002/anie.201305784.

131 Zhou, X., Lesiak, L., Lai, R. et al. (2017). Chemoselective alteration of fluorophore scaffolds as a strategy for the development of ratiometric chemodosimeters. *Angew. Chem. Int. Ed.* 56 (15): 4197–4200. doi: 10.1002/anie.201612628.

132 Panchuk-Voloshina, N., Haugland, R.P. et al. (1999). Alexa dyes, a series of new fluorescent dyes that yield exceptionally bright, photostable conjugates. *J Histochem. Cytochem.: Offic. J. Histochem. Soc.* 47 (9): 1179–1188. doi: 10.1177/002215549904700910.

133 Koide, Y., Urano, Y., Hanaoka, K. et al. (2012). Development of NIR fluorescent dyes based on Si-rhodamine for *in vivo* imaging. *J. Am. Chem. Soc.* 2012 (134): 5029–5031.

134 Kushida, Y., Hanaoka, K., Komatsu, T. et al. (2012). Red fluorescent scaffold for highly sensitive protease activity probes. *Bioorg. Med. Chem. Lett.* 22: 3908–3911.

135 Hanaoka, K., Kagami, Y., Piao, W. et al. (2018). Synthesis of unsymmetrical Si-rhodamine fluorophores and application to a far-red to near-infrared fluorescence probe for hypoxia. *Chem. Commun.* 54: 6939–6942. doi: 10.1039/C8CC02451K.

136 Zhou, X., Lai, R., Beck, J.R. et al. (2016). Nebraska Red: a phosphinate-based near-infrared fluorophore scaffold for chemical biology applications. *Chem. Commun.* 52 (83): 12290–12293. doi: 10.1039/c6cc05717a.

137 Chai, X., Cui, X., Wang, B. et al. (2015). Near-infrared phosphorus-substituted rhodamine with emission wavelength above 700 nm for bioimaging. *Chem. A Eur. J.* 21 (47): 16754–16758. doi: 10.1002/chem.201502921.

138 Koide, Y., Kawaguchi, M., Urano, Y. et al. (2012a). A reversible near-infrared fluorescence probe for reactive oxygen species based on Te-rhodamine. *Chem. Commun.* 48 (25): 3091–3093. doi: 10.1039/c2cc18011a.

139 Liu, J., Sun, Y.-Q., Zhang, H. et al. (2016). Sulfone-rhodamines: a new class of near-infrared fluorescent dyes for bioimaging. *ACS Appl. Mater. Interfaces* 8 (35): 22953–22962. doi: 10.1021/acsami.6b08338.

140 Hettie, K.S., Klockow, J.L., Glass, T.E., and Chin, F.T. (2019). Near-infrared fluorescent rosol dye tailored toward lymphatic mapping applications. *Anal. Chem.* 91: 3110–3117.

141 Vygranenko, K.V., Poronik, Y.M., Bousquet, M.H.E. et al. (2022). Direct transformation of coumarins into orange-red emitting rhodols. *Chem. Commun.* 58: 1542–1545. doi: 10.1039/D1CC02687A.

142 Vygranenko, K.V., Poronik, Y.M., Wrzosek, A. et al. (2021). Red emissive sulfone-rhodols as mitochondrial imaging agents. *Chem. Commun.* 57 (63): 7782–7785.

143 Miura, T., Urano, Y., Tanaka, K. et al. (2003). Rational design principle for modulating fluorescence properties of fluorescein-based probes by photoinduced electron transfer. *J. Am. Chem. Soc.* 125: 8666–8671. doi: 10.1021/JA035282S.

144 Bisballe, N. and Laursen, B.W. (2020). What is best strategy for water soluble fluorescence dyes? A case study using long fluorescence lifetime DAOTA dyes. *Chem. Eur. J.* 26: 15969–15976.

145 Albert, A. (1965). Acridine syntheses and reactions. Part VI. A new dehalogenation of 9-chloroacridine and its derivatives. Further acridine ionization constants and ultraviolet spectra. *J. Chem. Soc.* 1965: 4653–4657. doi: 10.1039/JR9650004653.

146 Strickler, S.J. and Berg, R.A. (1962). Relationship between absorption intensity and fluorescence lifetime of molecules. *J. Chem. Phys.* 37: 814–822. doi: 10.1063/1.1733166.

147 Du, H., Fuh, R.-C.A., Li, J. et al. (1998). Technical and software note photochemCAD: a computer-aided design and research tool. *Photochemistry* 68 (2): 141–142.

148 Melhuish, W.H. (1964). Measurement of quantum efficiencies of fluorescence and phosphorescence and some suggested luminescence standards. *J. Opt. Soc. Am.* 54 (2): 183–186. doi: Doi 10.1364/Josa.54.000183.

149 Aygün, S., Beşer, B.M., Acar, M., and Meral, K. (2020). Photophysical properties of some fluorescent dyes in SDS micellar solutions. *J. Fluoresc.* 30: 849–857. doi: 10.1007/s10895-020-02553-8.

150 Shaikh, M., Mohanty, J., Singh, P.K. et al. (2008). Complexation of acridine orange by cucurbit[7]uril and b-cyclodextrin: photophysical effects and pK_a shifts. *Photochem. Photobiol. Sci.* 7: 408–414.

151 Hinze, W.L., Riehl, T.E., Singh, N.H., and Baba, Y. (1984). Micelle-enhanced chemiluminescence and application to the determination of biological reductants using lucigenin. *Anal. Chem.* 56: 2180–2191.

152 Pereira, R.C., Pontinha, A.D.R., Pineiro, M., and de Melo, J.S.S. (2019). A comprehensive spectral, photophysical and electrochemical study of synthetic water-soluble acridones. A new class of pH and polarity sensitive fluorescent probes. *Dyes Pigm.* 166: 203–210.

153 Lowe, A.M., Ozer, B.H., Bai, Y. et al. (2007). Experimental and DFT studies: novel structural modifications greatly enhance the solvent sensitivity of live cell imaging dyes. *J. Phys. Chem. A* 111: 10849–10860.

154 Eicher, T., Hauptmann, S., and Speicher, A. (2012). *The Chemistry of Heterocycles: Structures, Reactions, Synthesis, and Applications,* 3e, completely revised and enlarged edition. Weinheim: Wiley-VCH.

155 Gvishi, R. and Reisfeld, R. (1989). An investigation of the equilibrium between various forms of oxazine-170 by means of absorption and fluorescence spectroscopy. *Chem. Phys. Lett.* 156 (2–3): 181–186. doi: 10.1016/S0009-2614(89)87116-7.

156 Baker, J.G., Middleton, R., Adams, L. et al. (2010). Influence of fluorophore and linker composition on the pharmacology of fluorescent adenosine A1 receptor ligands. *Br. J. Pharmacol.* 159: 772–786.

157 Frade, V.H.J., Coutinho, P.J.G., Moura, J.C.V.P. et al. (2007). Functionalised benzo[a]phenoxazine dyes as long-wavelength fluorescent probes for amino acids. *Tetrahedron* 63: 1654–1663.

158 Tasior, M., Kim, D., Singha, S. et al. (2015). π-Expanded coumarins: synthesis, optical properties and applications. *J. Mater. Chem. C* 3: 1421–1446.

159 Sun, W.-C., Gee, K.R., and Haugland, R.P. (1998). Synthesis of novel fluorinated coumarins: excellent UV-light excitable fluorescent dyes. *Bioorganic Med. Chem. Lett.* 8 (22): 3107–3110.

160 Taniguchi, M. and Lindsey, J.S. (2018). Database of absorption and fluorescence spectra of >300 common compounds for use in photochemCAD. *Photochem. Photobiol.* 94 (2): 290–327. doi: 10.1111/php.12860.

161 Kist, T.B.L. (2002). Fluorescence mean-lifetime measured using a time-correlated single photon counting instrument from PicoQuant (Berlin, Germany).

162 Păunescu, E., Louise, L., Jean, L. et al. (2011). A versatile access to new halogenated 7-azidocoumarins for photoaffinity labeling: synthesis and photophysical properties. *Dyes Pigm.* 91: 427–434.

163 Jones II, G., Jackson, W.R., Choi, C.-Y., and Bergmark, W.R. (1985). Solvent effects on emission yield and lifetime for coumarin laser dyes. Requirements for a rotatory decay mechanism. *J. Phys. Chem.* 89: 294–300.

164 Nizamov, S., Willig, K.I., Sednev, M.V. et al. (2012). Phosphorylated 3-heteroarylcoumarins and their use in fluorescence microscopy and nanoscopy. *Chem. Eur. J.* 18: 16339–16348.

165 Sednev, M. (2015). Fluorescent Dyes with Large Stokes Shifts of 80—200 nm for Optical Microscopy and Nanoscopy. PhD Thesis. Georg-August University, Göttingen.

166 Schill, H., Nizamov, S., Bottanelli, F. et al. (2013). 4-Trifluoromethyl-substituted coumarins with large Stokes shifts: synthesis, bioconjugates, and their use in super-resolution fluorescence microscopy. *Chem. – A Eur. J.* 19 (49): 16556–16565. doi: 10.1002/chem.201302037.

167 Sednev, M.V., Belov, V.N., and Hell, S.W. (2015). Fluorescent dyes with large Stokes shifts for super-resolution optical microscopy of biological objects: a review. *Methods Appl. Fluoresc.* 3: 042004.

168 Sun, X., Liu, T., Sun, J., and Wang, X. (2020). Synthesis and application of coumarin fluorescence probes. *RSC Adv.* 10: 10826–10847.

169 Tran-Thi, T.-H., Prayer, C., Millie, P. et al. (2002). Substituent and solvent effects on the nature of the transitions of pyrenol and pyranine. Identification of an intermediate in the excited-state proton-transfer reaction. *J. Phys. Chem. A* 106: 2244–2255.

170 Kist, T.B.L. (2021). https://youtu.be/X7ol2g-7BxE and https://youtu.be/OOn5ma12bpc.

171 Finkler, B., Spies, S., Vester, M. et al. (2014). Highly photostable "super"-photoacids for ultrasensitive fluorescence spectroscopy. *Photochem. Photobiol. Sci.* 13: 548–562.

172 Shang, X.F., Morris-Natschke, S.L., Yang, G.Z. et al. (2018). Biologically active quinoline and quinazoline alkaloids, Part II. *Med. Res. Rev.* 38: 1614–1660. doi: 10.1002/med.21492.

173 Stokes, G.G. (1852). On the change of refrangibility of light. *Phil. Trans.* 142: 463–562.

174 Eaton, D.F. (1988). Reference materials for fluorescence measurement. *Pure Appl. Chem.* 60 (7): 1107–1114.

175 Nawara, K., and Waluk, J. (2019). Goodbye to quinine in sulfuric acid solutions as a fluorescence quantum yield standard. *Anal. Chem.* 91Ç: 5389–5394.

176 Guo, Y., Cao, F., Qiu, P., and Wang, Z. (2019). Studies of the effect of halide ions on the fluorescence of quinine sulfate. *Luminescence* 34: 450–455. doi: 10.1002/bio.3627.

177 Uray, G., Strohmeier, G.A., and Fabian, W.M.F. (2000). sSrongly fluorescent push-pull substituted carbostyrils absorbing in the visible. *Proceedings of ECSOC-4, The Fourth International Electronic Conference on Synthetic Organic Chemistry*, 1–30 September. http://www.mdpi.org/ecsoc-4.htm.

178 Strohmeier, G.A., Fabian, W.M.F., and Uray, G. (2004). A combined experimental and theoretical approach toward the development of optimized luminescent carbostyrils. *Helv. Chim. Acta* 87: 215–226.

179 Kelterer, A.-M., Uray, G., and Fabian, W.M.F. (2015). Rational design of long-wavelength absorbing and emitting carbostyrils aided by time-dependent density functional calculations. *Comput. Theor. Chem.* 1055: 25–32.

180 Chou, P.T and Ching-Yen Wei, C.Y. (1996). Photophysics of 10-Hydroxybenzo[h]quinoline in aqueous solution. *J. Phys. Chem.* 100: 17059–17066.

181 Morton, R.A., Hopkinsj, T.A., and Seligerg, H.H. (1969). The spectroscopic properties of firefly luciferin and related compounds. An approach to product emission. *Biochemistry* 8 (4): 1598–1607. doi: 10.1021/bi00832a041.

182 Abrams, B., Diwu, Z., Guryev, O. et al. (2013). New violet-excitable reagents for multicolor flow applications. *Cytometry A* 83A: 752–762. doi: 10.1002/cyto.a.22309.

183 Barros, H.L., Marques, S.M.T., and Stefani, V. (2019). Water-soluble benzoxazole dyes fluorescent by ESIPT: structural characterization, photophysical properties and its application as a probe for direct staining of helminths. *J. Braz. Chem. Soc.* 30: 509–521. doi: 10.21577/0103-5053.20180193.

184 Barros, H.L. and Stefani, V. (2016). A new methodology for the visualization of latent fingermarks on the sticky side of adhesive tapes using novel fluorescent dyes. *Forensic Sci. Int.* 263: 83–91. doi: 10.1016/j.forsciint.2016.03.053.

185 Miao, J., Cui, H., Jin, J. et al. (2015). Development of 3-alkyl-6-methoxy-7-hydroxy-chromones (AMHCs) from natural isoflavones, a new class of fluorescent scaffolds for biological imaging. *Chem. Commun.* 51 (5): 881–884. doi: 10.1039/c4cc06762b.

186 Gaspar, A., Matos, M.J., Garrido, J. et al. (2014). Chromone: a valid scaffold in medicinal chemistry. *Chem. Rev.* 114: 4960–4992. doi: 10.1021/cr400265z.

187 Desfosses, B., Urios, P., Christeff, N. et al. (1986). The use of disulfonatonaphthalimide fluorescent dyes for the fluorescence polarization immunoassay of steroids. *Anal. Biochem.* 159 (1): 179–186. doi: 10.1016/0003-2697(86)90325-8.

188 Hanani, M. (2012). Lucifer yellow: an angel rather than the devil. *J. Cell. Mol. Med.* 16 (1): 22–31. doi: 10.1111/j.1582-4934.2011.01378.x.

189 Stewart, W.W. (1978). Functional connections between cells as revealed by dye-coupling with a highly fluorescent naphthalimide tracer. *Cell* 14: 741–759.

190 Aeby, P.G. (1998). Quantitative fluorescence imaging of tracer distributions in soil profiles. PhD Thesis. Swiss Federal Institute of Technology, Zurick.

191 Kelly, L.A., Roll, M., Joseph, J. et al. (2021). Solvent-dependent photophysics and reactivity of monomeric and dimeric 4-amino-1,8-naphthalimides. *J. Phys. Chem. A* 125: 2294–2307. doi: 10.1021/acs.jpca.0c11639.

192 Banerjee, S., Kitchen, J.A., Gunnlaugsson, T., and Kelly, J.M. (2012). Synthesis and photophysical evaluation of a pyridinium 4-amino-1,8-naphthalimide derivative that upon intercalation displays preference for AT-rich double-stranded DNA. *Org. Biomol. Chem.* 10: 3033. doi: 10.1039/c2ob06898b.

193 Rasse-Suriani, F.A.O., Denofrio, M.P., Yanuk, J.G. et al. (2016). Chemical and photochemical properties of chloroharmine derivatives in aqueous solutions. *Phys. Chem. Chem. Phys.* 18: 886–900. doi: 10.1039/C5CP05866J.

194 Nafisi, S., Panahyab, A., and Sadeghi, G.B. (2012). Interactions between β-carboline alkaloids and bovine serum albumin: investigation by spectroscopic approach. *J. Luminesc.* 132 (9): 2361–2366. doi: 10.1016/j.jlumin.2012.03.075.

195 Lakowicz, J.R. (2006). *Principles of Fluorescence Spectroscopy Principles of Fluorescence Spectroscopy, Principles of Fluorescence Spectroscopy*, 3e. Singapore: Springer. doi: 10.1007/978-0-387-46312-4.

196 Pardo, A., Reyman, D., Poyato, J.M.L., and Medina, F. (1992). Some β-carboline derivatives as fluorescence standards. *J. Luminesc.* 51: 269–274.

197 Savory, B. and Turnbull, J.H. (1984). Luminescence spectra of harmine, harmaline, reserpiline, and 3-dehydroreserpines. *J. Photochem.* 24: 355–371.

198 Booth, G. (2012). *Naphthalene derivatives.* In: *Ullmann's Encyclopedia of Industrial Chemistry*, 7e (ed. G. Bellussi). Hoboken: Wiley-VCH.

199 Weber, G. and Farris, F.J. (1979). Synthesis and spectral properties of a hydrophobic fluorescent probe: 6-propionyl-2-(dimethylamino)naphthalene. *Biochemistry* 18 (14): 3075–3078. doi: 10.1021/bi00581a025.

200 Kawski, A., Bojarski, P., and Kukliński, B. (2002). Excitation wavelength dependence of ACRYLODAN fluorescence spectra in some polar solvents. *Zeitschrift Fur Naturforschung – Sec A J. Phys. Sci.* 57 (1–2): 94–97. doi: 10.1515/zna-2002-1-215.

201 Prendergast, F.G., Meyer, M., Carlson, G.L. et al. (1983). Synthesis, spectral properties, and use of 6-acryloyl-2-dimethylaminonaphthalene (Acrylodan). *J. Biol. Chem.* 258: 7541–7544.

202 Haugland, R.P., Spence, M.T.Z., Johnson, I.D., and Basey, A. (2005). *The Handbook – A Guide to Fluorescent Probes and Labeling Technologies*, 10e. Oregon: Molecular Probes Inc.

203 Singh, P., Choudhury, S., Singha, S. et al. (2017). A sensitive fluorescent probe for the polar solvation dynamics at protein–surfactant interfaces. *Phys. Chem. Chem. Phys.* 19: 12237–12245. doi: 10.1039/c6cp08804j.

204 Vequi-Suplicy, C.C., Coutinho, K., and Lamy, M.T. (2013). Optical characterization of Prodan aggregates in water medium. *Phys. Chem. Chem. Phys.* 15: 11800–11807.

205 Ferreira, D.P., Conceição, D.S., Ferreira, V.R.A. et al. (2013). Photochemical properties of squarylium cyanine dyes. *Photochem. Photobiol. Sci.* 12: 1948–1959. doi: 10.1039/c3pp50132a.

206 He, J., Jo, Y.J., Sun, X. et al. (2021). Squaraine dyes for photovoltaic and biomedical applications. *Adv. Funct. Mater.* 31: 2008201. doi: 10.1002/adfm.202008201.

207 Xiao, S., Fu, N., Peckham, K., and Smith, B.D. (2010). Efficient synthesis of fluorescent squaraine rotaxane dendrimers. *Org. Lett.* 12: 140–143. doi: 10.1021/ol902546m.

208 Gassensmith, J.J., Baumes, J.M., and Smith, B.D. (2009). Discovery and early development of squaraine rotaxanes. *Chem. Commun.* 2009 (42): 6329–6338. doi: 10.1039/b911064j.

209 Tatarets, A.L., Fedyunyayeva, I.A., Dyubko, T.S. et al. (2006). Synthesis of water-soluble, ring-substituted squaraine dyes and their evaluation as fluorescent probes and labels. *Anal. Chim. Acta* 570: 214–223.

210 Whitby, O.S. (1953). A new method for preparing flavin-adenine dinucleotide. *Biochem. J.* 54: 437–442.

211 Wolfbeis, O.S. (1985). The fluorescence of organic natural products. In: *Molecular Luminescence: Methods and Applications – Part 1* (ed. S.G. Schulman). New York: Wiley.

212 Saleh, N., Graham, J., Afaneh, A. et al. (2012). Pteridine-based fluorescent pH sensors designed for physiological applications. *J. Photochem. Photobiol. A: Chem.* 247: 63–73. doi: 10.1016/j.jphotochem.2012.08.002.

213 Skalski, B., Paszyc, S., Adamiak, R.W. et al. (1989). Photophysical studies of luminarosine: a new, highly fluorescent ribonucleoside with pteridine-like betaine as the aglycone. *J. Chem. Soc. Perkin. Trans. II* 1691–1696. doi: 10.1039/P29890001691.

214 Visser, A.J.W.G. and Miiller, F. (1979). Absorption and fluorescence studies on neutral and cationic isoalloxazines. *Helv. Chim. Acta* 62: 593–608.

215 Belyy, A.Y., Platonov, D.N., Salikov, R.F. et al. (2021). Electron deficient 5-hydroxy-1,2-dihydroisoquinolin-1-ones: a new class of fluorescent dyes with large Stokes shifts. *Dyes Pigm.* 187: 109107.

216 Joshi, H.C., Gooijer, C., and Zwan, G.V.D. (2003). Substitution effects on the photophysical characteristics of the salicylic anion. *J. Fluoresc.* 13 (3): 227–234.

217 Beppu, T., Tomiguchi, K., Masuhara, A. et al. (2015). Single benzene green fluorophore: solid-state emissive, water-soluble, and solvent- and ph-independent fluorescence with large Stokes shifts. *Angew. Chem. Int. Ed.* 54 (25): 7332–7335. doi: 10.1002/anie.201502365.

218 Mochizuki, M., Hoshino, N., Sato, A. et al. (2022). BMeS-p-A succinimidyl ester as a sulfonyl aniline dye labeling reagent. *Tetrahedron Lett.* 90: 153625. doi: 10.1016/j.tetlet.2021.153625.

219 Weisstuch, A. and Testa, A.C. (1968). Fluorescence study of aminopyridines. *J. Phys. Chem.* 72 (6): 1982–1987. doi: 10.1021/j100852a021.

220 Ihlmes, H., Engels, B., Faulhaber, K., and Lennartz, C. (2000). New dyes based on amino-substituted acridizinium salts: synthesis and exceptional photochemical properties. *Chem. Eur. J.* 6 (15): 2854–2864.

221 Jagt, R.B.C., Kheibari, M.S., and Nitz, M. (2009). The synthesis and photophysical properties of halide sensitive acridizinium dyes. *Dyes Pigm.* 81: 161–165.

222 Ihmels, H., Faulhaber, K., Sturm, C. et al. (2001). Acridizinium salts as a novel class of DNA-binding and site-selective DNA-photodamaging chromophores. *Photochem. Photobiol.* 74 (4): 505–511.

223 Heyer, E., Lory, P., Leprince, J. et al. (2015). Highly fluorescent and water-soluble diketopyrrolopyrrole dyes for bioconjugation. *Angew. Chem. Int. Ed.* 54: 2995–2999. doi: 10.1002/anie.201411274.

224 Praetorius, A., Bailey, D.M., Schwarzlose, T., and Nau, W.M. (2008). Design of a fluorescent dye for indicator displacement from cucurbiturils: a macrocycle-responsive fluorescent switch operating through a pK_a shift. *Org. Lett.* 10 (18): 5089–4092. doi: 10.1021/ol8016275.

225 Peixoto, P.A., Boulangé, A., Ball, M. et al. (2014). Design and synthesis of epicocconone analogues with improved fluorescence properties. *J. Am. Chem. Soc.* 136: 15248–15256.

226 Shaya, J., Fontaine-Vive, F., Michel, B.Y., and Burger, A. (2016). Rational design of push-pull fluorene dyes: synthesis and structure–photophysics relationship. *Chem. Eur. J.* 22: 10627–10637. doi: 10.1002/chem.201600581.

227 Wang, C., Fukazawa, A., Tanabe, Y. et al. (2018). Water-soluble phospholo[3,2-b]phosphole-*P, P*'-dioxide-based fluorescent dyes with high photostability. *Chem. Asian J.* 13: 1616–1624. doi: 10.1002/asia.201800533.

228 Lei, Z., Li, X., Luo, X. et al. (2017). Bright, stable, and biocompatible organic fluorophores absorbing/emitting in the deep near-infrared spectral region. *Angew. Chem.* 129: 3025–3029.

229 Schlüter, F., Riehemann, K., Kehr, N.S. et al. (2018). A highly fluorescent water soluble spirobifluorene dye with a large Stokes shift: synthesis, characterization and bio-applications. *Chem. Commun.* 54: 642–645. doi: 10.1039/c7cc08761f.

230 Huang, K., Jiao, X., Liu, C.B. et al. (2017). Synthesis of a novel π-extended hybrid rhodamine dye with far-red fluorescence emission and its application in bioimaging. *Dyes Pigm.* 145: 561–569. doi: 10.1016/j.dyepig.2017.06.047.

231 Liu, X., Li, Y., Ren, X. et al. (2018). Methylated chromenoquinoline dyes: synthesis, optical properties, and application for mitochondrial labeling. *Chem. Commun.* 54: 1509–1512.

232 Modrzejewska, J., Szala, M., Grzelakowska, A. et al. (2021). Novel boronate probe based on 3-Benzothiazol-2-yl-7-hydroxychromen-2-one for the detection of peroxynitrite and hypochlorite. *Molecules* 26: 5940. doi: 10.3390/molecules26195940.

233 Kist, T.B.L. (2003). *Absorption Spectrum Measured Using a NanoDrop Spectrophotometer Model ND-1000 and Emission Spectra Measured Using Avantes Fiberoptic Spectrophotometer and Xenon Light Source.* XE–2000.

234 Thermo Fisher Scientific (2022). https://www.thermofisher.com/order/catalog/product/S6110 (accessed April 2022).

235 Majeed, S., Gao, W., Zholudov, Y., Muzyka, K., and Xu, G. (2016). Electrochemiluminescence of acridines. *Electroanalysis* 28: 2672–2679. doi: 0.1002/elan.201600209.

236 Gandioso, A., Bresolí-Obach, R., Nin-Hill, A. et al. (2018). Redesigning the coumarin scaffold into small bright fluorophores with far-red to near-infrared emission and large Stokes shifts useful for cell imaging. *J. Org. Chem.* 83: 1185–1195.

237 Biotium (2022). Biotium, Fremont, CA, USA. Datasheets accessed on April 2022 and personal communication.

238 Epstein, J.R., Biran, I., and Walt, D.R. (2002). Fluorescence-based nucleic acid detection and microarrays. *Anal. Chim. Acta* 469: 3–36.

239 Yang, Z., Cao, J., He, Y. et al. (2014). Macro-/micro-environment-sensitive chemosensing and biological imaging. *Chem. Soc. Rev.* 43: 4563–4601. doi: 10.1039/c4cs00051j.

240 Li, H., Cai, L., and Chen, Z. (2012). Coumarin-derived fluorescent chemosensors. In: *Advances in Chemical Sensors* (ed. W. Wang), Chapter 6, 121–150. IntechOpen Book Series. https://www.intechopen.com/chapters/26348 (accessed January 2022).

241 Lavis, L.D. (2017). Teaching old dyes new tricks: biological probes built from fluoresceins and rhodamines. *Annu Rev Biochem.* 86: 825–843. doi: 10.1146/annurev-biochem-061516-044839.

242 Jiang, C., Song, Z., Yu, L. et al. (2020). Fluorescent probes based on macrocyclic hosts: construction, mechanism and analytical applications. *TrAC – Trends Anal. Chem.* 133: 116086. doi: 10.1016/j.trac.2020.116086.

243 Zhang, X., Yin, J., and Yoon, J. (2014). Recent advances in development of chiral fluorescent and coloimetric sensors. *Chem. Rev.* 114: 4918–4959. doi: 10.1021/cr400568b.

244 Geddes, C.D. ed. (2010). *Metal-Enhanced Fluorescence*. Hoboken: Wiley.

245 Shaner, N.C., Steinbach, P.A., and Tsien, R.Y. (2005). A guide to choosing fluorescent proteins. *Nat. Methods* 2: 905–909. doi: 10.1038/NMETH819.

246 Fluhler, G.J., Burnham, E., and Loew, L.M. (1985). Spectra, membrane binding, and potentiometric reponses of new charge shift probes. *Biochemistry* 24: 5749–5755.

247 Bayer, M. and König, S. (2016). Pyrylium-based dye and charge tagging in proteomics. *Electrophoresis* 37: 2953–2958. doi: 10.1002/elps.201600318.

248 Höfelschweiger, B.K. (2005). The pyrylium dyes: a new class of biolabels. synthesis, spectroscopy, and application as labels and in general protein assay. PhD Thesis. Faculty of Chemistry and Pharmacy at the University of Regensburg.

249 Resch-Genger, U., Grabolle, M., Cavaliere-Jaricot, S. et al. (2008). Quantum dots versus organic dyes as fluorescent labels. *Nat. Methods* 5: 763–775 2008. doi: 10.1038/NMEtH.1248.

250 Hayashi, Y., Yamaguchi, S., Cha, W.Y. et al. (2011). Synthesis of directly connected BODIPY oligomers through Suzuki–Miyaura coupling. *Org. Lett.* 13: 2992–2995. doi: 10.1021/ol200799u.

251 Lin, W., Yuan, L., Cao, Z. et al. et al. (2010). Through-bond energy transfer cassettes with minimal spectral overlap between the donor emission and acceptor absorption: coumarin-rhodamine dyads with large pseudo-stokes shifts and emission shifts. *Angew. Chem. Int. Ed.* 49: 375–379. doi: 10.1002/anie.200904515.

252 Wu, L., Huang, C., Emery, B.P. et al. (2020). Förster resonance energy transfer (FRET)-based small-molecule sensors and imaging agents. *Chem. Soc. Rev.* 49: 5110–5139. doi: 10.1039/c9cs00318e.

5

Scattergrams of the Photophysical Properties

5.1 Introduction

In this chapter, the photophysical parameters of the fluorescent dyes of the current printed database are presented in scattergrams. This facilitates the quick identification of the best fluorescent dye labels and stains for each application or study. Dyes that have the smallest molecular size, largest molar absorption coefficients, largest quantum yield, largest brightness, largest Stokes shifts, largest (or smallest) fluorescence lifetime, and largest (or smallest) charge occurring in each range of the absorption or emission spectrum can easily be identified within these scattergrams.

Most of the available fluorophore data in the literature is related to fluorophores in organic solvents such as alcohols, dimethylsulfoxide, acetonitrile, alkanes, and even chloroform or dichloromethane. Worse still, many vendors do not report whether the parameters of their dyes were measured in an aqueous solution or organic solvent. Therefore, caution is advised when comparing data, as some photophysical parameters are significantly different when the fluorophore is in an organic solvent or an aqueous solution. For example, the fluorescence quantum yield of some dyes (e.g., NBD, Section 3.3.2) increases up to 50-fold when they are moved from aqueous solutions to organic media. Stokes shifts are, in general, much larger in aqueous solutions than in organic solvents. The molar absorption coefficients of most dyes are not expected to change very much, except when aggregation occurs in aqueous solutions.

Note that the entire database presented in this book contains photophysical parameters that were only taken from aqueous solutions such as acetate (A), borate (B), carbonate (C), phosphate (P), phosphate-buffered saline (PBS), tetraborate (T), tris(hydroxymethyl)aminomethane (Tris), and even pure water in some cases. This is important information for applications in life sciences, clinical analysis, food analysis, and pharmaceutical analysis, as most solutes in these fields occur in aqueous solutions.

It is important to consider that there are more discrepancies in data measured in aqueous solutions than in data measured in organic media. This occurs because water is a much more complex solvent with many complex solute–solvent interactions. For example, polyanionic buffers (B, P, PBS, and T) tend to aggregate with polycationic solutes (e.g., rhodamine 6G + 2, rhodamine 6G + 3, carborhodamine 16a, Si-9-iminopyronine 2f, DAOTA 4, AI3, and lucigenin, as well as some acridine, oxazine, and pyronine derivatives). Moreover, many dyes that have a poor solubility in water tend to form aggregates with themselves in aqueous solutions. This causes severe changes in the observed photophysical parameters. Additionally, some fluorophores have a flexible structure and are made up of a mixture of two or more conformational isomers in liquid solutions [1]. Others have a pK_a of around neutral pH, which makes the photophysical properties of these fluorophores very sensitive to pH (e.g., non-fluorinated fluoresceins, 7-hydroxycoumarins, and others). Finally, a few dyes are prone to easily change between their oxidized and reduced forms, each of which has its own distinct photophysical properties. These are

Fluorescent Dye Labels and Stains: A Database of Photophysical Properties, First Edition. Tarso B. Ledur Kist.

the main reasons why there are some discrepancies in the literature. Such differences are mainly seen in fluorescence quantum yields (Φ_f) (and consequently molecular brightness, B_{max}), but also in wavelength of absorption maxima ($\lambda_{a,max}$), wavelength of fluorescence emission maxima ($\lambda_{e,max}$), and molar absorption coefficients (ε_{max}). Nevertheless, these discrepancies are usually small (<10%) and do not compromise the scatterplot analyses and resulting conclusions of the present chapter. It is possible to resolve the discrepancies in the published data by remeasuring parameters with current instruments under controlled temperatures and sample conditions, but this would take many years to complete. Even if the analyses of this chapter were performed again with this updated dataset, the scatterplots produced and their conclusions would not change significantly. Therefore, data found in the literature was used as reported and only ~5% of the data was measured and re-measured by the author.

In the figures throughout this chapter, colored shapes are used to represent the different classes. The same colored shape is always used to represent the same class, details of which are presented in Table 5.1.

5.2 Photophysical Properties Along the Spectrum

This section examines how the photophysical parameters of fluorescent molecules change along the spectrum of $\lambda_{x,max}$ (or $\lambda_{a,max}$; they are used interchangeably within this chapter) in the 300–900 nm range. The goal is to answer the following questions: What are the $\lambda_{a,max}$ values of the smallest dyes? What are the brightest classes and in what range are their $\lambda_{a,max}$ values? What are the classes and $\lambda_{a,max}$ ranges of the dyes with the largest ε_{max}, Φ_f, Stokes shift (Δ_S), and fluorescence mean-lifetime (τ_f)? Are any of the above molecular and photophysical parameters correlated?

The scattergrams are limited to 800 nm because there are only two dyes in the whole database with $\lambda_{a,max}$ larger than 800 nm: ECXi, a pentacene with $\lambda_{a,max}$ = 887 nm and B_{max} = 1.3 mM^{-1}cm^{-1} [2], and ICG, a heptamethine cyanine with $\lambda_{a,max}$ = 807 nm and B_{max} = 12.2 mM^{-1}cm^{-1} [3, 4, 5].

5.2.1 Molecular Sizes vs. $\lambda_{a,max}$

Many methods and protocols require fluorescent labels or stains that are small and bright at the same time. Small sizes are important to avoid disturbances on the targets under study by the dyes. Small, organic, and bright fluorescent dyes also find interesting applications in the field of fluorescent imaging with genetically encoded non-canonical amino acids or any organic small bright structure with similar size; for example, the fluorescent dyes Acd, CouAA, Anap, and SansylA (see [6], and the references therein).

The size or hydrodynamic volume of a molecule in neutral aqueous solutions can be calculated from its structural formula, despite the difficulties in some cases. The presence of ionizable groups makes the calculations even more complex. The presence of charged groups on a fluorescent structure impacts its hydrodynamic volume significantly, as a few layers of water molecules will be attached to the fluorophore due to charge–dipole interactions (termed solvation). Therefore, uncharged molecules tend to be hydrodynamically smaller than molecules which have charged functional groups. Only a small number of the fluorophores of the current database (777 entries) do not carry a charged functional group at neutral pH (Section 5.3). The great majority have charged groups in neutral aqueous solutions. The charged groups play many roles, such as maintaining the resonance of the π-electron system (phenolates in fluoresceins and quaternary amines in rhodamines, rhodols, cyanines, squaraines, and merocyanines), making the fluorophore soluble in aqueous solutions (sulfonic, phosphonic, and sulfobetaine groups), and acting as a moiety to receive reactive groups for labeling (carboxylic acids).

In this section, the number of atoms (N_a) of the fluorescent structures will be used as an indication of molecular size. For dyes that occur as salts, the number of atoms of the fluorescent ion are counted while the number of atoms of the counter-ion(s) is/are ignored. The water molecules attached to the charged groups of the dye due to solvation will also be ignored. Common examples of counter-ions are Na^+ and K^+ (when the fluorescent structure is an anion) and Cl^-, Br^-, I^-, and ClO_4^- (when the fluorescent structure is a cation).

Table 5.1 Shapes used to represent over 40 classes of fluorescent dyes and stains.

Shapes	Classes Represented	Molecular Structure Section
○	Non-disclosed structures (the classes of some were revealed by the vendors but this cannot be confirmed without knowing the molecular structure of the dyes)	4.41
□	**Hybrid structures**	4.40
	Hybrid Structures: fusion of two existing dyes	4.40.1
	Hybrid Structures: single bond connected dyes	4.40.2
	Hybrid Structures: polymethine bridged dyes	4.40.3
◇	**Others**: which means classes that only have one or two fluorescent entities. Such classes were therefore grouped within a single section	4.39
	Small structures: aminopyridines, phenazines, acridiziniuns, diketopyrrolopyrroles, and carbazoles	4.39.1
	Medium-sized structures: epicocconones and aminofluorenes	4.39.2
	Large structures (N_a >80): phospholo[3,2-*b*]phospholes, pentacenes, and spirofluorenes	4.39.3
△	**Benzene derivatives**	4.38
●	**Isoquinolines and quinolines** are grouped together	
	Isoquinoliones	4.37
	Quinolines	4.29
■	**Pteridines** (a pyrimidine ring fused to a pyrazine ring)	4.36
◆	**Squaraines**	4.35
▲	**Naphthalenes**	4.34
○	**Indoles** (1*H*-indole or 2,3-benzopyrrole)	4.33
□	**Naphthalimides**	4.32
◇	**Chromones**	4.31
△	**Benzothiazoles** (1,3-thiazole ring fused to a benzene ring)	4.30
○	**Fluorones** and **HAS-fluorones**	
	Fluorones (fluoresceins with no carboxylic group at carbon-2′ on the pendant aryl group)	4.11
	HAS-fluorones (these are similar to HAS-fluoresceins, but with no carboxylic group at carbon-2′ on the pendant aryl group)	4.12
□	**Pyrenes**	4.28
◇	**Rhodols, HAS-rhodols, and sulforhodols** (these three classes are grouped under this empty green diamond shape)	
	First rhodos synthesized	4.15.1
	Recently synthesized rhodols	4.15.2
	Rhodol analogs	4.15.3
	HAS-rhodols	4.16
	Sulforhodols and sulforhodol analogs	4.27

(*Continued*)

Table 5.1 (Continued)

Shapes	Classes Represented	Molecular Structure Section
△	**Coumarins**	4.26
	7-Hydroxy coumarins	4.26.1
	Small 7-amino coumarins	4.26.2
	More elaborate 7-amino coumarins	4.26.3
○	**Phenoxazines** (an 2*H*-1,4-oxazine core flanked by two benzene rings)	4.25
□	**Merocyanines** (benzopyrylium ion connected to a pyrrole by a polymethine chain)	4.24
◇	**Acridines and acridones**	4.23
	Acridines	4.23.1
	Acridones	4.23.2
△	**Trianguleniums** (diazaoxatriangulenium derivatives are so far the brightest in water)	4.22
○	**Pyrodols** (6-amino-3*H*-xanthen-3-one) and **pyrodones** (3-oxo-3*H*-xanthen-6-olate)	4.21
□	**Rosol and HAS-rosol**	
	Rosol (rhodols with no carboxylic group at carbon-2′ on the pendant aryl group)	4.19
	HAS-rosol (HAS-rhodols with no carboxylic group at carbon-2′ on the pendant aryl group)	4.20
◇	**Cyanines**	
	Trimethine cyanines	4.13.1
	Pentamethine cyanines	4.13.2
	Heptamethine cyanines	4.13.3
△	**Borondipyrromethenes**	4.14
	Small water-soluble borondipyrromethenes	4.14.1
	Medium-sized, water-soluble borondipyrromenes	4.14.2
	Large water-soluble borondipyrromenes	4.14.3
	Other classes derived from borondipyrromethene	4.14.4
●	**Rosamines** (rhodamines with no carboxylic group at carbon-2′ on the pendant aryl group)	4.17
▲	**HAS-rosamines** (HAS-rhodamines with no carboxylic group at carbon-2′ on the pendant aryl group)	4.18
●	**Fluoresceins, mercaptofluoresceins, and sulfofluoresceins** (grouped as their spectral properties share some similarities)	
	Non-halogenated fluoresceins	4.8.1
	Halogenated fluoresceines	4.8.2
	Mercaptofluoresceins	4.8.3
	Sulfofluoresceins	4.10

Table 5.1 (Continued)

Shapes	Classes Represented	Molecular Structure Section
	**HAS-fluoresceins and fluorescein analogs** (grouped as their spectral properties share some similarities)	
	HAS-fluoresceins	4.9
	Fluorescein analogs	4.8.4
	**Pyronines**	4.4
	**HAS-pyronines**	4.5
	**Rhodamines and sulforhodamines**	
	Rhodamines	4.2
	Sulforhodamines	4.6
	**HAS-rhodamines, HAS-sulforhodamines, and sulforhodamine analogs** (grouped as their spectral properties share some similarities)	
	HAS-rhodamines	4.3
	HAS-sulforhodamines	4.7

It is important to know whether there is a correlation between $\lambda_{a,max}$ and the size (N_a) of the dyes of the present database (777 fluorescent dyes and stains with high brightness in aqueous solutions). This correlation does indeed exist and is shown in Figure 5.1, where N_a is the number of atoms of the fluorescent structure (an indication of molecular size) and $\lambda_{a,max}$ is the wavelength of absorption (or excitation) maximum of the longest wavelength band.

The smallest dyes in the database are two amino-pyridines and three benzene derivatives (4-methylsalicylic acid, 5-methoxysalicylic acid, and 5-methylsalicylic acid). They are made up of only 13 to 20 atoms and exhibit a $\lambda_{a,max}$ of nearly 300 nm. There is a staircase-like empty space below the points (shapes) in Figure 5.1, which is an important aspect of the relationship between minimum dye sizes (N_a) and $\lambda_{a,max}$ values. This data is presented in Table 5.2. This is caused by at least three reasons. First, according to quantum mechanics, the spacing between energy levels of an electron in a "potential well" decreases with one-dimensional well width (in the case of fluorescent dyes, this is related to the length of the π-electron system, or the number of conjugated bonds). An alternative statement would be that the energy difference between the ground and first excited state increases when the π-electron system is shortened as the number of conjugated bonds, and therefore atoms, is decreased. Therefore, according to quantum mechanics, there is no way around this. First, to increase $\lambda_{a,max}$, the length of the π-electron system (conjugated system) must be increased. Second, the π-electron system is usually made of carbons and hydrogens, yielding hydrophobic structures. In these cases, a few extra hydrophilic groups (e.g., sulfonic and carboxylic groups) must be bound to the fluorescent structure to confer water solubility. The longer the structure the larger the number of hydrophilic groups required to maintain at least some water solubility. Third, long and flexible structures are known to exhibit low fluorescence quantum yields. Therefore, extra $(CH_2)_n$ groups must be added to make the conjugated backbone more rigid and increase fluorescence quantum yield (see the discussion in Section 4.13). This effect can be partially mimicked by host/guest complexes (Section 2.7.10–1) or viscous solutions (Section 2.7.10.2).

There are 14 borondipyrromethenes made up of 120 atoms or more. They are exaggeratedly large because they are functionalized with branched oligo(ethylene glycol)methyl ether groups or with branched oligosaccharides (Section 4.14.3). Taking them out of the database, together with a coumarin and two others (which were also

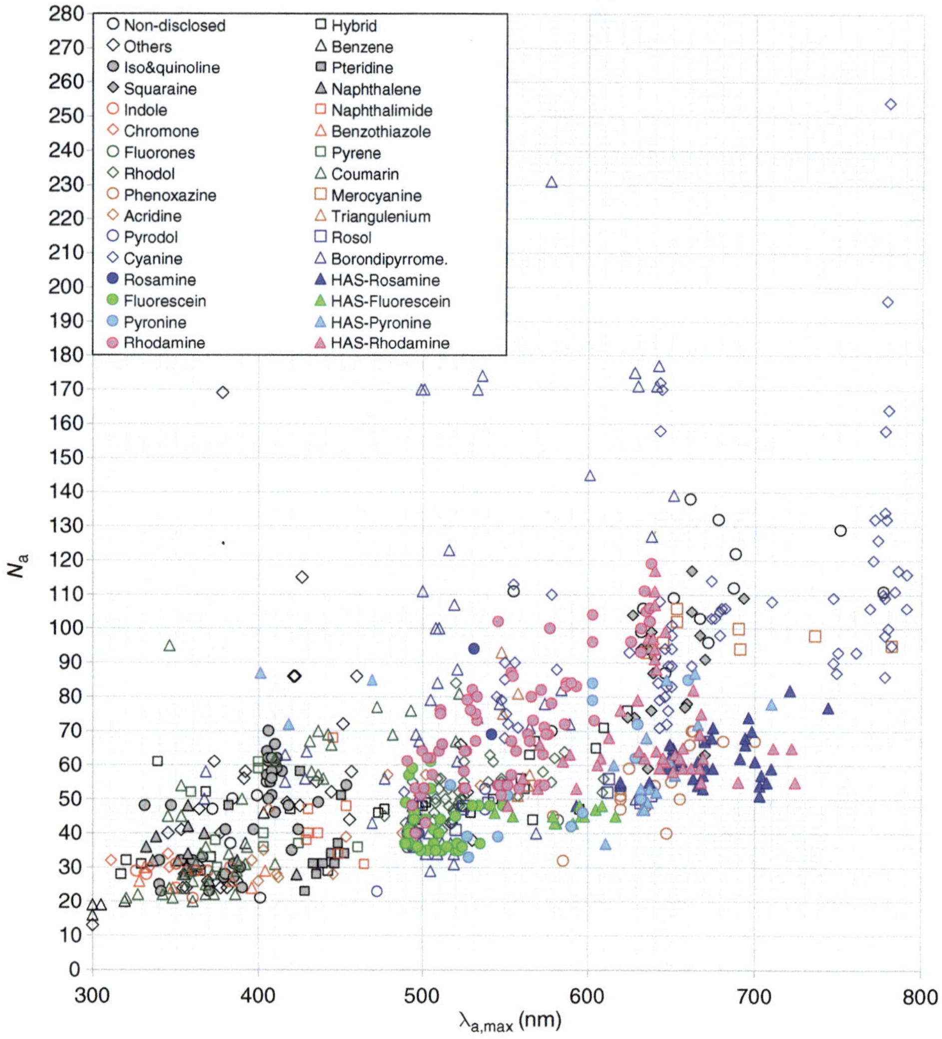

Figure 5.1 Scattergram of dye sizes (number of atoms of the fluorescent structures, N_a) plotted against the wavelength of absorption (or excitation) maximum of the longest wavelength band ($\lambda_{a,max}$).

Table 5.2 The minimum number of atoms required to synthesize a bright fluorescent dye with a given $\lambda_{a,max}$.

$\lambda_{a,max}$ (nm)	N_a Required	Notes
>300	>10	
>400	>20	
>500	>30	With the exception of one borondipyrromethene (meso-caboxyBODIPY J-K 1, see Section 4.14.1)
>600	>40	With the exception of one HAS-pyronine (JS-R, see Section 4.5)
>700	>50	
>800	>60	Dyes not shown in Figure 5.1

subjected to branched functionalization), results in a good correlation between N_a and $\lambda_{a,max}$. The linear regression produced from the resulting data, using the minimum squared method, gives $N_a = 0.155 \cdot \lambda_{a,max} - 22.7$, with $r^2 = 0.4822$ (where r-squared expresses goodness-of-fit in the regression model used).

Finally, it is worth noting the crowded rectangle in Figure 5.1, defined by $32 < N_a < 120$ and $\lambda_{a,max}$ >490 nm. The molecules in this corner are mainly bright xanthene derivatives (all classes derived from the xanthene super-class are shown in Figure 4.3). There are only three xanthene outliers: HAS-pyronines Si-9-iminopyronine 2c, Si-9-iminopyronine 2f, and Si-9-iminopyronine 2g, with $\lambda_{a,max}$ (N_a) = 401 nm (87), 418 nm (72), and 469 nm (85), respectively.

5.2.2 Molar Absorption Coefficients vs. $\lambda_{a,max}$

It is important to know molar absorption coefficients (ε_{max}), together with fluorescence quantum yields (Φ_f), to deeply understand brightness. Moreover, a high molar absorption coefficient is an important property of FRET donors and both methods and protocols involving absorbance measurements that use labels (bright-field microscopy, ELISA, and UV-Vis absorption detectors in HPLC, capillary, and microchip electrophoresis). Figure 5.2 shows which classes exhibit the largest molar absorption coefficients and the spectral ranges containing the entities which have the largest molar absorption coefficients.

Molar (decadic) absorption coefficients increase significantly from 300–630 nm (Figure 5.2). This, together with the results shown in Figure 5.1, shows the inability of small dyes to exhibit large molar absorption coefficients (at least for fluorescent dyes with medium to high brightness in aqueous solution). The excitation coupling term in the "molecule-photon Hamiltonian" is the dot product (or scalar product) of two vectors, the dipole moment of the dye, and the electric field of the photons of the excitation beam. Shorter wavelength photons (e.g., ~350 nm) have stronger electric fields than longer wavelength photons (e.g., ~650 nm). However, single small organic dyes which have a wavelength of excitation maximum of the longest wavelength band located at shorter wavelengths (e.g., ~350 nm) have much smaller dipole moments than longer dyes with a $\lambda_{a,max}$ at longer wavelengths (e.g., ~650 nm). Therefore, longer structures (with the π-electron system extended over many atoms) can have larger dipole moments and consequently exhibit much larger cross-sections of absorption and ε_{max} values when exposed to a flux of photons, as evidenced by Figure 5.2.

Below 350 nm, the four largest molar absorption coefficients are presented by three hybrid structures and one benzothiazole. There are five BOPAHY derivatives (Section 4.14.4) with $\lambda_{a,max}$ values of below 430 nm and ε_{max} values of ~50 $mM^{-1}cm^{-1}$. It is worth noting the presence of a solitary dye with $\lambda_{a,max}$ = 444 nm and ε_{max} = 69 $mM^{-1}cm^{-1}$, which is called epicocconone (Section 4.40.2). The neighboring empty orange diamond is an acridine (Atto 465 carboxylate, Section 4.23.1).

The bathochromic shift of the HAS-xanthenes (pink, green, light blue, and dark blue solid triangles) is evident when they are compared to their respective parent structures (pink, green, light blue, and blue solid circles). However, there are three exceptions: HAS-pyronines Si-9-iminopyronine 2c, Si-9-iminopyronine 2f, and Si-9-iminopyronine 2g [7], with $\lambda_{a,max}$ (ε_{max}) = 401 nm (18 $mM^{-1}cm^{-1}$), 418 nm (13 $mM^{-1}cm^{-1}$), and 469 nm (11 $mM^{-1}cm^{-1}$), respectively.

A column of cyanines (empty blue diamonds) can be seen around 550 nm (trimethine), around 650 nm (pentamethine), and between 750 and 790 nm (heptamethine). The positions of the non-disclosed structures (empty black circles) suggest that most of these molecules are pentamethine cyanines. The largest ε_{max} values are observed for pentamethine and heptamethine cyanines, followed by squaraines and both pentamethine and heptamethine merocyanines. As far as is known, no other classes have similarly high molar absorption coefficients. Out of all of the xanthenes, only two HAS-rosamines (solid blue triangles) have ε_{max} values at or above 160 $mM^{-1}cm^{-1}$.

5.2.3 Fluorescence Quantum Yield vs. $\lambda_{a,max}$

The fluorescence quantum yields (Φ_f) of dyes in aqueous solutions are plotted against their $\lambda_{a,max}$ in Figure 5.3. The entities can be identified by their class, $\lambda_{a,max}$, and Φ_f. Considering Figure 5.1 and the correlation between size

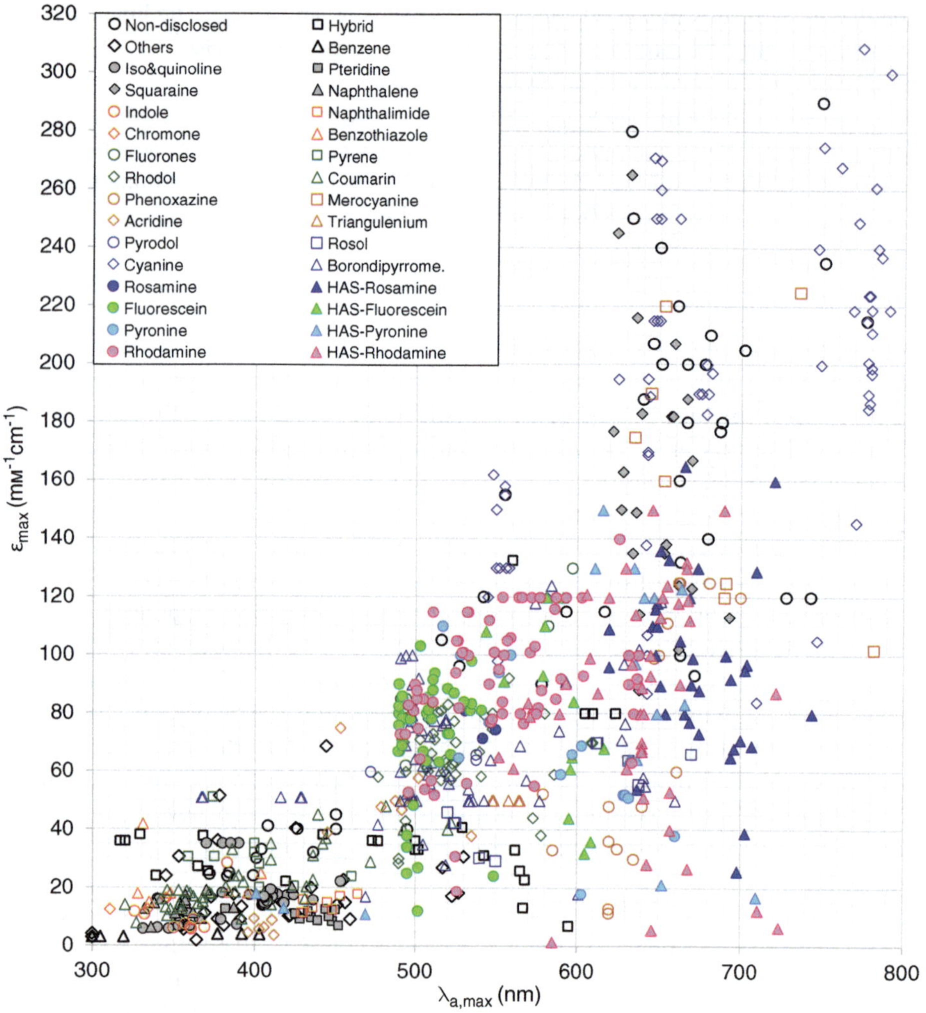

Figure 5.2 Scattergram of molar (decadic) absorption coefficients (ε_{max}) plotted against $\lambda_{a,max}$.

and $\lambda_{a,max}$, it is clear that smaller dyes have higher Φ_f values than larger dyes. This is attributed to their more rigid structures. Below 400 nm, the highest Φ_f values are presented by coumarins. Between 400 and 490 nm, the highest Φ_f values are from acridines and pyrenes. From 490–600 nm, there are many classes with Φ_f values of equal to or higher than 0.90: fluoresceins, rhodamones, rhodols, borondipyrromethenes, pyronines and one non-disclosed structure. Between 600 and 700 nm, only 24 xanthene derivatives and one non-disclosed structure (Atto 665 carboxylate) have a Φ_f value of higher than 0.50. With the exception of the dye HCDS (a rigidified heptamethine cyanine) [8], there are no other dyes with Φ_f>0.30 in the range 700 <$\lambda_{a,max}$ <900 nm (the current database covers $\lambda_{a,max}$ values up to 900 nm, even though figures of this chapter only show up to 800 nm). There is a corresponding large empty area in the upper right side of Figure 5.3 and beyond (up to 900 nm).

It is worth noting that long and flexible structures exhibit lower fluorescence quantum yields than small and rigid structures (see the discussion at the beginning of Section 4.13). Basically, upon excitation, the electronic energy of flexible and long π-electron systems is more easily converted into thermal energy as they bend and twist more easily. This bending and twisting is caused by the high collision rate of water molecules at room temperature,

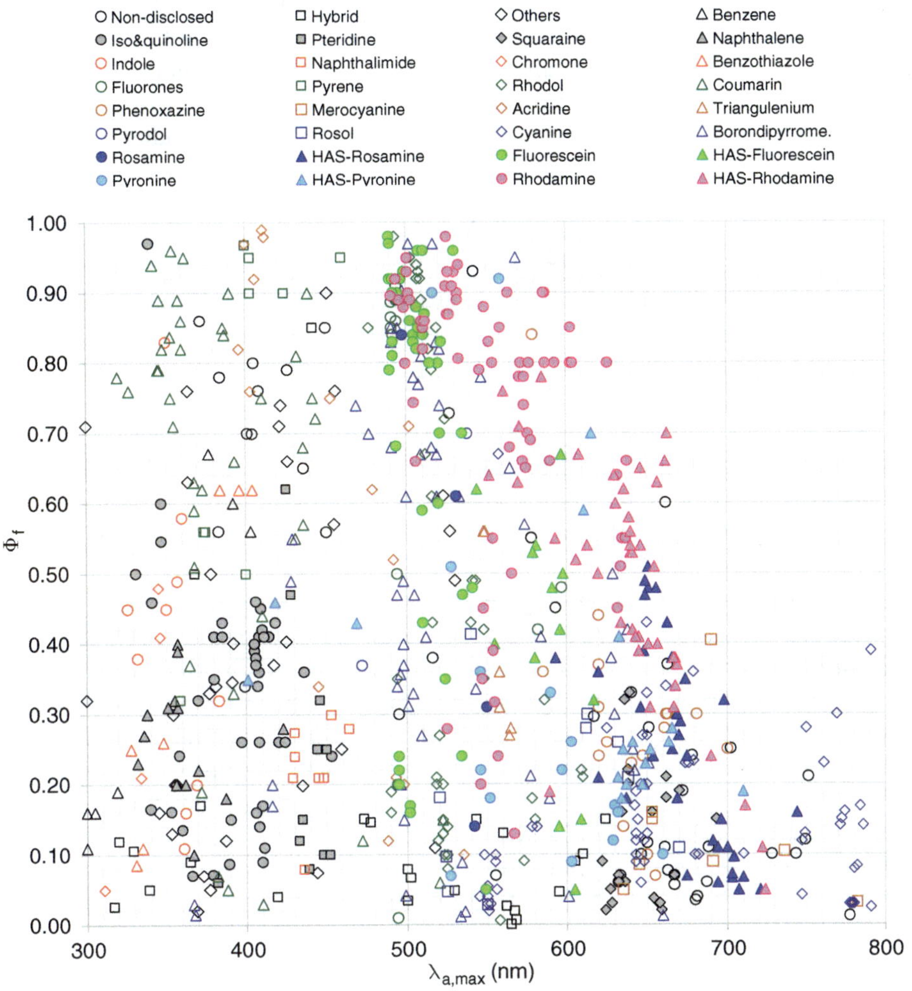

Figure 5.3 Scattergram of fluorescence quantum yields in aqueous media plotted against $\lambda_{a,max}$.

an effect which increases with temperature. Therefore, a negative correlation between fluorescence quantum yield and temperature is normally observed (Figure 2.19). Moreover, increasing the temperature also decreases the viscosity of aqueous solutions, which in turn also decreases fluorescence quantum yield (Section 2.7.10.2).

From a practical point of view there are a few measures that can be taken to increase the fluorescence quantum yields of small organic fluorescent dyes in general. This especially applies to cyanines, squaraines, and merocyanines. These measures are: i) working at lower temperatures. This has multiple positive effects, as already described; ii) increasing the dynamic viscosity of the samples when possible. For this, glycerol or another similar molecule must be used instead of diluted polymer solutions, which are widely used to increase dynamic viscosity. Diluted polymer solutions of the following high molecular mass polymers are unlikely to successfully increase the fluorescence quantum yield of small organic fluorescent dyes: linear poly(acrylamide), dextran, hydroxyethyl ether cellulose, poly(ethylene oxide), poly(*N*-vinylpyrrolidone), poly(*N*-vinylpyrrolidone), poly(vinyl acetate), poly(vinyl alcohol), or gels such as agarose and branched poly(acrylamide) (gel), to mention a few (see discussions in Section 2.7.10.1). These diluted polymer solutions increase the bulk viscosity measured; for example, by an

Ubbelohde viscometer, but not the local viscosity experienced by the dye in its microenvironment. In other words, in polymer solutions the "frictional coefficients involved in diffusional transport are different to the frictional coefficients involved in hydrodynamic transport;" and iii) adding a host/guest complexing agent or a protein. In these cases, the fluorophores are held inside the macrostructures that promote steric impediment and restrict stretching, bending, and twisting movements, thus minimizing the energy loss through non-radioactive decay channels after each excitation event. For a good review of fluorescent dyes and their supramolecular host/guest complexes with macrocycles in aqueous solution, see [9].

In summary, Figure 5.3 shows the opposite trend to Figure 5.2. The largest Φ_f are observed for dyes with $\lambda_{a,max}$ of <550 nm (Figure 5.3), while the largest ε_{max} are observed for dyes with $\lambda_{a,max}$ >600 nm (Figure 5.2). These opposite behaviors help explain why maximum brightness values are observed in the 480–680 nm range; i.e., in the middle of the whole spectra of the current database that goes from 300–900 nm (Section 5.2.4).

5.2.4 Brightness vs. $\lambda_{a,max}$

Molecular brightness (*B*) is one of the most important properties of a fluorescent dye. Its maximum value at the longest wavelength band is defined as $B_{max} = \varepsilon_{max}\,\Phi_f$, where ε_{max} is the molar (decadic) absorption coefficient (at $\lambda_{a,max}$) and Φ_f is the fluorescence quantum yield (see discussions in Section 2.7.8). Figure 5.4 shows the scattergram of B_{max} against $\lambda_{a,max}$.

The brightest fluorescent dyes in each spectral range, from $\lambda_{a,max}$ = 300–800 m, can be seen in Figure 5.4. There are only three dyes in the whole database (777 entries) which exhibit a brightness value of higher than 110 $\text{mM}^{-1}\text{cm}^{-1}$ in aqueous solutions: one non-disclosed structure (Atto 542 carboxylate), one rhodamine (Atto Rho14 carboxylate), and one cyanine (NIR4). There are five rhodamines (including rhodamine 6G), one sulforhodamine (sulforhodamine 101), and one HAS-pyronine (the $C(CH_3)_2$-pyronine Atto 610 carboxylate) with brightness values between 100 and 110 $\text{mM}^{-1}\text{cm}^{-1}$. In addition, eleven rhodamines (including JF559), two pyronines (pyronine 630 and Atto 520 carboxylate), one HAS-rhodamine (the $C(CH_3)_2$-rhodamine Atto 647N carboxylate), and one non-disclosed structure (Atto 665 carboxylate) have brightness values between 90 and 100 $\text{mM}^{-1}\text{cm}^{-1}$.

There are 164 dyes in the whole database that have brightness values of higher than 50 $\text{mM}^{-1}\text{cm}^{-1}$ in aqueous solutions. This is 21% of the total number of dyes (777 entries). The most impressive thing to note here is that all of these 164 dyes have $\lambda_{a,max}$ values of between 480 and 680 nm, with the exception of only the following 6 dyes ($\lambda_{a,max}$, B_{max}, class): Atto 465 carboxylate (453 nm, 56.2 $\text{mM}^{-1}\text{cm}^{-1}$, acridine); Alexa Fluor 700 (702 nm, 51.2 $\text{mM}^{-1}\text{cm}^{-1}$, non-disclosed); NIR3 (750 nm, 77 $\text{mM}^{-1}\text{cm}^{-1}$, cyanine); NIR1 (761 nm, 61.6 $\text{mM}^{-1}\text{cm}^{-1}$, cyanine); HCDO (769 nm, 65.6 $\text{mM}^{-1}\text{cm}^{-1}$, cyanine); and HCDS (791 nm, 85.3 $\text{mM}^{-1}\text{cm}^{-1}$, cyanine). There is a simple explanation for this: below 480 nm, the Φ_f of some dyes is high, but the ε_{max} of all of the dyes is small. The opposite happens above 680 nm as the ε_{max} of many dyes is high, but the Φ_f of almost all dyes is small. Luckily, between 480 and 680 nm there are a few dozen dyes with both high ε_{max} and high Φ_f values.

The scattergram in Figure 5.4 shows the predominance of fluorescent molecules with small brightness values at the beginning of the spectral range shown; i.e., for $\lambda_{a,max}$ in the UVA, violet, and purple ranges. When analyzed in conjunction with Figure 5.1, the absence of molecules that are both small and bright is noticed. Among the small molecules many have high fluorescence quantum yields, nonetheless all have relatively small molar absorption coefficients (Figure 5.2). This limitation may be difficult to overcome, because quantum mechanics show that it is not possible to have small molecules with molar absorption coefficients that are larger than a certain value (this is related to the staircase pattern shown in Table 5.2).

There is a sudden increase in the brightness within the blue range (~480 nm). This is mainly caused by the xanthene derivatives, represented by the solid light green shapes (fluoresceins), solid pink shapes (rhodamines), solid light blue shapes (pyronines), solid dark blue shapes (rosamines), empty green diamonds (rhodols), empty green circles (fluorones), empty blue circles (pyrodols and pyrodones), and empty blue squares (rosols) (all molecular structures are depicted in Figure 4.3). Rhodamines, pyronines, and cyanines outperform other dyes because

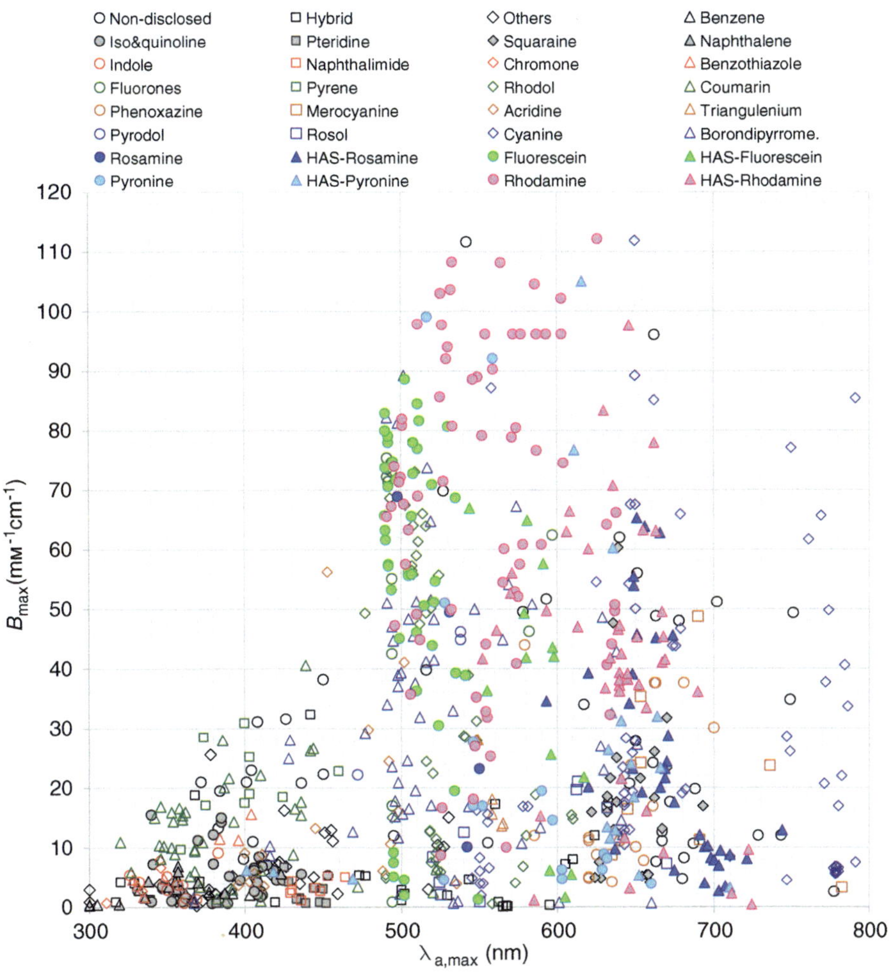

Figure 5.4 Scattergram of molecular fluorescence brightness (B_{max}) in aqueous media plotted against $\lambda_{a,max}$.

they have the highest brightness values, followed by fluoresceins and borondipyrromethenes. These classes contain the brightest entities among all currently available classes because their molar absorption coefficients are high for their range and their fluorescence quantum yields are also high. The low brightness and low number of bright dyes observed above 700 nm is due to the low Φ_f values exhibited by these large structures (cyanines, squaraines, and merocyanines, as shown in Figure 5.3); however, there are no fundamental principles that discourage the search for solutions to these poor performances. Indeed, the use of rotaxanes have proved to be one interesting way of increasing the brightness of single organic structures in the 675–900 nm spectral range (discussed in the beginning of Section 4.35).

5.2.5 Stokes Shift vs. $\lambda_{a,max}$

In addition to high brightness and photostability values, a large Stokes shift (Δ_S) is also a highly desired photophysical property. It facilitates the filtering of stray excitation radiation from the fluorescence radiation, rendering high signal-to-noise ratios and therefore high sensitivities (in detection schemes) and contrasts (in imaging systems). Sets of two, three, and even four dyes with different Stokes shifts are used for multicolor imaging in conventional confocal microscopy. Small, medium, and large Stokes shifts are required to allow spectral separation of different labels into two, three, or four excitation channels and one detection channel (or the opposite: one excitation channel and two, three, or four detection channels).

Dyes with extra-large Stokes shifts are also very important in two-color STED confocal nanoscopy, where two separated excitation bands and two coincident emission bands are usually used. The high-power doughnut-shaped STED beam (Figure 2.14B) is used to simultaneously deplete the excited state of the two fluorophores at the periphery of the two (double confocal) excitation spots by stimulated emission (Section 2.5.5). However, this high-power beam can re-excite the dye with the smallest Stokes shift (the dye whose excitation band is closer to the emission band). This is an undesired effect as it blurs the images generated. Therefore, photostable fluorescent dyes with high brightness values and large Stokes shifts are highly desired in the field of nanoscopy and others.

Stokes shift is defined as the energy difference (ΔE) between the excitation maximum and emission maximum. For practical reasons, the wavelength difference ($\Delta_S = \lambda_{e,max} - \lambda_{x,max}$) is used to represent the Stokes shift instead of ΔE or the wavenumber difference ($\Delta \nu$), which is linearly related to the energy difference. For a succinct description of the Stokes shift phenomenon, see Section 2.7.5 and Figure 2.15.

As shown in Figure 5.5, the largest fluorescence Stokes shifts are observed in the 300–470 nm range. The dyes with the largest Stokes shifts are benzothiazoles, naphthalenes, quinolines, isoquinolines, coumarins, and indoles. However, most dyes of these classes do not exhibit high brightness values, as shown in Figure 5.4. The Stokes shifts of most xanthenes is of the order of 20 nm (colored solid disks and triangles). The smallest Stokes shifts are those of borondipyrromethenes, but there are some exceptions (dye ($\lambda_{a,max}$, Δ_S, B_{max})): 8-trifluoromethyl-BODIPY 12 (477 nm, 107 nm, 29.2 $mM^{-1}cm^{-1}$), 8-trifluoromethyl-BODIPY 10 (469 nm, 73 nm, 12.6 $mM^{-1}cm^{-1}$), meso-caboxyBODIPY J-K 1 (505 nm, 49 nm, 48.3 $mM^{-1}cm^{-1}$), and distyryl-BODIPY 11 (601 nm, 54 nm, 0.7 $mM^{-1}cm^{-1}$).

Table 5.3 was created as an aid to Figure 5.5, and shows the average Stokes shift and standard deviation of each class of the whole database (777 entries). However, as shown in Section 5.2.6, the dyes with a large Stokes shift have a small brightness and the dyes with high brightness have a small Stokes shift.

5.2.6 Stokes Shift vs. Brightness

In Figure 5.6, the Stokes shifts of the dyes of the database are plotted against their respective brightnesses. This is very interesting from a practical point of view, as it shows that two very desired properties can be optimized by properly choosing the right dye. Only four dyes were not included in this scattergram. These were ECXi and ICG, which have $\lambda_{a,max}$ values of above 800 nm, and the following two because they have Δ_S values of above 220 nm: fluorene PP11, (S)-9,9-dimethyl-7-(2-methylaziridin-1-yl)-9*H*-fluorene-2-carbaldehyde, with $\lambda_{a,max}$ = 377 nm, $\lambda_{e,max}$ = 621 nm, Δ_S = 244 nm, ε_{max} = 36.0 $mM^{-1}cm^{-1}$, Φ_f = 0.05, and B_{max} = 1.8 $mM^{-1}cm^{-1}$ (all measured in water); and fluorene PP6, *N*-butyl-4-(7-(dimethylamino)-9,9-dimethyl-9*H*-fluoren-2-yl)-4-oxobutanamide, with $\lambda_{a,max}$ = 373 nm, $\lambda_{e,max}$ = 620 nm, Δ_S = 247 nm, ε_{max} = 16.5 $mM^{-1}cm^{-1}$, Φ_f = 0.07, and B_{max} = 1.1 $mM^{-1}cm^{-1}$ (all measured in water) (Section 4.39.3) [10].

Unfortunately, Figure 5.6 shows that it is possible to have either a high Stokes shift or a high brightness, but that it is not possible to have both together. The main reason for this is that a large Stokes shift is caused by the strong interactions (hydrogen bonds, dipole–dipole, and charge–dipole) of the groups belonging to the conjugated system of the dye with the water molecules of the aqueous solution (Section 2.7.5), and these increase the non-radiative decay rates, leading to smaller fluorescence quantum yields. As a result, there is a very large empty area in the upper right side of Figure 5.6. This means that there are currently no organic fluorescent dyes available which have both a high Stokes shift and high brightness. However, within the current best theories, there is no restriction that prohibits the existence of dyes that would occupy the empty space shown in the figure. It is only a matter of time before talented Chemists begin to fill that empty area.

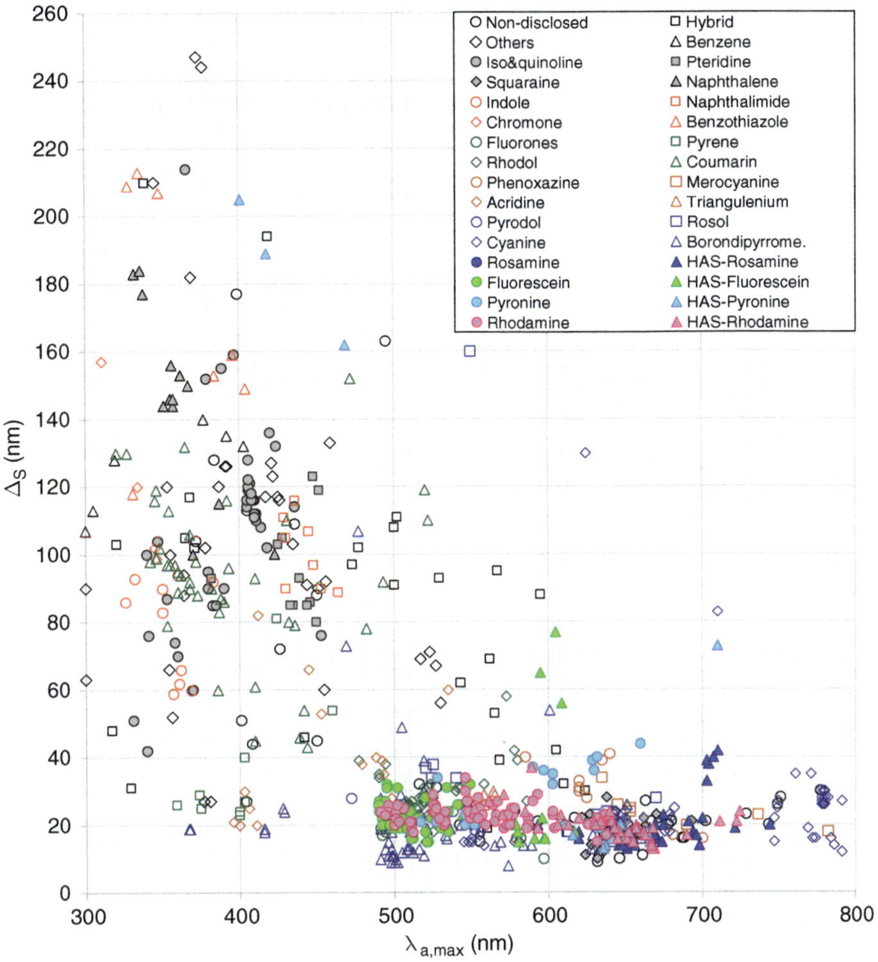

Figure 5.5 Stokes shifts of the 777 fluorescent dyes of the current database in aqueous solutions. Dyes can be identified by their respective class, $\lambda_{a,max}$, and Δ_S.

The heptamethine cyanine HCDN (empty blue diamond) stands out from the other dyes. It is a partially rigid cyanine with $\lambda_{a,max} = 625$ nm, $\lambda_{e,max} = 755$ nm, $\Delta_S = 130$ nm, $\varepsilon_{max} = 195.0$ mM^{-1}cm^{-1}, $\Phi_f = 0.28$, and $B_{max} = 54.5$ mM^{-1}cm^{-1} (all parameters measured in PBS pH 7.4) [8] (Section 4.13.3). Another dye with outstanding properties is Atto 465 carboxylate, an acridine (empty orange diamond) with Δ_S = 53 nm and B_{max} = 56.2 mM^{-1}cm^{-1} (all parameters measured in PBS pH 7.4 at 22°C) [11] Section 4.23).

Most of the xanthene derivatives exhibit a Stokes of ~20 nm, but there are three notorious exceptions: Si-9-Iminopyronine 2c, Si-9-Iminopyronine 2f, and Si-9-Iminopyronine 2g. Nonetheless, this comes at a cost, as their brightness is very low (only ~5 mM^{-1}cm^{-1}).

Borondipyrromethenes are known to exhibit small Stokes shifts. This can be noticed by the low-lying empty blue triangles along the bottom of Figure 5.6. However, there are a few that have large Stokes shifts and medium brightness values. For example, 8-trifluoromethyl-BODIPY 12 (Δ_S = 107 nm and B_{max} = 29.2 mM^{-1}cm^{-1}), meso-caboxyBODIPY J-K 1 (Δ_S = 49 nm and B_{max} = 48.3 mM^{-1}cm^{-1}), and meso-caboxyBODIPY J-K 3a (Δ_S = 107 nm and B_{max} = 29.2 mM^{-1}cm^{-1}); references are given in Section 4.14.1.

Table 5.3 Average Stokes shift and standard deviations of 777 fluorescent dyes and stains.

Class	Average Stokes Shift	Standard Deviation	Standard Deviation	Average Stokes Shift	Standard Deviation	Standard Deviation
	nm	nm	%	cm^{-1}	cm^{-1}	%
Rhodamines	23.2	3.0	13.1	745.5	148.6	19.9
Acridines and Acridones	39.2	17.8	45.3	1734.6	775.2	44.7
Benzenes	123.1	12.8	10.4	7886.7	1141.1	14.5
Benzothiazoles	172.6	34.3	19.9	9064.2	2072.9	22.9
Borondipyrromethenes	20.3	15.6	76.8	749.7	587.4	78.4
Chromones	119.5	23.1	19.3	7936.1	1742.4	22.0
Coumarins	93.5	23.3	24.9	5053.0	1661.1	32.9
Cyanines	22.1	9.0	40.7	461.2	154.5	33.5
Fluoresceins	22.1	4.2	18.9	826.5	158.4	19.2
Fluorones	19.3	1.2	6.2	764.3	43.5	5.7
HAS-Fluoresceins	27.7	19.8	71.5	748.2	469.8	62.8
HAS-Fluorones	19.4	6.4	33.0	509.8	141.9	27.8
HAS-Pyronines	17.8	2.9	16.5	424.4	76.4	18.0
HAS-Rhodamines	20.2	3.5	17.4	488.3	130.2	26.7
HAS-Rhodols	31.5	15.5	49.3	855.9	441.2	51.5
HAS-Rosamines	20.4	7.7	37.9	433.4	134.7	31.1
HAS-Rosols	22.2	3.8	17.2	531.0	54.8	10.3
HAS-Sulforhodamines	16.5	1.5	9.1	376.9	35.1	9.3
Indoles	78.5	14.1	18.0	5120.6	990.6	19.0
Isoquinolines	113.6	9.9	8.8	5255.8	547.2	10.4
Mercaptofluoresceins	30.0	2.5	8.4	1142.3	101.0	8.8
Merocyanines	23.4	5.0	21.6	495.5	154.4	31.2
Naphthalenes	147.4	25.2	17.1	8114.2	1646.3	20.3
Naphthalimides	100.6	9.9	9.8	4207.1	474.4	11.3
Phenoxazines	24.7	9.5	38.5	593.3	250.8	42.3
Pteridines	96.1	13.9	14.5	4169.5	563.6	13.5
Pyrenes	36.6	18.3	50.2	2012.4	698.8	34.7
Pyrodols and Pyrodones	25.3	1.9	7.4	924.7	185.1	20.0
Pyronines	30.5	8.4	27.4	842.7	147.3	17.5
Quinolines	102.9	40.7	39.5	5683.6	1522.5	26.8
Rhodols	27.6	4.4	15.9	992.2	183.0	18.4
Rosamines	22.8	2.2	9.5	778.5	85.4	11.0
Rosols	35.8	1.8	5.0	1205.7	277.7	23.0
Squaraines	18.2	5.0	27.3	418.8	110.7	26.4
Sulfluoresceins	22.5	2.2	9.7	797.6	30.4	3.8
Sulforhodamines	22.8	6.4	28.0	658.9	185.4	28.1
Triangulenium	28.2	1.3	4.8	863.9	32.8	3.8

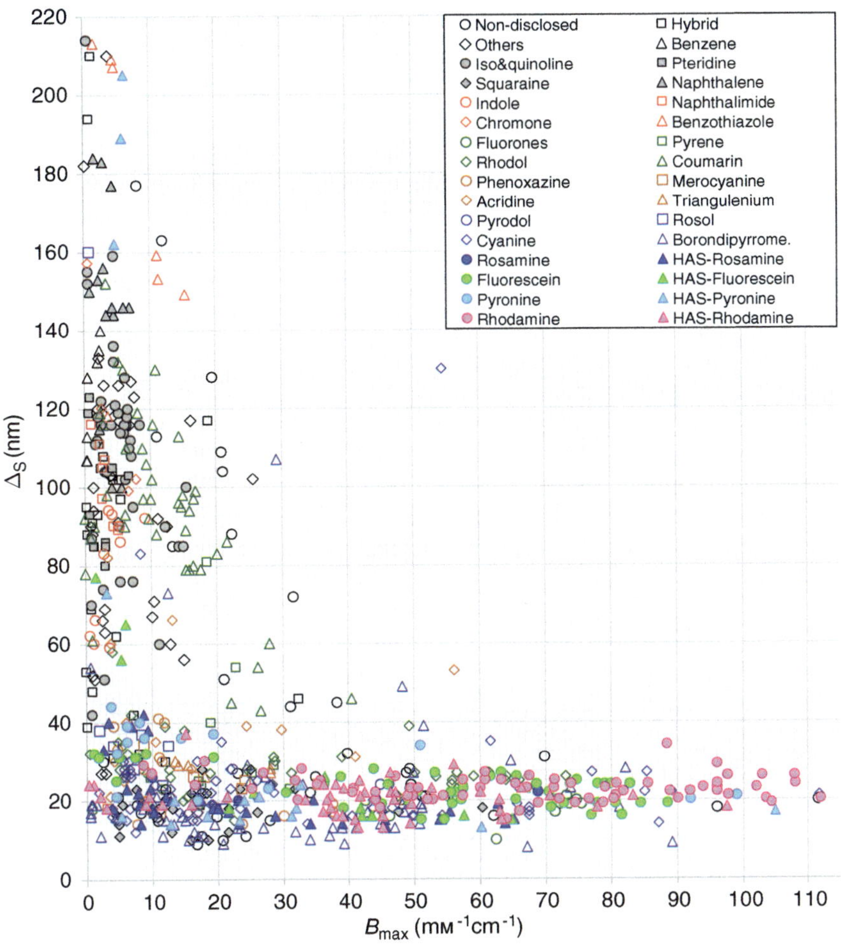

Figure 5.6 Stokes shifts of the 777 fluorescent dyes of the current database in aqueous solutions plotted against their respective brightnesses.

Coumarins are known to exhibit large Stokes shifts (empty green triangles in Figure 5.6). A total of seven coumarins have a Stokes shift that is larger than 40 nm and a brightness that is higher than 20 $mM^{-1}cm^{-1}$.

5.2.7 Fluorescence Mean-Lifetime vs. $\lambda_{a,max}$

Fluorescence mean-lifetime (τ_f) can be defined as the time required for a population of excited molecules to decrease to $1/e$ by radiative decay, where e is the Napier constant or natural logarithm base (Section 2.7). Fluorescence mean-lifetimes are crucial for applications in fields such as time-resolved fluorescence spectroscopy, STED confocal nanoscopy, fluorescence depolarization, fluorescence lifetime imaging microscopy, fluorescence lifetime sensing, and rotational relaxation time measurements, to mention but a few.

Only 295 dyes out of the whole database had their fluorescence mean-lifetime measured in aqueous solutions. These dyes are shown in Figure 5.7. As a dedicated instrument called "time-correlated single photon counting instrument" is required for the measurement of τ_f, many laboratories cannot measure this parameter for several reasons (cost of instrument, bench footprint of the instrument, and/or the necessity of skilled operators). The hybrid structure lucigenin (black empty square at the top left of Figure 5.7), is made of two 10-methylacridiniums held together by a single covalent bond. This dye has a τ_f of 25.5 ns and a B_{max} of 18.8 $mM^{-1}cm^{-1}$ (reference and molecular structure are given in Section 4.40.2). There are two indoles (red empty circles), harmane and nor-harmane (or β-carboline), with τ_f values of 20.0 and 21.2, respectively. There are six triangTuleniums (DAOTA 1 to 6, orange empty triangles) with τ_f values of between 12.0 and 19.0 ns, Stokes shifts between 26 and 30 nm, and brightnesses between 13 and 18 $mM^{-1}cm^{-1}$; this is remarkable (structures are given Figure 4.45) [12]. Five acridones and one acridine (9-aminoacridine) (orange empty diamonds) with τ_f values of between 15.5 and 17.0 ns are also worth noting. Finally, most xanthene derivatives have a τ_f of between 1 and 4.5 ns, while the τ_f of borondipyrromethenes falls between 2.5 and 14.1 ns, for coumarins it falls between 0.4 and 7 ns, and cyanines have very short mean-lifetimes (most are shorter than 1 ns).

5.2.8 Fluorescence Mean-Lifetime vs. Brightness

It is interesting to compare fluorescence mean-lifetime (τ_f) and brightness, as fluorescent dyes which have both long fluorescence mean-lifetimes and high brightness values would be useful for many niche applications. However, as shown in Figure 5.8, as brightness increases, the number of dyes with a long τ_f value decreases. Among the 295 dyes of the whole database that had a published τ_f measured in aqueous solution, there are only 3

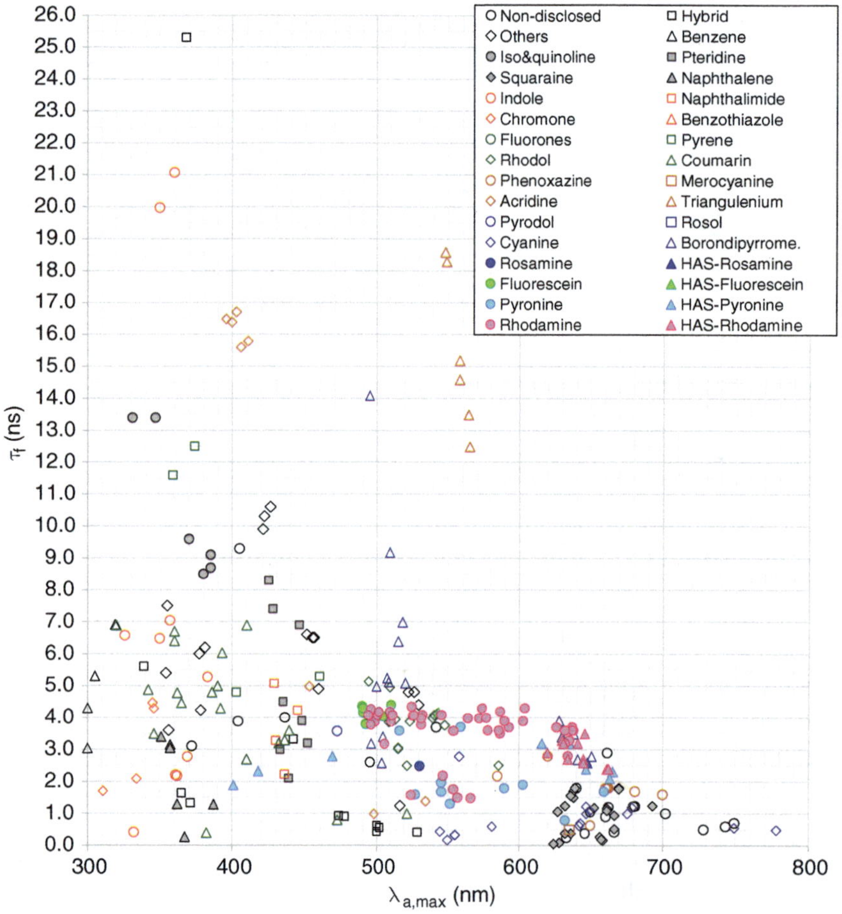

Figure 5.7 Scattergram of fluorescence mean-lifetimes vs. $\lambda_{a,max}$, both measured in aqueous solutions.

dyes with brightness values of higher than 30 $\text{mM}^{-1}\text{cm}^{-1}$ and a τ_f of longer than 5.5 ns. They are the following three borondipyrromethenes: B-Z 7b ($\lambda_{a,max}$ = 516 nm, B_{max} = 41.1 $\text{mM}^{-1}\text{cm}^{-1}$, and τ_f = 6.4 ns (Section 4.14.3)); Niu 27 ($\lambda_{a,max}$ = 510 nm, B_{max} = 51.2 $\text{mM}^{-1}\text{cm}^{-1}$, and τ_f = 9.2 ns (Section 4.14.2)); and 6-sulfonato-1,3,5,7-tetramethyl-BODIPY ($\lambda_{a,max}$ = 495 nm, B_{max} = 44.6 $\text{mM}^{-1}\text{cm}^{-1}$, and τ_f = 14.1 ns (Section 4.14.1)). Even setting the lower limit for brightness to only 20 $\text{mM}^{-1}\text{cm}^{-1}$, only 4 additional dyes will be found with τ_f >5.5 ns in aqueous solution: the coumarin Pacific blue[1] ($\lambda_{a,max}$ = 410 nm, B_{max} = 22.1 $\text{mM}^{-1}\text{cm}^{-1}$, and τ_f = 6.9 ns (Section 4.26.1)) [13]; the pyrene 1,3,6,8-pyrenetetrasulfonic acid ($\lambda_{a,max}$ = 374 nm, B_{max} = 28.5 $\text{mM}^{-1}\text{cm}^{-1}$, and τ_f = 12.5 ns (Section 4.28)); the 2 trianguleniums, DAOTA 4 ($\lambda_{a,max}$ = 549 nm, B_{max} = 28.0 $\text{mM}^{-1}\text{cm}^{-1}$, and τ_f = 18.5 ns (Section 4.22)); and DAOTA 5 ($\lambda_{a,max}$ = 548 nm, B_{max} = 28.0 $\text{mM}^{-1}\text{cm}^{-1}$, and τ_f = 18.6 ns (Section 4.22)) [12].

There is a large empty area in the upper right side of Figure 5.8 that represents the absence of dyes with both a high brightness and long mean-lifetime. It is known from the simulations of single dissipative quantum systems, which may actually correspond to concrete physical processes [14, 15], that the longer the mean-lifetime of a dye, the higher the probability of a non-radiative decay before the photoemission. This explains the observation that molecules with a high brightness do not, in general, have a very long mean-lifetime. However, there is nothing within current theories that would exclude the possibility of the existence of small organic fluorescent dyes

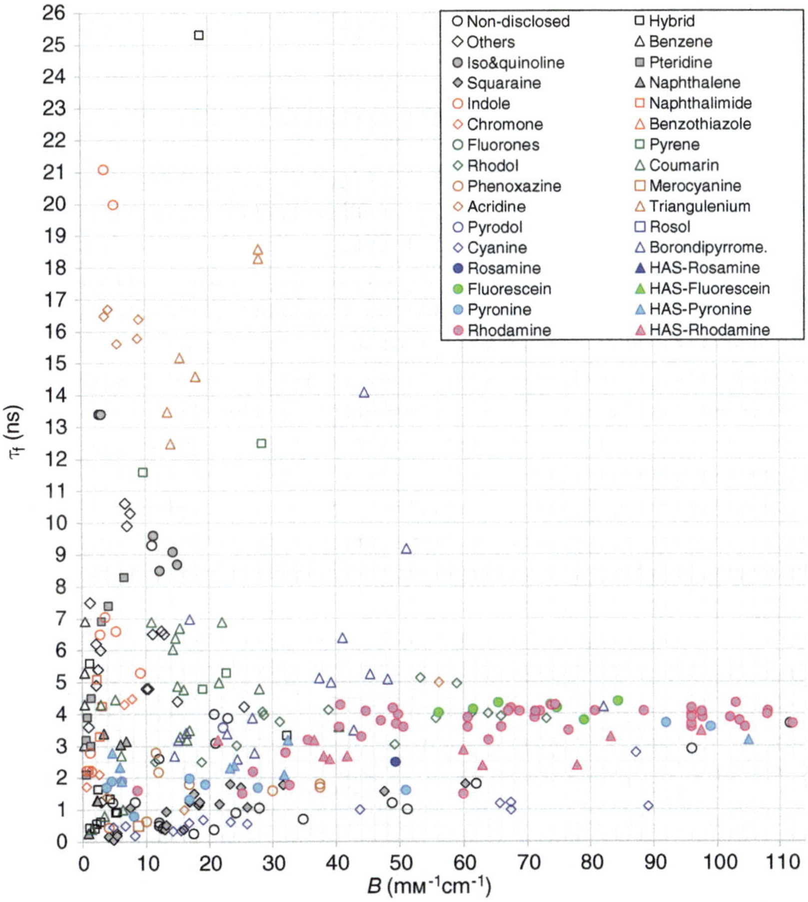

Figure 5.8 Scattergram of fluorescence mean-lifetimes vs. Brightness values, both measured in aqueous solutions.

1 Pacific blue is a trademark of Thermo Fisher Scientific (Waltham, MA, USA).

having both a high brightness and a long fluorescence mean-lifetime. The existence of such dyes could lead to new and interesting applications, reinforcing the potential and importance of this field of investigation.

5.3 Fluorophore Charges

Knowledge of the number of negatively and positively charged groups that a molecule possesses at its working pH is very important for many applications. For example, molecules with no charged groups are known to exhibit good cell permeability as they easily diffuse through membranes (plasma, nuclear) and into organelles. Small and uncharged fluorescent dyes are also good labels in liquid chromatography as they do not have a big effect on the partitioning coefficients of the analytes and thus produce better separations. Small, uncharged fluorescent dyes are highly desired for the labeling of immunoglobulin G and other antibodies as they do not affect the antibody-target affinity or ligation constant (immunostaining is discussed in Section 3.5). On the other hand, structures with a high net charge find important applications in many other fields. One such example is electrodriven separations such as electrophoresis along gel slabs, fused silica capillaries, and micro-channels in microchips. Fluorescent labeling increases the selectivity, sensitivity, and working range of calibration curves (usually up to six orders of magnitude). Moreover, selecting small and highly charged labels also increases separation efficiency (number of theoretical plates) and time efficiency (chapter 5 of [16]).

Figure 5.9 presents many important properties of the whole database in a single picture. The first column (from left to right) shows all dyes with no positive ionizable groups ($N_p = 0$) at the pH specified (usually neutral) (see Appendices for details). The second column shows all dyes possessing one positive ionizable group ($N_p = 1$), and so on. The first row (from bottom to top) shows all dyes with no negative ionizable groups ($N_n = 0$) at the pH specified. The second row shows all dyes with one negative ionizable group ($N_n = 1$), and so on. The numbers inside each cell show the net charge ($q = N_p - N_p$) of the dyes within each cell. Moreover, the abscissa within each cell indicates the Stokes shift (from 0–200 nm) and the ordinate the brightness (from 0–120 mM^{-1}cm^{-1}) (this is illustrated at the bottom right).

The rectangle or cell (C) corresponding to $N_n = 0$ and $N_p = 0$ ($C_{0,0}$) shows all dyes of the database with no ionizable groups at the pH indicated (usually neutral). The exact aqueous solution used to measure the photophysical parameters of each dye is given in the Appendices. The group of dyes belonging to $C_{0,0}$ contains many coumarins, quinolines, isoquinolines, and rosols (which are rhodols that lack the carboxylic group bound to carbon 2 of the pendant aryl group) (Figure 5.9). Note that the Stokes shifts are well distributed from 20–190 nm; however, the maximum brightness reached is only 20 mM^{-1}cm^{-1} (two rosols).

The cell (C) containing dyes with $N_n = 1$ and $N_p = 0$ ($C_{1,0}$) shows all dyes of the database for which the predominant ionic form carries only one negative ionizable group (one negative charge) at the pH indicated (usually neutral). This $C_{1,0}$ cell contains three fluoresceins and many rhodols (empty green diamonds), coumarins (empty green triangles), and fluorones (empty green circles).

The cell $C_{2,0}$ shows all dyes of the database for which the predominant ionic form carries two ionizable groups, both with a negative charge. This cell is dominated by fluorescein derivatives: one charge comes from the ionization of the carboxylic group (carboxylate) and the second from the ionization of the phenol group (phenolate).

The rectangle $C_{3,0}$ contains many carboxyfluorescein derivatives, where the third charge comes from the ionization of the second carboxylic group bound to the pendant aryl group. There is even a rarity: a fluorescein derivative occupying the rectangle $C_{5,0}$; i.e., it has five ionizable groups that are all negatively charged. These are in fact two superimposed solid green circles that contain the following two dyes: BCECF (2-(2,7-cis(2-carboxylatoethyl)-6-oxido-3-oxo-3*H*-xanthen-9-yl)terephthalate) and BCPCF (2-(2,7-bis(3-methoxy-3-oxopropyl)-6-oxido-3-oxo-3*H*-xanthen-9-yl)terephthalate). They have similar photophysical properties: both have $\lambda_{x,max} = 505$ nm and $\Delta_S = 22$ nm, but their B_{max} values differ ($B_{max} = 56.2$ mM^{-1}cm^{-1} and $B_{max} = 55.6$ mM^{-1}cm^{-1}, respectively). The molecular structures and references for these molecules are shown in Section 4.8.1.

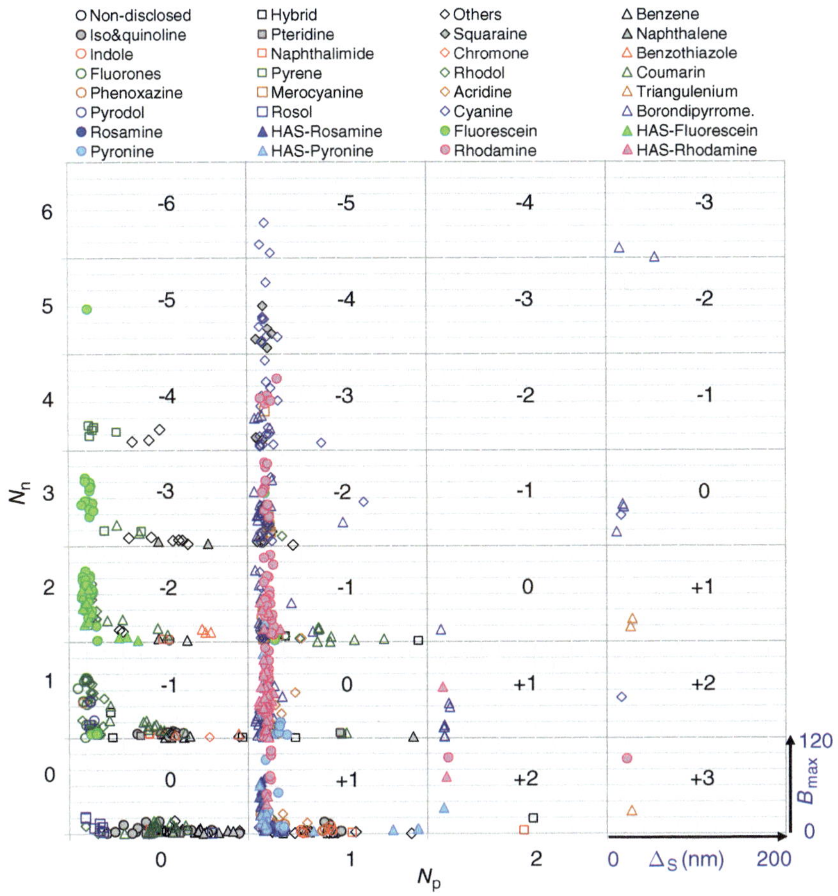

Figure 5.9 Scattergram showing the number of positive charges (N_p), number of negative charges (N_n), net charge, brightness, and Stokes shift of all dyes in the database.

The second column contains many rhodamines and pyronines from cell $C_{0,1}$ to $C_{4,1}$, cyanines from $C_{3,1}$ to $C_{5,1}$, and squaraines from $C_{4,1}$ to $C_{5,1}$. The only positive charge that the dyes of this column has comes from the iminium group found in rhodamines and pyronines and the 3*H*-pyrrol-1-ium heterocycle found in cyanines and squaraines. The large number of negative charges observed in cyanines comes from the presence of sulfonic groups and carboxylic groups, which are used to confer water solubility to these long structures. The same happens with squaraines, where the total negative charge comes from these groups plus one negative charge from the squaric acid located in the middle of the polymethine chain (Figures 4.1 and 4.61).

The two dyes in the cell $C_{0,3}$ are also worth noting. They have no negatively charged groups, only three positive charges. These dyes are rhodamine 6G + 3 and the triangulenium DAOTA4 [12] (Section 4.22).

Figure 5.10 was generated to identify dyes that are both the smallest and brightest within each cell denoted by $C_{Nn,Np}$, where N_n is the number of negative charges and N_p is the number of positive charges of the predominant ionic species at the pH specified. Now the abscissa within each cell indicates the number of atoms of each structure from 0–150 nm (dyes made of more than 150 atoms were excluded from this scatterplot) and the ordinate shows the brightness divided by the number of atoms (ranging from 0–2.62 $mM^{-1}cm^{-1}$). The numbers in the

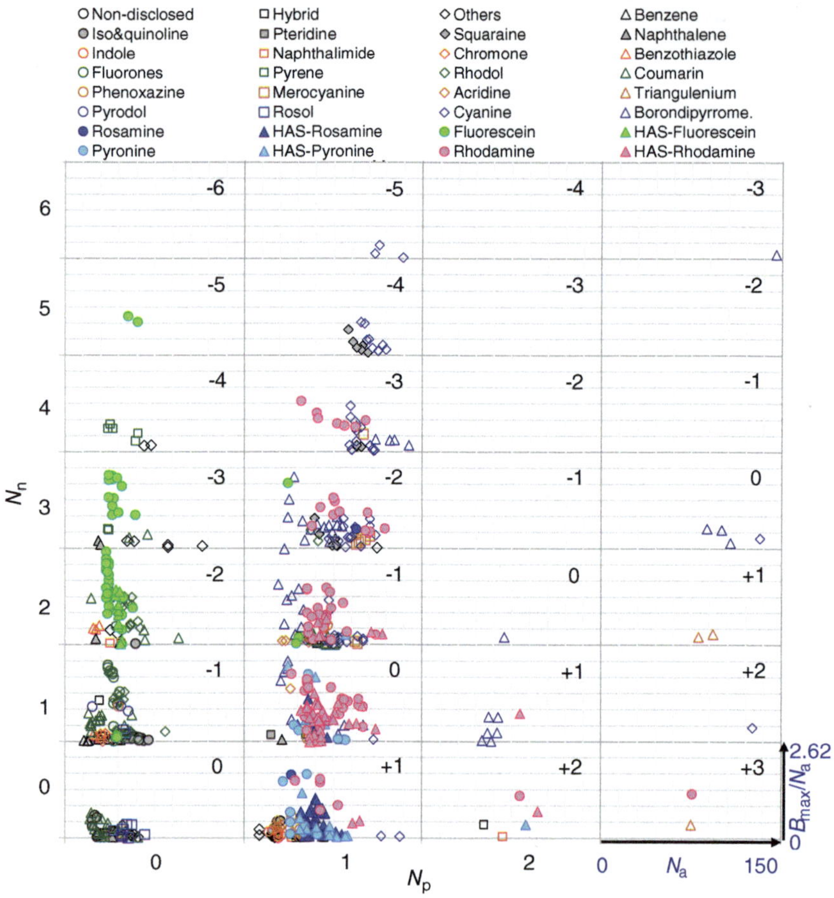

Figure 5.10 Scattergram showing the number of positive charges (N_p), number of negative charges (N_n), net charge, size (N_a), and brightness of molecules divided by their number of atoms (B_{max}/N_a).

upper right side of each cell indicate the net charge of the dyes. Note that after labeling, one negative charge (from the carboxylic group) is usually lost because it carries the reactive moiety that participates in the derivatization (or labeling) reaction.

In Figure 5.10, the smallest dyes are located on the left side of each cell and the brightest dyes (B_{max}/N_a) are located at the top of the groupings within each cell. The highest B_{max}/N_a ratio is shown by BODIPY FL (at the top of cell $C_{2,1}$), followed by chlorinated fluorescein G-C 1 (at the top of cell $C_{2,0}$), chlorinated fluorescein G-C 4, fluorescein S-H 1, fluorinated fluorescein S-H 19, chlorinated fluorescein G-C 5, and BODIPY (all at the top of cell $C_{2,0}$).

When labeling for electrodriven separations, the dye used must be small and exhibit a high B_{max}/N_a ratio at the same time. In this case the best dyes are usually, depending on the analyte, the two fluoresceins in cell $C_{5,0}$ and the rhodamine (rhodamine 6G + 3) in cell $C_{0,3}$. The triangulenium in $C_{0,3}$ (DAOTA4) would also be a good option, but it only has approximately one-third of the brightness of rhodamine 6G + 3. There are also three cyanines with a net charge of −5, shown in cell $C_{6,1}$. Many Dyomics dyes [17] have very high negative net charges; however, despite being highly water soluble, their fluorescence quantum yields and molar absorption coefficients in aqueous solutions are not known.

References

1 Bendig, J., Schedlera, U., Hardera, T. et al. (1995). Photophysical and photochemical deactivation behaviour of streptopolymethines. *J. Photochem. Photobiol. A Chem.* 91: 53–57.

2 Lei, Z., Li, X., Luo, X. et al. (2017). Bright, stable, and biocompatible organic fluorophores absorbing/emitting in the deep near-infrared spectral region. *Angew. Chem.* 129: 3025–3029.

3 Choi, H.S., Nasr, K., Alyabyev, S. et al. (2011). Synthesis and *in vivo* fate of zwitterionic near-infrared fluorophores. *J. Am. Chem. Soc.* 123: 2530–2536.

4 Pauli, J., Brehm, R., Spieles, M. et al. (2010). Novel fluorophores as building blocks for optical probes for *in vivo* near infrared fluorescence (NIRF) imaging. *J. Fluores.* 20 (3): 681–693. doi: 10.1007/s10895-010-0603-7.

5 Soper, S.A. and Mattingly, Q.L. (1994). Steady-state and picosecond laser fluorescence studies of nonradiative pathways in tricarbocyanine dyes: implications to the design of near-IR fluorochromes with high fluorescence efficiencies. *J. Am. Chem. Soc.* 116: 3744–3752.

6 Lee, S., Kim, J., and Koh, M. (2022). Recent advances in fluorescence imaging by genetically encoded non-canonical amino acids. *J. Mol. Biol.* 434 (8): 167248. doi: 10.1016/j.jmb.2021.167248.

7 Horváth, P., Šebej, P., Šolomek, T., and Klán, P. (2015). Small-molecule fluorophores with large stokes shifts: 9-iminopyronin analogues as clickable tags. *J. Org. Chem.* 80: 1299–1311.

8 Bertolino, C.A., Caputo, G., Barolo, C. et al. (2006). Novel heptamethine cyanine dyes with large Stokes shift for biological applications in the near infrared. *J. Fluoresc.* 16: 221–225. doi: 10.1007/s10895-006-0094-8.

9 Dsouza, R.N., Pischel, U., and Nau, W.M. (2011). Fluorescent dyes and their supramolecular host/guest complexes with macrocycles in aqueous solution. *Chem. Rev.* 111: 7941–7980.

10 Shaya, J., Fontaine-Vive, F., Michel, B.Y., and Burger, A. (2016). Rational design of push–pull fluorene dyes: synthesis and structure–photophysics relationship. *Chem. Eur. J.* 22: 10627–10637.

11 Atto-Tec (2020). *Fluorescent Labels and Dyes. Catalogue 2019–2020.* Siegen, Germany: ATTO-TEC GmbH.

12 Bisballe, N. and Laursen, B.W. (2020). What is best strategy for water soluble fluorescence dyes? a case study using long fluorescence lifetime DAOTA dyes. *Chem. Eur. J.* 26: 15969–15976.

13 Thermo Fisher Scientific (2022). https://www.thermofisher.com/order/catalog/product/S6110 (accessed on April 2022).

14 Kist, T.B.L. (1996). *Stochastic Schrödinger Equations Applied to Dissipative Quantum Systems in Optics.* PhD Thesis, Rio de Janeiro: PUC-RJ.

15 Kist, T.B.L., Orszag, M., Brun, T.A., and Davidovich, L. (1999). Stochastic Schrödinger equations in cavity QED: physical interpretation and localization. *Opt. B: Quantum Semiclass. Opt.* 1: 251–263.

16 Kist, T.B.L. (2020). *Open and Toroidal Electrophoresis: Ultra-High Separation Efficiencies in Capillaries, Microchips, and Slabs.* Chichester, UK: Wiley.

17 Dyomics (2017). *Fluorescent Dyes for Bioanalytical and Hightech Applications*, 8e, Jena, Germany: Dyomics GmbH.

6

Band Shapes and Excitation and Emission Ranges

6.1 Introduction

Many programs can be used to predict the spatial density probability profiles (or shapes) of the HOMO (highest occupied molecular orbital) and LUMO (lowest unoccupied molecular orbital). There are also techniques and pieces of software that can predict the energy required for the HOMO to LUMO transition. Many programs that are based on the time-dependent density functional theory, deep neural networks [1], and other theories, have been used to predict the electronic excitation energies of a great variety of fluorescent dyes and their derivatives. The electronic excitation energies of a few hundred BODIPY (borondipyrromethene) derivatives have been successfully tested [2]; however, almost all of the derivatives tested are hydrophobic and therefore the model is only expected to be accurate when these dyes are utilized in organic solvents. Predicting these excitation energies for hydrophilic structures dissolved in water (or aqueous solutions) is difficult due to the diversity, strength, and complexity of water–solute interactions (hydrogen bond, dipole–dipole, permanent charge–dipole, and transient "push-pull-charge"–dipole interactions).

It is even more difficult to predict the band shapes (excitation and emission bands) of fluorescent dyes [3]. This is especially true for flexible dyes in aqueous solutions. Knowledge of band shapes is very important as it serves as a guide to select the proper excitation sources (wavelengths), blocking filters (transmission ranges), and detectors (spectral responsivities) necessary to obtain a high effective brightness for a given fluorescent dye in a given experiment or protocol.

The first goal of this chapter is to show the typical excitation and emission characteristics (spectra, band shapes, and Stokes shifts) of the dyes of each class. When the excitation spectrum of a fluorescent dye is not available, then its absorption band is used instead. Note that from a practical point of view the differences between the excitation and absorption bands (of the longest wavelength absorption band) of a given dye in a given solvent are negligible in most cases. Therefore, no matter which spectrum is being considered, throughout this chapter the notation λ_a will be used instead of λ_x. The definitions of excitation (absorption) and emission spectra are discussed in Sections 2.7.3 and 2.7.4, respectively.

The absorption and emission spectra of many dyes in aqueous solutions (some of which were measured by the author) have already been published. The optimal excitation and emission ranges of these dyes are highlighted in gray and ranked in Appendices B and C, respectively. However, no reliable or quality absorption and/or emission spectra have been published for some dyes. For these dyes, a simple class-dependent calculation has been used to predict the approximate excitation and emission ranges (Section 6.3). Moreover, the presented simple calculation can be used for any new dye that is synthesized in the future and has an unknown or unpublished absorption and/or emission spectra. This coarse (or approximate) prediction only requires knowledge of the $\lambda_{a,max}$, $\lambda_{e,max}$, and class of the dye. Consequently, the importance of the detailed separation of the dyes into classes is again apparent.

Fluorescent Dye Labels and Stains: A Database of Photophysical Properties, First Edition. Tarso B. Ledur Kist.

Effective brightness is discussed in Section 2.7.8. Excitation wavelengths of $\lambda_{a,-0.25}$, $\lambda_{a,-0.50}$, and $\lambda_{a,+0.50}$ assure an effective brightness of $B_{max}/4$, $B_{max}/2$, and $B_{max}/2$, respectively (these parameters are depicted in Figure 6.1). Excitation involving a wavelength of between $\lambda_{a,-0.50}$ and $\lambda_{a,+0.50}$ assures a brightness of at least $B_{max}/2$. Excitation with any wavelength in the broader range of between $\lambda_{a,-0.25}$ and $\lambda_{a,+0.50}$ guarantees a brightness of at least $B_{max}/4$. This information is very important from a practical point of view. Unfortunately, the absorption spectrum and emission spectrum of many dyes remains unknown. Therefore, it is important to estimate approximate values for $\lambda_{a,-0.25}$, $\lambda_{a,-0.50}$, and $\lambda_{a,+0.50}$, as this data is very valuable for users of microscopy, analytical techniques, and many other applications.

In Section 6.3, a procedure to predict approximate values of $\lambda_{a,-0.25}$, $\lambda_{a,-0.50}$, and $\lambda_{a,+0.50}$ from only the $\lambda_{a,max}$ and class of a dye is proposed. The procedure is based on the average energy differences between $E(\lambda_{a,-0.25})-E(\lambda_{a,max})$, $E(\lambda_{a,-0.50})-E(\lambda_{a,max})$, and $E(\lambda_{a,max})-E(\lambda_{a,+0.50})$ of the dyes within a class. These energy differences, respectively termed a, b, and d (expressed in cm^{-1}, which is not a unit of energy, although it is linearly related to energy) (Figure 6.1), were assumed to be approximately constant for the dyes within a class. This is not totally true, as some dispersion around the mean values of a, b, and c is common. Such dispersion is a consequence of measurement errors and also real, small differences among the entities within a class. These differences result from modifications along the π-electron system of the scaffold of that class of dyes. Resonant structures (i.e., benzene and styryl) bound to one or more atoms belonging to the π-electron system affect $\lambda_{a,max}$ (which is common knowledge). They additionally affect, to some extent, the above cited energy differences or, equivalently, the band shape (shoulders in the wavenumber absorption bands). The large number of entities in the present database allows the average values of a, b, and c to be calculated, in addition to their dispersion values (standard deviations within a class).

A similar procedure is used to predict the optimal detection ranges for the emission bands. In this case, $\lambda_{e,max}$ (usually known) and the class of the dye are used to predict $\lambda_{e,-0.25}$, $\lambda_{e,-0.50}$, $\lambda_{e,+0.50}$, $\lambda_{e,+0.25}$, and $\lambda_{e,+0.10}$. Note that $\lambda_{e,+0.10}$ is very useful for efficient light harvesting in detection systems, as the emission bands usually exhibit long tails toward the longest wavelength of the spectrum (this phenomena, as well as the mirror symmetry between the

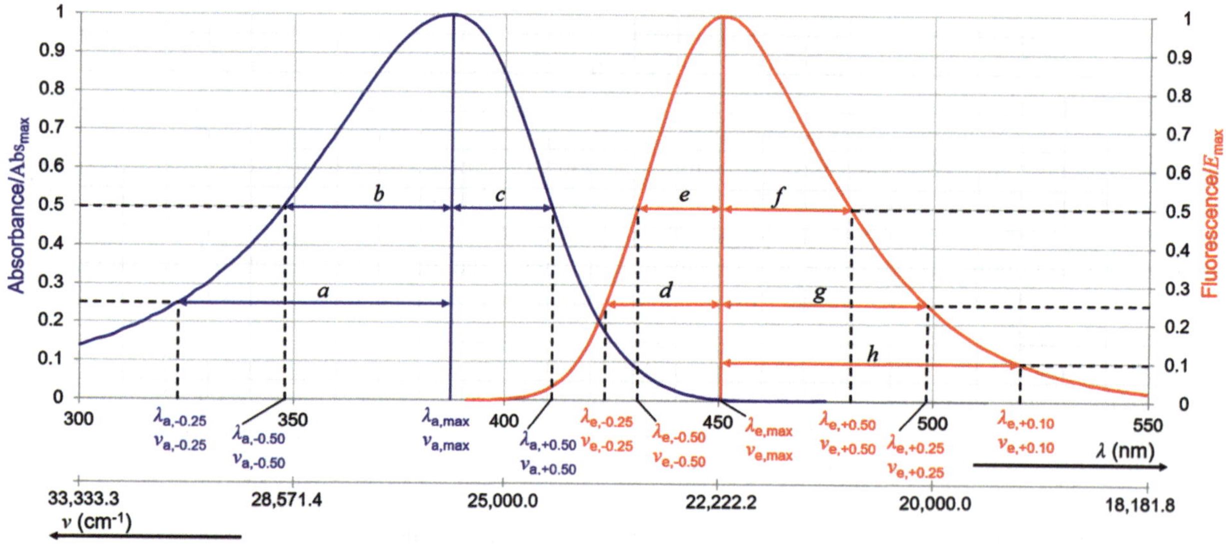

Figure 6.1 Normalized absorption and emission spectra of Pacific blue showing the parameters a, b, c, d, e, f, g, and h, as well as wavelengths $\lambda_{a,-0.25}$, $\lambda_{a,-0.50}$, $\lambda_{a,+0.50}$, $\lambda_{e,-0.25}$, $\lambda_{e,-0.50}$, $\lambda_{e,+0.50}$, $\lambda_{e,+0.25}$, and $\lambda_{e,+0.10}$. The subscript "a" stands for absorption, "e" for emission, "–" for the left side of the band, and "+" for the right side of the band. Numerals express the relative height of the band.

excitation and the fluorescence spectra of typical organic molecules, is explained in Section 2.7.5). The larger the amount of light collected the better. However, the opposite (related to excitation) is not true, i.e., knowledge of the value of $\lambda_{a,-0.10}$ is less useful because exciting at this wavelength results in a brightness of only $B_{max}/10$.

In summary, the use of any excitation wavelength between $\lambda_{a,-0.50}$ and $\lambda_{a,+0.50}$ assures a minimum brightness of at least $B_{max}/2$. The use of the full width at half maximum (FWHM), which is widely cited, is not a good guide for practitioners because the bands of almost all dyes are not symmetrical. Absorption bands always descend more sharply on the right side (the longest wavelength side) than on the left side. The opposite is observed with emission bands, as explained in Section 2.7 and Figure 2.15. Again, the importance of the knowledge of $\lambda_{e,+0.10}$ is apparent. Finally, the use of any excitation wavelength between $\lambda_{a,-0.25}$ and $\lambda_{a,+0.50}$ assures a minimum brightness of at least $B_{max}/4$, which is suitable for most applications. Note that exciting at wavelengths longer than $\lambda_{a,+0.50}$ is not recommended as it makes detection more difficult. In these cases it is harder to filter the scattered excitation light from the fluorescence light, especially when the absorption and emission bands of dyes significantly overlap (this happens for almost all dyes that have small Stokes shifts). For this reason the values of $\lambda_{a,+0.25}$ are not tabulated.

It is worth noting that there are at least a dozen spectra viewers on the Internet. They are useful tools, but generally have a few problems: i) they are restricted to only a handful of dyes; ii) they contain imprecise information (usually exaggeratedly high values of molar absorption coefficients and quantum yields in aqueous solutions); iii) the conditions used to measure photophysical parameters (solvent used, pH, buffer composition, whether the dyes are free or conjugated, among others) are not clearly shown; and iv) literature references are not presented or mentioned. In the present book, photophysical parameters are restricted exclusively to aqueous solutions with a maximum organic modifier or co-solvent(s) content of 1% (V/V), and all references are cited in Chapter 4 (~5% of the data was measured by the author).

6.2 Typical Absorption and Emission Spectra of Some Classes

In this section, the typical spectra of some classes are presented. Spectra are presented in terms of wavelength, not wavenumber, because wavelength is more commonly used in practical situations. The emission bands are characterized by a sharp ascendent on the left side of the band and a slow descendent (tail) on the right side of the band (Section 2.7.5). The sharpness of the left side is related to the Stokes shift. Dyes (classes) characterized by large Stokes shifts exhibit much wider bands and with smoother variations of the bands.

The right side (descending section) of an absorption band is usually a good mirror image of the left side (ascendent part) of the corresponding emission band. However, the left side of an absorption band usually exhibits shoulders. These shoulders can originate from many sources. Sometimes they are simply overtones; i.e., the second excited electronic state (S_2) lies close to the first excited electronic state (S_1). However, note that spontaneous fluorescence radiative transitions predominantly occur from the first excited state (S_1) to the ground state (S_0), which explains the simplicity of the emission bands (Section 2.7.5). The second reason why these shoulders may occur in the absorption bands is due to the existence of multiple π-electron systems. Let us take advantage of one extreme example: 1,3,6,8-pyrenetetrasulfonic acid is made by fusing four benzene rings with four sulfonic groups (at the 1, 3, 6, and 8 positions). This molecule consists of the superposition of many π-electron systems, as evidenced by the superposition of multiple absorption bands. The overtones of these transitions are deeper into the UV and should not be present in Figure 6.2F, because the spacings between adjacent electronic energy levels ($E(S_{n+1}) - E(S_n)$) increases for smaller values of $\lambda_{x,max}$ (nm) (red → green → blue → violet). Overtones are more easily observed for dyes with $\lambda_{x,max}$ values in the red range of the spectrum than in the violet. However, if this dye is modeled as the membrane of a drum (2D π-electron system), then the multiple transitions are the overtones themselves.

Figure 6.2 shows the typical absorption and emission spectra of one sample dye from each important class. Here, solutions were prepared in water and diluted with water until the maximum absorbance was ~0.050 in a

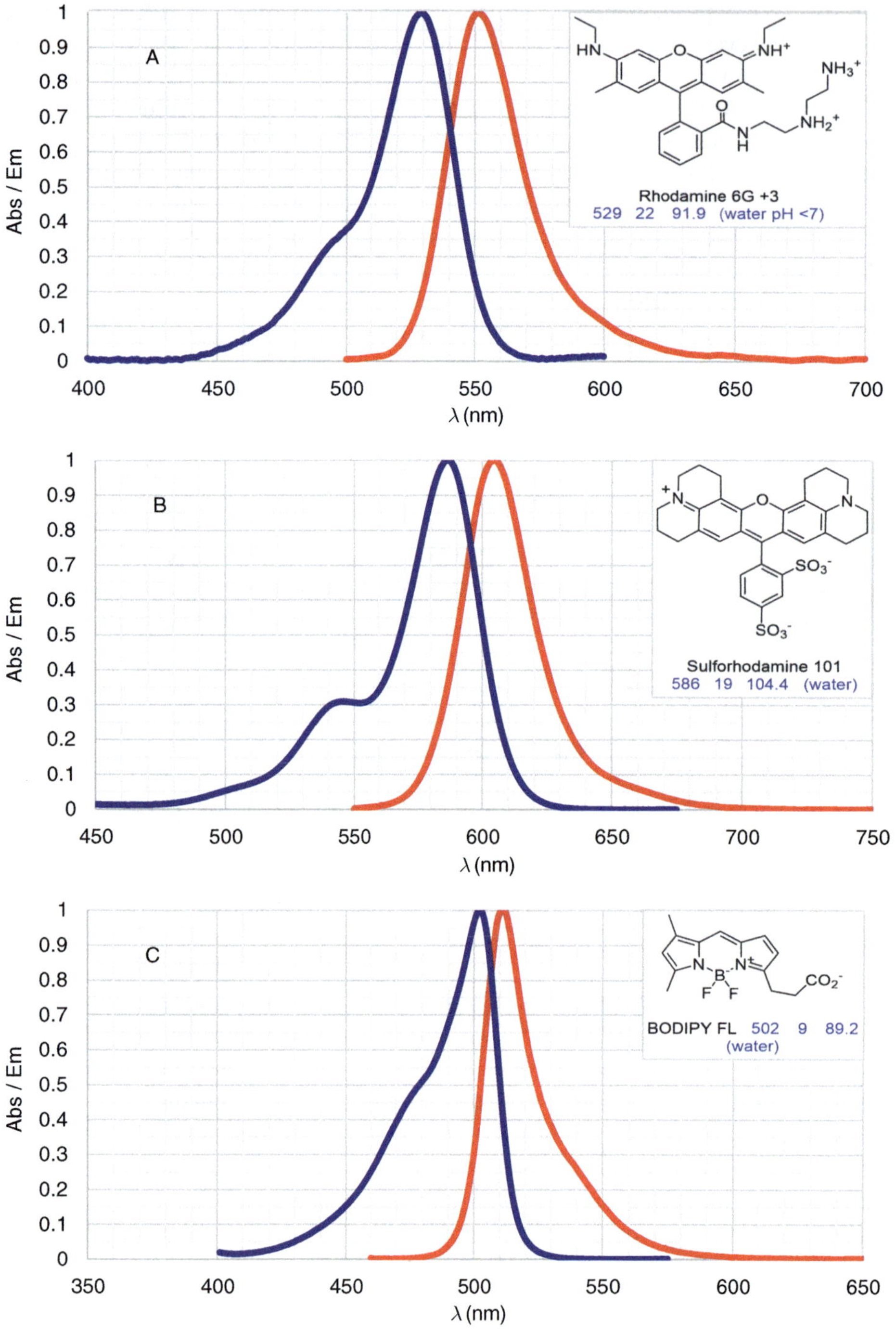

Figure 6.2 Typical normalized absorption and emission spectra of each class. Aqueous solutions were prepared with absorbance maximum of ~0.050 (with the exception of H and I that was ~0.500). The insets show the molecular structure and the following three photophysical parameters (in blue): $\lambda_{x,max}$ (nm); Δ_S (nm); and B_{max} (mM^{-1}cm^{-1}); in this order: (A) rhodamine 6G + 3 (rhodamine), (B) sulforhodamine 101 (sulforhodamine), (C) BODIPY FL (borondipyrromethene), (D) oxazine 170 (oxazine), (E) Pacific blue (coumarin), (F) 1,3,6,8-pyrenetetrasulfonic acid (pyrene), (G) lucifer yellow CH (naphthaliimide); (H) Cascade yellow (oxazole-pyridine), (I) Fluorescein (fluorescein) [5], and (J) sulfo-cyanine Cy3 [4].

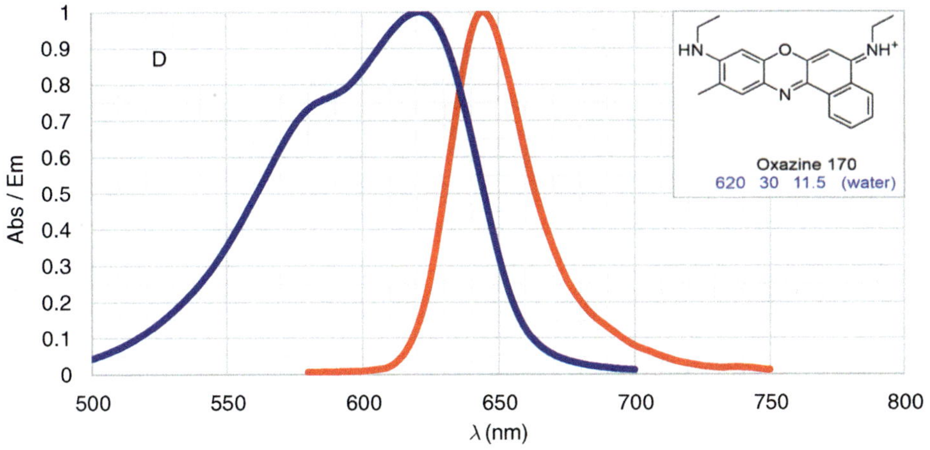

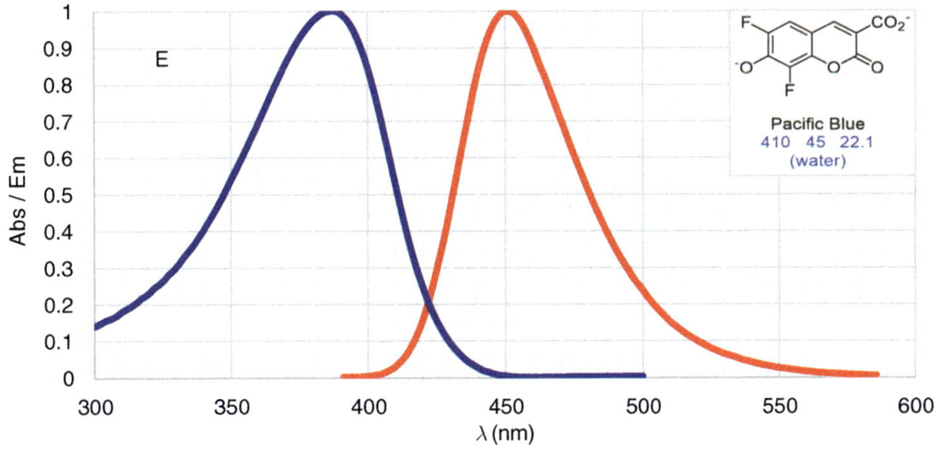

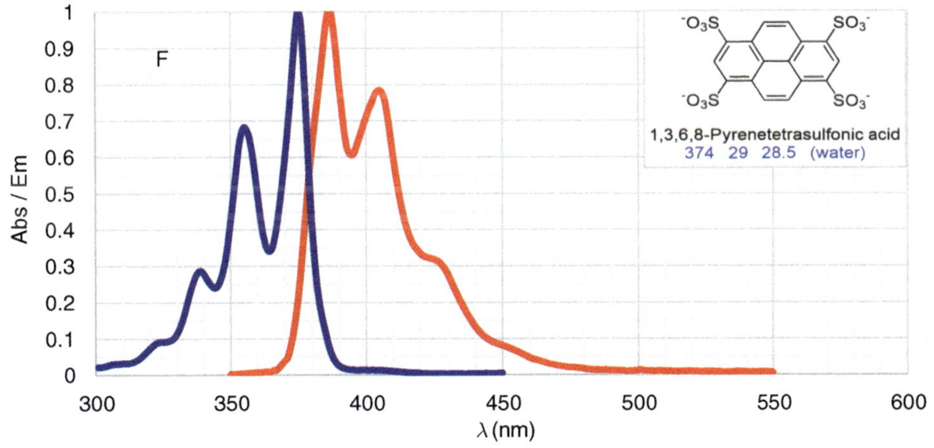

Figure 6.2 (Continued)

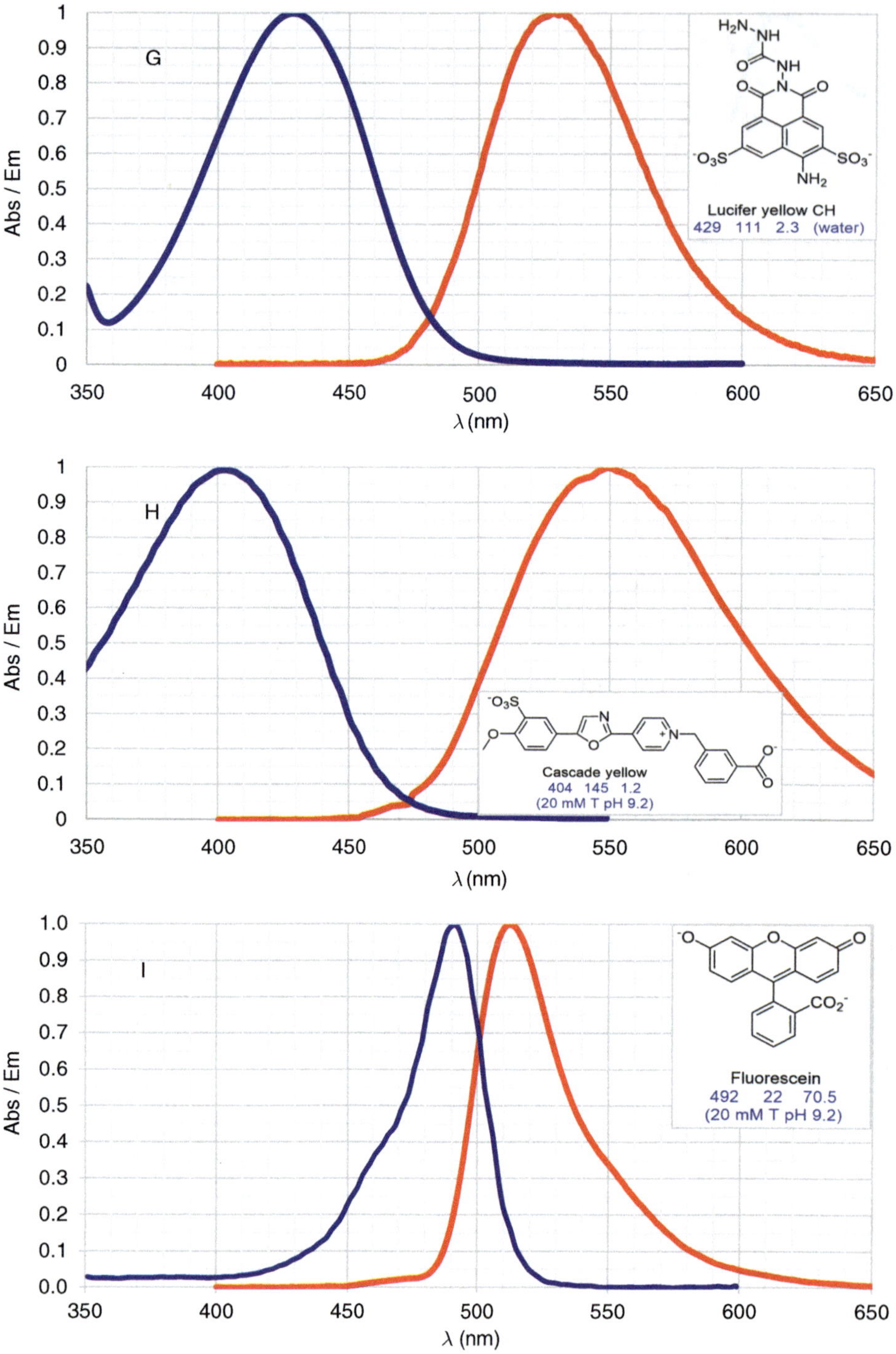

Figure 6.2 (Continued)

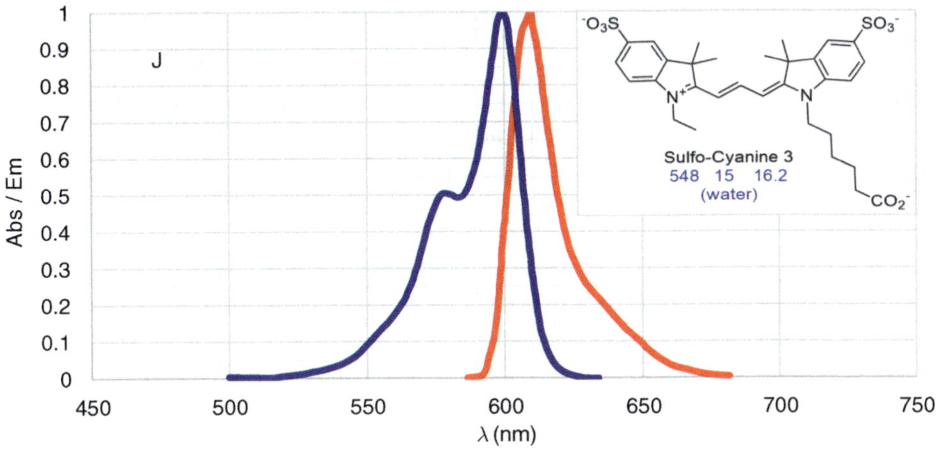

Figure 6.2 (Continued)

10 mm cuvette. Correct absorption bands (within the Lambert–Beer linearity range) and emission bands (without distortions due to self-absorption) were recorded in this way [4]. Moreover, the wavelength ranges of all spectra in Figure 6.2 were adjusted to 300 nm intervals to facilitate comparison of the spectra among themselves. Upon comparison, the following properties can be noticed. Some classes are characterized by wide bands while others are characterized by narrow bands. There are classes with symmetric bands and others with asymmetric bands, and some have large Stokes shifts while others have small strokes shifts, to mention but a few.

6.3 Coarse Prediction of Excitation and Emission Ranges

The mathematical relationships of $\lambda_{a,-0.25}$, $\lambda_{a,-0.50}$, $\lambda_{a,max}$, $\lambda_{a,+0.50}$, $\lambda_{e,-0.25}$, $\lambda_{e,-0.50}$, $\lambda_{e,max}$, $\lambda_{e,+0.50}$, $\lambda_{e,+0.25}$, and $\lambda_{e,+0.10}$ (in units of nm) with *a*, *b*, *c*, *d*, *e*, *f*, *g*, and *h* (in units of cm^{-1}) are shown in Table 6.1. The average values of *a*, *b*, *c*, *d*, *e*, *f*, *g*, and *h* for each class were determined using the present database. Note the importance of the definition of a class (discussed in detail in Section 4.1). The mathematical relationships shown in Table 6.1 were used to estimate the approximate values of the excitation ranges ($\lambda_{a,-0.25}$, $\lambda_{a,-0.50}$, and $\lambda_{a,+0.50}$) and emission ranges ($\lambda_{e,-0.25}$, $\lambda_{e,-0.50}$, $\lambda_{e,+0.50}$, $\lambda_{e,+0.25}$, and $\lambda_{e,+0.10}$) for each dye.

The average values of *a*, *b*, *c*, *d*, *e*, *f*, *g*, and *h* for each class are given in Table 6.2. These average values, which were extracted from the current database, can be used to predict the approximate (≈) excitation and emission ranges of any dye that already exists or will be synthesized in the near future. Note that in addition to the class of the dye, it is also necessary to know $\lambda_{a,max}$ to calculate $\lambda_{a,-0.25}$, $\lambda_{a,-0.50}$, and $\lambda_{a,+0.50}$, and $\lambda_{e,max}$ is needed to calculate $\lambda_{e,-0.25}$, $\lambda_{e,-0.50}$, $\lambda_{e,+0.50}$, $\lambda_{e,+0.25}$, and $\lambda_{e,+0.10}$ using the equations given in Table 6.1. The $\lambda_{a,max}$, $\lambda_{e,max}$, and class of almost all currently available dyes are readily available. For the convenience of the reader, excitation and emission ranges are given in the Appendices B and C, respectively. For some dyes they are based on the real spectra (Excitation and emission ranges highlighted with gray in appendices B and C. Excitation ranges highlighted with gray in appendices E, F and G), while for others the values were predicted (calculated) using the equations shown in Table 6.1 (Non-colored data of excitation and emission ranges of appendices B and C; non-colored data of excitation ranges of appendices E, F and G).

Table 6.1 Approximate (≈) relationships of the wavelengths $\lambda_{a,-0.25}$, $\lambda_{a,-0.50}$, $\lambda_{a,max}$, $\lambda_{a,+0.50}$, $\lambda_{e,-0.25}$, $\lambda_{e,-0.50}$, $\lambda_{e,max}$, $\lambda_{e,+0.50}$, $\lambda_{e,+0.25}$, and $\lambda_{e,+0.10}$ (in nm), with the parameters *a*, *b*, *c*, *d*, *e*, *f*, *g*, and *h* (in cm^{-1}). The constant 10^{-7} has units of cm^{-1}nm. Typical values of *a*, *b*, *c*, *d*, *e*, *f*, *g*, and *h* (parameters depicted in Figure 6.1) for each class are tabulated in Table 6.2.

Relationship	Definitions (All in cm^{-1})
$\lambda_{a,-0.25} \approx \left(\frac{1}{\lambda_{a,max}} + \frac{a}{10^7}\right)^{-1}$	$a = \nu_{a,-0.25} - \nu_{a,max}$
$\lambda_{a,-0.50} \approx \left(\frac{1}{\lambda_{a,max}} + \frac{b}{10^7}\right)^{-1}$	$b = \nu_{a,-0.50} - \nu_{a,max}$
$\lambda_{a,+0.50} \approx \left(\frac{1}{\lambda_{a,max}} - \frac{c}{10^7}\right)^{-1}$	$c = \nu_{a,max} - \nu_{a,+0.50}$
$\lambda_{e,-0.25} \approx \left(\frac{1}{\lambda_{e,max}} + \frac{d}{10^7}\right)^{-1}$	$d = \nu_{e,-0.25} - \nu_{e,max}$
$\lambda_{e,-0.50} \approx \left(\frac{1}{\lambda_{e,max}} + \frac{e}{10^7}\right)^{-1}$	$e = \nu_{e,-0.50} - \nu_{e,max}$
$\lambda_{e,+0.50} \approx \left(\frac{1}{\lambda_{e,max}} - \frac{f}{10^7}\right)^{-1}$	$f = \nu_{e,max} - \nu_{e,+0.50}$
$\lambda_{e,+0.25} \approx \left(\frac{1}{\lambda_{e,max}} - \frac{g}{10^7}\right)^{-1}$	$g = \nu_{e,max} - \nu_{e,+0.25}$
$\lambda_{e,+0.10} \approx \left(\frac{1}{\lambda_{e,max}} - \frac{h}{10^7}\right)^{-1}$	$h = \nu_{e,max} - \nu_{e,+010}$
Note: the parameter 10^7 is given in units of cm^{-1}nm.	Where ν represents the wavenumber (cm^{-1}).

Table 6.2 The average values of *a*, *b*, *c*, *d*, *e*, *f*, *g*, and *h* for each class, expressed in units of cm^{-1}. These values can be used to predict approximate excitation and emission ranges of the dyes (Table 6.1).

Class	*a*	*b*	*c*	*d*	*e*	*f*	*g*	*h*
Acridines	2677	1565	661.8	1462.6	1066.2	1607.9	2482.5	3384
$B(OH)_2$-Rosamines	1460.4	543	384.7	563.4	405.8	548.8	886.5	1457.9
Borondipyrromethenes	1818.8	956.5	331.7	470.6	311.2	485.5	1085.2	1616.2
$C(CH_3)_2$-Pyronines	1398.7	568.5	349.1	756.4	485.6	572.5	1005.3	1426.8
$C(CH_3)_2$-Rhodamines	1275.7	481.9	364.8	555.7	391.3	737.8	1363.1	2045.7

(Continued)

Table 6.2 (Continued)

Coumarins (7-amino)	3031.6	2024.5	1294.6	1391.6	1025.5	1523.8	2425.5	3323.9
$Ge(CH_3)_2$-Rhodamines	1013.9	478.1	427.4	573.5	377.6	490.2	827.3	1550.5
Naphthalimides	3356.6	2200.2	1665	1573.4	1120.8	1271.5	1845.4	2464.9
Phenoxazines	2019.2	1005.4	400.2	535.9	356.9	518.3	910.3	1677.2
Pyronines	1741.9	785.8	425.2	651.2	456.4	601.9	960.7	1683.9
Rhodamines	1670.1	640	454.2	668.6	479	704.3	1388.6	2051.3
Rosamines	1758.5	622.1	487.9	704.3	506.7	565.9	1014.9	1820.6
Rosols	1972.8	1052.3	454.9	1769	1227	1610	2370	3275
$Si(CH_3)_2$-Fluorones	1620.7	581.9	376.7	578.6	430.3	515	924.6	1455.2
$Si(CH_3)_2$-Rhodamines	1498.1	571.5	396.9	610	440.9	459	770	1446.5
$Si(CH_3)_2$-Rosamines	2068.4	938.9	397.4	565.2	393	457.9	849.3	1376.4
Squaraines	1176.3	408.4	411.9	584.3	409.3	410.8	715.4	1249.2
Sulforhodamines	1923.1	800.7	444.2	639.6	445.7	557.8	1093	1708.9
Sulforhodols	3095.2	1823.4	316.2	791.4	580.3	1567.9	2637.9	3317.3
Trianguleniums	3072.1	1704.1	578.5	734.4	521.5	876.1	1540.2	2107.4

References

1 Shao, J., Liu, Y., Yan, J. et al. (2022). Prediction of maximum absorption wavelength using deep neural networks. *J. Chem. Inf. Model.* 62: 1368–1375. doi: 10.1021/acs.jcim.1c01449.

2 Wang, J.-N., Jin, J.-L., Geng, Y. et al. (2013). An accurate and efficient method to predict the electronic excitation energies of BODIPY fluorescent dyes. *J. Comput. Chem.* 34: 566–575.

3 Zaleśny, R., Murugan, N.A., Gel'mukhanov, F. et al. (2015). Toward fully nonempirical simulations of optical band shapes of molecules in solution: a case study of heterocyclic ketoimine difluoroborates. *J. Phys. Chem.* 119 (21): 5145–5152. doi: 10.1021/jp5094417.

4 Kist, T.B.L. (2022). Absorption measured on an Agilent/Varian spectrophotometer, model Cary 5000, instrument version 2.24. fluorescence emission spectra measured on a Shimatzu spectrofluorometer model RF-5301PC.

5 Kist, T.B.L. (2003). Absorption spectrum measured using a nanodrop spectrophotometer model ND-1000 and emission spectra measured using Avantes Fiberoptic Spectrophotometer and Xenon light Source XE-2000.

7

Measuring Photostability and Mitigating Photobleaching

7.1 Introduction

The definitions of photostability, photobleaching, photoblinking, photoswitching, photocage compounds, and photochemical degradation are given in Section 2.8. An extensive list of photochemical phenomena and their definitions is detailed in the "Glossary of Terms used in Photochemistry" [1].

Photobleaching is one of the main limitations in fluorescence spectroscopy, fluorescence detection, and imaging techniques (especially in optical super-resolution microscopy, or nanoscopy). Fluorescence fading sometimes occurs too fast and results in short observation times, with poor contrasts, definitions, and/or detection limits. While some light detectors have a quantum efficiency that approaches 100%, and optical systems used for light harvesting have also almost reached their maximum possible performance, the current Achilles heel of the field is the photostability of the fluorescent dyes used. Fluorescent dye labels and stains that have high brightness values and bleaching numbers, defined as the average number of photons a molecule emits until photobleaching occurs, are highly desired.

The excitation wavelengths used in fluorescence nanoscopy and fluorescence detection schemes are usually close to $\lambda_{x,max}$; i.e., close to the excitation maximum of the longest excitation wavelength band. In almost all cases they are located between $\lambda_{x,-0.10}$ and $\lambda_{x,+0.10}$. When the "two photon excitation" technique is used, then the necessary wavelength is $\lambda \approx \lambda_{x,max}/2$. This is rarely used, as the brightness is much lower than the normal "one photon excitation."

Dyes are primarily "pumped" from the electronic ground state (S_0) to an energy level (electronic + vibrational state) that is located just above the first electronic excited state (S_1) and below the second electronic excited state (S_2) (Figure 2.15). Note that there are many quantized vibrational states just above each electronic state (S_n). For most of the good fluorescent dyes, this first excitation event is followed by a fast, non-radiative transition from the highly excited vibrational states where they first end up (above S_1) to a lower energy vibrational state (still above S_1) that is compatible with the temperature of the liquid solution. Then, after a mean-lifetime of a few nanoseconds, the dye will transit from S_1–S_0 through a photoemission event called fluorescence. However, just by chance, one of the following events may occur instead, some of which may lead to the bleaching of the dye:

1) *Photophysical event*: Upon excitation (from S_0–S_1) the fluorescent dyes may undergo a second excitation from S_1 to a higher electronic energy level (S_2 or even higher). The probability of this happening increases when dyes exhibit a long fluorescent lifetime (>20 ns) and when dyes are excited with high radiances (Wm^{-2}). These "two successive excitations" should not be confused with "two photon excitation" events. Two photon excitation events may also move the electron directly from S_0–S_2, but there is only a significant probability of this occurring when: i) the energy difference between S_0 and S_2 is similar to the sum of the energy of two photons with wavelength $\lambda_{x,max}$; ii) the excitation beam has high radiance; and iii) the fluorescent dyes exhibit a high

Fluorescent Dye Labels and Stains: A Database of Photophysical Properties, First Edition. Tarso B. Ledur Kist.

cross-sectional area for two-photon excitation. Fluorescent dyes at state S_2 or higher may undergo a chemical reaction to produce an adduct (sometimes a reversible intermediate, but further irreversible reactions including hydrolysis may occur) or end up in a triplet state (T_n) through an intersystem crossing event (see descriptions in events 2 and 3 below).

2) *Photochemistry event 1*: After excitation, the fluorescent dye ends up in the closest triplet state (T_1). This is a relatively long-lived, non-fluorescent excited state. The long mean-lifetime of this state is problematic because it causes blinking (reversible) and increases the probability of other follow-up chemical reactions. Eventually, this leads to irreversible bleaching through hydrolysis or other reactions.
3) *Photochemistry event 2*: After excitation, the fluorescent dye ends up in the closest triplet state (T_1) and is then deactivated by a triplet state quencher such as dissolved oxygen (O_2) or another ROS (reactive oxygen species). Solutions rich in dissolved O_2 and ROS are very effective quenchers, which is beneficial as quenching mechanisms reduce the mean-lifetime of the triplet and, consequently, the mean blinking duration time and photobleaching described in event 2 above. However, this does increase the concentration of the highly oxidizing singlet oxygen ($^1O_2$) and activated ROS species (ROS^*). These species are cytotoxic, are very efficient at oxidizing organic matter, and even contribute to the irreversible oxidation and bleaching of the dyes themselves.

From events 2 and 3 above, it is clear that dissolved oxygen has an ambivalent effect on photostability. However, it is effective in deactivating triplet states. This is important because it reduces the blinking time interval and the probability of other follow-up chemical reactions from these triplet states, which may eventually lead to bleaching (see event 2 above). On the other hand, this deactivation generates singlet oxygen ($^1O_2$) and other activated ROS compounds which are very effective at oxidizing CH-rich compounds. In CH-poor environments, such as plain water and buffer solutions made of inorganic acids, bases, and salts, dyes are oxidized at a higher rate and fluorescence fading is even more pronounced (see event 3 above) (Dobson et al.) [2]. In CH-rich environments, dyes are less affected because the other organic compounds also react with the $^1O_2$ and ROS compounds generated. This explains some discrepancies found in the literature, as some studies show the detrimental effect of dissolved oxygen in some experiments while others report the negligible or even opposite effect of degassed media. Some measures that can prevent these problematic outcomes are discussed in Section 7.3.

There are many comprehensive and important publications related to photostability. For a detailed description of the photooxidation processes of cyanines, see Gorka and Schnermann [3]. The photobleaching of four BODIPY and four Fluorescein dyes has been studied and compared by Hinkeldey et al. [4], using fluorescence correlation spectroscopy. They found BODIPY derivatives to be more stable than fluorescein derivatives. For a detailed chronological description of the landmark publications that lead to the current understanding of photobleaching, see Gong et al. [5]. Finally, for a concise and elegant description of the molecular mechanisms of fluorophore photobleaching, see Zheng and Lavis [6].

7.2 Measuring Photostability

There are well established testing procedures for the evaluation of the photostability of pharmaceutical drug substances and products [7]. These regulatory guidelines were harmonized in the 1980s and most are based on exposure to UV radiation over long time intervals. However, these testing protocols do not take into account the exact experimental conditions under which fluorescent dyes are commonly used as labels and strains. There is still a lack of consensus on the ideal protocol to test the photostability of fluorescent dyes, because many more variables affect the measurements. For example, illumination (or irradiation) may be with a laser line (Figure 7.1A) or a continuous spectrum (Figure 7.1B); radiance may be high or low, spatially homogeneous or not, and pulsed or continuous; temperature may be controlled or not; dye concentration may be higher or lower; buffer used; buffer pH; whether the solution is degassed or not; and both the additives and organic modifiers present (some have a protective effect and delay fluorophore fading). Additionally, some authors monitor absorbance

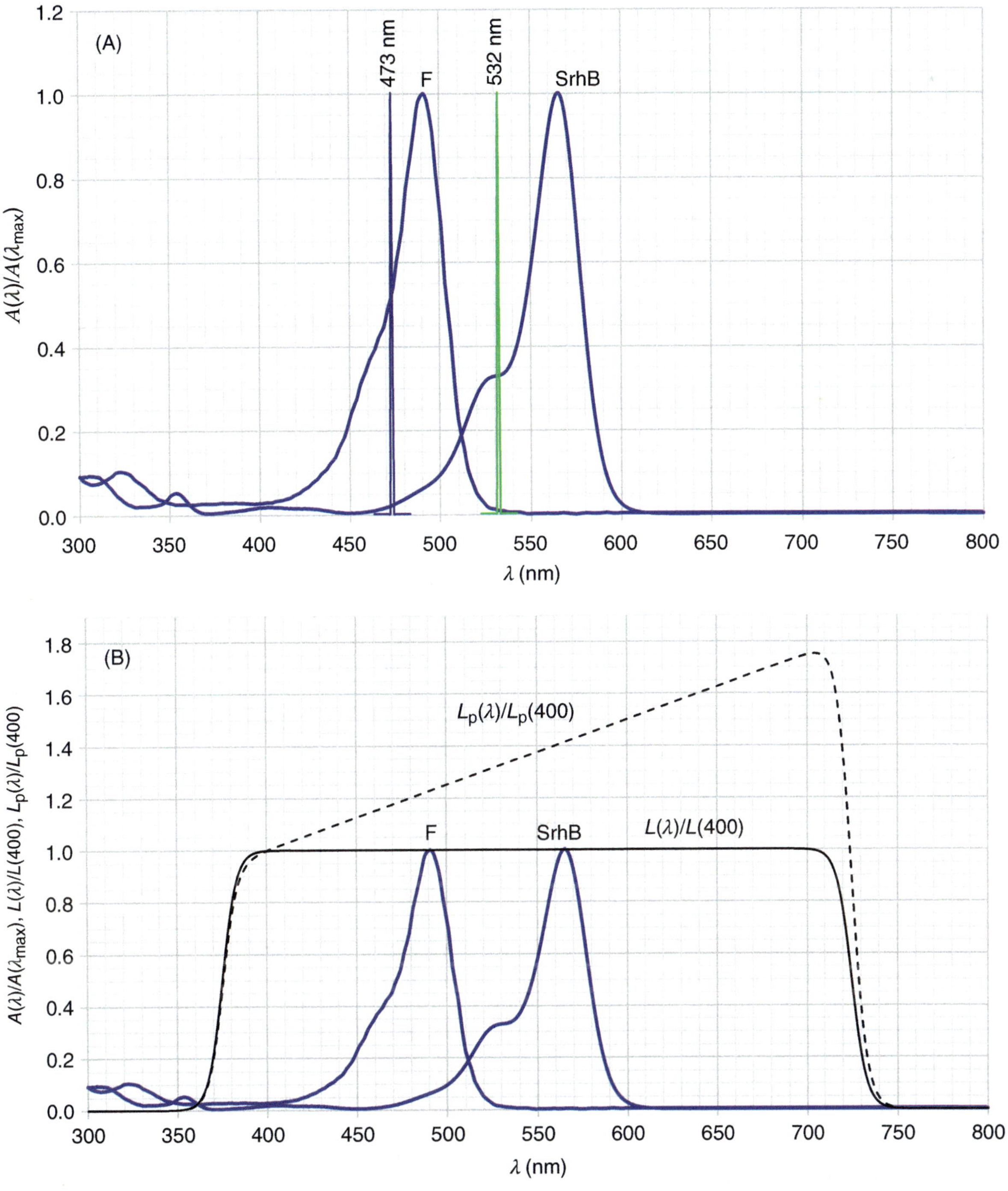

Figure 7.1 Possible irradiation conditions using (A) laser lines or (B) a lamp with a constant spectral radiance ($L(\lambda)$, continuous black line) and respective spectral photon radiance ($L_p(\lambda)$, dashed black curve) used in a photobleaching test of a dye, using fluorescein as a reference. Absorption band of fluorescein (F) and sulforhodamine B (SrhB).

fading over time, while others monitor fluorescence. The latter seems to be the more appropriate because fluorescence is the quantity of interest in the applications, not absorbance. Moreover, fluorescence fading usually precedes absorbance fading during the photobleaching process. Worse yet, an increase of absorption caused by the accumulation of colored photoproducts during irradiation can be observed [8].

Figure 7.1B shows another illumination system using a lamp that has a constant spectral radiance ($L(\lambda)$) from 400–700 nm, normalized to L(400 nm). Note that the corresponding spectral photon radiance ($L_p(\lambda)$, density probability distribution expressing the number of photons $s^{-1}m^{-2}nm^{-1}$) increases continuously with wavelength. This means that the flux of photons crossing a sample is 7/4 times larger at 700 nm than at 400 nm, which is unfair for dyes absorbing at longer wavelengths as they will be subjected to a higher rate of excitations and emissions. Nevertheless, the widely adopted condition of irradiation is a constant spectral radiance and not a constant spectral photon radiance. When lasers are used, care must be taken to ensure a homogenous illumination (of the whole sample).

There are many additional concepts that are peculiar to the field of fluorescent dyes. These include: excitation rates used (mean number of excitations per molecule per second); photoswitchable properties; photoblinking properties; mean number of photons emitted per single fluorophore before "photoburst;" photobleaching cross-section; and the quantum yield of photodegradation (Φ_{pd}). The Φ_{pd} expresses the number of molecules bleached per absorbed photon. It is a dimensionless quantity which lies in the 10^{-5}–10^{-7} range for most common dyes [9], and is dependent on temperature, radiance applied, dye concentration, and both the composition and pH of the buffer solution.

Fluorescent fading rates are usually presented in graphs which compare the fluorescent radiant power (output) of two or more fluorophores over time. Usually, the radiant power of one vanishes faster than the other (as decreasing exponentials) over dozens of minutes. Unfortunately, again there is no consistent data within the literature with which to compare, as most measurements were taken under conditions that are not fully detailed. Most of the available data compares the photostability of fluorophores to fluorescein. A consensus and reliable protocol for comparing photostability values also still needs to be sought and harmonized in this case.

Photobleaching mean-lifetime (τ_{pb}) (or survival mean-lifetime before photobleaching) is another interesting parameter and can be calculated from these fluorescence fading graphs. In its measurement, an ensemble of given fluorescent molecules are subjected to irradiation from constant radiance, and the fading of fluorescence is monitored over time. Note that the whole sample volume must be irradiated with homogenous radiance. A reference is always used (e.g., fluorescein) and the illumination conditions of both the dye and the reference must be identical (with fair spectral radiance, see Figure 7.1). Both are monitored simultaneously and from the generated graphs the photobleaching mean-lifetime relative to fluorescein ($\tau_{pb,r}$) can be extracted (Figure 7.2). Photobleaching mean-lifetime values are assumed to exponentially decay; however, this is not always the case as, among other factors, the consumption of dissolved oxygen in the medium changes the photobleaching rate over the course of the experiment.

The use of a reference (e.g., fluorescein) under identical illumination conditions (with fair spectral radiance, see Figure 7.2) gives more reproducible values of $\tau_{pb,r}$ compared to τ_{pb}. Moreover, most of the time, when lasers are used, then a different laser must be used for each dye (473 nm for fluorescein and 532 nm for sulforhodamine B, in the example given in Figure 7.1A). Therefore, the obtained τ_{pb} values must be divided by the radiant power of each laser to correct for laser power output differences. In addition, each τ_{pb} value must be corrected for the mismatch between the laser wavelength and λ_{max} of each dye. For example, in Figure 7.1A, the τ_{pb} of sulforhodamine B must be multiplied by $A_{SrhB}(532\ nm)/A_{SrhB}(\lambda_{max})$ and the τ_{pb} of fluorescein needs to be multiplied by $A_F(473\ nm)/A_F(\lambda_{max})$, where A represents the absorbances of the solutions used.

The photobleaching mean-lifetimes can be calculated directly from the original decaying curves (shown in Figure 7.2A) with the following equation:

$$\tau_{pb} = \frac{1}{n}\sum_{i=1}^{n} \frac{\text{Time}_i}{-\ln(\text{Emission intensity}_i)}, \tag{7.1}$$

where n is an arbitrarily chosen number that allows the calculation of an average value (and standard deviation or other statistical quantity). However, using the angular coefficients of the lines of Figure 7.2C produces more precise

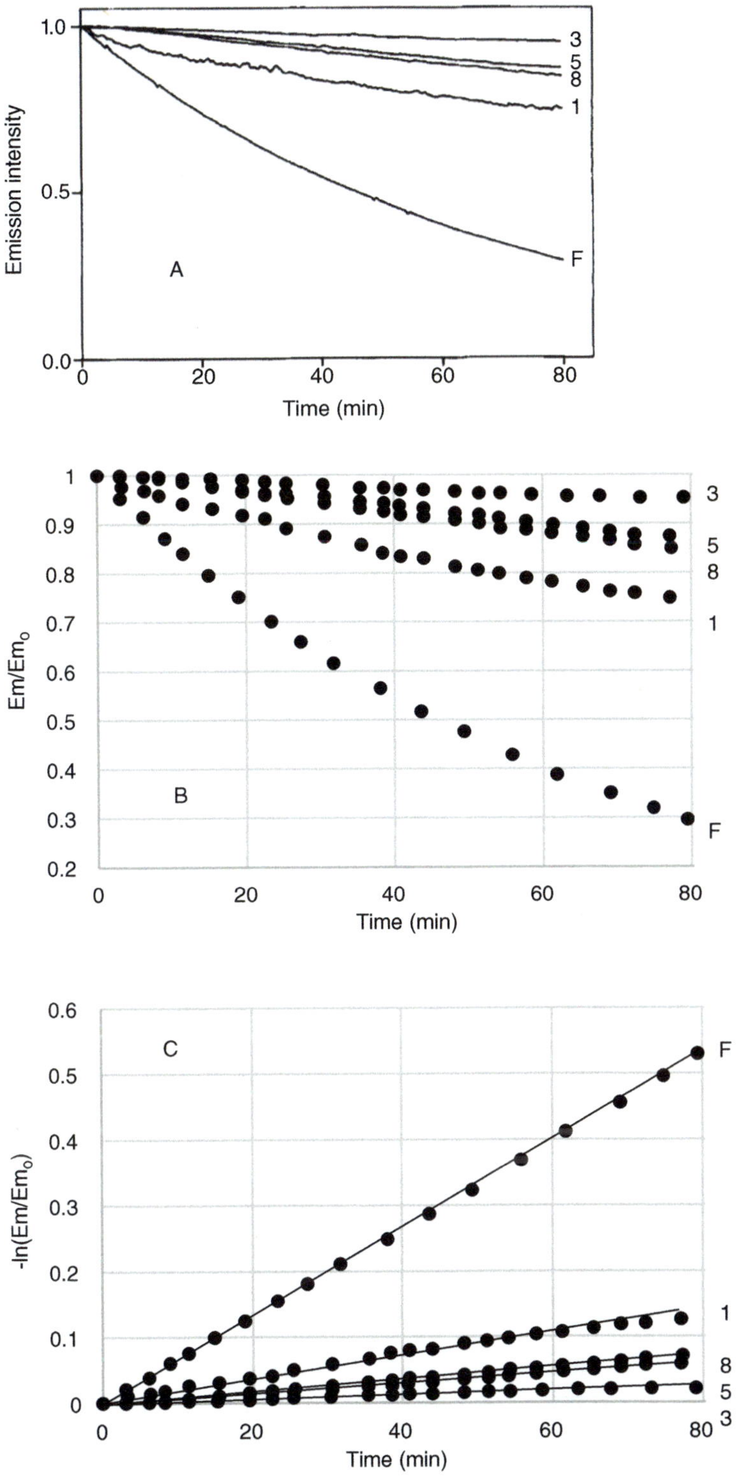

Figure 7.2 Photobleaching of a series of rhodols, using fluorescein as a reference. (A) Reproduced with permission from Whitaker et al. (1992) / Elsevier; (B) digitalized curves; (C) semi-log graph where the inverses of the angular coefficients are the respective photobleaching mean-lifetimes (τ_{pb}); and

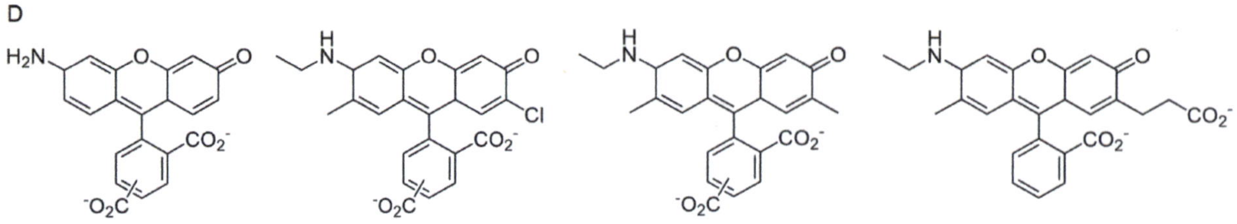

Figure 7.2 (Cont'd) (D) molecular structures of these dyes. From these, photobleaching relative mean-lifetimes ($\tau_{pb,r}$) can easily be calculated.

Table 7.1 Some relative photobleaching mean-lifetimes ($\tau_{pb,r}$) calculated from the decay curves found in the literature and using Equation (7.1). These references are indicated within this table.

Entry	Fluorescent Dye	$\tau_{pb,r}$
	Reference: fluorescein Solvent: 50 mM potassium phosphate buffer at pH 8.0 [10]. Decay curves shown in Figure 7.2A and dye structures in Figure 7.2D.	
1	3,2-Cl-7-Me-6-EtAm-carboxyrhodol	24.06
2	5,2,7-DiMe-6-EtAm-carboxyrhodol	9.81
3	8,2-PrAc-7-Me-6-EtAm-rhodol	8.48
4	1,5(6)-carboxyrhodol	3.46
Ref.	Fluorescein	1.00
	Reference: fluorescein Solvent: "aqueous buffer" at pH 8.0 (Mitronova et al. 2010) [11]. Decay curves shown in Figure 7.3A and dye structures in Figure 7.3B.	
5	3-H	121.07
6	Rhodamine green	74.15
7	1f	33.91
8	1e	26.18
9	Oregon Green 488[1]	2.72
Ref.	Fluorescein	1.00
	Reference: fluorescein Solvent: 1 mM NaOH [12], Figures 15 and 16 of page 71 of [12]. Structures are given in Sections 4.8.2 and 4.8.3.	
10	Mercaptofluorescein MK-69	68.55
11	Mercaptofluorescein MK-67	71.28
12	Mercaptofluorescein MK-75	16.09

1 AlexaFluor, BODIPY, and Oregon green are trademarks of Thermo Fisher Scientific (Waltham, MA, USA). Atto is a trademark of Atto-Tec GmbH (Siegen, Germany). Cy3, Cy3.5, Cy5, and Cy5.5 are trademarks of GE Healthcare (Chicago, IL, USA).

Table 7.1 (Continued)

Entry	Fluorescent Dye	$\tau_{pb,r}$
13	Mercaptofluorescein MK-43	9.21
14	Mercaptofluorescein MK-61	8.54
15	2′,7′-Dichlorofluorescein (DCFl)	2.23
Ref.	Fluorescein	1.00
	Reference: Cy5 Solvent: PBS pH 7.4 [13], PBS pH 7.2/7.5, and 22/20°C [14]. Structures are given in Chapter 4.	
16	Atto 740	359.16
17	Atto 725	355.61
18	Atto 655	126.11
19	Atto 647N	7.69
20	Alexa Fluor 660	1.69
Ref.	Cy5	1.00

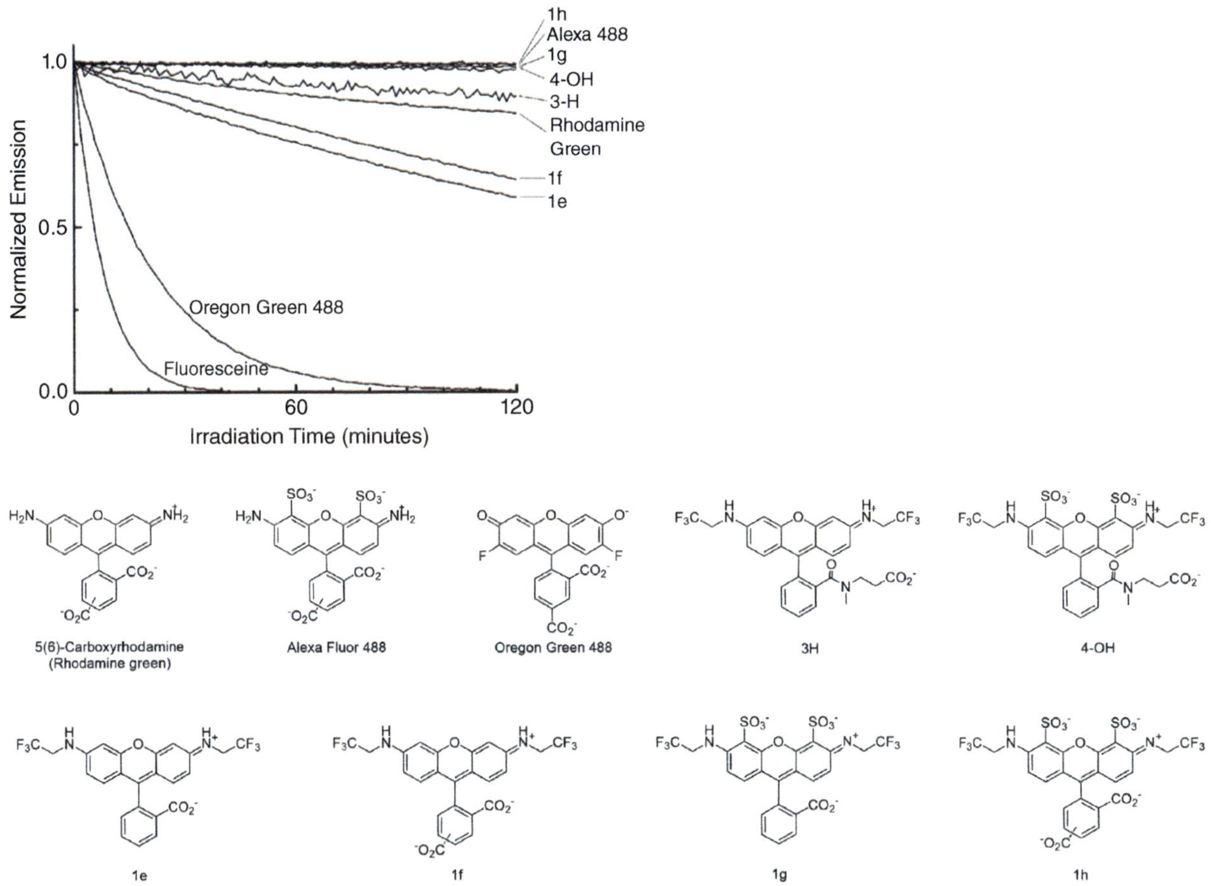

Figure 7.3 Photobleaching of a series of fluorinated rhodamines (and one fluorinated carboxyfluorescein) using fluorescein as the reference. (A) Raw data from Mitronova et al. (2010) [11] (reproduced with permission); and (B) molecular structures of the dyes. The photobleaching relative mean-lifetimes ($\tau_{pb,r}$) was calculated from the curves of A and shown in Table 7.1.

values of τ_{pb}, as they represent many more data points from across the whole monitoring time. This replaces the need of using a single point. The relative mean-lifetime is then calculated as $\tau_{pb,r} = (\tau_{pb}$ of the dye)/(τ_{pb} of the reference). Entries 1 to 5 of Table 7.1 show the results obtained from the digitalized points of Figure 7.2B, and the other entries were produced from the decay curves found in the literature (references are cited in Table 7.1).

7.3 Mitigating Photobleaching

Photoprotection systems are developed with the aim of creating more stable fluorophores by slowing down the photobleaching processes. The objective is to achieve longer observation times and with the use of higher radiances without flickering, photodegradation, and overall fluorescence fading.

Nani et al. [15], showed that photoinduced-oxidation is the main path that leads to the photobleaching of cyanines. Singlet oxygen ($^1O_2$), more than any other ROS, is involved in the regioselective photooxidation of heptamethine cyanines. The attack predominantly occurs on the polyene chain that ends up becoming an oxidative cleavage, i.e., a typical photodegradation process.

There are a great variety of proposed protocols and chemicals (for use in aqueous solutions) to increase photostability. Most of the proposed chemicals contain dissolved antioxidants, free radical scavengers, and triplet quenchers [16]. These special mounting media are used to preserve the fluorescence of tissue and cell smears, commonly known as antifading formulas. One such formula that contains reducing as well as oxidizing agents (called ROXS: reducing oxidizing system) has been successfully tested in the single-molecule fluorescence spectroscopy of oligonucleotides labeled with cyanines, (carbo-)rhodamines, and oxazines in aqueous solutions [17].

Handling the above-mentioned cocktails is demanding work, which is ultimately not very effective. This is because a concentration of up to a few hundred millimolar is required in some cases for them to work efficiently, and thus they have the potential to induce substantial changes in the samples under study.

Instead of using the photobleaching protection agents in solution (photostabilizing buffers), an alternative strategy is to use active agents that are directly covalently bound to the fluorescent dyes [18, 19]. Such dyes are then charmingly termed "fluorescent dyes with self-healing moieties." For example, a triplet state quencher that is directly attached to the dye [20] shortens the triplet mean-lifetime and, at the same time, prevents the formation of singlet oxygen and other activated ROS compounds. Such an approach was also used on commercially available and very popular fluorophores, using simple reactions to bind the photostabilizer or "self-healing" structure directly onto these commercially available dyes [21]. A large group of such triplet-state quenchers have been developed and attached to dyes with minor or no disturbance of dye brightness. These quenchers include 1,3,5,7-cyclooctatetraene (COT), 4-nitrobenzylalcohol, Trolox[2] (6-Hydroxy-2,5,7,8-tetramethylchroman-2-carboxylic acid, an analog of vitamin E), Trolox-quinone, ascorbic acid, ferrocene, nitrophenylalanine, nitrophenylacetic acid, methylviologen, and nitrobenzylalcohol, to mention only a few (Figure 7.4 shows two such structures). For a good and succinct description of recent advances in the strategies that prevent photobleaching, see Zheng and Lavis [6]. A detailed review of the use of triplet-state quenchers for many dyes is presented in Gong et al. [5]. For a review of the underlying concepts of triplet-state quenchers, state-of-the-art advances, and both the pros and cons of such quenchers, see Isselstein et al. [22]. A commercial dye modified with tris-*N*-nitrilotriacetic acid has been reported to have a 25-fold larger photon output [23], superseding the solution-based photostabilization by Ni(II) [23]. Pati et al. [24] tuned the Baird aromatic triplet-state energy of cyclooctatetraene to maximize its self-healing effect when attached to cyanines. They found that the efficiency of photostabilization scales directly with the contact rate (distance and geometry) of the fluorophore and photostabilizer pair. Finally, among photostabilizing buffers and self-healing moieties, Know et al. [25] examined a few dozen strategies that prevent photobleaching in super-resolution fluorescence microscopy.

---

2 A trademark of Roche Holding AG.

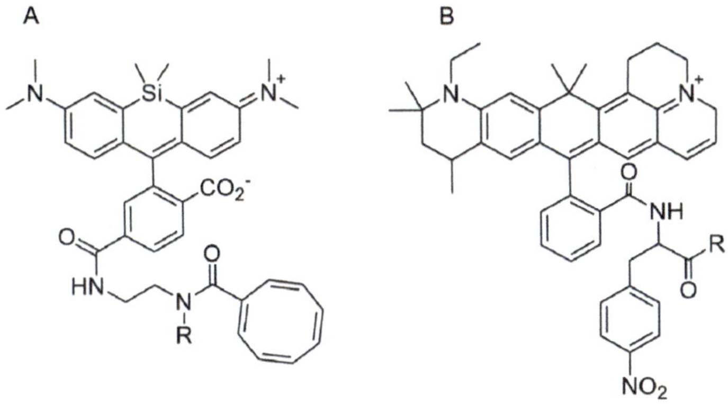

Figure 7.4 Examples of dyes with triplet-quenchers bound to them. (A) Si-rhodamine with a 1,3,5,7-cyclooctatetraene; and (B) a caborhodamine with a nitrobenzyl group. The radical R may be any reactive moiety used for labeling reactions (Section 3.4).

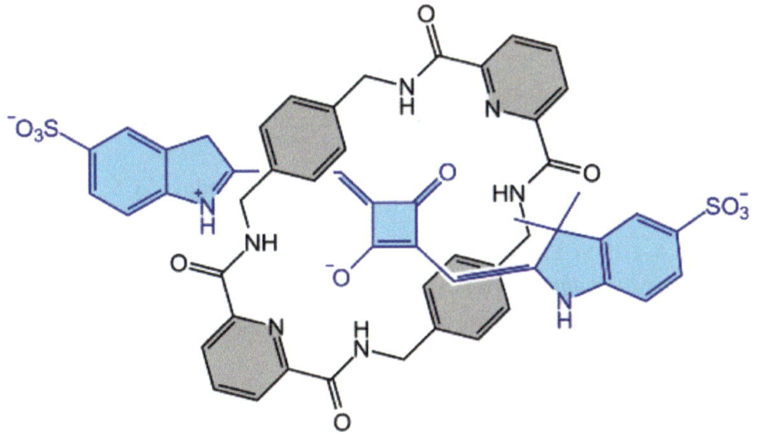

Figure 7.5 A squaraine (blue) with a macrocycle (a rotaxane, in gray).

Rotaxanes (Figure 7.5) are also used for such purposes. These are cyclic oligomers or macrocycles that encircle and interlock with the polymethine segment of squaraines, merocyanines, or cyanines. Initially they were used to mitigate the non-radiative relaxation mechanisms to increase fluorescence quantum yields. Tetralactam macrocycles were also used to improve the water solubility of squaraines [26]. Rotaxanes have the additional effect of preventing aggregation in aqueous solutions. Finally, they are also used to improve photostability by incorporating triplet-state quenchers and free radical scavengers onto the macrocycle. Moreover, this can function as a shield, protecting the squaraines and potentially cyanines, merocyanines, and other dyes from chemical attack [26]. The fluorescence quantum yield and photostability of tetraphenylethylene has been shown to increase with the use of a linked rotaxane structure made of permethylated α-cyclodextrins [27]. Additionally, Inouye et al. [28] observed circularly polarized luminescence (Section 2.5.6) in a double alkynylpyrene-threaded [4]rotaxane-type. This effect comes from the two stacked pyrenes that exist in an asymmetrically twisted manner inside the rotaxane, and it demonstrates new, additional, and unexpected potentials of rotaxanes. As the macrocycle moiety can be used for a great variety of functions, this has become a very active field of research.

Fluorescent dyes bound to DNA nanocubes (with sizes of ~6 × 6 × 6 nm), termed DNA FluoroCubes, is a completely different strategy. In these, the dye is located next to a larger, protective structure and has been claimed to emit many more photons (up to 43-fold more than an unbound dye) and exhibit longer mean-lifetimes (up to ~54-fold more than their unbound single counterparts) [29]. The mechanism behind this phenomenon is still not completely understood.

References

1 Braslavsky, S.E. (2007). Glossary of terms used in photochemistry, 3e. *Pure Appl. Chem.* 79: 293–465.
2 Dobson, D.E., Mahoney, E.R., Mach, T.P. et al. (2020). Pentamethine sulfobenzoindocyanine dyes with low net charge states and high photostability. *Photochem. Photobiol. Sci.* 19: 56–65.
3 Gorka, A.P. and Schnermann, M.J. (2016). Harnessing cyanine photooxidation: from slowing photobleaching to near-IR uncaging. *Curr. Opin. Chem. Biol.* 33: 117–125.
4 Hinkeldey, B., Schmitt, A., and Jung, G. (2008). Comparative photostability studies of BODIPY and fluorescein dyes by using fluorescence correlation spectroscopy. *Chem. Phys. Chem.* 9: 2019–2027. doi: 10.1002/cphc.200800299.
5 Gong, W., Das, P., Samanta, S. et al. (2019). Redefining the photo-stability of common fluorophores with triplet state quenchers: mechanistic insights and recent updates. *Chem. Commun.* 55: 8695–8704. doi: 10.1039/c9cc02616a.
6 Zheng, Q. and Lavis, L.D. (2017). Development of photostable fluorophores for molecular imaging. *Curr. Opin. Chem. Biol.* 39: 32–38. doi: 10.1016/j.cbpa.2017.04.017.
7 FDA (1996). Guidance for industry: Q1B photostability testing of new drugs substances and products. US Department of Health and Human Services (CDER).
8 Kuznetsova, N.A., Kaliya, O.L., and Luk'yanets, E.A. (1995). Photochemistry of laser dyes for the visible spectrum. *Proc. SPIE (International Conference on Atomic and Molecular Pulsed Laser)* 2619: 161–165. doi: 10.1117/12.216902.
9 Britt, A.D. and Moniz, W.B. (1972). The effect of pH on photobleaching of organbic laser dyes. *IEEE J. Quantum Electron.* 8(12): 913–914. doi: 10.1109/JQE.1972.1076894.
10 Whitaker, J.E., Haugland, R.P., Ryan, D. et al. (1992). Fluorescent rhodol derivatives: versatile, photostable labels and tracers. *Anal. Biochem.* 207 (1): 267–279.
11 Mitronova, G.Y. and Belov, V.N., et al. (2010). New fluorinated rhodamines for optical microscopy and nanoscopy. *Chem. Eur. J.* 16(15): 4477–4488. doi: 10.1002/chem.200903272.
12 Kotásková, M. (2012). Synthesis of new xanthene derivatives. PhD Thesis. Johannes Gutenberg University of Mainz.
13 Atto-Tec (2020). *Fluorescent Labels and Dyes. Catalogue 2019–2020.* Siegen, Germany: ATTO-TEC GmbH.
14 Thermo Fisher Scientific (2022). https://www.thermofisher.com/order/catalog/product/S6110 (accessed April 2022).
15 Nani, R.R., Kelley, J.A., Ivanic, J. et al. (2015). Reactive species involved in the regioselective photooxidation of heptamethine cyanines. *Chem. Sci.* 6: 6556–6563. doi: 10.1039/c5sc02396c.
16 Cooper, D., Uhm, H., Tauzin, L.J. et al. (2013). Photobleaching lifetimes of cyanine fluorophores used for single-molecule Förster resonance energy transfer in the presence of various photoprotection systems. *ChemBioChem* 14: 1075–1080. doi: 10.1002/cbic.201300030.
17 Vogelsang, J., Kasper, R., Steinhauer, C. et al. (2008). A reducing and oxidizing system minimizes photobleaching and blinking of fluorescent dyes. *Angew. Chem. Int. Ed.* 47: 5465–5469. doi: 10.1002/anie.200801518.
18 Liphardt, B., Liphardt, B., and Lüttke, W. (1981). Laser dyes with intramolecular triplet quenching. *Opt. Commun.* 38: 207–210.

19 Liphardt, B. and Lüttke, W. (1981). Laserfarbstoffe, I bifluorophore laserfarbstoffe zur steigerung des wirkungsgrades von Farbstoff-lasern. *Liebigs Ann. Chem.* 1981 (6): 1118–1138. doi: 10.1002/jlac.198119810617.

20 Zheng, Q., Juette, M.F., Jockusch, S. et al. (2014). Ultra-stable organic fluorophores for single-molecule research. *Chem. Soc. Rev.* 43: 1044–1056. doi: 10.1039/c3cs60237k.

21 van der Velde, J.H.M., Oelerich, J. et al. (2016). A simple and versatile design concept for fluorophore derivatives with intramolecular photostabilization. *Nat. Commun.* 7: 10144.

22 Isselstein, M., Zhang, L., Glembockyte, V. et al. (2020). Self-healing dyes-keeping the promise? *J. Phys. Chem. Lett.* 11: 4462–4480.

23 Glembockyte, V., Wieneke, R., Gatterdam, K. et al. (2018). Tris-N-nitrilotriacetic acid fluorophore as a self-healing dye for single-molecule fluorescence imaging. *J. Am. Chem. Soc.* 140: 11006–11012. doi: 10.1021/jacs.8b04681.

24 Pati, A.K., El Bakouri, O., Jockusch, S. et al. (2020). Tuning the Baird aromatic triplet-state energy of cyclooctatetraene to maximize the self-healing mechanism in organic fluorophores. *PNAS* 117 (39): 24305–24315. doi: 10.1073/pnas.2006517117.

25 Kwon, J., Elgawish, M.S., and Shim, S.-H. (2022). Bleaching-resistant super-resolution fluorescence microscopy. *Adv. Sci.* 9: 2101817. doi: 10.1002/advs.202101817.

26 Gassensmith, J.J., Baumes, J.M., and Smith, B.D. (2009). Discovery and early development of squaraine rotaxanes. *Chem. Commun.* 6329–6338. doi: 10.1039/b911064j.

27 Miyagishi, H.V., Masai, H., and Terao, J. (2022). Linked rotaxane structure restricts local molecular motions in solution to enhance fluorescence properties of tetraphenylethylene. *Chem. Eur. J.* 28: e202103175. (1 of 6) doi: 10.1002/chem.202103175.

28 Inouye, M., Hayashi, K., Yonenaga, Y. et al. (2014). A doubly alkynylpyrene-threaded [4] rotaxane that exhibits strong circularly polarized luminescence from the spatially restricted excimer. *Angew. Chem. Int. Ed.* 53: 14392–14396. doi: 10.1002/anie.201408193.

29 Niekamp, S., Stuurman, N., and Vale, R.D. (2020). A 6-nm ultra-photostable DNA FluoroCube for fluorescence imaging. *Nat. Methods* 17: 437–441.

Appendix A

A1. Short Name, Name, Class, Molecular Formula, and References

Appendix A provides information relating to the identity of over 700 fluorescent dyes with medium to high brightness in aqueous solutions in the present printed database. These dyes are used as fluorescent labels and stains and are listed in alphabetical order of their "short names." Note that many short names are registered names and trademarks (see footnote[1]).

Table A.1 shows the information given in the following Appendices as well as the photophysical and molecular parameters used to rank the database in each Appendix. The following photophysical parameters are detailed in the molecular structure figures of Chapter 4: wavelength of excitation maximum ($\lambda_{x,max}$ or $\lambda_{a,max}$), Stokes shift (Δ_S), brightness (B_{max}), and the aqueous solution used for the measurements. Chapter 5 compares the photophysical properties of the fluorescent dyes themselves, in addition to providing further molecular and photophysical data in scattergrams (one shape for each class).

Dye short names with a maximum of 30 characters were used throughout this book. Dyes with popular "short names" consisting of more than 30 characters have been shortened for practical reasons. The detailed information provided by the present Appendix is tabulated in Table A.2 as follows: entry number (first column); short name (second column); name, other names, the class to which the dye belongs, molecular formula, and CAS number (third column); and the sub-section where the molecular structures and references can be found (references from which the photophysical data was taken) (fourth column).

Most fluorescent dyes are found as salts. For example, sulfonated dyes may occur as salts of sodium, potassium, and other cations. In these cases, only the molecular formula of the fluorescent dye (anion) is given, and the counterion (cation) is ignored. Moreover, in most of these cases, the CAS numbers were omitted because each salt (of each cation) has its own CAS number, and providing the CAS number of only one of the salts may lead to some redundance and confusion. Furthermore, some suppliers sell a given dye as a particular salt, while others sell the same dye as another salt (with another cation). To be fair and equitable, when dyes occur as salts, the author omits the CAS numbers.

Table A.1 The information provided in each Appendix and the ranking system used.

	Appendix A	**Appendix B**	**Appendix C**	**Appendix D**	**Appendix E**	**Appendix F**	**Appendix G**
Ranked by:	Short names	Excitation maximum	Emission maximum	Stokes shift	Molecular brightness	Fluorescence mean-lifetime	Molecular net charge

1 Abberior STAR is a trademark of Abberior GmbH (Göttingen, Germany). Alexa Fluor, BODIPY, Cascade Blue, FAM, JOE, TET, HEX, Marina Blue, Oregon Green, Pacific Blue, ROX, TAMRA, and Texas Red are trademarks of Thermo Fisher Scientific (Waltham, MA, USA). ATTO is a trademark of ATTO-Tec GmbH (Siegen, Germany). CF Dye is a registered trademark of Biotium, Inc. (Fremont, CA, USA). Cy3, Cy3.5, Cy5, and Cy5.5 are trademarks of GE Healthcare (Chicago, IL, USA). Dy is a trademark of Dyomics GmbH (Jena, Germany). Seta, SeTau, and Square dyes are trademarks of SETA BioMedicals (Urbana, IL, USA).

Fluorescent Dye Labels and Stains: A Database of Photophysical Properties, First Edition. Tarso B. Ledur Kist.

Table A.2 Fluorescent dyes of the database ranked by their short names.

Ent.	Short Name	Name; Class; Molecular Formula; CAS Number	Ref.
1	(HO)-Dehydroluciferin	2-(6-hydroxybenzo[d]thiazol-2-yl)thiazole-4-carboxylate; Benzothiazole; $C_{11}H_6N_2O_3S_2$	4.30
2	(HO)-Luciferin	2-(6-hydroxybenzo[d]thiazol-2-yl)-4,5-dihydrothiazole-4-carboxylate; Benzothiazole; $C_{11}H_8N_2O_3S_2$; 2591–17-5	4.30
3	(HO)-Luciferyl adenylate	tl; Benzothiazole; tl	4.30
4	($^-$O)-Dehydroluciferin	2-(6-hydroxybenzo[d]thiazol-2-yl)thiazole-4-carboxylate; Benzothiazole; $C_{11}H_6N_2O_3S_2$	4.30
5	($^-$O)-Luciferin	2-(6-hydroxybenzo[d]thiazol-2-yl)-4,5-dihydrothiazole-4-carboxylate; Benzothiazole; $C_{11}H_8N_2O_3S_2$; 2591-17-5	4.30
6	1,2,7,8-Dibenzo-4,5-Br-fluorescein	6,8-Dibromo-14-(2-carboxyphenyl)-5-oxo-5H-dibenzo[a,j]xanthen-9-olate; Fluorescein (Halogenated); $C_{28}H_{13}Br_2O_5$	4.8.2
7	1,2,7,8-Dibenzofluorescein	14-(2-Carboxyphenyl)-5-oxo-5H-dibenzo[a,j]xanthen-9-olate; Fluorescein (Non-halogenated); $C_{28}H_{15}O_5$	4.8.1
8	1,2,7,8-Dibenzosulfofluorescein	2-(9-Oxido-5-oxo-5H-dibenzo[a,j]xanthen-14-yl)benzenesulfonate; Sulfofluorescein; $C_{27}H_{14}O_6S$	4.10
9	1,3,6,8-Pyrenetetrasulfonic acid	1,3,6,8-Pyrenetetrasulfonic acid tetrasodium salt; Pyrene; $C_{16}H_6O_{12}S_4$; 59572-10-0	4.28
10	10-Ethylacridine	10-Ethylacridin-10-ium; Acridine; $C_{14}H_{12}N$; 948-43-6	4.23.1
11	2′,4,5,6,7,7′-HexaF-Si-fluorescein	2-(2,8-Difluoro-5,5-dimethyl-7-oxido-3-oxo-3,5-dihydrodibenzo[b,e]silin-10-yl)-3,4,5,6-tetrafluorobenzoate; $Si(CH_3)_2$-Fluoresc.; $C_{22}H_{12}F_6O_4Si$	4.9
12	2,4,5,7-Tetra-Br-sulfofluorescein	2-(2,4,5,7-Tetrabromo-6-oxido-3-oxo-3H-xanthen-9-yl)benzenesulfonate; Sulfofluorescein; $C_{19}H_6Br_4O_6S$	4.10
13	2′,4′,5′,7′-TetraF-Si-fluorescein	2-(2,4,6,8-Tetrafluoro-5,5-dimethyl-7-oxido-3-oxo-3,5-dihydrodibenzo[b,e]silin-10-yl) benzoate; Maryland red; $Si(CH_3)_2$-Fluorescein; $C_{22}H_{14}F_4O_4Si$	4.9
14	2,4,5,7-Tetra-F-sulfofluorescein	2-(2,4,5,7-Tetrafluoro-6-oxido-3-oxo-3H-xanthen-9-yl)benzenesulfonate; Sulfofluorescein; $C_{19}H_6F_4O_6S$	4.10
15	2,4-DiMe TokyoGreen	9-(2,4-Dimethylphenyl)-3-oxo-3H-xanthen-6-olate; Fluorone; $C_{21}H_{16}O_3$	4.11
16	2,4-DiOMe TokyoGreen	9-(2,4-Dimethoxyphenyl)-3-oxo-3H-xanthen-6-olate; Fluorone; $C_{21}H_{15}O_5$	4.11
17	2,5-DiMe TokyoGreen	9-(2,5-Dimethylphenyl)-3-oxo-3H-xanthen-6-olate; Fluorone; $C_{21}H_{16}O_3$	4.11
18	2,5-DiOMe TokyoGreen	9-(2,5-Dimethoxyphenyl)-3-oxo-3H-xanthen-6-olate; Fluorone; $C_{21}H_{15}O_5$	4.11
19	2,6-Sulfon.-1,3,5,7,8-Me-BODIPY	PM556; 4,4-Difluoro-2,6-disulfonic acid-1,3,5,7,8-pentamethyl boron-dipyrromethene; Borondipyrromethene (Small); $C_{14}H_{15}BF_2N_2O_6S_2$; 121461-69-6	4.14.1
20	2,6-Sulfon.-1,3,5,7-Me-BODIPY	4,4-Difluoro-2,6-disulfonic acid-1,3,5,7-tetramethyl boron-dipyrromethene; BODIPY 491/510; Borondipyrromethene (Small); $C_{13}H_{12}BF_2N_2O_6S_2$	4.14.1

Ent.	Short Name	Name; Class; Molecular Formula; CAS Number	Ref.
21	2′,7′-Cl-4,5,6,7-F-fluorescein	2-(2,7-Dichloro-6-oxido-3-oxo-3H-xanthen-9-yl)-3,4,5,6-tetrafluorobenzoate; Fluorescein (Halogenated); $C_{20}H_4Cl_2F_4O_5$	4.8.2
22	2,7-DiCl-6-methylamino rhodol	2-(2,7-Dichloro-6-(methylamino)-3-oxo-3H-xanthen-9-yl)benzoic acid; compound 4q; Rhodol; $C_{21}H_{13}Cl_2NO_4$	4.15.1
23	2,7-DiCl-6-pyrrolidino sulforhodol	2-(2,7-Dichloro-3-oxo-6-(pyrrolidin-1-yl)-3H-xanthen-9-yl)-5-((Z)-4-((Z)-4-(dimethylamino) styryl)styryl)benzenesulfonate; Sulforhodol; $C_{41}H_{33}Cl_2N_2O_5S$	4.27
24	2′,7′-Difluoro-Si-Fluorescein	2-(2,8-Difluoro-5,5-dimethyl-7-oxido-3-oxo-3,5-dihydrodibenzo[b,e]silin-10-yl)benzoate; $Si(CH_3)_2$-Fluorescein; $C_{22}H_{16}F_2O_4Si$	4.9
25	2,7-DiMe-6-EtAm-carboxyrhodol	2-(6-(Ethylamino)-2,7-dimethyl-3-oxo-3H-xanthen-9-yl)-5-(6-)carboxybenzoic acid; compound 5; Rhodol; $C_{25}H_{21}NO_6$	4.15.1
26	2′,7′-Difluorocarbofluorescein	2-(2,7-Difluoro-10,10-dimethyl-6-oxido-3-oxo-3,10-dihydroanthracen-9-yl)benzoate; Virginia orange; $C(CH_3)_2$-Fluorescein; $C_{23}H_{16}F_2O_4$	4.9
27	2-Aminopyridine	Pyridin-2-amine; Others (Pyridine); $C_5H_6N_2$; 504-29-0	4.39.1
28	2-Br-7-Me-6-EtAm-carboxyrhodol	2-(2-Bromo-6-(ethylamino)-7-methyl-3-oxo-3H-xanthen-9-yl)-5-(6-)carboxybenzoic acid; compound 4; Rhodol; $C_{24}H_{16}BrNO_6$	4.15.1
29	2-Cl-6-Piperidinocarboxyrhodol	2-(8-Chloro-9-oxo-1,3,4,9-tetrahydro-2H-chromeno[3,2-g]quinolin-6-yl)-5-(6-) carboxybenzoic acid; compound 12; Rhodol; $C_{24}H_{16}ClNO_6$	4.15.1
30	2-Cl-7-Me-6-EtAm-carboxyrhodol	2-(2-Chloro-6-(ethylamino)-7-methyl-3-oxo-3H-xanthen-9-yl)-5-(6-)carboxybenzoic acid; compound 3; Rhodol; $C_{24}H_{18}ClNO_6$	4.15.1
31	2-Cloro-6-quinolinocarboxyrhodol	2-(11-chloro-12-oxo-2,3,6,7-tetrahydro-1H,5H,12H-chromeno[2,3-f]pyrido[3,2,1-ij] quinolin-9-yl)-5-(6-)carboxybenzoic acid; compound 14; Rhodol; $C_{27}H_{20}ClNO_6$	4.15.1
32	2-COOH DCTM	2-(4,6-Dichloro-5,5-dimethyl-7-oxido-3-oxo-3,5-dihydrodibenzo[b,e]silin-10-yl)benzoate; $Si(CH_3)_2$-Fluorescein; $C_{22}H_{14}Cl_2O_4Si$	4.9
33	2-COOH DFTM	2-(4,6-Difluoro-5,5-dimethyl-7-oxido-3-oxo-3,5-dihydrodibenzo[b,e]silin-10-yl)benzoate; $Si(CH_3)_2$-Fluorescein; $C_{22}H_{14}F_2O_4Si$	4.9
34	2-COOH TM	2-(5,5-Dimethyl-7-oxido-3-oxo-3,5-dihydrodibenzo[b,e]silin-10-yl)benzoate; $Si(CH_3)_2$-Fluorescein; $C_{22}H_{16}O_4Si$	4.9
35	2′COOH-4CH_2OH-Si-rhodol	7-(Dimethylamino)-4-(hydroxymethyl)-5,5-dimethyl-10-(o-tolyl)dibenzo[b,e]silin-3(5H)-one; $Si(CH_3)_2$-Rhodol; $C_{25}H_{27}NO_2Si$	4.16
36	2′COOH-Si-rhodol	7-(Dimethylamino)-5,5-dimethyl-10-(o-tolyl)dibenzo[b,e]silin-3(5H)-one; $Si(CH_3)_2$-Rhodol; $C_{24}H_{25}NOSi$	4.16
37	2-Et-7-Me-6-EtAm-carboxyrhodol	2-(2-Ethyl-6-(ethylamino)-7-methyl-3-oxo-3H-xanthen-9-yl)-5-(6-)carboxybenzoic acid; compound 6; Rhodol; $C_{26}H_{23}NO_6$	4.15.1

(Continued)

Ent.	Short Name	Name; Class; Molecular Formula; CAS Number	Ref.
38	2-Ethyl-6-quinolinocarboxyrhodol	4-(11-ethyl-12-oxo-2,3,41,6,7,9a-hexahydro-1H,5H,12H-chromeno[2,3-f]pyrido[3,2,1-ij] quinolin-9-yl)-5-(6-)carboxybenzoic acid; compound 17; Rhodol; $C_{29}H_{25}NO_6$	4.15.1
39	2-Me DCTM	4,6-Dichloro-5,5-dimethyl-7-oxo-10-(o-tolyl)-5,7-dihydrodibenzo[b,e]silin-3-olate; $Si(CH_3)_2$-Fluorone; $C_{22}H_{19}Cl_2O_2Si$	4.12
40	2-Me RG	6-Amino-9-(o-tolyl)-3H-xanthen-3-iminium; Rosamine; $C_{20}H_{17}N_2O$	4.17
41	2-Me SiR600	7-Amino-5,5-dimethyl-10-(o-tolyl)dibenzo[b,e]silin-3(5H)-iminium; $Si(CH_3)_2$-Rosamine; $C_{22}H_{23}N_2Si$	4.18.1
42	2-Me SiR620	N-(7-Amino-5,5-dimethyl-10-(o-tolyl)dibenzo[b,e]silin-3(5H)-ylidene)-N-methylmethanaminium; $Si(CH_3)_2$-Rosamine; $C_{24}H_{27}N_2Si$	4.18.1
43	2-Me SiR640	8-Amino-1,10,10-trimethyl-5-(o-tolyl)-3,10-dihydro-2H-benzo[5,6]silino[3,2-f]indol-1-ium; $Si(CH_3)_2$-Rosamine; $C_{25}H_{27}N_2Si$	4.18.1
44	2-Me SiR650	N-(7-(Dimethylamino)-5,5-dimethyl-10-(o-tolyl)dibenzo[b,e]silin-3(5H)-ylidene)-N-methylmethanaminium; $Si(CH_3)_2$-Rosamine; $C_{26}H_{31}N_2Si$	4.18.1
45	2-Me SiR654	8-(Ethylamino)-1,10,10-trimethyl-5-(o-tolyl)-3,10-dihydro-2H-benzo[5,6]silino[3,2-f] indol-1-ium; $Si(CH_3)_2$-Rosamine; $C_{27}H_{31}N_2Si$	4.18.1
46	2-Me SiR666	8-(Dimethylamino)-1,10,10-trimethyl-5-(o-tolyl)-3,10-dihydro-2H-benzo[5,6]silino[3,2-f] indol-1-ium; $Si(CH_3)_2$-Rosamine; $C_{27}H_{31}N_2Si$	4.18.1
47	2-Me SiR670	8-(Diethylamino)-1,10,10-trimethyl-5-(o-tolyl)-3,10-dihydro-2H-benzo[5,6]silino[3,2-f] indol-1-ium; $Si(CH_3)_2$-Rosamine; $C_{29}H_{35}N_2Si$	4.18.1
48	2-Me SiR671	1,11,11-Trimethyl-5-(o-tolyl)-2,3,7,8,9,11-hexahydrosilino[3,2-f:5,6-f']diindol-1-ium; $Si(CH_3)_2$-Rosamine; $C_{27}H_{29}N_2Si$	4.18.1
49	2-Me SiR675	9-(Ethylamino)-1,2,2,4,11,11-hexamethyl-6-(o-tolyl)-2,11-dihydrobenzo[5,6]silino[3,2-g] quinolin-1-ium; $Si(CH_3)_2$-Rosamine; $C_{31}H_{37}N_2Si$	4.18.1
50	2-Me SiR680	1,11,13,13-Tetramethyl-6-(o-tolyl)-2,3,4,8,9,10,11,13-octahydrosilino[3,2-g:5,6-g']diquinolin-1-ium; $Si(CH_3)_2$-Rosamine; $C_{30}H_{35}N_2Si$	4.18.1
51	2-Me SiR700	1,9,11,11-Tetramethyl-5-(o-tolyl)-2,3,7,8,9,11-hexahydrosilino[3,2-f:5,6-f']diindol-1-ium; $Si(CH_3)_2$-Rosamine; $C_{28}H_{31}N_2Si$	4.18.1
52	2-Me SiR720	1,2,2,4,8,10,10,11,13,13-Decamethyl-6-(o-tolyl)-2,10,11,13-tetrahydrosilino[3,2-g:5,6-g'] diquinolin-1-ium; $Si(CH_3)_2$-Rosamine; $C_{36}H_{43}N_2Si$	4.18.1
53	2-Me TeOR	N-(6-(Dimethylamino)-10-oxido-9-(o-tolyl)-3H-telluroxanthen-3-ylidene)-N-methylmethanaminium; TeO-Rosamine; $C_{24}H_{25}N_2OTe$	4.18.3
54	2-Me Tokyo Magenta	5,5-Dimethyl-7-oxo-10-(o-tolyl)-5,7-dihydrodibenzo[b,e]silin-3-olate; 2-methy-Si-fluorescein; $Si(CH_3)_2$-Fluorone; $C_{22}H_{19}O_2Si$	4.12
55	2-Me TokyoGreen	3-Oxo-9-(o-tolyl)-3H-xanthen-6-olate; Fluorone; $C_{20}H_{13}O_3$	4.11

Ent.	Short Name	Name; Class; Molecular Formula; CAS Number	Ref.
56	2′Me-4CH_2OH-Si-rosol	7-(Dimethylamino)-4-(hydroxymethyl)-5,5-dimethyl-10-(o-tolyl)dibenzo[b,e]silin-3(5H)-one; $Si(CH_3)_2$-Rosol; $C_{25}H_{27}NO_2Si$	4.20
57	2-Me-4-OMe TokyoGreen	9-(4-Methoxy-2-methylphenyl)-3-oxo-3H-xanthen-6-olate; Fluorone; $C_{21}H_{16}O_4$	4.11
58	2′Me-Si-rhosoL	7-(Dimethylamino)-5,5-dimethyl-10-(o-tolyl)dibenzo[b,e]silin-3(5H)-one (comp. 7 of SI); $Si(CH_3)_2$-Rosol; $C_{24}H_{25}NOSi$	4.20
59	2-Methyl harmine	8-Methoxy-1,2-dimethyl-2H-pyrido[3,4-b]indole; Indole; $C_{14}H_{14}N_2O$	4.33
60	2-OH-6-DiEtAm-benzo[c]xanthen	2-(10-(Diethyliminio)-5-oxido-10H-benzo[c]xanthen-7-yl)benzoate; Rhodol-analog; $C_{28}H_{22}NO_4$	4.15.3
61	2-OMe TokyoGreen	9-(2-Methoxyphenyl)-3-oxo-3H-xanthen-6-olate; Fluorone; $C_{20}H_{14}O_4$	4.11
62	2-OMe-5-Me TokyoGreen	9-(2-Methoxy-5-methylphenyl)-3-oxo-3H-xanthen-6-olate; Fluorone; $C_{21}H_{16}O_4$	4.11
63	2-PrAc-6-quinolinorhodol	2-(11-(2-carboxyethyl)-12-oxo-2,3,41,6,7,9a-hexahydro-1H,5H,12H-chromeno[2,3-f]pyrido[3,2,1-ij]quinolin-9-yl)benzoic acid; compound 19; Rhodol; $C_{29}H_{27}NO_6$	4.15.1
64	2-PrAc-7-Me-6-EtAm-rhodol	2-(7-(2-carboxylatoethyl)-3-(ethylamino)-2-methyl-6-oxo-6,8a-dihydro-3H-xanthen-9-yl)benzoate; compound 8; Rhodol; $C_{26}H_{23}NO_6$	4.15.1
65	3-(4-OH-Ph)-6-MeO-chromone	3-(4-Hydroxyphenyl)-6-methoxy-4H-chromen-4-one; Chromone; $C_{16}H_{12}O_4$	4.31
66	3,10-Dimethylisoalloxazine	3,10-Dimethylbenzo[g]pteridine-2,4(3H,10H)-dione; Pteridine; $C_{12}H_{10}N_4O_2$; 4074-59-3	4.36
67	3,6,10-Trimethylisoalloxazine	3,6,10-Trimethylbenzo[g]pteridine-2,4(3H,10H)-dione; Pteridine; $C_{13}H_{12}N_4O_2$; 70481-66-2	4.36
68	3,6,7,8,10-PentaMe-isoalloxazine	3,6,7,8,10-Pentamethylbenzo[g]pteridine-2,4(3H,10H)-dione; Pteridine; $C_{15}H_{16}N_4O_2$	4.36
69	3,6-Azetid.-9-CA.-Si-pyronine	3-(Azetidin-1-ium-1-ylidene)-7-(azetidin-1-yl)-5,5-dimethyl-3,5-dihydrodibenzo[b,e]siline-10-carboxylate; $Si(CH_3)_2$-Pyronine; $C_{22}H_{25}N_2O_2Si$	4.5
70	3,6-Azetid.-Si-pyronine	1-(7-(Azetidin-1-yl)-5,5-dimethyldibenzo[b,e]silin-3(5H)-ylidene)azetidin-1-ium; $Si(CH_3)_2$-Pyronine; $C_{21}H_{25}N_2Si$	4.5
71	3,6-Di(azetidin-1-yl)acridine	3,6-Di(azetidin-1-yl)acridine; Acridine; $C_{19}H_{19}N_3$	4.23.1
72	3,6-Diaminoacridine (Proflavine)	6-Aminoacridin-3(10H)-iminium; Acridine; $C_{13}H_{12}N_3$; 92-62-6	4.23.1
73	3,7,10-Trimethylisoalloxazine	3,7,10-Trimethylbenzo[g]pteridine-2,4(3H,10H)-dione; Pteridine; $C_{13}H_{12}N_4O_2$; 58537-70-5	4.36
74	3,7,8,10-TetraMe-isoalloxazine	3,7,8,10-tetramethylbenzo[g]pteridine-2,4(3H,10H)-dione; Pteridine; $C_{14}H_{14}N_4O_2$; 18636-32-3	4.36
75	3,7,9,10-TetraMe-isoalloxazine	3,7,9,10-Tetramethylbenzo[g]pteridine-2,4(3H,10H)-dione; Pteridine; $C_{14}H_{14}N_4O_2$; 1088-56-8	4.36
76	3,7-Di(azetidin)phenoxazine	3,7-Di(azetidin-1-yl)phenoxazin-5-ium; 1-(7-(azetidin-1-yl)-3H-phenoxazin-3-ylidene)azetidin-1-ium; Phenoxazine; $C_{18}H_{18}N_3O_1$	4.25
77	3,7-Di(ethylamino)phenoxazine	N-(7-(Diethylamino)-3H-phenoxazin-3-ylidene)-N-ethylethanaminium; Phenoxazine; $C_{20}H_{26}N_3O$	4.25

(Continued)

Ent.	Short Name	Name; Class; Molecular Formula; CAS Number	Ref.
78	3,7-Dimethyl-JF646	1-(7-(Azetidin-1-yl)-10-(2,6-dimethylphenyl)-5,5-dimethyldibenzo[b,e]silin-3(5H)-ylidene) azetidin-1-ium; $Si(CH_3)_2$-Rosamine; $C_{29}H_{33}N_2Si$	4.18.1
79	3,8,10-Trimethylisoalloxazine	3,8,10-Trimethylbenzo[g]pteridine-2,4(3H,10H)-dione; Pteridine; $C_{13}H_{12}N_4O_2$; 62738-33-4	4.36
80	3-AB-6-MeO-7-OH-chromone	3-(4-Azidobutyl)-6-methoxy-7-hydroxychromone; Comp. 83; Chromone; $C_{13}H_{14}O_3$	4.31
81	3-Amino-9-ethylcarbazole 1	9-Ethyl-9H-carbazol-3-amine; Others (Carbazole); $C_{14}H_{14}N_2$	4.39.1
82	3-Amino-9-ethylcarbazole 2	N1-(9-Ethyl-9H-carbazol-3-yl)-N4-methylbutane-1,4-diamine; Others (Carbazole); $C_{19}H_{25}N_3$	4.39.1
83	3-Aminopyridine	Pyridin-3-amine; Others (Pyridine); $C_5H_6N_2$	4.39.1
84	3-Bt-6-MeO-7-hydroxychromone	3-Buthyl-6-methoxy-7-hydroxychromone; Chromone; $C_{14}H_{16}O_4$	4.31
85	3-Buthyl-7-hydroxychromone	3-Butyl-7-hydroxy-4H-chromen-4-one; Chromone; $C_{13}H_{14}O_3$	4.31
86	3-Hydroxy-JF646	1-(7-(Azetidin-1-yl)-10-(2-hydroxyphenyl)-5,5-dimethyldibenzo[b,e]silin-3(5H)-ylidene) azetidin-1-ium; $Si(CH_3)_2$-Rosamine; $C_{27}H_{29}N_2OSi$	4.18.1
87	3-Methoxymethyl ether rhodol	2-(6-Amino-3-(methoxymethoxy)-3H-xanthen-9-yl)benzoate; compound 3p; Rhodol; $C_{22}H_{17}NO_5$	4.15.2
88	3-Methyl-5-Methoxy-JF646	1-(7-(Azetidin-1-yl)-10-(4-methoxy-2-methylphenyl)-5,5-dimethyldibenzo[b,e]silin-3(5H)-ylidene)azetidin-1-ium; $Si(CH_3)_2$-Rosamine; $C_{29}H_{33}N_2OSi$	4.18.1
89	3-Methyl-JF646	1-(7-(Azetidin-1-yl)-5,5-dimethyl-10-(o-tolyl)dibenzo[b,e]silin-3(5H)-ylidene)azetidin-1-ium; $Si(CH_3)_2$-Rosamine; $C_{28}H_{31}N_2Si$	4.18.1
90	3-MOMO-6-propylaminorhodol	2-(3-(Methoxymethoxy)-6-(propylamino)-3H-xanthen-9-yl)benzoate; compound 3h; Rhodol; $C_{25}H_{24}NO_5$	4.15.2
91	3-OH-4-nitroso-2,7-naphth-diSA	3-Hydroxy-4-nitroso-2,7-naphthalenedisulfonic acid disodium salt; HNNDS; Naphthalene; $C_{10}H_5NO_8S_2$; 525-05-3	4.34
92	3-OH-4-nitroso-2,7-naphth-diSA	3-Hydroxy-4-nitroso-2,7-naphthalenedisulfonic acid disodium salt; HNNDS; Naphthalene; $C_{12}H_5NO_8S_2$; 525-05-3	4.34
93	4,10-Dibromonaphthofluorescein	2-(4,10-Dibromo-11-oxido-3-oxo-3H-dibenzo[c,h]xanthen-7-yl)benzoate; Fluorescein-analog; $C_{28}H_{12}Br_2O_5$	4.8.4
94	4,5,6,7-Tetrafluoro-SiTMR	2-(7-(Dimethylamino)-3-(dimethyliminio)-5,5-dimethyl-3,5-dihydrodibenzo[b,e]silin-10-yl)-3,4,5,6-tetrafluorobenzoate; $Si(CH_3)_2$-Rhodamine; $C_{26}H_{24}F_4N_2O_2Si$	4.3.2
95	4,5,6,7-TetraF-Si-fluorescein	2-(5,5-Dimethyl-7-oxido-3-oxo-3,5-dihydrodibenzo[b,e]silin-10-yl)-3,4,5,6-tetrafluorobenzoate; $Si(CH_3)_2$-Fluorescein; $C_{22}H_{14}F_4O_4Si$	4.9
96	4,7-DiMe-6-EtAm-carboxyrhodol	2-(3-(ethylamino)-2,5-dimethyl-6-oxo-6,8a-dihydro-3H-xanthen-9-yl)terephthalate; compound 7; Rhodol; $C_{25}H_{21}NO_6$	4.15.1
97	4-Benzamidofluorescein	2-Benzamido-6-(6-oxido-3-oxo-3H-xanthen-9-yl)benzoate; Fluorescein (Non-halogenated); $C_{27}H_{15}NO_6$	4.8.1

Ent.	Short Name	Name; Class; Molecular Formula; CAS Number	Ref.
98	4-Carboxy Pennsylvania Green	4-(2,7-Difluoro-6-oxido-3-oxo-3H-xanthen-9-yl)-3-methylbenzoate; Fluorone; $C_{20}H_{11}F_2O_3$	4.11
99	4-Methylsalicylic acid	4-MeSA; 2-hydroxy-4-methylbenzoic acid; Benzene; $C_8H_8O_3$	4.38
100	5(6)-Carboxyfluorescein (FAM)	3′,6′-Dihydroxy-1-oxospiro[2-benzofuran-3,9′-xanthene]-5-carboxylic acid; 3′,6′-dihydroxy-3-oxospiro[2-benzofuran-1,9′-xanthene]-5-carboxylic acid; Fluorescein (Non-halogenated); $C_{21}H_{12}O_7$; 72088-94-9	4.8.1
101	5(6)-Carboxyrhodol	2-(3-amino-6-oxo-6,8a-dihydro-3H-xanthen-9-yl)terephthalate; compound 1; Rhodol; $C_{21}H_{13}NO_6$	4.15.1
102	5(6)-Carboxy-X-rhodamine (ROX)	tl; Rhodamine (550 < λa, max < 600 nm); $C_{33}H_{29}N_2O_5$; 194785-18-7	4.2.3
103	530RH	2-(3,9-dihydroxy-1,11-bis(2,2,2-trifluoroethyl)-3,4,8,9,10,11-hexahydro-2H-pyrano[3,2-g:5,6-g’]diquinolin-1-ium-6-yl)-4-hydroxybenzoate; Rhodamine (500 < λa, max < 550 nm); $C_{31}H_{21}F_6N_2O_7$	4.2.2
104	560CP	(Z)-2-(10,10-Dimethyl-6-((2,2,2-trifluoroethyl)amino)-3-((2,2,2-trifluoroethyl)iminio)-3,10-dihydroanthracen-9-yl)-4-((methylamino)oxy)benzoate; $C(CH_3)_2$-Rhodamine; $C_{28}H_{23}N_2O_3$	4.3.1
105	570CPH	(Z)-2-(6-((2,2-difluoro-3-hydroxypropyl)amino)-3-((2,2-difluoro-3-hydroxypropyl)iminio)-10,10-dimethyl-3,10-dihydroanthracen-9-yl)-4-hydroxybenzoate; $C(CH_3)_2$-Rhodamine; $C_{30}H_{29}N_2O_5$	4.3.1
106	575RH	tl; Rhodamine (550 < λa, max < 600 nm); $C_{33}H_{30}N_2O_6$	4.2.3
107	5-Aza-3-methyl-JF646	1-(7-(Azetidin-1-yl)-5,5-dimethyl-10-(3-methylpyridin-4-yl)dibenzo[b,e]silin-3(5H)-ylidene) azetidin-1-ium; $Si(CH_3)_2$-Rosamine; $C_{27}H_{30}N_3Si$	4.18.1
108	5-Carboxyfluorescein (S-H 2)	3′,6′-Dihydroxy-1-oxospiro[2-benzofuran-3,9′-xanthene]-5-carboxylic acid;3′,6′-dihydroxy-3-oxospiro[2-benzofuran-1,9′-xanthene]-5-carboxylic acid; Fluorescein (Non-halogenated); $C_{21}H_{12}O_7$; 3301-79-9	4.8.1
109	5-Methoxysalicylic acid	5-MeOSA; 2-hydroxy-5-methoxybenzoic acid; Benzene; $C_8H_8O_4$; 2612-02-4	4.38
110	5-Methylsalicylic acid	5-MeSA; 2-hydroxy-5-methylbenzoic acid; Benzene; $C_8H_8O_3$; 89-56-5	4.38
111	6,8-Dichloroharmine (6,8-DiCL-Ha)	6,8-Dichloro-7-methoxy-1-methyl-9H-pyrido[3,4-b]indol-2-ium; Indole; $C_{13}H_{11}Cl_2N_2O$	4.33
112	630GeRH	4-Hydroxy-2-(3-(3-hydroxyazetidin-1-ium-1-ylidene)-7-(3-hydroxyazetidin-1-yl)-5,5-dimethyl-3,5-dihydrodibenzo[b,e]germin-10-yl)benzoate; $Ge(C_{H3})_2$-Rhodamine; $C_{29}H_{27}GeN_2O_5$	4.3.3
113	640SiRH	4-Hydroxy-2-(3-(3-hydroxyazetidin-1-ium-1-ylidene)-7-(3-hydroxyazetidin-1-yl)-5,5-dimethyl-3,5-dihydrodibenzo[b,e]silin-10-yl)benzoate; $Si(CH_3)_2$-Rhodamine; $C_{28}H_{28}N_2O_5Si$	4.3.2
114	6-Acrylamidorhodol	2-(6-Acrylamido-3-oxo-3H-xanthen-9-yl)benzoate; compound 4n; Rhodol; $C_{23}H_{14}NO_5$	4.15.2
115	6-Allylaminorhodol	2-(6-(Allylamino)-3-oxo-3H-xanthen-9-yl)benzoate; compound 4j; Rhodol; $C_{23}H_{16}NO_4$	4.15.2
116	6-Aminorhodol (Rhodol)	2-(6-Amino-3-oxo-3H-xanthen-9-yl)benzoic acid; compound 4p; compound 4p; Rhodol; $C_{20}H_{12}NO_4$	4.15.2

(Continued)

Ent.	Short Name	Name; Class; Molecular Formula; CAS Number	Ref.
117	6-Azetidinyl naphthalimide	2-(6-(Azetidin-1-yl)-1,3-dioxo-1H-benzo[de]isoquinolin-2(3H)-yl)acetic acid; Naphthalimide; $C_{12}H_{12}N_3O_4$	4.32
118	6-Azetidinylaminorhodol	2-(6-(Azetidin-1-yl)-3-oxo-3H-xanthen-9-yl)benzoic acid; Rhodol; $C_{23}H_{18}NO_4$	4.15.2
119	6-Azido-4,5,7-trifluorofluorescein	4-Azido-2,3,5-trifluoro-6-(6-oxido-3-oxo-3H-xanthen-9-yl)benzoate; Fluorescein (Halogenated); $C_{20}H_6F_3N_3O_5$	4.8.2
120	6-Benzamidorhodol	2-(6-benzamido-3-oxo-3H-xanthen-9-yl)benzoate; compound 4o; Rhodol; $C_{27}H_{16}NO_5$	4.15.2
121	6-Benzylaminorhodol	2-(6-(benzylamino)-3-oxo-3H-xanthen-9-yl)benzoate; compound 4k; Rhodol; $C_{27}H_{18}NO_4$	4.15.2
122	6-Butylaminorhodol	2-(6-(butylamino)-3-oxo-3H-xanthen-9-yl)benzoate; compound 4i; Rhodol; $C_{24}H_{20}NO_4$	4.15.2
123	6-CA-MeThio-4,5,7-F-fluorescein	4-((Carboxylatomethyl)thio)-2,3,5-trifluoro-6-(6-oxido-3-oxo-3H-xanthen-9-yl)benzoate; 6-carboxymethylthio-4,5,7-trifluorofluorescein; Fluorescein (Halogenated); $C_{22}H_8F_3O_7S$	4.8.2
124	6-Chloro-harmine	6-Chloro-7-methoxy-1-methyl-9H-pyrido[3,4-b]indol-2-ium; Indole; $C_{13}H_{12}ClN_2O$	4.33
125	6-Dibutylaminorhodol	2-(6-(dibutylamino)-3-oxo-3H-xanthen-9-yl)benzoate; compound 4g; Rhodol; $C_{28}H_{28}NO_4$	4.15.2
126	6-Diethylaminorhodol	2-(6-(diethylamino)-3-oxo-3H-xanthen-9-yl)benzoate; compound 4e; Rhodol; $C_{24}H_{20}NO_4$	4.15.2
127	6-Dimethylaminorhodol	2-(6-(Dimethylamino)-3-oxo-3H-xanthen-9-yl)benzoate; compound 4d; Rhodol; $C_{22}H_{16}NO_4$	4.15.2
128	6-Dimethylaminorhodol	2-(6-(dimethylamino)-10,10-dimethyl-3-oxo-3,10-dihydroanthracen-9-yl)benzoic acid; compound 3; Rhodol; $C_{25}H_{23}NO_3$	4.15.1
129	6-Dimethylaminorhodol	2-(6-(N,N-dimethylamino)-3-oxo-3H-xanthen-9-yl)benzoic acid; Rhodol; $C_{21}H_{18}NO_4$	4.15.2
130	6-Dipropylaminorhodol	2-(3-Oxo-6-(propylamino)-3H-xanthen-9-yl)benzoic; compound 4f; Rhodol; $C_{26}H_{24}NO_4$	4.15.2
131	6-Morpholinorhodol	2-(6-morpholino-3-oxo-3H-xanthen-9-yl)benzoate; compound 4a; Rhodol; $C_{24}H_{18}NO_5$	4.15.2
132	6-Piperidinocarboxyrhodol	2-(9-Oxo-1,3,4,9-tetrahydro-2H-chromeno[3,2-g]quinolin-6-yl)-5-(6-)carboxybenzoate; compound 11; Rhodol; $C_{24}H_{17}NO_6$	4.15.1
133	6-Piperidinorhodol	2-(3-oxo-6-(piperidin-1-yl)-3H-xanthen-9-yl)benzoate; compound 4b; Rhodol; $C_{25}H_{20}NO_4$	4.15.2
134	6-Propylaminorhodol	2-(3-oxo-6-(propylamino)-3H-xanthen-9-yl)benzoate; compound 4h; Rhodol; $C_{23}H_{18}NO_4$	4.15.2
135	6-Pyrrolidinorhodol	2-(3-oxo-6-(pyrrolidin-1-yl)-3H-xanthen-9-yl)benzoate; compound 4c; Rhodol; $C_{24}H_{18}NO_4$	4.15.2
136	6-Quinoline-2-pentanoic-pyrodol	3-(12-Oxo-2,3,6,7-tetrahydro-1H,5H,12H-chromeno[2,3-f]pyrido[3,2,1-ij]quinolin-11-yl) pentanoic acid; Pyrodol; $C_{25}H_{24}NO_3$	4.21.1
137	6-Quinoline-2-propanoic-pyrodol	3-(12-Oxo-2,3,6,7-tetrahydro-1H,5H,12H-chromeno[2,3-f]pyrido[3,2,1-ij]quinolin-11-yl) propanoic acid; Pyrodol; $C_{23}H_{20}NO_3$	4.21.1
138	6-Quinolinocarboxyrhodol	2-(12-Oxo-2,3,6,7-tetrahydro-1H,5H,12H-chromeno[2,3-f]pyrido[3,2,1-ij]quinolin-9-yl)-5-(6-)carboxybenzoic acid; compound 13; Rhodol; $C_{27}H_{21}NO_6$	4.15.1
139	6-Sulfonato-1,3,5,7-Me-BODIPY	5,5-Difluoro-1,3,7,9-tetramethyl-5,9a-dihydro-3H-5l4,6l4-dipyrrolo[1,2-c:2′,1′-f][1,3,2] diazaborinine-8-sulfonate; Borondipyrromethene (Small); $C_{13}H_{14}BF_2N_2O_{3S}$	4.14.1

Ent.	Short Name	Name; Class; Molecular Formula; CAS Number	Ref.
140	7-(4-Hydroxy)cyclohexyl coum.	7-(((1s,4s)-4-Hydroxycyclohexyl)amino)-2H-chromen-2-one; Coumarin (7-Amino); $C_{15}H_{17}NO_3$	4.26.2
141	7-Am-4-Me-2-hydroxyquinoline	7-Amino-4-methylquinolin-2-ol; 7-amino-4-methyl-2-hydroxyquinoline; Quinoline; $C_{10}H_{12}N_2O$	4.29
142	7-Azetidinyl-3-CA coumarin	7-Azetidinyl-3-carboxylate coumarin; 7-(azetidin-1-yl)-2-oxo-2H-chromene-3-carboxylate; Coumarin (7-Amino); $C_{13}H_{11}NO_4$	4.26.2
143	7-Azetidinyl-4-Me-coumarin	7-(Azetidin-1-yl)-4-methyl-2H-chromen-2-one; Coumarin (7-Amino); $C_{13}H_{13}N_1O_1$	4.26.2
144	7-Diethylamino-3-CA coumarin	7-Diethylamino-4-methyl-2H-chromen-2-one; Coumarin (7-Amino); $C_{26}H_{25}N_2O_7S$	4.26.2
145	7-Diethylamino-coumarin	7-(Diethylamino)-2H-chromen-2-one; DEAC; Coumarin (7-Amino); $C_{13}H_{15}NO_2$	4.26.2
146	7-Dimethylamino-4-Me-coumarin	7-(Dimethylamino)-4-methyl-2H-chromen-2-one; Coumarin (7-Amino); $C_{12}H_{13}NO_2$	4.26.2
147	7-Ethylamino-coumarin	7-(Ethylamino)-2H-chromen-2-one; Coumarin (7-Amino); $C_{11}H_{11}NO_2$	4.26.2
148	7-Hydroxy-4-Me-coumarin-3-CA	7-Hydroxy-4-methyl-2-oxo-2H-chromene-3-carboxylate; Coumarin (7-Hydroxy); $C_{12}H_{10}O_4$	4.26.1
149	7-Hydroxy-4-Me-coumarin-3-CA	4-Methyl-7-oxido-2-oxo-2H-chromene-3-carboxylate; Coumarin (7-Hydroxy); $C_{11}H_6O_5$	4.26.1
150	7-Hydroxy-4-methylcoumarin	7-Hydroxy-4-methyl-2H-chromen-2-one; Coumarin (7-Hydroxy); $C_{10}H_8O_3$	4.26.1
151	7-Hydroxy-4-methylcoumarin	4-Methyl-2-oxo-2H-chromen-7-olate; Coumarin (7-Hydroxy); $C_{10}H_7O_3$	4.26.1
152	7-Me-6-EtAm-carboxyrhodol	2-(6-(ethylamino)-7-methyl-3-oxo-3H-xanthen-9-yl)-5-(6-)carboxybenzoic acid; compound 2; Rhodol; $C_{24}H_{19}NO_6$	4.15.1
153	8-Aminonaphthalene-1,3,6-triSA	8-Aminonaphthalene-1,3,6-trisulfonic acid; ANTS; Naphthalene; $C_{10}H_7NO_9S_3$; 5398-34-5	4.34
154	8-Aminopyrene-1,3,6-triSA	8-Aminopyrene-1,3,6-trisulfonic acid trisodium salt; HPTS; pyranine; APTS; Pyrene; $C_{16}H_8NO_9S_3$; 6358-69-6	4.28
155	8-Chloro-harmine	8-Chloro-7-methoxy-1-methyl-9H-pyrido[3,4-b]indol-2-ium; Indole; $C_{13}H_{12}ClN_2O$	4.33
156	8-Hydroxypyrene-1,3,6-triSA	8-Hydroxypyrene-1,3,6-trisulfonic acid trisodium salt; HPTS; pyranine; HPTS; Pyrene; $C_{16}H_7O_{10}S_3$; 6358-69-6	4.28
157	8-Hydroxypyrene-1,3,6-triSA	8-Hydroxypyrene-1,3,6-trisulfonic acid trisodium salt; HPTS; pyranine; HPTS; Pyrene; $C_{16}H_7O_{10}S_3$; 6358-69-6	4.28
158	8-Trifluoromethyl-BODIPY 10	6-((7-Chloro-5,5-difluoro-10-(trifluoromethyl)-5H-4l4,5l4-dipyrrolo[1,2-c:2′,1′-f][1,3,2]diazaborinin-3-yl)amino)hexanoate; Borondipyrromethene (Small); $C_{16}H_{15}BClF_5N_3O_2$	4.14.1
159	8-Trifluoromethyl-BODIPY 11	3-(((7-Chloro-5,5-difluoro-10-(trifluoromethyl)-5H-4l4,5l4-dipyrrolo[1,2-c:2′,1′-f][1,3,2]diazaborinin-3-yl)methyl)thio)propane-1-sulfonate; Borondipyrromethene (Small); $C_{14}H_{12}BClF_5N_2O_3S_2$	4.14.1
160	8-Trifluoromethyl-BODIPY 12	6-((5,5-Difluoro-7-(((2-sulfonatoethyl)thio)methyl)-10-(trifluoromethyl)-5H-4l4,5l4-dipyrrolo[1,2-c:2′,1′-f][1,3,2]diazaborinin-3-yl)amino)hexanoate; Borondipyrromethene (Small); $C_{19}H_{21}BF_5N_3O_5S_2$	4.14.1

(Continued)

Ent.	Short Name	Name; Class; Molecular Formula; CAS Number	Ref.
161	9-Aminoacridine	Acridin-9-amine; Acridine; $C_{13}H_{11}N_2$; 134-50-9	4.23.1
162	Abberior Star 440SXP	4-(6-(((Hydroxyoxidophosphoryl)oxy)methyl)-8,8-dimethyl-2-oxo-3-(pyridin-1-ium-4-yl)-2H-pyrano[3,2-g]quinolin-9(8H)-yl)butanoate; Coumarin (More elaborated 7-Amino); $C_{28}H_{28}N_3PO_{10}$	4.26.3
163	Abberior Star 470SXP	(E)-4-(6-(((Hydroxyoxidophosphoryl)oxy)methyl)-8,8-dimethyl-2-oxo-3-(2-(pyridin-1-ium-2-yl)vinyl)-4-(trifluoromethyl)-2H-pyrano[3,2-g]quinolin-9(8H)-yl)butanoate; Coumarin (More elaborated 7-Amino); $C_{31}H_{29}F_3N_3O_{10}P$	4.26.3
164	Abberior Star 488 carboxylate	(E)-4-(4,5-Disulfonato-6-((2,2,2-trifluoroethyl)amino)-3-((2,2,2-trifluoroethyl)iminio)-3H-xanthen-9-yl)isophthalate; Rhodamine (Highly charged); $C_{25}H_{16}F_6N_2O_{11}S_2$	4.2.5
165	Abberior Star 512 carboxylate	(E)-4-(2-(4,5-Disulfonato-6-((2,2,2-trifluoroethyl)amino)-3-((2,2,2-trifluoroethyl)iminio)-3H-xanthen-9-yl)-N-methylbenzamido)butanoate; Rhodamine (500 < λa, max < 550 nm); $C_{29}H_{25}F_6N_3O_{10}S_2$	4.2.2
166	Abberior Star 520SXP	4-(6-(((Hydroxyoxidophosphoryl)oxy)methyl)-8,8-dimethyl-2-oxo-3-(1-(3,3,4,4-tetrafluoro-2-oxido-5-oxocyclopent-1-en-1-yl)pyridin-1-ium-4-yl)-2H-pyrano[3,2-g]quinolin-9(8H)-yl) butanoate; Coumarin (More elaborated 7-Amino); $C_{33}H_{28}F_4N_3O_{12}P$	4.26.3
167	Abberior Star 580 carboxylate	tl; Rhodamine (550 < λa, max < 600 nm); $C_{37}H_{39}N_3O_4$	4.2.3
168	Abberior Star 600 carboxylate	tl; Rhodamine (λa, max > 600 nm); $C_{34}H_{26}F_3N_2O_6S_2$	4.2.4
169	Abberior Star 635 carboxylate	tl; Rhodamine (λa, max > 600 nm); $C_{45}H_{49}F_3N_3O_9S_2$	4.2.4
170	Abberior Star 635P carboxylate	tl; Rhodamine (λa, max > 600 nm); $C_{43}H_{41}F_4N_3O_{12}P_2$	4.2.4
171	Abberior Star Red carboxylate	tl; Rhodamine (λa, max >600 nm); $C_{48}H_{48}F_4N_4O_{13}S_2$	4.2.4
172	Acedan	1-(6-(Dimethylamino)naphthalen-2-yl)ethan-1-one; Naphthalene; $C_{14}H_{15}NO$	4.34
173	Acedan 2	1-(6-((1s,4s)-7-Azabicyclo[2.2.1]heptan-7-yl)naphthalen-2-yl)ethan-1-one; Naphthalene; $C_{18}H_{19}NO$	4.34
174	Acedan 3	1-(6-(Isopropylamino)naphthalen-2-yl)ethan-1-one; Naphthalene; $C_{15}H_{17}NO$	4.34
175	Acedan 4	1-(6-(Methylamino)naphthalen-2-yl)ethan-1-one; Naphthalene; $C_{13}H_{13}NO$	4.34
176	Acedan 5	1-(6-((4-hydroxycyclohexyl)amino)naphthalen-2-yl)ethan-1-one; Naphthalene; $C_{18}H_{21}NO_2$	4.34
177	Acedan 6	N-(4-((6-Acetylnaphthalen-2-yl)amino)cyclohexyl)acetamide; Naphthalene; $C_{20}H_{24}N_2O_2$	4.34
178	Acedan 7	1-(6-((2-Hydroxyethyl)amino)naphthalen-2-yl)ethan-1-one; Naphthalene; $C_{14}H_{15}NO_2$	4.34
179	Acedan 8	1-(6-Morpholinonaphthalen-2-yl)ethan-1-one; Naphthalene; $C_{16}H_{17}NO_2$	4.34
180	Acridine orange	N-(6-(Dimethylamino)acridin-3(10H)-ylidene)-N-methylmethanaminium; Acridine; $C_{17}H_{20}N_3$	4.23.1
181	Acridine orange	N-(6-(Dimethylamino)acridin-3(10H)-ylidene)-N-methylmethanaminium; Acridine; $C_{17}H_{19}N_2$	4.23.1

Ent.	Short Name	Name; Class; Molecular Formula; CAS Number	Ref.
182	Acridine red	(E)-N-(6-(methylamino)-3H-xanthen-3-ylidene)methanaminium; Pyronine; $C_{15}H_{15}N_2O$	4.4
183	Acridizinium I-L 5a	9-Aminopyrido[1,2-b]isoquinolin-5-ium; Others (Acridizinium); $C_{13}H_{11}N_2$	4.39.1
184	Acridizinium I-L 5b	8-Aminopyrido[1,2-b]isoquinolin-5-ium; Others (Acridizinium); $C_{13}H_{11}N_2$	4.39.1
185	Acridizinium J-N 1a	Pyrido[1,2-b]isoquinolin-5-ium bromide; Others (Acridizinium); $C_{13}H_{10}N$	4.39.1
186	Acridizinium J-N 1b	8-Fluoropyrido[1,2-b]isoquinolin-5-ium bromide; Others (Acridizinium); $C_{13}H_9FN$	4.39.1
187	Acridizinium J-N 1c	8,10-Difluoropyrido[1,2-b]isoquinolin-5-ium bromide; Others (Acridizinium); $C_{13}H_8F_2N$	4.39.1
188	Acridizinium J-N 1d	8-Chloropyrido[1,2-b]isoquinolin-5-ium bromide; Others (Acridizinium); $C_{13}H_9ClN$	4.39.1
189	Acridizinium J-N 1e	8-Methoxypyrido[1,2-b]isoquinolin-5-ium chloride; Others (Acridizinium); $C_{14}H_{12}NO$	4.39.1
190	Acrylodan	6-Acryloyl-2-dimethylaminonaphthalene; Naphthalene; $C_{15}H_{15}NO$; 86636-92-2	4.34
191	$ACSO_3H$	9-Oxo-4a,9,9a,10-tetrahydroacridine-2,7-disulfonate; Acridones; $C_{13}H_9NO_7S_2$	4.23.2
192	Acyman	(1r,4r)-4-(Methyl(naphthalen-2-yl)amino)cyclohexan-1-ol; Naphthalene; $C_{17}H_{21}NO$	4.34
193	AF555	6-(3,3-Dimethyl-2-((E)-3-((E)-3-methyl-5-sulfonato-1,3-bis(3-sulfonatopropyl)indolin-2-ylidene)prop-1-en-1-yl)-5-sulfonato-3H-indol-1-ium-1-yl)hexanoate; Cyanine (Trimethine); $C_{40}H_{50}N_4O_{15}S_4$	4.13.1
194	AF647	6-(2-((1E,3E)-5-((E)-3,3-Dimethyl-5-sulfonato-1-(3-sulfonatopropyl)indolin-2-ylidene) penta-1,3-dien-1-yl)-3-methyl-5-sulfonato-3-(4-sulfonatobutyl)-3H-indol-1-ium-1-yl) hexanoate; Cyanine (Pentamethine); $C_{37}H_{44}N_2O_{14}S_4$	4.13.2
195	AFD647	6-(2-((1E,3E)-5-((E)-3,3-Dimethyl-5-sulfonato-1-(3-sulfonatopropyl)indolin-2-ylidene) penta-1,3-dien-1-yl)-3-methyl-5-sulfonato-3-(3-sulfonatopropyl)-3H-indol-1-ium-1-yl) hexanoate; Cyanine (Pentamethine); $C_{36}H_{42}N_2O_{14}S_4$	4.13.2
196	AI3	2-(2-(2-(2-Aminoethoxy)ethoxy)ethyl)-6-((2-(2-(2-aminoethoxy)ethoxy)ethyl)amino)-1H-benzo[de]isoquinoline-1,3(2H)-dione; Naphthalimide; $C_{24}H_{34}N_4O_6$	4.32
197	Alexa Fluor 350	2-(7-Amino-4-methyl-2-oxo-6-sulfonato-2H-chromen-3-yl)acetate; Coumarin (More elaborated 7-Amino); $C_{16}H_{14}N_2O_6S_2$	4.26.3
198	Alexa Fluor 405	1-(2-((3,6,8-Trisulfonatopyren-1-yl)oxy)acetyl)piperidine-4-carboxylate; Pyrene; $C_{24}H_{18}NO_{13}S_3$	4.28
199	Alexa Fluor 430	6-(8,8-Dimethyl-2-oxo-6-(sulfonatomethyl)-4-(trifluoromethyl)-2H-pyrano[3,2-g]quinolin-9(8H)-yl)hexanoate; Coumarin (More elaborated 7-Amino); $C_{26}H_{26}F_3N_2O_9S$	4.26.3
200	Alexa Fluor 488	2-(6-Amino-3-iminio-4,5-disulfonato-3H-xanthen-9-yl)terephthalate; Rhodamine (Highly charged); $C_{22}H_{11}N_2O_{11}S_2$	4.2.5
201	Alexa Fluor 532	4-((2R,8S)-2,3,3,7,7,8-Hexamethyl-10,12-disulfonato-2,3,7,8-tetrahydro-1H-pyrano[3,2-f:5,6-f']diindol-9-ium-5-yl)benzoate; Rosamine; $C_{34}H_{40}N_2O_{14}S_4$; 247145-11-5	4.17

(Continued)

Ent.	Short Name	Name; Class; Molecular Formula; CAS Number	Ref.
202	Alexa Fluor 546	4-((2-((5-Carb oxylatopentyl)amino)-2-oxoethyl)thio)-2,3,5-trichloro-6-((4R,8S)-2,2,4,8,10,10-hexamethyl-12,14-disulfonato-3,4,8,9,10,11-hexahydro-2H-pyrano[3,2-g:5,6-g'] diquinolin-1-ium-6-yl)benzoate; Rhodamine (Highly charged); $C_{40}H_{41}Cl_3N_3O_{12}S_3$	4.2.5
203	Alexa Fluor 555	6-(2-((E)-3-((E)-3,3-Dimethyl-5-sulfonato-1-(3-sulfonatopropyl)indolin-2-ylidene)prop-1-en-1-yl)-3-methyl-5-sulfonato-1-(3-sulfonatopropyl)-3H-indol-1-ium-3-yl)hexanoate; Cyanine (Trimethine); $C_{40}H_{50}N_4O_{15}S_4$	4.13.1
204	Alexa Fluor 568	2-(2,2,10,10-Tetramethyl-4,8-bis(sulfonatomethyl)-10,11-dihydro-2H-pyrano[3,2-g:5,6-g'] diquinolin-1-ium-6-yl)terephthalate; Rhodamine (Highly charged); $C_{33}H_{30}N_2O_{11}S_2$; 247145-38-6	4.2.5
205	Alexa Fluor 594	2-(1,2,2,10,10,11-Hexamethyl-4,8-bis(sulfonatomethyl)-10,11-dihydro-2H-pyrano[3,2-g:5,6-g']diquinolin-1-ium-6-yl)terephthalate; Rhodamine (Highly charged); $C_{35}H_{34}N_2O_{11}S_2$	4.2.5
206	Alexa Fluor 633	nd; Sulforhodamine; $C_{38}H_{33}F_2N_2O_{12}S_6$; 1132773-86-4	4.6
207	Alexa Fluor 647	6-(2-((1E,3E)-5-((E)-3,3-Dimethyl-5-sulfonato-1-(3-sulfonatopropyl)indolin-2-ylidene) penta-1,3-dien-1-yl)-3-methyl-5-sulfonato-1-(3-sulfonatopropyl)-3H-indol-1-ium-3-yl) hexanoate; Cyanine (Pentamethine); $C_{36}H_{42}N_2O_{14}S_4$	4.13.2
208	Alexa Fluor 660	nd; nd; nd; 422309-89-5	4.41
209	Alexa Fluor 680	(E)-2-((2E,4E)-5-(5-Bromo-3,3-dimethyl-7-(3-sulfonatopropyl)-3H-pyrrolo[2,3-b]pyridin-7-ium-2-yl)penta-2,4-dien-1-ylidene)-3-(6-((2-(2,5-dioxo-2,5-dihydro-1H-pyrrol-1-yl)ethyl) amino)-6-oxohexyl)-3-methyl-1-(3-sulfonatopropyl)indoline-5-sulfonate; Cyanine (Pentamethine); $C_{35}H_{45}BrN_3O_{11}S_3$; 948558-33-6	4.13.2
210	Alexa Fluor 700	nd; nd; nd; 1246956-22-8	4.41
211	Alexa Fluor 750	nd; nd; nd; 697795-06-5	4.41
212	AR116	6-(Diethylamino)-3H-xanthen-3-iminium; Pyronine; $C_{17}H_{19}N_2O$	4.4
213	Atto 390 carboxylate	4-(4,6,8,8-Tetramethyl-2-oxo-7,8-dihydro-2H-pyrano[3,2-g]quinolin-9(6H)-yl)butanoate; Coumarin (More elaborated 7-Amino); $C_{20}H_{25}NO_4$	4.26.3
214	Atto 425 carboxylate	4-(3-(Ethoxycarbonyl)-6,8,8-trimethyl-2-oxo-2H-pyrano[3,2-g]quinolin-9(8H)-yl)butanoate; Coumarin (More elaborated 7-Amino); $C_{22}H_{27}NO_6$	4.26.3
215	Atto 430LS carboxylate	nd; nd (pb Coumarin); nd	4.41
216	Atto 465 carboxylate	4-(3,6-Diaminoacridin-10-ium-10-yl)butanoate; Acridine; $C_{17}H_{17}N_3O_2$	4.23.1
217	Atto 488 carboxylate	4-(2-(6-Amino-3-iminio-4,5-disulfonato-3H-xanthen-9-yl)-N-methylbenzamido)butanoate; Rhodamine (λa, max < 500 nm); $C_{25}H_{24}N_3O_{10}S_2$	4.2.1
218	Atto 490LS carboxylate	nd; nd (pb Coumarin); nd	4.41
219	Atto 495 carboxylate	4-(3,6-Bis(dimethylamino)acridin-10-ium-10-yl)butanoate; Acridine; $C_{21}H_{25}N_3O_2$	4.23.1

Ent.	Short Name	Name; Class; Molecular Formula; CAS Number	Ref.
220	Atto 514 carboxylate	(Z)-4-(2-(4,5-Disulfonato-6-((2,2,2-trifluoroethyl)amino)-3-((2,2,2-trifluoroethyl)iminio)-3H-xanthen-9-yl)-N-methylbenzamido)butanoate; Rhodamine (500 < λa, max < 550 nm); $C_{29}H_{26}F_6N_3O_{10}S_2$	4.2.2
221	Atto 520 carboxylate	(E)-3-(6-(ethylamino)-3-(ethyliminio)-2,7-dimethyl-3H-xanthen-9-yl)propanoate; Pyronine; $C_{22}H_{27}N_2O_3$	4.4
222	Atto 532 carboxylate	(Z)-4-(2-(6-(Ethylamino)-3-(ethyliminio)-4,5-disulfonato-3H-xanthen-9-yl)-N-methylbenzamido)butanoate; Rhodamine (500 < λa, max < 550 nm); $C_{29}H_{31}N_3O_{10}S_2$	4.2.2
223	Atto 542 carboxylate	nd; nd (pb Rhodamine); nd	4.41
224	Atto 550 carboxylate	4-(2-(2,2,4,8,10,10-hexamethyl-3,4,8,9,10,11-hexahydro-2H-pyrano[3,2-g:5,6-g']diquinolin-1-ium-6-yl)-N-methylbenzamido)butanoate; Rhodamine (550 < λa, max < 600 nm); $C_{37}H_{43}N_3O_4$	4.2.3
225	Atto 565 carboxylate	2-(1,11-Diethyl-3,4,8,9,10,11-hexahydro-2H-pyrano[3,2-g:5,6-g']diquinolin-1-ium-6-yl) terephthalate; Rhodamine (550 < λa, max < 600 nm); $C_{31}H_{31}N_2O_5$	4.2.3
226	Atto 590 carboxylate	2-(1,11-Diethyl-2,2,4,8,10,10-hexamethyl-10,11-dihydro-2H-pyrano[3,2-g:5,6-g']diquinolin-1-ium-6-yl)terephthalate; Rhodamine (550 < λa, max < 600 nm); $C_{37}H_{39}N_2O_5$	4.2.3
227	Atto 594 carboxylate	4-(2-(1,11-Diethyl-2,2,10,10-tetramethyl-4,8-bis(sulfonatomethyl)-10,11-dihydro-2H-pyrano[3,2-g:5,6-g']diquinolin-1-ium-6-yl)-N-methylbenzamido)butanoate; Rhodamine (λa, max > 600 nm); $C_{41}H_{48}N_3O_{10}S_2$	4.2.4
228	Atto 610 carboxylate	4-(9-(Dimethyl-l4-azanylidene)-11,11-dimethyl-3,4,9,11-tetrahydronaphtho[2,3-g]quinolin-1(2H)-yl)butanoate; $C(CH_3)_2$-Pyronine; $C_{25}H_{31}N_2O_2$	4.5
229	Atto 620 carboxylate	4-(2-(6-(Dimethylamino)-3-(dimethyliminio)-10,10-dimethyl-3,10-dihydroanthracen-9-yl)-N-methylbenzamido)butanoate; $C(CH_3)_2$-Rhodamine; $C_{32}H_{38}N_3O_3$	4.3.1
230	Atto 633 carboxylate	4-(2-(9-(Dimethyliminio)-1,11,11-trimethyl-1,2,3,4,9,11-hexahydronaphtho[2,3-g]quinolin-6-yl)-N-methylbenzamido)butanoate; $C(CH_3)_2$-Rhodamine; $C_{34}H_{39}N_3O_3$	4.3.1
231	Atto 635 carboxylate	4-(9-(Dimethylamino)-2,2,4,11,11-pentamethyl-2,11-dihydronaphtho[2,3-g]quinolin-1-ium-1-yl)butanoate; $C(CH_3)_2$-Pyronine; $C_{29}H_{35}N_2O_2$	4.5
232	Atto 647 carboxylate	(R)-4-(1-Ethyl-2,2,8,10,10,13,13-heptamethyl-4-(sulfonatomethyl)-8,9,10,13-tetrahydrobenzo[1,2-g:5,4-g']diquinolin-1-ium-11(2H)-yl)butanoate; $C(CH_3)_2$-Pyronine; $C_{34}H_{43}N_2O_5S$	4.5
233	Atto 647N carboxylate	tl; $C(CH_3)_2$-Rhodamine; $C_{42}H_{51}N_3O_3$	4.3.1
234	Atto 655 carboxylate	4-(1-Ethyl-10,10-dimethyl-8-(sulfonatomethyl)-3,4,9,10-tetrahydro-2H-dipyrido[3,2-b:2′,3′-i]phenoxazin-1-ium-11(8H)-yl)butanoate; Phenoxazine; $C_{27}H_{33}N_3O_6S$	4.25
235	Atto 665 carboxylate	nd; nd (pb $C(CH_3)_2$-Pyronine); nd	4.41
236	Atto 680 carboxylate	4-(1-Ethyl-10,10-dimethyl-8-(sulfonatomethyl)-3,4-dihydro-2H-dipyrido[3,2-b:2′,3′-i] phenoxazin-1-ium-11(10H)-yl)butanoate; Phenoxazine; $C_{27}H_{28}N_5O_6S_1$	4.25

(Continued)

Ent.	Short Name	Name; Class; Molecular Formula; CAS Number	Ref.
237	Atto 700 carboxylate	4-(1-Ethyl-2,2,4,10,10-pentamethyl-8-(sulfonatomethyl)-2H-dipyrido[3,2-b:2′,3′-i] phenoxazin-1-ium-11(10H)-yl)butanoate; Phenoxazine; $C_{27}H_{28}N_5O_6S1$	4.25
238	Atto 725 carboxylate	nd; nd; nd	4.41
239	Atto 740 carboxylate	nd; nd; nd	4.41
240	Atto Oxa12 carboxylate	nd; Phenoxazine; $C_{27}H_{33}N_3O_6S$	4.25
241	Atto Rho101 carboxylate	tl; Rhodamine ($550 < \lambda a$, max < 600 nm); $C_{37}H_{40}N_3O_4$	4.2.3
242	Atto Rho11 carboxylate	4-(2-(1,11-Diethyl-3,4,8,9,10,11-hexahydro-2H-pyrano[3,2-g:5,6-g']diquinolin-1-ium-6-yl)-N-methylbenzamido)butanoate; Rhodamine ($550 < \lambda a$, max < 600 nm); $C_{35}H_{40}N_3O_4$	4.2.3
243	Atto Rho12 carboxylate	4-(2-((4R,8S)-1,11-Diethyl-2,2,4,8,10,10-hexamethyl-3,4,8,9,10,11-hexahydro-2H-pyrano[3,2-g:5,6-g']diquinolin-1-ium-6-yl)-N-methylbenzamido)butanoate; Rhodamine (550 λa, max < 600 nm); $C_{41}H_{52}N_3O_4$	4.2.3
244	Atto Rho13 carboxylate	4-(2-(1,11-Diethyl-2,2,4,8,10,10-hexamethyl-10,11-dihydro-2H-pyrano[3,2-g:5,6-g']diquinolin-1-ium-6-yl)-N-methylbenzamido)butanoate; Rhodamine (λa, max > 600 nm); $C_{41}H_{48}N_3O_4$	4.2.4
245	Atto Rho14 carboxylate	4-(2,3,4,5-Tetrachloro-6-(1,11-diethyl-2,2,4,8,10,10-hexamethyl-10,11-dihydro-2H-pyrano[3,2-g:5,6-g']diquinolin-1-ium-6-yl)-N-methylbenzamido)butanoate; Rhodamine (λa, max > 600 nm); $C_{41}H_{44}Cl_4N_3O_4$	4.2.4
246	Atto Rho3B carboxylate	4-(2-(6-(diethylamino)-3-(diethyliminio)-3H-xanthen-9-yl)-N-methylbenzamido)butanoate; Rhodamine ($550 < \lambda a$, max < 600 nm); $C_{33}H_{40}N_3O_4$	4.2.3
247	Atto Rho6G carboxylate	(E)-4-(2-(6-(Ethylamino)-3-(ethyliminio)-2,7-dimethyl-3H-xanthen-9-yl)-N-methylbenzamido)butanoate; Rhodamine ($500 < \lambda a$, max < 550 nm); $C_{31}H_{35}N_3O_4$	4.2.2
248	BC-BA	(3-(Benzo[d]thiazol-2-yl)-2-oxo-2H-chromen-7-yl)boronic acid; Hybrid – Single bond connected dye; $C_{16}H_{10}BNO_4S$	4.40.2
249	BCECF	2-(2,7-Bis(2-carboxylatoethyl)-6-oxido-3-oxo-3H-xanthen-9-yl)terephthalate; Fluorescein (Non-halogenated); $C_{27}H_{15}O_{11}$	4.8.1
250	BC-H	3-(Benzo[d]thiazol-2-yl)-2H-chromen-2-one; Hybrid – Single bond connected dye; $C_{16}H_9NO_2S$	4.40.2
251	BC-OH	3-(Benzo[d]thiazol-2-yl)-2-oxo-2H-chromen-7-olate; Hybrid – Single bond connected dye; $C_{16}H_8NO_3S$	4.40.2
252	BCPCF	2-(2,7-Bis(3-methoxy-3-oxopropyl)-6-oxido-3-oxo-3H-xanthen-9-yl)terephthalate; Fluorescein (Non-halogenated); $C_{29}H_{21}O_{11}$	4.8.1
253	Benzo[h]quinoline (BQ)	Benzo[h]quinoline; Quinoline; $C_{13}H_9N$	4.29
254	BMeS-*p*-A	2,5-Bis(methylsulfonyl)-1,4-diaminobenzene; Benzene; $C_8H_{12}N_2O_4S_2$; 1678513-97-7	4.38
255	BMeS-*p*-A-2COOH	3,3′-((2,5-bis(methylsulfonyl)-1,4-phenylene)bis(azanediyl))dipropionic acid; Benzene; $C_{14}H_{20}N_2O_8S_2$	4.38

Ent.	Short Name	Name; Class; Molecular Formula; CAS Number	Ref.
256	BMeS-*p*-A-COOH	3-((4-amino-2,5-bis(methylsulfonyl)phenyl)amino)propanoic acid; Benzene; $C_{11}H_{16}N_2O_6S_2$	4.38
257	BODIPY	4,4-Difluoro-4-bora-3a,4a-diaza-s-indacene; Borondipyrromethene (Small); $C_{15}H_{17}BF_2N_2$	4.14.1
258	BODIPY 630/650	(E)-3-(4-(2-(5,5-Difluoro-7-(thiophen-2-yl)-5H-4l4,5l4-dipyrrolo[1,2-c:2′,1′-f][1,3,2] diazaborinin-3-yl)vinyl)phenoxy)-2-oxopropanoate; Borondipyrromethene (Small); $C_{24}H_{16}BF_2N_2O_4S$	4.14.1
259	BODIPY FL	3-(5,5-Difluoro-7,9-dimethyl-5H-4l4,5l4-dipyrrolo[1,2-c:2′,1′-f][1,3,2]diazaborinin-3-yl) propanoate; Borondipyrromethene (Small); $C_{14}H_{14}BF_2N_2O_2$; 165599-63-3	4.14.1
260	BOPAHY H-D 18ab’	7,7,11,11-Tetrafluoro-1-hexyl-3,14-dimethyl-9-(1H-pyrrol-2-yl)-1,4,5,11-tetrahydro-7H-3l4,7l4,8l4,11l4,12l4-[1,3,4,2]oxadiazaborolo[3′,4′:1,2][1,2,4,3]triazaborinino[4,5-a][1,2,3] triazolo[4,5-e]indole; Borondipyrromethene-analog; $C_{22}H_{28}B_2F_4N_7O$	4.14.5
261	BOPAHY H-D 18bb’	7,7,11,11-Tetrafluoro-1-(2-methoxyethyl)-3,14-dimethyl-9-(1H-pyrrol-2-yl)-1,4,5,11-tetrahydro-7H-3l4,7l4,8l4,11l4,12l4-[1,3,4,2]oxadiazaborolo[3′,4′:1,2][1,2,4,3] triazaborinino[4,5-a][1,2,3]triazolo[4,5-e]indole; Borondipyrromethene-analog; $C_{19}H_{22}B_2F_4N_7O_2$	4.14.5
262	BOPAHY H-D 19aa’	1,1,5,5-Tetrafluoro-8-(1-hexyl-3-methyl-1H-1,2,3l4-triazol-5-yl)-3,7,9-trimethyl-1H,5H-1l4,4l4,5l4,11l4-[1,3,4,2]oxadiazaborolo[3,4-a]pyrrolo[1,2-d][1,2,4,3]triazaborinine; Borondipyrromethene-analog; $C_{18}H_{27}B_2F_4N_6O$	4.14.5
263	BOPAHY H-D 19ab’	1,1,5,5-Tetrafluoro-8-(1-hexyl-3-methyl-1H-1,2,3l4-triazol-5-yl)-7,9-dimethyl-3-(1H-pyrrol-2-yl)-1H,5H-1l4,4l4,5l4,11l4-[1,3,4,2]oxadiazaborolo[3,4-a]pyrrolo[1,2-d][1,2,4,3] triazaborinine; Borondipyrromethene-analog; $C_{21}H_{28}B_2F_4N_7O$	4.14.5
264	BOPAHY H-D 19ba’	1,1,5,5-Tetrafluoro-8-(1-(2-methoxyethyl)-3-methyl-1H-1,2,3l4-triazol-5-yl)-3,7,9-trimethyl-1H,5H-1l4,4l4,5l4,11l4-[1,3,4,2]oxadiazaborolo[3,4-a]pyrrolo[1,2-d][1,2,4,3]triazaborinine; Borondipyrromethene-analog; $C_{15}H_{21}B_2F_4N_6O_2$	4.14.5
265	BOPAHY H-D 19bb’	1,1,5,5-Tetrafluoro-8-(1-(2-methoxyethyl)-3-methyl-1H-1,2,3l4-triazol-5-yl)-7,9-dimethyl-3-(1H-pyrrol-2-yl)-1H,5H-1l4,4l4,5l4,11l4-[1,3,4,2]oxadiazaborolo[3,4-a]pyrrolo[1,2-d][1,2,4,3] triazaborinine; Borondipyrromethene-analog; $C_{18}H_{22}B_2F_4N_7O_2$	4.14.5
266	BOPAHY H-D 20bb’	1,1,5,5-Tetrafluoro-8-(1-(2-methoxyethyl)-3-methyl-1H-1,2,3l4-triazol-5-yl)-7,9-dimethyl-3-(1H-pyrrol-2-yl)-1H,5H-1l4,4l4,5l4,11l4-[1,3,4,2]oxadiazaborolo[3,4-a]pyrrolo[1,2-d][1,2,4,3] triazaborinine; Borondipyrromethene-analog; $C_{18}H_{22}B_2F_4N_7O_2$	4.14.5
267	Bpe-H	N-Ethyl-N-(5-((3-hydroxypropyl)amino)-9H-benzo[a]phenoxazin-9-ylidene)ethanaminium; Phenoxazine; $C_{23}H_{26}N_3O_2$	4.25
268	Briliant sulfaflavine	6-Amino-2-(2,4-dimethylphenyl)-1,3-dioxo-2,3-dihydro-1H-benzo[de]isoquinoline-5-sulfonate; Naphthalimide; $C_{19}H_{13}N_2O_5S$; 2391-30-2	4.32
269	B-Z 14	4-(3,7-Bis((E)-4-((ethoxyoxidophosphoryl)methoxy)styryl)-5,5-bis(3-(2-methoxyethoxy) prop-1-yn-1-yl)-1,9-dimethyl-5H-4l4,5l4-dipyrrolo[1,2-c:2′,1′-f][1,3,2]diazaborinin-10-yl) benzoate; Borondipyrromethene (Large); $C_{52}H_{56}BN_2O_{14}P_2$	4.14.3

(Continued)

Ent.	Short Name	Name; Class; Molecular Formula; CAS Number	Ref.
270	B-Z 19	4-(3,7-Bis((E)-2-(6-((ethoxyoxidophosphoryl)methoxy)naphthalen-2-yl)vinyl)-5,5-bis(3-(2-methoxyethoxy)prop-1-yn-1-yl)-1,9-dimethyl-5H-4l4,5l4-dipyrrolo[1,2-c:2′,1′-f][1,3,2]diazaborinin-10-yl)benzoate; Borondipyrromethene (Large); $C_{60}H_{60}BN_2O_{14}P_2$	4.14.3
271	B-Z 7a	4-(5,5-Bis(3-(2-(2-((ethoxyoxidophosphoryl)methoxy)ethoxy)ethoxy)prop-1-yn-1-yl)-1,3,7,9-tetramethyl-5H-4l4,5l4-dipyrrolo[1,2-c:2′,1′-f][1,3,2]diazaborinin-10-yl)benzoate; Borondipyrromethene (Large); $C_{40}H_{52}BN_2O_{14}P_2$	4.14.3
272	B-Z 7b	4-(5,5-Bis(3-(2-(2-((ethoxyoxidophosphoryl)methoxy)ethoxy)ethoxy)prop-1-yn-1-yl)-2,8-diethyl-1,3,7,9-tetramethyl-5H-4l4,5l4-dipyrrolo[1,2-c:2′,1′-f][1,3,2]diazaborinin-10-yl)benzoate; Borondipyrromethene (Large); $C_{44}H_{60}BN_2O_{14}P_2$	4.14.3
273	Bza-Oet	(E)-N-(5-((4-Ethoxy-4-oxobutyl)amino)-10-methyl-9H-benzo[a]phenoxazin-9-ylidene)ethanaminium; Phenoxazine; $C_{25}H_{28}N_3O_3$	4.25
274	Bza-OH	(E)-N-(5-((3-Carboxypropyl)amino)-10-methyl-9H-benzo[a]phenoxazin-9-ylidene)ethanaminium; Phenoxazine; $C_{23}H_{24}N_3O_3$	4.25
275	Bzh-Cl	(E)-N-(5-((3-Hydroxypropyl)amino)-10-methyl-9H-benzo[a]phenoxazin-9-ylidene)ethanaminium; Phenoxazine; $C_{22}H_{24}N_3O_2$	4.25
276	Bzh-H	(E)-N-(5-((3-Chloropropyl)amino)-10-methyl-9H-benzo[a]phenoxazin-9-ylidene)ethanaminium; Phenoxazine; $C_{22}H_{23}ClN_3O$	4.25
277	Bzm-H	N-(5-((3-Hydroxypropyl)amino)-9H-benzo[a]phenoxazin-9-ylidene)-N-methylmethanaminium; Phenoxazine; $C_{21}H_{22}N_3O_2$	4.25
278	Carbofluorescein	2-(6-Hydroxy-10,10-dimethyl-3-oxo-3,10-dihydroanthracen-9-yl)benzoate; $C(CH_3)_2$-Fluorescein; $C_{23}H_{19}O_4$	4.9
279	Carborhodamine	2-(6-Amino-3-iminio-10,10-dimethyl-3,10-dihydroanthracen-9-yl)benzoate; $C(CH_3)_2$-Rhodamine; $C_{23}H_{21}N_2O_2$	4.3.1
280	Carborhodamine 16a	6-(2-((2-Ammonio-2-methylpropoxy)carbonyl)phenyl)-1,11-bis(2-methoxyethyl)-13,13-dimethyl-2,3,4,8,9,10,11,13-octahydrobenzo[1,2-g:5,4-g']diquinolin-1-ium; $C(CH_3)_2$-Rhodamine; $C_{39}H_{51}N_3O_4$	4.3.1
281	Carborhodamine 1b	1,11-Bis(2-methoxyethyl)-13,13-dimethyl-6-(2-(methyl(3-oxobutyl)carbamoyl)phenyl)-2,3,4,8,9,10,11,13-octahydrobenzo[1,2-g:5,4-g']diquinolin-1-ium; $C(CH_3)_2$-Rhodamine; $C_{40}H_{50}N_3O_4$	4.3.1
282	Carborhodamine 1c	3-(2-(1,11-Bis(2-methoxyethyl)-13,13-dimethyl-2,3,4,8,9,10,11,13-octahydrobenzo[1,2-g:5,4-g']diquinolin-1-ium-6-yl)-N-methylbenzamido)propanoate; $C(CH_3)_2$-Rhodamine; $C_{39}H_{47}N_3O_5$	4.3.1
283	Carborhodamine 20a	1,11-Bis(2-hydroxyethyl)-13,13-dimethyl-6-(2-(methyl(3-oxobutyl)carbamoyl)phenyl)-2,3,4,8,9,10,11,13-octahydrobenzo[1,2-g:5,4-g']diquinolin-1-ium; $C(CH_3)_2$-Rhodamine; $C_{38}H_{46}N_3O_4$	4.3.1
284	Carborhodamine 20b	3-(2-(1,11-Bis(2-hydroxyethyl)-13,13-dimethyl-2,3,4,8,9,10,11,13-octahydrobenzo[1,2-g:5,4-g']diquinolin-1-ium-6-yl)-N-methylbenzamido)propanoate; $C(CH_3)_2$-Rhodamine; $C_{37}H_{43}N_3O_5$	4.3.1

Ent.	Short Name	Name; Class; Molecular Formula; CAS Number	Ref.
285	Carborhodamine 21a	2-(3-(2-(1,11-Bis(2-methoxyethyl)-13,13-dimethyl-2,3,4,8,9,10,11,13-octahydrobenzo[1,2-g:5,4-g']diquinolin-1-ium-6-yl)-N-methylbenzamido)propanamido)-3-methoxy-3-oxopropane-1-sulfonate; $C(CH_3)_2$-Rhodamine; $C_{43}H_{54}N_4O_9S$	4.3.1
286	Carborhodamine 21b	(3-(2-(1,11-Bis(2-methoxyethyl)-13,13-dimethyl-2,3,4,8,9,10,11,13-octahydrobenzo[1,2-g:5,4-g']diquinolin-1-ium-6-yl)-N-methylbenzamido)propanoyl)(sulfonato)alaninate; $C(CH_3)_2$-Rhodamine; $C_{42}H_{51}N_4O_9S$	4.3.1
287	Carborhodamine 22	3-(2-(3-(2-(1,11-Bis(2-methoxyethyl)-13,13-dimethyl-2,3,4,8,9,10,11,13-octahydrobenzo[1,2-g:5,4-g']diquinolin-1-ium-6-yl)-N-methylbenzamido)propanamido)-3-sulfonatopropanamido)propanoate; $C(CH_3)_2$-Rhodamine; $C_{45}H_{56}N_5O_{10}S$	4.3.1
288	Carborhodol S-H 13-H	4-(2-(6-(Dimethylamino)-10,10-dimethyl-3-oxo-3,10-dihydroanthracen-9-yl)benzamido) butanoate; $C(CH_3)_2$-Rhodol; $C_{29}H_{29}N_2O_4$	4.16
289	Carbostyril 1	6,7-Dimethoxy-4-(trifluoromethyl)quinolin-2(1H)-one; Quinoline; $C_{12}H_{10}F_3NO_3$	4.29
290	Carbostyril 124	7-Amino-4-(trifluoromethyl)quinolin-2(1H)-one; compound 10; Quinoline; $C_{10}H_7F_3N_2O$	4.29
291	Carbostyril 13	1,4-Dimethyl-9-(trifluoromethyl)-1,3,4,6-tetrahydropyrido[2,3-g]quinoxalin-7(2H)-one; Quinoline; $C_{14}H_{14}F_3N_3O$	4.29
292	Carbostyril 2	6,7-Dimethoxy-1-methyl-4-(trifluoromethyl)quinolin-2(1H)-one; Quinoline; $C_{13}H_{12}F_3NO_3$	4.29
293	Carbostyril 23b	1,4-Diethyl-9-(trifluoromethyl)-1,3,4,6-tetrahydropyrido[2,3-g]quinoxalin-7(2H)-one; Quinoline; $C_{16}H_{18}F_3N_3O$	4.29
294	Carbostyril 23b	1,4-Diethyl-9-(trifluoromethyl)-1,3,4,6-tetrahydropyrido[2,3-g]quinoxalin-7(2H)-one; Quinoline; $C_{16}H_{18}F_3N_3O$	4.29
295	Carbostyril 23b	1,4-Diethyl-9-(trifluoromethyl)-1,3,4,6-tetrahydropyrido[2,3-g]quinoxalin-7(2H)-one; Quinoline; $C_{16}H_{18}F_3N_3O$	4.29
296	Carbostyril 4	6-Amino-7-methoxy-4-(trifluoromethyl)quinolin-2(1H)-one; Quinoline; $C_{11}H_9F_3N_2O_2$	4.29
297	Carbostyril 5	N-(7-Methoxy-2-oxo-4-(trifluoromethyl)-1,2-dihydroquinolin-6-yl)acetamide; Quinoline; $C_{13}H_{11}F_3N_2O_3$	4.29
298	Carbostyril 6	7-Methoxy-6-(methylamino)-4-(trifluoromethyl)quinolin-2(1H)-one; Quinoline; $C_{12}H_{11}F_3N_2O_2$	4.29
299	Carbostyril 7	6-(Dimethylamino)-7-methoxy-4-(trifluoromethyl)quinolin-2(1H)-one; Quinoline; $C_{13}H_{13}F_3N_2O_2$	4.29
300	Carbostyril 9	N-(6-Methoxy-2-oxo-4-(trifluoromethyl)-1,2-dihydroquinolin-7-yl)acetamide; Quinoline; $C_{13}H_{11}F_3N_2O_3$	4.29
301	Cascade Blue	2-((3,6,8-Trisulfonatopyren-1-yl)oxy)acetate; Pyrene; $C_{18}H_9O_{10}S_3$; 138039-52-8	4.28
302	Cascade yellow	5-[2-[1-[[3-(2,5-dioxopyrrolidin-1-yl)oxycarbonylphenyl]methyl]pyridin-1-ium-4-yl]-1,3-oxazol-5-yl]-2-methoxybenzenesulfonate; PyMPO; Hybrid – Single bond connected dye; $C_{27}H_{21}N_3O_9S$	4.40.2

(Continued)

Ent.	Short Name	Name; Class; Molecular Formula; CAS Number	Ref.
303	CF405M	nd; nd (Coumarin); nd	4.41
304	CF405S	nd; nd (Pyrene); nd	4.41
305	CF430	nd; nd (Cyanine); nd	4.41
306	CF450	nd; nd (Cyanine); nd	4.41
307	CF514	nd; nd (Rhodamine); nd	4.41
308	CF532	nd; nd (Rhodamine); nd	4.41
309	CF594	nd; nd (Rhodamine); nd	4.41
310	CF620R	nd; nd (Rhodamine); nd	4.41
311	CF647	nd; nd (Cyanine); nd	4.41
312	CF660C	nd; nd (Cyanine); nd	4.41
313	CF660R	nd; nd (Rhodamine); nd	4.41
314	CF680	nd; nd (Cyanine); nd	4.41
315	CF680R	nd; nd (Rhodamine); nd	4.41
316	Chlorinated fluorescein G-C 1	2-(2,7-Dichloro-6-oxido-3-oxo-3H-xanthen-9-yl)benzoate; Fluorescein (Halogenated); $C_{20}H_8Cl_2O_5$	4.8.2
317	Chlorinated fluorescein G-C 2(3)	2(5)-Chloro-6-(6-oxido-3-oxo-3H-xanthen-9-yl)benzoate; Fluorescein (Halogenated); $C_{20}H_9ClO_5$	4.8.2
318	Chlorinated fluorescein G-C 4	2(3)-Chloro-6-(6-oxido-3-oxo-3H-xanthen-9-yl)benzoate; Fluorescein (Halogenated); $C_{20}H_9ClO_5$	4.8.2
319	Chlorinated fluorescein G-C 5	2,3,4,5-Tetrachloro-6-(6-oxido-3-oxo-3H-xanthen-9-yl)benzoate; Fluorescein (Halogenated); $C_{20}H_6Cl_4O_5$	4.8.2
320	Chlorinated fluorescein G-C 6	2,3,4,5-Tetrachloro-6-(2,7-dichloro-6-oxido-3-oxo-3H-xanthen-9-yl)benzoate; Fluorescein (Halogenated); $C_{20}H_4Cl_6O_5$	4.8.2
321	Chromenoquinoline L-S 4a	N,N-Diethyl-9-methyl-6H-chromeno[4,3-b]quinolin-3-amine; Hybrid – Fusion of two existing dye; $C_{21}H_{22}N_2O$	4.40.1
322	Chromenoquinoline L-S 4b	N,N-Diethyl-9-methoxy-6H-chromeno[4,3-b]quinolin-3-amine; Hybrid – Fusion of two existing dye; $C_{21}H_{22}N_2O_2$	4.40.1
323	Chromenoquinoline L-S 4c	3-(Diethylamino)-6H-chromeno[4,3-b]quinoline-9-carbonitrile; Hybrid – Fusion of two existing dye; $C_{21}H_{19}N_3O$	4.40.1
324	Chromenoquinoline L-S 4d	tl; Hybrid – Fusion of two existing dye; $C_{23}H_{22}N_2O$	4.40.1
325	Chromenoquinoline L-S 4e	tl; Hybrid – Fusion of two existing dye; $C_{23}H_{22}N_2O_2$	4.40.1
326	Chromenoquinoline L-S 4f	tl; Hybrid – Fusion of two existing dye; $C_{23}H_{19}N_3O$	4.40.1

Ent.	Short Name	Name; Class; Molecular Formula; CAS Number	Ref.
327	cis-PO-PB430	3-(5,10-Dioxido-5,10-bis(4-(3-sulfonatopropoxy)phenyl)phosphindolo[3,2-b]phosphindol-2-yl)benzoate; Others (Phosphole); $C_{39}H_{31}O_{12}P_2S_2$	4.39.3
328	Coumarin 1	7-(Diethylamino)-4-methyl-coumarin; 7-(diethylamino)-4-methyl-2H-chromen-2-one; Coumarin (7-Amino); $C_{14}H_{17}NO_2$	4.26.3
329	Coumarin 102	9-Methyl-2,3,6,7-tetrahydro-1H,5H,11H-pyrano[2,3-f]pyrido[3,2,1-ij]quinolin-11-one; Coumarin (7-Amino); $C_{16}H_{17}NO_2$	4.26.2
330	Coumarin 120	7-Diamino-4-methyl-2H-chromen-2-one; Coumarin (7-Amino); $C_{10}H_9NO_2$; 26093-31-2	4.26.2
331	Coumarin 152	7-(Diethylamino)-4-(trifluoromethyl)-2H-chromen-2-one; Coumarin (7-Amino); $C_{14}H_{14}F_3NO_2$	4.26.2
332	Coumarin 2	7-(Ethylamino)-4,5-dimethyl-2H-chromen-2-one; Coumarin (7-Amino); $C_{13}H_{15}NO_2$; 26078-25-1	4.26.2
333	Coumarin 307	7-(Ethylamino)-5-methyl-4-(trifluoromethyl)-2H-chromen-2-one; Coumarin (7-Amino); $C_{13}H_{12}F_3NO_2$	4.26.2
334	Coumarin P-R 12	3-Acetyl-2-oxo-2H-chromen-7-olate; Coumarin (7-Hydroxy); $C_{11}H_7O_4$	4.26.1
335	Coumarin P-R 13	3-Acetyl-8-fluoro-2-oxo-2H-chromen-7-olate; Coumarin (7-Hydroxy); $C_{11}H_6FO_4$	4.26.1
336	Coumarin P-R 14	3-Acetyl-6-chloro-2-oxo-2H-chromen-7-olate; Coumarin (7-Hydroxy); $C_{11}H_6ClO_4$	4.26.1
337	Coumarin P-R 21	8-Fluoro-7-(pyrrolidin-1-ylamino)-3-(pyrrolidine-1-carbonyl)-2H-chromen-2-one; Coumarin (7-Amino); $C_{18}H_{20}FN_3O_3$	4.26.2
338	Coumarin P-R 22	6-Chloro-7-(pyrrolidin-1-ylamino)-3-(pyrrolidine-1-carbonyl)-2H-chromen-2-one; Coumarin (7-Amino); $C_{18}H_{20}ClN_3O_3$	4.26.2
339	Coumarin P-R 3	7-Amino-2-oxo-2H-chromene-3-carboxylate; Coumarin (7-Amino); $C_{10}H_6NO_4$	4.26.2
340	Coumarin P-R 5	3-Acetyl-7-amino-8-fluoro-2H-chromen-2-one; Coumarin (7-Amino); $C_{11}H_8FNO_3$	4.26.2
341	Coumarin P-R 6	3-Acetyl-7-amino-6-chloro-2H-chromen-2-one; Coumarin (7-Amino); $C_{11}H_8ClNO_3$	4.26.2
342	Coumarin-pyronine 9	6-(Diethylamino)-9-(4,7-dihydroxy-2-oxo-2H-chromen-3-yl)-3H-xanthen-3-iminium; Hybrid – Single bond connected dye; $C_{26}H_{23}N_2O_5$	4.40.2
343	COUPY 10	4-(Cyano(7-(diethylamino)-4-methyl-2H-chromen-2-ylidene)methyl)-1-(2,2,2-trifluoroethyl)pyridin-1-ium; Hybrid – Polymethine bridged dye; $C_{23}H_{23}F_3N_3O$	4.40.3
344	COUPY 11	(Z)-4-(Cyano(7-(dimethylamino)-4-(trifluoromethyl)-2H-chromen-2-ylidene)methyl)-1-methylpyridin-1-ium; Hybrid – Polymethine bridged dye; $C_{20}H_{17}F_3N_3O$	4.40.3
345	COUPY 12	(Z)-4-(Cyano(7-(dimethylamino)-4-(trifluoromethyl)-2H-chromen-2-ylidene)methyl)-1-(2,2,2-trifluoroethyl)pyridin-1-ium; Hybrid – Polymethine bridged dye; $C_{21}H_{16}F_6N_3O$	4.40.3
346	COUPY 9	(Z)-4-(Cyano(7-(diethylamino)-4-methyl-2H-chromen-2-ylidene)methyl)-1-methylpyridin-1-ium; Hybrid – Polymethine bridged dye; $C_{22}H_{24}N_3O$	4.40.3
347	Cresyl violet (CV)	9-Amino-5H-benzo[a]phenoxazin-5-iminium; Phenoxazine; $C_{16}H_{12}N_3O$	4.25

(Continued)

Ent.	Short Name	Name; Class; Molecular Formula; CAS Number	Ref.
348	CW800	6-((E)-2-((E)-2-(3-((E)-2-(3,3-dimethyl-5-sulfonato-1-(4-sulfonatobutyl)-3H-indol-1-ium-2-yl)vinyl)-2-(4-sulfonatophenoxy)cyclohex-2-en-1-ylidene)ethylidene)-3,3-dimethyl-5-sulfonatoindolin-1-yl)hexanoate; Cyanine (Heptamethine); $C_{46}H_{50}N_2O_{15}S_4$; 1239619-02-3	4.13.3
349	Cy3	1-Ethyl-2-((E)-3-((E)-1-ethyl-3,3-dimethyl-5-sulfonatoindolin-2-ylidene)prop-1-en-1-yl)-3,3-dimethyl-3H-indol-1-ium-5-sulfonate; Cyanine (Trimethine); $C_{27}H_{31}N_2O_6S_2$	4.13.1
350	Cy3'	4-((E)-2-((E)-3-(3,3-Dimethyl-1-(4-sulfonatobutyl)-3H-indol-1-ium-2-yl)allylidene)-3,3-dimethylindolin-1-yl)butane-1-sulfonate; Cyanine (Trimethine); $C_{31}H_{39}N_2O_6S_2$	4.13.1
351	Cy3.10-OH	2-((E)-2-((E)-3-(5-(Carboxylatomethyl)-1-ethyl-3,3-dimethyl-3H-indol-1-ium-2-yl)allylidene)-1-ethyl-3,3-dimethylindolin-5-yl)acetate; Cyanine (Trimethine); $C_{31}H_{35}N_2O_4$	4.13.1
352	Cy3.12-OH	2-((E)-2-((E)-3-(5-(Carboxylatomethyl)-3,3-dimethyl-1-(4-sulfonatobutyl)-3H-indol-1-ium-2-yl)allylidene)-3,3-dimethyl-1-(4-sulfonatobutyl)indolin-5-yl)acetate; Cyanine (Trimethine); $C_{35}H_{41}N_2O_{10}S_2$	4.13.1
353	Cy3.13-OH	2-((E)-2-((E)-3-(3,3-Dimethyl-1-(4-sulfonatobutyl)-3H-indol-1-ium-2-yl)allylidene)-3,3-dimethyl-1-(4-sulfonatobutyl)indolin-5-yl)acetate; Cyanine (Trimethine); $C_{33}H_{40}N_2O_8S_2$	4.13.1
354	Cy3.18.OH	6-((E)-2-((E)-3-(1-(5-Carboxylatopentyl)-3,3-dimethyl-5-sulfonato-3H-indol-1-ium-2-yl)allylidene)-3,3-dimethyl-5-sulfonatoindolin-1-yl)hexanoate; Cyanine (Trimethine); $C_{35}H_{41}N_2O_{10}S_2$	4.13.1
355	Cy3.205	6-(2-((1E,3E)-3-(3-(5-c1arboxylatopentyl)-1,1-dimethyl-6,8-disulfonato-1,3,6,7-tetrahydro-2H-benzo[e]indol-2-ylidene)prop-1-en-1-yl)-1,1-dimethyl-6,8-disulfonato-1H-benzo[e]indol-3-ium-3-yl)hexanoate; Cyanine (Trimethine); $C_{43}H_{45}N_2O_{16}S_4$	4.13.1
356	Cy3.5	3-Ethyl-2-((1E,3E)-3-(3-ethyl-1,1-dimethyl-6,8-disulfonato-1,3,6,7-tetrahydro-2H-benzo[e]indol-2-ylidene)prop-1-en-1-yl)-1,1-dimethyl-1H-benzo[e]indol-3-ium-6,8-disulfonate; Cyanine (Trimethine); $C_{35}H_{35}N_2O_{12}S_4$	4.13.1
357	Cy3B	24-Carboxy-5,5,27,27-tetramethyl-16-oxa-12-aza-20-azoniaheptacy clo[15.11.0.03,15.04,12.06,11.020,28.021,26]octacosa-1,3,6(11),7,9,20(28),21(26),22,24-nonaene-8-sulfonate; Cyanine (Trimethine); $C_{31}H_{31}N_2O_6S$	4.13.1
358	Cy5	1-Ethyl-2-((1E,3E)-5-((E)-1-ethyl-3,3-dimethyl-5-sulfonatoindolin-2-ylidene)penta-1,3-dien-1-yl)-3,3-dimethyl-3H-indol-1-ium-5-sulfonate; Cyanine (Pentamethine); $C_{29}H_{33}N_2O_6S_2$	4.13.2
359	Cy5'	4-((E)-2-((2E,4E)-5-(3,3-Dimethyl-1-(4-sulfonatobutyl)-3H-indol-1-ium-2-yl)penta-2,4-dien-1-ylidene)-3,3-dimethylindolin-1-yl)butane-1-sulfonate; Cyanine (Pentamethine); $C_{33}H_{41}N_2O_6S_2$	4.13.2
360	Cy5.10-OH	2-((E)-2-((2E,4E)-5-(5-(Carboxylatomethyl)-1-ethyl-3,3-dimethyl-3H-indol-1-ium-2-yl)penta-2,4-dien-1-ylidene)-1-ethyl-3,3-dimethylindolin-5-yl)acetate; Cyanine (Pentamethine); $C_{33}H_{37}N_2O_4$	4.13.2
361	Cy5.11-OH	2-((E)-2-((2E,4E)-5-(3,3-Dimethyl-1-(4-sulfonatobutyl)-3H-indol-1-ium-2-yl)penta-2,4-dien-1-ylidene)-1-ethyl-3,3-dimethylindolin-5-yl)acetate; Cyanine (Pentamethine); $C_{33}H_{39}N_2O_5S$	4.13.2

Ent.	Short Name	Name; Class; Molecular Formula; CAS Number	Ref.
362	Cy5.12-OH	2-((E)-2-((2E,4E)-5-(5-(Carboxylatomethyl)-3,3-dimethyl-1-(4-sulfonatobutyl)-3H-indol-1-ium-2-yl)penta-2,4-dien-1-ylidene)-3,3-dimethyl-1-(4-sulfonatobutyl)indolin-5-yl)acetate; Cyanine (Pentamethine); $C_{37}H_{43}N_2O_{10}S_2$	4.13.2
363	Cy5.13-OH	2-((E)-2-((2E,4E)-5-(3,3-Dimethyl-1-(4-sulfonatobutyl)-3H-indol-1-ium-2-yl)penta-2,4-dien-1-ylidene)-3,3-dimethyl-1-(4-sulfonatobutyl)indolin-5-yl)acetate; Cyanine (Pentamethine); $C_{35}H_{42}N_2O_8S_2$	4.13.2
364	Cy5.18.OH	6-((E)-2-((2E,4E)-5-(1-(5-C1arboxylatopentyl)-3,3-dimethyl-5-sulfonato-3H-indol-1-ium-2-yl)penta-2,4-dien-1-ylidene)-3,3-dimethyl-5-sulfonatoindolin-1-yl)hexanoate; Cyanine (Pentamethine); $C_{37}H_{43}N_2O_{10}S_2$	4.13.2
365	Cy5.203	6-(2-((1E,3E,5E)-5-(3-(5-Carboxylatopentyl)-1,1-dimethyl-7-sulfonato-1,3-dihydro-2H-benzo[e]indol-2-ylidene)penta-1,3-dien-1-yl)-1,1-dimethyl-7-sulfonato-1H-benzo[e]indol-3-ium-3-yl)hexanoate; Cyanine (Pentamethine); $C_{45}H_{47}N_2O_{10}S_2$	4.13.2
366	Cy5.204	6-(2-((1E,3E,5E)-5-(1-(5-Carboxylatopentyl)-3,3-dimethyl-6-sulfonato-1,3-dihydro-2H-benzo[g]indol-2-ylidene)penta-1,3-dien-1-yl)-3,3-dimethyl-6-sulfonato-3H-benzo[g]indol-1-ium-1-yl)hexanoate; Cyanine (Pentamethine); $C_{45}H_{47}N_2O_{10}S_2$	4.13.2
367	Cy5.205	6-(2-((1E,3E,5E)-5-(3-(5-Carboxylatopentyl)-1,1-dimethyl-6,8-disulfonato-1,3,6,7-tetrahydro-2H-benzo[e]indol-2-ylidene)penta-1,3-dien-1-yl)-1,1-dimethyl-6,8-disulfonato-1H-benzo[e]indol-3-ium-3-yl)hexanoate; Cyanine (Pentamethine); $C_{45}H_{47}N_2O_{16}S_4$	4.13.2
368	Cy5.29.OH	6-(2-((1E,3E)-5-((E)-1-Ethyl-3,3-dimethyl-5-sulfonatoindolin-2-ylidene)penta-1,3-dien-1-yl)-3,3-dimethyl-5-sulfonato-3H-indol-1-ium-1-yl)hexanoate; Cyanine (Pentamethine); $C_{33}H_{38}N_2O_8S_2$	4.13.2
369	Cy5.5	6-(2-((1E,3E,5E)-5-(3-Ethyl-1,1-dimethyl-6,8-disulfonato-1,3,6,7-tetrahydro-2H-benzo[e]indol-2-ylidene)penta-1,3-dien-1-yl)-1,1-dimethyl-6,8-disulfonato-1H-benzo[e]indol-3-ium-3-yl)hexanoate; Cyanine (Pentamethine); $C_{41}H_{42}N_2O_{14}S_4$	4.13.2
370	Cyan 500 carboxylate	nd; nd (pb Coumarin); nd	4.41
371	CypHer5	4-(2-((1E,3E)-5-((E)-3,3-Dimethyl-5-sulfonatoindolin-2-ylidene)penta-1,3-dien-1-yl)-3,3-dimethyl-5-sulfonato-3H-indol-1-ium-1-yl)butanoate; Cyanine (Pentamethine); $C_{29}H_{30}N_2O_8S_2$	4.13.2
372	DAMBO-P^H-T	8-(5-Benzotriazolyl)-2,6-bis(2-carboxyethyl)-4,4-difluoro-1,3,5,7-tetramethyl-4-bora-3a,4a-diaza-s-indacene; Borondipyrromethene (Small); $C_{25}H_{24}BF_2N_5O_4$	4.14.1
373	DAMBO-T	8-(5-Benzotriazolyl)-4,4-difluoro-1,3,5,7-tetramethyl-4-bora-3a,4a-diaza-s-indacene; Borondipyrromethene (Small); $C_{19}H_{18}BF_2N_5$	4.14.1
374	DAOTA 1	8,12-Dipropyl-8,12-dihydro-3a2H-benzo[ij]xantheno[1,9,8-cdef][2,7]naphthyridin-3a2-ylium; Triangulenium; $C_{25}H_{23}N_2O$	4.22
375	DAOTA 2	8,12-Dipropyl-8,12-dihydrobenzo[ij]xantheno[1,9,8-cdef][2,7]naphthyridin-3a2-ylium-3-sulfonate; Triangulenium; $C_{25}H_{22}N_2O_4S$	4.22

(Continued)

Ent.	Short Name	Name; Class; Molecular Formula; CAS Number	Ref.
376	DAOTA 3	8,12-Dipropyl-8,12-dihydrobenzo[ij]xantheno[1,9,8-cdef][2,7]naphthyridin-3a2-ylium-3,5-disulfonate; Triangulenium; $C_{25}H_{21}N_2O_7S_2$	4.22
377	DAOTA 4	8,12-Bis(3-(trimethylammonio)propyl)-8,12-dihydro-3a2H-benzo[ij]xantheno[1,9,8-cdef][2,7]naphthyridin-3a2-ylium; Triangulenium; $C_{31}H_{39}N_4O$	4.22
378	DAOTA 5	3,3′-((Benzo[ij]xantheno[1,9,8-cdef][2,7]naphthyridine-3a2-ylium-8,12-diylbis(propane-3,1-diyl))bis(dimethylammoniumdiyl))bis(propane-1-sulfonate); Triangulenium; $C_{35}H_{45}N_4O_7S_2$	4.22
379	DAOTA 6	8,12-Bis(3-(trimethylammonio)propyl)-8,12-dihydrobenzo[ij]xantheno[1,9,8-cdef][2,7]naphthyridin-3a2-ylium-3,5-disulfonate; Triangulenium; $C_{31}H_{37}N_4O_7S_2$	4.22
380	DEAC-SR101	2-(13-(Diethylamino)-10-oxo-2,3,6,7-tetrahydro-1H,5H,10H-chromeno[3′,4′:5,6]pyrano[2,3-f]pyrido[3,2,1-ij]quinolin-4-ium-9-yl)benzenesulfonate; Hybrid – Fusion of two existing dye; $C_{32}H_{30}N_2O_6S$	4.40.1
381	Des-carboxy-JF646	1-(7-(Azetidin-1-yl)-5,5-dimethyl-10-phenyldibenzo[b,e]silin-3(5H)-ylidene)azetidin-1-ium; $Si(CH_3)_2$-Rosamine; $C_{27}H_{29}N_2Si$	4.18.1
382	Distyryl-BODIPY 11	3-(3-(4-(3,7-Bis((E)-4-(3-(dimethyl(3-sulfonatopropyl)ammonio)prop-1-yn-1-yl)styryl)-5,5-difluoro-1,9-dimethyl-5H-4l4,5l4-dipyrrolo[1,2-c:2′,1′-f][1,3,2]diazaborinin-10-yl)benzamido)-2-sulfonatopropanamido)-2-sulfonatopropanoate; Borondipyrromethene (Medium sized); $C_{56}H_{60}BF_2N_6O_{16}S_4$	4.14.2
383	Distyryl-BODIPY 12	3-(3-(4-(3,7-Bis((E)-4-(3-(dimethyl(3-sulfonatopropyl)ammonio)prop-1-yn-1-yl)styryl)-5,5-bis(3-(2-methoxyethoxy)prop-1-yn-1-yl)-1,9-dimethyl-5H-4l4,5l4-dipyrrolo[1,2-c:2′,1′-f][1,3,2]diazaborinin-10-yl)benzamido)-2-sulfonatopropanamido)-2-sulfonatopropanoate; Borondipyrromethene (Medium sized); $C_{68}H_{78}BN_6O_{20}S_4$	4.14.2
384	Distyryl-BODIPY 13	4-(5,5-Bis(3-(dimethyl(3-sulfonatopropyl)ammonio)prop-1-yn-1-yl)-3,7-bis((E)-4-(3-(dimethyl(3-sulfonatopropyl)ammonio)prop-1-yn-1-yl)styryl)-1,9-dimethyl-5H-4l4,5l4-dipyrrolo[1,2-c:2′,1′-f][1,3,2]diazaborinin-10-yl)benzoate; Borondipyrromethene (Medium sized); $C_{66}H_{80}BN_6O_{14}S_4$	4.14.2
385	Distyryl-BODIPY 17	4-(5,5-Bis(3-(dimethyl(3-sulfonatopropyl)ammonio)prop-1-yn-1-yl)-3,7-bis((E)-4-(3-(dimethyl(3-sulfonatopropyl)ammonio)prop-1-yn-1-yl)styryl)-1,9-dimethyl-5H-4l4,5l4-dipyrrolo[1,2-c:2′,1′-f][1,3,2]diazaborinin-10-yl)benzoate; Borondipyrromethene (Medium sized); $C_{66}H_{80}BN_6O_{14}S_4$	4.14.2
386	Distyryl-BODIPY 18	3,3′-(((3,7-Bis((E)-4-(3-(dimethyl(3-sulfonatopropyl)ammonio)prop-1-yn-1-yl)styryl)-10-(4-(methoxycarbonyl)phenyl)-1,9-dimethyl-5H-4l4,5l4-dipyrrolo[1,2-c:2′,1′-f][1,3,2]diazaborinine-5,5-diyl)bis(prop-2-yne-3,1-diyl))bis(dimethylammoniumdiyl))bis(propane-1-sulfonate; Borondipyrromethene (Medium sized); $C_{67}H_{83}BN_6O_{14}S_4$	4.14.2
387	DMAC-SR101	2-(13-(Dimethylamino)-10-oxo-2,3,6,7-tetrahydro-1H,5H,10H-chromeno[3′,4′:5,6]pyrano[2,3-f]pyrido[3,2,1-ij]quinolin-4-ium-9-yl)benzenesulfonate; Hybrid – Fusion of two existing dye; $C_{30}H_{26}N_2O_6S$	4.40.1

Ent.	Short Name	Name; Class; Molecular Formula; CAS Number	Ref.
388	DMAX-EP	9,10-Dimethyl-3-(6-oxido-3-oxo-3H-xanthen-9-yl)-8a,9,10,10a-tetrahydro-9,10-epidioxyanthracene-2-carboxylate; Fluorescein (Non-halogenated); $C_{30}H_{20}O_7$	4.8.1
389	DMPT	6,7-Diphenyl-2-morpholinylpterin; Pteridine; $C_{22}H_{19}N_5O_2$	4.36
390	DPP 10	2,2′-((2,2′-(1,4-dioxo-3,6-di(thiophen-2-yl)pyrrolo[3,4-c]pyrrole-2,5(1H,4H)-diyl)bis(acetyl))bis(azanediyl))bis(ethane-1,1-disulfonate); Others (Diketopyrrolopyrrole); $C_{22}H_{18}N_4O_{16}S_6$	4.39.1
391	DPP 5	2,2′-(1,4-Dioxo-3,6-diphenylpyrrolo[3,4-c]pyrrole-2,5(1H,4H)-diyl)diacetate; Others (Diketopyrrolopyrrole); $C_{22}H_{14}N_2O_6$	4.39.1
392	DPP 6	2,2′-(1,4-Dioxo-3,6-di(thiophen-2-yl)pyrrolo[3,4-c]pyrrole-2,5(1H,4H)-diyl)diacetate; Others (Diketopyrrolopyrrole); $C_{18}H_{10}N_2O_6S_2$	4.39.1
393	DPP 7	2-(5-(2-((2,2-Disulfonatoethyl)amino)-2-oxoethyl)-1,4-dioxo-3,6-diphenyl-4,5-dihydropyrrolo[3,4-c]pyrrol-2(1H)-yl)acetate; Others (Diketopyrrolopyrrole); $C_{24}H_{18}N_3O_{11}S_2$	4.39.1
394	DPP 8	2-(5-(2-((2,2-Disulfonatoethyl)amino)-2-oxoethyl)-1,4-dioxo-3,6-di(thiophen-2-yl)-4,5-dihydropyrrolo[3,4-c]pyrrol-2(1H)-yl)acetate; Others (Diketopyrrolopyrrole); $C_{20}H_{14}N_3O_{11}S_4$	4.39.1
395	DPP 9	2,2′-((2,2′-(1,4-Dioxo-3,6-diphenylpyrrolo[3,4-c]pyrrole-2,5(1H,4H)-diyl)bis(acetyl))bis(azanediyl))bis(ethane-1,1-disulfonate); Others (Diketopyrrolopyrrole); $C_{26}H_{22}N_4O_{16}S_4$	4.39.1
396	DY-350 carboxylate	4-(3-(7-Amino-4-oxo-6-sulfonato-1,2,3,4-tetrahydrocyclopenta[c]chromen-1-yl)-3-sulfonatopropanamido)butanoate; Coumarin (More elaborated 7-Amino); $C_{19}H_{20}N_2O_{11}S_2$	4.26.3
397	DY-405 carboxylate	6-(2-((3,6,8-Trisulfonatopyren-1-yl)oxy)acetamido)hexanoate; Pyrene; $C_{24}H_{20}NO_{13}S_3$	4.28
398	DY-430 carboxylate	6-(6,8,8-Trimethyl-2-oxo-9-(3-sulfonatopropyl)-8,9-dihydro-2H-pyrano[3,2-g]quinoline-3-carboxamido)hexanoate; Coumarin (More elaborated 7-Amino); $C_{25}H_{30}N_2O_8S$	4.26.3
399	DY-431 carboxylate	6-(8,8-Dimethyl-2-oxo-6-(sulfonatomethyl)-9-(3-sulfonatopropyl)-8,9-dihydro-2H-pyrano[3,2-g]quinoline-3-carboxamido)hexanoate; Coumarin (More elaborated 7-Amino); $C_{25}H_{29}N_2O_{11}S_2$	4.26.3
400	DY-485XL carboxylate	6-(Ethyl(2-oxo-3-(1-(3-sulfonatopropyl)pyridin-1-ium-4-yl)-2H-chromen-7-yl)amino)hexanoate; Coumarin (More elaborated 7-Amino); $C_{27}H_{32}N_2O_7S$	4.26.3
401	DY-490 carboxylate	4-(6-Amino-3-iminio-5-sulfonato-4-(N-(3-sulfonatopropyl)sulfamoyl)-3H-xanthen-9-yl)isophthalate; Rhodamine (Highly charged); $C_{24}H_{18}N_3O_{13}S_3$	4.2.5
402	DY-495 carboxylate	5-((5-Carboxylatopentyl)carbamoyl)-2-(6-oxido-3-oxo-3H-xanthen-9-yl)benzoate; Fluorescein (Non-halogenated); $C_{27}H_{23}NO_8$	4.8.1
403	DY-505 carboxylate	2-(6-Amino-3-iminio-3H-xanthen-9-yl)-5-((5-carboxylatopentyl)carbamoyl)benzoate; Rhodamine (500 < λa, max < 550 nm); $C_{27}H_{26}N_3O_6$	4.2.2
404	DY-510XL carboxylate	6-(4-(6,8,8-Trimethyl-2-oxo-9-(3-sulfonatopropyl)-8,9-dihydro-2H-pyrano[3,2-g]quinolin-3-yl)pyridin-1-ium-1-yl)hexanoate; Coumarin (More elaborated 7-Amino); $C_{31}H_{35}N_2O_7S$	4.26.3

(Continued)

Ent.	Short Name	Name; Class; Molecular Formula; CAS Number	Ref.
405	DY-530 carboxylate	(Z)-4-(Nethyl-2-(6-((3-sulfonatopropyl)amino)-3-((3-sulfonatopropyl)iminio)-3H-xanthen-9-yl)benzamido)butanoate; Rhodamine (500 < λa, max < 550 nm); $C_{31}H_{34}N_3O_{10}S_2$	4.2.2
406	DY-555 carboxylate	5-((5-carboxylatopentyl)carbamoyl)-2-(6-(diethylamino)-3-(diethyliminio)-3H-xanthen-9-yl) benzoate; Rhodamine (550 < λa, max < 600 nm); $C_{35}H_{42}N_3O_6$	4.2.3
407	DY-635 carboxylate	6-((Z)-2-((E)-3-(11-(Tert-butyl)-2,3,6,7-tetrahydro-1H,5H-pyrano[2,3-f]pyrido[3,2,1-ij] quinolin-12-ium-9-yl)allylidene)-3,3-dimethyl-5-sulfonatoindolin-1-yl)hexanoate; Merocyanine; $C_{38}H_{46}N_2O_6S$	4.24
408	DY-636 carboxylate	4-((Z)-2-((E)-3-(11-(Tert-butyl)-2,3,6,7-tetrahydro-1H,5H-pyrano[2,3-f]pyrido[3,2,1-ij] quinolin-12-ium-9-yl)allylidene)-3-methyl-5-sulfonato-1-(3-sulfonatopropyl)indolin-3-yl) butanoate; Merocyanine; $C_{38}H_{45}N_2O_9S_2$	4.24
409	DY-651 carboxylate	4-((E)-2-((Z)-3-(2-(Tert-butyl)-9-ethyl-6,8,8-trimethyl-8,9-dihydropyrano[3,2-g]quinolin-1-ium-4-yl)allylidene)-3-methyl-5-sulfonato-1-(3-sulfonatopropyl)indolin-3-yl)butanoate; Merocyanine; $C_{40}H_{49}N_2O_9S_2$	4.24
410	DY-654 carboxylate	4-((Z)-2-((E)-3-(2-(Tert-butyl)-8,8-dimethyl-6-(sulfonatomethyl)-9-(3-sulfonatopropyl)-8,9-dihydropyrano[3,2-g]quinolin-1-ium-4-yl)allylidene)-3-methyl-5-sulfonato-1-(3-sulfonatopropyl)indolin-3-yl)butanoate; Merocyanine; $C_{41}H_{43}N_2O_{15}S_5$	4.24
411	DY-679P1	6-((E)-2-((2E,4E)-5-(3-(2-Methoxyethyl)-1,1-dimethyl-6,8-disulfonato-1H-benzo[e] indol-3-ium-2-yl)penta-2,4-dien-1-ylidene)-1,1-dimethyl-6,8-disulfonato-1,2-dihydro-3H-benzo[e]indol-3-yl)hexanoate; Cyanine (Pentamethine); $C_{42}H_{42}N_2O_{15}S_4$	4.13.2
412	DY-681 carboxylate	4-((E)-2-((E)-3-(4-(Tert-butyl)-7-(diethylamino)chromenylium-2-yl)allylidene)-3-methyl-5-sulfonato-1-(3-sulfonatopropyl)indolin-3-yl)butanoate; Merocyanine; $C_{36}H_{45}N_2O_9S_2$	4.24
413	DY-682 carboxylate	4-((E)-2-((E)-3-(4-(Tert-butyl)-7-(ethyl(3-sulfonatopropyl)amino)chromenylium-2-yl) allylidene)-3-methyl-5-sulfonato-1-(3-sulfonatopropyl)indolin-3-yl)butanoate; Merocyanine; $C_{37}H_{46}N_2O_{12}S_3$	4.24
414	DY-731 carboxylate	5-((Z)-2-((2E,4E)-5-(2-(Tert-butyl)-7-(diethylamino)chromenylium-4-yl)penta-2,4-dien-1-ylidene)-3-methyl-5-sulfonato-1-(3-sulfonatopropyl)indolin-3-yl)pentanoate; Merocyanine; $C_{38}H_{47}N_2O_9S_2$	4.24
415	DY-747P1	(E)-1-(5-Carboxylatopentyl)-2-((2E,4E,6E)-7-(1-(2-methoxyethyl)-3,3-dimethyl-5-sulfonato-3H-indol-1-ium-2-yl)hepta-2,4,6-trien-1-ylidene)-3,3-dimethylindoline-5-carboxylate; Cyanine (Heptamethine); $C_{37}H_{42}N_2O_8S$	4.13.3
416	DY-780 carboxylate	6-((E)-2-((2E,4E)-5-(4-(Tert-butyl)-7-(diethylamino)chromenylium-2-yl)penta-2,4-dien-1-ylidene)-3,3-dimethyl-5-sulfonatoindolin-1-yl)hexanoate; Merocyanine; $C_{38}H_{48}N_2O_6S$	4.24
417	ECXi	tl; Others (Pentacene); $C_{53}H_{44}N_2O_{10}S_3$	4.39.2
418	Edans	5-((2-Ammonioethyl)amino)naphthalene-1-sulfonate; Naphthalene; $C_{12}H_{14}N_2O_3S$; 146-17-8	4.34

Ent.	Short Name	Name; Class; Molecular Formula; CAS Number	Ref.
419	Epicocconone	(6S,9aS)-3-((1Z,4E,6E,8E)-1-Hydroxy-3-oxodeca-1,4,6,8-tetraen-1-yl)-6-(hydroxymethyl)-9a-methyl-5,6-dihydro-2H-furo[3,2-g]isochromene-2,9(9aH)-dione; Others (Epicocconone); $C_{23}H_{22}O_7$	4.39.2
420	Epicocconone P-F 17a	(9aS)-3-(3-(4-Methoxyphenyl)-2-methyl-3-oxopropanoyl)-6,6,9a-trimethyl-5,6-dihydro-2H-furo[3,2-g]isochromene-2,9(9aH)-dione; Others (Epicocconone); $C_{25}H_{24}O_7$	4.39.2
421	Epicocconone P-F 17d	(9aS)-6,6,9a-Trimethyl-3-(2-methyl-3-(naphthalen-2-yl)-3-oxopropanoyl)-5,6-dihydro-2H-furo[3,2-g]isochromene-2,9(9aH)-dione; Others (Epicocconone); $C_{28}H_{24}O_6$	4.39.2
422	Epicocconone P-F 4b	(S,Z)-3-(1-Hydroxy-3-(naphthalen-2-yl)-3-oxoprop-1-en-1-yl)-6,6,9a-trimethyl-5,6-dihydro-2H-furo[3,2-g]isochromene-2,9(9aH)-dione; Others (Epicocconone); $C_{27}H_{22}O_6$	4.39.2
423	Epicocconone P-F 4c	(S,Z)-3-(3-(6-Chloropyridin-3-yl)-1-hydroxy-3-oxoprop-1-en-1-yl)-6,6,9a-trimethyl-5,6-dihydro-2H-furo[3,2-g]isochromene-2,9(9aH)-dione; Others (Epicocconone); $C_{22}H_{18}ClNO_6$	4.39.2
424	Epicocconone P-F 4d	(S,Z)-3-(1-Hydroxy-3-oxo-3-phenylprop-1-en-1-yl)-6,6,9a-trimethyl-5,6-dihydro-2H-furo[3,2-g]isochromene-2,9(9aH)-dione; Others (Epicocconone); $C_{23}H_{20}O_6$	4.39.2
425	Et-rosol	6-(Diethylamino)-9-phenyl-3H-xanthen-3-one; Rosol; $C_{23}H_{21}NO_2$	4.19
426	EVOblue 30	(Z)-3-((3-(Ethyl(3-sulfonatopropyl)iminio)-3H-phenoxazin-7-yl)(methyl)amino)propanoate; Phenoxazine; $C_{21}H_{24}N_3O_6S$	4.25
427	Flavin mononucleotide (FMN)	(2S,3S,4R)-5-(7,8-Dimethyl-2,4-dioxo-3,4-dihydrobenzo[g]pteridin-10(2H)-yl)-2,3,4-trihydroxypentyl hydrogen phosphate; riboflavin-5′-phosphate; Pteridine; $C_{17}H_{20}N_4O_9P$	4.36
428	Fluorene I-6	N,N,9,9-Tetramethyl-7-vinyl-9H-fluoren-2-amine; Others (Fluorene); $C_{19}H_{21}N$	4.39.2
429	Fluorene PP11	(S)-9,9-Dimethyl-7-(2-methylaziridin-1-yl)-9H-fluorene-2-carbaldehyde; Others (Fluorene); $C_{19}H_{19}NO$	4.39.2
430	Fluorene PP12	7-(Azetidin-1-yl)-9,9-dimethyl-9H-fluorene-2-carbaldehyde; Others (Fluorene); $C_{19}H_{19}NO$	4.39.2
431	Fluorene PP6	N-Butyl-4-(7-(dimethylamino)-9,9-dimethyl-9H-fluoren-2-yl)-4-oxobutanamide; Others (Fluorene); $C_{25}H_{32}N_2O_2$	4.39.2
432	Fluorescein	2-(6-Oxido-3-oxo-3H-xanthen-9-yl)benzoate; Fluorescein (Non-halogenated); $C_{20}H_{12}O_5$; 2321-07-5	4.8.1
433	Fluorescein	2-(6-Oxido-3-oxo-3H-xanthen-9-yl)benzoate; Fluorescein (Non-halogenated); $C_{20}H_{12}O_5$; 2321-07-5	4.8.1
434	Fluorescein (G-C 1)	2-(6-Oxido-3-oxo-3H-xanthen-9-yl)benzoate; Fluorescein (Non-halogenated); $C_{20}H_{12}O_5$	4.8.1
435	Fluorescein (S-H 1)	2-(6-Oxido-3-oxo-3H-xanthen-9-yl)benzoate; Fluorescein (Non-halogenated); $C_{20}H_{12}O_5$	4.8.1
436	Fluorinated fluorescein S-H 14a,b	5(6)-Carboxy-2′,7′-difluorofluorescein; Fluorescein (Halogenated); $C_{21}H_{10}Cl_2O_7$; 111843-78-8	4.8.2
437	Fluorinated fluorescein S-H 15a,b	5(6)-Carboxy-4′,5′-difluorofluorescein; Fluorescein (Halogenated); $C_{21}H_{10}F_2O_7$	4.8.2
438	Fluorinated fluorescein S-H 16a,b	5(6)-Carboxy-2′,4′,5′,7′-tetrafluorofluorescein; Fluorescein (Halogenated); $C_{21}H_8F_4O_7$	4.8.2

(Continued)

Ent.	Short Name	Name; Class; Molecular Formula; CAS Number	Ref.
439	Fluorinated fluorescein S-H 17	2-(2,7-Difluoro-6-oxido-3-oxo-3H-xanthen-9-yl)benzoate; Fluorescein (Halogenated); $C_{20}H_{10}Cl_2O_5$; 76-54-0	4.8.2
440	Fluorinated fluorescein S-H 18	2,3,4,5-Tetrafluoro-6-(6-oxido-3-oxo-3H-xanthen-9-yl)benzoate; 4,5,6,7-tetrafluorofluorescein; Fluorescein (Halogenated); $C_{20}H_6F_4O_5$; 185318-74-5	4.8.2
441	Fluorinated fluorescein S-H 19	2-(2,7-Difluoro-6-oxido-3-oxo-3H-xanthen-9-yl)-3,4,5,6-tetrafluorobenzoate; 2′,4,5,6,7,7′-hexafluorofluorescein; Fluorescein (Halogenated); $C_{20}H_4F_6O_5$	4.8.2
442	Fluorinated fluorescein S-H 20	2′,4,4′,5,5′,6,7,7′-Octafluorofluorescein; Fluorescein (Halogenated); $C_{20}H_4F_8O_5$	4.8.2
443	Fluorinated rhodam. 1f-5′-COOH	(E)-4-(6-((2,2,2-Trifluoroethyl)amino)-3-((2,2,2-trifluoroethyl)iminio)-3H-xanthen-9-yl) isophthalate; Rhodamine (500 < λa, max < 550 nm); $C_{25}H_{15}F_6N_2O_5$	4.2.2
444	Fluorinated rhodam. 1f-6′-COOH	(E)-2-(6-((2,2,2-Trifluoroethyl)amino)-3-((2,2,2-trifluoroethyl)iminio)-3H-xanthen-9-yl) terephthalate; Rhodamine (500 < λa, max < 550 nm); $C_{25}H_{15}F_6N_2O_5$	4.2.2
445	Fluorinated rhodamine 1e	(E)-2-(6-((2,2,2-Trifluoroethyl)amino)-3-((2,2,2-trifluoroethyl)iminio)-3H-xanthen-9-yl) benzoate; Rhodamine (λa, max < 500 nm); $C_{24}H_{16}F_6N_2O_3$	4.2.1
446	Fluorinated rhodamine 1j	(Z)-4-(2,7-Difluoro-6-((2,2,2-trifluoroethyl)amino)-3-((2,2,2-trifluoroethyl)iminio)-3H-xanthen-9-yl)isophthalate; Rhodamine (λa, max < 500 nm); $C_{25}H_{13}F_8N_2O_5$	4.2.1
447	Fluorinated rhodamine 1k	(E)-2-(2,7-Difluoro-4,5-disulfonato-6-((2,2,2-trifluoroethyl)amino)-3-((2,2,2-trifluoroethyl) iminio)-3H-xanthen-9-yl)benzoate; Rhodamine (500 < λa, max < 550 nm); $C_{24}H_{12}F_8N_2O_9S_2$	4.2.2
448	Fluorinated rhodamine 1r	2-(6-(Dimethylamino)-3-(dimethyliminio)-2,7-difluoro-3H-xanthen-9-yl)terephthalate; Rhodamine (500 < λa, max < 550 nm); $C_{26}H_{23}F_2N_2O_5$	4.2.2
449	Fluorinated rhodamine 3-H	(E)-3-(N-Methyl-2-(6-((2,2,2-trifluoroethyl)amino)-3-((2,2,2-trifluoroethyl)iminio)-3H-xanthen-9-yl)benzamido)propanoate; Rhodamine (500 < λa, max < 550 nm); $C_{27}H_{24}F_6N_3O_4$	4.2.2
450	Fluorinated rhodamine 5-SO_3H	(E)-3-(2-(4,5-Disulfonato-6-((2,2,2-trifluoroethyl)amino)-3-((2,2,2-trifluoroethyl)iminio)-3H-xanthen-9-yl)-N-methylbenzamido)propanoate; Rhodamine (500 < λa, max < 550 nm); $C_{26}H_{20}F_6N_3O_9S_2$	4.2.2
451	GeR	2-(7-(dimethylamino)-3-(dimethyliminio)-5,5-dimethyl-3,5-dihydrodibenzo[b,e]germin-10-yl)-4-((methylamino)oxy)benzoate; $Ge(CH_3)_2$-Rhodamine; $C_{27}H_{29}GeN_2O_3$	4.3.3
452	G-L 13a	2,6-Bis(2,3-dideoxy-β-D-erythro-hex-2-enopyranosyl)-8-phenyl-1,3,5,7-tetramethyl-4,4-difluoro-4-bora-3a,4a-diaza-s-indacene; Borondipyrromethene (Medium sized); $C_{31}H_{37}BF_2N_2O_6$	4.14.2
453	G-L 13b	2,6-Bis(2,3-dideoxy-β-D-erythro-hex-2-enopyranosyl)-8-(2-azidomethyl)-phenyl-1,3,5,7-tetramethyl-4,4-difluoro-4-bora-3a,4adiaza-s-indacene; Borondipyrromethene (Medium sized); $C_{32}H_{38}BF_2N_5O_6$	4.14.2
454	G-L 14	2,6-Bis(2,3-dideoxy-β-D-erythro-hex-2-enopyranosyl)-8-[(2,3-dideoxy-α-D-erythro-hex-2-enopyranosyl)-2-methyl-phenyl]-1,3,5,7-tetramethyl-4,4-difluoro-4-bora-3a,4a-diaza-s-indacene; Borondipyrromethene (Medium sized); $C_{38}H_{47}BF_2N_2O_{10}$	4.14.2
455	Harmaline	8-Methoxy-1-methyl-4,9-dihydro-3H-pyrido[3,4-b]indole; Indole; $C_{13}H_{14}N_2O$; 304-21-2	4.33

Ent.	Short Name	Name; Class; Molecular Formula; CAS Number	Ref.
456	Harmane	1-Methyl-9H-pyrido[3,4-b]indole; Indole; $C_{12}H_{10}N_2$; 486-84-0	4.33
457	Harmine	7-Methoxy-1-methyl-9H-pyrido[3,4-b]indole; Indole; $C_{14}H_{12}N_2O$	4.33
458	Harmine	7-Methoxy-1-methyl-9H-pyrido[3,4-b]indole; Indole; $C_{14}H_{12}N_2O$	4.33
459	Harmine	7-Methoxy-1-methyl-9H-pyrido[3,4-b]indole; Indole; $C_{13}H_{12}N_2O$; 442-51-3	4.33
460	HBQ	Benzo[h]quinolin-10-ol; Quinoline; $C_{13}H_9NO$	4.29
461	HC6H-SR101	tl; Hybrid – Fusion of two existing dye; $C_{34}H_{29}N_2O_9S_2$	4.40.1
462	HCDN	3-(((E)-2-((E)-2-(1-Ethyl-3,3-dimethyl-5-sulfonato-3H-indol-1-ium-2-yl)vinyl)-6-(2-((E)-1-ethyl-3,3-dimethyl-5-sulfonatoindolin-2-ylidene)ethylidene)cyclohex-1-en-1-yl)amino) propanoate; Cyanine (Heptamethine); $C_{37}H_{43}N_3O_8S_2$	4.13.3
463	HCDO	4-((4-(((E)-2-((E)-2-(1-Ethyl-3,3-dimethyl-5-sulfonato-3H-indol-1-ium-2-yl)vinyl)-6-(2-((E)-1-ethyl-3,3-dimethyl-5-sulfonatoindolin-2-ylidene)ethylidene)cyclohex-1-en-1-yl)oxy) phenyl)amino)-4-oxobutanoate; Cyanine (Heptamethine); $C_{44}H_{47}N_3O_{10}S_2$	4.13.3
464	HCDS	4-((4-(((E)-2-((E)-2-(1-Ethyl-3,3-dimethyl-5-sulfonato-3H-indol-1-ium-2-yl)vinyl)-6-(2-((E)-1-ethyl-3,3-dimethyl-5-sulfonatoindolin-2-ylidene)ethylidene)cyclohex-1-en-1-yl)thio) phenyl)amino)-4-oxobutanoate; Cyanine (Heptamethine); $C_{44}H_{47}N_3O_9S_3$	4.13.3
465	HEX	2,5-Dichloro-3-(2,4,5,7-tetrachloro-6-oxido-3-oxo-3H-xanthen-9-yl)terephthalate; Fluorescein (Halogenated); $C_{21}H_3Cl_6O_7$; 155911-15-2	4.8.2
466	HMDER	6-(Diethylamino)-9-(2-(hydroxymethyl)phenyl)-3H-xanthen-3-one; Rhodol; $C_{24}H_{23}NO_3$	4.15.2
467	HMDSeR	6-(Diethylamino)-9-(2-(hydroxymethyl)phenyl)-3H-selenoxanthen-3-one; Se-Rhodol; $C_{24}H_{23}NO_2Se$	4.16
468	ICG	4-(2-((1E,3E,5E,7E)-7-(1,1-Dimethyl-3-(4-sulfonatobutyl)-1,3-dihydro-2H-benzo[e] indol-2-ylidene)hepta-1,3,5-trien-1-yl)-1,1-dimethyl-1H-benzo[e]indol-3-ium-3-yl)butane-1-sulfonate; Cyanine (Heptamethine); $C_{43}H_{47}N_2O_6S_2$	4.13.3
469	ICG	4-[2-[(1E,3E,5E,7Z)-7-[1,1-dimethyl-3-(4-sulfonatobutyl)benzo[e]indol-2-ylidene]hepta-1,3,5-trienyl]-1,1-dimethylbenzo[e]indol-3-ium-3-yl]butane-1-sulfonate; Cyanine (Heptamethine); $C_{43}H_{47}N_2O_6S_2$	4.13.3
470	IR-783	4-((E)-2-((E)-2-(2-Chloro-3-((E)-2-(3,3-dimethyl-1-(4-sulfonatobutyl)-3H-indol-1-ium-2-yl) vinyl)cyclohex-2-en-1-ylidene)ethylidene)-3,3-dimethylindolin-1-yl)butane-1-sulfonate; Cyanine (Heptamethine); $C_{38}H_{46}ClN_2O_6S_2$; 115970-66-6	4.13.3
471	IRDye78-CA	3-(4-(((E)-2-((E)-2-(3,3-Dimethyl-5-sulfonato-1-(4-sulfonatobutyl)-3H-indol-1-ium-2-yl) vinyl)-6-(2-((E)-3,3-dimethyl-5-sulfonato-1-(4-sulfonatobutyl)indolin-2-ylidene)ethylidene) cyclohex-1-en-1-yl)oxy)phenyl)propanoate; Cyanine (Heptamethine); $C_{47}H_{52}N_2O_{15}S_4$	4.13.3
472	Isoquinoline 10b	3,4,6,7,8-Pentamethoxy-1-oxo-2-(1H-pyrrol-1-yl)-1,2-dihydroisoquinolin-5-olate; Isoquinoline; $C_{18}H_{19}N_2O_7$	4.37

(Continued)

Ent.	Short Name	Name; Class; Molecular Formula; CAS Number	Ref.
473	Isoquinoline 3d	3,4,6,7,8-Pentakis(methoxycarbonyl)-1-oxo-1,2-dihydroisoquinolin-5-olate; Isoquinoline; $C_{19}H_{16}NO_{12}$	4.37
474	Isoquinoline 3e	3,4,6,7,8-Pentakis(methoxycarbonyl)-2-methyl-1-oxo-1,2-dihydroisoquinolin-5-olate; Isoquinoline; $C_{20}H_{18}NO_{12}$	4.37
475	Isoquinoline 3f	2-Isopropyl-3,4,6,7,8-pentakis(methoxycarbonyl)-1-oxo-1,2-dihydroisoquinolin-5-olate; Isoquinoline; $C_{22}H_{22}NO_{12}$	4.37
476	Isoquinoline 3g	2-Cyclobutyl-3,4,6,7,8-pentakis(methoxycarbonyl)-1-oxo-1,2-dihydroisoquinolin-5-olate; Isoquinoline; $C_{23}H_{22}NO_{12}$	4.37
477	Isoquinoline 3h	2-Cyclopentyl-3,4,6,7,8-pentakis(methoxycarbonyl)-1-oxo-1,2-dihydroisoquinolin-5-olate; Isoquinoline; $C_{24}H_{24}NO_{12}$	4.37
478	Isoquinoline 3i	2-Cyclohexyl-3,4,6,7,8-pentakis(methoxycarbonyl)-1-oxo-1,2-dihydroisoquinolin-5-olate; Isoquinoline; $C_{25}H_{26}NO_{12}$	4.37
479	Isoquinoline 3j	3,4,6,7,8-Pentakis(methoxycarbonyl)-1-oxo-2-phenyl-1,2-dihydroisoquinolin-5-olate; Isoquinoline; $C_{25}H_{20}NO_{12}$	4.37
480	Isoquinoline 3k	2-(4-Hydroxyphenyl)-3,4,6,7,8-pentakis(methoxycarbonyl)-1-oxo-1,2-dihydroisoquinolin-5-olate; Isoquinoline; $C_{25}H_{20}NO_{13}$	4.37
481	Isoquinoline 3l	3,4,6,7,8-pentakis(methoxycarbonyl)-1-oxo-2-(p-tolyl)-1,2-dihydroisoquinolin-5-olate; Isoquinoline; $C_{26}H_{22}NO_{12}$	4.37
482	Isoquinoline 3m	2-(4-Bromophenyl)-3,4,6,7,8-pentakis(methoxycarbonyl)-1-oxo-1,2-dihydroisoquinolin-5-olate; Isoquinoline; $C_{25}H_{19}BrNO_{12}$	4.37
483	Isoquinoline 3n	3,4,6,7,8-Pentakis(methoxycarbonyl)-2-(4-methoxyphenyl)-1-oxo-1,2-dihydroisoquinolin-5-olate; Isoquinoline; $C_{26}H_{22}NO_{13}$	4.37
484	Isoquinoline 3o	2-(2,4-Dimethoxyphenyl)-3,4,6,7,8-pentakis(methoxycarbonyl)-1-oxo-1,2-dihydroisoquinolin-5-olate; Isoquinoline; $C_{27}H_{24}NO_{14}$	4.37
485	Isoquinoline 4b	7,8,9,11-tetrakis(methoxycarbonyl)-1,6-dioxo-1,3,4,6-tetrahydro-[1,4]oxazino[4,3-b] isoquinolin-10-olate; Isoquinoline; $C_{20}H_{16}NO_{12}$	4.37
486	Isoquinoline 4c	7,9,10,11-tetrakis(methoxycarbonyl)-6,12-dioxo-6,12-dihydro-5H-isoquinolino[2,3-a] quinoxalin-8-olate; Isoquinoline; $C_{24}H_{17}N_2O_{11}$	4.37
487	Isoquinoline 7a	(3,4,6,7,8-Pentakis(methoxycarbonyl)-5-oxido-1-oxoisoquinolin-2(1H)-yl)glycinate; Isoquinoline; $C_{21}H_{18}N_2O_{14}$	4.37
488	Isoquinoline 7b	2-(tert-Butylamino)-3,4,6,7,8-pentakis(methoxycarbonyl)-1-oxo-1,2-dihydroisoquinolin-5-olate; Isoquinoline; $C_{23}H_{25}N_2O_{12}$	4.37
489	Isoquinoline 7c	2-(2-Hydroxyacetamido)-3,4,6,7,8-pentakis(methoxycarbonyl)-1-oxo-1,2-dihydroisoquinolin-5-olate; Isoquinoline; $C_{21}H_{19}N_2O_{14}$	4.37

Ent.	Short Name	Name; Class; Molecular Formula; CAS Number	Ref.
490	Isoquinoline 7d	2-Benzamido-3,4,6,7,8-pentakis(methoxycarbonyl)-1-oxo-1,2-dihydroisoquinolin-5-olate; Isoquinoline; $C_{26}H_{21}N_2O_{13}$	4.37
491	Isoquinoline 7f	2-(3,4-Dimethoxybenzamido)-3,4,6,7,8-pentakis(methoxycarbonyl)-1-oxo-1,2-dihydroisoquinolin-5-olate; Isoquinoline; $C_{28}H_{25}N_2O_{15}$	4.37
492	Isoquinoline 8	2-Amino-3,4,6,7,8-pentakis(methoxycarbonyl)-1-oxo-1,2-dihydroisoquinolin-5-olate; Isoquinoline; $C_{19}H_{17}N_2O_{12}$	4.37
493	JF479	2-(3-(3,3-Difluoroazetidin-1-ium-1-ylidene)-6-(3,3-difluoroazetidin-1-yl)-10-methyl-3,10-dihydroacridin-9-yl)benzoate; Acridine; $C_{27}H_{21}F_4N_3O_2$	4.23.1
494	JF502	2-(3-(Azetidin-1-ium-1-ylidene)-6-(azetidin-1-yl)-10-methyl-3,10-dihydroacridin-9-yl) benzoate; Acridine; $C_{27}H_{25}N_3O_2$	4.23.1
495	JF525	2-(3-(3,3-Difluoroazetidin-1-ium-1-ylidene)-6-(3,3-difluoroazetidin-1-yl)-3H-xanthen-9-yl) benzoate; Rhodamine (500 < λa, max < 550 nm); $C_{26}H_{18}F_4N_2O_3$	4.2.2
496	JF526	2-(3-(3,3-Difluoroazetidin-1-ium-1-ylidene)-6-(3,3-difluoroazetidin-1-yl)-2,7-difluoro-3H-xanthen-9-yl)benzoate; Rhodamine (500 < λa, max < 550 nm); $C_{26}H_{16}F_6N_2O_3$	4.2.2
497	JF549	2-(3-(Azetidin-1-ium-1-ylidene)-6-(azetidin-1-yl)-3H-xanthen-9-yl)benzoate; Rhodamine (500 < λa, max < 550 nm); $C_{26}H_{23}N_2O_3$	4.2.2
498	JF552	2-(3-(Azetidin-1-ium-1-ylidene)-6-(azetidin-1-yl)-2,7-difluoro-3H-xanthen-9-yl)benzoate; Rhodamine (550 < λa, max < 600 nm); $C_{26}H_{21}F_2N_2O_3$	4.2.3
499	JF559	2,3,4,5-Tetrafluoro-6-(3-(3-fluoroazetidin-1-ium-1-ylidene)-6-(3-fluoroazetidin-1-yl)-3H-xanthen-9-yl)benzoate; Rhodamine (550 < λa, max < 600 nm); $C_{26}H_{16}F_6N_2O_3$	4.2.3
500	JF570	2-(3-(Azetidin-1-ium-1-ylidene)-6-(azetidin-1-yl)-3H-thioxanthen-9-yl)benzoate; S-Rhodamine; $C_{26}H_{22}N_2O_2S$	4.3.3
501	JF571	2-(3-(azetidin-1-ium-1-ylidene)-6-(azetidin-1-yl)-3H-xanthen-9-yl)-3,4,5,6-tetrafluorobenzoate; Rhodamine (550 < λa, max < 600 nm); $C_{26}H_{19}F_4N_2O_3$	4.2.3
502	JF574	2-(3-(Azetidin-1-ium-1-ylidene)-6-(azetidin-1-yl)-2,7-difluoro-3H-xanthen-9-yl)-3,4,5,6-tetrafluorobenzoate; Rhodamine (550 < λa, max < 600 nm); $C_{26}H_{17}F_6N_2O_3$	4.2.3
503	JF585	2-(3-(3,3-Difluoroazetidin-1-ium-1-ylidene)-6-(3,3-difluoroazetidin-1-yl)-10,10-dimethyl-3,10-dihydroanthracen-9-yl)benzoate; $C(CH_3)_2$-Rhodamine; $C_{29}H_{24}F_4N_2O_2$	4.3.1
504	JF593	2-(3-(Azetidin-1-ium-1-ylidene)-6-(azetidin-1-yl)-3H-thioxanthen-9-yl)-3,4,5,6-tetrafluorobenzoate; S-Rhodamine; $C_{26}H_{18}F_4N_2O_2S$	4.3.3
505	JF608	2-(3-(Azetidin-1-ium-1-ylidene)-6-(azetidin-1-yl)-10,10-dimethyl-3,10-dihydroanthracen-9-yl)benzoate; $C(CH_3)_2$-Rhodamine; $C_{29}H_{29}N_2O_2$	4.3.1
506	JF646	2-(3-(Azetidin-1-ium-1-ylidene)-7-(azetidin-1-yl)-5,5-dimethyl-3,5-dihydrodibenzo[b,e]silin-10-yl)benzoate; $Si(CH_3)_2$-Rhodamine; $C_{28}H_{28}N_2O_2Si$	4.3.2

(Continued)

Ent.	Short Name	Name; Class; Molecular Formula; CAS Number	Ref.
507	JF668	2-(3-(Azetidin-1-ium-1-ylidene)-7-(azetidin-1-yl)-5,5-dioxido-3H-acridophosphin-10-yl) benzoate; POOH-Rhodamine; $C_{26}H_{22}N_2O_4P$	4.3.3
508	JF669	2-(3-(Azetidin-1-ium-1-ylidene)-7-(azetidin-1-yl)-5,5-dimethyl-3,5-dihydrodibenzo[b,e] silin-10-yl)-3,4,5,6-tetrafluorobenzoate; $Si(CH_3)_2$-Rhodamine; $C_{28}H_{25}F_4N_2O_2Si$	4.3.2
509	JF669-6-SCH2CO2Et	2-(3-(Azetidin-1-ium-1-ylidene)-7-(azetidin-1-yl)-5,5-dimethyl-3,5-dihydrodibenzo[b,e] silin-10-yl)-3,5,6-trifluoro-4-((2-oxopropyl)thio)benzoate; $Si(CH_3)_2$-Rhodamine; $C_{32}H_{32}F_3N_2O_4SSi$	4.3.2
510	JF669-6-SCH2CO2H	2-(3-(Azetidin-1-ium-1-ylidene)-7-(azetidin-1-yl)-5,5-dimethyl-3,5-dihydrodibenzo[b,e] silin-10-yl)-4-((carboxylatomethyl)thio)-3,5,6-trifluorobenzoate; $Si(CH_3)_2$-Rhodamine; $C_{30}H_{28}F_3N_2O_4SSi$	4.3.2
511	JF690	2-(3-(Azetidin-1-ium-1-ylidene)-7-(azetidin-1-yl)-5,5-dioxido-3H-acridophosphin-10-yl)-3,4,5,6-tetrafluorobenzoate; POOH-Rhodamine; $C_{26}H_{18}F_4N_2O_4P$	4.3.3
512	JF711	2,3,4,5-Tetrafluoro-6-(3-(3-fluoroazetidin-1-ium-1-ylidene)-7-(3-fluoroazetidin-1-yl)-5-oxido-5-phenyl-3H-acridophosphin-10-yl)benzoate; POPh-Rhodamine; $C_{32}H_{21}F_6N_2O_3P$	4.3.3
513	JF722	2-(3-(Azetidin-1-ium-1-ylidene)-7-(azetidin-1-yl)-5-oxido-5-phenyl-3H-acridophosphin-10-yl)-3,4,5,6-tetrafluorobenzoate; POPh-Rhodamine; $C_{32}H_{23}F_4N_2O_3P$	4.3.3
514	JF724	2-(3-(Azetidin-1-ium-1-ylidene)-6-(azetidin-1-yl)-10,10-dioxido-3H-thioxanthen-9-yl)-3,4,5,6-tetrafluorobenzoate; S-Rhodamine; $C_{26}H_{18}F_4N_2O_4S$	4.3.3
515	JOE	2-(4,5-Dichloro-2,7-dimethoxy-6-oxido-3-oxo-3H-xanthen-9-yl)terephthalate; Fluorescein (Halogenated); $C_{23}H_{11}Cl_2O_9$; 113394-23-3	4.8.2
516	JS-R	7-(Dimethylamino)-3-(dimethyliminio)-5,5-dihydroxy-3,5-dihydrodibenzo[b,e]borinin-5-uide; $B(OH)_2$-pyronine; $B(OH)_2$-Pyronine; $C_{11}H_{21}BN_2O_2$	4.5
517	Jul-rosol	9-Phenyl-2,3,6,7-tetrahydro-1H,5H,12H-chromeno[2,3-f]pyrido[3,2,1-ij]quinolin-12-one; Rosol; $C_{25}H_{21}NO_2$	4.19
518	KK1119	tl; Rhodamine (λa, max > 600 nm); $C_{43}H_{44}F_4N_3O_{10}S_2$	4.2.4
519	KK114	Similar to Abberior Star Red; Rhodamine (λa, max > 600 nm); $C_{43}H_{40}F_4N_3O_{10}S_2$	4.2.4
520	KK9046	tl; Rhodamine (λa, max > 600 nm); $C_{43}H_{44}F_4N_3O_6$	4.2.4
521	L-B 1a	5,5-Difluoro-10-(4-iodophenyl)-1,3,7,9-tetramethyl-5H-4l4,5l4-dipyrrolo[1,2-c:2′,1′-f][1,3,2] diazaborinine-2-sulfonate; Borondipyrromethene (Small); $C_{19}H_{17}BF_2IN_2O_3S$	4.14.1
522	L-B 1b	5,5-difluoro-10-(4-iodophenyl)-1,3,7,9-tetramethyl-5H-4l4,5l4-dipyrrolo[1,2-c:2′,1′-f][1,3,2] diazaborinine-2,8-disulfonate; Borondipyrromethene (Small); $C_{19}H_{16}BF_2IN_2O_6S_2$	4.14.1
523	L-B 2a	10-(4-Bromophenyl)-3,7-dichloro-5,5-difluoro-5H-4l4,5l4-dipyrrolo[1,2-c:2′,1′-f][1,3,2] diazaborinine-2-sulfonate; Borondipyrromethene (Small); $C_{15}H_7BBrCl_2F_2N_2O_3S$	4.14.1
524	L-B 2b	10-(4-Bromophenyl)-3,7-dichloro-5,5-difluoro-5H-4l4,5l4-dipyrrolo[1,2-c:2′,1′-f][1,3,2] diazaborinine-2,8-disulfonate; Borondipyrromethene (Small); $C_{15}H_6BBrCl_2F_2N_2O_6S_2$	4.14.1

Ent.	Short Name	Name; Class; Molecular Formula; CAS Number	Ref.
525	L-B 4a	10-(4-Azidophenyl)-5,5-difluoro-1,3,7,9-tetramethyl-5H-4l4,5l4-dipyrrolo[1,2-c:2′,1′-f][1,3,2]diazaborinine-2-sulfonate; Borondipyrromethene (Small); $C_{19}H_{17}BF_2N_5O_3S$	4.14.1
526	L-B 4b	10-(4-Azidophenyl)-5,5-difluoro-1,3,7,9-tetramethyl-5H-4l4,5l4-dipyrrolo[1,2-c:2′,1′-f][1,3,2]diazaborinine-2,8-disulfonate; Borondipyrromethene (Small); $C_{19}H_{16}BF_2N_5O_6S_2$	4.14.1
527	L-B 5	10-(4-(4-(3-Carboxypropyl)-1H-1,2,3-triazol-1-yl)phenyl)-5,5-difluoro-1,3,7,9-tetramethyl-5H-4l4,5l4-dipyrrolo[1,2-c:2′,1′-f][1,3,2]diazaborinine-2,8-disulfonate; Borondipyrromethene (Small); $C_{25}H_{24}BF_2N_5O_8S_2$	4.14.1
528	LS479	2,2′,2″,2‴-((3-((E)-1-ethyl-2-((2E,4E,6E)-7-(1-Ethyl-3,3-dimethyl-3H-indol-1-ium-2-yl)hepta-2,4,6-trien-1-ylidene)-3,3-dimethylindolin-5-yl)pentane-1,5-diyl)bis(azanetriyl))tetraacetate; Cyanine (Heptamethine); $C_{44}H_{53}N_4O_8$	4.13.3
529	LS482	2,2′,2″,2‴-(((((1E,3Z,5E)-1-(1-Ethyl-3,3-dimethyl-3H-indol-1-ium-2-yl)-7-((E)-1-ethyl-3,3-dimethylindolin-2-ylidene)hepta-1,3,5-trien-4-yl)azanediyl)bis(ethane-2,1-diyl))bis(azanetriyl))tetraacetate; Cyanine (Heptamethine); $C_{43}H_{52}N_5O_8$	4.13.3
530	Lucifer yellow CH	6-Amino-2-(hydrazinecarboxamido)-1,3-dioxo-2,3-dihydro-1H-benzo[de]isoquinoline-5,8-disulfonate; Naphthalimide; $C_{13}H_{10}N_4O_9S_2$; 67769-47-5	4.32
531	Lucifer yellow VS	6-Amino-1,3-dioxo-2-(3-(vinylsulfonyl)phenyl)-2,3-dihydro-1H-benzo[de]isoquinoline-5,8-disulfonate; Naphthalimide; $C_{20}H_{12}N_2O_{10}S_3$	4.32
532	Lucigenin	10,10′-Dimethyl-[9,9′-biacridine]-10,10′-diium; Hybrid – Single bond connected dye; $C_{28}H_{22}N_2$	4.40.2
533	Luminarine	4-Aminopyrido[2,1-h]pteridin-11-ium-6-olate; Pteridine; $C_{10}H_7N_5O$	4.36
534	Marina Blue	2-(6,8-Difluoro-7-hydroxy-4-methyl-2-oxo-2H-chromen-3-yl)acetic acid; Coumarin (7-Hydroxy); $C_{12}H_{10}F_2O_5$	4.26.1
535	Me-PR	(R)-N-(7-(Diethylamino)-5-methyl-5-oxido-10-(o-tolyl)-3H-acridophosphin-3-ylidene)-N-ethylethanaminium; POR-Rosamine; $C_{29}H_{35}N_2OP$	4.18.2
536	Mercaptofluorescein MK-43	2-(6-((2-Hydroxyethyl)thio)-3-oxo-3H-xanthen-9-yl)benzoate; Mercaptofluorescein; $C_{22}H_{16}O_5S$	4.8.3
537	Mercaptofluorescein MK-61	2-(6-((2-Azidoethyl)thio)-3-oxo-3H-xanthen-9-yl)benzoate; Mercaptofluorescein; $C_{22}H_{16}N_3O_4S$	4.8.3
538	Mercaptofluorescein MK-67	2-(2,7-Dichloro-6-((2-hydroxyethyl)thio)-3-oxo-3H-xanthen-9-yl)benzoate; Mercaptofluorescein; $C_{22}H_{13}Cl_2O_5S$	4.8.3
539	Mercaptofluorescein MK-69	2-(6-((2-Aminoethyl)thio)-3-oxo-3H-xanthen-9-yl)benzoate; Mercaptofluorescein; $C_{22}H_{16}NO_4S$	4.8.3
540	Mercaptofluorescein MK-75	2-(6-((2-Azidoethyl)thio)-2,7-dichloro-3-oxo-3H-xanthen-9-yl)benzoate; Mercaptofluorescein; $C_{22}H_{12}Cl_2N_3O_4S$	4.8.3
541	Me-rosol	6-(Dimethylamino)-9-phenyl-3H-xanthen-3-one; Rosol; $C_{21}H_{17}NO_2$	4.19

(Continued)

Ent.	Short Name	Name; Class; Molecular Formula; CAS Number	Ref.
542	*meso*-CaboxyBODIPY J-K 1	5,5-Difluoro-3,7-dimethyl-5H-4l4,5l4-dipyrrolo[1,2-c:2′,1′-f][1,3,2]diazaborinine-10-carboxylate; Borondipyrromethene (Small); $C_{12}H_{10}BF_2N_2O_2$	4.14.1
543	*meso*-CaboxyBODIPY J-K 3a	5,5-Difluoro-3,7-dimethyl-5H-4l4,5l4-dipyrrolo[1,2-c:2′,1′-f][1,3,2]diazaborinine-10-carboxamide; Borondipyrromethene (Small); $C_{12}H_{12}BF_2N_3O$	4.14.1
544	*meso*-CaboxyBODIPY J-K 3b	5,5-Difluoro-N,3,7-trimethyl-5H-4l4,5l4-dipyrrolo[1,2-c:2′,1′-f][1,3,2]diazaborinine-10-carboxamide; Borondipyrromethene (Small); $C_{13}H_{14}BF_2N_3O$	4.14.1
545	*meso*-CaboxyBODIPY J-K 3c	5,5-Difluoro-N,N,3,7-tetramethyl-5H-4l4,5l4-dipyrrolo[1,2-c:2′,1′-f][1,3,2]diazaborinine-10-carboxamide; Borondipyrromethene (Small); $C_{14}H_{16}BF_2N_3O$	4.14.1
546	MetACSO3H	10-Methyl-9-oxo-4a,9,9a,10-tetrahydroacridine-2,7-disulfonate; Acridones; $C_{14}H_{11}NO_7S_2$	4.23.2
547	M-S 10a	4-(3-(5,5-Difluoro-1,3,10-trimethyl-7-(thiophen-2-yl)-5H-4l4,5l4-dipyrrolo[1,2-c:2′,1′-f][1,3,2]diazaborinin-2-yl)propanamido)-5-oxo-5-((2-sulfonatoethyl)amino)pentanoate; Borondipyrromethene (Small); $C_{26}H_{29}BF_2N_4O_7S_2$	4.14.1
548	M-S 10b	4-(3-(5,5-Difluoro-7-(4-methoxyphenyl)-1,3,10-trimethyl-5H-4l4,5l4-dipyrrolo[1,2-c:2′,1′-f][1,3,2]diazaborinin-2-yl)propanamido)-5-oxo-5-((2-sulfonatoethyl)amino)pentanoate; Borondipyrromethene (Small); $C_{29}H_{33}BF_2N_4O_8S$	4.14.1
549	M-S 10c	4-(3-(5,5-Difluoro-1,3,10-trimethyl-7-phenyl-5H-4l4,5l4-dipyrrolo[1,2-c:2′,1′-f][1,3,2]diazaborinin-2-yl)propanamido)-5-oxo-5-((2-sulfonatoethyl)amino)pentanoate; Borondipyrromethene (Small); $C_{28}H_{31}BF_2N_4O_7S$	4.14.1
550	M-S 10d	(E)-4-(3-(5,5-Difluoro-1,3,10-trimethyl-7-styryl-5H-4l4,5l4-dipyrrolo[1,2-c:2′,1′-f][1,3,2]diazaborinin-2-yl)propanamido)-5-oxo-5-((2-sulfonatoethyl)amino)pentanoate; Borondipyrromethene (Small); $C_{30}H_{33}BF_2N_4O_7S$	4.14.1
551	M-S 10e	(E)-4-(3-(5,5-Difluoro-7-(4-methoxystyryl)-1,3,10-trimethyl-5H-4l4,5l4-dipyrrolo[1,2-c:2′,1′-f][1,3,2]diazaborinin-2-yl)propanamido)-5-oxo-5-((2-sulfonatoethyl)amino)pentanoate; Borondipyrromethene (Small); $C_{31}H_{35}BF_2N_4O_8S$	4.14.1
552	M-S 10f	4-(3-(5,5-Difluoro-1,3,10-trimethyl-7-(1H-pyrrol-2-yl)-5H-4l4,5l4-dipyrrolo[1,2-c:2′,1′-f][1,3,2]diazaborinin-2-yl)propanamido)-5-oxo-5-((2-sulfonatoethyl)amino)pentanoate; Borondipyrromethene (Small); $C_{26}H_{30}BF_2N_5O_7S$	4.14.1
553	Naphtahlimide 10b	1-(2-Hydroxyethyl)-4-(methylamino)-1H-benzo[de]isoquinoline-1,3(2H)-dione; Naphthalimide; $C_{15}H_{14}N_2O_3$	4.32
554	Naphtahlimide 10c	4-((4-Hydroxycyclohexyl)(methyl)amino)-1-(2-hydroxyethyl)-1H-benzo[de]isoquinoline-1,3(2H)-dione; Naphthalimide; $C_{20}H_{22}N_2O_4$	4.32
555	Naphthalimide B-K 1	1-(2-(6-Amino-1,3-dioxo-1H-benzo[de]isoquinolin-2(3H)-yl)ethyl)pyridin-1-ium; Naphthalimide; $C_{19}H_{16}N_3O_2$	4.32
556	Naphthofluorescein	7′,19′-Dihydroxyspiro[2-benzofuran-3,13′-2-oxapentacyclo[12.8.0.03,12.04,9.017,22]docosa-1(14),3(12),4(9),5,7,10,15,17(22),18,20-decaene]-1-one; Fluorescein-analog; $C_{28}H_{15}O_5$	4.8.4

Ent.	Short Name	Name; Class; Molecular Formula; CAS Number	Ref.
557	NBDY	4,4-Difluoro-1,3,5,7-tetramethyl-2-iodoacetamide boron-dipyrromethene; Borondipyrromethene (Small); $C_{15}H_{17}BF_2IN_3O$	4.14.1
558	NBQ	10-Hydroxy-1-methylbenzo[h]quinolin-1-ium; Quinoline; $C_{14}H_{12}NO$	4.29
559	NCR	2-(10-(Diethyliminio)-1-methyl-3-oxo-3,10-dihydrochromeno[8,7-c]xanthen-7-yl)benzoate; Hybrid – Fusion of two existing dye; $C_{32}H_{25}NO_5$	4.40.1
560	NCRMe	N-Ethyl-N-(7-(2-(methoxycarbonyl)phenyl)-1-methyl-3-oxochromeno[8,7-c]xanthen-10(3H)-ylidene)ethanaminium; Hybrid – Fusion of two existing dye; $C_{33}H_{28}NO_5$	4.40.1
561	Neo-Cy3	(E)-2-((E)-3-(5-(2-Iodoacetamido)-1,3,3-trimethyl-3H-indol-1-ium-2-yl)allylidene)-3,3-dimethyl-1-(3-sulfonatopropyl)indoline-6-sulfonate; Cyanine (Trimethine); $C_{29}H_{33}IN_3O_7S_2$	4.13.1
562	Neo-Cy5	(E)-2-((2E,4E)-5-(5-(2-I1odoacetamido)-1,3,3-trimethyl-3H-indol-1-ium-2-yl)penta-2,4-dien-1-ylidene)-3,3-dimethyl-1-(3-sulfonatopropyl)indoline-6-sulfonate; Cyanine (Pentamethine); $C_{31}H_{35}IN_3O_7S_2$	4.13.2
563	NIR1	(E)-2-((2E,4E,6E)-7-(3-Ethyl-1,1-dimethyl-6,8-disulfonato-1H-benzo[e]indol-3-ium-2-yl) hepta-2,4,6-trien-1-ylidene)-3,3-dimethyl-1-(4-sulfonatobutyl)indoline-5-carboxylate; Cyanine (Heptamethine); $C_{38}H_{39}N_2O_{11}S_3$	4.13.3
564	NIR2	(E)-2-((2E,4E)-5-(3-Ethyl-1,1-dimethyl-6,8-disulfonato-1H-benzo[e]indol-3-ium-2-yl) penta-2,4-dien-1-ylidene)-3,3-dimethyl-1-(4-sulfonatobutyl)indoline-5-carboxylate; Cyanine (Pentamethine); $C_{36}H_{37}N_2O_{11}S_3$	4.13.2
565	NIR3	(E)-2-((2E,4E,6E)-7-(3,3-Dimethyl-5-sulfonato-1-(4-sulfonatobutyl)-3H-indol-1-ium-2-yl) hepta-2,4,6-trien-1-ylidene)-3,3-dimethyl-1-(4-sulfonatobutyl)indoline-5-carboxylate; Cyanine (Heptamethine); $C_{36}H_{41}N_2O_{11}S_3$	4.13.3
566	NIR4	(E)-2-((2E,4E)-5-(3,3-Dimethyl-5-sulfonato-1-(4-sulfonatobutyl)-3H-indol-1-ium-2-yl) penta-2,4-dien-1-ylidene)-3,3-dimethyl-1-(4-sulfonatobutyl)indoline-5-carboxylate; Cyanine (Pentamethine); $C_{34}H_{39}N_2O_{11}S_3$	4.13.2
567	Niu 18	3-((3-(4-(5,5-Difluoro-1,3,7,9-tetramethyl-5H-4l4,5l4-dipyrrolo[1,2-c:2′,1′-f][1,3,2] diazaborinin-10-yl)phenyl)prop-2-yn-1-yl)dimethylammonio)propane-1-sulfonate; Borondipyrromethene (Medium sized); $C_{27}H_{32}BF_2N_3O_3S$	4.14.2
568	Niu 25	3,3′-(((5,5-Difluoro-1,3,7,9,10-pentamethyl-5H-4l4,5l4-dipyrrolo[1,2-c:2′,1′-f][1,3,2] diazaborinine-2,8-diyl)bis(prop-2-yne-3,1-diyl))bis(dimethylammoniumdiyl))bis(propane-1-sulfonate); Borondipyrromethene (Medium sized); $C_{30}H_{43}BF_2N_4O_6S_2$	4.14.2
569	Niu 27	3,3′-(((2,8-Diethyl-1,3,7,9,10-pentamethyl-5H-4l4,5l4-dipyrrolo[1,2-c:2′,1′-f][1,3,2] diazaborinine-5,5-diyl)bis(prop-2-yne-3,1-diyl))bis(dimethylammoniumdiyl))bis(propane-1-sulfonate); Borondipyrromethene (Medium sized); $C_{34}H_{53}BN_4O_6S_2$	4.14.2
570	Niu 29	3,3′-(((2,8-Diethyl-10-(4-iodophenyl)-1,3,7,9-tetramethyl-5H-4l4,5l4-dipyrrolo[1,2-c:2′,1′-f] [1,3,2]diazaborinine-5,5-diyl)bis(prop-2-yne-3,1-diyl))bis(dimethylammoniumdiyl)) bis(propane-1-sulfonate); Borondipyrromethene (Medium sized); $C_{39}H_{54}BIN_4O_6S_2$	4.14.2

(Continued)

Ent.	Short Name	Name; Class; Molecular Formula; CAS Number	Ref.
571	N-Methylacridone	10-Methyl-9-oxo-8a,9-dihydroacridin-10-ium; Acridones; $C_{14}H_{12}NO$	4.23.2
572	NOBQ	10-Methoxy-1-methylbenzo[h]quinolin-1-ium; Quinoline; $C_{15}H_{14}NO$	4.29
573	nor-Harmane (β-carboline)	9H-Pyrido[3,4-b]indole; Indole; $C_{11}H_8N_2$	4.33
574	nor-SqB1	2-((3-((3,3-Dimethyl-5-sulfonatoindolin-2-ylidene)methyl)-2-oxido-4-(2,4,6-trioxotetrahydropyrimidin-5(2H)-ylidene)cyclobut-2-en-1-ylidene)methyl)-3,3-dimethyl-3H-1l4-indole-5-sulfonate; Squaraine; $C_{31}H_{29}N_4O_{10}S_2$	4.35
575	nor-SqB2	6-(2-((3-((3-(5-Carboxylatopentyl)-3-methyl-5-sulfonato-3H-1l4-indol-2-yl)methylene)-2-oxido-4-(2,4,6-trioxotetrahydropyrimidin-5(2H)-ylidene)cyclobut-1-en-1-yl)methylene)-3-methyl-5-sulfonatoindolin-3-yl)hexanoate; Squaraine; $C_{40}H_{39}N_4O_{14}S_2$	4.35
576	nor-SqB3	6-(2-((3-((1-(5-Carboxylatopentyl)-1-methyl-6,8-disulfonato-1,3-dihydro-2H-benzo[e]indol-2-ylidene)methyl)-2-oxido-4-(2,4,6-trioxotetrahydropyrimidin-5(2H)-ylidene)cyclobut-2-en-1-ylidene)methyl)-1-methyl-6,8-disulfonato-1H-3l4-benzo[e]indol-1-yl)hexanoate; Squaraine; $C_{48}H_{41}N_4O_{20}S_4$	4.35
577	nor-SqCN1	2-((4-(Dicyanomethylene)-3-((3,3-dimethyl-5-sulfonatoindolin-2-ylidene)methyl)-2-oxidocyclobut-2-en-1-ylidene)methyl)-3,3-dimethyl-3H-1l4-indole-5-sulfonate; Squaraine; $C_{29}H_{21}N_4O_7S_2$	4.35
578	nor-SqCN2	6-(2-((3-((3-(5-Carboxylatopentyl)-3-methyl-5-sulfonato-3H-1l4-indol-2-yl)methylene)-4-(dicyanomethylene)-2-oxidocyclobut-1-en-1-yl)methylene)-3-methyl-5-sulfonatoindolin-3-yl)hexanoate; Squaraine; $C_{39}H_{35}N_4O_{11}S_2$	4.35
579	nor-SqCN3	6-(2-((3-((1-(5-Carboxylatopentyl)-1-methyl-6,8-disulfonato-1,3-dihydro-2H-benzo[e]indol-2-ylidene)methyl)-4-(dicyanomethylene)-2-oxidocyclobut-2-en-1-ylidene)methyl)-1-methyl-6,8-disulfonato-1H-3l4-benzo[e]indol-1-yl)hexanoate; Squaraine; $C_{47}H_{37}N_4O_{17}S_4$	4.35
580	nor-SqO1	2-((3-((3,3-Dimethyl-5-sulfonatoindolin-2-ylidene)methyl)-2-oxido-4-oxocyclobut-2-en-1-ylidene)methyl)-3,3-dimethyl-3H-1l4-indole-5-sulfonate; Squaraine; $C_{26}H_{21}N_2O_8S_2$	4.35
581	nor-SqO2	6-(2-((3-((3-(5-Carboxylatopentyl)-3-methyl-5-sulfonato-3H-1l4-indol-2-yl)methylene)-2-oxido-4-oxocyclobut-1-en-1-yl)methylene)-3-methyl-5-sulfonatoindolin-3-yl)hexanoate; Squaraine; $C_{36}H_{35}N_2O_{12}S_2$	4.35
582	nor-SqO3	6-(2-((3-((1-(5-Carboxylatopentyl)-1-methyl-6,8-disulfonato-1,3-dihydro-2H-benzo[e]indol-2-ylidene)methyl)-2-oxido-4-oxocyclobut-2-en-1-ylidene)methyl)-1-methyl-6,8-disulfonato-1H-3l4-benzo[e]indol-1-yl)hexanoate; Squaraine; $C_{44}H_{37}N_2O_{18}S_4$	4.35
583	NR600	10-(2,6-Dimethylphenyl)-3-oxo-3H-acridophosphine-5,7-bis(olate) 5-oxide; POOH-Fluorone; $C_{21}H_{15}O_4P$	4.12
584	NR604	10-(2,6-Dimethoxyphenyl)-3-oxo-3H-acridophosphine-5,7-bis(olate) 5-oxide; POOH-Fluorone; $C_{21}H_{15}O_6P$	4.12
585	NR632	7-(Dimethylamino)-10-(2,6-dimethylphenyl)-3-oxo-3H-acridophosphin-5-olate 5-oxide; POPh-Rosol; $C_{23}H_{21}NO_3P$	4.20

Ent.	Short Name	Name; Class; Molecular Formula; CAS Number	Ref.
586	NR638	10-(2,6-Dimethoxyphenyl)-7-(dimethylamino)-3-oxo-3H-acridophosphin-5-olate 5-oxide; POPh-Rosol; $C_{23}H_{21}NO_5P$	4.20
587	NR666	7-(Dimethylamino)-3-(dimethyliminio)-10-(o-tolyl)-3H-acridophosphin-5-olate 5-oxide; PO_2H-Rosamine; $C_{24}H_{25}N_2O_2P$	4.18.2
588	NR669	7-(Dimethylamino)-3-(dimethyliminio)-10-(2,6-dimethylphenyl)-3H-acridophosphin-5-olate 5-oxide; PO_2H-Rosamine; $C_{25}H_{27}N_2O_2P$	4.18.2
589	NR675	10-(2,6-Dimethoxyphenyl)-7-(dimethylamino)-3-(dimethyliminio)-3H-acridophosphin-5-olate 5-oxide; PO_2H-Rosamine; $C_{25}H_{27}N_2O_4P$	4.18.2
590	NR698	tl; PO_2H-Rosamine; $C_{32}H_{33}N_2O_2P$	4.18.2
591	NR700	(S)-N-(7-(Dimethylamino)-5-ethoxy-5-oxido-10-(o-tolyl)-3H-acridophosphin-3-ylidene)-N-methylmethanaminium; PO_2R-Rosamine; $C_{26}H_{30}N_2O_2P$	4.18.2
592	NR744	tl; PO_2R-Rosamine; $C_{34}H_{38}N_2O_2P$	4.18.2
593	NX	3-(6-Oxido-3-oxo-3H-xanthen-9-yl)-2-naphthoate; Fluorescein (Non-halogenated); $C_{24}H_{12}O_5$	4.8.1
594	OBQ	10-Methoxybenzo[h]quinoline; Quinoline; $C_{14}H_{11}NO$	4.29
595	Oregon Green 488	4-(2,7-Difluoro-6-oxido-3-oxo-3H-xanthen-9-yl)isophthalate; Fluorescein (Halogenated); $C_{21}H_7F_2O_7$	4.8.2
596	Oregon Green 514	4-((Carboxylatomethyl)thio)-2-(2,7-difluoro-6-oxido-3-oxo-3H-xanthen-9-yl)-3,5,6-trifluorobenzoate; Fluorescein (Halogenated); $C_{22}H_6F_{5O7}S$	4.8.2
597	Oxazine 170	(E)-N-(9-(Ethylamino)-10-methyl-5H-benzo[a]phenoxazin-5-ylidene)ethanaminium; Phenoxazine; $C_{21}H_{22}N_3O$	4.25
598	Pacific Blue	6,8-Difluoro-7-hydroxy-2-oxo-2H-chromene-3-carboxylate; Coumarin (7-Hydroxy); $C_{14}H_7F_2NO_7$; 215868-33-0	4.26.1
599	Pacific orange	2-(5,7-Difluoro-6-hydroxybenzo[d]thiazol-2-yl)-4-methyl-4,5-dihydrothiazole-4-carboxylate; Benzothiazole; $C_{12}H_8F_2N_2O_3S_2$	4.30
600	PB430	(R)-4-(3-(5-Oxido-5-phenyl-12,12-bis(4-(3-sulfonatopropoxy)phenyl)-12H-benzo[f]indeno[1,2-b]phosphindol-2-yl)benzamido)butanoate; Others (Phosphole); $C_{54}H_{45}NO_{12}PS_2$	4.39.3
601	Pennsylvania Green	2,7-Difluoro-3-oxo-9-(o-tolyl)-3H-xanthen-6-olate; Fluorone; $C_{20}H_{11}F_2O_3$	4.11
602	PetACSO3H	9-Oxo-10-pentyl-4a,9,9a,10-tetrahydroacridine-2,7-disulfonate; Acridones; $C_{18}H_{19}NO_7S_2$	4.23.2
603	Phosphorilated coum. N-H 21c	4-(6-(((Hydroxyoxidophosphoryl)oxy)methyl)-8,8-dimethyl-2-oxo-3-(pyridin-1-ium-2-yl)-2H-pyrano[3,2-g]quinolin-9(8H)-yl)butanoate; Coumarin (More elaborated 7-Amino); $C_{24}H_{23}N_2PO_8$	4.26.3

(Continued)

Ent.	Short Name	Name; Class; Molecular Formula; CAS Number	Ref.
604	Phosphorilated coum. N-H 22c	4-(6-(((Hydroxyoxidophosphoryl)oxy)methyl)-4,8,8-trimethyl-2-oxo-3-(pyridin-1-ium-4-yl)-2H-pyrano[3,2-g]quinolin-9(8H)-yl)butanoate; Coumarin (More elaborated 7-Amino); $C_{25}H_{26}N_2O_8P$	4.26.3
605	Phosphorilated oxazine 2-H	4-(1-(3-((Hydroxyoxidophosphoryl)oxy)propyl)-3,4,9,10-tetrahydro-2H-dipyrido[3,2-b:2′,3′-i]phenoxazin-1-ium-11(8H)-yl)butanoate; Phenoxazine; $C_{25}H_{30}N_3O_7P$	4.25
606	Pip-rosol	9-Phenyl-6-(piperidin-1-yl)-3H-xanthen-3-one; Rosol; $C_{24}H_{21}NO_2$	4.19
607	POF	(R)-10-(2,6-Dimethoxyphenyl)-3-oxo-5-phenyl-3H-acridophosphin-7-olate 5-oxide; POPh-Fluorone; $C_{27}H_{21}O_4P$	4.12
608	PONy A1	4-(6-(Dimethylamino)-3-(dimethyliminio)-9-(oxidohydrophosphoryl)acridin-10(3H)-yl) butanoate; Acridine; $C_{21}H_{25}N_3O_4P$	4.23.1
609	PONy C7	3-(Benzo[d]thiazol-2-yl)-7-(diethylamino)-2-oxo-2H-chromen-4-yl]phosphonic acid; Hybrid – Single bond connected dye; $C_{20}H_{18}N_2O_5PS$	4.40.2
610	PONy CP1	9-[(Allyloxy)(diisopropylamino)phosphoryl]-3,6-bis(dimethylamino)-10,10-dimethyl-9,10-dihydroanthracenylium; $C(CH_3)_2$-Pyronine; $C_{29}H_{43}N_3O_2P$	4.5
611	PONy P1	(6-(Dimethylamino)-3-(dimethyliminio)-3H-xanthen-9-yl)phosphinate; Pyronine; $C_{17}H_{19}N_2O_3P$	4.4
612	PONy P10	9-(Diisopropoxyphosphoryl)-3,6-bis(dimethylamino)-9H-xanthenylium trifluoroacetate; Pyronine; $C_{23}H_{32}N_2O_4P$	4.4
613	PONy P13	tl; Pyronine; $C_{29}H_{36}N_2O_4P$	4.4
614	PONy P1-Halo	(3-((2-(2-((6-Chlorohexyl)oxy)ethoxy)ethyl)amino)-3-oxopropyl)(6-(dimethylamino)-3-(dimethyliminio)-3H-xanthen-9-yl)phosphinate; Pyronine; $C_{30}H_{43}ClN_3O_6P$	4.4
615	PONy P2	9-(Dimethoxyphosphoryl)-3,6-bis(dimethylamino)-9H-xanthenylium; Pyronine; $C_{19}H_{24}N_2O_4P$	4.4
616	PONy P5	tl; Pyronine; $C_{32}H_{42}N_3O_7P$	4.4
617	PONy P6	2-[[3,6-Bis(dimethylamino)-9H-xanthenylium-9-yl](dimethylamino)phosphoryloxy]-benzoate; Pyronine; $C_{34}H_{36}N_3O_5P$	4.4
618	PONy P9	7-(Dimethoxyphosphoryl)-3,10-bis(dimethylamino)-5,6-dihydrobenzo[c]xanthen-12-ium; Pyronine; $C_{18}H_{21}N_2O_4P$	4.4
619	PORL	(R)-7-(Diethylamino)-10-(2,6-dimethoxyphenyl)-5-phenyl-3H-acridophosphin-3-one 5-oxide; POPh-Rosol; $C_{31}H_{30}NO_4P$	4.20
620	PR	(R)-N-(7-(Diethylamino)-5-methyl-5-oxido-10-phenyl-3H-acridophosphin-3-ylidene)-N-ethylethanaminium; POR-Rosamine; $C_{28}H_{34}N_2OP$	4.18.2
621	Prodan	6-Propionyl-2-(dimethy-1-amino)naphthalene; Naphthalene; $C_{15}H_{17}NO$; 70504-01-7	4.34
622	Pyrene-1-sulfonate	Pyrene-1-sulfonate; Pyrene; $C_{16}H_{10}O_3S$	4.28

Ent.	Short Name	Name; Class; Molecular Formula; CAS Number	Ref.
623	Pyrodone	3-Oxo-3H-xanthen-6-olate; Pyrodone; $C_{13}H_7O_3$	4.21.2
624	Pyronine 630	1,11-Diethyl-3,4,8,9,10,11-hexahydro-2H-pyrano[3,2-g:5,6-g']diquinolin-1-ium; Pyronine; $C_{23}H_{27}N_2O$	4.4
625	Pyronine B	N-(6-(diethylamino)-3H-xanthen-3-ylidene)-N-ethylethanaminium; Pyronine; $C_{21}H_{27}N_2O$	4.4
626	Pyronine Y (OP)	N-(6-(dimethylamino)-3H-xanthen-3-ylidene)-N-methylmethanaminium; Pyronine; $C_{17}H_{19}N_2O$	4.4
627	Pyronine Y (OP)	N-(6-(dimethylamino)-3H-xanthen-3-ylidene)-N-methylmethanaminium; Pyronine; $C_{17}H_{19}N_2O$	4.4
628	Pyrrolidinylcarborhodamine	2-(10,10-dimethyl-3-(pyrrolidin-1-ium-1-ylidene)-6-(pyrrolidin-1-yl)-3,10-dihydroanthracen-9-yl)benzoate; $C(CH_3)_2$-Rhodamine; $C_{31}H_{33}N_2O_2$	4.3.1
629	Quinine	(6-methoxyquinolin-4-yl)((5R)-5-vinylquinuclidin-2-yl)methanol; Quinoline; $C_{20}H_{24}N_2O_2$; 130-95-0	4.29
630	Quinine	(6-methoxyquinolin-4-yl)((5R)-5-vinylquinuclidin-2-yl)methanol; Quinoline; $C_{20}H_{24}N_2O_2$; 130-95-0	4.29
631	Quinine	(6-methoxyquinolin-4-yl)((5R)-5-vinylquinuclidin-2-yl)methanol; Quinoline; $C_{20}H_{24}N_2O_2$; 130-95-0	4.29
632	Redmond Red	3-(3-Oxo-7-(pivaloyloxy)-3H-phenoxazin-2-yl)propanoate; Phenoxazine; $C_{20}H_{18}NO_6$	4.25
633	RF620	7-(Dimethylamino)-3-(dimethyliminio)-5,5-dihydroxy-10-(o-tolyl)-3,5-dihydrodibenzo[b,e]borinin-5-uide; $B(OH)_2$-Rosamine; $C_{24}H_{27}N_2O_2$	4.18.3
634	Rhodamine 110	2-(6-Amino-3-(l4-azanylidene)-3H-xanthen-9-yl)benzoate; Rhodamine (λa, max < 500 nm); $C_{21}H_{14}N_2O_3$; 13558-31-1	4.2.1
635	Rhodamine 123	6-Amino-9-(2-(methoxycarbonyl)phenyl)-3H-xanthen-3-iminium; Rhodamine (500 < λa, max < 550 nm); $C_{21}H_{17}N_2O_3$; 62669-70-9	4.2.2
636	Rhodamine 6G	(E)-N-(9-(2-(ethoxycarbonyl)phenyl)-6-(ethylamino)-2,7-dimethyl-3H-xanthen-3-ylidene)ethanaminium; Rhodamine (500 < λa, max < 550 nm); $C_{28}H_{31}N_2O_3$; 989-38-8	4.2.2
637	Rhodamine 6G	(E)-N-(9-(2-(ethoxycarbonyl)phenyl)-6-(ethylamino)-2,7-dimethyl-3H-xanthen-3-ylidene)ethanaminium; Rhodamine (500 < λa, max < 550 nm); $C_{28}H_{31}N_2O_3$; 989-38-8	4.2.2
638	Rhodamine 6G +2	(E)-N-(9-(2-((2-(2-(2-Ammonioethoxy)ethoxy)ethyl)carbamoyl)phenyl)-6-(ethylamino)-2,7-dimethyl-3H-xanthen-3-ylidene)ethanaminium; N-(8-amino-3,6-dioxaoctyl)-rhodamin 6G-amid-bis-(trifluoracetat); Rhodamine (500 < λa, max < 550 nm); $C_{32}H_{42}N_4O_4$; 591742-76-6	4.2.2
639	Rhodamine 6G +3	(E)-N-(9-(2-((2-((2-Ammonioethyl)amino)ethyl)carbamoyl)phenyl)-6-(ethylamino)-2,7-dimethyl-3H-xanthen-3-ylidene)ethanaminium; N-[2-(2-aminoethylamino)-ethyl]-rhodamin 6G-amid-bis-(trifluoracetat); Rhodamine (Highly charged); $C_{30}H_{39}N_5O_2$	4.2.5

(Continued)

Ent.	Short Name	Name; Class; Molecular Formula; CAS Number	Ref.
640	Rhodamine RBzET	(E)-N-(9-(2-(2-(4-(Ethoxycarbonyl)phenyl)acetyl)phenyl)-6-(ethylamino)-2,7-dimethyl-3H-xanthen-3-ylidene)ethanaminium; Rhodamine (500 < λa, max < 550 nm); $C_{36}H_{37}N_2O_4$	4.2.2
641	Rosamine W-B 5h	1-(9-(2,6-Dimethoxyphenyl)-6-(piperazin-1-yl)-3H-xanthen-3-ylidene)piperazin-1-ium; Rosamine; $C_{29}H_{33}N_4O_3$	4.17
642	RS800	6-((E)-2-((E)-2-(3-((E)-2-(3,3-Dimethyl-1-(4-sulfonatobutyl)-3H-indol-1-ium-2-yl)vinyl)-2-(4-sulfonatophenoxy)cyclohex-2-en-1-ylidene)ethylidene)-3,3-dimethylindolin-1-yl) hexanoate; Cyanine (Heptamethine); $C_{46}H_{52}N_2O_9S_2$	4.13.3
643	Safranin T	3,7-diamino-2,8-dimethyl-5-phenylphenazin-5-ium; Others (Phenazine); $C_{20}H_{19}N_4$	4.39.1
644	Salicylic acid	2-hydroxybenzoic acid; Benzene; $C_7H_6O_3$; 69-72-7	4.38
645	Seta-375-carboxylate	nd; nd; $C_{20}H_{21}BrN_2O_3$	4.41
646	Seta-385-carboxylate	nd; nd; $C_{21}H_{23}BrN_2O_3$	4.41
647	Seta-400-carboxylate	nd; nd; $C_{21}H_{23}BrN_2O_4$	4.41
648	Seta-405-carboxylate	nd; nd; $C_{10}H_4F_2O_5$	4.41
649	Seta-555-carboxylate	nd; nd; $C_{38}H_{53}N_2O_{14}PS_3$	4.41
650	Seta-580-carboxylate	nd; nd; $C_{33}H_{30}N_2O_5$	4.41
651	Seta-632-carboxylate	nd; nd; $C_{37}H_{44}N_2O_{13}S_3$	4.41
652	Seta-633-carboxylate	nd; nd; $C_{38}H_{46}N_2O_{16}S_4$	4.41
653	Seta-635-pH-di-carboxylate	nd; nd; $C_{36}H_{40}N_2O_{12}S_2$	4.41
654	Seta-646-carboxylate	nd; nd; $C_{42}H_{38}N_2S_5$	4.41
655	Seta-650-carboxylate	nd; nd; $C_{40}H_{49}N_2O_{14}PS_3$	4.41
656	Seta-660-di-carboxylate	nd; nd; $C_{50}H_{56}N_2O_{24}S_6$	4.41
657	Seta-670-carboxylate	nd; nd; $C_{40}H_{44}N_4O_{12}S_3$	4.41
658	Seta-670-pH-di-carboxylate	nd; nd; $C_{39}H_{40}N_4O_{11}S_2$	4.41
659	Seta-680-carboxylate	nd; nd; $C_{47}H_{57}N_2O_{20}PS_5$	4.41
660	Seta-700-carboxylate	nd; nd; $C_{47}H_{46}N_4O_{12}S_3$	4.41
661	Seta-700-di-carboxylate	nd; nd; $C_{49}H_{48}N_4O_{17}S_4$	4.41
662	Seta-750-carboxylate	nd; nd; $C_{44}H_{61}N_2O_{17}PS_4$	4.41
663	Seta-780-carboxylate	nd; nd; $C_{45}H_{50}N_2O_{11}S_3$	4.41
664	SeTau-380-carboxylate	nd; nd; $C_{13}H_{13}N_1O_8S_2$	4.41
665	SeTau-405-carboxylate	nd; nd; $C_{24}H_{23}ClN_3O_4$	4.41

Ent.	Short Name	Name; Class; Molecular Formula; CAS Number	Ref.
666	Si-9-Iminopyronine 2c	(E)-N-(7-(Dimethylamino)-10-((2-(2-(2-methoxyethoxy)ethoxy)acetyl)imino)-5,5-dimethyl-10,10a-dihydrodibenzo[b,e]silin-3(5H)-ylidene)-N-methylmethanaminium; $Si(CH_3)_2$-Pyronine; $C_{26}H_{38}N_3O_4Si$	4.5
667	Si-9-Iminopyronine 2c	(E)-N-(7-(Dimethylamino)-10-((2-(2-(2-methoxyethoxy)ethoxy)acetyl)imino)-5,5-dimethyl-10,10a-dihydrodibenzo[b,e]silin-3(5H)-ylidene)-N-methylmethanaminium; $Si(CH_3)_2$-Pyronine; $C_{26}H_{38}N_3O_4Si$	4.5
668	Si-9-Iminopyronine 2f	N-((E)-7-(Dimethylamino)-5,5-dimethyl-10-((4-(1-((3R,4R,5S)-2,4,5-trihydroxy-6-(hydroxymethyl)tetrahydro-2H-pyran-3-yl)-1H-1,2,3-triazol-4-yl)butanoyl)imino)-10,10a-dihydrodibenzo[b,e]silin-3(5H)-ylidene)-N-methylmethanaminium; $Si(CH_3)_2$-Pyronine; $C_{31}H_{43}N_6O_6Si$	4.5
669	Si-9-Iminopyronine 2f	N-((E)-7-(Dimethylamino)-5,5-dimethyl-10-((4-(1-((3R,4R,5S)-2,4,5-trihydroxy-6-(hydroxymethyl)tetrahydro-2H-pyran-3-yl)-1H-1,2,3-triazol-4-yl)butanoyl)imino)-10,10a-dihydrodibenzo[b,e]silin-3(5H)-ylidene)-N-methylmethanaminium; $Si(CH_3)_2$-Pyronine; $C_{31}H_{43}N_6O_6Si$	4.5
670	Si-9-Iminopyronine 2g	N-((E)-7-(Dimethylamino)-5,5-dimethyl-10-(((2-(1-((3S,4S,5R)-2,4,5-trihydroxy-6-(hydroxymethyl)tetrahydro-2H-pyran-3-yl)-1H-1,2,3-triazol-4-yl)ethyl)sulfonyl)imino)-10,10a-dihydrodibenzo[b,e]silin-3(5H)-ylidene)-N-methylmethanaminium; $Si(CH_3)_2$-Pyronine; $C_{29}H_{41}N_6O_7SSi$	4.5
671	Si-Fluorescein	2-(5,5-Dimethyl-7-oxido-3-oxo-3,5-dihydrodibenzo[b,e]silin-10-yl)benzoate; $Si(CH_3)_2$-Fluorescein; $C_{22}H_{18}O_4Si$	4.9
672	$SiOH_2R$	N-(7-(dimethylamino)-5,5-dihydroxy-10-(o-tolyl)dibenzo[b,e]silin-3(5H)-ylidene)-N-methylmethanaminium; $Si(OH)_2$-Rosamine; $C_{24}H_{27}N_2O_2Si$	4.18.1
673	SiP-F	(R)-N-(7-(Dimethylamino)-5-methyl-5-(3,3,3-trifluoropropyl)dibenzo[b,e]silin-3(5H)-ylidene)-N-methylmethanaminium; $C(CH_3)_2$-Pyronine; $C_{21}H_{26}F_3N_2Si$	4.5
674	SiP-Ph	N-(7-(Dimethylamino)-5,5-diphenyldibenzo[b,e]silin-3(5H)-ylidene)-N-methylmethanaminium; $C(CH_3)_2$-Pyronine; $C_{29}H_{29}N_2Si$	4.5
675	SiP-Pr	N-(7-(Dimethylamino)-5,5-diisopropyldibenzo[b,e]silin-3(5H)-ylidene)-N-methylmethanaminium; $C(CH_3)_2$-Pyronine; $C_{21}H_{29}N_2Si$	4.5
676	SiR	2-(7-(Dimethylamino)-3-(dimethyliminio)-5,5-dimethyl-3,5-dihydrodibenzo[b,e]silin-10-yl) benzoate; $Si(CH_3)_2$-Rhodamine; $C_{27}H_{29}N_2O_3Si$	4.3.2
677	SiR-COOH	2-(7-(Dimethylamino)-3-(dimethyliminio)-5,5-dimethyl-3,5-dihydrodibenzo[b,e]silin-10-yl) terephthalate; $Si(CH_3)_2$-Rhodamine; $C_{27}H_{28}N_2O_4Si$	4.3.2
678	SiRF 13a	6-(7-(Dimethylamino)-3-(dimethyliminio)-5,5-dimethyl-3,5-dihydrodibenzo[b,e]silin-10-yl)-2,3,4-trifluorobenzoate; $Si(CH_3)_2$-Rhodamine; $C_{26}H_{25}F_3N_2O_2Si$	4.3.2
679	SiRF 15a	2-((Carboxylatomethyl)thio)-6-(7-(dimethylamino)-3-(dimethyliminio)-5,5-dimethyl-3,5-dihydrodibenzo[b,e]silin-10-yl)-3,4-difluorobenzoate; $Si(CH_3)_2$-Rhodamine; $C_{28}H_{27}F_2N_2O_4SSi$	4.3.2

(Continued)

Ent.	Short Name	Name; Class; Molecular Formula; CAS Number	Ref.
680	SiRF 1a-H	N-(7-(Dimethylamino)-10-(2-formylphenyl)-5,5-dimethyldibenzo[b,e]silin-3(5H)-ylidene)-N-methylmethanaminium; $Si(CH_3)_2$-Rhodamine; $C_{26}H_{28}N_2O_2Si$	4.3.2
681	SiRF 1b-H	5-(2-(2-(7-(Dimethylamino)-3-(dimethyliminio)-5,5-dimethyl-3,5-dihydrodibenzo[b,e]silin-10-yl)phenyl)-2-oxoacetamido)-4-oxo-6-sulfonatohexanoate; $Si(CH_3)_2$-Rhodamine; $C_{33}H_{36}N_3O_8SSi$	4.3.2
682	SiR-methyl	3-(7-(Dimethylamino)-3-(dimethyliminio)-5,5-dimethyl-3,5-dihydrodibenzo[b,e]silin-10-yl)-4-methylbenzoate; $Si(CH_3)_2$-Rosam.; $C_{27}H_{30}N_2O_1Si$	4.18.1
683	SiTMR	Si-Tetramethylrhodamine; $Si(CH_3)_2$-Rhodamine; $C_{26}H_{28}N_2O_2Si$	4.3.2
684	S-JF646	3,6-Bis(azetidinyl)-9-propoxy-Si-pyronine; 3-(3-(azetidin-1-ium-1-ylidene)-7-(azetidin-1-yl)-5,5-dimethyl-3,5-dihydrodibenzo[b,e]silin-10-yl)propanoate; $Si(CH_3)_2$-Pyronine; $C_{24}H_{28}N_2O_2Si$	4.5
685	SMSiR	(2-(7-(Dimethylamino)-3-(dimethyliminio)-5,5-dimethyl-3,5-dihydrodibenzo[b,e]silin-10-yl)phenyl)methanesulfonate; $Si(CH_3)_2$-Sulforhodamine; $C_{26}H_{30}N_2O_3SSi$	4.7
686	SNARF-1	2-(10-(Diethyliminio)-3-oxo-6a,10-dihydro-3H-benzo[c]xanthen-7-yl)terephthalate; Rhodol-analog; $C_{29}H_{22}NO_6$	4.15.3
687	SO_2R1	N-(6-(dimethylamino)-10,10-dioxido-9-phenyl-3H-thioxanthen-3-ylidene)-N-methylmethanaminium; SO_2-Rosamine; $C_{23}H_{23}N_2O_2S$	4.18.3
688	SO_2R2	N-(6-(dimethylamino)-10,10-dioxido-9-(o-tolyl)-3H-thioxanthen-3-ylidene)-N-methylmethanaminium; SO_2-Rosamine; $C_{24}H_{25}N_2O_2S$	4.18.3
689	SO_2R3	N-(6-(dimethylamino)-9-(2-methoxyphenyl)-10,10-dioxido-3H-thioxanthen-3-ylidene)-N-methylmethanaminium; SO_2-Rosamine; $C_{24}H_{25}N_2O_3S$	4.18.3
690	SO_2R4	N-(6-(dimethylamino)-9-(2,6-dimethylphenyl)-10,10-dioxido-3H-thioxanthen-3-ylidene)-N-methylmethanaminium; SO_2-Rosamine; $C_{25}H_{27}N_2O_2S$	4.18.3
691	SO_2R5	N-(9-(2,6-dimethoxyphenyl)-6-(dimethylamino)-10,10-dioxido-3H-thioxanthen-3-ylidene)-N-methylmethanaminium; SO_2-Rosamine; $C_{25}H_{27}N_2O_4S$	4.18.3
692	Spirobifluorene 1	4,4′,4′′,4′′′-((((9,9′-Spirobi[fluorene]-2,7-diylbis(4,1-phenylene))bis(azanetriyl))tetrakis(benzenene-4,1-diyl))tetrakis(oxy))tetrakis(butane-1-sulfonate); Others (Spirobifluorene); $C_{77}H_{70}N_2O_{16}S_4$	4.39.3
693	SqB1	6-(2-((3-((1-(5-Carboxylatopentyl)-3,3-dimethyl-5-sulfonato-3H-1l4-indol-2-yl)methylene)-2-oxido-4-(2,4,6-trioxotetrahydropyrimidin-5(2H)-ylidene)cyclobut-1-en-1-yl)methylene)-3,3-dimethyl-5-sulfonatoindolin-1-yl)hexanoate; Squaraine; $C_{42}H_{42}N_4O_{14}S_2$	4.35
694	SqCN1	6-(2-((3-((1-(5-Carboxylatopentyl)-3,3-dimethyl-5-sulfonato-3H-1l4-indol-2-yl)methylene)-4-(dicyanomethylene)-2-oxidocyclobut-1-en-1-yl)methylene)-3,3-dimethyl-5-sulfonatoindolin-1-yl)hexanoate; Squaraine; $C_{41}H_{40}N_4O_{11}S_2$	4.35

Ent.	Short Name	Name; Class; Molecular Formula; CAS Number	Ref.
695	SqO1	6-(2-((3-((1-(5-Carboxylatopentyl)-3,3-dimethyl-5-sulfonato-3H-1l4-indol-2-yl)methylene)-2-oxido-4-oxocyclobut-1-en-1-yl)methylene)-3,3-dimethyl-5-sulfonatoindolin-1-yl)hexanoate; Squaraine; $C_{38}H_{40}N_2O_{12}S_2$	4.35
696	Squaraine T-P 4a	SO_3-CH_3-hexanoate S-Squaraine; Squaraine; $C_{32}H_{32}N_2O_6S_2$	4.35
697	Squaraine T-P 4b	SO_3-CH_3-hexanoate $C(CN)_2$-Sq.; Squaraine; $C_{35}H_{32}N_4O_6S$	4.35
698	Squaraine T-P 4c	SO_3-CH_3-hexanoate O-Squaraine; Squaraine; $C_{32}H_{32}N_2O_7S$	4.35
699	Squaraine T-P 4d	SO_3-hexanoate-Et $C(CN)_2$-Sq.; Squaraine; $C_{33}H_{34}N_2O_7S$	4.35
700	Squaraine T-P 5a	Di-SO_3-di-hexanoate S-Squaraine; Squaraine; $C_{38}H_{40}N_2O_{11}S_3$	4.35
701	Squaraine T-P 5b	Di-SO_3-di-hexanoate $C(CN)_2$-Sq.; Squaraine; $C_{41}H_{40}N_4O_{11}S_2$	4.35
702	Squaraine T-P 5c	Di-SO_3-di-hexanoate O-Squaraine; Squaraine; $C_{38}H_{40}N_2O_{12}S_2$	4.35
703	Square-635-carboxylate	nd; Squaraine; nd	4.35
704	Square-650-carboxylate	nd; Squaraine; nd	4.35
705	Square-655-carboxylate	nd; Squaraine; nd	4.35
706	Square-660-carboxylate	nd; Squaraine; nd	4.35
707	Square-680-carboxylate	nd; Squaraine; nd	4.35
708	SR101-110	4-(12-Amino-1,2,3,5,6,7-hexahydrochromeno[2,3-f]pyrido[3,2,1-ij]quinolin-4-ium-9-yl)benzene-1,3-disulfonate; Sulforhodamine; $C_{24}H_{22}N_2O_7S_2$	4.6
709	SR101-6G	4-(12-(Ethylamino)-11-methyl-1,2,3,5,6,7-hexahydrochromeno[2,3-f]pyrido[3,2,1-ij]quinolin-4-ium-9-yl)benzene-1,3-disulfonate; Sulforhodamine; $C_{27}H_{29}N_2O_7S_2$	4.6
710	SR101-B	4-(12-(Diethylamino)-1,2,3,5,6,7-hexahydrochromeno[2,3-f]pyrido[3,2,1-ij]quinolin-4-ium-9-yl)benzene-1,3-disulfonate; Sulforhodamine; $C_{28}H_{30}N_2O_7S_2$	4.6
711	SR101-Me_2	4-(12-(Dimethylamino)-1,2,3,5,6,7-hexahydrochromeno[2,3-f]pyrido[3,2,1-ij]quinolin-4-ium-9-yl)benzene-1,3-disulfonate; Sulforhodamine; $C_{26}H_{26}N_2O_7S_2$	4.6
712	SR101-NaphtNH_2	4-(13-Amino-1,2,3,5,6,7-hexahydrobenzo[7,8]chromeno[2,3-f]pyrido[3,2,1-ij]quinolin-4-ium-9-yl)benzene-1,3-disulfonate; Sulforhodamine-analog; $C_{28}H_{24}N_2O_7S_2$	4.6
713	SR101-NaphtOH	4-(13-Hydroxy-1,2,3,5,6,7-hexahydrobenzo[7,8]chromeno[2,3-f]pyrido[3,2,1-ij]quinolin-4-ium-9-yl)benzene-1,3-disulfonate; Sulforhodol-analog; $C_{28}H_{23}N_1O_8S_2$	4.27
714	SR101-OH	4-(12-Oxido-1,2,3,5,6,7-hexahydrochromeno[2,3-f]pyrido[3,2,1-ij]quinolin-4-ium-9-yl)benzene-1,3-disulfonate; Sulforhodol; $C_{24}H_{21}N_1O_8S_2$	4.27
715	Stefani HB-11	2-(5-Amino-2-hydroxyphenyl)benzo[d]oxazole-5-sulfonate; Hybrid – Single bond connected dye; $C_{13}H_{10}N_2O_5S$	4.40.2
716	Stefani HB-6	3-(Benzo[d]oxazol-2-yl)-4-hydroxybenzenesulfonate; Hybrid – Single bond connected dye; $C_{13}H_8NO_5S$	4.40.2

(Continued)

Ent.	Short Name	Name; Class; Molecular Formula; CAS Number	Ref.
717	Stefani HB-7	3-(Benzo[d]thiazol-2-yl)-4-hydroxybenzenesulfonic acid; Benzothiazole; $C_{13}H_{10}NO_4S_2$	4.30
718	Stefani HB-9	4-Hydroxy-3-(6-methylbenzo[d]oxazol-2-yl)benzenesulfonate; Hybrid – Single bond connected dye; $C_{14}H_{11}NO_5S$	4.40.2
719	Sulfobetaine pyrene 21	3-(Dimethyl(3-(pyren-1-yl)prop-2-yn-1-yl)ammonio)propane-1-sulfonate; Pyrene; $C_{24}H_{23}NO_3S$	4.28
720	Sulfo-cyanine3	6-((E)-2-((E)-3-(1-Ethyl-3,3-dimethyl-5-sulfonato-3H-indol-1-ium-2-yl)allylidene)-3,3-dimethyl-5-sulfonatoindolin-1-yl)hexanoate; Cyanine (Trimethine); $C_{31}H_{36}N_2O_8S_2$	4.13.1
721	Sulfo-cyanine5	6-((E)-2-((2E,4E)-5-(1-(5-Carboxylatopentyl)-3,3-dimethyl-5-sulfonato-3H-indol-1-ium-2-yl)penta-2,4-dien-1-ylidene)-3,3-dimethyl-5-sulfonatoindolin-1-yl)hexanoate; Cyanine (Pentamethine); $C_{37}H_{43}N_2O_{10}S_2$	4.13.2
722	Sulfofluorescein	2-(6-Oxido-3-oxo-3H-xanthen-9-yl)benzenesulfonate; Sulfofluorescein; $C_{19}H_{10}O_6S$	4.10
723	Sulfo-JF646	Si-Tetramethylsulforhodamine; 2-(7-(dimethylamino)-3-(dimethyliminio)-5,5-dimethyl-3,5-dihydrodibenzo[b,e]silin-10-yl)benzenesulfonate; $Si(CH_3)_2$-Sulforhodamine; $C_{27}H_{28}N_2O_3SSi$	4.7
724	Sulfonated coumarin N-H 6-H	4-(8,8-Dimethyl-2-oxo-3-(pyridin-1-ium-4-yl)-6-(sulfonatomethyl)-2H-pyrano[3,2-g]quinolin-9(8H)-yl)butanoate; Coumarin (More elaborated 7-Amino); $C_{24}H_{23}N_2O_7S$	4.26.3
725	Sulfonated coumarin N-H 7-H	4-(8,8-Dimethyl-2-oxo-6-(sulfonatomethyl)-3-(1-(3,3,4,4-tetrafluoro-2-oxido-5-oxocyclopent-1-en-1-yl)pyridin-1-ium-4-yl)-2H-pyrano[3,2-g]quinolin-9(8H)-yl)butanoate; Coumarin (More elaborated 7-Amino); $C_{29}H_{22}F_4N_2O_9S$	4.26.3
726	Sulforhodamine 101	2-(3-Oxa-23-aza-9-azoniaheptacyclo[17.7.1.15,9.02,17.04,15.023,27.013,28]octacosa-1(27),2(17),4,9(28),13,15,18-heptaen-16-yl)-5-sulfobenzenesulfonate; Texas red; Sulforhodamine; $C_{31}H_{30}N_2O_7S_2$; 60311-02-6	4.6
727	Sulforhodamine B	4-(12-(Diethylamino)-1,2,3,5,6,7-hexahydrochromeno[2,3-f]pyrido[3,2,1-ij]quinolin-4-ium-9-yl)benzene-1,3-disulfonate; Sulforhodamine; $C_{27}H_{29}N_2O_7S_2$; 3520-42-1	4.6
728	TAMRA	5(6)-Carboxytetramethylrhodamine; Rhodamine (500 < λa, max < 550 nm); $C_{25}H_{22}N_2O_5$; 98181-63-6	4.2.2
729	TAMRA	5(6)-Carboxytetramethylrhodamine; Rhodamine (500 < λa, max < 550 nm); $C_{25}H_{22}N_2O_5$; 98181-63-6	4.2.2
730	TET	2-(2,4,5,7-Tetrachloro-6-oxido-3-oxo-3H-xanthen-9-yl)terephthalate; Fluorescein (Halogenated); $C_{21}H_5Cl_4O_7$; 155911-14-1	4.8.2
731	Tetraethylrhodamine	2-(6-(Diethylamino)-3-(diethyliminio)-3H-xanthen-9-yl)benzoate; Rhodamine B; Rhodamine (550 < λa, max < 600 nm); $C_{28}H_{31}N_2O_3$; 81-88-9	4.2.3
732	Tetramethycarborhodamine	2-(6-(Dimethylamino)-3-(dimethyliminio)-10,10-dimethyl-3,10-dihydroanthracen-9-yl)benzoate; $C(CH_3)_2$-Rhodamine; $C_{27}H_{29}N_2O_2$	4.3.1
733	Tetramethylrhodamine	2-(6-(Dimethylamino)-3-(dimethyliminio)-3H-xanthen-9-yl)benzoate; TMR; Rhodamine (500 < λa, max < 550 nm); $C_{24}H_{22}N_2O_3$	4.2.2

Ent.	Short Name	Name; Class; Molecular Formula; CAS Number	Ref.
734	Tetramethylrosamine	N-(6-dimethylamino)-9-(o-tolyl)-3H-xanthen-3-ylidene)-N-methylmethanaminium; TMRosamine; Rosamine; $C_{24}H_{25}N_2O$	4.17
735	TG170	1-Ethyl-2-((E)-2-((E)-3-(2-((E)-1-ethyl-3,3-dimethylindolin-2-ylidene)ethylidene)-2-((2,3,5,6,8,9,11,12-octahydrobenzo[b][1,4,7,10,13]pentaoxacyclopentadecin-15-yl)amino)cyclohex-1-en-1-yl)vinyl)-3,3-dimethyl-3H-indol-1-ium; Cyanine (Heptamethine); $C_{48}H_{60}N_3O_5$	4.13.3
736	Thienyl-JF646	4-(3-(Azetidin-1-ium-1-ylidene)-7-(azetidin-1-yl)-5,5-dimethyl-3,5-dihydrodibenzo[b,e]silin-10-yl)thiophene-3-carboxylate; $Si(CH_3)_2$-Rhodamine; $C_{26}H_{26}N_2O_2SSi$	4.3.2
737	THQ-rosol	1,4-Diethyl-11-phenyl-1,2,3,4-tetrahydro-8H-chromeno[2,3-g]quinoxalin-8-one; Rosol; $C_{25}H_{24}N_2O_2$	4.19
738	TMDHS or SiP-Me	N-(7-(Dimethylamino)-5,5-dimethyldibenzo[b,e]silin-3(5H)-ylidene)-N-methylmethanaminium; $Si(CH_3)_2$-Pyronine; $C_{19}H_{25}N_2Si$	4.5
739	TMDHS or SiP-Me	N-(7-(Dimethylamino)-5,5-dimethyldibenzo[b,e]silin-3(5H)-ylidene)-N-methylmethanaminium; $Si(CH_3)_2$-Pyronine; $C_{19}H_{25}N_2Si$	4.5
740	tMe-PR	(R)-N-(7-(Diethylamino)-10-mesityl-5-methyl-5-oxido-3H-acridophosphin-3-ylidene)-N-ethylethanaminium; POR-Rosamine; $C_{31}H_{39}N_2OP$	4.18.2
741	trans-PO-PB430	3-(5,10-Dioxido-5,10-bis(4-(3-sulfonatopropoxy)phenyl)phosphindolo[3,2-b]phosphindol-2-yl)benzoate; Others (Phosphole); $C_{39}H_{31}O_{12}P_2S_2$	4.39.3
742	trans-PO-PB460	3-(5,10-Dioxido-5,10-bis(4-(3-sulfonatopropoxy)phenyl)phosphindolo[3,2-b]phosphindol-2-yl)benzoate; Others (Phosphole); $C_{39}H_{31}O_{12}P_2S_2$	4.39.3
743	Tri-*O*-acetyl-P-luminoarosin	4-(((2R,3S,4S,5R)-3,4-Diethoxy-5-(ethoxymethyl)tetrahydrofuran-2-yl)amino)pyrido[2,1-h]pteridin-11-ium-6-olate; Pteridine; $C_{21}H_{27}N_5O_5$	4.36
744	Vita Blue	2-(11-Oxido-3-oxo-3H-dibenzo[c,h]xanthen-7-yl)benzenesulfonate; Sulfofluorescein-analog; $C_{27}H_{14}O_6S$	4.10
745	W-B 15	2,2′-((2-(3-(Piperazin-1-ium-1-ylidene)-6-(piperazin-1-yl)-3H-xanthen-9-yl)-1,3-phenylene)bis(oxy))diacetate; Rhodamine ($550 < \lambda a, max < 600$ nm); $C_{31}H_{31}N_4O_7$	4.2.3
746	W-L 8a	3-(4-((1E,3Z)-1-(1-(2-((1,3-bis(2,3-Dihydroxypropoxy)propan-2-yl)oxy)ethyl)-3,3-dimethyl-5-sulfonato-3H-indol-1-ium-2-yl)-5-((E)-1-(2-((1,3-bis(2,3-dihydroxypropoxy)propan-2-yl)oxy)ethyl)-3,3-dimethyl-5-sulfonatoindolin-2-ylidene)penta-1,3-dien-3-yl)phenyl)propanoate; Cyanine (Pentamethine); $C_{56}H_{76}N_2O_{22}S_2$	4.13.2
747	W-L 8b	3-(4-((1E,3Z)-1-(1-(4-((1,3-bis(2,3-Dihydroxypropoxy)propan-2-yl)oxy)butyl)-3,3-dimethyl-5-sulfonato-3H-indol-1-ium-2-yl)-5-((E)-1-(4-((1,3-bis(2,3-dihydroxypropoxy)propan-2-yl)oxy)butyl)-3,3-dimethyl-5-sulfonatoindolin-2-ylidene)penta-1,3-dien-3-yl)phenyl)propanoate; Cyanine (Pentamethine); $C_{60}H_{84}N_2O_{22}S_2$	4.13.2

(Continued)

Ent.	Short Name	Name; Class; Molecular Formula; CAS Number	Ref.
748	W-L 8c	3-(4-((1E,3Z)-1-(1-(2-(2-((1,3-Bis(2,3-dihydroxypropoxy)propan-2-yl)oxy)ethoxy)ethyl)-3,3-dimethyl-5-sulfonato-3H-indol-1-ium-2-yl)-5-((E)-1-(2-(2-((1,3-bis(2,3-dihydroxypropoxy)propan-2-yl)oxy)ethoxy)ethyl)-3,3-dimethyl-5-sulfonatoindolin-2-ylidene)penta-1,3-dien-3-yl)phenyl)propanoate; Cyanine (Pentamethine); $C_{60}H_{84}N_2O_{24}S_2$	4.13.2
749	W-L 8d	3-(4-((1E,3Z)-1-(3,3-Dimethyl-5-sulfonato-1-(4-sulfonatobutyl)-3H-indol-1-ium-2-yl)-5-((E)-3,3-dimethyl-5-sulfonato-1-(4-sulfonatobutyl)indolin-2-ylidene)penta-1,3-dien-3-yl)phenyl)propanoate; Cyanine (Pentamethine); $C_{42}H_{46}N_2O_{14}S_4$	4.13.2
750	W-L 8e	tl; Cyanine (Pentamethine); tl	4.13.2
751	Y-A 1 (Cypate)	3-(2-((1E,3E,5E,7E)-7-(3-(2-Carboxylatoethyl)-1,1-dimethyl-1,3-dihydro-2H-benzo[e]indol-2-ylidene)hepta-1,3,5-trien-1-yl)-1,1-dimethyl-1H-benzo[e]indol-3-ium-3-yl)propanoate; Cyanine (Heptamethine); $C_{41}H_{39}N_2O_4$	4.13.3
752	Y-A 10	tl; Cyanine (Heptamethine); $C_{87}H_{115}N_{12}O_{40}$	4.13.3
753	Y-A 11	tl; Cyanine (Heptamethine); $C_{105}H_{143}N_{16}O_{54}$	4.13.3
754	Y-A 2	2,2′-((3-(2-((1E,3E,5E,7E)-7-(3-(2-Carboxylatoethyl)-1,1-dimethyl-1,3-dihydro-2H-benzo[e]indol-2-ylidene)hepta-1,3,5-trien-1-yl)-1,1-dimethyl-1H-benzo[e]indol-3-ium-3-yl)propanoyl)azanediyl)diacetate; Cyanine (Heptamethine); $C_{45}H_{43}N_3O_7$	4.13.3
755	Y-A 3	2,2′-((3-(2-((1E,3E,5E,7E)-7-(3-(3-(Bis(carboxylatomethyl)amino)-3-oxopropyl)-1,1-dimethyl-1,3-dihydro-2H-benzo[e]indol-2-ylidene)hepta-1,3,5-trien-1-yl)-1,1-dimethyl-1H-benzo[e]indol-3-ium-3-yl)propanoyl)azanediyl)diacetate; Cyanine (Heptamethine); $C_{49}H_{47}N_4O_{10}$	4.13.3
756	Y-A 4	2,2′,2′′,2′′′-((2,2′-((3-(2-((1E,3E,5E,7E)-7-(3-(3-(Bis(carboxylatomethyl)amino)-3-oxopropyl)-1,1-dimethyl-1,3-dihydro-2H-benzo[e]indol-2-ylidene)hepta-1,3,5-trien-1-yl)-1,1-dimethyl-1H-benzo[e]indol-3-ium-3-yl)propanoyl)azanediyl)bis(acetyl))bis(azanetriyl))tetraacetate; Cyanine (Heptamethine); $C_{57}H_{55}N_6O_{16}$	4.13.3
757	Y-A 5	2,2′,2′′,2′′′-((2,2′-((3-(2-((1E,3E,5E,7E)-7-(3-(3-Bis(2-(bis(carboxylatomethyl)amino)-2-oxoethyl)amino)-3-oxopropyl)-1,1-dimethyl-1,3-dihydro-2H-benzo[e]indol-2-ylidene)hepta-1,3,5-trien-1-yl)-1,1-dimethyl-1H-benzo[e]indol-3-ium-3-yl)propanoyl)azanediyl)bis(acetyl))bis(azanetriyl))tetraacetate; Cyanine (Heptamethine); $C_{65}H_{63}N_8O_{22}$	4.13.3
758	Y-A 6	3-((E)-2-((2E,4E,6E)-7-(1,1-Dimethyl-3-(3-oxo-3-(((2R,3R,4R,5S,6R)-2,4,5-trihydroxy-6-(hydroxymethyl)tetrahydro-2H-pyran-3-yl)amino)propyl)-1H-benzo[e]indol-3-ium-2-yl)hepta-2,4,6-trien-1-ylidene)-1,1-dimethyl-1,2-dihydro-3H-benzo[e]indol-3-yl)propanoate; Cyanine (Heptamethine); $C_{47}H_{51}N_3O_8$	4.13.3
759	Y-A 7	2-((1E,3E,5E,7E)-7-(1,1-Dimethyl-3-(3-oxo-3-(((2R,3R,4R,5S,6R)-2,4,5-trihydroxy-6-(hydroxymethyl)tetrahydro-2H-pyran-3-yl)amino)propyl)-1,3-dihydro-2H-benzo[e]indol-2-ylidene)hepta-1,3,5-trien-1-yl)-1,1-dimethyl-3-(3-oxo-3-(((2R,3R,4R,5S,6R)-2,4,5-trihydroxy-6-(hydroxymethyl)tetrahydro-2H-pyran-3-yl)amino)propyl)-1H-benzo[e]indol-3-ium; Cyanine (Heptamethine); $C_{53}H_{63}N_4O_{12}$	4.13.3

Ent.	Short Name	Name; Class; Molecular Formula; CAS Number	Ref.
760	Y-A 8	2-((1E,3E,5E,7E)-7-(1,1-Dimethyl-3-(3-oxo-3-(((2R,3R,4R,5S,6R)-2,4,5-trihydroxy-6-(hydroxymethyl)tetrahydro-2H-pyran-3-yl)amino)propyl)-1,3-dihydro-2H-benzo[e]indol-2-ylidene)hepta-1,3,5-trien-1-yl)-3-(3-((1,3-dioxo-1,3-bis(((2R,3R,5S,6R)-2,4,5-trihydroxy-6-(hydroxymethyl)tetrahydro-2H-pyran-3-yl)amino)propan-2-yl)amino)-3-oxopropyl)-1,1-dimethyl-1H-benzo[e]indol-3-ium; Cyanine (Heptamethine); $C_{62}H_{77}N_6O_{19}$	4.13.3
761	Y-A 9	3-(3-((1,3-Dioxo-1,3-bis(((2R,3R,5S,6R)-2,4,5-trihydroxy-6-(hydroxymethyl)tetrahydro-2H-pyran-3-yl)amino)propan-2-yl)amino)-3-oxopropyl)-2-((1E,3E,5E,7E)-7-(3-(3-((1,3-dioxo-1,3-bis(((2R,3R,5S,6R)-2,4,5-trihydroxy-6-(hydroxymethyl)tetrahydro-2H-pyran-3-yl)amino)propan-2-yl)amino)-3-oxopropyl)-1,1-dimethyl-1,3-dihydro-2H-benzo[e]indol-2-ylidene)hepta-1,3,5-trien-1-yl)-1,1-dimethyl-1H-benzo[e]indol-3-ium; Cyanine (Heptamethine); $C_{71}H_{91}N_8O_{26}$	4.13.3
762	Yakima Yellow carboxylate	3-(2′,5,5′,6-Tetrachloro-4′-methyl-3-oxo-3′,6′-bis(pivaloyloxy)-3H-spiro[isobenzofuran-1,9′-xanthen]-7′-yl)propanoate; Fluorescein (Halogenated); $C_{24}H_{13}Cl_4O_7$	4.8.2
763	Z-L A	4,4-Difluoro-8-[3,4-bis(3-{2-[2-(2-methoxyethoxy)ethoxy]ethoxy}-2-{2-[2-(2-methoxyethoxy)ethoxy]ethoxymethyl}propoxy)phenyl]-1,3,5,7-tetramethyl-4-bora3a,4adiaza-s-indacene; Borondipyrromethene (Large); $C_{55}H_{92}BF_2N_2O_{18}$	4.14.3
764	Z-L B	4,4-Bis(3-{2-[2-(2-methoxyethoxy)ethoxy]ethoxy}propynyl)-8-[3,4-bis(3-{2-[2-(2-methoxyethoxy)ethoxy]ethoxy}-2-{2-[2-(2-methoxyethoxy)ethoxy]ethoxymethyl}-propoxy)-phenyl]-1,3,5,7-tetramethyl-4-bora-3a,4a-diaza-s-indacene; Borondipyrromethene (Large); $C_{75}H_{126}BN_2O_{26}$	4.14.3
765	Z-L C	4,4-Difluoro-8-[3,4-bis(3-{2-[2-(2-methoxyethoxy)ethoxy]ethoxy}-2-{2-[2-(2-methoxyethoxy)ethoxy]ethoxymethyl}propoxy)phenyl]-1,3,5,7-tetramethyl-2,6-diiodo-4-bora-3a,4a-diaza-s-indacene; Borondipyrromethene (Large); $C_{55}H_{90}BF_2I_2N_2O_{18}$	4.14.3
766	Z-L D	10-(3,4-bis((13-(2,5,8,11-tetraoxadodecyl)-2,5,8,11-tetraoxatetradecan-14-yl)oxy)phenyl)-2,8-diethynyl-5,5-difluoro-1,3,7,9-tetramethyl-3H,5H-5l4-dipyrrolo[1,2-c:2′,1′-f][1,3,2]diazaborinine; Borondipyrromethene (Large); $C_{59}H_{92}BF_2N_2O_{18}$	4.14.3
767	Z-L E	4,4-Dimethoxy-8-[3,4-bis(3-{2-[2-(2-methoxyethoxy)ethoxy]ethoxy}-2-{2-[2-(2-methoxyethoxy)ethoxy]ethoxymethyl)propoxy)phenyl]-1,3,5,7-tetramethyl-2,6-diethynyl-4-bora-3a,4a-diaza-s-indacene; Borondipyrromethene (Large); $C_{61}H_{98}BN_2O_{20}$	4.14.3
768	Z-L F	4,4-Difluoro-8-[2,4-bis-(3-{2-[2-(2-methoxyethoxy)ethoxy]ethoxy}-2-{2-[2-(2-methoxy-ethoxy)ethoxy]ethoxymethyl}propoxy)phenyl]-1,3,5,7-tetramethyl-4-bora3a,4a-diaza-s-indacene; Borondipyrromethene (Large); $C_{55}H_{92}BF_2N_2O_{18}$	4.14.3
769	Z-L G	4,4-Difluoro-8-[2,4-bis-(3-{2-[2-(2-methoxyethoxy)ethoxy]ethoxy}-2-{2-[2-(2-methoxy-ethoxy)ethoxy]ethoxymethyl}propoxy)phenyl]-1,5,7-trimethyl-3-[(1E)-2-(2,4-bis-{2-[2-(2-methoxyethoxy)ethoxy]ethoxy}phenyl)ethenyl]-4-bora-3a,4a-diaza-sindacene; Borondipyrromethene (Large); $C_{76}H_{124}BF_2N_2O_{26}$	4.14.3

(Continued)

Ent.	Short Name	Name; Class; Molecular Formula; CAS Number	Ref.
770	Z-L H	4,4-Difluoro-8-[2,4-bis-(3-{2-[2-(2-methoxyethoxy)ethoxy]ethoxy}-2-{2-[2-(2-methoxy-ethoxy)ethoxy]ethoxymethyl}propoxy)phenyl]-1,7-dimethyl-3,5-bis-[(1E)-2-(2,4-bis-{2-[2-(2-methoxyethoxy)ethoxy]ethoxy}phenyl) ethenyl]-4-bora-3a,4a-diaza-sindacene-4-bora-3a,4a-diaza-s-indacene; Borondipyrromethene (Large); tl	4.14.3
771	Z-L I	4,4-Bis[1-(3-{2-[2-(2-methoxyethoxy)ethoxy]ethoxy}-2-{2-[2-(2-methoxyethoxy)-ethox] ethoxymethyl}propoxy)-3,6,9,12-tetraoxapentadec-14-ynyl]-1,3,5,7,8-pentamethyl-4-bora-3a,4a-diaza-s-indacene; Borondipyrromethene (Large); tl	4.14.3
772	Z-L J	4,4-Difluoro-8-(4-methoxyphenyl)-1,3,5,7-tetramethyl-2,6-di[1-(3-{2-[2-(2-methoxyetho-xy) ethoxy]ethoxy}-2-{2-[2-(2-methoxyethoxy)ethoxy]ethoxymethyl}propoxy)-3,6,9,12-tetraoxapentadec-14-ynyl]-4- bora-3a,4a-diaza-s-indacene; Borondipyrromethene (Large); tl	4.14.3
773	ZW800-1	3-(4-(((E)-2-((E)-2-(3,3-dimethyl-5-sulfonato-1-(3-(trimethylammonio)propyl)-3H-indol-1-ium-2-yl)vinyl)-6-(2-((E)-3,3-dimethyl-5-sulfonato-1-(3-(trimethylammonio)propyl) indolin-2-ylidene)ethylidene)cyclohex-1-en-1-yl)oxy)phenyl)propanoate; Cyanine (Heptamethine); $C_{51}H_{66}N_4O_9S_2$	4.13.3
774	ZW800-3a	3-(4-(((E)-2-((E)-2-(3,3-dimethyl-1-(3-(trimethylammonio)propyl)-3H-indol-1-ium-2-yl) vinyl)-6-(2-((E)-3,3-dimethyl-1-(3-(trimethylammonio)propyl)indolin-2-ylidene)ethylidene) cyclohex-1-en-1-yl)oxy)phenyl)propanoate; Cyanine (Heptamethine); $C_{51}H_{68}N_4O_3$	4.13.3
775	Zy5	6-(2-((1E,3E)-5-((E)-1-Ethyl-3,3-dimethyl-5-sulfonatoindolin-2-ylidene)penta-1,3-dien-1-yl)-3,3-dimethyl-5-sulfonato-3H-indol-1-ium-1-yl)hexanoate; Cyanine (Pentamethine); $C_{33}H_{38}N_2O_8S_2$; 146368-14-1	4.13.2
776	Zy5.5	6-(2-((1E,3E,5E)-5-(3-Ethyl-1,1-dimethyl-6,8-disulfonato-1,3,6,7-tetrahydro-2H-benzo[e] indol-2-ylidene)penta-1,3-dien-1-yl)-1,1-dimethyl-6,8-disulfonato-1H-benzo[e]indol-3-ium-3-yl)hexanoate; Cyanine (Pentamethine); $C_{41}H_{42}N_2O_{14}S_4$	4.13.2
777	Zy7 carboxylate	6-((E)-2-((2E,4E,6E)-7-(1-ethyl-3,3-dimethyl-5-sulfonato-3H-indol-1-ium-2-yl)hepta-2,4,6-trien-1-ylidene)-3,3-dimethyl-5-sulfonatoindolin-1-yl)hexanoate; Cyanine (Heptamethine); $C_{35}H_{40}N_2O_8S_2$	4.13.3

Abbreviations: AC, acetic acid; BODIPY, borondipyrromethene; CA, carboxylic acid; nd, non-disclosed; pb, probably; SA, sulfonic acid; and tl, too large. The column "Ref." indicates the sub-section where the bibliographical references can be found together with the molecular structures and some photophysical data.

Appendix B

B1. Ranked by Excitation Maximum

Appendix B contains the database of over 700 fluorescent dyes with medium to high brightness in aqueous solutions, ranked by ascending values of the excitation maximum (ranked by $\lambda_{x,max}$). The dyes are identified by their short names. Note that many short names are registered names and trademarks (see footnote[1]). The exact identity of each dye is given in Appendix A. Only the longest wavelength excitation band of each dye is considered. The spectra of the absorption bands were used when the spectra of the excitation bands were not available. The differences between the absorption bands and excitation bands (at the longest wavelength band) are assumed to be small. The values of excitation maximum ($\lambda_{x,max}$) (or absorption maximum when $\lambda_{x,max}$ was not available) for most dyes were taken from the literature, while some were measured by the author. The parameters highlighted in gray are from actual spectra measured by the author or taken from the literature (column "Ref." shows the subsections where the references can be found). However, the "excitation spectrum," and even the "absorption spectrum," of some dyes is currently not available. In these cases (cells without color fill), the approximated excitation ranges were predicted using the equations and parameters (for each class) detailed in Chapter 6.

There are more precise ways to predict the "wavelength at height α of the left side of the longest wavelength excitation band" ($\lambda_{x,-\alpha}$) or the "wavelength at height α of the right side of the longest wavelength excitation band" ($\lambda_{x,+\alpha}$). However, this would excessively increase the calculations. In addition, such a refined calculation is not justified here since the changes in the photophysical properties caused by the chemical micro-environment and other factors are, in many cases, much larger than the imprecisions of the calculations used here. For example, the values of $\lambda_{x,-0.25}$, $\lambda_{x,-0.5}$, $\lambda_{x,max}$, and $\lambda_{x,+0.5}$ may experience small differences if they are measured in plain water or biological fluids containing some proteins. Moreover, if the dyes are bound to proteins they may change even further. Finally, some authors took their measurements in aqueous solutions containing 1% (v/v) DMSO or other water-mixable organic solvents (Section 3.2.2). These solvents usually come from the stock solutions of the labels and stains. These small amounts of remaining organic modifiers may also cause some changes to some photophysical parameters.

Acronyms used in Table B.1: A, acetate; AC, acetic acid; ACN, acetonitrile; BODIPY, borondipyrromethene; BSA, bovine serum albumin; C, carbonate; CA, carboxylic acid; FBS, fetal bovine serum; Et, ethanol; nd, non-disclosed; ni, not informed; P, phosphate; pb, probably; pH >7, basic medium; pH < 7, acidic medium (when the exact value of the pH is not given in the literature); SA, sulfonic acid; and TX-100, triton X-100.

1 Abberior STAR is a trademark of Abberior GmbH (Göttingen, Germany). AlexaFluor, BODIPY, Cascade Blue, FAM, JOE, TET, HEX, Marina Blue, Oregon Green, Pacific Blue, ROX, TAMRA, and Texas Red are trademarks of Thermo Fisher Scientific (Waltham, MA, USA). ATTO is a trademark of ATTO-Tec GmbH (Siegen, Germany). CF Dye is a registered trademark of Biotium, Inc. (Fremont, CA, USA). Cy3, Cy3.5, Cy5, and Cy5.5 are trademarks of GE Healthcare (Chicago, IL, USA). Dy is a trademark of Dyomics GmbH (Jena, Germany). Seta, SeTau, and Square dyes are trademarks of SETA BioMedicals (Urbana, IL, USA).

Fluorescent Dye Labels and Stains: A Database of Photophysical Properties, First Edition. Tarso B. Ledur Kist.

Table B.1 Photophysical properties of fluorescent labels and stains in aqueous solutions and ranked by the wavelength of excitation maximum of the longest wavelength band ($\lambda_{x,max}$).

Ent.	Short Names	Aqueous Solutions	$\lambda_{x,-0.25}$ nm	$\lambda_{x,-0.50}$ nm	$\lambda_{x,max}$ nm	$\lambda_{x,+0.50}$ nm	$\lambda_{e,-0.25}$ nm	$\lambda_{e,-0.50}$ nm	$\lambda_{e,max}$ nm	$\lambda_{e,+0.50}$ nm	$\lambda_{e,+0.25}$ nm	$\lambda_{e,+0.10}$ nm	Δ_S nm	ε_{max} $mM^{-1}cm^{-1}$	Φ_f	B_{max} $mM^{-1}cm^{-1}$	Class	Ref.
1	4-Methylsalicylic acid	H_2O pH 7	252	272	300	317	370	379	407	446	467	491	107	3.0	0.11	0.3	Benzene	4.38
2	Salicylic acid	H_2O pH 7	272	281	300	316	370	377	407	441	461	482	107	3.0	0.16	0.4	Benzene	4.38
3	2-Aminopyridine	H_2O			300				363				63	4.1	0.71	2.9	Pyridine	4.39.1
4	3-Aminopyridine	H_2O			300				390				90	2.8	0.32	0.9	Pyridine	4.39.1
5	5-Methylsalicylic acid	H_2O pH 7	268	281	305	323	379	388	418	456	477	497	113	3.0	0.16	0.4	Benzene	4.38
6	3-(4-OH-Ph)-6-MeO-chromone	100 mM Tris pH 8			311				468				157	12.6	0.05	0.6	Chromone	4.31
7	Stefani HB-6	H_2O pH 6			317				365				48	35.9	0.03	0.9	Hybrid-Bonded	4.40.2
8	5-Methoxysalicylic acid	H_2O pH 7	285	295	319	339	403	414	447	489	511	526	128	3.0	0.19	0.5	Benzene	4.38
9	7-Hydroxy-4-methylcoumarin	H_2O pH 5	276	295	320	341	415	424	450	483	501	523	130	14.0	0.78	10.9	Coumarin 7-OH	4.26.1
10	Stefani HB-9	H_2O pH 6			320				423				103	35.9	0.12	4.2	Hybrid-Bonded	4.40.2
11	Harmine	50 mM H_2SO_4	304	309	326	343	383	389	412	445	464	483	86	11.9	0.45	5.3	Indole	4.33
12	Coumarin P-R 12	PBS pH 7.5	282	301	327	348	421	430	457	491	510	533	130	7.9	0.76	5.9	Coumarin 7-OH	4.26.1
13	(HO)-Luciferin	H_2O pH 5			328				537				209	18.2	0.25	4.5	Benzothiazole	4.30
14	Stefani HB-11	H_2O pH 6			329				360				31	38.1	0.11	4.0	Hybrid-Bonded	4.40.2
15	Stefani HB-7	H_2O			331				449				118	42.0	0.09	3.6	Benzothiazole	4.30
16	Quinine	H_2O pH 7			331				382				51	5.7	0.50	2.8	Quinoline	4.29
17	Acedan 8	H_2O	278	296	332	365	468	482	515	551	568	595	183	12.6	0.23	2.9	Naphthalene	4.34
18	Harmine	H_2O pH 9	309	315	332	350	393	400	425	458	477	498	93	11.0	0.38	4.1	Indole	4.33
19	3-Buthyl-7-hydroxychromone	100 mM Tris pH 8			334				454				120	12.7	0.21	2.6	Chromone	4.31
20	(HO)-Luciferyl adenylate	H_2O pH 5			335				548				213	15.1	0.11	1.6	Benzothiazole	4.30
21	Edans	H_2O	281	300	336	370	472	487	520	557	574	601	184	6.2	0.27	1.6	Naphthalene	4.34
22	Acedan 2	H_2O	282	301	338	372	468	482	515	551	568	595	177	14.6	0.30	4.3	Naphthalene	4.34
23	Carbostyril 9	H_2O			340				382				42	6.0	0.17	0.9	Quinoline	4.29
24	7-Am-4-Me-2-hydroxyquinoline	H_2O pH 5-9			340				440				100	16.0	0.97	15.5	Quinoline	4.29

Ent.	Short Names	Aqueous Solutions	$\lambda_{x,-0.25}$ nm	$\lambda_{x,-0.50}$ nm	$\lambda_{x,max}$ nm	$\lambda_{x,+0.50}$ nm	$\lambda_{e,-0.25}$ nm	$\lambda_{e,-0.50}$ nm	$\lambda_{e,max}$ nm	$\lambda_{e,+0.50}$ nm	$\lambda_{e,+0.25}$ nm	$\lambda_{e,+0.10}$ nm	Δ_S nm	ε_{max} mM^{-1}cm^{-1}	Φ_f	B_{max} mM^{-1}cm^{-1}	Class	Ref.
25	Carbostyril 124	H_2O			341				417				76	15.8	0.46	7.2	Quinoline	4.29
26	Coumarin 120	H_2O	293	314	342	366	407	415	440	471	489	510	98	16.0	0.94	15.0	Coumarin 7-Am	4.26.2
27	Coumarin P-R 21	PBS pH 7.5	295	316	345	369	425	434	461	495	515	538	116	13.2	0.79	10.4	Coumarin 7-Am	4.26.2
28	3-Bt-6-MeO-7-hydroxychromone	100 mM Tris pH 8			345				447				102	16.4	0.48	7.8	Chromone	4.31
29	Fluorene PP12	H_2O			345				555				210	24.0	0.16	3.8	Fluorene	4.39.2
30	Alexa Fluor 350	H_2O pH 7	296	317	346	370	411	420	445	477	495	516	99	19.0	0.89	16.9	Coumarin 7-Am	4.26.3
31	Coumarin P-R 5	PBS pH 7.5	296	317	346	370	428	437	465	500	520	544	119	10.3	0.79	8.1	Coumarin 7-Am	4.26.2
32	3-AB-6-MeO-7-OH-chromone	100 mM Tris pH 8			346				445				99	16.3	0.41	6.6	Chromone	4.31
33	Quinine	100 mM $HClO_4$			347				451				104	5.7	0.60	3.4	Quinoline	4.29
34	Quinine	50 mM H_2SO_4			347				451				104	5.7	0.55	3.1	Quinoline	4.29
35	Coumarin P-R 3	PBS pH 7.5	297	319	348	372	415	424	450	483	501	523	102	12.5	0.82	10.2	Coumarin 7-Am	4.26.2
36	(HO)-Dehydroluciferin	H_2O pH 4.5			348				555				207	18.2	0.26	4.7	Benzothiazole	4.30
37	Harmane	50 mM H_2SO_4	324	331	350	370	405	412	440	474	494	517	90	6.3	0.83	5.1	Indole	4.33
38	2-Methyl harmine	50 mM H_2SO_4	324	331	350	370	399	406	433	466	485	509	83	6.3	0.45	2.8	Indole	4.33
39	Acedan 7	H_2O	291	311	351	388	452	465	495	528	544	568	144	11.0	0.31	3.4	Naphthalene	4.34
40	DY-350 carboxylate	PBS pH 7.4	301	323	353	378	400	408	432	462	479	499	79	19.0	0.84	15.9	Coumarin 7-Am	4.26.3
41	Coumarin P-R 22	PBS pH 7.5	307	324	353	376	417	425	450	484	504	527	97	13.4	0.75	10.0	Coumarin 7-Am	4.26.2
42	Carbostyril 5	H_2O			353				440				87	6.6	0.16	1.0	Quinoline	4.29
43	Fluorene I-6	H_2O			353				473				120	30.5	0.13	3.9	Fluorene	4.39.2
44	7-Azetidinyl-4-Me-coumarin	10 mM HEPES pH 7.3	308	325	354	377	432	441	467	504	525	551	113	15.0	0.96	14.4	Coumarin 7-Am	4.26.2

(Continued)

Ent.	Short Names	Aqueous Solutions	$\lambda_{x,-0.25}$ nm	$\lambda_{x,-0.50}$ nm	$\lambda_{x,max}$ nm	$\lambda_{x,+0.50}$ nm	$\lambda_{e,-0.25}$ nm	$\lambda_{e,-0.50}$ nm	$\lambda_{e,max}$ nm	$\lambda_{e,+0.50}$ nm	$\lambda_{e,+0.25}$ nm	$\lambda_{e,+0.10}$ nm	Δ_S nm	ε_{max} mм$^{-1}$cm^{-1}	Φ_f	B_{max} mм$^{-1}$cm^{-1}	Class	Ref.
45	Acridizinium J-N 1c	H_2O			354				420				66	8.4	0.30	2.5	Acridizinium	4.39.1
46	Acedan 4	H_2O	294	315	355	393	457	470	501	535	551	576	146	13.5	0.32	4.3	Naphthalene	4.34
47	Coumarin P-R 6	PBS pH 7.5	309	326	355	378	419	427	452	486	507	530	97	12.6	0.71	8.9	Coumarin 7-Am	4.26.2
48	Acridizinium J-N 1e	H_2O			355				455				100	6.7	0.20	1.3	Acridizinium	4.39.1
49	8-Aminonaphthalene-1,3,6-triSA	100 mм P pH 7	295	315	356	394	466	479	512	547	564	591	156	15.0	0.20	3.0	Naphthalene	4.34
50	Acridizinium J-N 1d	H_2O			356				408				52	5.4	0.20	1.0	Acridizinium	4.39.1
51	Acedan 3	H_2O	295	316	357	396	457	470	501	535	551	576	144	14.5	0.31	4.5	Naphthalene	4.34
52	Acedan 5	H_2O	295	316	357	396	458	472	503	537	553	579	146	17.4	0.40	6.9	Naphthalene	4.34
53	Acedan 6	H_2O	295	316	357	396	458	472	503	537	553	579	146	15.4	0.39	6.0	Naphthalene	4.34
54	Harmine	H_2O pH 4	330	337	357	378	387	393	416	449	465	484	59	7.3	0.49	3.6	Indole	4.33
55	Marina Blue	100 mм P pH 10/8	311	328	358	381	421	430	455	490	510	534	97	18.7	0.89	16.6	Coumarin 7-OH	4.26.1
56	Carbostyril 1	H_2O			358				432				74	11.6	0.24	2.7	Quinoline	4.29
57	Sulfobetaine pyrene 21	H_2O	314	327	359	365	371	375	385	412	428	441	26	30.6	0.32	9.7	Pyrene	4.28
58	7-Hydroxy-4-methylcoumarin	H_2O pH 10	313	330	360	384	416	424	449	483	503	526	89	18.0	0.86	15.4	Coumarin 7-OH	4.26.1
59	7-Hydroxy-4-Me-coumarin-3-CA	H_2O pH 5	313	330	360	384	421	430	455	490	510	534	95	18.0	0.82	14.7	Coumarin 7-OH	4.26.1
60	nor-Harmane (β-carboline)	50 mм H_2SO_4	333	340	360	381	418	425	454	492	514	538	94	6.3	0.58	3.6	Indole	4.33
61	Carbostyril 2	H_2O			360				430				70	6.7	0.14	0.9	Quinoline	4.29
62	8-Chloro-harmine	H_2O pH 4	334	341	361	382	392	398	423	456	474	495	62	5.9	0.11	0.6	Indole	4.33
63	Acedan	H_2O	299	320	362	402	468	482	515	551	568	595	153	11.2	0.20	2.2	Naphthalene	4.34
64	Coumarin 2	H_2O	314	332	362	386	422	431	456	491	512	535	94	16.9	0.95	16.0	Coumarin 7-Am	4.26.2
65	6-Chloro-harmine	H_2O pH 4	335	341	362	383	396	402	428	462	481	502	66	9.0	0.16	1.4	Indole	4.33
66	Prodan	H_2O	300	322	364	404	481	496	531	569	587	616	167	14.5	0.10	1.4	Naphthalene	4.34
67	3-Amino-9-ethylcarbazole 1	H_2O			364				452				88	1.8	0.63	1.1	Carbazole	4.39.1
68	3-Amino-9-ethylcarbazole 2	H_2O			364				458				94	1.8	0.76	1.3	Carbazole	4.39.1

Ent.	Short Names	Aqueous Solutions	$\lambda_{x,-0.25}$ nm	$\lambda_{x,-0.50}$ nm	$\lambda_{x,max}$ nm	$\lambda_{x,+0.50}$ nm	$\lambda_{e,-0.25}$ nm	$\lambda_{e,-0.50}$ nm	$\lambda_{e,max}$ nm	$\lambda_{e,+0.50}$ nm	$\lambda_{e,+0.25}$ nm	$\lambda_{e,+0.10}$ nm	Δ_S nm	ε_{max} mм$^{-1}$cm^{-1}	Φ_f	B_{max} mм$^{-1}$cm^{-1}	Class	Ref.
69	Coumarin 152	H_2O	316	334	365	389	457	467	497	539	564	593	132	14.3	0.37	5.2	Coumarin 7-Am	4.26.2
70	BC-H	100 mм P, 20% Et pH 7.4			365				470				105	27.2	0.09	2.4	Hybrid-Bonded	4.40.2
71	Carbostyril 7	H_2O			366				580				214	9.0	0.07	0.6	Quinoline	4.29
72	Acyman	H_2O	302	324	367	408	470	484	517	553	570	597	150	9.1	0.10	0.9	Naphthalene	4.34
73	BOPAHY H-D 19ba′	H_2O	344	355	367	372	379	381	386	393	403	412	19	51	0.03	1.5	Bodipy-analog	4.14.5
74	7-(4-Hydroxy) cyclohexyl coum.	H_2O	319	337	368	393	426	434	460	496	517	541	92	16.5	0.59	9.7	Coumarin 7-Am	4.26.2
75	7-Ethylamino-coumarin	H_2O	319	337	368	393	424	432	458	493	514	538	90	11.9	0.51	6.0	Coumarin 7-Am	4.26.2
76	Coumarin P-R 13	PBS pH 7.5	319	337	368	393	438	447	474	512	534	560	106	15.0	0.63	9.4	Coumarin 7-OH	4.26.1
77	BOPAHY H-D 19aa′	H_2O	345	355	368	373	380	382	387	394	404	413	19	51	0.02	0.7	Bodipy-analog	4.14.5
78	Lucigenin	H_2O			368				485				117	37.6	0.50	18.8	Hybrid-Bonded	4.40.2
79	6,8-Dichloroharmine (6,8-DiCL-Ha)	H_2O pH 4	341	348	369	391	397	403	429	463	482	503	60	6.3	0.20	1.2	Indole	4.33
80	Acridizinium I-L 5b	H_2O			369				551				182	11.0	0.02	0.2	Acridizinium	4.39.1
81	3-OH-4-nitroso-2, 7-naphth-diSA	A pH 4	304	326	370	412	431	442	470	500	514	535	100	19.0	0.22	4.1	Naphthalene	4.34
82	Benzo[h]quinoline (BQ)	H_2O pH 2-5			370				430				60	35.0	0.32	11.2	Quinoline	4.29
83	BC-BA	100 mм P, 20% Et pH 7.4			371				473				102	25.5	0.17	4.3	Hybrid-Bonded	4.40.2
84	7-Dimethylamino-4-Me-coumarin	10 mм HEPES pH 7.3	322	340	372	397	434	443	470	507	529	555	98	18.0	0.19	3.4	Coumarin 7-Am	4.26.2
85	Seta-375-carboxylate	P pH 7.4			372				476				104	24.5	0.86	21.0	nd	4.41
86	Coumarin P-R 14	PBS pH 7.5	322	341	373	398	426	435	461	497	518	542	88	17.7	0.62	10.9	Coumarin 7-OH	4.26.1
87	Fluorene PP6	H_2O			373				620				247	16.5	0.07	1.1	Fluorene	4.39.2
88	1,3,6,8-Pyrenete-trasulfonic acid	H_2O pH 7	335	350	374	378	391	394	403	430	451	465	29	51.0	0.56	28.5	Pyrene	4.28
89	Pyrene-1-sulfonate	H_2O	327	340	375	382	385	389	400	429	446	461	25	30.6	0.56	17.1	Pyrene	4.28

(Continued)

Ent.	Short Names	Aqueous Solutions	$\lambda_{x,-0.25}$ nm	$\lambda_{x,-0.50}$ nm	$\lambda_{x,max}$ nm	$\lambda_{x,+0.50}$ nm	$\lambda_{e,-0.25}$ nm	$\lambda_{e,-0.50}$ nm	$\lambda_{e,max}$ nm	$\lambda_{e,+0.50}$ nm	$\lambda_{e,+0.25}$ nm	$\lambda_{e,+0.10}$ nm	Δ_S nm	ε_{max} mM^{-1}cm^{-1}	Φ_f	B_{max} mM^{-1}cm^{-1}	Class	Ref.
90	BMeS-*p*-A	H_2O	322	341	377	404	458	472	517	576	610	643	140	3.8	0.67	2.5	Benzene	4.38
91	Fluorene PP11	H_2O			377				621				244	36.0	0.05	1.8	Fluorene	4.39.2
92	Acridizinium J-N 1a	H_2O			377				404				27	8.9	0.33	2.9	Acridizinium	4.39.1
93	Spirobifluorene 1	H_2O			378				480				102	51	0.50	25.6	Spirobifluorene	4.39.3
94	Carbostyril 4	H_2O			379				531				152	8.7	0.07	0.6	Quinoline	4.29
95	Carbostyril 23b	H2O pH 2.8			380				475				95	18.0	0.41	7.3	Quinoline	4.29
96	NBQ	H_2O pH 2-5			380				470				90	35.0	0.35	12.2	Quinoline	4.29
97	Acridizinium J-N 1b	H_2O			381				408				27	6.7	0.34	2.2	Acridizinium	4.39.1
98	Coumarin 1	H_2O	329	349	382	409	436	445	472	510	532	558	90	20.1	0.07	1.4	Coumarin 7-Am	4.26.2
99	DMPT	H_2O pH 7			382				475				93	12.4	0.06	0.7	Pteridine	4.36
100	Harmaline	5 mM H_2SO_4	352	360	383	407	436	444	475	517	541	568	92	28.6	0.32	9.1	Indole	4.33
101	SeTau-380-carboxylate	P pH 7.3			383				468				85	23.8	0.56	13.3	nd	4.41
102	($^-$O)-Luciferin	H_2O pH 11			384				537				153	18.5	0.62	11.4	Benzothiazole	4.30
103	Seta-385-carboxylate	P pH 7.4			384				512				128	25.0	0.78	19.5	nd	4.41
104	NOBQ	H_2O pH 2-5			385				470				85	35.0	0.41	14.3	Quinoline	4.29
105	OBQ	H_2O pH 2-5			385				470				85	35.0	0.43	15.0	Quinoline	4.29
106	7-Hydroxy-4-Me-coumarin-3-CA	H_2O pH 10	332	352	386	413	414	422	446	479	499	522	60	33.0	0.85	28.0	Coumarin 7-OH	4.26.1
107	Acrylodan	H_2O	315	339	387	433	457	471	502	536	552	577	115	12.9	0.18	2.3	Naphthalene	4.34
108	7-Azetidinyl-3-CA coumarin	10 mM HEPES pH 7.3	333	353	387	414	434	443	470	507	529	555	83	24.0	0.84	20.1	Coumarin 7-Am	4.26.2
109	Acridizinium I-L 5a	H_2O			387				507				120	17.4	0.12	2.0	Acridizinium	4.39.1
110	7-Diethylamino-coumarin	H_2O	333	354	388	416	438	448	475	513	536	562	87	18.3	0.05	0.9	Coumarin 7-Am	4.26.2
111	Carbostyril 6	H_2O			389				544				155	7.0	0.09	0.6	Quinoline	4.29
112	Atto 390 carboxylate	PBS pH 7.4, 22 °C	343	357	390	415	441	450	476	516	540	568	86	24.0	0.90	21.6	Coumarin 7-Am	4.26.3

Ent.	Short Names	Aqueous Solutions	$\lambda_{x,-0.25}$ nm	$\lambda_{x,-0.50}$ nm	$\lambda_{x,max}$ nm	$\lambda_{x,+0.50}$ nm	$\lambda_{e,-0.25}$ nm	$\lambda_{e,-0.50}$ nm	$\lambda_{e,max}$ nm	$\lambda_{e,+0.50}$ nm	$\lambda_{e,+0.25}$ nm	$\lambda_{e,+0.10}$ nm	Δ_S nm	ε_{max} mM^{-1}cm^{-1}	Φ_f	B_{max} mM^{-1}cm^{-1}	Class	Ref.
113	HBQ	H_2O pH 2-5			390				480				90	35.0	0.15	5.2	Quinoline	4.29
114	Epicocconone P-F 17a	H_2O, 0.1% SDS pH 2.5			391				517				126	8.8	0.35	3.0	Epicocconone	4.39.2
115	BMeS-*p*-A-COOH	H_2O	333	353	392	421	466	481	527	589	624	659	135	3.9	0.60	2.3	Benzene	4.38
116	Coumarin 307	H_2O	335	355	392	417	471	481	508	550	576	606	116	9.0	0.33	2.9	Coumarin 7-Am	4.26.2
117	Epicocconone P-F 17d	H_2O, 0.1% SDS pH 2.5			392				518				126	13.2	0.40	5.2	Epicocconone	4.39.2
118	Coumarin 102	H_2O	336	356	393	418	455	464	489	528	552	579	96	21.8	0.66	14.3	Coumarin 7-Am	4.26.2
119	$ACSO_3H$	H_2O pH 5.5	353	366	396	404	405	408	417	449	458	472	21	4.4	0.82	3.6	Acridones	4.23.2
120	($^-$O)-Dehydroluciferin	H_2O pH 11			396				555				159	18.2	0.62	11.2	Benzothiazole	4.30
121	Carbostyril 23b	H_2O pH 12			397				556				159	17.3	0.26	4.5	Quinoline	4.29
122	Seta-400-carboxylate	P pH 7.4			399				576				177	24.0	0.34	8.1	nd	4.41
123	DY-405 carboxylate	PBS pH 7.4	336	349	400	410	404	408	423	458	474	492	23	32.0	0.97	30.9	Pyrene	4.28
124	Alexa Fluor 405	PBS pH 7.2, 22 °C	346	360	400	408	408	411	424	457	476	494	24	35.0	0.50	17.5	Pyrene	4.28
125	9-Aminoacridine	P pH 7	366	379	400	410	400	405	420	449	467	487	20	9.3	0.97	9.0	Acridine	4.23.1
126	Si-9-Iminopyronine 2f	~1 mM NaOH, 1% DMSO pH 11	375	389	401	408	583	590	606	629	643	675	205	17.8	0.35	6.2	Si(CH3)2-Pyron	4.5
127	Seta-405-carboxylate	P pH 7.4			401				452				51	30.0	0.70	21.0	nd	4.41
128	BMeS-*p*-A-2COOH	H_2O	341	362	403	434	473	487	535	599	635	671	132	3.7	0.56	2.0	Benzene	4.38
129	Cascade Blue	50 mM P pH 8	348	363	403	411	413	417	430	464	484	502	27	28.0	0.90	25.2	Pyrene	4.28
130	8-Hydroxypyrene-1, 3,6-triSA	H_2O pH 2	348	363	403	411	425	429	443	479	501	519	40	20.0	0.95	19.0	Pyrene	4.28
131	MetACSO3H	H_2O pH 5.5	364	376	403	411	418	423	433	470	487	495	30	5.6	0.76	4.2	Acridones	4.23.2
132	Cascade yellow	H_2O	329	356	404	440	491	505	549	603	657	663	145	24.0	0.05	1.2	Hybrid-Bonded	4.40.2

(Continued)

Ent.	Short Names	Aqueous Solutions	$\lambda_{x,-0.25}$ nm	$\lambda_{x,-0.50}$ nm	$\lambda_{x,max}$ nm	$\lambda_{x,+0.50}$ nm	$\lambda_{e,-0.25}$ nm	$\lambda_{e,-0.50}$ nm	$\lambda_{e,max}$ nm	$\lambda_{e,+0.50}$ nm	$\lambda_{e,+0.25}$ nm	$\lambda_{e,+0.10}$ nm	Δ_S nm	ε_{max} $mM^{-1}cm^{-1}$	Φ_f	B_{max} $mM^{-1}cm^{-1}$	Class	Ref.
133	Pacific orange	C pH 9.5			404				553				149	25.0	0.62	15.5	Benzothiazole	4.30
134	CF405S	H_2O			404				431				27	33.0	0.70	23.0	nd (Pyrene)	4.41
135	SeTau-405-carboxylate	P pH 7.4			405				518				113	13.8	0.80	11.0	nd	4.41
136	Isoquinoline 3f	PBS pH 7.4			405				519				114	15.4	0.39	6.0	Isoquinoline	4.37
137	Isoquinoline 3i	PBS pH 7.4			405				521				116	15.4	0.40	6.1	Isoquinoline	4.37
138	PetACSO3H	H_2O pH 5.5	367	379	406	414	417	421	431	467	480	495	25	6.2	0.92	5.6	Acridones	4.23.2
139	Isoquinoline 3h	PBS pH 7.4			406				525				119	13.7	0.38	5.2	Isoquinoline	4.37
140	Isoquinoline 7b	PBS pH 7.4			406				526				120	14.4	0.46	6.6	Isoquinoline	4.37
141	Isoquinoline 7f	PBS pH 7.4			406				528				122	16.3	0.16	2.6	Isoquinoline	4.37
142	Isoquinoline 8	PBS pH 7.4			406				534				128	16.8	0.37	6.2	Isoquinoline	4.37
143	Isoquinoline 7a	PBS pH 7.4			407				528				121	14.1	0.34	4.7	Isoquinoline	4.37
144	CF405M	H_2O			408				452				44	41.0	0.76	31.1	nd (Coumarin)	4.41
145	Isoquinoline 3e	PBS pH 7.4			408				524				116	15.8	0.41	6.4	Isoquinoline	4.37
146	Isoquinoline 3m	PBS pH 7.4			408				526				118	18.1	0.36	6.5	Isoquinoline	4.37
147	Isoquinoline 7c	PBS pH 7.4			408				524				116	15.8	0.26	4.1	Isoquinoline	4.37
148	Isoquinoline 7d	PBS pH 7.4			408				526				118	15.0	0.14	2.1	Isoquinoline	4.37
149	Isoquinoline 3j	PBS pH 7.4			409				525				116	18.8	0.45	8.4	Isoquinoline	4.37
150	Pacific Blue	PBS pH 7.2, 22 °C	339	366	410	437	428	436	455	486	505	526	45	29.5	0.75	22.1	Coumarin 7-OH	4.26.1
151	Phosphorilated coum. N-H 22c	PBS pH 7.4	348	370	410	437	467	477	503	545	570	599	93	14.0	0.44	6.1	Coumarin 7-Am	4.26.3

Ent.	Short Names	Aqueous Solutions	$\lambda_{x,-0.25}$ nm	$\lambda_{x,-0.50}$ nm	$\lambda_{x,max}$ nm	$\lambda_{x,+0.50}$ nm	$\lambda_{e,-0.25}$ nm	$\lambda_{e,-0.50}$ nm	$\lambda_{e,max}$ nm	$\lambda_{e,+0.50}$ nm	$\lambda_{e,+0.25}$ nm	$\lambda_{e,+0.10}$ nm	Δ_S nm	ε_{max} $mM^{-1}cm^{-1}$	Φ_f	B_{max} $mM^{-1}cm^{-1}$	Class	Ref.
152	7-Diethylamino-3-CA coumarin	10 mM HEPES pH 7.3	357	376	410	432	446	452	471	503	525	548	61	35.0	0.03	1.0	Coumarin 7-Am	4.26.2
153	Isoquinoline 10b	PBS pH 7.4			410				526				116	15.5	0.17	2.6	Isoquinoline	4.37
154	Isoquinoline 3l	PBS pH 7.4			410				522				112	16.8	0.42	7.0	Isoquinoline	4.37
155	Isoquinoline 3o	PBS pH 7.4			410				521				111	19.2	0.09	1.7	Isoquinoline	4.37
156	N-Methylacridone	H_2O	369	382	411	419	417	421	431	467	480	493	20	8.8	0.99	8.7	Acridones	4.23.2
157	Isoquinoline 3k	PBS pH 7.4			411				523				112	18.6	0.11	2.0	Isoquinoline	4.37
158	Isoquinoline 3n	PBS pH 7.4			411				521				110	16.6	0.41	6.8	Isoquinoline	4.37
159	10-Ethylacridine	P pH 7	376	390	412	423	467	474	494	534	560	589	82	3.5	0.98	3.4	Acridine	4.23.1
160	Isoquinoline 3g	PBS pH 7.4			414				522				108	17.5	0.41	7.1	Isoquinoline	4.37
161	BOPAHY H-D 19ab′	H_2O	387	400	416	422	426	429	435	444	457	468	19	51	0.20	10.2	Bodipy-analog	4.14.5
162	BOPAHY H-D 19bb′	H_2O	387	400	416	422	425	428	434	443	455	467	18	51	0.17	8.6	Bodipy-analog	4.14.5
163	BOPAHY H-D 20bb′	H_2O	387	400	416	422	426	429	435	444	457	468	19	51	0.17	8.6	Bodipy-analog	4.14.5
164	Epicocconone P-F 4d	H_2O, 0.1% SDS pH 2.5			417				534				117	14.5	0.37	5.3	Epicocconone	4.39.2
165	Si-9-Iminopyronine 2c	~1 mM NaOH, 1% DMSO pH 11	390	405	418	426	584	591	607	630	645	676	189	12.9	0.46	5.9	Si(CH3)2-Pyron	4.5
166	Isoquinoline 3d	PBS pH 7.4			418				520				102	14.5	0.43	6.2	Isoquinoline	4.37
167	PONy C7	PBS pH 7.4			419				613				194	22.0	0.04	0.8	Hybrid-Bonded	4.40.2
168	Carbostyril 13	H_2O pH 7			420				556				136	17.3	0.26	4.5	Quinoline	4.29
169	cis-PO-PB430	PBS pH 7.4			421				548				127	10.0	0.71	7.1	Phosphole	4.39.3
170	trans-PO-PB430	PBS pH 7.4			422				545				123	10.3	0.74	7.6	Phosphole	4.39.3

(Continued)

Ent.	Short Names	Aqueous Solutions	$\lambda_{x,-0.25}$ nm	$\lambda_{x,-0.50}$ nm	$\lambda_{x,max}$ nm	$\lambda_{x,+0.50}$ nm	$\lambda_{e,-0.25}$ nm	$\lambda_{e,-0.50}$ nm	$\lambda_{e,max}$ nm	$\lambda_{e,+0.50}$ nm	$\lambda_{e,+0.25}$ nm	$\lambda_{e,+0.10}$ nm	Δ_S nm	ε_{max} mM^{-1}cm^{-1}	Φ_f	B_{max} mM^{-1}cm^{-1}	Class	Ref.
171	3-OH-4-nitroso-2, 7-naphth-diSA	T pH 9	339	367	423	478	475	489	523	560	578	605	100	20.0	0.28	5.6	Naphthalene	4.34
172	8-Aminopyrene-1, 3,6-triSA	H_2O	363	380	424	433	482	487	505	552	581	607	81	20.6	0.90	18.5	Pyrene	4.28
173	Carbostyril 23b	H_2O pH 7			424				556				132	17.3	0.26	4.5	Quinoline	4.29
174	Tri-*O*-acetyl-P-luminoarosin	H_2O pH 7			425				528				103	10.8	0.62	6.7	Pteridine	4.36
175	Epicocconone P-F 4c	H_2O, 0.1% SDS pH 2.5			425				542				117	40.3	0.40	16.2	Epicocconone	4.39.2
176	CF430	H_2O			426				498				72	40.0	0.79	31.6	nd (Cyanine)	4.41
177	PB430	PBS pH 7.4			426				542				116	10.4	0.66	6.8	Phosphole	4.39.3
178	BOPAHY H-D 18bb′	H_2O	397	411	428	434	444	447	453	463	476	489	25	51	0.49	24.9	Bodipy-analog	4.14.5
179	Luminarine	H_2O pH 7			428				533				105	9.0	0.47	4.2	Pteridine	4.36
180	Lucifer yellow CH	H_2O pH 7	375	392	429	462	498	509	540	580	600	623	111	11.3	0.21	2.3	Naphthalimide	4.32
181	BOPAHY H-D 18ab′	H_2O	398	412	429	435	444	447	453	463	476	489	24	51	0.55	28.0	Bodipy-analog	4.14.5
182	Briliant sulfaflavine	H_2O	376	393	430	463	481	491	520	557	575	596	90	16.0	0.27	4.3	Naphthalimide	4.32
183	Lucifer yellow VS	H_2O pH 7	376	393	430	463	493	505	535	574	594	616	105	11.3	0.24	2.7	Naphthalimide	4.32
184	Alexa Fluor 430	PBS pH 7.4	373	393	431	456	508	517	541	584	613	645	110	16.0	0.55	8.8	Coumarin 7-Am	4.26.3
185	Phosphorilated coum. N-H 21c	PBS pH 7.4	374	394	432	457	482	490	512	550	576	604	80	20.4	0.81	16.5	Coumarin 7-Am	4.26.3
186	3,8,10-Trimethyli-soalloxazine	H_2O pH 7			433				518				85	11.5	0.12	1.3	Pteridine	4.36
187	3,10-Dimethylis-oalloxazine	H_2O pH 7			435				520				85	9.5	0.15	1.4	Pteridine	4.36
188	Epicocconone P-F 4b	H_2O, 0.1% SDS pH 2.5			435				538				103	19.8	0.20	3.9	Epicocconone	4.39.2
189	Abberior Star 440SXP	PBS pH 7.4	377	398	436	461	485	493	515	554	580	608	79	22.7	0.68	15.4	Coumarin 7-Am	4.26.3
190	Sulfonated coumarin N-H 6-H	PBS pH 7.4	377	398	436	461	485	493	515	554	580	608	79	31.0	0.57	17.6	Coumarin 7-Am	4.26.3

Ent.	Short Names	Aqueous Solutions	$\lambda_{x,-0.25}$ nm	$\lambda_{x,-0.50}$ nm	$\lambda_{x,max}$ nm	$\lambda_{x,+0.50}$ nm	$\lambda_{e,-0.25}$ nm	$\lambda_{e,-0.50}$ nm	$\lambda_{e,max}$ nm	$\lambda_{e,+0.50}$ nm	$\lambda_{e,+0.25}$ nm	$\lambda_{e,+0.10}$ nm	Δ_S nm	ε_{max} mM^{-1}cm^{-1}	Φ_f	B_{max} mM^{-1}cm^{-1}	Class	Ref.
191	Naphthalimide B-K 1	10 mM P pH 7.0	380	398	436	470	508	520	552	594	615	639	116	13.2	0.08	1.0	Naphthalimide	4.32
192	Isoquinoline 4b	PBS pH 7.4			436				550				114	15.5	0.36	5.5	Isoquinoline	4.37
193	Atto 430LS carboxylate	PBS pH 7.4, 22 °C			436				545				109	32.0	0.65	20.8	nd (pb Coum.)	4.41
194	Atto 425 carboxylate	PBS pH 7.4, 22 °C	395	409	439	460	460	466	485	521	548	576	46	45.0	0.90	40.5	Coumarin 7-Am	4.26.3
195	3,6,10-Trimethy-lisoalloxazine	H_2O pH 7			439				532				93	8.5	0.08	0.6	Pteridine	4.36
196	DY-431 carboxylate	PBS pH 7.4	388	405	442	467	470	477	496	529	550	573	54	35.0	0.75	26.2	Coumarin 7-Am	4.26.3
197	BC-OH	100 mM P, 20% Et pH 7.4			442				488				46	38.1	0.85	32.3	Hybrid-Bonded	4.40.2
198	DY-430 carboxylate	PBS pH 7.4	394	410	444	467	462	468	487	521	545	570	43	37.0	0.72	26.6	Coumarin 7-Am	4.26.3
199	3,7,8,10-TetraMe-isoalloxazine	H_2O pH 7			444				529				85	12.4	0.25	3.1	Pteridine	4.36
200	Epicocconone	H_2O, 0.1% SDS pH 2.5			444				535				91	68.6	0.07	5.0	Epicocconone	4.39.2
201	AI3	H_2O pH 7.4	387	405	445	481	508	520	552	594	615	639	107	15.0	0.21	3.1	Naphthalimide	4.32
202	3,6-Diaminoacridine (Proflavine)	H_2O pH 7	404	420	445	457	482	490	511	554	582	614	66	38.9	0.34	13.2	Acridine	4.23.1
203	3,7,10-Trimethy-lisoalloxazine	H_2O pH 7			446				532				86	9.8	0.32	3.1	Pteridine	4.36
204	Naphtahlimide 10b	H_2O	389	408	448	484	502	514	545	586	606	630	97	12.4	0.21	2.6	Naphthalimide	4.32
205	3,6,7,8,10-PentaMe-isoalloxazine	H_2O pH 7			448				571				123	8.3	0.10	0.8	Pteridine	4.36
206	Flavin mononucleotide (FMN)	H_2O			450				530				80	12.2	0.25	3.0	Pteridine	4.36
207	CF450	H_2O			450				538				88	40.0	0.56	22.3	nd (Cyanine)	4.41
208	Cyan 500 carboxylate	PBS pH 7.4, 22 °C			450				495				45	45.0	0.85	38.2	nd (pb Coum.)	4.41

(Continued)

Ent.	Short Names	Aqueous Solutions	$\lambda_{x,-0.25}$ nm	$\lambda_{x,-0.50}$ nm	$\lambda_{x,max}$ nm	$\lambda_{x,+0.50}$ nm	$\lambda_{e,-0.25}$ nm	$\lambda_{e,-0.50}$ nm	$\lambda_{e,max}$ nm	$\lambda_{e,+0.50}$ nm	$\lambda_{e,+0.25}$ nm	$\lambda_{e,+0.10}$ nm	Δ_S nm	ε_{max} mM^{-1}cm^{-1}	Φ_f	B_{max} mM^{-1}cm^{-1}	Class	Ref.
209	DPP 9	PBS			451				541				90	13.9	0.90	12.5	Diketopyrrolop	4.39.1
210	3,7,9,10-TetraMe-isoalloxazine	H_2O pH 7			452				571				119	6.8	0.10	0.6	Pteridine	4.36
211	Naphtahlimide 10c	H_2O	393	412	453	490	500	512	543	583	603	627	90	17.3	0.30	5.1	Naphthalimide	4.32
212	Atto 465 carboxylate	PBS pH 7.4, 22 °C	404	423	453	467	471	480	506	551	579	611	53	75.0	0.75	56.2	Acridine	4.23.1
213	Isoquinoline 4c	PBS pH 7.4			453				529				76	22.1	0.24	5.3	Isoquinoline	4.37
214	DPP 5	PBS			455				515				60	22.7	0.57	12.9	Diketopyrrolop	4.39.1
215	DPP 7	PBS			456				548				92	14.7	0.76	11.1	Diketopyrrolop	4.39.1
216	trans-PO-PB460	PBS pH 7.4			459				592				133	9.0	0.25	2.2	Phosphole	4.39.3
217	8-Hydroxypyrene-1, 3,6-triSA	H_2O pH 10	389	408	460	470	490	495	514	563	593	620	54	24.0	0.95	22.8	Pyrene	4.28
218	6-Azetidinyl naphthalimide	10 mM HEPES pH 7.3	401	421	464	503	509	521	553	595	616	640	89	18.0	0.28	5.0	Naphthalimide	4.32
219	8-Trifluoromethyl-BODIPY 10	100 mM P pH 7.4	432	449	469	476	529	533	542	557	576	594	73	17.1	0.74	12.6	Bodipy-Small	4.14.1
220	Si-9-Iminopyronine 2g	PBS pH 7.4	434	452	469	479	606	613	631	656	672	706	162	11	0.43	4.7	Si(CH3)2-Pyron	4.5
221	Abberior Star 470SXP	PBS pH 7.4	416	434	472	499	583	594	624	681	722	767	152	29.0	0.12	3.4	Coumarin 7-Am	4.26.3
222	Pyrodone	100 mM P pH 7.4	436	454	472	481	486	490	500	513	523	539	28	60.0	0.37	22.2	Pyrodone	4.21.2
223	Chromenoquinoline L-S 4a	H_2O			473				570				97	36.19	0.15	5.4	Hybrid-Fused	4.40.1
224	8-Trifluoromethyl-BODIPY 12	100 mM P pH 7.4	439	456	477	485	568	574	584	601	624	645	107	41.8	0.70	29.2	Bodipy-Small	4.14.1
225	3-Methoxymethyl ether rhodol	50 mM P, <0.1% ACN pH 8	439	457	477	488	492	500	516	539	563	587	39	58.0	0.85	49.3	Rhodol	4.15.2
226	Chromenoquinoline L-S 4b	H_2O			477				579				102	36.05	0.15	5.2	Hybrid-Fused	4.40.1

Ent.	Short Names	Aqueous Solutions	$\lambda_{x,-0.25}$ nm	$\lambda_{x,-0.50}$ nm	$\lambda_{x,max}$ nm	$\lambda_{x,+0.50}$ nm	$\lambda_{e,-0.25}$ nm	$\lambda_{e,-0.50}$ nm	$\lambda_{e,max}$ nm	$\lambda_{e,+0.50}$ nm	$\lambda_{e,+0.25}$ nm	$\lambda_{e,+0.10}$ nm	Δ_S nm	ε_{max} mм$^{-1}$cm^{-1}	Φ_f	B_{max} mм$^{-1}$cm^{-1}	Class	Ref.
227	JF479	10 mм HEPES pH 7.3	432	450	479	493	488	495	517	561	589	623	38	47.9	0.62	29.7	Acridine	4.23.1
228	DY-485XL carboxylate	H_2O	424	442	482	510	527	535	560	605	638	672	78	48.0		0.0	Coumarin 7-Am	4.26.3
229	Acridine orange	H_2O	439	458	488	503	497	505	528	574	604	639	40	50.0	0.12	6.0	Acridine	4.23.1
230	Fluorescein	100 mм NaOH in D_2O	453	472	489.7	502	495	501	516.3	542	564	587	26	67.0	0.98	65.6	Fluorescein-nh	4.8.1
231	6-Benzamidorhodol	50 mм P, <0.1% ACN pH 8	450	469	490	502	501	508	525	549	573	599	35	30.1	0.16	4.8	Rhodol	4.15.2
232	6-Acrylamidorhodol	50 mм P, <0.1% ACN pH 8	450	469	490	502	500	507	524	548	572	598	34	28.5	0.19	5.4	Rhodol	4.15.2
233	4-Benzamidofluorescein	100 mм P pH 7.5	453	472	490	502	496	502	517	543	565	588	27	80.0	0.79	63.2	Fluorescein-nh	4.8.1
234	Oregon Green 488	H_2O pH 9	453	472	490	502	493	499	514	540	561	584	24	76.0	0.97	73.7	Fluorescein-ha	4.8.2
235	Fluorescein (S-H 1)	P pH 9	453	472	490	502	493	499	514	540	561	584	24	90.0	0.92	82.8	Fluorescein-nh	4.8.1
236	Fluorinated fluorescein S-H 17	P pH 9	453	472	490	502	493	499	514	540	561	584	24	82.4	0.97	79.9	Fluorescein-ha	4.8.2
237	Fluorescein	100 mм NaOH in H_2O	453	472	490.2	503	495	501	516.3	542	564	587	26	67.0	0.92	61.6	Fluorescein-nh	4.8.1
238	Z-L I	0.5 м PBS pH 7.4	451	469	491	499	489	493	501	513	530	545	10	50	0.68	34.0	Bodipy-Large	4.14.3
239	2,6-Sulfon.-1,3,5,7,8-Me-BODIPY	H_2O	451	469	491	499	507	511	519	532	550	567	28	99.0	0.83	82.1	Bodipy-Small	4.14.1
240	2,6-Sulfon.-1,3,5,7-Me-BODIPY	H_2O	451	469	491	499	498	502	510	523	540	556	19	60.0	0.85	51.0	Bodipy-Small	4.14.1
241	DY-490 carboxylate	PBS pH 7.4	452	468	491	504	495	500	515	536	553	572	24	73.0	0.90	65.4	Rhodam Hi Ch	4.2.5
242	2,4-DiMe TokyoGreen	100 mм NaOH	453	475	491	503	493	497	510	548	563	582	19	85.0	0.87	73.5	Fluorone	4.11

(Continued)

Ent.	Short Names	Aqueous Solutions	$\lambda_{x,-0.25}$ nm	$\lambda_{x,-0.50}$ nm	$\lambda_{x,max}$ nm	$\lambda_{x,+0.50}$ nm	$\lambda_{e,-0.25}$ nm	$\lambda_{e,-0.50}$ nm	$\lambda_{e,max}$ nm	$\lambda_{e,+0.50}$ nm	$\lambda_{e,+0.25}$ nm	$\lambda_{e,+0.10}$ nm	Δ_S nm	ε_{max} mM^{-1}cm^{-1}	Φ_f	B_{max} mM^{-1}cm^{-1}	Class	Ref.
243	2,5-DiMe TokyoGreen	100 mM NaOH	453	475	491	503	493	497	510	548	563	582	19	85.0	0.89	75.4	Fluorone	4.11
244	2-Me TokyoGreen	H_2O pH 9	453	475	491	503	493	497	510	548	563	582	19	85.0	0.85	72.2	Fluorone	4.11
245	3,6-Di(azetidin-1-yl) acridine	10 mM HEPES pH 7.3	442	461	492	507	500	508	531	578	608	643	39	47.0	0.52	24.4	Acridine	4.23.1
246	2-Me-4-OMe TokyoGreen	100 mM NaOH	454	476	492	504	492	496	509	547	562	581	17	85.0	0.84	71.4	Fluorone	4.11
247	DMAX-EP	100 mM P pH 7.5	454	474	492	504	494	500	515	541	563	585	23	71.0	0.81	57.5	Fluorescein-nh	4.8.1
248	NX	100 mM P pH 7.5	454	474	492	504	491	497	512	537	559	582	20	69.0	0.83	57.2	Fluorescein-nh	4.8.1
249	Fluorescein (G-C 1)	H_2O pH 9.18, 25 °C	454	474	492	504	493	499	514	540	561	584	22	78.4	0.90	70.5	Fluorescein-nh	4.8.1
250	5-Carboxyfluorescein (S-H 2)	P pH 9	454	474	492	504	495	501	516	542	564	587	24	84.7	0.92	77.9	Fluorescein-nh	4.8.1
251	Fluorinated fluorescein S-H 14a,b	P pH 9	454	474	492	504	495	501	516	542	564	587	24	85.9	0.92	79.0	Fluorescein-ha	4.8.2
252	DY-510XL carboxylate	H_2O	432	451	493	522	549	558	585	635	670	709	92	40.0		0.0	Coumarin 7-Am	4.26.3
253	Acridine orange	HEPES pH 7.3	443	462	493	508	497	505	528	574	604	639	35	50.0	0.21	10.5	Acridine	4.23.1
254	6-Aminorhodol (Rhodol)	50 mM P, <0.1% ACN pH 8	452	471	493	505	492	500	516	539	563	587	23	70.0	0.98	68.6	Rhodol	4.15.2
255	5(6)-Carboxyrhodol	50 mM P pH 8-9	451	472	494	503	495	501	520	539	562	584	26	59.9	0.89	53.3	Rhodol	4.15.1
256	L-B 1a	H_2O	453	472	494	502	495	499	507	520	537	552	13	100.0	0.47	47.0	Bodipy-Small	4.14.1
257	L-B 4a	H_2O	453	472	494	502	495	499	507	520	537	552	13	69.2	0.34	23.5	Bodipy-Small	4.14.1
258	3-MOMO-6-propylaminorhodol	50 mM P, <0.1% ACN pH 8	453	472	494	506	507	515	532	557	582	608	38	43.0	0.35	15.0	Rhodol	4.15.2
259	2,5-DiOMe TokyoGreen	100 mM NaOH	456	478	494	506	495	499	512	550	566	585	18	85.0	0.01	0.8	Fluorone	4.11

Ent.	Short Names	Aqueous Solutions	$\lambda_{x,-0.25}$ nm	$\lambda_{x,-0.50}$ nm	$\lambda_{x,max}$ nm	$\lambda_{x,+0.50}$ nm	$\lambda_{e,-0.25}$ nm	$\lambda_{e,-0.50}$ nm	$\lambda_{e,max}$ nm	$\lambda_{e,+0.50}$ nm	$\lambda_{e,+0.25}$ nm	$\lambda_{e,+0.10}$ nm	Δ_S nm	ε_{max} $mM^{-1}cm^{-1}$	Φ_f	B_{max} $mM^{-1}cm^{-1}$	Class	Ref.
260	2-OMe TokyoGreen	100 mM NaOH	456	478	494	506	497	502	515	554	569	589	21	85.0	0.86	73.1	Fluorone	4.11
261	2-OMe-5-Me TokyoGreen	100 mM NaOH	456	478	494	506	497	501	514	553	568	587	20	85.0	0.50	42.5	Fluorone	4.11
262	4-Carboxy Pennsylvania Green	PBS pH 7.4	456	478	494	506	497	502	515	554	569	589	21	62.0	0.89	55.1	Fluorone	4.11
263	Pennsylvania Green	PBS pH 7.4	456	478	494	506	497	501	514	553	568	587	20	82.0	0.91	74.6	Fluorone	4.11
264	2,4-DiOMe TokyoGreen	100 mM NaOH	456	478	494	506	496	500	513	551	567	586	19	85.0	0.20	17.0	Fluorone	4.11
265	DY-495 carboxylate	PBS pH 7.4	456	476	494	507	499	505	521	547	570	593	27	78.0	0.68	53.2	Fluorescein-nh	4.8.1
266	Alexa Fluor 488	PBS pH 7.2, 22 °C	457	477	494	506	499	504	517	538	555	576	23	73.0	0.92	67.1	Rhodam Hi Ch	4.2.5
267	6-Sulfonato-1,3,5,7-Me-BODIPY	H_2O	454	473	495	503	503	507	515	528	545	562	20	52.5	0.85	44.6	Bodipy-Small	4.14.1
268	Mercaptofluorescein MK-43	1 mM NaOH	457	477	495	508	504	510	526	553	576	600	31	34.0	0.22	7.4	Fluorescein-SH	4.8.3
269	Mercaptofluorescein MK-61	1 mM NaOH	457	477	495	508	505	511	527	554	577	601	32	38.0	0.24	9.1	Fluorescein-SH	4.8.3
270	Mercaptofluorescein MK-69	1 mM NaOH	457	477	495	508	504	510	526	553	576	600	31	25.0	0.20	5.0	Fluorescein-SH	4.8.3
271	5(6)-Carboxyfluorescein (FAM)	100 mM PBS pH 9	457	477	495	508	499	504	520	546	569	592	25	83.0	0.90	74.7	Fluorescein-nh	4.8.1
272	Atto 490LS carboxylate	PBS pH 7.4, 22 °C			495				658				163	40.0	0.30	12.0	nd (pb Coum.)	4.41
273	Z-L B	0.5 M PBS pH 7.4	455	474	496	504	495	499	507	520	537	552	11	50	0.36	17.8	Bodipy-Large	4.14.3
274	Niu 18	H_2O	455	474	496	504	499	503	511	524	541	557	15	75.5	0.20	15.1	Bodipy-Med.	4.14.2
275	Fluorinated rhodamine 1j	Buffer pH 7	458	482	496	507	504	509	522	542	566	588	26	53.0	0.89	47.1	Rhodam <500	4.2.1
276	Rhodamine 110	H_2O pH <7	458	482	496	507	503	507	520	540	564	586	24	83.0	0.89	73.8	Rhodam <500	4.2.1
277	Sulfofluorescein	C pH 9.5	464	482	497	508	499	504	517	556	572	591	20	78.0	0.92	71.7	Sulfofluoresc.	4.10

(Continued)

Ent.	Short Names	Aqueous Solutions	$\lambda_{x,-0.25}$ nm	$\lambda_{x,-0.50}$ nm	$\lambda_{x,max}$ nm	$\lambda_{x,+0.50}$ nm	$\lambda_{e,-0.25}$ nm	$\lambda_{e,-0.50}$ nm	$\lambda_{e,max}$ nm	$\lambda_{e,+0.50}$ nm	$\lambda_{e,+0.25}$ nm	$\lambda_{e,+0.10}$ nm	Δ_S nm	ε_{max} mм$^{-1}$cm^{-1}	Φ_f	B_{max} mм$^{-1}$cm^{-1}	Class	Ref.
278	Atto 495 carboxylate	PBS pH 7.4, 22 °C	455	471	498	512	503	509	526	569	597	631	28	80.0	0.20	16.0	Acridine	4.23.1
279	DAMBO-T	100 mм P pH 7.4	457	475	498	506	495	499	507	520	537	552	9	52.0	0.40	20.8	Bodipy-Small	4.14.1
280	BODIPY	H_2O	457	475	498	506	504	508	516	529	547	563	18	90.0	0.90	81.2	Bodipy-Small	4.14.1
281	L-B 1b	H_2O	457	475	498	506	497	501	509	522	539	555	11	100.0	0.37	37.0	Bodipy-Small	4.14.1
282	L-B 4b	H_2O	457	475	498	506	497	501	509	522	539	555	11	77.6	0.15	11.6	Bodipy-Small	4.14.1
283	L-B 5	H_2O	457	475	498	506	499	503	511	524	541	557	13	79.4	0.49	38.9	Bodipy-Small	4.14.1
284	2-Me RG	PBS pH 7.4	458	483	498	510	502	507	520	536	549	574	22	82.0	0.84	68.8	Rosamine	4.17
285	Z-L A	0.5 м PBS pH 7.4	457	476	499	507	498	502	510	523	540	556	11	50	0.04	2.1	Bodipy-Large	4.14.3
286	Chlorinated fluorescein G-C 2(3)	H_2O pH 9.18, 25 °C	460	480	499	512	497	503	518	544	566	589	19	48.4	0.93	45.0	Fluorescein-ha	4.8.2
287	Fluorinated rhodamine 1e	Buffer pH 7	461	485	499	510	504	509	522	542	566	588	23	81.0	0.88	71.2	Rhodam <500	4.2.1
288	B-Z 7a	H_2O	458	477	500	508	497	501	509	522	539	555	9	64.5	0.61	39.3	Bodipy-Large	4.14.3
289	Atto 488 carboxylate	PBS pH 7.4, 22 °C	461	486	500	511	503	507	520	540	564	586	20	90.0	0.80	72.0	Rhodam <500	4.2.1
290	Chromenoquinoline L-S 4c	H_2O			500				591				91	33.16	0.03	1.1	Hybrid-Fused	4.40.1
291	Chromenoquinoline L-S 4d	H_2O			500				608				108	36.06	0.08	2.8	Hybrid-Fused	4.40.1
292	Z-L F	0.5 м PBS pH 7.4	459	478	501	509	499	503	511	524	541	557	10	50	0.31	15.6	Bodipy-Large	4.14.3
293	Fluorinated rhodam. 1f-5′-COOH	Buffer pH 7	463	484	501	513	506	512	525	546	565	586	24	85.0	0.95	80.7	Rhoda 500-550	4.2.2
294	Fluorinated rhodam. 1f-6′-COOH	Buffer pH 7	463	484	501	513	505	510	523	544	562	584	22	88.0	0.93	81.8	Rhoda 500-550	4.2.2
295	JF502	10 mм HEPES pH 7.3	450	470	502	518	502	510	533	580	610	646	31	57.8	0.71	41.0	Acridine	4.23.1
296	BODIPY FL	H_2O	460	479	502	511	499	503	511	524	541	557	9	92.0	0.97	89.2	Bodipy-Small	4.14.1
297	Mercaptofluorescein MK-67	1 mм NaOH	463	483	502	515	505	511	527	554	577	601	25	27.0	0.17	4.5	Fluorescein-SH	4.8.3

Ent.	Short Names	Aqueous Solutions	$\lambda_{x,-0.25}$ nm	$\lambda_{x,-0.50}$ nm	$\lambda_{x,max}$ nm	$\lambda_{x,+0.50}$ nm	$\lambda_{e,-0.25}$ nm	$\lambda_{e,-0.50}$ nm	$\lambda_{e,max}$ nm	$\lambda_{e,+0.50}$ nm	$\lambda_{e,+0.25}$ nm	$\lambda_{e,+0.10}$ nm	Δ_S nm	ε_{max} mM^{-1}cm^{-1}	Φ_f	B_{max} mM^{-1}cm^{-1}	Class	Ref.
298	Mercaptofluorescein MK-75	1 mM NaOH	463	483	502	515	511	517	533	561	584	609	31	12.0	0.16	1.9	Fluorescein-SH	4.8.3
299	Rhodamine 123	H_2O pH <7	464	485	502	514	509	514	528	549	568	590	26	75.0	0.90	67.5	Rhoda 500-550	4.2.2
300	Chromenoquinoline L-S 4e	H_2O			502				613				111	33.16	0.07	2.2	Hybrid-Fused	4.40.1
301	6-Benzylaminorhodol	50 mM P, <0.1% ACN pH 8	461	480	503	515	504	512	529	554	578	604	26	71.0	0.95	67.4	Rhodol	4.15.2
302	Chlorinated fluorescein G-C 1	H_2O pH 9.18, 25 °C	464	484	503	516	500	506	522	548	571	595	19	103.0	0.86	88.5	Fluorescein-ha	4.8.2
303	Abberior Star 488 carboxylate	PBS pH 7.4	465	485	503	515	505	511	524	545	564	585	21	64.5	0.89	57.4	Rhodam Hi Ch	4.2.5
304	NBDY	H_2O	462	481	504	513	520	524	533	547	566	583	29	74.4	0.33	24.5	Bodipy-Small	4.14.1
305	DY-505 carboxylate	PBS pH 7.4	459	478	505	515	506	512	530	552	570	590	25	85.0	0.74	63.2	Rhoda 500-550	4.2.2
306	*meso*-CaboxyBODIPY J-K 1	10 mM PBS pH 7.4	463	482	505	514	540	545	554	569	589	608	49	62.0	0.78	48.3	Bodipy-Small	4.14.1
307	G-L 13a	H_2O	463	482	505	514	505	509	517	530	548	564	12	35.0	0.47	16.4	Bodipy-Med.	4.14.2
308	BCECF	100 mM NaOH	466	486	505	518	505	511	527	554	577	601	22	67.0	0.84	56.2	Fluorescein-nh	4.8.1
309	BCPCF	100 mM NaOH	466	486	505	518	505	511	527	554	577	601	22	67.0	0.83	55.6	Fluorescein-nh	4.8.1
310	Fluorinated rhodamine 1k	Buffer pH 7	467	488	506	518	512	517	531	553	572	594	25	54.0	0.66	35.6	Rhoda 500-550	4.2.2
311	6-Allylaminorhodol	50 mM P, <0.1% ACN pH 8	464	484	507	519	509	517	534	559	584	611	27	61.5	0.92	56.5	Rhodol	4.15.2
312	6-Propylaminorhodol	50 mM P, <0.1% ACN pH 8	464	484	507	519	506	514	531	556	581	607	24	61.0	0.94	57.3	Rhodol	4.15.2
313	6-Azido-4,5, 7-trifluorofluorescein	100 mM NaOH	467	488	507	520	505	511	527	554	577	601	20	79.9	0.82	65.5	Fluorescein-ha	4.8.2

(Continued)

Ent.	Short Names	Aqueous Solutions	$\lambda_{x,-0.25}$ nm	$\lambda_{x,-0.50}$ nm	$\lambda_{x,max}$ nm	$\lambda_{x,+0.50}$ nm	$\lambda_{e,-0.25}$ nm	$\lambda_{e,-0.50}$ nm	$\lambda_{e,max}$ nm	$\lambda_{e,+0.50}$ nm	$\lambda_{e,+0.25}$ nm	$\lambda_{e,+0.10}$ nm	Δ_S nm	ε_{max} mM^{-1}cm^{-1}	Φ_f	B_{max} mM^{-1}cm^{-1}	Class	Ref.
314	6-CA-MeThio-4, 5,7-F-fluorescein	100 mM NaOH	467	488	507	520	504	510	526	553	576	600	19	63.5	0.88	55.8	Fluorescein-ha	4.8.2
315	7-Me-6-EtAm-carboxyrhodol	50 mM P pH 8-9	465	485	508	521	509	517	534	559	584	611	26	78.1	0.82	64.0	Rhodol	4.15.1
316	6-Butylaminorhodol	50 mM P, <0.1% ACN pH 8	465	485	508	521	508	516	533	558	583	609	25	60.0	0.93	55.8	Rhodol	4.15.2
317	Fluorinated fluorescein S-H 18	P pH 9	468	489	508	521	505	511	527	554	577	601	19	85.6	0.85	72.7	Fluorescein-ha	4.8.2
318	Fluorinated fluorescein S-H 19	P pH 9	468	489	508	521	505	511	527	554	577	601	19	81.2	0.96	77.9	Fluorescein-ha	4.8.2
319	G-L 14	H_2O	472	492	508	518	501	505	521	555	575	603	13	59.0	0.77	45.4	Bodipy-Med.	4.14.2
320	2-PrAc-7-Me-6-EtAm-rhodol	50 mM P pH 8-9	466	486	509	522	510	518	535	560	585	612	26	79.4	0.92	73.0	Rhodol	4.15.1
321	L-B 2a	H_2O	473	493	509	519	503	507	523	557	577	606	14	72.4	0.27	19.5	Bodipy-Small	4.14.1
322	G-L 13b	H_2O	473	493	509	519	502	506	522	556	576	604	13	56.0	0.67	37.5	Bodipy-Med.	4.14.2
323	2,7-DiMe-6-EtAm-carboxyrhodol	50 mM P pH 8-9	467	487	510	522	511	519	536	561	586	612	26	66.3	0.89	59.0	Rhodol	4.15.1
324	Fluorinated fluorescein S-H 15a,b	P pH 9	470	490	510	523	511	518	534	562	585	610	24	84.2	0.43	36.2	Fluorescein-ha	4.8.2
325	Fluorinated fluorescein S-H 16a,b	P pH 9	470	490	510	523	511	518	534	562	585	610	24	78.1	0.59	46.0	Fluorescein-ha	4.8.2
326	Fluorinated rhodamine 3-H	Buffer pH 7	471	492	510	523	512	517	531	553	572	594	21	57.0	0.86	49.0	Rhoda 500-550	4.2.2
327	Niu 27	H_2O	474	494	510	520	508	512	528	563	583	612	18	63.3	0.81	51.2	Bodipy-Med.	4.14.2
328	2-Et-7-Me-6-EtAm-carboxyrhodol	50 mM P pH 8-9	468	488	511	524	512	520	538	563	589	616	27	73.0	0.84	61.3	Rhodol	4.15.1
329	Chlorinated fluorescein G-C 5	H_2O pH 9.18, 25 °C	471	491	511	524	505	511	527	554	577	601	16	91.6	0.84	76.9	Fluorescein-ha	4.8.2
330	Oregon Green 514	PBS pH 7.2, 22 °C	471	491	511	524	508	514	530	557	581	605	19	88.0	0.96	84.4	Fluorescein-ha	4.8.2
331	Abberior Star 512 carboxylate	PBS pH 7.4	472	493	511	524	511	516	530	552	571	593	19	84.0	0.82	68.8	Rhoda 500-550	4.2.2
332	Atto 514 carboxylate	PBS pH 7.4, 22 °C	474	494	511	522	512	517	532	553	575	599	21	115.0	0.85	97.7	Rhoda 500-550	4.2.2

Ent.	Short Names	Aqueous Solutions	$\lambda_{x,-0.25}$ nm	$\lambda_{x,-0.50}$ nm	$\lambda_{x,max}$ nm	$\lambda_{x,+0.50}$ nm	$\lambda_{e,-0.25}$ nm	$\lambda_{e,-0.50}$ nm	$\lambda_{e,max}$ nm	$\lambda_{e,+0.50}$ nm	$\lambda_{e,+0.25}$ nm	$\lambda_{e,+0.10}$ nm	Δ_S nm	ε_{max} $mM^{-1}cm^{-1}$	Φ_f	B_{max} $mM^{-1}cm^{-1}$	Class	Ref.
333	2,7-DiCl-6-methylamino rhodol	50 mM P pH 8-9	469	488	512	525	507	515	533	558	583	611	21	71.0	0.67	47.5	Rhodol	4.15.1
334	Chlorinated fluorescein G-C 4	H_2O pH 9.18, 25 °C	471	492	512	526	506	512	528	555	578	602	16	93.8	0.87	81.6	Fluorescein-ha	4.8.2
335	Fluorinated rhodamine 5-SO_3H	Buffer pH 7	472	494	512	525	511	516	530	552	571	593	18	52.0	0.86	44.7	Rhoda 500-550	4.2.2
336	L-B 2b	H_2O	476	496	512	522	504	508	524	559	578	607	12	77.6	0.41	31.8	Bodipy-Small	4.14.1
337	2-Cl-7-Me-6-EtAm-carboxyrhodol	50 mM P pH 8-9	479	495	514	538	514	525	539	573	600	634	25	80.5	0.82	66.0	Rhodol	4.15.1
338	2,4,5,7-Tetra-F-sulfofluorescein	P pH 9.18	480	499	515	527	518	523	537	579	596	618	22	63.2	0.80	50.5	Sulfofluoresc.	4.10
339	4,7-DiMe-6-EtAm-carboxyrhodol	50 mM P pH 8-9	472	492	516	529	518	526	544	570	596	624	28	56.9	0.43	24.4	Rhodol	4.15.1
340	6-Piperidino-carboxyrhodol	50 mM P pH 8-9	472	492	516	529	515	523	541	567	593	620	25	80.9	0.79	63.9	Rhodol	4.15.1
341	2-Br-7-Me-6-EtAm-carboxyrhodol	50 mM P pH 8-9	472	492	516	529	513	521	539	565	590	618	23	80.9	0.61	49.3	Rhodol	4.15.1
342	B-Z 7b	H_2O	479	500	516	526	509	513	529	564	584	614	13	60.5	0.68	41.1	Bodipy-Large	4.14.3
343	CF514	H_2O			516				548				32	105.0	0.38	39.8	nd (Rhodam.)	4.41
344	Atto 520 carboxylate	PBS pH 7.4, 22 °C	474	497	517	529	520	525	538	556	567	592	21	110.0	0.90	99.0	Pyronine	4.4
345	*meso*-Caboxy-BODIPY J-K 3c	10 mM PBS pH 7.4	480	501	517	527	519	523	540	577	598	629	23	76.0	0.97	73.7	Bodipy-Small	4.14.1
346	Safranin T	H_2O			517				586				69	27.0	0.11	2.9	Phenazine	4.39.1
347	6-Dimethylaminorhodol	50 mM P, <0.1% ACN pH 8	473	494	518	531	518	526	544	570	596	624	26	64.0	0.20	12.8	Rhodol	4.15.2
348	6-Dimethylaminorhodol	10 mM HEPES pH 7.3	473	494	518	531	520	528	546	572	599	626	28	60.0	0.21	12.6	Rhodol	4.15.2
349	6-Azetidinylaminor-hodol	10 mM HEPES pH 7.3	474	495	519	532	520	528	546	572	599	626	27	59.0	0.85	50.1	Rhodol	4.15.2

(Continued)

Ent.	Short Names	Aqueous Solutions	$\lambda_{x,-0.25}$ nm	$\lambda_{x,-0.50}$ nm	$\lambda_{x,max}$ nm	$\lambda_{x,+0.50}$ nm	$\lambda_{e,-0.25}$ nm	$\lambda_{e,-0.50}$ nm	$\lambda_{e,max}$ nm	$\lambda_{e,+0.50}$ nm	$\lambda_{e,+0.25}$ nm	$\lambda_{e,+0.10}$ nm	Δ_S nm	ε_{max} $mM^{-1}cm^{-1}$	Φ_f	B_{max} $mM^{-1}cm^{-1}$	Class	Ref.
350	6-Pyrrolidinorhodol	50 mM P, <0.1% ACN pH 8	474	495	519	532	523	532	550	577	603	632	31	63.0	0.12	7.5	Rhodol	4.15.2
351	Niu 29	H_2O	477	497	519	531	511	515	530	561	584	612	11	28.0	0.61	17.0	Bodipy-Med.	4.14.2
352	*meso*-CaboxyBODIPY J-K 3a	10 mM PBS pH 7.4	477	497	519	531	537	541	558	592	618	649	39	77.0	0.67	51.5	Bodipy-Small	4.14.1
353	*meso*-CaboxyBODIPY J-K 3b	10 mM PBS pH 7.4	477	497	519	531	528	532	549	582	607	637	30	78.0	0.83	64.7	Bodipy-Small	4.14.1
354	Sulfonated coumarin N-H 7-H	PBS pH 7.4	453	474	520	552	596	607	639	699	742	789	119	40.0	0.06	2.4	Coumarin 7-Am	4.26.3
355	2,7-DiCl-6-pyrrolidino sulforhodol	PBS pH 7.4	470	491	520	534	516	521	535	555	584	604	15	83.0	0.27	22.4	Sulforhodol	4.27
356	Me-rosol	50 mM P pH 7.4	475	496	520	533	536	542	557	591	624	653	37	46.1	0.18	8.4	Rosol	4.19
357	2′,7′-Cl-4,5,6, 7-F-fluorescein	100 mM NaOH	478	500	520	534	515	521	538	566	590	615	18	88.5	0.80	70.8	Fluorescein-ha	4.8.2
358	JOE	H_2O pH 9	478	500	520	534	524	531	548	577	602	628	28	73.0	0.60	43.8	Fluorescein-ha	4.8.2
359	DAMBO-P^{H}-T	100 mM P pH 7.4	479	499	521	533	517	521	537	569	592	621	16	56.0	0.74	41.4	Bodipy-Small	4.14.1
360	Niu 25	PBS pH 7.4	479	499	521	533	520	524	540	572	596	625	19	58.8	0.82	48.2	Bodipy-Med.	4.14.2
361	TET	H_2O pH 9	479	501	521	535	513	519	536	564	588	613	15	73.0	0.70	51.1	Fluorescein-ha	4.8.2
362	Abberior Star 520SXP	PBS pH 7.4	454	475	522	555	590	601	632	691	733	779	110	42.5	0.15	6.3	Coumarin 7-Am	4.26.3
363	6-Morpholinorhodol	50 mM P, <0.1% ACN pH 8	477	498	522	535	526	534	553	580	607	636	31	61.0	0.13	7.9	Rhodol	4.15.2
364	6-Diethylaminorhodol	50 mM P, <0.1% ACN pH 8	477	498	522	535	522	531	549	576	602	630	27	61.0	0.15	9.1	Rhodol	4.15.2
365	6-Dimethy-laminorhodol	C pH 9.5	477	498	522	535	519	527	545	571	597	625	23	57.0	0.20	11.4	Rhodol	4.15.1
366	Chlorinated fluorescein G-C 6	H_2O pH 9.18, 25 °C	480	501	522	536	514	520	537	565	589	614	15	65.8	0.83	54.6	Fluorescein-ha	4.8.2

Ent.	Short Names	Aqueous Solutions	$\lambda_{x,-0.25}$ nm	$\lambda_{x,-0.50}$ nm	$\lambda_{x,max}$ nm	$\lambda_{x,+0.50}$ nm	$\lambda_{e,-0.25}$ nm	$\lambda_{e,-0.50}$ nm	$\lambda_{e,max}$ nm	$\lambda_{e,+0.50}$ nm	$\lambda_{e,+0.25}$ nm	$\lambda_{e,+0.10}$ nm	Δ_S nm	ε_{max} mм$^{-1}$cm^{-1}	Φ_f	B_{max} mм$^{-1}$cm^{-1}	Class	Ref.
367	6-Piperidinorhodol	50 mм P, <0.1% ACN pH 8	478	499	523	536	528	536	555	582	609	638	32	62.5	0.10	6.2	Rhodol	4.15.2
368	DPP 10	PBS			523				594				71	17.1	0.61	10.4	Diketopyrrolop	4.39.1
369	2-Cl-6-Piperidino-carboxyrhodol	50 mм P pH 8-9	478	500	524	537	522	530	548	574	601	629	24	77.4	0.72	55.7	Rhodol	4.15.1
370	Et-rosol	50 mм P pH 7.4	478	500	524	537	538	544	558	589	622	649	34	42.6	0.10	4.1	Rosol	4.19
371	2,4,5,7-Tetra-Br-sulfofluorescein	100 mм NaOH pH 12	488	507	524	536	526	531	546	590	607	630	22	87.0	0.35	30.4	Sulfofluoresc.	4.10
372	6-Dibutylaminorhodol	50 mм P, <0.1% ACN pH 8	479	500	525	538	526	534	553	580	607	636	28	63.0	0.08	5.0	Rhodol	4.15.2
373	6-Dipropylaminorhodol	50 mм P, <0.1% ACN pH 8	479	500	525	538	525	533	552	579	606	634	27	59.0	0.10	5.9	Rhodol	4.15.2
374	HMDER	PBS pH 7.4	479	500	525	538	517	525	543	569	595	622	18	70.0	0.14	9.8	Rhodol	4.15.2
375	Pip-rosol	50 mм P pH 7.4	479	500	525	538	537	544	563	599	630	669	38	42.5	0.05	2.0	Rosol	4.19
376	JF525	10 mм HEPES pH 7.3	483	506	525	538	529	534	549	572	593	617	24	94.0	0.91	85.5	Rhoda 500-550	4.2.2
377	Fluorinated rhodamine 1r	Buffer pH 7	483	506	525	538	533	539	554	578	598	623	29	31.0	0.28	8.6	Rhoda 500-550	4.2.2
378	Rhodamine 6G	D_2O pH <7	484	506	525.5	539	531	537	552.0	575	596	620	26	105.0	0.98	102.9	Rhoda 500-550	4.2.2
379	JF526	10 mм HEPES pH 7.3	484	507	526	539	530	535	550	573	594	618	24	19.0	0.87	16.5	Rhoda 500-550	4.2.2
380	Rhodamine 6G	H_2O pH <7	485	507	526.3	540	532	537	552.3	576	596	621	26	105.0	0.93	97.6	Rhoda 500-550	4.2.2
381	AR116	100 mм P pH 7.5	483	506	527	539	529	535	548	567	578	604	21	64.8	0.07	4.5	Pyronine	4.4

(Continued)

Ent.	Short Names	Aqueous Solutions	$\lambda_{x,-0.25}$ nm	$\lambda_{x,-0.50}$ nm	$\lambda_{x,max}$ nm	$\lambda_{x,+0.50}$ nm	$\lambda_{e,-0.25}$ nm	$\lambda_{e,-0.50}$ nm	$\lambda_{e,max}$ nm	$\lambda_{e,+0.50}$ nm	$\lambda_{e,+0.25}$ nm	$\lambda_{e,+0.10}$ nm	Δ_S nm	ε_{max} mM^{-1}cm^{-1}	Φ_f	B_{max} mM^{-1}cm^{-1}	Class	Ref.
382	Rhodamine RBzET	H_2O	485	508	527	541	531	537	552	575	596	620	25	82.0	0.87	71.3	Rhoda 500-550	4.2.2
383	CF532	H_2O			527				558				31	96.0	0.73	69.8	nd (Rhodam.)	4.41
384	DPP 8	PBS			527				594				67	18.1	0.56	10.1	Diketopyrrolop	4.39.1
385	Acridine red	H_2O	484	507	528	540	542	548	562	582	594	621	34	100.0	0.51	51.0	Pyronine	4.4
386	Rhodamine 6G +3	H_2O pH <7	486	510	529	544	532	537	551	570	582	602	22	101.0	0.91	91.9	Rhodam Hi Ch	4.2.5
387	Chromenoquinoline L-S 4f	H_2O			529				622				93	40.64	0.05	1.9	Hybrid-Fused	4.40.1
388	Yakima Yellow carboxylate	100 mM P/ Tris pH 7.0/7.4	487	509	530	545	525	532	549	578	603	630	19	84.0	0.96	80.6	Fluorescein-ha	4.8.2
389	Rhodamine 6G +2	H_2O pH <7	487	511	530	544	533	539	553	571	583	603	23	101.0	0.93	93.9	Rhoda 500-550	4.2.2
390	DPP 6	PBS			530				586				56	30.7	0.49	15.0	Diketopyrrolop	4.39.1
391	Alexa Fluor 532	PBS pH 7.2, 22 °C	486	514	531	545	533	539	554	572	587	616	23	81.0	0.61	49.4	Rosamine	4.17
392	530RH	PBS pH 7.4	491	514	532	547	533	538	553	574	592	618	21	56.0	0.89	49.8	Rhoda 500-550	4.2.2
393	Atto 532 carboxylate	PBS pH 7.4, 22 °C	491	515	532	544	533	538	552	575	600	625	20	115.0	0.90	103.5	Rhoda 500-550	4.2.2
394	Atto Rho6G carboxylate	PBS pH 7.4, 22 °C	489	513	533	546	534	540	557	583	609	636	24	115.0	0.94	108.1	Rhoda 500-550	4.2.2
395	Z-L C	0.5 M PBS pH 7.4	489	510	533	546	529	533	550	583	608	638	17	50	0.01	0.6	Bodipy-Large	4.14.3
396	M-S 10c	H_2O, 0.1% TX-100	489	510	533	546	536	540	557	591	617	648	24	54.0	0.61	32.9	Bodipy-Small	4.14.1
397	DY-530 carboxylate	PBS pH 7.4	490	513	533	547	533	539	554	578	598	623	21	100.0	0.81	80.6	Rhoda 500-550	4.2.2
398	Z-L E	0.5 M PBS pH 7.4	490	511	534	547	529	533	550	583	608	638	16	50	0.09	4.4	Bodipy-Large	4.14.3
399	1,2,7,8-Dibenzo-fluorescein	C pH 9.5	490	513	534	549	540	547	565	596	623	651	31	81.0	0.24	19.4	Fluorescein-nh	4.8.1
400	PONy A1	PBS pH 7.4	476	499	535	553	556	566	595	654	693	739	60	38.0	0.10	3.8	Acridine	4.23.1

Ent.	Short Names	Aqueous Solutions	$\lambda_{x,-0.25}$ nm	$\lambda_{x,-0.50}$ nm	$\lambda_{x,max}$ nm	$\lambda_{x,+0.50}$ nm	$\lambda_{e,-0.25}$ nm	$\lambda_{e,-0.50}$ nm	$\lambda_{e,max}$ nm	$\lambda_{e,+0.50}$ nm	$\lambda_{e,+0.25}$ nm	$\lambda_{e,+0.10}$ nm	Δ_S nm	ε_{max} mM^{-1}cm^{-1}	Φ_f	B_{max} mM^{-1}cm^{-1}	Class	Ref.
401	HEX	H_2O pH 9	491	513	535	550	534	541	559	589	616	643	24	98.0	0.70	68.6	Fluorescein-ha	4.8.2
402	Fluorinated fluorescein S-H 20	P pH 9	491	513	535	550	529	535	553	583	608	635	18	83.4	0.47	39.2	Fluorescein-ha	4.8.2
403	Z-L D	0.5 M PBS pH 7.4	492	513	536	549	534	538	555	589	614	645	19	50	0.02	0.9	Bodipy-Large	4.14.3
404	6-Quinoline-2-pentanoic-pyrodol	50 mM P pH 8-9	492	515	538	550	544	549	562	579	591	611	24	64.1	0.70	44.8	Pyrodol	4.21.1
405	6-Quinoline-2-propanoic-pyrodol	50 mM P pH 8-9	492	515	538	550	544	549	562	579	591	611	24	65.9	0.70	46.1	Pyrodol	4.21.1
406	Jul-rosol	50 mM P pH 7.4	492	514	540	554	548	554	574	614	648	686	34	30.4	0.41	12.5	Rosol	4.19
407	6-Quinolinocarbo-xyrhodol	50 mM P pH 8-9	492	514	540	554	541	550	570	599	627	658	30	66.8	0.43	28.7	Rhodol	4.15.1
408	2-PrAc-6-quinoli-norhodol	50 mM P pH 8-9	493	515	541	555	543	552	572	601	630	661	31	58.2	0.49	28.5	Rhodol	4.15.1
409	1,2,7,8-Dibenzosul-fofluorescein	C pH 9.5	502	523	541	554	546	551	567	614	633	658	26	81.0	0.48	38.8	Sulfofluoresc.	4.10
410	Rosamine W-B 5h	100 mM P pH 7.4	495	524	542	557	546	552	568	587	603	634	26	71.5	0.14	10.0	Rosamine	4.17
411	Atto 542 carboxylate	PBS pH 7.4, 22 °C			542				562				20	120.0	0.93	111.6	nd (pb Rhoda)	4.41
412	2-Ethyl-6-quinolinocarboxyrhodol	50 mM P pH 8-9	494	517	543	557	545	554	574	603	632	664	31	79.4	0.49	38.9	Rhodol	4.15.1
413	Z-L J	0.5 M PBS pH 7.4	498	519	543	556	541	546	563	598	624	656	20	50	0.34	16.8	Bodipy-Large	4.14.3
414	COUPY 9	H_2O			543				605				62	31.0	0.15	4.6	Hybrid-Bridged	4.40.3
415	Carbofluorescein	100 mM NaOH	505	526	544	557	546	551	567	614	633	658	23	108.0	0.62	66.9	C(CH3)2-Fluore	4.9
416	Cy3′	H_2O	483	504	545	556	543	547	560	575	606	624	15	120.0	0.04	4.8	Cyanine-Trimet	4.13.1
417	Pyronine Y (OP)	H_2O	498	523	546	559	546	550	566	584	598	631	20	77.0	0.22	16.9	Pyronine	4.4
418	Pyronine Y (OP)	PBS pH 7.4	500	525	546	560	519	526	569	557	567	586	23	77.0	0.36	27.7	Pyronine	4.4
419	TAMRA	H_2O pH 7	501	525	546	561	554	560	576	602	624	651	30	90.0	0.20	18.0	Rhoda 500-550	4.2.2

(Continued)

Ent.	Short Names	Aqueous Solutions	$\lambda_{x,-0.25}$ nm	$\lambda_{x,-0.50}$ nm	$\lambda_{x,max}$ nm	$\lambda_{x,+0.50}$ nm	$\lambda_{e,-0.25}$ nm	$\lambda_{e,-0.50}$ nm	$\lambda_{e,max}$ nm	$\lambda_{e,+0.50}$ nm	$\lambda_{e,+0.25}$ nm	$\lambda_{e,+0.10}$ nm	Δ_S nm	ε_{max} mм$^{-1}$cm^{-1}	Φ_f	B_{max} mм$^{-1}$cm^{-1}	Class	Ref.
420	Alexa Fluor 546	PBS pH 7.2, 22 °C	506	537	546	567	553	558	580	592	621	648	34	112.0	0.79	88.4	Rhodam Hi Ch	4.2.5
421	M-S 10b	H_2O, 0.1% TX-100	501	523	547	560	551	556	574	610	638	671	27	64.0	0.78	49.9	Bodipy-Small	4.14.1
422	TAMRA	50 mм PBS pH 8-9	502	526	547	562	552	558	574	599	622	648	27	77.0	0.35	26.9	Rhoda 500-550	4.2.2
423	DAOTA 5	H_2O	479	505	548	567	551	558	576	606	630	656	28	50.0	0.56	28.0	Triangulenium	4.22
424	Sulfo-cyanine3	H_2O	486	506	548	559	546	550	563	578	610	628	15	162.0	0.10	16.2	Cyanine-Trimet	4.13.1
425	SR101-OH	H_2O	493	516	548	564	550	556	572	595	628	652	24	80.0	0.35	28.0	Sulforhodol	4.27
426	2-Cloro-6-quinolinocarboxyrhodol	50 mм P pH 8-9	498	521	548	563	546	555	575	604	634	665	27	74.5	0.42	31.2	Rhodol	4.15.1
427	Tetramethylrhodamine	10 mм HEPES pH 7.3	503	527	548	563	550	556	572	597	619	646	24	78.0	0.45	35.1	Rhoda 500-550	4.2.2
428	DAOTA 4	H_2O	481	502	549	567	553	559	576	607	632	655	27	50.0	0.56	28.0	Triangulenium	4.22
429	1,2,7,8-Dibenzo-4,5-Br-fluorescein	C pH 9.5	503	526	549	565	554	562	581	614	642	672	32	24.0	0.05	1.2	Fluorescein-ha	4.8.2
430	JF549	10 mм HEPES pH 7.3	504	528	549	564	549	555	571	596	618	644	22	101	0.88	88.8	Rhoda 500-550	4.2.2
431	Cy3	H_2O	487	508	550	561	553	557	570	585	618	636	20	150.0	0.06	8.2	Cyanine-Trimet	4.13.1
432	Cy3.13-OH	H_2O/PBS	487	508	550	561	549	553	566	581	613	631	16	130.0	0.03	3.9	Cyanine-Trimet	4.13.1
433	Cy3.18.OH	PBS	487	508	550	561	548	552	565	580	612	630	15	150.0	0.04	6.0	Cyanine-Trimet	4.13.1
434	THQ-rosol	50 mм P pH 7.4	500	523	550	565	631	653	710	802	854	925	160	29.4	0.03	0.8	Rosol	4.19
435	Tetramethylrosamine	10 mм PBS pH 7.4	501	532	550	565	548	554	570	589	605	636	20	74.5	0.31	23.1	Rosamine	4.17
436	Neo-Cy3	PBS	488	509	551	562	555	560	573	588	622	640	22	98.0	0.02	1.9	Cyanine-Trimet	4.13.1
437	Cy3.10-OH	H_2O/PBS	489	510	552	563	553	557	570	585	618	636	18	130.0	0.03	3.9	Cyanine-Trimet	4.13.1
438	Pyronine B	H_2O	504	529	552	565	551	557	572	592	605	633	20	94.1	0.18	16.9	Pyronine	4.4
439	JF552	10 mм HEPES pH 7.3	505	535	552	566	554	560	575	600	629	657	23	95.2	0.83	79.0	Rhoda 550-600	4.2.3

Ent.	Short Names	Aqueous Solutions	$\lambda_{x,-0.25}$ nm	$\lambda_{x,-0.50}$ nm	$\lambda_{x,max}$ nm	$\lambda_{x,+0.50}$ nm	$\lambda_{e,-0.25}$ nm	$\lambda_{e,-0.50}$ nm	$\lambda_{e,max}$ nm	$\lambda_{e,+0.50}$ nm	$\lambda_{e,+0.25}$ nm	$\lambda_{e,+0.10}$ nm	Δ_S nm	ε_{max} $mM^{-1}cm^{-1}$	Φ_f	B_{max} $mM^{-1}cm^{-1}$	Class	Ref.
440	Carborhodamine	10 mM HEPES pH 7.3	515	537	552	564	558	564	577	605	628	654	25	65.0	0.64	41.6	C(CH3)2-Rhoda	4.3.1
441	SR101-110	PBS	485	507	554	570	555	560	576	600	638	661	22	80.0	0.55	44.0	Sulforhodam.	4.6
442	Atto 550 carboxylate	PBS pH 7.4, 22 °C	507	535	554	567	554	560	576	598	626	654	22	120.0	0.80	96.0	Rhoda 550-600	4.2.3
443	Sulforhodamine B	H_2O pH 7	508	537	554	567	561	568	582	599	609	628	28	84.0	0.39	32.7	Sulforhodam.	4.6
444	AF555	PBS pH 7.2, 22 °C	491	512	555	566	555	559	572	587	620	639	17	158.0	0.09	14.2	Cyanine-Trimet	4.13.1
445	Alexa Fluor 555	PBS pH 7.2, 22 °C	491	512	555	566	555	559	572	587	620	639	17	155.0	0.10	15.5	Cyanine-Trimet	4.13.1
446	DY-555 carboxylate	PBS pH 7.4	508	537	555	569	558	564	580	605	635	663	25	100.0	0.32	31.7	Rhoda 550-600	4.2.3
447	2′,7′-Difluorocarbo-fluorescein	100 mM NaOH	514	536	555	569	559	564	581	631	651	677	26	90.9	0.40	36.3	C(CH3)2-Fluore	4.9
448	Seta-555-carboxylate	P pH 7.4			555				570				15	155.0	0.07	10.8	nd	4.41
449	Cy3.12-OH	H_2O/PBS	492	513	556	567	556	561	574	590	623	641	18	130.0	0.05	6.5	Cyanine-Trimet	4.13.1
450	Tetraethylrhodamine	H_2O pH 7	510	539	557	571	558	564	580	605	635	663	23	105.0	0.24	25.2	Rhoda 550-600	4.2.3
451	DAOTA 1	H_2O	489	511	558	578	563	570	587	621	650	674	29	50.0	0.31	15.5	Triangulenium	4.22
452	DAOTA 6	H_2O	489	509	558	575	560	567	584	614	640	660	26	50.0	0.36	18.0	Triangulenium	4.22
453	Cy3B	H_2O	493	515	558	569	555	559	572	587	620	639	14	130.0	0.67	87.1	Cyanine-Trimet	4.13.1
454	HMDSeR	100 mM P pH 7.4	516	540	558	571	568	574	590	631	656	680	32	92.0	0.01	0.5	Se-Rhodol	4.16
455	Pyronine 630	H_2O, 25 °C	509	535	559	573	558	564	579	600	613	642	20	100.0	0.92	92.0	Pyronine	4.4
456	JF559	10 mM HEPES pH 7.3	511	541	559	573	557	563	579	604	633	662	20	106.0	0.85	90.1	Rhoda 550-600	4.2.3
457	Coumarin-pyronine 9	100 mM P pH 7.5			560				579				19	132.5	0.13	17.2	Hybrid-Bonded	4.40.2
458	560CP	PBS pH 7.4	522	541	561	576	565	572	588	625	646	669	27	61.0	0.76	46.3	C(CH3)2-Rhoda	4.3.1
459	COUPY 10	H_2O			562				631				69	33.0	0.03	0.8	Hybrid-Bridged	4.40.3
460	DAOTA 2	H_2O	493	515	564	583	569	576	594	627	654	678	30	50.0	0.27	13.5	Triangulenium	4.22

(Continued)

Ent.	Short Names	Aqueous Solutions	$\lambda_{x,-0.25}$ nm	$\lambda_{x,-0.50}$ nm	$\lambda_{x,max}$ nm	$\lambda_{x,+0.50}$ nm	$\lambda_{e,-0.25}$ nm	$\lambda_{e,-0.50}$ nm	$\lambda_{e,max}$ nm	$\lambda_{e,+0.50}$ nm	$\lambda_{e,+0.25}$ nm	$\lambda_{e,+0.10}$ nm	Δ_S nm	ε_{max} mM^{-1}cm^{-1}	Φ_f	B_{max} mM^{-1}cm^{-1}	Class	Ref.
461	Atto 565 carboxylate	PBS pH 7.4, 22 °C	513	543	564	578	565	572	590	617	648	676	26	120.0	0.90	108.0	Rhoda 550-600	4.2.3
462	DAOTA 3	H_2O	449	511	565	583	571	577	594	626	654	681	29	50.0	0.28	14.0	Triangulenium	4.22
463	SR101-6G	PBS	508	540	565	579	563	571	586	608	638	666	21	80.0	0.68	54.4	Sulforhodam.	4.6
464	M-S 10a	H_2O, 0.1% TX-100	516	539	565	579	556	561	579	616	644	678	14	69.0	0.65	44.8	Bodipy-Small	4.14.1
465	NCR	H_2O			565				618				53	25.8	0.00	0.0	Hybrid-Fused	4.40.1
466	Atto Rho3B carboxylate	PBS pH 7.4, 22 °C	517	548	566	579	566	572	589	613	645	675	23	120.0	0.50	60.0	Rhoda 550-600	4.2.3
467	W-B 15	100 mM P pH 7.4	518	549	567	582	571	578	594	620	651	681	27	76.5	0.13	9.9	Rhoda 550-600	4.2.3
468	COUPY 11	H_2O			567				662				95	13.0	0.02	0.2	Hybrid-Bridged	4.40.3
469	NCRMe	H_2O			568				607				39	22.8	0.01	0.1	Hybrid-Fused	4.40.1
470	8-Trifluoromethyl-BODIPY 11	100 mM P pH 7.4	519	543	569	583	561	565	584	622	650	685	15	57.0	0.95	54.1	Bodipy-Small	4.14.1
471	JF570	10 mM HEPES pH 7.3	521	548	570	586	569	576	593	620	644	673	23	83.6	0.63	52.6	S-Rhodamine	4.3.3
472	JF571	10 mM HEPES pH 7.3	521	552	571	586	568	574	590	616	647	676	19	101	0.78	78.7	Rhoda 550-600	4.2.3
473	570CPH	PBS pH 7.4	580	602	571	633	617	623	600	673	707	740	29	79.0	0.71	56.0	C(CH3)2-Rhoda	4.3.1
474	Atto Rho11 carboxylate	PBS pH 7.4, 22 °C	522	553	572	585	571	577	595	619	650	680	23	120.0	0.80	96.0	Rhoda 550-600	4.2.3
475	SR101-Me_2	PBS	517	553	573	587	572	578	593	613	632	668	20	80.0	0.66	52.8	Sulforhodam.	4.6
476	SNARF-1	P pH 10	519	543	573	590	594	606	631	667	703	744	58	44.0	0.09	4.0	Rhodol-analog	4.15.3
477	M-S 10d	H_2O, 0.1% TX-100	523	547	574	589	559	563	582	619	648	682	8	118.0	0.57	67.2	Bodipy-Small	4.14.1
478	JF574	10 mM HEPES pH 7.3	524	555	574	589	571	578	594	620	651	681	20	103	0.78	80.3	Rhoda 550-600	4.2.3
479	575RH	PBS pH 7.4	524	555	574	589	574	580	597	624	655	685	23	55.0	0.74	40.7	Rhoda 550-600	4.2.3
480	SR101-B	PBS	522	557	575	590	574	580	595	615	631	665	20	80.0	0.65	52.0	Sulforhodam.	4.6

Ent.	Short Names	Aqueous Solutions	$\lambda_{x,-0.25}$ nm	$\lambda_{x,-0.50}$ nm	$\lambda_{x,max}$ nm	$\lambda_{x,+0.50}$ nm	$\lambda_{e,-0.25}$ nm	$\lambda_{e,-0.50}$ nm	$\lambda_{e,max}$ nm	$\lambda_{e,+0.50}$ nm	$\lambda_{e,+0.25}$ nm	$\lambda_{e,+0.10}$ nm	Δ_S nm	ε_{max} mM^{-1}cm^{-1}	Φ_f	B_{max} mM^{-1}cm^{-1}	Class	Ref.
481	5(6)-Carboxy-X-rhodamine (ROX)	H_2O pH <7	525	557	576	591	578	584	601	628	660	691	25	82.0	0.70	57.4	Rhoda 550-600	4.2.3
482	Z-L G	0.5 M PBS pH 7.4	526	550	577	592	574	578	598	637	667	704	21	50	0.21	10.6	Bodipy-Large	4.14.3
483	Atto Rho12 carboxylate	PBS pH 7.4, 22 °C	526	557	577	590	577	583	600	627	659	689	23	120.0	0.80	96.0	Rhoda 550-600	4.2.3
484	Cy3.205	PBS pH 7.4	509	532	578	590	574	579	593	610	645	665	15	120.0	0.14	16.8	Cyanine-Trimet	4.13.1
485	2-OH-6-DiEtAm-benzo[c]xanthen	H_2O	523	547	578	595	584	596	620	654	689	728	42	38.1	0.18	6.8	Rhodol-analog	4.15.3
486	Alexa Fluor 568	PBS pH 7.2/7.0, 20/22 °C	527	559	578	593	580	586	603	630	662	693	25	88.0	0.69	60.7	Rhodam Hi Ch	4.2.5
487	Seta-580-carboxylate	H_2O			578				602				24	90.0	0.55	49.5	nd	4.41
488	Redmond Red	100 mM P pH 7.1	518	547	579	593	577	583	595	614	629	661	16	52.3	0.84	43.9	Phenoxazine	4.25
489	Si-Fluorescein	100 mM NaOH	535	559	579	594	575	581	599	652	674	701	20	93.0	0.53	49.2	Si(CH3)2-Fluore	4.9
490	SR101-NaphtOH	PBS	481	509	580	607	580	588	619	676	700	735	39	80.0	0.15	12.0	Sulforhodol-an	4.27
491	2-COOH TM	100 mM P, 1% DMSO pH 10	536	560	580	595	575	580	598	651	672	700	18	110.0	0.38	41.8	Si(CH3)2-Fluore	4.9
492	Cy3.5	H_2O	511	534	581	593	577	582	596	613	649	669	15	120.0	0.14	16.8	Cyanine-Trimet	4.13.1
493	2-COOH DFTM	100 mM P, 1% DMSO pH 10	537	560	581	596	573	578	596	649	670	697	15	120.0	0.54	64.8	Si(CH3)2-Fluore	4.9
494	2-Me Tokyo Magenta	100 mM P pH 9	532	563	582	595	578	583	598	617	633	655	16	110.0	0.42	46.2	Si(CH3)2-Fluoron	4.12
495	M-S 10e	H_2O, 0.1% TX-100	532	556	584	599	574	578	598	637	667	704	14	124.0	0.41	50.8	Bodipy-Small	4.14.1
496	Cresyl violet (CV)	H_2O	523	553	585	599	605	611	625	646	663	698	40	33.1	0.36	11.9	Phenoxazine	4.25
497	JF585	10 mM HEPES pH 7.3	544	568	585	598	588	594	609	640	666	696	24	1.5	0.78	1.1	C(CH3)2-Rhoda	4.3.1

(Continued)

Ent.	Short Names	Aqueous Solutions	$\lambda_{x,-0.25}$ nm	$\lambda_{x,-0.50}$ nm	$\lambda_{x,max}$ nm	$\lambda_{x,+0.50}$ nm	$\lambda_{e,-0.25}$ nm	$\lambda_{e,-0.50}$ nm	$\lambda_{e,max}$ nm	$\lambda_{e,+0.50}$ nm	$\lambda_{e,+0.25}$ nm	$\lambda_{e,+0.10}$ nm	Δ_S nm	ε_{max} $mM^{-1}cm^{-1}$	Φ_f	B_{max} $mM^{-1}cm^{-1}$	Class	Ref.
498	Sulforhodamine 101	H_2O pH 7	534	569	586	602	584	590	605	622	632	647	19	116.0	0.90	104.4	Sulforhodam.	4.6
499	Carborhodol S-H 13-H	PBS pH 7.4	539	567	586	601	589	595	613	657	685	711	27	59.0	0.32	18.8	C(CH3)2-Rhod	4.16
500	Abberior Star 580 carboxylate	PBS pH 7.4	535	567	587	603	583	590	607	635	667	699	20	85.0	0.90	76.5	Rhoda 550-600	4.2.3
501	Atto Rho101 carboxylate	PBS pH 7.4, 22 °C	535	569	587	601	586	592	609	635	664	698	22	120.0	0.80	96.0	Rhoda 550-600	4.2.3
502	SR101-NaphtNH_2	PBS	493	526	589	610	593	602	626	666	691	722	37	80.0	0.19	15.2	Srhod.-analog	4.6
503	M-S 10f	H_2O, 0.1% TX-100	536	561	589	604	578	583	603	643	674	711	14	74.0	0.18	13.3	Bodipy-Small	4.14.1
504	PONy P1	PBS pH 7.4	535	564	590	605	602	610	627	652	667	701	37	59.0	0.33	19.4	Pyronine	4.4
505	Alexa Fluor 594	PBS pH 7.2/7.0, 20/22 °C	537	570	590	606	593	599	617	646	679	712	27	92.0	0.66	60.7	Rhodam Hi Ch	4.2.5
506	2-COOH DCTM	100 mм P, 1% DMSO pH 10	545	570	591	606	583	589	607	662	684	712	16	120.0	0.48	57.6	Si(CH3)2-Fluore	4.9
507	Atto 590 carboxylate	PBS pH 7.4, 22 °C	535	569	593	609	594	601	622	651	687	718	29	120.0	0.80	96.0	Rhoda 550-600	4.2.3
508	JF593	10 mм HEPES pH 7.3	541	569	593	610	587	594	612	641	667	697	19	90.3	0.55	49.6	S-Rhodamine	4.3.3
509	2-Me SiR600	H_2O	542	574	593	607	593	599	613	633	650	674	20	91.0	0.38	34.5	Si(CH3)2-Rosa	4.18.1
510	CF594	H_2O			593				614				21	115.0	0.45	51.7	nd (Rhodam.)	4.41
511	Naphthofluorescein	C pH 9.5	541	568	595	613	626	635	660	703	740	780	65	44.0	0.14	6.1	Fluores-analog	4.8.4
512	COUPY 12	H_2O			595				683				88	6.9	0.05	0.3	Hybrid-Bridged	4.40.3
513	2′,7′-Difluoro-Si-Fluorescein	100 mм NaOH	549	574	596	612	593	599	618	675	698	727	22	60.9	0.42	25.5	Si(CH3)2-Fluore	4.9
514	PONy P9	PBS pH 7.4	541	570	597	613	608	615	633	658	674	709	36	66.0	0.22	14.5	Pyronine	4.4
515	2-Me DCTM	100 mм P pH 9	544	577	597	611	586	592	607	627	643	666	10	130.0	0.48	62.4	Si(CH3)2-Fluoron	4.12
516	2′,4′,5′,7′-TetraF-Si-fluorescein	100 mм NaOH	550	575	597	613	588	594	613	669	691	720	16	65.0	0.67	43.5	Si(CH3)2-Fluore	4.9

Ent.	Short Names	Aqueous Solutions	$\lambda_{x,-0.25}$ nm	$\lambda_{x,-0.50}$ nm	$\lambda_{x,max}$ nm	$\lambda_{x,+0.50}$ nm	$\lambda_{e,-0.25}$ nm	$\lambda_{e,-0.50}$ nm	$\lambda_{e,max}$ nm	$\lambda_{e,+0.50}$ nm	$\lambda_{e,+0.25}$ nm	$\lambda_{e,+0.10}$ nm	Δ_S nm	ε_{max} mM^{-1}cm^{-1}	Φ_f	B_{max} mM^{-1}cm^{-1}	Class	Ref.
517	4,5,6,7-TetraF-Si-fluorescein	100 mM NaOH	551	576	598	614	589	595	614	670	693	722	16	84.1	0.50	42.0	Si(CH3)2-Fluore	4.9
518	NR600	PBS, 1% DMSO pH 7.4	551	580	600	613	597	603	619	646	665	689	19	67.6	0.68	45.9	POOH-Fluoron	4.12
519	Distyryl-BODIPY 11	PBS pH 7.5	546	572	601	617	626	632	655	703	739	784	54	17.9	0.04	0.7	Bodipy-Med.	4.14.2
520	PONy P1-Halo	PBS pH 7.4	546	576	603	619	610	617	635	660	676	711	32	18.0	0.26	4.6	Pyronine	4.4
521	PONy P6	PBS pH 7.4	546	576	603	619	613	620	638	663	680	715	35	69.0	0.09	6.2	Pyronine	4.4
522	Atto 594 carboxylate	PBS pH 7.4, 22 °C	547	583	603	617	601	608	626	652	688	718	23	120.0	0.85	102.0	Rhodam >600	4.2.4
523	Atto Rho13 carboxylate	PBS pH 7.4, 22 °C	547	584	603	618	602	609	627	654	690	723	24	120.0	0.80	96.0	Rhodam >600	4.2.4
524	Abberior Star 600 carboxylate	PBS pH 7.4	550	587	604	619	600	606	623	647	677	714	19	93.0	0.80	74.4	Rhodam >600	4.2.4
525	NR604	PBS, 1% DMSO pH 7.4	554	584	604	618	604	610	627	654	672	697	23	45.7	0.48	21.9	POOH-Fluoron	4.12
526	4,10-Dibromona-phthofluorescein	C pH 9.5	549	578	605	624	646	655	682	728	768	811	77	32.0	0.05	1.6	Fluores-analog	4.8.4
527	DMAC-SR101	PBS			605				647				42	80.0	0.09	7.2	Hybrid-Fused	4.40.1
528	Tetramethycar-borhodamine	10 mM HEPES pH 7.3	562	588	606	620	604	610	626	659	686	718	20	121.0	0.52	62.9	C(CH3)2-Rhoda	4.3.1
529	JF608	10 mM HEPES pH 7.3	564	589	608	623	609	615	631	664	692	725	23	99.0	0.67	66.3	C(CH3)2-Rhoda	4.3.1
530	2′COOH-Si-rhodol	100 mM P pH 7.4	559	588	609	625	605	611	630	677	706	734	21	70.0	0.22	15.4	Si(CH3)2-Rhod	4.16
531	Vita Blue	C pH 9.5	560	586	609	625	636	643	665	731	758	793	56	36.0	0.15	5.4	S-Fluore-analog	4.10
532	2′COOH-4CH_2OH-Si-rhodol	100 mM P pH 7.4	560	589	610	626	605	611	630	677	706	734	20	70.0	0.21	14.7	Si(CH3)2-Rhod	4.16
533	DEAC-SR101	PBS			610				642				32	80.0	0.10	8.0	Hybrid-Fused	4.40.1

(Continued)

Ent.	Short Names	Aqueous Solutions	$\lambda_{x,-0.25}$ nm	$\lambda_{x,-0.50}$ nm	$\lambda_{x,max}$ nm	$\lambda_{x,+0.50}$ nm	$\lambda_{e,-0.25}$ nm	$\lambda_{e,-0.50}$ nm	$\lambda_{e,max}$ nm	$\lambda_{e,+0.50}$ nm	$\lambda_{e,+0.25}$ nm	$\lambda_{e,+0.10}$ nm	Δ_S nm	ε_{max} mM^{-1}cm^{-1}	Φ_f	B_{max} mM^{-1}cm^{-1}	Class	Ref.
534	JS-R	10 mM HEPES pH 7.4	558	589	611	625	609	617	631	648	659	671	20	130.0	0.59	76.7	B(OH)2-Pyron	4.5
535	2′Me-Si-rhosoL	100 mM P pH 7.4	553	582	612	627	608	614	630	651	667	693	18	70.0	0.28	19.6	Si(CH3)2-Rosol	4.20
536	2′Me-4CH$_2$OH-Si-rosol	100 mM P pH 7.4	554	583	613	628	609	615	631	652	668	694	18	70.0	0.30	21.0	Si(CH3)2-Rosol	4.20
537	Pyrrolidinylcar-borhodamine	10 mM HEPES pH 7.3	568	594	613	628	610	617	633	666	694	727	20	87.0	0.54	46.9	C(CH3)2-Rhoda	4.3.1
538	Atto 610 carboxylate	PBS pH 7.4, 22 °C	529	554	616	581	568	573	633	608	634	659	17	150.0	0.70	105.0	C(CH3)2-Pyron	4.5
539	2′,4,5,6,7,7′-HexaF-Si-fluorescein	100 mM NaOH	567	594	617	634	609	615	635	695	720	751	18	67.9	0.32	21.7	Si(CH3)2-Fluore	4.9
540	CF620R	H_2O			617				639				22	115.0	0.30	34.0	nd (Rhodam.)	4.41
541	Oxazine 170	H_2O pH 2-7	539	560	620	643	624	632	650	668	679	692	30	48.0	0.24	11.5	Phenoxazine	4.25
542	2-Me SiR620	100 mM P, 1% DMSO pH 7.4	550	586	620	636	616	622	638	657	675	699	18	96.0	0.21	20.1	Si(CH3)2-Rosa	4.18.1
543	Bza-OH	H_2O pH 7.4	551	584	620	636	631	638	653	676	694	733	33	12.8	0.37	4.7	Phenoxazine	4.25
544	Bzh-Cl	H_2O pH 7.4	551	584	620	636	629	636	651	674	692	731	31	36.1	0.31	11.1	Phenoxazine	4.25
545	Bzh-H	H_2O pH 7.4	551	584	620	636	629	636	651	674	692	731	31	11.3	0.44	4.9	Phenoxazine	4.25
546	RF620	10 mM PBS pH 7.4	569	600	620	635	614	620	636	659	674	701	16	109.0	0.36	39.2	B(OH)2-Rosa	4.18.3
547	Atto 620 carboxylate	PBS pH 7.4, 22 °C	580	602	620	633	617	623	642	673	707	740	22	120.0	0.50	60.0	C(CH3)2-Rhoda	4.3.1
548	Square-655-carboxylate	P pH 7.4	576	604	622	638	623	630	644	664	682	720	22	177.0	0.09	15.9	Squaraine	4.35
549	Squaraine T-P 4c	67 mM P pH 7.4	577	606	624	639	615	621	635	655	672	709	11	245.0	0.02	4.9	Squaraine	4.35
550	HC6H-SR101	PBS, 5% BSA			624				654				30	80.0	0.15	12.0	Hybrid-Fused	4.40.1

Ent.	Short Names	Aqueous Solutions	$\lambda_{x,-0.25}$ nm	$\lambda_{x,-0.50}$ nm	$\lambda_{x,max}$ nm	$\lambda_{x,+0.50}$ nm	$\lambda_{e,-0.25}$ nm	$\lambda_{e,-0.50}$ nm	$\lambda_{e,max}$ nm	$\lambda_{e,+0.50}$ nm	$\lambda_{e,+0.25}$ nm	$\lambda_{e,+0.10}$ nm	Δ_S nm	ε_{max} mM^{-1}cm^{-1}	Φ_f	B_{max} mM^{-1}cm^{-1}	Class	Ref.
551	Bza-Oet	H_2O pH 7.4	555	588	625	641	631	638	653	676	694	733	28	33.5	0.26	8.7	Phenoxazine	4.25
552	HCDN	PBS pH 7.4	564	603	625	641	725	735	755	778	789	802	130	195.0	0.28	54.5	Cyanine-Hepta	4.13.3
553	Atto Rho14 carboxylate	PBS pH 7.4, 22 °C	564	606	626	639	622	628	646	668	693	738	20	140.0	0.80	112.0	Rhodam >600	4.2.4
554	SqB1	P pH 7.4	580	609	627	642	629	636	650	671	688	728	23	150.0	0.05	7.5	Squaraine	4.35
555	Distyryl-BODIPY 18	H_2O	568	596	628	646	613	619	641	687	721	764	13	71.0	0.38	26.9	Bodipy-Med.	4.14.2
556	Squaraine T-P 4a	67 mM P pH 7.4	581	610	628	644	621	628	642	662	679	718	14	163.0	0.03	4.8	Squaraine	4.35
557	PONy P10	PBS pH 7.4	567	599	629	646	640	648	668	696	714	753	39	52.0	0.12	6.2	Pyronine	4.4
558	BODIPY 630/650	H_2O	569	597	629	647	618	623	646	692	728	772	17	97.0	0.50	48.5	Bodipy-Small	4.14.1
559	PONy P13	PBS pH 7.4	568	600	630	647	638	646	666	694	712	750	36	52.0	0.17	8.8	Pyronine	4.4
560	Distyryl-BODIPY 17	H_2O	570	598	630	648	615	620	643	689	724	767	13	76.6	0.30	22.9	Bodipy-Med.	4.14.2
561	Atto 633 carboxylate	PBS pH 7.4, 22 °C	575	610	630	646	627	635	651	687	724	756	21	130.0	0.64	83.2	C(CH3)2-Rhoda	4.3.1
562	630GeRH	PBS pH 7.4	582	612	631	647	629	635	651	671	683	722	20	61.0	0.60	36.6	Ge(CH3)2-Rhoda	4.3.3
563	POF	PBS pH 7.4	561	591	632	652	635	643	661	687	714	750	29	54.5	0.24	13.0	POPh-Fluorone	4.12
564	Alexa Fluor 633	PBS/H_2O pH 7.2, 20/22 °C	568	608	632	650	621	629	647	670	691	722	15	100.0	0.64	64.0	Sulforhodam.	4.6
565	PONy P2	PBS pH 7.4	569	602	632	649	644	652	672	700	718	758	40	51.0	0.16	8.1	Pyronine	4.4
566	KK9046	H_2O	573	613	632	648	629	636	654	681	714	755	22	90.0	0.45	40.5	Rhodam >600	4.2.4
567	NR632	10 mM PBS pH 7.4	578	608	632	646	631	638	655	678	695	723	23	64.0	0.26	16.6	POPh-Rosol	4.20
568	SiP-Pr	50 mM PBS pH 7.4	579	613	632	647	607	619	652	661	675	691	20	64.2	0.21	13.4	C(CH3)2-Pyron	4.5

(Continued)

Ent.	Short Names	Aqueous Solutions	$\lambda_{x,-0.25}$ nm	$\lambda_{x,-0.50}$ nm	$\lambda_{x,max}$ nm	$\lambda_{x,+0.50}$ nm	$\lambda_{e,-0.25}$ nm	$\lambda_{e,-0.50}$ nm	$\lambda_{e,max}$ nm	$\lambda_{e,+0.50}$ nm	$\lambda_{e,+0.25}$ nm	$\lambda_{e,+0.10}$ nm	Δ_S nm	ε_{max} mм$^{-1}$cm^{-1}	Φ_f	B_{max} mм$^{-1}$cm^{-1}	Class	Ref.
569	Squaraine T-P 5c	67 mм P pH 7.4	584	614	632	648	621	628	642	662	679	718	10	265.0	0.06	15.9	Squaraine	4.35
570	SqO1	P pH 7.4	584	614	632	648	621	628	642	662	679	718	10	265.0	0.07	18.5	Squaraine	4.35
571	Seta-632-carboxylate	P pH 7.4			632				641				9	280.0	0.06	16.8	nd	4.41
572	TMDHS or SiP-Me	50 mм PBS pH 7.4	570	603	633	651	627	635	654	681	698	735	21	64.2	0.41	26.3	Si(CH3)2-Pyron	4.5
573	Seta-633-carboxylate	P pH 7.4			633				644				11	250.0	0.07	17.5	nd	4.41
574	TMDHS or SiP-Me	H_2O	571	604	634	652	626	634	653	680	697	734	19	64.2	0.18	11.5	Si(CH3)2-Pyron	4.5
575	Abberior Star 635 carboxylate	PBS pH 7.4	574	615	634	650	629	636	654	681	714	755	20	63.0	0.51	32.1	Rhodam >600	4.2.4
576	Square-635-carboxylate	P pH 5.6	591	617	634	648	624	633	646	665	679	725	12	135.0	0.16	21.6	Squaraine	4.35
577	GeR	PBS pH 7.4	601	621	634	658	635	643	655	681	697	734	21	97.0	0.43	41.7	Ge(CH3)2-Rhoda	4.3.3
578	DY-635 carboxylate	PBS pH 7.4	550	583	635	658	630	641	669	729	766	808	34	175.0	0.05	8.7	Merocyanine	4.24
579	Bzm-H	H_2O pH 7.4	563	597	635	652	651	658	674	698	718	760	39	29.9	0.14	4.1	Phenoxazine	4.25
580	Abberior Star 635P carboxylate	PBS pH 7.4	575	616	635	652	630	637	655	682	715	756	20	80.0	0.55	44.0	Rhodam >600	4.2.4
581	Atto 635 carboxylate	PBS pH 7.4, 22 °C	579	612	635	649	626	638	659	683	701	720	24	130.0	0.25	32.5	C(CH3)2-Pyron	4.5
582	3,6-Azetid.-Si-pyronine	10 mм HEPES pH 7.3	573	606	636	654	623	630	649	675	692	729	13	97.1	0.62	60.2	Si(CH3)2-Pyron	4.5
583	nor-SqO1	P pH 7.4	588	618	636	652	631	639	653	674	692	731	17	149.0	0.32	47.6	Squaraine	4.35
584	Carborhodamine 16a	H_2O	588	616	636	652	633	640	657	693	723	759	21	114.0	0.62	70.6	C(CH3)2-Rhoda	4.3.1
585	Squaraine T-P 5a	67 mм P pH 7.4	592	620	636	653	624	631	648	666	680	705	12	216.0	0.06	12.9	Squaraine	4.35
586	2-Me SiR640	100 mм P, 1% DMSO pH 7.4	509	548	637	658	635	643	660	683	707	733	23	54.0	0.18	9.7	Si(CH3)2-Rosa	4.18.1
587	KK1119	H_2O	577	618	637	654	634	641	660	687	721	763	23	90.0	0.55	49.5	Rhodam >600	4.2.4

Ent.	Short Names	Aqueous Solutions	$\lambda_{x,-0.25}$ nm	$\lambda_{x,-0.50}$ nm	$\lambda_{x,max}$ nm	$\lambda_{x,+0.50}$ nm	$\lambda_{e,-0.25}$ nm	$\lambda_{e,-0.50}$ nm	$\lambda_{e,max}$ nm	$\lambda_{e,+0.50}$ nm	$\lambda_{e,+0.25}$ nm	$\lambda_{e,+0.10}$ nm	Δ_S nm	ε_{max} mм$^{-1}$cm^{-1}	Φ_f	B_{max} mм$^{-1}$cm^{-1}	Class	Ref.
588	SiP-F	50 mм PBS pH 7.4	577	611	637	653	602	612	651	655	667	679	14	64.2	0.20	12.8	C(CH3)2-Pyron	4.5
589	KK114	H_2O	580	619	637	652	635	642	660	685	713	762	23	92.0	0.55	50.6	Rhodam >600	4.2.4
590	B-Z 14	H_2O	576	605	638	656	626	632	655	703	739	784	17	102.0	0.42	42.8	Bodipy-Large	4.14.3
591	Abberior Star Red carboxylate	PBS pH 7.4	578	619	638	655	630	637	655	682	715	756	17	100.0	0.66	66.0	Rhodam >600	4.2.4
592	NR638	10 mм PBS pH 7.4	583	613	638	652	637	645	662	686	703	732	24	55.6	0.21	11.6	POPh-Rosol	4.20
593	nor-SqB1	P pH 7.4	589	619	638	654	644	651	666	688	706	748	28	88.0	0.20	17.6	Squaraine	4.35
594	nor-SqB2	P pH 7.4	589	619	638	654	644	651	666	688	706	748	28	114.0	0.22	25.0	Squaraine	4.35
595	nor-SqO2	P pH 7.4	590	620	639	655	635	642	657	678	696	736	18	183.0	0.33	60.3	Squaraine	4.35
596	Carborhodamine 1b	H_2O	590	619	639	655	636	644	661	698	728	764	22	80.0	0.58	46.4	C(CH3)2-Rhoda	4.3.1
597	Bpe-H	H_2O pH 7.4	567	601	640	657	657	665	681	706	726	769	41	48.1	0.23	11.0	Phenoxazine	4.25
598	Carborhodamine 1c	H_2O	591	619	640	656	637	645	662	699	730	766	22	89.0	0.53	47.1	C(CH3)2-Rhoda	4.3.1
599	Carborhodamine 20a	H_2O	591	619	640	656	636	644	661	698	728	764	21	70.0	0.53	37.1	C(CH3)2-Rhoda	4.3.1
600	Carborhodamine 21a	H_2O	591	619	640	656	638	646	663	700	731	767	23	70.0	0.56	39.2	C(CH3)2-Rhoda	4.3.1
601	Carborhodamine 21b	H_2O	591	619	640	656	639	646	664	701	732	768	24	68.0	0.56	38.0	C(CH3)2-Rhoda	4.3.1
602	Carborhodamine 22	H_2O	591	619	640	656	639	646	664	701	732	768	24	67.0	0.54	36.1	C(CH3)2-Rhoda	4.3.1
603	Seta-635-pH-di-carboxylate	P pH 9			640				656				16	188.0	0.33	62.0	nd	4.41
604	3,6-Azetid.-9-CA.-Si-pyronine	10 mм HEPES pH 7.3	577	610	641	659	630	638	657	684	701	739	16	120.0	0.26	31.2	Si(CH3)2-Pyron	4.5
605	Distyryl-BODIPY 13	PBS pH 7.5	579	608	641	659	628	633	657	705	742	787	16	58.0	0.25	14.5	Bodipy-Med.	4.14.2
606	640SiRH	PBS pH 7.4	584	519	641	658	639	646	662	681	694	726	21	51.0	0.42	21.4	Si(CH3)2-Rhoda	4.3.2
607	Carborhodamine 20b	H_2O	592	620	641	657	638	646	663	700	731	767	22	80.0	0.53	42.4	C(CH3)2-Rhoda	4.3.1
608	Distyryl-BODIPY 12	PBS pH 7.5	579	609	642	660	628	633	657	705	742	787	15	55.1	0.22	12.1	Bodipy-Med.	4.14.2
609	Cy5′	H_2O	583	621	642	656	639	645	660	677	688	716	18	138.0	0.17	23.4	Cyanine-Penta	4.13.2

(Continued)

Ent.	Short Names	Aqueous Solutions	$\lambda_{x,-0.25}$ nm	$\lambda_{x,-0.50}$ nm	$\lambda_{x,max}$ nm	$\lambda_{x,+0.50}$ nm	$\lambda_{e,-0.25}$ nm	$\lambda_{e,-0.50}$ nm	$\lambda_{e,max}$ nm	$\lambda_{e,+0.50}$ nm	$\lambda_{e,+0.25}$ nm	$\lambda_{e,+0.10}$ nm	Δ_S nm	ε_{max} $mM^{-1}cm^{-1}$	Φ_f	B_{max} $mM^{-1}cm^{-1}$	Class	Ref.
610	CypHer5	A pH 4.6	583	621	643	657	641	646	662	679	690	719	19	100.0	0.19	19.0	Cyanine-Penta	4.13.2
611	W-L 8a	PBS	583	621	643	657	645	651	667	684	696	725	24	87.0	0.12	10.4	Cyanine-Penta	4.13.2
612	W-L 8c	PBS	583	621	643	657	645	650	666	683	695	724	23	107.0	0.12	12.8	Cyanine-Penta	4.13.2
613	W-L 8d	PBS	583	621	643	657	645	650	666	683	695	724	23	169.0	0.13	21.9	Cyanine-Penta	4.13.2
614	W-L 8e	PBS	583	621	643	657	644	649	665	682	694	722	22	195.0	0.09	17.5	Cyanine-Penta	4.13.2
615	Neo-Cy5	PBS pH 7.4	583	621	643	657	639	645	660	677	688	716	17	170.0	0.10	17.0	Cyanine-Penta	4.13.2
616	SiTMR	10 mM HEPES pH 7.3	586	566	643	660	638	645	662	682	696	729	19	28.2	0.41	11.5	Si(CH3)2-Rhoda	4.3.2
617	W-L 8b	PBS	584	622	644	659	646	652	668	685	697	726	24	189.0	0.15	28.3	Cyanine-Penta	4.13.2
618	DY-636 carboxylate	PBS pH 7.4	558	591	645	669	632	643	671	732	769	811	26	190.0	0.09	16.3	Merocyanine	4.24
619	SiR-COOH	H_2O	588	567	645	662	637	644	661	681	695	728	16	100.0	0.39	39.0	Si(CH3)2-Rhoda	4.3.2
620	SiR	PBS pH 7.4	589	623	645	663	640	647	661	687	702	737	16	93.0	0.41	38.1	Si(CH3)2-Rhoda	4.3.2
621	Sulfo-cyanine5	H_2O	586	624	646	661	641	646	662	679	690	719	16	271.0	0.20	54.2	Cyanine-Penta	4.13.2
622	Cy5.11-OH	H_2O/PBS	586	624	646	661	645	651	667	684	696	725	21	215.0	0.06	12.9	Cyanine-Penta	4.13.2
623	JF646	10 mM HEPES pH 7.3	589	568	646	663	639	646	664	684	698	732	18	5.6	0.54	3.0	Si(CH3)2-Rhoda	4.3.2
624	2-Me SiR650	H_2O	591	628	646	661	636	642	660	681	707	749	14	110.0	0.31	34.1	Si(CH3)2-Rosa	4.18.1
625	Atto 647N carboxylate	PBS pH 7.4, 22 °C	592	626	646	661	642	648	664	690	706	753	18	150.0	0.65	97.5	C(CH3)2-Rhoda	4.3.1
626	Seta-646-carboxylate	P pH 7.4			646				656				10	207.0	0.10	20.7	nd	4.41
627	3,7-Di(azetidin) phenoxazine	10 mM HEPES pH 7.3	572	607	647	664	638	646	661	684	703	743	14	99.0	0.24	23.7	Phenoxazine	4.25
628	Cy5	H_2O	587	625	647	662	642	647	663	680	691	720	16	250.0	0.27	67.5	Cyanine-Penta	4.13.2
629	Zy5	P pH 7.4	587	625	647	662	644	649	665	682	694	722	18	250.0	0.27	67.5	Cyanine-Penta	4.13.2
630	Atto 647 carboxylate	PBS pH 7.4, 22 °C	609	624	647	659	639	646	667	700	734	767	20	120.0	0.20	24.0	C(CH3)2-Pyron	4.5

Ent.	Short Names	Aqueous Solutions	$\lambda_{x,-0.25}$ nm	$\lambda_{x,-0.50}$ nm	$\lambda_{x,max}$ nm	$\lambda_{x,+0.50}$ nm	$\lambda_{e,-0.25}$ nm	$\lambda_{e,-0.50}$ nm	$\lambda_{e,max}$ nm	$\lambda_{e,+0.50}$ nm	$\lambda_{e,+0.25}$ nm	$\lambda_{e,+0.10}$ nm	Δ_S nm	ε_{max} $mM^{-1}cm^{-1}$	Φ_f	B_{max} $mM^{-1}cm^{-1}$	Class	Ref.
631	Des-carboxy-JF646	10 mM HEPES pH 7.3	571	611	648	665	638	645	662	683	701	728	14	116	0.20	23.2	Si(CH3)2-Rosa	4.18.1
632	SiR-methyl	H_2O	571	611	648	665	638	645	662	683	701	728	14	100.0	0.39	39.0	Si(CH3)2-Rosa	4.18.1
633	Cy5.10-OH	H_2O/PBS	588	626	648	663	645	651	667	684	696	725	19	215.0	0.09	19.3	Cyanine-Penta	4.13.2
634	Cy5.13-OH	H_2O/PBS	588	626	648	663	647	653	669	686	698	727	21	215.0	0.12	25.8	Cyanine-Penta	4.13.2
635	3-Methyl-5-Methoxy-JF646	10 mM HEPES pH 7.3	572	612	649	666	640	647	664	685	704	731	15	110	0.49	53.9	Si(CH3)2-Rosa	4.18.1
636	3-Methyl-JF646	10 mM HEPES pH 7.3	572	612	649	666	642	649	666	687	706	733	17	118	0.47	55.4	Si(CH3)2-Rosa	4.18.1
637	SiP-Ph	50 mM PBS pH 7.4	588	624	649	665	613	630	665	672	683	698	16	80.0	0.23	18.4	C(CH3)2-Pyron	4.5
638	EVOblue 30	H_2O	575	610	650	667	644	651	667	691	710	751	17	100.0	0.10	10.0	Phenoxazine	4.25
639	NIR4	H_2O	589	628	650	665	649	655	671	688	700	729	21	260.0	0.43	111.8	Cyanine-Penta	4.13.2
640	Cy5.12-OH	H_2O/PBS	589	628	650	665	652	658	674	692	703	733	24	215.0	0.13	27.9	Cyanine-Penta	4.13.2
641	Cy5.29.OH	PBS	589	628	650	665	645	650	666	683	695	724	16	250.0	0.20	50.0	Cyanine-Penta	4.13.2
642	AF647	PBS pH 7.2, 20/22 °C	589	628	650	665	650	656	672	690	701	731	22	270.0	0.33	89.1	Cyanine-Penta	4.13.2
643	AFD647	PBS pH 7.2, 20/22 °C	589	628	650	665	650	656	672	690	701	731	22	270.0	0.33	89.1	Cyanine-Penta	4.13.2
644	Alexa Fluor 647	PBS pH 7.2, 20/22 °C	589	628	650	665	650	656	672	690	701	731	22	270.0	0.33	89.1	Cyanine-Penta	4.13.2
645	Cy5.18.OH	PBS	590	628	650	664	646	651	667	685	698	739	17	250.0	0.27	67.5	Cyanine-Penta	4.13.2
646	CF647	H_2O			650				665				15	240.0	0.12	27.9	nd (Cyanine)	4.41
647	3,7-Dimethyl-JF646	10 mM HEPES pH 7.3	574	614	651	668	640	647	664	685	704	731	13	89.6	0.51	45.7	Si(CH3)2-Rosa	4.18.1
648	3-Hydroxy-JF646	10 mM HEPES pH 7.3	574	614	651	668	642	649	666	687	706	733	15	136	0.48	65.2	Si(CH3)2-Rosa	4.18.1

(Continued)

Ent.	Short Names	Aqueous Solutions	$\lambda_{x,-0.25}$ nm	$\lambda_{x,-0.50}$ nm	$\lambda_{x,max}$ nm	$\lambda_{x,+0.50}$ nm	$\lambda_{e,-0.25}$ nm	$\lambda_{e,-0.50}$ nm	$\lambda_{e,max}$ nm	$\lambda_{e,+0.50}$ nm	$\lambda_{e,+0.25}$ nm	$\lambda_{e,+0.10}$ nm	Δ_S nm	ε_{max} mM^{-1}cm^{-1}	Φ_f	B_{max} mM^{-1}cm^{-1}	Class	Ref.
649	B-Z 19	H_2O	587	617	651	670	646	652	677	728	767	816	26	118.5	0.23	27.2	Bodipy-Large	4.14.3
650	Thienyl-JF646	10 mM HEPES pH 7.3	593	572	651	668	642	649	667	687	701	735	16	113	0.40	45.2	Si(CH3)2-Rhoda	4.3.2
651	Seta-650-carboxylate	P pH 7.4			651				671				20	200.0	0.28	56.0	nd	4.41
652	SMSiR	H_2O	584	626	652	671	643	651	670	695	717	751	18	120.0	0.31	37.2	Si(CH3)2-Srhod	4.7
653	S-JF646	10 mM HEPES pH 7.3	586	620	652	671	640	648	668	696	714	753	16	21.2	0.25	5.3	Si(CH3)2-Pyron	4.5
654	DY-654 carboxylate	PBS pH 7.4	594	627	653	673	651	659	677	699	711	732	24	220.0	0.16	35.2	Merocyanine	4.24
655	Square-650-carboxylate	P pH 5.6	598	632	653	668	651	658	671	694	712	765	18	135.0	0.16	21.6	Squaraine	4.35
656	DY-651 carboxylate	PBS pH 7.4	599	631	653	672	652	659	678	699	711	731	25	160.0	0.15	24.1	Merocyanine	4.24
657	2-Me SiR654	100 mM P, 1% DMSO pH 7.4	576	616	654	671	647	655	672	693	713	740	18	80.0	0.24	19.2	Si(CH3)2-Rosa	4.18.1
658	Square-680-carboxylate	P pH 7.4	603	634	654	671	647	655	670	692	711	753	16	138.0	0.04	5.1	Squaraine	4.35
659	3,7-Di(ethylamino) phenoxazine	10 mM HEPES pH 7.3	578	615	655	673	646	653	669	693	712	754	14	111.0	0.07	7.7	Phenoxazine	4.25
660	Sulfo-JF646	10 mM HEPES pH 7.3	586	629	655	674	643	651	670	695	717	751	15	124.0	0.51	63.2	Si(CH3)2-Srhod	4.7
661	5-Aza-3-methyl-JF646	10 mM HEPES pH 7.3	578	618	656	674	646	653	670	691	710	738	14	133	0.48	63.8	Si(CH3)2-Rosa	4.18.1
662	SiRF 13a	H_2O	598	577	657	675	651	659	677	698	712	747	20	53.0	0.63	33.3	Si(CH3)2-Rhoda	4.3.2
663	SiRF 15a	H_2O	598	577	657	675	651	658	676	697	711	746	19	40.0	0.40	16.0	Si(CH3)2-Rhoda	4.3.2
664	Square-660-carboxylate	P pH 7.4	600	635	657	675	654	660	676	702	730	769	19	182.0	0.03	5.4	Squaraine	4.35
665	Squaraine T-P 4d	67 mM P pH 7.4	606	638	658	675	654	662	677	700	719	762	19	182.0	0.03	5.4	Squaraine	4.35
666	Squaraine T-P 4b	67 mM P pH 7.4	607	639	659	676	653	661	676	698	718	760	17	207.0	0.02	4.1	Squaraine	4.35
667	PONy P5	PBS pH 7.4	592	627	660	679	673	682	704	735	755	799	44	38.0	0.10	3.8	Pyronine	4.4

Ent.	Short Names	Aqueous Solutions	$\lambda_{x,-0.25}$ nm	$\lambda_{x,-0.50}$ nm	$\lambda_{x,max}$ nm	$\lambda_{x,+0.50}$ nm	$\lambda_{e,-0.25}$ nm	$\lambda_{e,-0.50}$ nm	$\lambda_{e,max}$ nm	$\lambda_{e,+0.50}$ nm	$\lambda_{e,+0.25}$ nm	$\lambda_{e,+0.10}$ nm	Δ_S nm	ε_{max} mм$^{-1}$cm^{-1}	Φ_f	B_{max} mм$^{-1}$cm^{-1}	Class	Ref.
668	Z-L H	0.5 м PBS pH 7.4	594	625	660	679	645	651	676	727	766	815	16	50	0.01	0.6	Bodipy-Large	4.14.3
669	Phosphorilated oxazine 2-H	H_2O pH ~7	583	620	661	679	654	662	678	703	723	765	17	60.0	0.28	16.8	Phenoxazine	4.25
670	Seta-660-di-carboxylate	P pH 7.4			661				672				11	220.0	0.11	24.2	nd	4.41
671	Atto Oxa12 carboxylate	PBS pH 7.4, 22 °C	584	621	662	680	657	665	681	706	726	769	19	125.0	0.30	37.5	Phenoxazine	4.25
672	NIR2	H_2O	599	639	662	677	661	667	684	702	714	745	22	250.0	0.34	85.0	Cyanine-Penta	4.13.2
673	SiRF 1a-H	H_2O	602	581	662	680	654	662	680	701	716	751	18	118.0	0.66	77.8	Si(CH3)2-Rhoda	4.3.2
674	nor-SqB3	P pH 7.4	610	642	662	679	660	668	684	707	727	770	22	102.0	0.18	18.3	Squaraine	4.35
675	nor-SqO3	P pH 7.4	610	642	662	679	656	663	679	702	721	764	17	124.0	0.21	26.0	Squaraine	4.35
676	Atto 665 carboxylate	PBS pH 7.4, 22 °C			662				680				18	160.0	0.60	96.0	nd (pb Pyron.)	4.41
677	Atto 655 carboxylate	PBS pH 7.4, 22 °C	587	629	663	679	656	664	680	707	733	784	17	125.0	0.30	37.5	Phenoxazine	4.25
678	$SiOH_2R$	10 mм PBS pH 7.4	599	639	663	679	656	662	681	700	713	738	18	105.0	0.43	45.1	Si(OH)2-Rosa	4.18.1
679	SiRF 1b-H	H_2O	603	581	663	681	654	662	680	701	716	751	17	90.0	0.70	63.0	Si(CH3)2-Rhoda	4.3.2
680	CF660R	H_2O			663				682				19	100	0.08	7.6	nd (Rhodam.)	4.41
681	Alexa Fluor 660	PBS pH 7.2/7.5, 20/22 °C			663				690				27	132.0	0.37	48.8	nd	4.41
682	Si-9-Iminopyronine 2f	~1 mм HCl, 1% DMSO pH 3	595	631	664	683	653	661	682	711	730	770	18	123	0.26	31.9	Si(CH3)2-Pyron	4.5
683	2-Me SiR666	100 mм P, 1% DMSO pH 7.4	585	627	666	684	659	666	684	706	726	755	18	80.0	0.25	20.0	Si(CH3)2-Rosa	4.18.1
684	Si-9-Iminopyronine 2c	~1 mм HCl, 1% DMSO pH 3	597	633	666	685	652	660	681	710	729	769	15	83.2	0.28	23.3	Si(CH3)2-Pyron	4.5

(Continued)

Ent.	Short Names	Aqueous Solutions	$\lambda_{x,-0.25}$ nm	$\lambda_{x,-0.50}$ nm	$\lambda_{x,max}$ nm	$\lambda_{x,+0.50}$ nm	$\lambda_{e,-0.25}$ nm	$\lambda_{e,-0.50}$ nm	$\lambda_{e,max}$ nm	$\lambda_{e,+0.50}$ nm	$\lambda_{e,+0.25}$ nm	$\lambda_{e,+0.10}$ nm	Δ_S nm	ε_{max} mM^{-1}cm^{-1}	Φ_f	B_{max} mM^{-1}cm^{-1}	Class	Ref.
685	NR666	PBS, 1% DMSO pH 7.4	608	646	666	683	653	661	685	719	741	778	19	165.0	0.38	62.7	PO2H-Rosam.	4.18.2
686	JF669-6-SCH2CO2H	10 mM HEPES pH 7.3	606	584	667	685	655	663	681	702	717	752	14	130.0	0.38	49.4	Si(CH3)2-Rhoda	4.3.2
687	4,5,6,7-Tetrafluoro-SiTMR	10 mM HEPES pH 7.3	606	584	667	685	656	664	682	703	718	754	15	132.0	0.31	40.9	Si(CH3)2-Rhoda	4.3.2
688	Squaraine T-P 5b	67 mM P pH 7.4	614	647	667	685	661	669	685	708	728	772	18	188.0	0.07	13.1	Squaraine	4.35
689	SqCN1	P pH 7.4	614	647	667	685	661	669	685	708	728	772	18	188.0	0.07	13.1	Squaraine	4.35
690	CF660C	H_2O			667				685				18	200	0.06	11.1	nd (Cyanine)	4.41
691	Seta-670-carboxylate	P pH 7.4			667				688				21	180.0	0.07	12.6	nd	4.41
692	JF669-6-SCH2CO2Et	10 mM HEPES pH 7.3	607	585	668	686	655	663	681	702	717	752	13	119.0	0.38	45.2	Si(CH3)2-Rhoda	4.3.2
693	JF668	10 mM HEPES pH 7.3	607	585	668	686	661	668	687	708	724	760	19	26.7	0.34	9.0	POOH-Rhoda	4.3.3
694	2-Me TeOR	PBS pH 7.4	599	642	669	692	654	663	686	714	737	784	17	120.0	0.18	21.6	TeO-Rosamine	4.18.3
695	JF669	10 mM HEPES pH 7.3	608	586	669	687	656	664	682	703	718	754	13	112.0	0.37	41.4	Si(CH3)2-Rhoda	4.3.2
696	NR669	10 mM PBS pH 7.4	619	653	669	684	664	672	689	712	726	757	20	77.2	0.30	23.1	PO2H-Rosam.	4.18.2
697	2-Me SiR670	100 mM P, 1% DMSO pH 7.4	588	630	670	688	661	669	687	709	730	759	17	90.0	0.27	24.3	Si(CH3)2-Rosa	4.18.1
698	PORL	PBS, 1% DMSO pH 7.4	609	643	670	686	671	679	698	724	744	776	28	66.1	0.11	7.2	POPh-Rosol	4.20
699	nor-SqCN1	P pH 7.4	616	650	670	687	669	677	693	717	737	782	23	123.0	0.19	23.3	Squaraine	4.35
700	nor-SqCN2	P pH 7.4	616	650	670	687	669	677	693	717	737	782	23	167.0	0.19	31.7	Squaraine	4.35

Ent.	Short Names	Aqueous Solutions	$\lambda_{x,-0.25}$ nm	$\lambda_{x,-0.50}$ nm	$\lambda_{x,max}$ nm	$\lambda_{x,+0.50}$ nm	$\lambda_{e,-0.25}$ nm	$\lambda_{e,-0.50}$ nm	$\lambda_{e,max}$ nm	$\lambda_{e,+0.50}$ nm	$\lambda_{e,+0.25}$ nm	$\lambda_{e,+0.10}$ nm	Δ_S nm	ε_{max} mм$^{-1}$cm^{-1}	Φ_f	B_{max} mм$^{-1}$cm^{-1}	Class	Ref.
701	2-Me SiR671	100 mм P, 1% DMSO pH 7.4	589	631	671	689	661	669	687	709	730	759	16	99.0	0.29	28.7	Si(CH3)2-Rosa	4.18.1
702	Seta-670-pH-di-carboxylate	P pH 6			672				694				22	93.0	0.19	17.6	nd	4.41
703	2-Me SiR680	H_2O	592	634	674	693	663	671	689	711	732	761	15	130.0	0.35	45.5	Si(CH3)2-Rosa	4.18.1
704	Cy5.5	PBS	609	650	674	690	671	677	694	713	725	757	20	190.0	0.23	43.7	Cyanine-Penta	4.13.2
705	Cy5.205	PBS pH 7.4	609	650	674	690	671	677	694	713	725	757	20	190.0	0.23	43.7	Cyanine-Penta	4.13.2
706	2-Me SiR675	100 mм P, 1% DMSO pH 7.4	592	635	675	694	672	679	698	721	742	772	23	88.0	0.07	6.1	Si(CH3)2-Rosa	4.18.1
707	NR675	10 mм PBS pH 7.4	625	657	675	689	668	675	695	718	733	762	20	73.2	0.24	17.5	PO2H-Rosam.	4.18.2
708	Zy5.5	P pH 7.4	610	652	676	692	671	677	694	713	725	757	18	190.0	0.23	43.7	Cyanine-Penta	4.13.2
709	Seta-680-carboxylate	P pH 7.4			678				698				20	200.0	0.24	48.0	nd	4.41
710	DY-679P1	PBS pH 7.4	612	655	679	696	673	680	697	715	726	745	18	200.0	0.23	46.6	Cyanine-Penta	4.13.2
711	Alexa Fluor 680	PBS pH 7.2/7.5, 20/22 °C	613	655	679	695	678	685	702	721	734	766	23	183.0	0.36	65.8	Cyanine-Penta	4.13.2
712	Cy5.203	PBS pH 7.4	614	656	680	696	678	685	702	721	734	766	22	190.0	0.10	19.0	Cyanine-Penta	4.13.2
713	CF680R	H_2O			680				701				21	140	0.03	4.7	nd (Rhodam.)	4.41
714	Atto 680 carboxylate	PBS pH 7.4, 22 °C	603	648	681	696	675	682	698	727	749	807	17	125.0	0.30	37.5	Phenoxazine	4.25
715	CF680	H_2O			681				698				17	210	0.04	8.2	nd (Cyanine)	4.41
716	Cy5.204	PBS pH 7.4	615	658	682	698	683	689	707	726	739	772	25	197.0	0.10	19.7	Cyanine-Penta	4.13.2
717	Seta-700-carboxylate	P pH 7.4			687				703				16	177.0	0.06	10.6	nd	4.41
718	Seta-700-di-carboxylate	P pH 7.4			688				704				16	180.0	0.11	19.8	nd	4.41
719	DY-682 carboxylate	PBS pH 7.4	617	654	690	708	671	685	710	733	745	764	20	120.0	0.41	48.6	Merocyanine	4.24

(Continued)

Ent.	Short Names	Aqueous Solutions	$\lambda_{x,-0.25}$ nm	$\lambda_{x,-0.50}$ nm	$\lambda_{x,max}$ nm	$\lambda_{x,+0.50}$ nm	$\lambda_{e,-0.25}$ nm	$\lambda_{e,-0.50}$ nm	$\lambda_{e,max}$ nm	$\lambda_{e,+0.50}$ nm	$\lambda_{e,+0.25}$ nm	$\lambda_{e,+0.10}$ nm	Δ_S nm	ε_{max} mM^{-1}cm^{-1}	Φ_f	B_{max} mM^{-1}cm^{-1}	Class	Ref.
720	JF690	10 mM HEPES pH 7.3	625	602	690	710	679	687	707	730	746	784	17	150.0	0.24	36.0	POOH-Rhoda	4.3.3
721	DY-681 carboxylate	PBS pH 7.4	592	630	691	718	665	676	708	776	818	866	17	125.0	0.09	11.2	Merocyanine	4.24
722	2-Me SiR700	H_2O	617	665	691	710	686	695	712	732	746	764	21	100.0	0.12	12.0	Si(CH3)2-Rosa	4.18.1
723	nor-SqCN3	P pH 7.4	636	671	693	712	688	697	714	739	761	809	21	113.0	0.15	16.9	Squaraine	4.35
724	Me-PR	PBS, 1% DMSO pH 7.4	632	672	694	711	682	690	712	740	757	789	18	92.0	0.11	10.1	POR-Rosam.	4.18.2
725	PR	PBS, 1% DMSO pH 7.4	632	672	694	711	682	690	712	740	757	789	18	65.0	0.06	3.9	POR-Rosam.	4.18.2
726	tMe-PR	PBS, 1% DMSO pH 7.4	634	674	696	714	682	691	713	741	759	790	17	68.0	0.15	10.2	POR-Rosam.	4.18.2
727	NR698	PBS, 1% DMSO pH 7.4	631	676	698	715	686	693	712	738	751	773	14	25.9	0.32	8.2	PO2H-Rosam.	4.18.2
728	Atto 700 carboxylate	PBS pH 7.4, 22 °C	622	666	700	717	693	701	716	744	768	828	16	120.0	0.25	30.0	Phenoxazine	4.25
729	NR700	PBS, 1% DMSO pH 7.4	630	674	700	720	685	696	722	752	772	808	22	71.0	0.11	7.8	PO2R-Rosam.	4.18.2
730	Alexa Fluor 700	PBS/H_2O pH 7.2, 20/22 °C			702				723				21	205.0	0.25	51.2	nd	4.41
731	SO_2R1	10 mM PBS, 0.25% ACN pH 7.4	633	678	703	724	701	711	736	772	798	828	33	39	0.07	2.6	SO2-Rosamine	4.18.3
732	SO_2R2	10 mM PBS, 0.25% ACN pH 7.4	633	678	703	724	706	717	742	779	805	836	39	95	0.07	6.9	SO2-Rosamine	4.18.3

Ent.	Short Names	Aqueous Solutions	$\lambda_{x,-0.25}$ nm	$\lambda_{x,-0.50}$ nm	$\lambda_{x,max}$ nm	$\lambda_{x,+0.50}$ nm	$\lambda_{e,-0.25}$ nm	$\lambda_{e,-0.50}$ nm	$\lambda_{e,max}$ nm	$\lambda_{e,+0.50}$ nm	$\lambda_{e,+0.25}$ nm	$\lambda_{e,+0.10}$ nm	Δ_S nm	ε_{max} $mM^{-1}cm^{-1}$	Φ_f	B_{max} $mM^{-1}cm^{-1}$	Class	Ref.
733	SO_2R4	10 mM PBS, 0.25% ACN pH 7.4	634	679	704	725	706	717	742	779	805	836	38	97.0	0.10	9.4	SO2-Rosamine	4.18.3
734	SO_2R3	10 mM PBS, 0.25% ACN pH 7.4	636	681	707	728	711	721	747	784	811	842	40	69	0.05	3.4	SO2-Rosamine	4.18.3
735	LS482	Buffer pH 8	632	682	710	731	760	771	793	819	830	845	83	84.0	0.10	8.4	Cyanine-Hepta	4.13.3
736	SO_2R5	10 mM PBS, 0.25% ACN pH 7.4	639	684	710	732	715	726	752	790	817	848	42	129.0	0.07	8.7	SO2-Rosamine	4.18.3
737	PONy CP1	PBS pH 7.4	640	681	710	728	737	754	783	817	843	870	73	17.0	0.19	3.2	C(CH3)2-Pyron	4.5
738	JF711	10 mM HEPES pH 7.3	642	618	711	732	702	711	732	756	774	815	21	12.4	0.17	2.1	POPh-Rhodam	4.3.3
739	2-Me SiR720	H_2O	627	675	721	742	710	719	740	766	790	824	19	160.0	0.05	8.0	Si(CH3)2-Rosa	4.18.1
740	JF722	10 mM HEPES pH 7.3	651	626	722	743	712	721	743	768	786	829	21	87.2	0.11	9.5	POPh-Rhodam	4.3.3
741	JF724	10 mM HEPES pH 7.3	647	688	724	750	711	721	748	792	831	879	24	6.6	0.05	0.3	SO2-Rhodam	4.3.3
742	Atto 725 carboxylate	PBS pH 7.4, 22 °C			728				751				23	120.0	0.10	12.0	nd	4.41
743	DY-731 carboxylate	PBS pH 7.4	678	709	736	761	725	735	759	783	794	811	23	225.0	0.11	23.6	Merocyanine	4.24
744	Atto 740 carboxylate	PBS pH 7.4, 22 °C			743				763				20	120.0	0.10	12.0	nd	4.41
745	NR744	PBS, 1% DMSO pH 7.4	672	719	744	763	725	737	764	795	812	841	20	80.0	0.16	12.8	PO2R-Rosam.	4.18.2
746	LS479	Buffer pH 8	661	716	747	770	731	741	762	786	796	810	15	105.0	0.04	4.4	Cyanine-Hepta	4.13.3

(Continued)

Ent.	Short Names	Aqueous Solutions	$\lambda_{x,-0.25}$ nm	$\lambda_{x,-0.50}$ nm	$\lambda_{x,max}$ nm	$\lambda_{x,+0.50}$ nm	$\lambda_{e,-0.25}$ nm	$\lambda_{e,-0.50}$ nm	$\lambda_{e,max}$ nm	$\lambda_{e,+0.50}$ nm	$\lambda_{e,+0.25}$ nm	$\lambda_{e,+0.10}$ nm	Δ_S nm	ε_{max} $mM^{-1}cm^{-1}$	Φ_f	B_{max} $mM^{-1}cm^{-1}$	Class	Ref.
747	DY-747P1	PBS pH 7.4	661	716	747	770	738	748	769	793	804	818	22	240.0	0.12	28.5	Cyanine-Hepta	4.13.3
748	Zy7 carboxylate	P pH 7.4	663	718	749	772	743	753	774	798	810	824	25	200.0	0.13	26.0	Cyanine-Hepta	4.13.3
749	Alexa Fluor 750	PBS/H_2O pH 7.2, 20/22 °C			749				775				26	290.0	0.12	34.8	nd	4.41
750	NIR3	H_2O	663	719	750	773	745	756	777	802	813	827	27	275.0	0.28	77.0	Cyanine-Hepta	4.13.3
751	Seta-750-carboxylate	P pH 7.4			751				779				28	235.0	0.21	49.3	nd	4.41
752	NIR1	H_2O	672	729	761	785	763	773	796	822	834	849	35	268.0	0.23	61.6	Cyanine-Hepta	4.13.3
753	HCDO	PBS pH 7.4	678	736	769	793	755	766	788	813	825	840	19	218.8	0.30	65.6	Cyanine-Hepta	4.13.3
754	IRDye78-CA	10 mM HEPES pH 7.4	680	738	771	796	772	783	806	832	845	860	35	145.7	0.14	20.6	Cyanine-Hepta	4.13.3
755	ZW800-1	FBS	680	739	772	797	755	766	788	813	825	840	16	249.0	0.15	37.6	Cyanine-Hepta	4.13.3
756	ZW800-3a	FBS	682	741	774	799	757	768	790	815	827	842	16	309.0	0.16	49.7	Cyanine-Hepta	4.13.3
757	Y-A 3	PBS pH 7.2	684	744	777	802	773	784	807	834	846	861	30	215.0	0.03	6.4	Cyanine-Hepta	4.13.3
758	Seta-780-carboxylate	P pH 7.4			777				807				30	215.0	0.01	2.5	nd	4.41
759	Y-A 1 (Cypate)	PBS pH 7.2	685	744	778	803	771	782	805	831	844	859	27	224.0	0.03	6.7	Cyanine-Hepta	4.13.3
760	Y-A 2	PBS pH 7.2	685	744	778	803	772	783	806	832	845	860	28	190.0	0.03	5.7	Cyanine-Hepta	4.13.3
761	Y-A 4	PBS pH 7.2	685	744	778	803	771	782	805	831	844	859	27	201.0	0.03	6.0	Cyanine-Hepta	4.13.3
762	Y-A 5	PBS pH 7.2	685	744	778	803	770	781	804	830	842	858	26	185.0	0.03	5.5	Cyanine-Hepta	4.13.3
763	Y-A 6	PBS pH 7.2	685	744	778	803	771	782	805	831	844	859	27	224.0	0.03	6.7	Cyanine-Hepta	4.13.3
764	Y-A 7	PBS pH 7.2	686	745	779	804	771	782	805	831	844	859	26	224.0	0.03	6.7	Cyanine-Hepta	4.13.3

Ent.	Short Names	Aqueous Solutions	$\lambda_{x,-0.25}$ nm	$\lambda_{x,-0.50}$ nm	$\lambda_{x,max}$ nm	$\lambda_{x,+0.50}$ nm	$\lambda_{e,-0.25}$ nm	$\lambda_{e,-0.50}$ nm	$\lambda_{e,max}$ nm	$\lambda_{e,+0.50}$ nm	$\lambda_{e,+0.25}$ nm	$\lambda_{e,+0.10}$ nm	Δ_S nm	ε_{max} mM^{-1}cm^{-1}	Φ_f	B_{max} mM^{-1}cm^{-1}	Class	Ref.
765	Y-A 9	PBS pH 7.2	686	745	779	804	775	786	809	836	848	864	30	187.0	0.03	5.6	Cyanine-Hepta	4.13.3
766	ICG	PBS	687	746	780	805	776	787	810	837	849	865	30	211.2	0.08	16.9	Cyanine-Hepta	4.13.3
767	Y-A 10	PBS pH 7.2	687	746	780	805	774	785	808	835	847	862	28	219.0	0.03	6.5	Cyanine-Hepta	4.13.3
768	Y-A 11	PBS pH 7.2	687	746	780	805	773	784	807	834	846	861	27	199.0	0.03	5.9	Cyanine-Hepta	4.13.3
769	Y-A 8	PBS pH 7.2	687	746	780	805	772	783	806	832	845	860	26	197.0	0.03	5.9	Cyanine-Hepta	4.13.3
770	DY-780 carboxylate	PBS pH 7.4	657	705	782	817	745	760	800	888	943	1007	18	102.0	0.03	3.2	Merocyanine	4.24
771	IR-783	PBS	688	748	782	807	776	787	810	837	849	865	28	261.0	0.08	21.9	Cyanine-Hepta	4.13.3
772	RS800	FBS	690	750	784	809	766	777	800	826	838	853	16	240.0	0.17	40.5	Cyanine-Hepta	4.13.3
773	CW800	FBS	691	752	786	812	766	777	800	826	838	853	14	237.0	0.14	33.6	Cyanine-Hepta	4.13.3
774	TG170	H_2O	695	756	791	817	769	780	803	829	841	857	12	300.0	0.02	7.4	Cyanine-Hepta	4.13.3
775	HCDS	PBS pH 7.4	695	756	791	817	783	794	818	845	858	874	27	218.8	0.39	85.3	Cyanine-Hepta	4.13.3
776	ICG	FBS	708	771	807	834	787	798	822	850	862	878	15	121.0	0.09	11.2	Cyanine-Hepta	4.13.3
777	ECXi	H_2O			887				938				51	93.3	0.01	1.3	Pentacene	4.39.2

Appendix C

C1. Ranked by Emission Maximum

Appendix C contains the over 700 fluorescent dyes with medium to high brightness in aqueous solutions, ranked by ascending values of their emission range (Maximum by $\lambda_{e,max}$) (Table C.1). The fluorescent dyes of this database are identified by their short names. Some of the names are registered trademarks (listed in footnote[1]). The parameters highlighted in gray were taken from actual spectra measured by the author or from the literature (column "Ref." shows the sub-sections where they can be found). The emission spectra of some dyes were not found. In these cases, the emission ranges (cells without color fill) were predicted using the equation and parameters (for each class) explained and shown in Chapter 6.

As explained in Appendix B for excitation ranges, there are more precise ways to predict the wavelength at height $\alpha(0 < \alpha < 1)$ of the left side of the emission bands ($\lambda_{e,-\alpha}$) and right side of the emission bands ($\lambda_{e,+\alpha}$); however, they would excessively increase the calculations. Moreover, such a refined calculation would not be justified here, since the changes in the photophysical properties caused by the chemical micro-environment and other factors are, in many cases, much larger than the imprecisions of the calculations used. For example, the values of $\lambda_{e,-0.5}$, $\lambda_{e,max}$, and $\lambda_{e,+0.1}$ may experience small differences if they are measured in plain water or dissolved in either intracellular or extracellular fluids. Moreover, if the dyes are bound to proteins or nucleic acids, for instance, they may change even more. Finally, some authors took measurements in aqueous solutions containing 1% (v/v) DMSO or other water-mixable organic solvents (Section 3.2.2). Usually, these solvents originate from the stock solutions used. The small amounts of remaining organic modifiers are also known to cause small changes to some photophysical parameters.

Molecular structures and all references are shown in Chapter 4. Panoramic views of the photophysical properties are given in Chapter 5 with the aid of scattergrams. The photostability (relative photobleaching mean-lifetime) of some dyes is shown in Chapter 7.

Acronyms used in Table C.1: A, acetate; AC, acetic acid; ACN, acetonitrile; BODIPY, borondipyrromethene; BSA, bovine serum albumin; C, carbonate; CA, carboxylic acid; FBS, fetal bovine serum; Et, ethanol; nd, nondisclosed; ni, not informed; P, phosphate; pb, probably; pH >7, basic medium; pH < 7, acidic medium (when the exact value of the pH is not given in the literature); SA, sulfonic acid; and TX-100, triton X-100.

1 Abberior STAR is a trademark of Abberior GmbH (Göttingen, Germany). Alexa Fluor, BODIPY, Cascade Blue, FAM, JOE, TET, HEX, Marina Blue, Oregon Green, Pacific Blue, ROX, TAMRA, and Texas Red are trademarks of Thermo Fisher Scientific (Waltham, MA, USA). ATTO is a trademark of ATTO-Tec GmbH (Siegen, Germany). CF Dye is a registered trademark of Biotium, Inc. (Fremont, CA, USA). Cy3, Cy3.5, Cy5, and Cy5.5 are trademarks of GE Healthcare (Chicago, IL, USA). Dy is a trademark of Dyomics GmbH (Jena, Germany). Seta, SeTau, and Square dyes are trademarks of SETA BioMedicals (Urbana, IL, USA).

Fluorescent Dye Labels and Stains: A Database of Photophysical Properties, First Edition. Tarso B. Ledur Kist.

Table C.1 Photophysical properties of fluorescent labels and stains in aqueous solutions ranked by the wavelenght of emission maximum ($\lambda_{e,max}$).

Ent.	Short Names	Aqueous Solutions	$\lambda_{x,-0.25}$ nm	$\lambda_{x,-0.50}$ nm	$\lambda_{x,max}$ nm	$\lambda_{x,+0.50}$ nm	$\lambda_{e,-0.25}$ nm	$\lambda_{e,-0.50}$ nm	$\lambda_{e,max}$ nm	$\lambda_{e,+0.50}$ nm	$\lambda_{e,+0.25}$ nm	$\lambda_{e,+0.10}$ nm	Δ_S nm	ε_{max} mM^{-1}cm^{-1}	Φ_f	B_{max} mM^{-1}cm^{-1}	Class	Ref.
1	Stefani HB-11	H_2O pH 6			329				360				31	38.1	0.11	4.0	Hybrid-Bonded	4.40.2
2	2-Aminopyridine	H_2O			300				363				63	4.1	0.71	2.9	Pyridine	4.39.1
3	Stefani HB-6	H_2O pH 6			317				365				48	35.9	0.03	0.9	Hybrid-Bonded	4.40.2
4	Carbostyril 9	H_2O			340				382				42	6.0	0.17	0.9	Quinoline	4.29
5	Quinine	H_2O pH 7			331				382				51	5.7	0.50	2.8	Quinoline	4.29
6	Sulfobetaine pyrene 21	H_2O	314	327	359	365	371	375	385	412	428	441	26	30.6	0.32	9.7	Pyrene	4.28
7	BOPAHY H-D 19ba′	H_2O	344	355	367	372	379	381	386	393	403	412	19	51	0.03	1.5	Bodipy-analog	4.14.5
8	BOPAHY H-D 19aa′	H_2O	345	355	368	373	380	382	387	394	404	413	19	51	0.02	0.7	Bodipy-analog	4.14.5
9	3-Aminopyridine	H_2O			300				390				90	2.8	0.32	0.9	Pyridine	4.39.1
10	Pyrene-1-sulfonate	H_2O	327	340	375	382	385	389	400	429	446	461	25	30.6	0.56	17.1	Pyrene	4.28
11	1,3,6,8-Pyrenetetra-sulfonic acid	H_2O pH 7	335	350	374	378	391	394	403	430	451	465	29	51.0	0.56	28.5	Pyrene	4.28
12	Acridizinium J-N 1a	H_2O			377				404				27	8.9	0.33	2.9	Acridizinium	4.39.1
13	Salicylic acid	H_2O pH 7	272	281	300	316	370	377	407	441	461	482	107	3.0	0.16	0.4	Benzene	4.38
14	4-Methylsalicylic acid	H_2O pH 7	252	272	300	317	370	379	407	446	467	491	107	3.0	0.11	0.3	Benzene	4.38
15	Acridizinium J-N 1b	H_2O			381				408				27	6.7	0.34	2.2	Acridizinium	4.39.1
16	Acridizinium J-N 1d	H_2O			356				408				52	5.4	0.20	1.0	Acridizinium	4.39.1
17	Harmine	50 mM H_2SO_4	304	309	326	343	383	389	412	445	464	483	86	11.9	0.45	5.3	Indole	4.33
18	Harmine	H_2O pH 4	330	337	357	378	387	393	416	449	465	484	59	7.3	0.49	3.6	Indole	4.33
19	$ACSO_3H$	H_2O pH 5.5	353	366	396	404	405	408	417	449	458	472	21	4.4	0.82	3.6	Acridones	4.23.2
20	Carbostyril 124	H_2O			341				417				76	15.8	0.46	7.2	Quinoline	4.29
21	5-Methylsalicylic acid	H_2O pH 7	268	281	305	323	379	388	418	456	477	497	113	3.0	0.16	0.4	Benzene	4.38
22	9-Aminoacridine	P pH 7	366	379	400	410	400	405	420	449	467	487	20	9.3	0.97	9.0	Acridine	4.23.1
23	Acridizinium J-N 1c	H_2O			354				420				66	8.4	0.30	2.5	Acridizinium	4.39.1
24	8-Chloro-harmine	H_2O pH 4	334	341	361	382	392	398	423	456	474	495	62	5.9	0.11	0.6	Indole	4.33
25	DY-405 carboxylate	PBS pH 7.4	336	349	400	410	404	408	423	458	474	492	23	32.0	0.97	30.9	Pyrene	4.28
26	Stefani HB-9	H_2O pH 6			320				423				103	35.9	0.12	4.2	Hybrid-Bonded	4.40.2

Ent.	Short Names	Aqueous Solutions	$\lambda_{x,-0.25}$ nm	$\lambda_{x,-0.50}$ nm	$\lambda_{x,max}$ nm	$\lambda_{x,+0.50}$ nm	$\lambda_{e,-0.25}$ nm	$\lambda_{e,-0.50}$ nm	$\lambda_{e,max}$ nm	$\lambda_{e,+0.50}$ nm	$\lambda_{e,+0.25}$ nm	$\lambda_{e,+0.10}$ nm	Δ_S nm	ε_{max} mM^{-1}cm^{-1}	Φ_f	B_{max} mM^{-1}cm^{-1}	Class	Ref.
27	Alexa Fluor 405	PBS pH 7.2, 22 °C	346	360	400	408	408	411	424	457	476	494	24	35.0	0.50	17.5	Pyrene	4.28
28	Harmine	H_2O pH 9	309	315	332	350	393	400	425	458	477	498	93	11.0	0.38	4.1	Indole	4.33
29	6-Chloro-harmine	H_2O pH 4	335	341	362	383	396	402	428	462	481	502	66	9.0	0.16	1.4	Indole	4.33
30	6,8-Dichloroharmine (6,8-DiCL-Ha)	H_2O pH 4	341	348	369	391	397	403	429	463	482	503	60	6.3	0.20	1.2	Indole	4.33
31	Cascade Blue	50 mM P pH 8	348	363	403	411	413	417	430	464	484	502	27	28.0	0.90	25.2	Pyrene	4.28
32	Benzo[h]quinoline (BQ)	H_2O pH 2-5			370				430				60	35.0	0.32	11.2	Quinoline	4.29
33	Carbostyril 2	H_2O			360				430				70	6.7	0.14	0.9	Quinoline	4.29
34	N-Methylacridone	H_2O	369	382	411	419	417	421	431	467	480	493	20	8.8	0.99	8.7	Acridones	4.23.2
35	PetACSO3H	H_2O pH 5.5	367	379	406	414	417	421	431	467	480	495	25	6.2	0.92	5.6	Acridones	4.23.2
36	CF405S	H_2O			404				431				27	33.0	0.70	23.0	nd (Pyrene)	4.41
37	DY-350 carboxylate	PBS pH 7.4	301	323	353	378	400	408	432	462	479	499	79	19.0	0.84	15.9	Coumarin 7-Am	4.26.3
38	Carbostyril 1	H_2O			358				432				74	11.6	0.24	2.7	Quinoline	4.29
39	2-Methyl harmine	50 mM H_2SO_4	324	331	350	370	399	406	433	466	485	509	83	6.3	0.45	2.8	Indole	4.33
40	MetACSO3H	H_2O pH 5.5	364	376	403	411	418	423	433	470	487	495	30	5.6	0.76	4.2	Acridones	4.23.2
41	BOPAHY H-D 19bb′	H_2O	387	400	416	422	425	428	434	443	455	467	18	51	0.17	8.6	Bodipy-analog	4.14.5
42	BOPAHY H-D 19ab′	H_2O	387	400	416	422	426	429	435	444	457	468	19	51	0.20	10.2	Bodipy-analog	4.14.5
43	BOPAHY H-D 20bb′	H_2O	387	400	416	422	426	429	435	444	457	468	19	51	0.17	8.6	Bodipy-analog	4.14.5
44	Harmane	50 mM H_2SO_4	324	331	350	370	405	412	440	474	494	517	90	6.3	0.83	5.1	Indole	4.33
45	Coumarin 120	H_2O	293	314	342	366	407	415	440	471	489	510	98	16.0	0.94	15.0	Coumarin 7-Am	4.26.2
46	7-Am-4-Me-2-hydroxyquinoline	H_2O pH 5-9			340				440				100	16.0	0.97	15.5	Quinoline	4.29
47	Carbostyril 5	H_2O			353				440				87	6.6	0.16	1.0	Quinoline	4.29
48	8-Hydroxypyrene-1,3,6-triSA	H_2O pH 2	348	363	403	411	425	429	443	479	501	519	40	20.0	0.95	19.0	Pyrene	4.28
49	Alexa Fluor 350	H_2O pH 7	296	317	346	370	411	420	445	477	495	516	99	19.0	0.89	16.9	Coumarin 7-Am	4.26.3

(Continued)

Ent.	Short Names	Aqueous Solutions	$\lambda_{x,-0.25}$ nm	$\lambda_{x,-0.50}$ nm	$\lambda_{x,max}$ nm	$\lambda_{x,+0.50}$ nm	$\lambda_{e,-0.25}$ nm	$\lambda_{e,-0.50}$ nm	$\lambda_{e,max}$ nm	$\lambda_{e,+0.50}$ nm	$\lambda_{e,+0.25}$ nm	$\lambda_{e,+0.10}$ nm	Δ_S nm	ε_{max} mM^{-1}cm^{-1}	Φ_f	B_{max} mM^{-1}cm^{-1}	Class	Ref.
50	3-AB-6-MeO-7-OH-chromone	100 mM Tris pH 8			346				445				99	16.3	0.41	6.6	Chromone	4.31
51	7-Hydroxy-4-Me-coumarin-3-CA	H_2O pH 10	332	352	386	413	414	422	446	479	499	522	60	33.0	0.85	28.0	Coumarin 7-OH	4.26.1
52	5-Methoxysalicylic acid	H_2O pH 7	285	295	319	339	403	414	447	489	511	526	128	3.0	0.19	0.5	Benzene	4.38
53	3-Bt-6-MeO-7-hydroxychromone	100 mM Tris pH 8			345				447				102	16.4	0.48	7.8	Chromone	4.31
54	7-Hydroxy-4-methylcoumarin	H_2O pH 10	313	330	360	384	416	424	449	483	503	526	89	18.0	0.86	15.4	Coumarin 7-OH	4.26.1
55	Stefani HB-7	H_2O			331				449				118	42.0	0.09	3.6	Benzothiazole	4.30
56	7-Hydroxy-4-methylcoumarin	H_2O pH 5	276	295	320	341	415	424	450	483	501	523	130	14.0	0.78	10.9	Coumarin 7-OH	4.26.1
57	Coumarin P-R 3	PBS pH 7.5	297	319	348	372	415	424	450	483	501	523	102	12.5	0.82	10.2	Coumarin 7-Am	4.26.2
58	Coumarin P-R 22	PBS pH 7.5	307	324	353	376	417	425	450	484	504	527	97	13.4	0.75	10.0	Coumarin 7-Am	4.26.2
59	Quinine	100 mM $HClO_4$			347				451				104	5.7	0.60	3.4	Quinoline	4.29
60	Quinine	50 mM H_2SO_4			347				451				104	5.7	0.55	3.1	Quinoline	4.29
61	Coumarin P-R 6	PBS pH 7.5	309	326	355	378	419	427	452	486	507	530	97	12.6	0.71	8.9	Coumarin 7-Am	4.26.2
62	3-Amino-9-ethylcarbazole 1	H_2O			364				452				88	1.8	0.63	1.1	Carbazole	4.39.1
63	CF405M	H_2O			408				452				44	41.0	0.76	31.1	nd (Coumarin)	4.41
64	Seta-405-carboxylate	P pH 7.4			401				452				51	30.0	0.70	21.0	nd	4.41
65	BOPAHY H-D 18ab′	H_2O	398	412	429	435	444	447	453	463	476	489	24	51	0.55	28.0	Bodipy-analog	4.14.5
66	BOPAHY H-D 18bb′	H_2O	397	411	428	434	444	447	453	463	476	489	25	51	0.49	24.9	Bodipy-analog	4.14.5
67	nor-Harmane (β-carboline)	50 mM H_2SO_4	333	340	360	381	418	425	454	492	514	538	94	6.3	0.58	3.6	Indole	4.33
68	3-Buthyl-7-hydroxychromone	100 mM Tris pH 8			334				454				120	12.7	0.21	2.6	Chromone	4.31
69	7-Hydroxy-4-Me-coumarin-3-CA	H_2O pH 5	313	330	360	384	421	430	455	490	510	534	95	18.0	0.82	14.7	Coumarin 7-OH	4.26.1
70	Marina Blue	100 mM P pH 10/8	311	328	358	381	421	430	455	490	510	534	97	18.7	0.89	16.6	Coumarin 7-OH	4.26.1

Ent.	Short Names	Aqueous Solutions	$\lambda_{x,-0.25}$ nm	$\lambda_{x,-0.50}$ nm	$\lambda_{x,max}$ nm	$\lambda_{x,+0.50}$ nm	$\lambda_{e,-0.25}$ nm	$\lambda_{e,-0.50}$ nm	$\lambda_{e,max}$ nm	$\lambda_{e,+0.50}$ nm	$\lambda_{e,+0.25}$ nm	$\lambda_{e,+0.10}$ nm	Δ_S nm	ε_{max} mм$^{-1}$cm^{-1}	Φ_f	B_{max} mм$^{-1}$cm^{-1}	Class	Ref.
71	Pacific Blue	PBS pH 7.2, 22 °C	339	366	410	437	428	436	455	486	505	526	45	29.5	0.75	22.1	Coumarin 7-OH	4.26.1
72	Acridizinium J-N 1e	H_2O			355				455				100	6.7	0.20	1.3	Acridizinium	4.39.1
73	Coumarin 2	H_2O	314	332	362	386	422	431	456	491	512	535	94	16.9	0.95	16.0	Coumarin 7-Am	4.26.2
74	Coumarin P-R 12	PBS pH 7.5	282	301	327	348	421	430	457	491	510	533	130	7.9	0.76	5.9	Coumarin 7-OH	4.26.1
75	7-Ethylamino-coumarin	H_2O	319	337	368	393	424	432	458	493	514	538	90	11.9	0.51	6.0	Coumarin 7-Am	4.26.2
76	3-Amino-9-sethylcarbazole 2	H_2O			364				458				94	1.8	0.76	1.3	Carbazole	4.39.1
77	7-(4-Hydroxy)cyclohexyl coum.	H_2O	319	337	368	393	426	434	460	496	517	541	92	16.5	0.59	9.7	Coumarin 7-Am	4.26.2
78	Coumarin P-R 21	PBS pH 7.5	295	316	345	369	425	434	461	495	515	538	116	13.2	0.79	10.4	Coumarin 7-Am	4.26.2
79	Coumarin P-R 14	PBS pH 7.5	322	341	373	398	426	435	461	497	518	542	88	17.7	0.62	10.9	Coumarin 7-OH	4.26.1
80	Coumarin P-R 5	PBS pH 7.5	296	317	346	370	428	437	465	500	520	544	119	10.3	0.79	8.1	Coumarin 7-Am	4.26.2
81	7-Azetidinyl-4-Me-coumarin	10 mм HEPES pH 7.3	308	325	354	377	432	441	467	504	525	551	113	15.0	0.96	14.4	Coumarin 7-Am	4.26.2
82	3-(4-OH-Ph)-6-MeO-chromone	100 mм Tris pH 8			311				468				157	12.6	0.05	0.6	Chromone	4.31
83	SeTau-380-carboxylate	P pH 7.3			383				468				85	23.8	0.56	13.3	nd	4.41
84	3-OH-4-nitroso-2,7-naphth-diSA	A pH 4	304	326	370	412	431	442	470	500	514	535	100	19.0	0.22	4.1	Naphthalene	4.34
85	7-Azetidinyl-3-CA coumarin	10 mм HEPES pH 7.3	333	353	387	414	434	443	470	507	529	555	83	24.0	0.84	20.1	Coumarin 7-Am	4.26.2
86	7-Dimethylamino-4-Me-coumarin	10 mм HEPES pH 7.3	322	340	372	397	434	443	470	507	529	555	98	18.0	0.19	3.4	Coumarin 7-Am	4.26.2
87	BC-H	100 mм P, 20% Et pH 7.4			365				470				105	27.2	0.09	2.4	Hybrid-Bonded	4.40.2
88	NBQ	H_2O pH 2-5			380				470				90	35.0	0.35	12.2	Quinoline	4.29

(Continued)

Ent.	Short Names	Aqueous Solutions	$\lambda_{x,-0.25}$ nm	$\lambda_{x,-0.50}$ nm	$\lambda_{x,max}$ nm	$\lambda_{x,+0.50}$ nm	$\lambda_{e,-0.25}$ nm	$\lambda_{e,-0.50}$ nm	$\lambda_{e,max}$ nm	$\lambda_{e,+0.50}$ nm	$\lambda_{e,+0.25}$ nm	$\lambda_{e,+0.10}$ nm	Δ_S nm	ε_{max} $mM^{-1}cm^{-1}$	Φ_f	B_{max} $mM^{-1}cm^{-1}$	Class	Ref.
89	NOBQ	H_2O pH 2-5			385				470				85	35.0	0.41	14.3	Quinoline	4.29
90	OBQ	H_2O pH 2-5			385				470				85	35.0	0.43	15.0	Quinoline	4.29
91	7-Diethylamino-3-CA coumarin	10 mM HEPES pH 7.3	357	376	410	432	446	452	471	503	525	548	61	35.0	0.03	1.0	Coumarin 7-Am	4.26.2
92	Coumarin 1	H_2O	329	349	382	409	436	445	472	510	532	558	90	20.1	0.07	1.4	Coumarin 7-Am	4.26.2
93	BC-BA	100 mM P, 20% Et pH 7.4			371				473				102	25.5	0.17	4.3	Hybrid-Bonded	4.40.2
94	Fluorene I-6	H_2O			353				473				120	30.5	0.13	3.9	Fluorene	4.39.2
95	Coumarin P-R 13	PBS pH 7.5	319	337	368	393	438	447	474	512	534	560	106	15.0	0.63	9.4	Coumarin 7-OH	4.26.1
96	Harmaline	5 mM H_2SO_4	352	360	383	407	436	444	475	517	541	568	92	28.6	0.32	9.1	Indole	4.33
97	7-Diethylamino-coumarin	H_2O	333	354	388	416	438	448	475	513	536	562	87	18.3	0.05	0.9	Coumarin 7-Am	4.26.2
98	Carbostyril 23b	H2O pH 2.8			380				475				95	18.0	0.41	7.3	Quinoline	4.29
99	DMPT	H_2O pH 7			382				475				93	12.4	0.06	0.7	Pteridine	4.36
100	Atto 390 carboxylate	PBS pH 7.4, 22 °C	343	357	390	415	441	450	476	516	540	568	86	24.0	0.90	21.6	Coumarin 7-Am	4.26.3
101	Seta-375-carboxylate	P pH 7.4			372				476				104	24.5	0.86	21.0	nd	4.41
102	HBQ	H_2O pH 2-5			390				480				90	35.0	0.15	5.2	Quinoline	4.29
103	Spirobifluorene 1	H_2O			378				480				102	51	0.50	25.6	Spiro-bifluorene	4.39.3
104	Atto 425 carboxylate	PBS pH 7.4, 22 °C	395	409	439	460	460	466	485	521	548	576	46	45.0	0.90	40.5	Coumarin 7-Am	4.26.3
105	Lucigenin	H_2O			368				485				117	37.6	0.50	18.8	Hybrid-Bonded	4.40.2
106	DY-430 carboxylate	PBS pH 7.4	394	410	444	467	462	468	487	521	545	570	43	37.0	0.72	26.6	Coumarin 7-Am	4.26.3
107	BC-OH	100 mM P, 20% Et pH 7.4			442				488				46	38.1	0.85	32.3	Hybrid-Bonded	4.40.2
108	Coumarin 102	H_2O	336	356	393	418	455	464	489	528	552	579	96	21.8	0.66	14.3	Coumarin 7-Am	4.26.2
109	10-Ethylacridine	P pH 7	376	390	412	423	467	474	494	534	560	589	82	3.5	0.98	3.4	Acridine	4.23.1

Ent.	Short Names	Aqueous Solutions	$\lambda_{x,-0.25}$ nm	$\lambda_{x,-0.50}$ nm	$\lambda_{x,max}$ nm	$\lambda_{x,+0.50}$ nm	$\lambda_{e,-0.25}$ nm	$\lambda_{e,-0.50}$ nm	$\lambda_{e,max}$ nm	$\lambda_{e,+0.50}$ nm	$\lambda_{e,+0.25}$ nm	$\lambda_{e,+0.10}$ nm	Δ_S nm	ε_{max} mм$^{-1}$cm^{-1}	Φ_f	B_{max} mм$^{-1}$cm^{-1}	Class	Ref.
110	Acedan 7	H_2O	291	311	351	388	452	465	495	528	544	568	144	11.0	0.31	3.4	Naphthalene	4.34
111	Cyan 500 carboxylate	PBS pH 7.4, 22 °C			450				495				45	45.0	0.85	38.2	nd (pb Coum.)	4.41
112	DY-431 carboxylate	PBS pH 7.4	388	405	442	467	470	477	496	529	550	573	54	35.0	0.75	26.2	Coumarin 7-Am	4.26.3
113	Coumarin 152	H_2O	316	334	365	389	457	467	497	539	564	593	132	14.3	0.37	5.2	Coumarin 7-Am	4.26.2
114	CF430	H_2O			426				498				72	40.0	0.79	31.6	nd (Cyanine)	4.41
115	Pyrodone	100 mм P pH 7.4	436	454	472	481	486	490	500	513	523	539	28	60.0	0.37	22.2	Pyrodone	4.21.2
116	Acedan 3	H_2O	295	316	357	396	457	470	501	535	551	576	144	14.5	0.31	4.5	Naphthalene	4.34
117	Acedan 4	H_2O	294	315	355	393	457	470	501	535	551	576	146	13.5	0.32	4.3	Naphthalene	4.34
118	Z-L I	0.5 м PBS pH 7.4	451	469	491	499	489	493	501	513	530	545	10	50	0.68	34.0	Bodipy-Large	4.14.3
119	Acrylodan	H_2O	315	339	387	433	457	471	502	536	552	577	115	12.9	0.18	2.3	Naphthalene	4.34
120	Acedan 5	H_2O	295	316	357	396	458	472	503	537	553	579	146	17.4	0.40	6.9	Naphthalene	4.34
121	Acedan 6	H_2O	295	316	357	396	458	472	503	537	553	579	146	15.4	0.39	6.0	Naphthalene	4.34
122	Phosphorilated coum. N-H 22c	PBS pH 7.4	348	370	410	437	467	477	503	545	570	599	93	14.0	0.44	6.1	Coumarin 7-Am	4.26.3
123	8-Aminopyrene-1,3,6-triSA	H_2O	363	380	424	433	482	487	505	552	581	607	81	20.6	0.90	18.5	Pyrene	4.28
124	Atto 465 carboxylate	PBS pH 7.4, 22 °C	404	423	453	467	471	480	506	551	579	611	53	75.0	0.75	56.2	Acridine	4.23.1
125	DAMBO-T	100 mм P pH 7.4	457	475	498	506	495	499	507	520	537	552	9	52.0	0.40	20.8	Bodipy-Small	4.14.1
126	L-B 1a	H_2O	453	472	494	502	495	499	507	520	537	552	13	100.0	0.47	47.0	Bodipy-Small	4.14.1
127	L-B 4a	H_2O	453	472	494	502	495	499	507	520	537	552	13	69.2	0.34	23.5	Bodipy-Small	4.14.1
128	Z-L B	0.5 м PBS pH 7.4	455	474	496	504	495	499	507	520	537	552	11	50	0.36	17.8	Bodipy-Large	4.14.3
129	Acridizinium I-L 5a	H_2O			387				507				120	17.4	0.12	2.0	Acridizinium	4.39.1
130	Coumarin 307	H_2O	335	355	392	417	471	481	508	550	576	606	116	9.0	0.33	2.9	Coumarin 7-Am	4.26.2
131	2-Me-4-OMe TokyoGreen	100 mм NaOH	454	476	492	504	492	496	509	547	562	581	17	85.0	0.84	71.4	Fluorone	4.11
132	B-Z 7a	H_2O	458	477	500	508	497	501	509	522	539	555	9	64.5	0.61	39.3	Bodipy-Large	4.14.3
133	L-B 1b	H_2O	457	475	498	506	497	501	509	522	539	555	11	100.0	0.37	37.0	Bodipy-Small	4.14.1

(Continued)

Ent.	Short Names	Aqueous Solutions	$\lambda_{x,-0.25}$ nm	$\lambda_{x,-0.50}$ nm	$\lambda_{x,max}$ nm	$\lambda_{x,+0.50}$ nm	$\lambda_{e,-0.25}$ nm	$\lambda_{e,-0.50}$ nm	$\lambda_{e,max}$ nm	$\lambda_{e,+0.50}$ nm	$\lambda_{e,+0.25}$ nm	$\lambda_{e,+0.10}$ nm	Δ_S nm	ε_{max} mм$^{-1}$cm^{-1}	Φ_f	B_{max} mм$^{-1}$cm^{-1}	Class	Ref.
134	L-B 4b	H_2O	457	475	498	506	497	501	509	522	539	555	11	77.6	0.15	11.6	Bodipy-Small	4.14.1
135	2,4-DiMe TokyoGreen	100 mм NaOH	453	475	491	503	493	497	510	548	563	582	19	85.0	0.87	73.5	Fluorone	4.11
136	2,5-DiMe TokyoGreen	100 mм NaOH	453	475	491	503	493	497	510	548	563	582	19	85.0	0.89	75.4	Fluorone	4.11
137	2-Me TokyoGreen	H_2O pH 9	453	475	491	503	493	497	510	548	563	582	19	85.0	0.85	72.2	Fluorone	4.11
138	2,6-Sulfon.-1,3,5,7-Me-BODIPY	H_2O	451	469	491	499	498	502	510	523	540	556	19	60.0	0.85	51.0	Bodipy-Small	4.14.1
139	Z-L A	0.5 м PBS pH 7.4	457	476	499	507	498	502	510	523	540	556	11	50	0.04	2.1	Bodipy-Large	4.14.3
140	3,6-Diaminoacridine (Proflavine)	H_2O pH 7	404	420	445	457	482	490	511	554	582	614	66	38.9	0.34	13.2	Acridine	4.23.1
141	BODIPY FL	H_2O	460	479	502	511	499	503	511	524	541	557	9	92.0	0.97	89.2	Bodipy-Small	4.14.1
142	L-B 5	H_2O	457	475	498	506	499	503	511	524	541	557	13	79.4	0.49	38.9	Bodipy-Small	4.14.1
143	Niu 18	H_2O	455	474	496	504	499	503	511	524	541	557	15	75.5	0.20	15.1	Bodipy-Med.	4.14.2
144	Z-L F	0.5 м PBS pH 7.4	459	478	501	509	499	503	511	524	541	557	10	50	0.31	15.6	Bodipy-Large	4.14.3
145	8-Aminonaphthalene-1,3,6-triSA	100 mм P pH 7	295	315	356	394	466	479	512	547	564	591	156	15.0	0.20	3.0	Naphthalene	4.34
146	Phosphorilated coum. N-H 21c	PBS pH 7.4	374	394	432	457	482	490	512	550	576	604	80	20.4	0.81	16.5	Coumarin 7-Am	4.26.3
147	NX	100 mм P pH 7.5	454	474	492	504	491	497	512	537	559	582	20	69.0	0.83	57.2	Fluorescein-nh	4.8.1
148	2,5-DiOMe TokyoGreen	100 mм NaOH	456	478	494	506	495	499	512	550	566	585	18	85.0	0.01	0.8	Fluorone	4.11
149	Seta-385-carboxylate	P pH 7.4			384				512				128	25.0	0.78	19.5	nd	4.41
150	2,4-DiOMe TokyoGreen	100 mм NaOH	456	478	494	506	496	500	513	551	567	586	19	85.0	0.20	17.0	Fluorone	4.11
151	8-Hydroxypyrene-1,3,6-triSA	H_2O pH 10	389	408	460	470	490	495	514	563	593	620	54	24.0	0.95	22.8	Pyrene	4.28
152	Fluorescein (G-C 1)	H_2O pH 9.18, 25 °C	454	474	492	504	493	499	514	540	561	584	22	78.4	0.90	70.5	Fluorescein-nh	4.8.1
153	Fluorescein (S-H 1)	P pH 9	453	472	490	502	493	499	514	540	561	584	24	90.0	0.92	82.8	Fluorescein-nh	4.8.1
154	Fluorinated fluorescein S-H 17	P pH 9	453	472	490	502	493	499	514	540	561	584	24	82.4	0.97	79.9	Fluorescein-ha	4.8.2
155	Oregon Green 488	H_2O pH 9	453	472	490	502	493	499	514	540	561	584	24	76.0	0.97	73.7	Fluorescein-ha	4.8.2
156	2-OMe-5-Me TokyoGreen	100 mм NaOH	456	478	494	506	497	501	514	553	568	587	20	85.0	0.50	42.5	Fluorone	4.11

Ent.	Short Names	Aqueous Solutions	$\lambda_{x,-0.25}$ nm	$\lambda_{x,-0.50}$ nm	$\lambda_{x,max}$ nm	$\lambda_{x,+0.50}$ nm	$\lambda_{e,-0.25}$ nm	$\lambda_{e,-0.50}$ nm	$\lambda_{e,max}$ nm	$\lambda_{e,+0.50}$ nm	$\lambda_{e,+0.25}$ nm	$\lambda_{e,+0.10}$ nm	Δ_S nm	ε_{max} mM^{-1}cm^{-1}	Φ_f	B_{max} mM^{-1}cm^{-1}	Class	Ref.
157	Pennsylvania Green	PBS pH 7.4	456	478	494	506	497	501	514	553	568	587	20	82.0	0.91	74.6	Fluorone	4.11
158	Acedan	H_2O	299	320	362	402	468	482	515	551	568	595	153	11.2	0.20	2.2	Naphthalene	4.34
159	Acedan 2	H_2O	282	301	338	372	468	482	515	551	568	595	177	14.6	0.30	4.3	Naphthalene	4.34
160	Acedan 8	H_2O	278	296	332	365	468	482	515	551	568	595	183	12.6	0.23	2.9	Naphthalene	4.34
161	Abberior Star 440SXP	PBS pH 7.4	377	398	436	461	485	493	515	554	580	608	79	22.7	0.68	15.4	Coumarin 7-Am	4.26.3
162	Sulfonated coumarin N-H 6-H	PBS pH 7.4	377	398	436	461	485	493	515	554	580	608	79	31.0	0.57	17.6	Coumarin 7-Am	4.26.3
163	DMAX-EP	100 mM P pH 7.5	454	474	492	504	494	500	515	541	563	585	23	71.0	0.81	57.5	Fluorescein-nh	4.8.1
164	DY-490 carboxylate	PBS pH 7.4	452	468	491	504	495	500	515	536	553	572	24	73.0	0.90	65.4	Rhodam Hi Ch	4.2.5
165	2-OMe TokyoGreen	100 mM NaOH	456	478	494	506	497	502	515	554	569	589	21	85.0	0.86	73.1	Fluorone	4.11
166	4-Carboxy Pennsylvania Green	PBS pH 7.4	456	478	494	506	497	502	515	554	569	589	21	62.0	0.89	55.1	Fluorone	4.11
167	6-Sulfonato-1,3,5,7-Me-BODIPY	H_2O	454	473	495	503	503	507	515	528	545	562	20	52.5	0.85	44.6	Bodipy-Small	4.14.1
168	DPP 5	PBS			455				515				60	22.7	0.57	12.9	Diketopyrrolop	4.39.1
169	3-Methoxymethyl ether rhodol	50 mM P, <0.1% ACN pH 8	439	457	477	488	492	500	516	539	563	587	39	58.0	0.85	49.3	Rhodol	4.15.2
170	6-Aminorhodol (Rhodol)	50 mM P, <0.1% ACN pH 8	452	471	493	505	492	500	516	539	563	587	23	70.0	0.98	68.6	Rhodol	4.15.2
171	5-Carboxyfluorescein (S-H 2)	P pH 9	454	474	492	504	495	501	516	542	564	587	24	84.7	0.92	77.9	Fluorescein-nh	4.8.1
172	Fluorinated fluorescein S-H 14a,b	P pH 9	454	474	492	504	495	501	516	542	564	587	24	85.9	0.92	79.0	Fluorescein-ha	4.8.2
173	BODIPY	H_2O	457	475	498	506	504	508	516	529	547	563	18	90.0	0.90	81.2	Bodipy-Small	4.14.1
174	Fluorescein	100 mM NaOH in D_2O	453	472	489.7	502	495	501	516.3	542	564	587	26	67.0	0.98	65.6	Fluorescein-nh	4.8.1
175	Fluorescein	100 mM NaOH in H_2O	453	472	490.2	503	495	501	516.3	542	564	587	26	67.0	0.92	61.6	Fluorescein-nh	4.8.1
176	BMeS-*p*-A	H_2O	322	341	377	404	458	472	517	576	610	643	140	3.8	0.67	2.5	Benzene	4.38
177	Acyman	H_2O	302	324	367	408	470	484	517	553	570	597	150	9.1	0.10	0.9	Naphthalene	4.34

(Continued)

Ent.	Short Names	Aqueous Solutions	$\lambda_{x,-0.25}$ nm	$\lambda_{x,-0.50}$ nm	$\lambda_{x,max}$ nm	$\lambda_{x,+0.50}$ nm	$\lambda_{e,-0.25}$ nm	$\lambda_{e,-0.50}$ nm	$\lambda_{e,max}$ nm	$\lambda_{e,+0.50}$ nm	$\lambda_{e,+0.25}$ nm	$\lambda_{e,+0.10}$ nm	Δ_S nm	ε_{max} $mM^{-1}cm^{-1}$	Φ_f	B_{max} $mM^{-1}cm^{-1}$	Class	Ref.
178	JF479	10 mM HEPES pH 7.3	432	450	479	493	488	495	517	561	589	623	38	47.9	0.62	29.7	Acridine	4.23.1
179	4-Benzamidofluorescein	100 mM P pH 7.5	453	472	490	502	496	502	517	543	565	588	27	80.0	0.79	63.2	Fluorescein-nh	4.8.1
180	Alexa Fluor 488	PBS pH 7.2, 22 °C	457	477	494	506	499	504	517	538	555	576	23	73.0	0.92	67.1	Rhodam Hi Ch	4.2.5
181	Sulfofluorescein	C pH 9.5	464	482	497	508	499	504	517	556	572	591	20	78.0	0.92	71.7	Sulfofluoresc.	4.10
182	G-L 13a	H_2O	463	482	505	514	505	509	517	530	548	564	12	35.0	0.47	16.4	Bodipy-Med.	4.14.2
183	Epicocconone P-F 17a	H_2O, 0.1% SDS pH 2.5			391				517				126	8.8	0.35	3.0	Epicocconone	4.39.2
184	Chlorinated fluorescein G-C 2(3)	H_2O pH 9.18, 25 °C	460	480	499	512	497	503	518	544	566	589	19	48.4	0.93	45.0	Fluorescein-ha	4.8.2
185	3,8,10-Trimethylisoallox-azine	H_2O pH 7			433				518				85	11.5	0.12	1.3	Pteridine	4.36
186	Epicocconone P-F 17d	H_2O, 0.1% SDS pH 2.5			392				518				126	13.2	0.40	5.2	Epicocconone	4.39.2
187	SeTau-405-carboxylate	P pH 7.4			405				518				113	13.8	0.80	11.0	nd	4.41
188	2,6-Sulfon.-1,3,5,7,8-Me-BODIPY	H_2O	451	469	491	499	507	511	519	532	550	567	28	99.0	0.83	82.1	Bodipy-Small	4.14.1
189	Isoquinoline 3f	PBS pH 7.4			405				519				114	15.4	0.39	6.0	Isoquinoline	4.37
190	Edans	H_2O	281	300	336	370	472	487	520	557	574	601	184	6.2	0.27	1.6	Naphthalene	4.34
191	Briliant sulfaflavine	H_2O	376	393	430	463	481	491	520	557	575	596	90	16.0	0.27	4.3	Naphthalimide	4.32
192	5(6)-Carboxyrhodol	50 mM P pH 8-9	451	472	494	503	495	501	520	539	562	584	26	59.9	0.89	53.3	Rhodol	4.15.1
193	5(6)-Carboxyfluorescein (FAM)	100 mM PBS pH 9	457	477	495	508	499	504	520	546	569	592	25	83.0	0.90	74.7	Fluorescein-nh	4.8.1
194	2-Me RG	PBS pH 7.4	458	483	498	510	502	507	520	536	549	574	22	82.0	0.84	68.8	Rosamine	4.17
195	Atto 488 carboxylate	PBS pH 7.4, 22 °C	461	486	500	511	503	507	520	540	564	586	20	90.0	0.80	72.0	Rhodam <500	4.2.1
196	Rhodamine 110	H_2O pH <7	458	482	496	507	503	507	520	540	564	586	24	83.0	0.89	73.8	Rhodam <500	4.2.1
197	3,10-Dimethylisoal-loxazine	H_2O pH 7			435				520				85	9.5	0.15	1.4	Pteridine	4.36
198	Isoquinoline 3d	PBS pH 7.4			418				520				102	14.5	0.43	6.2	Isoquinoline	4.37
199	DY-495 carboxylate	PBS pH 7.4	456	476	494	507	499	505	521	547	570	593	27	78.0	0.68	53.2	Fluorescein-nh	4.8.1

Ent.	Short Names	Aqueous Solutions	$\lambda_{x,-0.25}$ nm	$\lambda_{x,-0.50}$ nm	$\lambda_{x,max}$ nm	$\lambda_{x,+0.50}$ nm	$\lambda_{e,-0.25}$ nm	$\lambda_{e,-0.50}$ nm	$\lambda_{e,max}$ nm	$\lambda_{e,+0.50}$ nm	$\lambda_{e,+0.25}$ nm	$\lambda_{e,+0.10}$ nm	Δ_S nm	ε_{max} $mM^{-1}cm^{-1}$	Φ_f	B_{max} $mM^{-1}cm^{-1}$	Class	Ref.
200	G-L 14	H_2O	472	492	508	518	501	505	521	555	575	603	13	59.0	0.77	45.4	Bodipy-Med.	4.14.2
201	Isoquinoline 3i	PBS pH 7.4			405				521				116	15.4	0.40	6.1	Isoquinoline	4.37
202	Isoquinoline 3n	PBS pH 7.4			411				521				110	16.6	0.41	6.8	Isoquinoline	4.37
203	Isoquinoline 3o	PBS pH 7.4			410				521				111	19.2	0.09	1.7	Isoquinoline	4.37
204	Chlorinated fluorescein G-C 1	H_2O pH 9.18, 25 °C	464	484	503	516	500	506	522	548	571	595	19	103.0	0.86	88.5	Fluorescein-ha	4.8.2
205	G-L 13b	H_2O	473	493	509	519	502	506	522	556	576	604	13	56.0	0.67	37.5	Bodipy-Med.	4.14.2
206	Fluorinated rhodamine 1e	Buffer pH 7	461	485	499	510	504	509	522	542	566	588	23	81.0	0.88	71.2	Rhodam <500	4.2.1
207	Fluorinated rhodamine 1j	Buffer pH 7	458	482	496	507	504	509	522	542	566	588	26	53.0	0.89	47.1	Rhodam <500	4.2.1
208	Isoquinoline 3g	PBS pH 7.4			414				522				108	17.5	0.41	7.1	Isoquinoline	4.37
209	Isoquinoline 3l	PBS pH 7.4			410				522				112	16.8	0.42	7.0	Isoquinoline	4.37
210	3-OH-4-nitroso-2,7-naphth-diSA	T pH 9	339	367	423	478	475	489	523	560	578	605	100	20.0	0.28	5.6	Naphthalene	4.34
211	L-B 2a	H_2O	473	493	509	519	503	507	523	557	577	606	14	72.4	0.27	19.5	Bodipy-Small	4.14.1
212	Fluorinated rhodam. 1f-6′-COOH	Buffer pH 7	463	484	501	513	505	510	523	544	562	584	22	88.0	0.93	81.8	Rhoda 500-550	4.2.2
213	Isoquinoline 3k	PBS pH 7.4			411				523				112	18.6	0.11	2.0	Isoquinoline	4.37
214	6-Acrylamidorhodol	50 mм P, <0.1% ACN pH 8	450	469	490	502	500	507	524	548	572	598	34	28.5	0.19	5.4	Rhodol	4.15.2
215	L-B 2b	H_2O	476	496	512	522	504	508	524	559	578	607	12	77.6	0.41	31.8	Bodipy-Small	4.14.1
216	Abberior Star 488 carboxylate	PBS pH 7.4	465	485	503	515	505	511	524	545	564	585	21	64.5	0.89	57.4	Rhodam Hi Ch	4.2.5
217	Isoquinoline 3e	PBS pH 7.4			408				524				116	15.8	0.41	6.4	Isoquinoline	4.37
218	Isoquinoline 7c	PBS pH 7.4			408				524				116	15.8	0.26	4.1	Isoquinoline	4.37
219	6-Benzamidorhodol	50 mм P, <0.1% ACN pH 8	450	469	490	502	501	508	525	549	573	599	35	30.1	0.16	4.8	Rhodol	4.15.2
220	Fluorinated rhodam. 1f-5′-COOH	Buffer pH 7	463	484	501	513	506	512	525	546	565	586	24	85.0	0.95	80.7	Rhoda 500-550	4.2.2
221	Isoquinoline 3h	PBS pH 7.4			406				525				119	13.7	0.38	5.2	Isoquinoline	4.37
222	Isoquinoline 3j	PBS pH 7.4			409				525				116	18.8	0.45	8.4	Isoquinoline	4.37
223	Atto 495 carboxylate	PBS pH 7.4, 22 °C	455	471	498	512	503	509	526	569	597	631	28	80.0	0.20	16.0	Acridine	4.23.1

(Continued)

Ent.	Short Names	Aqueous Solutions	$\lambda_{x,-0.25}$ nm	$\lambda_{x,-0.50}$ nm	$\lambda_{x,max}$ nm	$\lambda_{x,+0.50}$ nm	$\lambda_{e,-0.25}$ nm	$\lambda_{e,-0.50}$ nm	$\lambda_{e,max}$ nm	$\lambda_{e,+0.50}$ nm	$\lambda_{e,+0.25}$ nm	$\lambda_{e,+0.10}$ nm	Δ_S nm	ε_{max} $mM^{-1}cm^{-1}$	Φ_f	B_{max} $mM^{-1}cm^{-1}$	Class	Ref.
224	6-CA-MeThio-4,5,7-F-fluorescein	100 mM NaOH	467	488	507	520	504	510	526	553	576	600	19	63.5	0.88	55.8	Fluorescein-ha	4.8.2
225	Mercaptofluorescein MK-43	1 mM NaOH	457	477	495	508	504	510	526	553	576	600	31	34.0	0.22	7.4	Fluorescein-SH	4.8.3
226	Mercaptofluorescein MK-69	1 mM NaOH	457	477	495	508	504	510	526	553	576	600	31	25.0	0.20	5.0	Fluorescein-SH	4.8.3
227	Isoquinoline 10b	PBS pH 7.4			410				526				116	15.5	0.17	2.6	Isoquinoline	4.37
228	Isoquinoline 3m	PBS pH 7.4			408				526				118	18.1	0.36	6.5	Isoquinoline	4.37
229	Isoquinoline 7b	PBS pH 7.4			406				526				120	14.4	0.46	6.6	Isoquinoline	4.37
230	Isoquinoline 7d	PBS pH 7.4			408				526				118	15.0	0.14	2.1	Isoquinoline	4.37
231	BMeS-*p*-A-COOH	H_2O	333	353	392	421	466	481	527	589	624	659	135	3.9	0.60	2.3	Benzene	4.38
232	6-Azido-4,5,7-trifluorofluorescein	100 mM NaOH	467	488	507	520	505	511	527	554	577	601	20	79.9	0.82	65.5	Fluorescein-ha	4.8.2
233	BCECF	100 mM NaOH	466	486	505	518	505	511	527	554	577	601	22	67.0	0.84	56.2	Fluorescein-nh	4.8.1
234	BCPCF	100 mM NaOH	466	486	505	518	505	511	527	554	577	601	22	67.0	0.83	55.6	Fluorescein-nh	4.8.1
235	Chlorinated fluorescein G-C 5	H_2O pH 9.18, 25 °C	471	491	511	524	505	511	527	554	577	601	16	91.6	0.84	76.9	Fluorescein-ha	4.8.2
236	Fluorinated fluorescein S-H 18	P pH 9	468	489	508	521	505	511	527	554	577	601	19	85.6	0.85	72.7	Fluorescein-ha	4.8.2
237	Fluorinated fluorescein S-H 19	P pH 9	468	489	508	521	505	511	527	554	577	601	19	81.2	0.96	77.9	Fluorescein-ha	4.8.2
238	Mercaptofluorescein MK-61	1 mM NaOH	457	477	495	508	505	511	527	554	577	601	32	38.0	0.24	9.1	Fluorescein-SH	4.8.3
239	Mercaptofluorescein MK-67	1 mM NaOH	463	483	502	515	505	511	527	554	577	601	25	27.0	0.17	4.5	Fluorescein-SH	4.8.3
240	Acridine orange	H_2O	439	458	488	503	497	505	528	574	604	639	40	50.0	0.12	6.0	Acridine	4.23.1
241	Acridine orange	HEPES pH 7.3	443	462	493	508	497	505	528	574	604	639	35	50.0	0.21	10.5	Acridine	4.23.1
242	Chlorinated fluorescein G-C 4	H_2O pH 9.18, 25 °C	471	492	512	526	506	512	528	555	578	602	16	93.8	0.87	81.6	Fluorescein-ha	4.8.2
243	Niu 27	H_2O	474	494	510	520	508	512	528	563	583	612	18	63.3	0.81	51.2	Bodipy-Med.	4.14.2
244	Rhodamine 123	H_2O pH <7	464	485	502	514	509	514	528	549	568	590	26	75.0	0.90	67.5	Rhoda 500-550	4.2.2
245	Isoquinoline 7a	PBS pH 7.4			407				528				121	14.1	0.34	4.7	Isoquinoline	4.37
246	Isoquinoline 7f	PBS pH 7.4			406				528				122	16.3	0.16	2.6	Isoquinoline	4.37
247	Tri-*O*-acetyl-P-luminoarosin	H_2O pH 7			425				528				103	10.8	0.62	6.7	Pteridine	4.36

Ent.	Short Names	Aqueous Solutions	$\lambda_{x,-0.25}$ nm	$\lambda_{x,-0.50}$ nm	$\lambda_{x,max}$ nm	$\lambda_{x,+0.50}$ nm	$\lambda_{e,-0.25}$ nm	$\lambda_{e,-0.50}$ nm	$\lambda_{e,max}$ nm	$\lambda_{e,+0.50}$ nm	$\lambda_{e,+0.25}$ nm	$\lambda_{e,+0.10}$ nm	Δ_S nm	ε_{max} mM^{-1}cm^{-1}	Φ_f	B_{max} mM^{-1}cm^{-1}	Class	Ref.
248	6-Benzylaminorhodol	50 mM P, <0.1% ACN pH 8	461	480	503	515	504	512	529	554	578	604	26	71.0	0.95	67.4	Rhodol	4.15.2
249	B-Z 7b	H_2O	479	500	516	526	509	513	529	564	584	614	13	60.5	0.68	41.1	Bodipy-Large	4.14.3
250	3,7,8,10-TetraMe-isoalloxazine	H_2O pH 7			444				529				85	12.4	0.25	3.1	Pteridine	4.36
251	Isoquinoline 4c	PBS pH 7.4			453				529				76	22.1	0.24	5.3	Isoquinoline	4.37
252	DY-505 carboxylate	PBS pH 7.4	459	478	505	515	506	512	530	552	570	590	25	85.0	0.74	63.2	Rhoda 500-550	4.2.2
253	Oregon Green 514	PBS pH 7.2, 22 °C	471	491	511	524	508	514	530	557	581	605	19	88.0	0.96	84.4	Fluorescein-ha	4.8.2
254	Niu 29	H_2O	477	497	519	531	511	515	530	561	584	612	11	28.0	0.61	17.0	Bodipy-Med.	4.14.2
255	Fluorinated rhodamine 5-SO_3H	Buffer pH 7	472	494	512	525	511	516	530	552	571	593	18	52.0	0.86	44.7	Rhoda 500-550	4.2.2
256	Abberior Star 512 carboxylate	PBS pH 7.4	472	493	511	524	511	516	530	552	571	593	19	84.0	0.82	68.8	Rhoda 500-550	4.2.2
257	Flavin mononucleotide (FMN)	H_2O			450				530				80	12.2	0.25	3.0	Pteridine	4.36
258	Prodan	H_2O	300	322	364	404	481	496	531	569	587	616	167	14.5	0.10	1.4	Naphthalene	4.34
259	3,6-Di(azetidin-1-yl)acridine	10 mM HEPES pH 7.3	442	461	492	507	500	508	531	578	608	643	39	47.0	0.52	24.4	Acridine	4.23.1
260	6-Propylaminorhodol	50 mM P, <0.1% ACN pH 8	464	484	507	519	506	514	531	556	581	607	24	61.0	0.94	57.3	Rhodol	4.15.2
261	Fluorinated rhodamine 1k	Buffer pH 7	467	488	506	518	512	517	531	553	572	594	25	54.0	0.66	35.6	Rhoda 500-550	4.2.2
262	Fluorinated rhodamine 3-H	Buffer pH 7	471	492	510	523	512	517	531	553	572	594	21	57.0	0.86	49.0	Rhoda 500-550	4.2.2
263	Carbostyril 4	H_2O			379				531				152	8.7	0.07	0.6	Quinoline	4.29
264	3-MOMO-6-propylaminorhodol	50 mM P, <0.1% ACN pH 8	453	472	494	506	507	515	532	557	582	608	38	43.0	0.35	15.0	Rhodol	4.15.2
265	Atto 514 carboxylate	PBS pH 7.4, 22 °C	474	494	511	522	512	517	532	553	575	599	21	115.0	0.85	97.7	Rhoda 500-550	4.2.2
266	3,6,10-Trimethylisoalloxazine	H_2O pH 7			439				532				93	8.5	0.08	0.6	Pteridine	4.36
267	3,7,10-Trimethylisoalloxazine	H_2O pH 7			446				532				86	9.8	0.32	3.1	Pteridine	4.36
268	JF502	10 mM HEPES pH 7.3	450	470	502	518	502	510	533	580	610	646	31	57.8	0.71	41.0	Acridine	4.23.1
269	2,7-DiCl-6-methylamino rhodol	50 mM P pH 8-9	469	488	512	525	507	515	533	558	583	611	21	71.0	0.67	47.5	Rhodol	4.15.1

(Continued)

Ent.	Short Names	Aqueous Solutions	$\lambda_{x,-0.25}$ nm	$\lambda_{x,-0.50}$ nm	$\lambda_{x,max}$ nm	$\lambda_{x,+0.50}$ nm	$\lambda_{e,-0.25}$ nm	$\lambda_{e,-0.50}$ nm	$\lambda_{e,max}$ nm	$\lambda_{e,+0.50}$ nm	$\lambda_{e,+0.25}$ nm	$\lambda_{e,+0.10}$ nm	Δ_S nm	ε_{max} mм$^{-1}$cm^{-1}	Φ_f	B_{max} mм$^{-1}$cm^{-1}	Class	Ref.
270	6-Butylaminorhodol	50 mм P, <0.1% ACN pH 8	465	485	508	521	508	516	533	558	583	609	25	60.0	0.93	55.8	Rhodol	4.15.2
271	Mercaptofluorescein MK-75	1 mм NaOH	463	483	502	515	511	517	533	561	584	609	31	12.0	0.16	1.9	Fluorescein-SH	4.8.3
272	NBDY	H_2O	462	481	504	513	520	524	533	547	566	583	29	74.4	0.33	24.5	Bodipy-Small	4.14.1
273	Luminarine	H_2O pH 7			428				533				105	9.0	0.47	4.2	Pteridine	4.36
274	6-Allylaminorhodol	50 mм P, <0.1% ACN pH 8	464	484	507	519	509	517	534	559	584	611	27	61.5	0.92	56.5	Rhodol	4.15.2
275	7-Me-6-EtAm-carboxyrhodol	50 mм P pH 8-9	465	485	508	521	509	517	534	559	584	611	26	78.1	0.82	64.0	Rhodol	4.15.1
276	Fluorinated fluorescein S-H 15a,b	P pH 9	470	490	510	523	511	518	534	562	585	610	24	84.2	0.43	36.2	Fluorescein-ha	4.8.2
277	Fluorinated fluorescein S-H 16a,b	P pH 9	470	490	510	523	511	518	534	562	585	610	24	78.1	0.59	46.0	Fluorescein-ha	4.8.2
278	Epicocconone P-F 4d	H_2O, 0.1% SDS pH 2.5			417				534				117	14.5	0.37	5.3	Epicocconone	4.39.2
279	Isoquinoline 8	PBS pH 7.4			406				534				128	16.8	0.37	6.2	Isoquinoline	4.37
280	BMeS-*p*-A-2COOH	H_2O	341	362	403	434	473	487	535	599	635	671	132	3.7	0.56	2.0	Benzene	4.38
281	Lucifer yellow VS	H_2O pH 7	376	393	430	463	493	505	535	574	594	616	105	11.3	0.24	2.7	Naphthalimide	4.32
282	2-PrAc-7-Me-6-EtAm-rhodol	50 mм P pH 8-9	466	486	509	522	510	518	535	560	585	612	26	79.4	0.92	73.0	Rhodol	4.15.1
283	2,7-DiCl-6-pyrrolidino sulforhodol	PBS pH 7.4	470	491	520	534	516	521	535	555	584	604	15	83.0	0.27	22.4	Sulforhodol	4.27
284	Epicocconone	H_2O, 0.1% SDS pH 2.5			444				535				91	68.6	0.07	5.0	Epicocconone	4.39.2
285	2,7-DiMe-6-EtAm-carboxyrhodol	50 mм P pH 8-9	467	487	510	522	511	519	536	561	586	612	26	66.3	0.89	59.0	Rhodol	4.15.1
286	TET	H_2O pH 9	479	501	521	535	513	519	536	564	588	613	15	73.0	0.70	51.1	Fluorescein-ha	4.8.2
287	Chlorinated fluorescein G-C 6	H_2O pH 9.18, 25 °C	480	501	522	536	514	520	537	565	589	614	15	65.8	0.83	54.6	Fluorescein-ha	4.8.2
288	DAMBO-P^{H}-T	100 mм P pH 7.4	479	499	521	533	517	521	537	569	592	621	16	56.0	0.74	41.4	Bodipy-Small	4.14.1
289	2,4,5,7-Tetra-F-sulfofluorescein	P pH 9.18	480	499	515	527	518	523	537	579	596	618	22	63.2	0.80	50.5	Sulfofluoresc.	4.10

Ent.	Short Names	Aqueous Solutions	$\lambda_{x,-0.25}$ nm	$\lambda_{x,-0.50}$ nm	$\lambda_{x,max}$ nm	$\lambda_{x,+0.50}$ nm	$\lambda_{e,-0.25}$ nm	$\lambda_{e,-0.50}$ nm	$\lambda_{e,max}$ nm	$\lambda_{e,+0.50}$ nm	$\lambda_{e,+0.25}$ nm	$\lambda_{e,+0.10}$ nm	Δ_S nm	ε_{max} mM^{-1}cm^{-1}	Φ_f	B_{max} mM^{-1}cm^{-1}	Class	Ref.
290	(HO)-Luciferin	H_2O pH 5			328				537				209	18.2	0.25	4.5	Benzothiazole	4.30
291	($^-$O)-Luciferin	H_2O pH 11			384				537				153	18.5	0.62	11.4	Benzothiazole	4.30
292	2-Et-7-Me-6-EtAm-carboxyrhodol	50 mM P pH 8-9	468	488	511	524	512	520	538	563	589	616	27	73.0	0.84	61.3	Rhodol	4.15.1
293	2′,7′-Cl-4,5,6,7-F-fluorescein	100 mM NaOH	478	500	520	534	515	521	538	566	590	615	18	88.5	0.80	70.8	Fluorescein-ha	4.8.2
294	Atto 520 carboxylate	PBS pH 7.4, 22 °C	474	497	517	529	520	525	538	556	567	592	21	110.0	0.90	99.0	Pyronine	4.4
295	CF450	H_2O			450				538				88	40.0	0.56	22.3	nd (Cyanine)	4.41
296	Epicocconone P-F 4b	H_2O, 0.1% SDS pH 2.5			435				538				103	19.8	0.20	3.9	Epicocconone	4.39.2
297	2-Br-7-Me-6-EtAm-carboxyrhodol	50 mM P pH 8-9	472	492	516	529	513	521	539	565	590	618	23	80.9	0.61	49.3	Rhodol	4.15.1
298	2-Cl-7-Me-6-EtAm-carboxyrhodol	50 mM P pH 8-9	479	495	514	538	514	525	539	573	600	634	25	80.5	0.82	66.0	Rhodol	4.15.1
299	Lucifer yellow CH	H_2O pH 7	375	392	429	462	498	509	540	580	600	623	111	11.3	0.21	2.3	Naphthalimide	4.32
300	*meso*-CaboxyBODIPY J-K 3c	10 mM PBS pH 7.4	480	501	517	527	519	523	540	577	598	629	23	76.0	0.97	73.7	Bodipy-Small	4.14.1
301	Niu 25	PBS pH 7.4	479	499	521	533	520	524	540	572	596	625	19	58.8	0.82	48.2	Bodipy-Med.	4.14.2
302	Alexa Fluor 430	PBS pH 7.4	373	393	431	456	508	517	541	584	613	645	110	16.0	0.55	8.8	Coumarin 7-Am	4.26.3
303	6-Piperidinocarboxyrhodol	50 mM P pH 8-9	472	492	516	529	515	523	541	567	593	620	25	80.9	0.79	63.9	Rhodol	4.15.1
304	DPP 9	PBS			451				541				90	13.9	0.90	12.5	Diketopyrrolop	4.39.1
305	8-Trifluoromethyl-BODIPY 10	100 mM P pH 7.4	432	449	469	476	529	533	542	557	576	594	73	17.1	0.74	12.6	Bodipy-Small	4.14.1
306	Epicocconone P-F 4c	H_2O, 0.1% SDS pH 2.5			425				542				117	40.3	0.40	16.2	Epicocconone	4.39.2
307	PB430	PBS pH 7.4			426				542				116	10.4	0.66	6.8	Phosphole	4.39.3
308	Naphtahlimide 10c	H_2O	393	412	453	490	500	512	543	583	603	627	90	17.3	0.30	5.1	Naphthalimide	4.32
309	HMDER	PBS pH 7.4	479	500	525	538	517	525	543	569	595	622	18	70.0	0.14	9.8	Rhodol	4.15.2
310	6-Dimethylaminorhodol	50 mM P, <0.1% ACN pH 8	473	494	518	531	518	526	544	570	596	624	26	64.0	0.20	12.8	Rhodol	4.15.2

(Continued)

Ent.	Short Names	Aqueous Solutions	$\lambda_{x,-0.25}$ nm	$\lambda_{x,-0.50}$ nm	$\lambda_{x,max}$ nm	$\lambda_{x,+0.50}$ nm	$\lambda_{e,-0.25}$ nm	$\lambda_{e,-0.50}$ nm	$\lambda_{e,max}$ nm	$\lambda_{e,+0.50}$ nm	$\lambda_{e,+0.25}$ nm	$\lambda_{e,+0.10}$ nm	Δ_S nm	ε_{max} $mM^{-1}cm^{-1}$	Φ_f	B_{max} $mM^{-1}cm^{-1}$	Class	Ref.
311	4,7-DiMe-6-EtAm-carboxyrhodol	50 mM P pH 8-9	472	492	516	529	518	526	544	570	596	624	28	56.9	0.43	24.4	Rhodol	4.15.1
312	Carbostyril 6	H_2O			389				544				155	7.0	0.09	0.6	Quinoline	4.29
313	Naphtahlimide 10b	H_2O	389	408	448	484	502	514	545	586	606	630	97	12.4	0.21	2.6	Naphthalimide	4.32
314	6-Dimethylaminorhodol	C pH 9.5	477	498	522	535	519	527	545	571	597	625	23	57.0	0.20	11.4	Rhodol	4.15.1
315	Atto 430LS carboxylate	PBS pH 7.4, 22 °C			436				545				109	32.0	0.65	20.8	nd (pb Coum.)	4.41
316	trans-PO-PB430	PBS pH 7.4			422				545				123	10.3	0.74	7.6	Phosphole	4.39.3
317	6-Dimethylaminorhodol	10 mM HEPES pH 7.3	473	494	518	531	520	528	546	572	599	626	28	60.0	0.21	12.6	Rhodol	4.15.2
318	6-Azetidinylaminorhodol	10 mM HEPES pH 7.3	474	495	519	532	520	528	546	572	599	626	27	59.0	0.85	50.1	Rhodol	4.15.2
319	2,4,5,7-Tetra-Br-sulfofluorescein	100 mM NaOH pH 12	488	507	524	536	526	531	546	590	607	630	22	87.0	0.35	30.4	Sulfofluoresc.	4.10
320	2-Cl-6-Piperidinocarboxyrhodol	50 mM P pH 8-9	478	500	524	537	522	530	548	574	601	629	24	77.4	0.72	55.7	Rhodol	4.15.1
321	JOE	H_2O pH 9	478	500	520	534	524	531	548	577	602	628	28	73.0	0.60	43.8	Fluorescein-ha	4.8.2
322	AR116	100 mM P pH 7.5	483	506	527	539	529	535	548	567	578	604	21	64.8	0.07	4.5	Pyronine	4.4
323	(HO)-Luciferyl adenylate	H_2O pH 5			335				548				213	15.1	0.11	1.6	Benzothiazole	4.30
324	CF514	H_2O			516				548				32	105.0	0.38	39.8	nd (Rhodam.)	4.41
325	cis-PO-PB430	PBS pH 7.4			421				548				127	10.0	0.71	7.1	Phosphole	4.39.3
326	DPP 7	PBS			456				548				92	14.7	0.76	11.1	Diketopyrrolop	4.39.1
327	Cascade yellow	H_2O	329	356	404	440	491	505	549	603	657	663	145	24.0	0.05	1.2	Hybrid-Bonded	4.40.2
328	6-Diethylaminorhodol	50 mM P, <0.1% ACN pH 8	477	498	522	535	522	531	549	576	602	630	27	61.0	0.15	9.1	Rhodol	4.15.2
329	Yakima Yellow carboxylate	100 mM P/Tris pH 7.0/7.4	487	509	530	545	525	532	549	578	603	630	19	84.0	0.96	80.6	Fluorescein-ha	4.8.2
330	*meso*-CaboxyBODIPY J-K 3b	10 mM PBS pH 7.4	477	497	519	531	528	532	549	582	607	637	30	78.0	0.83	64.7	Bodipy-Small	4.14.1

Ent.	Short Names	Aqueous Solutions	$\lambda_{x,-0.25}$ nm	$\lambda_{x,-0.50}$ nm	$\lambda_{x,max}$ nm	$\lambda_{x,+0.50}$ nm	$\lambda_{e,-0.25}$ nm	$\lambda_{e,-0.50}$ nm	$\lambda_{e,max}$ nm	$\lambda_{e,+0.50}$ nm	$\lambda_{e,+0.25}$ nm	$\lambda_{e,+0.10}$ nm	Δ_S nm	ε_{max} mм$^{-1}$cm^{-1}	Φ_f	B_{max} mм$^{-1}$cm^{-1}	Class	Ref.
331	JF525	10 mм HEPES pH 7.3	483	506	525	538	529	534	549	572	593	617	24	94.0	0.91	85.5	Rhoda 500-550	4.2.2
332	6-Pyrrolidinorhodol	50 mм P, <0.1% ACN pH 8	474	495	519	532	523	532	550	577	603	632	31	63.0	0.12	7.5	Rhodol	4.15.2
333	Z-L C	0.5 м PBS pH 7.4	489	510	533	546	529	533	550	583	608	638	17	50	0.01	0.6	Bodipy-Large	4.14.3
334	Z-L E	0.5 м PBS pH 7.4	490	511	534	547	529	533	550	583	608	638	16	50	0.09	4.4	Bodipy-Large	4.14.3
335	JF526	10 mм HEPES pH 7.3	484	507	526	539	530	535	550	573	594	618	24	19.0	0.87	16.5	Rhoda 500-550	4.2.2
336	Isoquinoline 4b	PBS pH 7.4			436				550				114	15.5	0.36	5.5	Isoquinoline	4.37
337	Rhodamine 6G +3	H_2O pH <7	486	510	529	544	532	537	551	570	582	602	22	101.0	0.91	91.9	Rhodam Hi Ch	4.2.5
338	Acridizinium I-L 5b	H_2O			369				551				182	11.0	0.02	0.2	Acridizinium	4.39.1
339	AI3	H_2O pH 7.4	387	405	445	481	508	520	552	594	615	639	107	15.0	0.21	3.1	Naphthalimide	4.32
340	Naphthalimide B-K 1	10 mм P pH 7.0	380	398	436	470	508	520	552	594	615	639	116	13.2	0.08	1.0	Naphthalimide	4.32
341	6-Dipropylaminorhodol	50 mм P, <0.1% ACN pH 8	479	500	525	538	525	533	552	579	606	634	27	59.0	0.10	5.9	Rhodol	4.15.2
342	Rhodamine 6G	D_2O pH <7	484	506	525.5	539	531	537	552.0	575	596	620	26	105.0	0.98	102.9	Rhoda 500-550	4.2.2
343	Rhodamine RBzET	H_2O	485	508	527	541	531	537	552	575	596	620	25	82.0	0.87	71.3	Rhoda 500-550	4.2.2
344	Atto 532 carboxylate	PBS pH 7.4, 22 °C	491	515	532	544	533	538	552	575	600	625	20	115.0	0.90	103.5	Rhoda 500-550	4.2.2
345	Rhodamine 6G	H_2O pH <7	485	507	526.3	540	532	537	552.3	576	596	621	26	105.0	0.93	97.6	Rhoda 500-550	4.2.2
346	6-Azetidinyl naphthalimide	10 mм HEPES pH 7.3	401	421	464	503	509	521	553	595	616	640	89	18.0	0.28	5.0	Naphthalimide	4.32
347	6-Morpholinorhodol	50 mм P, <0.1% ACN pH 8	477	498	522	535	526	534	553	580	607	636	31	61.0	0.13	7.9	Rhodol	4.15.2
348	6-Dibutylaminorhodol	50 mм P, <0.1% ACN pH 8	479	500	525	538	526	534	553	580	607	636	28	63.0	0.08	5.0	Rhodol	4.15.2

(Continued)

Ent.	Short Names	Aqueous Solutions	$\lambda_{x,-0.25}$ nm	$\lambda_{x,-0.50}$ nm	$\lambda_{x,max}$ nm	$\lambda_{x,+0.50}$ nm	$\lambda_{e,-0.25}$ nm	$\lambda_{e,-0.50}$ nm	$\lambda_{e,max}$ nm	$\lambda_{e,+0.50}$ nm	$\lambda_{e,+0.25}$ nm	$\lambda_{e,+0.10}$ nm	Δ_S nm	ε_{max} $mM^{-1}cm^{-1}$	Φ_f	B_{max} $mM^{-1}cm^{-1}$	Class	Ref.
349	Fluorinated fluorescein S-H 20	P pH 9	491	513	535	550	529	535	553	583	608	635	18	83.4	0.47	39.2	Fluorescein-ha	4.8.2
350	530RH	PBS pH 7.4	491	514	532	547	533	538	553	574	592	618	21	56.0	0.89	49.8	Rhoda 500-550	4.2.2
351	Rhodamine 6G +2	H_2O pH <7	487	511	530	544	533	539	553	571	583	603	23	101.0	0.93	93.9	Rhoda 500-550	4.2.2
352	Pacific orange	C pH 9.5			404				553				149	25.0	0.62	15.5	Benzothiazole	4.30
353	Alexa Fluor 532	PBS pH 7.2, 22 °C	486	514	531	545	533	539	554	572	587	616	23	81.0	0.61	49.4	Rosamine	4.17
354	DY-530 carboxylate	PBS pH 7.4	490	513	533	547	533	539	554	578	598	623	21	100.0	0.81	80.6	Rhoda 500-550	4.2.2
355	Fluorinated rhodamine 1r	Buffer pH 7	483	506	525	538	533	539	554	578	598	623	29	31.0	0.28	8.6	Rhoda 500-550	4.2.2
356	*meso*-CaboxyBODIPY J-K 1	10 mᴍ PBS pH 7.4	463	482	505	514	540	545	554	569	589	608	49	62.0	0.78	48.3	Bodipy-Small	4.14.1
357	6-Piperidinorhodol	50 mᴍ P, <0.1% ACN pH 8	478	499	523	536	528	536	555	582	609	638	32	62.5	0.10	6.2	Rhodol	4.15.2
358	Z-L D	0.5 ᴍ PBS pH 7.4	492	513	536	549	534	538	555	589	614	645	19	50	0.02	0.9	Bodipy-Large	4.14.3
359	(HO)-Dehydroluciferin	H_2O pH 4.5			348				555				207	18.2	0.26	4.7	Benzothiazole	4.30
360	($^-$O)-Dehydroluciferin	H_2O pH 11			396				555				159	18.2	0.62	11.2	Benzothiazole	4.30
361	Fluorene PP12	H_2O			345				555				210	24.0	0.16	3.8	Fluorene	4.39.2
362	Carbostyril 13	H_2O pH 7			420				556				136	17.3	0.26	4.5	Quinoline	4.29
363	Carbostyril 23b	H_2O pH 12			397				556				159	17.3	0.26	4.5	Quinoline	4.29
364	Carbostyril 23b	H_2O pH 7			424				556				132	17.3	0.26	4.5	Quinoline	4.29
365	Atto Rho6G carboxylate	PBS pH 7.4, 22 °C	489	513	533	546	534	540	557	583	609	636	24	115.0	0.94	108.1	Rhoda 500-550	4.2.2
366	M-S 10c	H_2O, 0.1% TX-100	489	510	533	546	536	540	557	591	617	648	24	54.0	0.61	32.9	Bodipy-Small	4.14.1
367	Me-rosol	50 mᴍ P pH 7.4	475	496	520	533	536	542	557	591	624	653	37	46.1	0.18	8.4	Rosol	4.19
368	*meso*-CaboxyBODIPY J-K 3a	10 mᴍ PBS pH 7.4	477	497	519	531	537	541	558	592	618	649	39	77.0	0.67	51.5	Bodipy-Small	4.14.1
369	Et-rosol	50 mᴍ P pH 7.4	478	500	524	537	538	544	558	589	622	649	34	42.6	0.10	4.1	Rosol	4.19
370	CF532	H_2O			527				558				31	96.0	0.73	69.8	nd (Rhodam.)	4.41
371	HEX	H_2O pH 9	491	513	535	550	534	541	559	589	616	643	24	98.0	0.70	68.6	Fluorescein-ha	4.8.2
372	DY-485XL carboxylate	H_2O	424	442	482	510	527	535	560	605	638	672	78	48.0		0.0	Coumarin 7-Am	4.26.3

Ent.	Short Names	Aqueous Solutions	$\lambda_{x,-0.25}$ nm	$\lambda_{x,-0.50}$ nm	$\lambda_{x,max}$ nm	$\lambda_{x,+0.50}$ nm	$\lambda_{e,-0.25}$ nm	$\lambda_{e,-0.50}$ nm	$\lambda_{e,max}$ nm	$\lambda_{e,+0.50}$ nm	$\lambda_{e,+0.25}$ nm	$\lambda_{e,+0.10}$ nm	Δ_S nm	ε_{max} mM^{-1}cm^{-1}	Φ_f	B_{max} mM^{-1}cm^{-1}	Class	Ref.
373	Cy3′	H_2O	483	504	545	556	543	547	560	575	606	624	15	120.0	0.04	4.8	Cyanine-Trimet	4.13.1
374	Acridine red	H_2O	484	507	528	540	542	548	562	582	594	621	34	100.0	0.51	51.0	Pyronine	4.4
375	6-Quinoline-2-pentanoic-pyrodol	50 mM P pH 8-9	492	515	538	550	544	549	562	579	591	611	24	64.1	0.70	44.8	Pyrodol	4.21.1
376	6-Quinoline-2-propanoic-pyrodol	50 mM P pH 8-9	492	515	538	550	544	549	562	579	591	611	24	65.9	0.70	46.1	Pyrodol	4.21.1
377	Atto 542 carboxylate	PBS pH 7.4, 22 °C			542				562				20	120.0	0.93	111.6	nd (pb Rhoda)	4.41
378	Pip-rosol	50 mM P pH 7.4	479	500	525	538	537	544	563	599	630	669	38	42.5	0.05	2.0	Rosol	4.19
379	Z-L J	0.5 M PBS pH 7.4	498	519	543	556	541	546	563	598	624	656	20	50	0.34	16.8	Bodipy-Large	4.14.3
380	Sulfo-cyanine3	H_2O	486	506	548	559	546	550	563	578	610	628	15	162.0	0.10	16.2	Cyanine-Trimet	4.13.1
381	1,2,7,8-Dibenzofluorescein	C pH 9.5	490	513	534	549	540	547	565	596	623	651	31	81.0	0.24	19.4	Fluorescein-nh	4.8.1
382	Cy3.18.OH	PBS	487	508	550	561	548	552	565	580	612	630	15	150.0	0.04	6.0	Cyanine-Trimet	4.13.1
383	Pyronine Y (OP)	H_2O	498	523	546	559	546	550	566	584	598	631	20	77.0	0.22	16.9	Pyronine	4.4
384	Cy3.13-OH	H_2O/PBS	487	508	550	561	549	553	566	581	613	631	16	130.0	0.03	3.9	Cyanine-Trimet	4.13.1
385	1,2,7,8-Dibenzosulfo-fluorescein	C pH 9.5	502	523	541	554	546	551	567	614	633	658	26	81.0	0.48	38.8	Sulfofluoresc.	4.10
386	Carbofluorescein	100 mM NaOH	505	526	544	557	546	551	567	614	633	658	23	108.0	0.62	66.9	C(CH3) 2-Fluore	4.9
387	Rosamine W-B 5h	100 mM P pH 7.4	495	524	542	557	546	552	568	587	603	634	26	71.5	0.14	10.0	Rosamine	4.17
388	Pyronine Y (OP)	PBS pH 7.4	500	525	546	560	519	526	569	557	567	586	23	77.0	0.36	27.7	Pyronine	4.4
389	6-Quinolinocarboxyrhodol	50 mM P pH 8-9	492	514	540	554	541	550	570	599	627	658	30	66.8	0.43	28.7	Rhodol	4.15.1
390	Tetramethylrosamine	10 mM PBS pH 7.4	501	532	550	565	548	554	570	589	605	636	20	74.5	0.31	23.1	Rosamine	4.17
391	Cy3	H_2O	487	508	550	561	553	557	570	585	618	636	20	150.0	0.06	8.2	Cyanine-Trimet	4.13.1
392	Cy3.10-OH	H_2O/PBS	489	510	552	563	553	557	570	585	618	636	18	130.0	0.03	3.9	Cyanine-Trimet	4.13.1
393	Chromenoquinoline L-S 4a	H_2O			473				570				97	36.19	0.15	5.4	Hybrid-Fused	4.40.1
394	Seta-555-carboxylate	P pH 7.4			555				570				15	155.0	0.07	10.8	nd	4.41
395	JF549	10 mM HEPES pH 7.3	504	528	549	564	549	555	571	596	618	644	22	101	0.88	88.8	Rhoda 500-550	4.2.2

(Continued)

Ent.	Short Names	Aqueous Solutions	$\lambda_{x,-0.25}$ nm	$\lambda_{x,-0.50}$ nm	$\lambda_{x,max}$ nm	$\lambda_{x,+0.50}$ nm	$\lambda_{e,-0.25}$ nm	$\lambda_{e,-0.50}$ nm	$\lambda_{e,max}$ nm	$\lambda_{e,+0.50}$ nm	$\lambda_{e,+0.25}$ nm	$\lambda_{e,+0.10}$ nm	Δ_S nm	ε_{max} mM^{-1}cm^{-1}	Φ_f	B_{max} mM^{-1}cm^{-1}	Class	Ref.
396	3,6,7,8,10-PentaMe-isoalloxazine	H_2O pH 7			448				571				123	8.3	0.10	0.8	Pteridine	4.36
397	3,7,9,10-TetraMe-isoalloxazine	H_2O pH 7			452				571				119	6.8	0.10	0.6	Pteridine	4.36
398	2-PrAc-6-quinolinorhodol	50 mM P pH 8-9	493	515	541	555	543	552	572	601	630	661	31	58.2	0.49	28.5	Rhodol	4.15.1
399	Tetramethylrhodamine	10 mM HEPES pH 7.3	503	527	548	563	550	556	572	597	619	646	24	78.0	0.45	35.1	Rhoda 500-550	4.2.2
400	SR101-OH	H_2O	493	516	548	564	550	556	572	595	628	652	24	80.0	0.35	28.0	Sulforhodol	4.27
401	Pyronine B	H_2O	504	529	552	565	551	557	572	592	605	633	20	94.1	0.18	16.9	Pyronine	4.4
402	AF555	PBS pH 7.2, 22 °C	491	512	555	566	555	559	572	587	620	639	17	158.0	0.09	14.2	Cyanine-Trimet	4.13.1
403	Alexa Fluor 555	PBS pH 7.2, 22 °C	491	512	555	566	555	559	572	587	620	639	17	155.0	0.10	15.5	Cyanine-Trimet	4.13.1
404	Cy3B	H_2O	493	515	558	569	555	559	572	587	620	639	14	130.0	0.67	87.1	Cyanine-Trimet	4.13.1
405	Neo-Cy3	PBS	488	509	551	562	555	560	573	588	622	640	22	98.0	0.02	1.9	Cyanine-Trimet	4.13.1
406	2-Ethyl-6-quinolinocarbo-xyrhodol	50 mM P pH 8-9	494	517	543	557	545	554	574	603	632	664	31	79.4	0.49	38.9	Rhodol	4.15.1
407	Jul-rosol	50 mM P pH 7.4	492	514	540	554	548	554	574	614	648	686	34	30.4	0.41	12.5	Rosol	4.19
408	M-S 10b	H_2O, 0.1% TX-100	501	523	547	560	551	556	574	610	638	671	27	64.0	0.78	49.9	Bodipy-Small	4.14.1
409	TAMRA	50 mM PBS pH 8-9	502	526	547	562	552	558	574	599	622	648	27	77.0	0.35	26.9	Rhoda 500-550	4.2.2
410	Cy3.12-OH	H_2O/PBS	492	513	556	567	556	561	574	590	623	641	18	130.0	0.05	6.5	Cyanine-Trimet	4.13.1
411	2-Cloro-6-quinolinocarbo-xyrhodol	50 mM P pH 8-9	498	521	548	563	546	555	575	604	634	665	27	74.5	0.42	31.2	Rhodol	4.15.1
412	JF552	10 mM HEPES pH 7.3	505	535	552	566	554	560	575	600	629	657	23	95.2	0.83	79.0	Rhoda 550-600	4.2.3
413	DAOTA 5	H_2O	479	505	548	567	551	558	576	606	630	656	28	50.0	0.56	28.0	Triangulenium	4.22
414	DAOTA 4	H_2O	481	502	549	567	553	559	576	607	632	655	27	50.0	0.56	28.0	Triangulenium	4.22
415	TAMRA	H_2O pH 7	501	525	546	561	554	560	576	602	624	651	30	90.0	0.20	18.0	Rhoda 500-550	4.2.2
416	Atto 550 carboxylate	PBS pH 7.4, 22 °C	507	535	554	567	554	560	576	598	626	654	22	120.0	0.80	96.0	Rhoda 550-600	4.2.3

Ent.	Short Names	Aqueous Solutions	$\lambda_{x,-0.25}$ nm	$\lambda_{x,-0.50}$ nm	$\lambda_{x,max}$ nm	$\lambda_{x,+0.50}$ nm	$\lambda_{e,-0.25}$ nm	$\lambda_{e,-0.50}$ nm	$\lambda_{e,max}$ nm	$\lambda_{e,+0.50}$ nm	$\lambda_{e,+0.25}$ nm	$\lambda_{e,+0.10}$ nm	Δ_S nm	ε_{max} mM^{-1}cm^{-1}	Φ_f	B_{max} mM^{-1}cm^{-1}	Class	Ref.
417	SR101-110	PBS	485	507	554	570	555	560	576	600	638	661	22	80.0	0.55	44.0	Sulforhodam.	4.6
418	Seta-400-carboxylate	P pH 7.4			399				576				177	24.0	0.34	8.1	nd	4.41
419	Carborhodamine	10 mM HEPES pH 7.3	515	537	552	564	558	564	577	605	628	654	25	65.0	0.64	41.6	C(CH3)2-Rhoda	4.3.1
420	M-S 10a	H_2O, 0.1% TX-100	516	539	565	579	556	561	579	616	644	678	14	69.0	0.65	44.8	Bodipy-Small	4.14.1
421	JF559	10 mM HEPES pH 7.3	511	541	559	573	557	563	579	604	633	662	20	106.0	0.85	90.1	Rhoda 550-600	4.2.3
422	Pyronine 630	H_2O, 25 °C	509	535	559	573	558	564	579	600	613	642	20	100.0	0.92	92.0	Pyronine	4.4
423	Chromenoquinoline L-S 4b	H_2O			477				579				102	36.05	0.15	5.2	Hybrid-Fused	4.40.1
424	Coumarin-pyronine 9	100 mM P pH 7.5			560				579				19	132.5	0.13	17.2	Hybrid-Bonded	4.40.2
425	Alexa Fluor 546	PBS pH 7.2, 22 °C	506	537	546	567	553	558	580	592	621	648	34	112.0	0.79	88.4	Rhodam Hi Ch	4.2.5
426	DY-555 carboxylate	PBS pH 7.4	508	537	555	569	558	564	580	605	635	663	25	100.0	0.32	31.7	Rhoda 550-600	4.2.3
427	Tetraethylrhodamine	H_2O pH 7	510	539	557	571	558	564	580	605	635	663	23	105.0	0.24	25.2	Rhoda 550-600	4.2.3
428	Carbostyril 7	H_2O			366				580				214	9.0	0.07	0.6	Quinoline	4.29
429	1,2,7,8-Dibenzo-4,5-Br-fluorescein	C pH 9.5	503	526	549	565	554	562	581	614	642	672	32	24.0	0.05	1.2	Fluorescein-ha	4.8.2
430	2′,7′-Difluorocarbofluorescein	100 mM NaOH	514	536	555	569	559	564	581	631	651	677	26	90.9	0.40	36.3	C(CH3)2-Fluore	4.9
431	M-S 10d	H_2O, 0.1% TX-100	523	547	574	589	559	563	582	619	648	682	8	118.0	0.57	67.2	Bodipy-Small	4.14.1
432	Sulforhodamine B	H_2O pH 7	508	537	554	567	561	568	582	599	609	628	28	84.0	0.39	32.7	Sulforhodam.	4.6
433	DAOTA 6	H_2O	489	509	558	575	560	567	584	614	640	660	26	50.0	0.36	18.0	Triangulenium	4.22
434	8-Trifluoromethyl-BODIPY 11	100 mM P pH 7.4	519	543	569	583	561	565	584	622	650	685	15	57.0	0.95	54.1	Bodipy-Small	4.14.1
435	8-Trifluoromethyl-BODIPY 12	100 mM P pH 7.4	439	456	477	485	568	574	584	601	624	645	107	41.8	0.70	29.2	Bodipy-Small	4.14.1
436	DY-510XL carboxylate	H_2O	432	451	493	522	549	558	585	635	670	709	92	40.0		0.0	Coumarin 7-Am	4.26.3
437	SR101-6G	PBS	508	540	565	579	563	571	586	608	638	666	21	80.0	0.68	54.4	Sulforhodam.	4.6
438	DPP 6	PBS			530				586				56	30.7	0.49	15.0	Diketopyrrolop	4.39.1
439	Safranin T	H_2O			517				586				69	27.0	0.11	2.9	Phenazine	4.39.1

(Continued)

Ent.	Short Names	Aqueous Solutions	$\lambda_{x,-0.25}$ nm	$\lambda_{x,-0.50}$ nm	$\lambda_{x,max}$ nm	$\lambda_{x,+0.50}$ nm	$\lambda_{e,-0.25}$ nm	$\lambda_{e,-0.50}$ nm	$\lambda_{e,max}$ nm	$\lambda_{e,+0.50}$ nm	$\lambda_{e,+0.25}$ nm	$\lambda_{e,+0.10}$ nm	Δ_S nm	ε_{max} mM^{-1}cm^{-1}	Φ_f	B_{max} mM^{-1}cm^{-1}	Class	Ref.
440	DAOTA 1	H_2O	489	511	558	578	563	570	587	621	650	674	29	50.0	0.31	15.5	Triangulenium	4.22
441	560CP	PBS pH 7.4	522	541	561	576	565	572	588	625	646	669	27	61.0	0.76	46.3	C(CH3)2-Rhoda	4.3.1
442	Atto Rho3B carboxylate	PBS pH 7.4, 22 °C	517	548	566	579	566	572	589	613	645	675	23	120.0	0.50	60.0	Rhoda 550-600	4.2.3
443	Atto 565 carboxylate	PBS pH 7.4, 22 °C	513	543	564	578	565	572	590	617	648	676	26	120.0	0.90	108.0	Rhoda 550-600	4.2.3
444	JF571	10 mM HEPES pH 7.3	521	552	571	586	568	574	590	616	647	676	19	101	0.78	78.7	Rhoda 550-600	4.2.3
445	HMDSeR	100 mM P pH 7.4	516	540	558	571	568	574	590	631	656	680	32	92.0	0.01	0.5	Se-Rhodol	4.16
446	Chromenoquinoline L-S 4c	H_2O			500				591				91	33.16	0.03	1.1	Hybrid-Fused	4.40.1
447	trans-PO-PB460	PBS pH 7.4			459				592				133	9.0	0.25	2.2	Phosphole	4.39.3
448	JF570	10 mM HEPES pH 7.3	521	548	570	586	569	576	593	620	644	673	23	83.6	0.63	52.6	S-Rhodamine	4.3.3
449	SR101-Me$_2$	PBS	517	553	573	587	572	578	593	613	632	668	20	80.0	0.66	52.8	Sulforhodam.	4.6
450	Cy3.205	PBS pH 7.4	509	532	578	590	574	579	593	610	645	665	15	120.0	0.14	16.8	Cyanine-Trimet	4.13.1
451	DAOTA 2	H_2O	493	515	564	583	569	576	594	627	654	678	30	50.0	0.27	13.5	Triangulenium	4.22
452	DAOTA 3	H_2O	449	511	565	583	571	577	594	626	654	681	29	50.0	0.28	14.0	Triangulenium	4.22
453	JF574	10 mM HEPES pH 7.3	524	555	574	589	571	578	594	620	651	681	20	103	0.78	80.3	Rhoda 550-600	4.2.3
454	W-B 15	100 mM P pH 7.4	518	549	567	582	571	578	594	620	651	681	27	76.5	0.13	9.9	Rhoda 550-600	4.2.3
455	DPP 10	PBS			523				594				71	17.1	0.61	10.4	Diketopyrrolop	4.39.1
456	DPP 8	PBS			527				594				67	18.1	0.56	10.1	Diketopyrrolop	4.39.1
457	PONy A1	PBS pH 7.4	476	499	535	553	556	566	595	654	693	739	60	38.0	0.10	3.8	Acridine	4.23.1
458	Atto Rho11 carboxylate	PBS pH 7.4, 22 °C	522	553	572	585	571	577	595	619	650	680	23	120.0	0.80	96.0	Rhoda 550-600	4.2.3
459	SR101-B	PBS	522	557	575	590	574	580	595	615	631	665	20	80.0	0.65	52.0	Sulforhodam.	4.6
460	Redmond Red	100 mM P pH 7.1	518	547	579	593	577	583	595	614	629	661	16	52.3	0.84	43.9	Phenoxazine	4.25

Ent.	Short Names	Aqueous Solutions	$\lambda_{x,-0.25}$ nm	$\lambda_{x,-0.50}$ nm	$\lambda_{x,max}$ nm	$\lambda_{x,+0.50}$ nm	$\lambda_{e,-0.25}$ nm	$\lambda_{e,-0.50}$ nm	$\lambda_{e,max}$ nm	$\lambda_{e,+0.50}$ nm	$\lambda_{e,+0.25}$ nm	$\lambda_{e,+0.10}$ nm	Δ_S nm	ε_{max} mм$^{-1}$cm^{-1}	Φ_f	B_{max} mм$^{-1}$cm^{-1}	Class	Ref.
461	2-COOH DFTM	100 mм P, 1% DMSO pH 10	537	560	581	596	573	578	596	649	670	697	15	120.0	0.54	64.8	Si(CH3)2-Fluore	4.9
462	Cy3.5	H_2O	511	534	581	593	577	582	596	613	649	669	15	120.0	0.14	16.8	Cyanine-Trimet	4.13.1
463	575RH	PBS pH 7.4	524	555	574	589	574	580	597	624	655	685	23	55.0	0.74	40.7	Rhoda 550-600	4.2.3
464	M-S 10e	H_2O, 0.1% TX-100	532	556	584	599	574	578	598	637	667	704	14	124.0	0.41	50.8	Bodipy-Small	4.14.1
465	Z-L G	0.5 м PBS pH 7.4	526	550	577	592	574	578	598	637	667	704	21	50	0.21	10.6	Bodipy-Large	4.14.3
466	2-COOH TM	100 mм P, 1% DMSO pH 10	536	560	580	595	575	580	598	651	672	700	18	110.0	0.38	41.8	Si(CH3)2-Fluore	4.9
467	2-Me Tokyo Magenta	100 mм P pH 9	532	563	582	595	578	583	598	617	633	655	16	110.0	0.42	46.2	Si(CH3)2-Fluoron	4.12
468	Si-Fluorescein	100 mм NaOH	535	559	579	594	575	581	599	652	674	701	20	93.0	0.53	49.2	Si(CH3)2-Fluore	4.9
469	Atto Rho12 carboxylate	PBS pH 7.4, 22 °C	526	557	577	590	577	583	600	627	659	689	23	120.0	0.80	96.0	Rhoda 550-600	4.2.3
470	570CPH	PBS pH 7.4	580	602	571	633	617	623	600	673	707	740	29	79.0	0.71	56.0	C(CH3)2-Rhoda	4.3.1
471	5(6)-Carboxy-X-rhodamine (ROX)	H_2O pH <7	525	557	576	591	578	584	601	628	660	691	25	82.0	0.70	57.4	Rhoda 550-600	4.2.3
472	Seta-580-carboxylate	H_2O			578				602				24	90.0	0.55	49.5	nd	4.41
473	M-S 10f	H_2O, 0.1% TX-100	536	561	589	604	578	583	603	643	674	711	14	74.0	0.18	13.3	Bodipy-Small	4.14.1
474	Alexa Fluor 568	PBS pH 7.2/7.0, 20/22 °C	527	559	578	593	580	586	603	630	662	693	25	88.0	0.69	60.7	Rhodam Hi Ch	4.2.5
475	Sulforhodamine 101	H_2O pH 7	534	569	586	602	584	590	605	622	632	647	19	116.0	0.90	104.4	Sulforhodam.	4.6
476	COUPY 9	H_2O			543				605				62	31.0	0.15	4.6	Hybrid-Bridged	4.40.3
477	Si-9-Iminopyronine 2f	~1 mм NaOH, 1% DMSO pH 11	375	389	401	408	583	590	606	629	643	675	205	17.8	0.35	6.2	Si(CH3)2-Pyron	4.5
478	2-COOH DCTM	100 mм P, 1% DMSO pH 10	545	570	591	606	583	589	607	662	684	712	16	120.0	0.48	57.6	Si(CH3)2-Fluore	4.9
479	Abberior Star 580 carboxylate	PBS pH 7.4	535	567	587	603	583	590	607	635	667	699	20	85.0	0.90	76.5	Rhoda 550-600	4.2.3

(Continued)

Ent.	Short Names	Aqueous Solutions	$\lambda_{x,-0.25}$ nm	$\lambda_{x,-0.50}$ nm	$\lambda_{x,max}$ nm	$\lambda_{x,+0.50}$ nm	$\lambda_{e,-0.25}$ nm	$\lambda_{e,-0.50}$ nm	$\lambda_{e,max}$ nm	$\lambda_{e,+0.50}$ nm	$\lambda_{e,+0.25}$ nm	$\lambda_{e,+0.10}$ nm	Δ_S nm	ε_{max} $mM^{-1}cm^{-1}$	Φ_f	B_{max} $mM^{-1}cm^{-1}$	Class	Ref.
480	Si-9-Iminopyronine 2c	~1 mM NaOH, 1% DMSO pH 11	390	405	418	426	584	591	607	630	645	676	189	12.9	0.46	5.9	Si(CH3)2-Pyron	4.5
481	2-Me DCTM	100 mM P pH 9	544	577	597	611	586	592	607	627	643	666	10	130.0	0.48	62.4	Si(CH3)2-Fluoron	4.12
482	NCRMe	H_2O			568				607				39	22.8	0.01	0.1	Hybrid-Fused	4.40.1
483	Chromenoquinoline L-S 4d	H_2O			500				608				108	36.06	0.08	2.8	Hybrid-Fused	4.40.1
484	Atto Rho101 carboxylate	PBS pH 7.4, 22 °C	535	569	587	601	586	592	609	635	664	698	22	120.0	0.80	96.0	Rhoda 550-600	4.2.3
485	JF585	10 mM HEPES pH 7.3	544	568	585	598	588	594	609	640	666	696	24	1.5	0.78	1.1	C(CH3)2-Rhoda	4.3.1
486	JF593	10 mM HEPES pH 7.3	541	569	593	610	587	594	612	641	667	697	19	90.3	0.55	49.6	S-Rhodamine	4.3.3
487	2′,4′,5′,7′-TetraF-Si-fluorescein	100 mM NaOH	550	575	597	613	588	594	613	669	691	720	16	65.0	0.67	43.5	Si(CH3)2-Fluore	4.9
488	Carborhodol S-H 13-H	PBS pH 7.4	539	567	586	601	589	595	613	657	685	711	27	59.0	0.32	18.8	C(CH3)2-Rhod	4.16
489	2-Me SiR600	H_2O	542	574	593	607	593	599	613	633	650	674	20	91.0	0.38	34.5	Si(CH3)2-Rosa	4.18.1
490	Chromenoquinoline L-S 4e	H_2O			502				613				111	33.16	0.07	2.2	Hybrid-Fused	4.40.1
491	PONy C7	PBS pH 7.4			419				613				194	22.0	0.04	0.8	Hybrid-Bonded	4.40.2
492	4,5,6,7-TetraF-Si-fluorescein	100 mM NaOH	551	576	598	614	589	595	614	670	693	722	16	84.1	0.50	42.0	Si(CH3)2-Fluore	4.9
493	CF594	H_2O			593				614				21	115.0	0.45	51.7	nd (Rhodam.)	4.41
494	Alexa Fluor 594	PBS pH 7.2/7.0, 20/22 °C	537	570	590	606	593	599	617	646	679	712	27	92.0	0.66	60.7	Rhodam Hi Ch	4.2.5
495	2′,7′-Difluoro-Si-Fluorescein	100 mM NaOH	549	574	596	612	593	599	618	675	698	727	22	60.9	0.42	25.5	Si(CH3)2-Fluore	4.9
496	NCR	H_2O			565				618				53	25.8	0.00	0.0	Hybrid-Fused	4.40.1
497	SR101-NaphtOH	PBS	481	509	580	607	580	588	619	676	700	735	39	80.0	0.15	12.0	Sulforhodol-an	4.27
498	NR600	PBS, 1% DMSO pH 7.4	551	580	600	613	597	603	619	646	665	689	19	67.6	0.68	45.9	POOH-Fluoron	4.12

Ent.	Short Names	Aqueous Solutions	$\lambda_{x,-0.25}$ nm	$\lambda_{x,-0.50}$ nm	$\lambda_{x,max}$ nm	$\lambda_{x,+0.50}$ nm	$\lambda_{e,-0.25}$ nm	$\lambda_{e,-0.50}$ nm	$\lambda_{e,max}$ nm	$\lambda_{e,+0.50}$ nm	$\lambda_{e,+0.25}$ nm	$\lambda_{e,+0.10}$ nm	Δ_S nm	ε_{max} $mM^{-1}cm^{-1}$	Φ_f	B_{max} $mM^{-1}cm^{-1}$	Class	Ref.
499	2-OH-6-DiEtAm-benzo[c] xanthen	H_2O	523	547	578	595	584	596	620	654	689	728	42	38.1	0.18	6.8	Rhodol-analog	4.15.3
500	Fluorene PP6	H_2O			373				620				247	16.5	0.07	1.1	Fluorene	4.39.2
501	Fluorene PP11	H_2O			377				621				244	36.0	0.05	1.8	Fluorene	4.39.2
502	Atto 590 carboxylate	PBS pH 7.4, 22 °C	535	569	593	609	594	601	622	651	687	718	29	120.0	0.80	96.0	Rhoda 550-600	4.2.3
503	Chromenoquinoline L-S 4f	H_2O			529				622				93	40.64	0.05	1.9	Hybrid-Fused	4.40.1
504	Abberior Star 600 carboxylate	PBS pH 7.4	550	587	604	619	600	606	623	647	677	714	19	93.0	0.80	74.4	Rhodam >600	4.2.4
505	Abberior Star 470SXP	PBS pH 7.4	416	434	472	499	583	594	624	681	722	767	152	29.0	0.12	3.4	Coumarin 7-Am	4.26.3
506	Cresyl violet (CV)	H_2O	523	553	585	599	605	611	625	646	663	698	40	33.1	0.36	11.9	Phenoxazine	4.25
507	SR101-NaphtNH_2	PBS	493	526	589	610	593	602	626	666	691	722	37	80.0	0.19	15.2	Srhod.-analog	4.6
508	Atto 594 carboxylate	PBS pH 7.4, 22 °C	547	583	603	617	601	608	626	652	638	718	23	120.0	0.85	102.0	Rhodam >600	4.2.4
509	Tetramethycarborhodamine	10 mM HEPES pH 7.3	562	588	606	620	604	610	626	659	686	718	20	121.0	0.52	62.9	C(CH3)2-Rhoda	4.3.1
510	Atto Rho13 carboxylate	PBS pH 7.4, 22 °C	547	584	603	618	602	609	627	654	690	723	24	120.0	0.80	96.0	Rhodam >600	4.2.4
511	PONy P1	PBS pH 7.4	535	564	590	605	602	610	627	652	667	701	37	59.0	0.33	19.4	Pyronine	4.4
512	NR604	PBS, 1% DMSO pH 7.4	554	584	604	618	604	610	627	654	672	697	23	45.7	0.48	21.9	POOH-Fluoron	4.12
513	2′COOH-4CH_2OH-Si-rhodol	100 mM P pH 7.4	560	589	610	626	605	611	630	677	706	734	20	70.0	0.21	14.7	Si(CH3)2-Rhod	4.16
514	2′COOH-Si-rhodol	100 mM P pH 7.4	559	588	609	625	605	611	630	677	706	734	21	70.0	0.22	15.4	Si(CH3)2-Rhod	4.16
515	2′Me-Si-rhosoL	100 mM P pH 7.4	553	582	612	627	608	614	630	651	667	693	18	70.0	0.28	19.6	Si(CH3)2-Rosol	4.20
516	SNARF-1	P pH 10	519	543	573	590	594	606	631	667	703	744	58	44.0	0.09	4.0	Rhodol-analog	4.15.3
517	Si-9-Iminopyronine 2g	PBS pH 7.4	434	452	469	479	606	613	631	656	672	706	162	11	0.43	4.7	Si(CH3)2-Pyron	4.5
518	JF608	10 mM HEPES pH 7.3	564	589	608	623	609	615	631	664	692	725	23	99.0	0.67	66.3	C(CH3)2-Rhoda	4.3.1
519	2′Me-4CH_2OH-Si-rosol	100 mM P pH 7.4	554	583	613	628	609	615	631	652	668	694	18	70.0	0.30	21.0	Si(CH3)2-Rosol	4.20

(Continued)

Ent.	Short Names	Aqueous Solutions	$\lambda_{x,-0.25}$ nm	$\lambda_{x,-0.50}$ nm	$\lambda_{x,max}$ nm	$\lambda_{x,+0.50}$ nm	$\lambda_{e,-0.25}$ nm	$\lambda_{e,-0.50}$ nm	$\lambda_{e,max}$ nm	$\lambda_{e,+0.50}$ nm	$\lambda_{e,+0.25}$ nm	$\lambda_{e,+0.10}$ nm	Δ_S nm	ε_{max} mM^{-1}cm^{-1}	Φ_f	B_{max} mM^{-1}cm^{-1}	Class	Ref.
520	JS-R	10 mM HEPES pH 7.4	558	589	611	625	609	617	631	648	659	671	20	130.0	0.59	76.7	B(OH)2-Pyron	4.5
521	COUPY 10	H_2O			562				631				69	33.0	0.03	0.8	Hybrid-Bridged	4.40.3
522	Abberior Star 520SXP	PBS pH 7.4	454	475	522	555	590	601	632	691	733	779	110	42.5	0.15	6.3	Coumarin 7-Am	4.26.3
523	Atto 610 carboxylate	PBS pH 7.4, 22 °C	529	554	616	581	568	573	633	608	634	659	17	150.0	0.70	105.0	C(CH3)2-Pyron	4.5
524	PONy P9	PBS pH 7.4	541	570	597	613	608	615	633	658	674	709	36	66.0	0.22	14.5	Pyronine	4.4
525	Pyrrolidinylcarborhodamine	10 mM HEPES pH 7.3	568	594	613	628	610	617	633	666	694	727	20	87.0	0.54	46.9	C(CH3)2-Rhoda	4.3.1
526	2′,4,5,6,7,7′-HexaF-Si-fluorescein	100 mM NaOH	567	594	617	634	609	615	635	695	720	751	18	67.9	0.32	21.7	Si(CH3)2-Fluore	4.9
527	PONy P1-Halo	PBS pH 7.4	546	576	603	619	610	617	635	660	676	711	32	18.0	0.26	4.6	Pyronine	4.4
528	Squaraine T-P 4c	67 mM P pH 7.4	577	606	624	639	615	621	635	655	672	709	11	245.0	0.02	4.9	Squaraine	4.35
529	RF620	10 mM PBS pH 7.4	569	600	620	635	614	620	636	659	674	701	16	109.0	0.36	39.2	B(OH)2-Rosa	4.18.3
530	PONy P6	PBS pH 7.4	546	576	603	619	613	620	638	663	680	715	35	69.0	0.09	6.2	Pyronine	4.4
531	2-Me SiR620	100 mM P, 1% DMSO pH 7.4	550	586	620	636	616	622	638	657	675	699	18	96.0	0.21	20.1	Si(CH3)2-Rosa	4.18.1
532	Sulfonated coumarin N-H 7-H	PBS pH 7.4	453	474	520	552	596	607	639	699	742	789	119	40.0	0.06	2.4	Coumarin 7-Am	4.26.3
533	CF620R	H_2O			617				639				22	115.0	0.30	34.0	nd (Rhodam.)	4.41
534	Distyryl-BODIPY 18	H_2O	568	596	628	646	613	619	641	687	721	764	13	71.0	0.38	26.9	Bodipy-Med.	4.14.2
535	Seta-632-carboxylate	P pH 7.4			632				641				9	280.0	0.06	16.8	nd	4.41
536	Atto 620 carboxylate	PBS pH 7.4, 22 °C	580	602	620	633	617	623	642	673	707	740	22	120.0	0.50	60.0	C(CH3)2-Rhoda	4.3.1
537	SqO1	P pH 7.4	584	614	632	648	621	628	642	662	679	718	10	265.0	0.07	18.5	Squaraine	4.35
538	Squaraine T-P 4a	67 mM P pH 7.4	581	610	628	644	621	628	642	662	679	718	14	163.0	0.03	4.8	Squaraine	4.35
539	Squaraine T-P 5c	67 mM P pH 7.4	584	614	632	648	621	628	642	662	679	718	10	265.0	0.06	15.9	Squaraine	4.35
540	DEAC-SR101	PBS			610				642				32	80.0	0.10	8.0	Hybrid-Fused	4.40.1

Ent.	Short Names	Aqueous Solutions	$\lambda_{x,-0.25}$ nm	$\lambda_{x,-0.50}$ nm	$\lambda_{x,max}$ nm	$\lambda_{x,+0.50}$ nm	$\lambda_{e,-0.25}$ nm	$\lambda_{e,-0.50}$ nm	$\lambda_{e,max}$ nm	$\lambda_{e,+0.50}$ nm	$\lambda_{e,+0.25}$ nm	$\lambda_{e,+0.10}$ nm	Δ_S nm	ε_{max} $mM^{-1}cm^{-1}$	Φ_f	B_{max} $mM^{-1}cm^{-1}$	Class	Ref.
541	Distyryl-BODIPY 17	H_2O	570	598	630	648	615	620	643	689	724	767	13	76.6	0.30	22.9	Bodipy-Med.	4.14.2
542	Square-655-carboxylate	P pH 7.4	576	604	622	638	623	630	644	664	682	720	22	177.0	0.09	15.9	Squaraine	4.35
543	Seta-633-carboxylate	P pH 7.4			633				644				11	250.0	0.07	17.5	nd	4.41
544	BODIPY 630/650	H_2O	569	597	629	647	618	623	646	692	728	772	17	97.0	0.50	48.5	Bodipy-Small	4.14.1
545	Atto Rho14 carboxylate	PBS pH 7.4, 22 °C	564	606	626	639	622	628	646	668	693	738	20	140.0	0.80	112.0	Rhodam >600	4.2.4
546	Square-635-carboxylate	P pH 5.6	591	617	634	648	624	633	646	665	679	725	12	135.0	0.16	21.6	Squaraine	4.35
547	Alexa Fluor 633	PBS/H_2O pH 7.2, 20/22 °C	568	608	632	650	621	629	647	670	691	722	15	100.0	0.64	64.0	Sulforhodam.	4.6
548	DMAC-SR101	PBS			605				647				42	80.0	0.09	7.2	Hybrid-Fused	4.40.1
549	Squaraine T-P 5a	67 mM P pH 7.4	592	620	636	653	624	631	648	666	680	705	12	216.0	0.06	12.9	Squaraine	4.35
550	3,6-Azetid.-Si-pyronine	10 mM HEPES pH 7.3	573	606	636	654	623	630	649	675	692	729	13	97.1	0.62	60.2	Si(CH3)2-Pyron	4.5
551	Oxazine 170	H_2O pH 2-7	539	560	620	643	624	632	650	668	679	692	30	48.0	0.24	11.5	Phenoxazine	4.25
552	SqB1	P pH 7.4	580	609	627	642	629	636	650	671	688	728	23	150.0	0.05	7.5	Squaraine	4.35
553	SiP-F	50 mM PBS pH 7.4	577	611	637	653	602	612	651	655	667	679	14	64.2	0.20	12.8	C(CH3)2-Pyron	4.5
554	Atto 633 carboxylate	PBS pH 7.4, 22 °C	575	610	630	646	627	635	651	687	724	756	21	130.0	0.64	83.2	C(CH3)2-Rhoda	4.3.1
555	630GeRH	PBS pH 7.4	582	612	631	647	629	635	651	671	683	722	20	61.0	0.60	36.6	Ge(CH3)2-Rhoda	4.3.3
556	Bzh-Cl	H_2O pH 7.4	551	584	620	636	629	636	651	674	692	731	31	36.1	0.31	11.1	Phenoxazine	4.25
557	Bzh-H	H_2O pH 7.4	551	584	620	636	629	636	651	674	692	731	31	11.3	0.44	4.9	Phenoxazine	4.25
558	SiP-Pr	50 mM PBS pH 7.4	579	613	632	647	607	619	652	661	675	691	20	64.2	0.21	13.4	C(CH3)2-Pyron	4.5
559	TMDHS or SiP-Me	H_2O	571	604	634	652	626	634	653	680	697	734	19	64.2	0.18	11.5	Si(CH3)2-Pyron	4.5
560	Bza-Oet	H_2O pH 7.4	555	588	625	641	631	638	653	676	694	733	28	33.5	0.26	8.7	Phenoxazine	4.25

(Continued)

Ent.	Short Names	Aqueous Solutions	$\lambda_{x,-0.25}$ nm	$\lambda_{x,-0.50}$ nm	$\lambda_{x,max}$ nm	$\lambda_{x,+0.50}$ nm	$\lambda_{e,-0.25}$ nm	$\lambda_{e,-0.50}$ nm	$\lambda_{e,max}$ nm	$\lambda_{e,+0.50}$ nm	$\lambda_{e,+0.25}$ nm	$\lambda_{e,+0.10}$ nm	Δ_S nm	ε_{max} mм$^{-1}$cm^{-1}	Φ_f	B_{max} mм$^{-1}$cm^{-1}	Class	Ref.
561	Bza-OH	H_2O pH 7.4	551	584	620	636	631	638	653	676	694	733	33	12.8	0.37	4.7	Phenoxazine	4.25
562	nor-SqO1	P pH 7.4	588	618	636	652	631	639	653	674	692	731	17	149.0	0.32	47.6	Squaraine	4.35
563	TMDHS or SiP-Me	50 mм PBS pH 7.4	570	603	633	651	627	635	654	681	698	735	21	64.2	0.41	26.3	Si(CH3)2-Pyron	4.5
564	Abberior Star 635 carboxylate	PBS pH 7.4	574	615	634	650	629	636	654	681	714	755	20	63.0	0.51	32.1	Rhodam >600	4.2.4
565	KK9046	H_2O	573	613	632	648	629	636	654	681	714	755	22	90.0	0.45	40.5	Rhodam >600	4.2.4
566	HC6H-SR101	PBS, 5% BSA			624				654				30	80.0	0.15	12.0	Hybrid-Fused	4.40.1
567	B-Z 14	H_2O	576	605	638	656	626	632	655	703	739	784	17	102.0	0.42	42.8	Bodipy-Large	4.14.3
568	Distyryl-BODIPY 11	PBS pH 7.5	546	572	601	617	626	632	655	703	739	784	54	17.9	0.04	0.7	Bodipy-Med.	4.14.2
569	Abberior Star 635P carboxylate	PBS pH 7.4	575	616	635	652	630	637	655	682	715	756	20	80.0	0.55	44.0	Rhodam >600	4.2.4
570	Abberior Star Red carboxylate	PBS pH 7.4	578	619	638	655	630	637	655	682	715	756	17	100.0	0.66	66.0	Rhodam >600	4.2.4
571	NR632	10 mм PBS pH 7.4	578	608	632	646	631	638	655	678	695	723	23	64.0	0.26	16.6	POPh-Rosol	4.20
572	GeR	PBS pH 7.4	601	621	634	658	635	643	655	681	697	734	21	97.0	0.43	41.7	Ge(CH3) 2-Rhoda	4.3.3
573	Seta-635-pH-di-carboxylate	P pH 9			640				656				16	188.0	0.33	62.0	nd	4.41
574	Seta-646-carboxylate	P pH 7.4			646				656				10	207.0	0.10	20.7	nd	4.41
575	Distyryl-BODIPY 12	PBS pH 7.5	579	609	642	660	628	633	657	705	742	787	15	55.1	0.22	12.1	Bodipy-Med.	4.14.2
576	Distyryl-BODIPY 13	PBS pH 7.5	579	608	641	659	628	633	657	705	742	787	16	58.0	0.25	14.5	Bodipy-Med.	4.14.2
577	3,6-Azetid.-9-CA.-Si-pyronine	10 mм HEPES pH 7.3	577	610	641	659	630	638	657	684	701	739	16	120.0	0.26	31.2	Si(CH3)2-Pyron	4.5
578	Carborhodamine 16a	H_2O	588	616	636	652	633	640	657	693	723	759	21	114.0	0.62	70.6	C(CH3)2-Rhoda	4.3.1
579	nor-SqO2	P pH 7.4	590	620	639	655	635	642	657	678	696	736	18	183.0	0.33	60.3	Squaraine	4.35
580	Atto 490LS carboxylate	PBS pH 7.4, 22 °C			495				658				163	40.0	0.30	12.0	nd (pb Coum.)	4.41
581	Atto 635 carboxylate	PBS pH 7.4, 22 °C	579	612	635	649	626	638	659	683	701	720	24	130.0	0.25	32.5	C(CH3)2-Pyron	4.5
582	Naphthofluorescein	C pH 9.5	541	568	595	613	626	635	660	703	740	780	65	44.0	0.14	6.1	Fluores-analog	4.8.4
583	KK1119	H_2O	577	618	637	654	634	641	660	687	721	763	23	90.0	0.55	49.5	Rhodam >600	4.2.4

Ent.	Short Names	Aqueous Solutions	$\lambda_{x,-0.25}$ nm	$\lambda_{x,-0.50}$ nm	$\lambda_{x,max}$ nm	$\lambda_{x,+0.50}$ nm	$\lambda_{e,-0.25}$ nm	$\lambda_{e,-0.50}$ nm	$\lambda_{e,max}$ nm	$\lambda_{e,+0.50}$ nm	$\lambda_{e,+0.25}$ nm	$\lambda_{e,+0.10}$ nm	Δ_S nm	ε_{max} mM^{-1}cm^{-1}	Φ_f	B_{max} mM^{-1}cm^{-1}	Class	Ref.
584	2-Me SiR640	100 mM P, 1% DMSO pH 7.4	509	548	637	658	635	643	660	683	707	733	23	54.0	0.18	9.7	Si(CH3)2-Rosa	4.18.1
585	KK114	H_2O	580	619	637	652	635	642	660	685	713	762	23	92.0	0.55	50.6	Rhodam >600	4.2.4
586	2-Me SiR650	H_2O	591	628	646	661	636	642	660	681	707	749	14	110.0	0.31	34.1	Si(CH3)2-Rosa	4.18.1
587	Cy5′	H_2O	583	621	642	656	639	645	660	677	688	716	18	138.0	0.17	23.4	Cyanine-Penta	4.13.2
588	Neo-Cy5	PBS pH 7.4	583	621	643	657	639	645	660	677	688	716	17	170.0	0.10	17.0	Cyanine-Penta	4.13.2
589	POF	PBS pH 7.4	561	591	632	652	635	643	661	687	714	750	29	54.5	0.24	13.0	POPh-Fluorone	4.12
590	Carborhodamine 1b	H_2O	590	619	639	655	636	644	661	698	728	764	22	80.0	0.58	46.4	C(CH3)2-Rhoda	4.3.1
591	Carborhodamine 20a	H_2O	591	619	640	656	636	644	661	698	728	764	21	70.0	0.53	37.1	C(CH3)2-Rhoda	4.3.1
592	SiR-COOH	H_2O	588	567	645	662	637	644	661	681	695	728	16	100.0	0.39	39.0	Si(CH3)2-Rhoda	4.3.2
593	3,7-Di(azetidin)phenoxazine	10 mM HEPES pH 7.3	572	607	647	664	638	646	661	684	703	743	14	99.0	0.24	23.7	Phenoxazine	4.25
594	SiR	PBS pH 7.4	589	623	645	663	640	647	661	687	702	737	16	93.0	0.41	38.1	Si(CH3)2-Rhoda	4.3.2
595	Carborhodamine 1c	H_2O	591	619	640	656	637	645	662	699	730	766	22	89.0	0.53	47.1	C(CH3)2-Rhoda	4.3.1
596	NR638	10 mM PBS pH 7.4	583	613	638	652	637	645	662	686	703	732	24	55.6	0.21	11.6	POPh-Rosol	4.20
597	SiTMR	10 mM HEPES pH 7.3	586	566	643	660	638	645	662	682	696	729	19	28.2	0.41	11.5	Si(CH3)2-Rhoda	4.3.2
598	Des-carboxy-JF646	10 mM HEPES pH 7.3	571	611	648	665	638	645	662	683	701	728	14	116	0.20	23.2	Si(CH3)2-Rosa	4.18.1
599	SiR-methyl	H_2O	571	611	648	665	638	645	662	683	701	728	14	100.0	0.39	39.0	Si(CH3)2-Rosa	4.18.1
600	640SiRH	PBS pH 7.4	584	519	641	658	639	646	662	681	694	726	21	51.0	0.42	21.4	Si(CH3)2-Rhoda	4.3.2
601	CypHer5	A pH 4.6	583	621	643	657	641	646	662	679	690	719	19	100.0	0.19	19.0	Cyanine-Penta	4.13.2
602	Sulfo-cyanine5	H_2O	586	624	646	661	641	646	662	679	690	719	16	271.0	0.20	54.2	Cyanine-Penta	4.13.2
603	COUPY 11	H_2O			567				662				95	13.0	0.02	0.2	Hybrid-Bridged	4.40.3
604	Carborhodamine 20b	H_2O	592	620	641	657	638	646	663	700	731	767	22	80.0	0.53	42.4	C(CH3)2-Rhoda	4.3.1
605	Carborhodamine 21a	H_2O	591	619	640	656	638	646	663	700	731	767	23	70.0	0.56	39.2	C(CH3)2-Rhoda	4.3.1
606	Cy5	H_2O	587	625	647	662	642	647	663	680	691	720	16	250.0	0.27	67.5	Cyanine-Penta	4.13.2
607	Carborhodamine 21b	H_2O	591	619	640	656	639	646	664	701	732	768	24	68.0	0.56	38.0	C(CH3)2-Rhoda	4.3.1
608	Carborhodamine 22	H_2O	591	619	640	656	639	646	664	701	732	768	24	67.0	0.54	36.1	C(CH3)2-Rhoda	4.3.1

(Continued)

Ent.	Short Names	Aqueous Solutions	$\lambda_{x,-0.25}$ nm	$\lambda_{x,-0.50}$ nm	$\lambda_{x,max}$ nm	$\lambda_{x,+0.50}$ nm	$\lambda_{e,-0.25}$ nm	$\lambda_{e,-0.50}$ nm	$\lambda_{e,max}$ nm	$\lambda_{e,+0.50}$ nm	$\lambda_{e,+0.25}$ nm	$\lambda_{e,+0.10}$ nm	Δ_S nm	ε_{max} $mM^{-1}cm^{-1}$	Φ_f	B_{max} $mM^{-1}cm^{-1}$	Class	Ref.
609	JF646	10 mM HEPES pH 7.3	589	568	646	663	639	646	664	684	698	732	18	5.6	0.54	3.0	Si(CH3)2-Rhoda	4.3.2
610	3,7-Dimethyl-JF646	10 mM HEPES pH 7.3	574	614	651	668	640	647	664	685	704	731	13	89.6	0.51	45.7	Si(CH3)2-Rosa	4.18.1
611	3-Methyl-5-Methoxy-JF646	10 mM HEPES pH 7.3	572	612	649	666	640	647	664	685	704	731	15	110	0.49	53.9	Si(CH3)2-Rosa	4.18.1
612	Atto 647N carboxylate	PBS pH 7.4, 22 °C	592	626	646	661	642	648	664	690	706	753	18	150.0	0.65	97.5	C(CH3)2-Rhoda	4.3.1
613	SiP-Ph	50 mM PBS pH 7.4	588	624	649	665	613	630	665	672	683	698	16	80.0	0.23	18.4	C(CH3)2-Pyron	4.5
614	Vita Blue	C pH 9.5	560	586	609	625	636	643	665	731	758	793	56	36.0	0.15	5.4	S-Fluore-analog	4.10
615	W-L 8e	PBS	583	621	643	657	644	649	665	682	694	722	22	195.0	0.09	17.5	Cyanine-Penta	4.13.2
616	Zy5	P pH 7.4	587	625	647	662	644	649	665	682	694	722	18	250.0	0.27	67.5	Cyanine-Penta	4.13.2
617	CF647	H_2O			650				665				15	240.0	0.12	27.9	nd (Cyanine)	4.41
618	PONy P13	PBS pH 7.4	568	600	630	647	638	646	666	694	712	750	36	52.0	0.17	8.8	Pyronine	4.4
619	3-Hydroxy-JF646	10 mM HEPES pH 7.3	574	614	651	668	642	649	666	687	706	733	15	136	0.48	65.2	Si(CH3)2-Rosa	4.18.1
620	3-Methyl-JF646	10 mM HEPES pH 7.3	572	612	649	666	642	649	666	687	706	733	17	118	0.47	55.4	Si(CH3)2-Rosa	4.18.1
621	nor-SqB1	P pH 7.4	589	619	638	654	644	651	666	688	706	748	28	88.0	0.20	17.6	Squaraine	4.35
622	nor-SqB2	P pH 7.4	589	619	638	654	644	651	666	688	706	748	28	114.0	0.22	25.0	Squaraine	4.35
623	Cy5.29.OH	PBS	589	628	650	665	645	650	666	683	695	724	16	250.0	0.20	50.0	Cyanine-Penta	4.13.2
624	W-L 8c	PBS	583	621	643	657	645	650	666	683	695	724	23	107.0	0.12	12.8	Cyanine-Penta	4.13.2
625	W-L 8d	PBS	583	621	643	657	645	650	666	683	695	724	23	169.0	0.13	21.9	Cyanine-Penta	4.13.2
626	Atto 647 carboxylate	PBS pH 7.4, 22 °C	609	624	647	659	639	646	667	700	734	767	20	120.0	0.20	24.0	C(CH3)2-Pyron	4.5
627	Thienyl-JF646	10 mM HEPES pH 7.3	593	572	651	668	642	649	667	687	701	735	16	113	0.40	45.2	Si(CH3)2-Rhoda	4.3.2

Ent.	Short Names	Aqueous Solutions	$\lambda_{x,-0.25}$ nm	$\lambda_{x,-0.50}$ nm	$\lambda_{x,max}$ nm	$\lambda_{x,+0.50}$ nm	$\lambda_{e,-0.25}$ nm	$\lambda_{e,-0.50}$ nm	$\lambda_{e,max}$ nm	$\lambda_{e,+0.50}$ nm	$\lambda_{e,+0.25}$ nm	$\lambda_{e,+0.10}$ nm	Δ_S nm	ε_{max} mM^{-1}cm^{-1}	Φ_f	B_{max} mM^{-1}cm^{-1}	Class	Ref.
628	EVOblue 30	H_2O	575	610	650	667	644	651	667	691	710	751	17	100.0	0.10	10.0	Phenoxazine	4.25
629	Cy5.10-OH	H_2O/PBS	588	626	648	663	645	651	667	684	696	725	19	215.0	0.09	19.3	Cyanine-Penta	4.13.2
630	Cy5.11-OH	H_2O/PBS	586	624	646	661	645	651	667	684	696	725	21	215.0	0.06	12.9	Cyanine-Penta	4.13.2
631	W-L 8a	PBS	583	621	643	657	645	651	667	684	696	725	24	87.0	0.12	10.4	Cyanine-Penta	4.13.2
632	Cy5.18.OH	PBS	590	628	650	664	646	651	667	685	698	739	17	250.0	0.27	67.5	Cyanine-Penta	4.13.2
633	PONy P10	PBS pH 7.4	567	599	629	646	640	648	668	696	714	753	39	52.0	0.12	6.2	Pyronine	4.4
634	S-JF646	10 mM HEPES pH 7.3	586	620	652	671	640	648	668	696	714	753	16	21.2	0.25	5.3	Si(CH3)2-Pyron	4.5
635	W-L 8b	PBS	584	622	644	659	646	652	668	685	697	726	24	189.0	0.15	28.3	Cyanine-Penta	4.13.2
636	DY-635 carboxylate	PBS pH 7.4	550	583	635	658	630	641	669	729	766	808	34	175.0	0.05	8.7	Merocyanine	4.24
637	3,7-Di(ethylamino) phenoxazine	10 mM HEPES pH 7.3	578	615	655	673	646	653	669	693	712	754	14	111.0	0.07	7.7	Phenoxazine	4.25
638	Cy5.13-OH	H_2O/PBS	588	626	648	663	647	653	669	686	698	727	21	215.0	0.12	25.8	Cyanine-Penta	4.13.2
639	SMSiR	H_2O	584	626	652	671	643	651	670	695	717	751	18	120.0	0.31	37.2	Si(CH3)2-Srhod	4.7
640	Sulfo-JF646	10 mM HEPES pH 7.3	586	629	655	674	643	651	670	695	717	751	15	124.0	0.51	63.2	Si(CH3)2-Srhod	4.7
641	5-Aza-3-methyl-JF646	10 mM HEPES pH 7.3	578	618	656	674	646	653	670	691	710	738	14	133	0.48	63.8	Si(CH3)2-Rosa	4.18.1
642	Square-680-carboxylate	P pH 7.4	603	634	654	671	647	655	670	692	711	753	16	138.0	0.04	5.1	Squaraine	4.35
643	DY-636 carboxylate	PBS pH 7.4	558	591	645	669	632	643	671	732	769	811	26	190.0	0.09	16.3	Merocyanine	4.24
644	NIR4	H_2O	589	628	650	665	649	655	671	688	700	729	21	260.0	0.43	111.8	Cyanine-Penta	4.13.2
645	Square-650-carboxylate	P pH 5.6	598	632	653	668	651	658	671	694	712	765	18	135.0	0.16	21.6	Squaraine	4.35
646	Seta-650-carboxylate	P pH 7.4			651				671				20	200.0	0.28	56.0	nd	4.41
647	PONy P2	PBS pH 7.4	569	602	632	649	644	652	672	700	718	758	40	51.0	0.16	8.1	Pyronine	4.4
648	2-Me SiR654	100 mM P, 1% DMSO pH 7.4	576	616	654	671	647	655	672	693	713	740	18	80.0	0.24	19.2	Si(CH3)2-Rosa	4.18.1
649	AF647	PBS pH 7.2, 20/22 °C	589	628	650	665	650	656	672	690	701	731	22	270.0	0.33	89.1	Cyanine-Penta	4.13.2
650	AFD647	PBS pH 7.2, 20/22 °C	589	628	650	665	650	656	672	690	701	731	22	270.0	0.33	89.1	Cyanine-Penta	4.13.2

(Continued)

Ent.	Short Names	Aqueous Solutions	$\lambda_{x,-0.25}$ nm	$\lambda_{x,-0.50}$ nm	$\lambda_{x,max}$ nm	$\lambda_{x,+0.50}$ nm	$\lambda_{e,-0.25}$ nm	$\lambda_{e,-0.50}$ nm	$\lambda_{e,max}$ nm	$\lambda_{e,+0.50}$ nm	$\lambda_{e,+0.25}$ nm	$\lambda_{e,+0.10}$ nm	Δ_S nm	ε_{max} $mM^{-1}cm^{-1}$	Φ_f	B_{max} $mM^{-1}cm^{-1}$	Class	Ref.
651	Alexa Fluor 647	PBS pH 7.2, 20/22 °C	589	628	650	665	650	656	672	690	701	731	22	270.0	0.33	89.1	Cyanine-Penta	4.13.2
652	Seta-660-di-carboxylate	P pH 7.4			661				672				11	220.0	0.11	24.2	nd	4.41
653	Bzm-H	H_2O pH 7.4	563	597	635	652	651	658	674	698	718	760	39	29.9	0.14	4.1	Phenoxazine	4.25
654	Cy5.12-OH	H_2O/PBS	589	628	650	665	652	658	674	692	703	733	24	215.0	0.13	27.9	Cyanine-Penta	4.13.2
655	Z-L H	0.5 м PBS pH 7.4	594	625	660	679	645	651	676	727	766	815	16	50	0.01	0.6	Bodipy-Large	4.14.3
656	SiRF 15a	H_2O	598	577	657	675	651	658	676	697	711	746	19	40.0	0.40	16.0	Si(CH3)2-Rhoda	4.3.2
657	Squaraine T-P 4b	67 mм P pH 7.4	607	639	659	676	653	661	676	698	718	760	17	207.0	0.02	4.1	Squaraine	4.35
658	Square-660-carboxylate	P pH 7.4	600	635	657	675	654	660	676	702	730	769	19	182.0	0.03	5.4	Squaraine	4.35
659	B-Z 19	H_2O	587	617	651	670	646	652	677	728	767	816	26	118.5	0.23	27.2	Bodipy-Large	4.14.3
660	DY-654 carboxylate	PBS pH 7.4	594	627	653	673	651	659	677	699	711	732	24	220.0	0.16	35.2	Merocyanine	4.24
661	SiRF 13a	H_2O	598	577	657	675	651	659	677	698	712	747	20	53.0	0.63	33.3	Si(CH3)2-Rhoda	4.3.2
662	Squaraine T-P 4d	67 mм P pH 7.4	606	638	658	675	654	662	677	700	719	762	19	182.0	0.03	5.4	Squaraine	4.35
663	DY-651 carboxylate	PBS pH 7.4	599	631	653	672	652	659	678	699	711	731	25	160.0	0.15	24.1	Merocyanine	4.24
664	Phosphorilated oxazine 2-H	H_2O pH ~7	583	620	661	679	654	662	678	703	723	765	17	60.0	0.28	16.8	Phenoxazine	4.25
665	nor-SqO3	P pH 7.4	610	642	662	679	656	663	679	702	721	764	17	124.0	0.21	26.0	Squaraine	4.35
666	SiRF 1a-H	H_2O	602	581	662	680	654	662	680	701	716	751	18	118.0	0.66	77.8	Si(CH3)2-Rhoda	4.3.2
667	SiRF 1b-H	H_2O	603	581	663	681	654	662	680	701	716	751	17	90.0	0.70	63.0	Si(CH3)2-Rhoda	4.3.2
668	Atto 655 carboxylate	PBS pH 7.4, 22 °C	587	629	663	679	656	664	680	707	733	784	17	125.0	0.30	37.5	Phenoxazine	4.25
669	Atto 665 carboxylate	PBS pH 7.4, 22 °C			662				680				18	160.0	0.60	96.0	nd (pb Pyron.)	4.41
670	Si-9-Iminopyronine 2c	~1 mм HCl, 1% DMSO pH 3	597	633	666	685	652	660	681	710	729	769	15	83.2	0.28	23.3	Si(CH3)2-Pyron	4.5
671	JF669-6-SCH2CO2Et	10 mм HEPES pH 7.3	607	585	668	686	655	663	681	702	717	752	13	119.0	0.38	45.2	Si(CH3)2-Rhoda	4.3.2
672	JF669-6-SCH2CO2H	10 mм HEPES pH 7.3	606	584	667	685	655	663	681	702	717	752	14	130.0	0.38	49.4	Si(CH3)2-Rhoda	4.3.2

Ent.	Short Names	Aqueous Solutions	$\lambda_{x,-0.25}$ nm	$\lambda_{x,-0.50}$ nm	$\lambda_{x,max}$ nm	$\lambda_{x,+0.50}$ nm	$\lambda_{e,-0.25}$ nm	$\lambda_{e,-0.50}$ nm	$\lambda_{e,max}$ nm	$\lambda_{e,+0.50}$ nm	$\lambda_{e,+0.25}$ nm	$\lambda_{e,+0.10}$ nm	Δ_S nm	ε_{max} mм$^{-1}$cm^{-1}	Φ_f	B_{max} mм$^{-1}$cm^{-1}	Class	Ref.
673	$SiOH_2R$	10 mм PBS pH 7.4	599	639	663	679	656	662	681	700	713	738	18	105.0	0.43	45.1	Si(OH)2-Rosa	4.18.1
674	Atto Oxa12 carboxylate	PBS pH 7.4, 22 °C	584	621	662	680	657	665	681	706	726	769	19	125.0	0.30	37.5	Phenoxazine	4.25
675	Bpe-H	H_2O pH 7.4	567	601	640	657	657	665	681	706	726	769	41	48.1	0.23	11.0	Phenoxazine	4.25
676	4,10-Dibromonaphtho-fluorescein	C pH 9.5	549	578	605	624	646	655	682	728	768	811	77	32.0	0.05	1.6	Fluores-analog	4.8.4
677	Si-9-Iminopyronine 2f	~1 mм HCl, 1% DMSO pH 3	595	631	664	683	653	661	682	711	730	770	18	123	0.26	31.9	Si(CH3)2-Pyron	4.5
678	4,5,6,7-Tetrafluoro-SiTMR	10 mм HEPES pH 7.3	606	584	667	685	656	664	682	703	718	754	15	132.0	0.31	40.9	Si(CH3)2-Rhoda	4.3.2
679	JF669	10 mм HEPES pH 7.3	608	586	669	687	656	664	682	703	718	754	13	112.0	0.37	41.4	Si(CH3)2-Rhoda	4.3.2
680	CF660R	H_2O			663				682				19	100	0.08	7.6	nd (Rhodam.)	4.41
681	COUPY 12	H_2O			595				683				88	6.9	0.05	0.3	Hybrid-Bridged	4.40.3
682	2-Me SiR666	100 mм P, 1% DMSO pH 7.4	585	627	666	684	659	666	684	706	726	755	18	80.0	0.25	20.0	Si(CH3)2-Rosa	4.18.1
683	nor-SqB3	P pH 7.4	610	642	662	679	660	668	684	707	727	770	22	102.0	0.18	18.3	Squaraine	4.35
684	NIR2	H_2O	599	639	662	677	661	667	684	702	714	745	22	250.0	0.34	85.0	Cyanine-Penta	4.13.2
685	NR666	PBS, 1% DMSO pH 7.4	608	646	666	683	653	661	685	719	741	778	19	165.0	0.38	62.7	PO2H-Rosam.	4.18.2
686	SqCN1	P pH 7.4	614	647	667	685	661	669	685	708	728	772	18	188.0	0.07	13.1	Squaraine	4.35
687	Squaraine T-P 5b	67 mм P pH 7.4	614	647	667	685	661	669	685	708	728	772	18	188.0	0.07	13.1	Squaraine	4.35
688	CF660C	H_2O			667				685				18	200	0.06	11.1	nd (Cyanine)	4.41
689	2-Me TeOR	PBS pH 7.4	599	642	669	692	654	663	686	714	737	784	17	120.0	0.18	21.6	TeO-Rosamine	4.18.3
690	JF668	10 mм HEPES pH 7.3	607	585	668	686	661	668	687	708	724	760	19	26.7	0.34	9.0	POOH-Rhoda	4.3.3

(Continued)

Ent.	Short Names	Aqueous Solutions	$\lambda_{x,-0.25}$ nm	$\lambda_{x,-0.50}$ nm	$\lambda_{x,max}$ nm	$\lambda_{x,+0.50}$ nm	$\lambda_{e,-0.25}$ nm	$\lambda_{e,-0.50}$ nm	$\lambda_{e,max}$ nm	$\lambda_{e,+0.50}$ nm	$\lambda_{e,+0.25}$ nm	$\lambda_{e,+0.10}$ nm	Δ_S nm	ε_{max} mM^{-1}cm^{-1}	Φ_f	B_{max} mM^{-1}cm^{-1}	Class	Ref.
691	2-Me SiR670	100 mM P, 1% DMSO pH 7.4	588	630	670	688	661	669	687	709	730	759	17	90.0	0.27	24.3	Si(CH3)2-Rosa	4.18.1
692	2-Me SiR671	100 mM P, 1% DMSO pH 7.4	589	631	671	689	661	669	687	709	730	759	16	99.0	0.29	28.7	Si(CH3)2-Rosa	4.18.1
693	Seta-670-carboxylate	P pH 7.4			667				688				21	180.0	0.07	12.6	nd	4.41
694	2-Me SiR680	H_2O	592	634	674	693	663	671	689	711	732	761	15	130.0	0.35	45.5	Si(CH3)2-Rosa	4.18.1
695	NR669	10 mM PBS pH 7.4	619	653	669	684	664	672	689	712	726	757	20	77.2	0.30	23.1	PO2H-Rosam.	4.18.2
696	Alexa Fluor 660	PBS pH 7.2/7.5, 20/22 °C			663				690				27	132.0	0.37	48.8	nd	4.41
697	nor-SqCN1	P pH 7.4	616	650	670	687	669	677	693	717	737	782	23	123.0	0.19	23.3	Squaraine	4.35
698	nor-SqCN2	P pH 7.4	616	650	670	687	669	677	693	717	737	782	23	167.0	0.19	31.7	Squaraine	4.35
699	Cy5.205	PBS pH 7.4	609	650	674	690	671	677	694	713	725	757	20	190.0	0.23	43.7	Cyanine-Penta	4.13.2
700	Cy5.5	PBS	609	650	674	690	671	677	694	713	725	757	20	190.0	0.23	43.7	Cyanine-Penta	4.13.2
701	Zy5.5	P pH 7.4	610	652	676	692	671	677	694	713	725	757	18	190.0	0.23	43.7	Cyanine-Penta	4.13.2
702	Seta-670-pH-di-carboxylate	P pH 6			672				694				22	93.0	0.19	17.6	nd	4.41
703	NR675	10 mM PBS pH 7.4	625	657	675	689	668	675	695	718	733	762	20	73.2	0.24	17.5	PO2H-Rosam.	4.18.2
704	DY-679P1	PBS pH 7.4	612	655	679	696	673	680	697	715	726	745	18	200.0	0.23	46.6	Cyanine-Penta	4.13.2
705	PORL	PBS, 1% DMSO pH 7.4	609	643	670	686	671	679	698	724	744	776	28	66.1	0.11	7.2	POPh-Rosol	4.20
706	2-Me SiR675	100 mM P, 1% DMSO pH 7.4	592	635	675	694	672	679	698	721	742	772	23	88.0	0.07	6.1	Si(CH3)2-Rosa	4.18.1
707	Atto 680 carboxylate	PBS pH 7.4, 22 °C	603	648	681	696	675	682	698	727	749	807	17	125.0	0.30	37.5	Phenoxazine	4.25
708	CF680	H_2O			681				698				17	210	0.04	8.2	nd (Cyanine)	4.41
709	Seta-680-carboxylate	P pH 7.4			678				698				20	200.0	0.24	48.0	nd	4.41
710	CF680R	H_2O			680				701				21	140	0.03	4.7	nd (Rhodam.)	4.41
711	Alexa Fluor 680	PBS pH 7.2/7.5, 20/22 °C	613	655	679	695	678	685	702	721	734	766	23	183.0	0.36	65.8	Cyanine-Penta	4.13.2

Ent.	Short Names	Aqueous Solutions	$\lambda_{x,-0.25}$ nm	$\lambda_{x,-0.50}$ nm	$\lambda_{x,max}$ nm	$\lambda_{x,+0.50}$ nm	$\lambda_{e,-0.25}$ nm	$\lambda_{e,-0.50}$ nm	$\lambda_{e,max}$ nm	$\lambda_{e,+0.50}$ nm	$\lambda_{e,+0.25}$ nm	$\lambda_{e,+0.10}$ nm	Δ_S nm	ε_{max} mM^{-1}cm^{-1}	Φ_f	B_{max} mM^{-1}cm^{-1}	Class	Ref.
712	Cy5.203	PBS pH 7.4	614	656	680	696	678	685	702	721	734	766	22	190.0	0.10	19.0	Cyanine-Penta	4.13.2
713	Seta-700-carboxylate	P pH 7.4			687				703				16	177.0	0.06	10.6	nd	4.41
714	PONy P5	PBS pH 7.4	592	627	660	679	673	682	704	735	755	799	44	38.0	0.10	3.8	Pyronine	4.4
715	Seta-700-di-carboxylate	P pH 7.4			688				704				16	180.0	0.11	19.8	nd	4.41
716	JF690	10 mM HEPES pH 7.3	625	602	690	710	679	687	707	730	746	784	17	150.0	0.24	36.0	POOH-Rhoda	4.3.3
717	Cy5.204	PBS pH 7.4	615	658	682	698	683	689	707	726	739	772	25	197.0	0.10	19.7	Cyanine-Penta	4.13.2
718	DY-681 carboxylate	PBS pH 7.4	592	630	691	718	665	676	708	776	818	866	17	125.0	0.09	11.2	Merocyanine	4.24
719	THQ-rosol	50 mM P pH 7.4	500	523	550	565	631	653	710	802	854	925	160	29.4	0.03	0.8	Rosol	4.19
720	DY-682 carboxylate	PBS pH 7.4	617	654	690	708	671	685	710	733	745	764	20	120.0	0.41	48.6	Merocyanine	4.24
721	Me-PR	PBS, 1% DMSO pH 7.4	632	672	694	711	682	690	712	740	757	789	18	92.0	0.11	10.1	POR-Rosam.	4.18.2
722	PR	PBS, 1% DMSO pH 7.4	632	672	694	711	682	690	712	740	757	789	18	65.0	0.06	3.9	POR-Rosam.	4.18.2
723	NR698	PBS, 1% DMSO pH 7.4	631	676	698	715	686	693	712	738	751	773	14	25.9	0.32	8.2	PO2H-Rosam.	4.18.2
724	2-Me SiR700	H_2O	617	665	691	710	686	695	712	732	746	764	21	100.0	0.12	12.0	Si(CH3)2-Rosa	4.18.1
725	tMe-PR	PBS, 1% DMSO pH 7.4	634	674	696	714	682	691	713	741	759	790	17	68.0	0.15	10.2	POR-Rosam.	4.18.2
726	nor-SqCN3	P pH 7.4	636	671	693	712	688	697	714	739	761	809	21	113.0	0.15	16.9	Squaraine	4.35
727	Atto 700 carboxylate	PBS pH 7.4, 22 °C	622	666	700	717	693	701	716	744	768	828	16	120.0	0.25	30.0	Phenoxazine	4.25
728	NR700	PBS, 1% DMSO pH 7.4	630	674	700	720	685	696	722	752	772	808	22	71.0	0.11	7.8	PO2R-Rosam.	4.18.2
729	Alexa Fluor 700	PBS/H_2O pH 7.2, 20/22 °C			702				723				21	205.0	0.25	51.2	nd	4.41

(Continued)

Ent.	Short Names	Aqueous Solutions	$\lambda_{x,-0.25}$ nm	$\lambda_{x,-0.50}$ nm	$\lambda_{x,max}$ nm	$\lambda_{x,+0.50}$ nm	$\lambda_{e,-0.25}$ nm	$\lambda_{e,-0.50}$ nm	$\lambda_{e,max}$ nm	$\lambda_{e,+0.50}$ nm	$\lambda_{e,+0.25}$ nm	$\lambda_{e,+0.10}$ nm	Δ_S nm	ε_{max} mM^{-1}cm^{-1}	Φ_f	B_{max} mM^{-1}cm^{-1}	Class	Ref.
730	JF711	10 mM HEPES pH 7.3	642	618	711	732	702	711	732	756	774	815	21	12.4	0.17	2.1	POPh-Rhodam	4.3.3
731	SO_2R1	10 mM PBS, 0.25% ACN pH 7.4	633	678	703	724	701	711	736	772	798	828	33	39	0.07	2.6	SO2-Rosamine	4.18.3
732	2-Me SiR720	H_2O	627	675	721	742	710	719	740	766	790	824	19	160.0	0.05	8.0	Si(CH3)2-Rosa	4.18.1
733	SO_2R2	10 mM PBS, 0.25% ACN pH 7.4	633	678	703	724	706	717	742	779	805	836	39	95	0.07	6.9	SO2-Rosamine	4.18.3
734	SO_2R4	10 mM PBS, 0.25% ACN pH 7.4	634	679	704	725	706	717	742	779	805	836	38	97.0	0.10	9.4	SO2-Rosamine	4.18.3
735	JF722	10 mM HEPES pH 7.3	651	626	722	743	712	721	743	768	786	829	21	87.2	0.11	9.5	POPh-Rhodam	4.3.3
736	SO_2R3	10 mM PBS, 0.25% ACN pH 7.4	636	681	707	728	711	721	747	784	811	842	40	69	0.05	3.4	SO2-Rosamine	4.18.3
737	JF724	10 mM HEPES pH 7.3	647	688	724	750	711	721	748	792	831	879	24	6.6	0.05	0.3	SO2-Rhodam	4.3.3
738	Atto 725 carboxylate	PBS pH 7.4, 22 °C			728				751				23	120.0	0.10	12.0	nd	4.41
739	SO_2R5	10 mM PBS, 0.25% ACN pH 7.4	639	684	710	732	715	726	752	790	817	848	42	129.0	0.07	8.7	SO2-Rosamine	4.18.3
740	HCDN	PBS pH 7.4	564	603	625	641	725	735	755	778	789	802	130	195.0	0.28	54.5	Cyanine-Hepta	4.13.3
741	DY-731 carboxylate	PBS pH 7.4	678	709	736	761	725	735	759	783	794	811	23	225.0	0.11	23.6	Merocyanine	4.24
742	LS479	Buffer pH 8	661	716	747	770	731	741	762	786	796	810	15	105.0	0.04	4.4	Cyanine-Hepta	4.13.3
743	Atto 740 carboxylate	PBS pH 7.4, 22 °C			743				763				20	120.0	0.10	12.0	nd	4.41
744	NR744	PBS, 1% DMSO pH 7.4	672	719	744	763	725	737	764	795	812	841	20	80.0	0.16	12.8	PO2R-Rosam.	4.18.2

Ent.	Short Names	Aqueous Solutions	$\lambda_{x,-0.25}$ nm	$\lambda_{x,-0.50}$ nm	$\lambda_{x,max}$ nm	$\lambda_{x,+0.50}$ nm	$\lambda_{e,-0.25}$ nm	$\lambda_{e,-0.50}$ nm	$\lambda_{e,max}$ nm	$\lambda_{e,+0.50}$ nm	$\lambda_{e,+0.25}$ nm	$\lambda_{e,+0.10}$ nm	Δ_S nm	ε_{max} mM^{-1}cm^{-1}	Φ_f	B_{max} mM^{-1}cm^{-1}	Class	Ref.
745	DY-747P1	PBS pH 7.4	661	716	747	770	738	748	769	793	804	818	22	240.0	0.12	28.5	Cyanine-Hepta	4.13.3
746	Zy7 carboxylate	P pH 7.4	663	718	749	772	743	753	774	798	810	824	25	200.0	0.13	26.0	Cyanine-Hepta	4.13.3
747	Alexa Fluor 750	PBS/H_2O pH 7.2, 20/22 °C			749				775				26	290.0	0.12	34.8	nd	4.41
748	NIR3	H_2O	663	719	750	773	745	756	777	802	813	827	27	275.0	0.28	77.0	Cyanine-Hepta	4.13.3
749	Seta-750-carboxylate	P pH 7.4			751				779				28	235.0	0.21	49.3	nd	4.41
750	PONy CP1	PBS pH 7.4	640	681	710	728	737	754	783	817	843	870	73	17.0	0.19	3.2	C(CH3)2-Pyron	4.5
751	HCDO	PBS pH 7.4	678	736	769	793	755	766	788	813	825	840	19	218.8	0.30	65.6	Cyanine-Hepta	4.13.3
752	ZW800-1	FBS	680	739	772	797	755	766	788	813	825	840	16	249.0	0.15	37.6	Cyanine-Hepta	4.13.3
753	ZW800-3a	FBS	682	741	774	799	757	768	790	815	827	842	16	309.0	0.16	49.7	Cyanine-Hepta	4.13.3
754	LS482	Buffer pH 8	632	682	710	731	760	771	793	819	830	845	83	84.0	0.10	8.4	Cyanine-Hepta	4.13.3
755	NIR1	H_2O	672	729	761	785	763	773	796	822	834	849	35	268.0	0.23	61.6	Cyanine-Hepta	4.13.3
756	DY-780 carboxylate	PBS pH 7.4	657	705	782	817	745	760	800	888	943	1007	18	102.0	0.03	3.2	Merocyanine	4.24
757	CW800	FBS	691	752	786	812	766	777	800	826	838	853	14	237.0	0.14	33.6	Cyanine-Hepta	4.13.3
758	RS800	FBS	690	750	784	809	766	777	800	826	838	853	16	240.0	0.17	40.5	Cyanine-Hepta	4.13.3
759	TG170	H_2O	695	756	791	817	769	780	803	829	841	857	12	300.0	0.02	7.4	Cyanine-Hepta	4.13.3
760	Y-A 5	PBS pH 7.2	685	744	778	803	770	781	804	830	842	858	26	185.0	0.03	5.5	Cyanine-Hepta	4.13.3
761	Y-A 1 (Cypate)	PBS pH 7.2	685	744	778	803	771	782	805	831	844	859	27	224.0	0.03	6.7	Cyanine-Hepta	4.13.3
762	Y-A 4	PBS pH 7.2	685	744	778	803	771	782	805	831	844	859	27	201.0	0.03	6.0	Cyanine-Hepta	4.13.3
763	Y-A 6	PBS pH 7.2	685	744	778	803	771	782	805	831	844	859	27	224.0	0.03	6.7	Cyanine-Hepta	4.13.3
764	Y-A 7	PBS pH 7.2	686	745	779	804	771	782	805	831	844	859	26	224.0	0.03	6.7	Cyanine-Hepta	4.13.3
765	IRDye78-CA	10 mM HEPES pH 7.4	680	738	771	796	772	783	806	832	845	860	35	145.7	0.14	20.6	Cyanine-Hepta	4.13.3
766	Y-A 2	PBS pH 7.2	685	744	778	803	772	783	806	832	845	860	28	190.0	0.03	5.7	Cyanine-Hepta	4.13.3
767	Y-A 8	PBS pH 7.2	687	746	780	805	772	783	806	832	845	860	26	197.0	0.03	5.9	Cyanine-Hepta	4.13.3
768	Y-A 11	PBS pH 7.2	687	746	780	805	773	784	807	834	846	861	27	199.0	0.03	5.9	Cyanine-Hepta	4.13.3
769	Y-A 3	PBS pH 7.2	684	744	777	802	773	784	807	834	846	861	30	215.0	0.03	6.4	Cyanine-Hepta	4.13.3
770	Seta-780-carboxylate	P pH 7.4			777				807				30	215.0	0.01	2.5	nd	4.41
771	Y-A 10	PBS pH 7.2	687	746	780	805	774	785	808	835	847	862	28	219.0	0.03	6.5	Cyanine-Hepta	4.13.3

(Continued)

Ent.	Short Names	Aqueous Solutions	$\lambda_{x,-0.25}$ nm	$\lambda_{x,-0.50}$ nm	$\lambda_{x,max}$ nm	$\lambda_{x,+0.50}$ nm	$\lambda_{e,-0.25}$ nm	$\lambda_{e,-0.50}$ nm	$\lambda_{e,max}$ nm	$\lambda_{e,+0.50}$ nm	$\lambda_{e,+0.25}$ nm	$\lambda_{e,+0.10}$ nm	Δ_S nm	ε_{max} $mM^{-1}cm^{-1}$	Φ_f	B_{max} $mM^{-1}cm^{-1}$	Class	Ref.
772	Y-A 9	PBS pH 7.2	686	745	779	804	775	786	809	836	848	864	30	187.0	0.03	5.6	Cyanine-Hepta	4.13.3
773	ICG	PBS	687	746	780	805	776	787	810	837	849	865	30	211.2	0.08	16.9	Cyanine-Hepta	4.13.3
774	IR-783	PBS	688	748	782	807	776	787	810	837	849	865	28	261.0	0.08	21.9	Cyanine-Hepta	4.13.3
775	HCDS	PBS pH 7.4	695	756	791	817	783	794	818	845	858	874	27	218.8	0.39	85.3	Cyanine-Hepta	4.13.3
776	ICG	FBS	708	771	807	834	787	798	822	850	862	878	15	121.0	0.09	11.2	Cyanine-Hepta	4.13.3
777	ECXi	H_2O			887				938				51	93.3	0.01	1.3	Pentacene	4.39.2

Appendix D

D1. Ranked by Stokes Shifts

Appendix D contains the over 700 fluorescent dyes with medium to high brightness in aqueous solutions, ranked by descending values of Stokes shifts (Δ_S). Some of the names are registered trademarks (listed in footnote[1]). Only the longest excitation (or absorption) wavelength bands of each dye are considered in the Stokes shift calculations. The values of emission maximum ($\lambda_{e,max}$) and excitation maximum ($\lambda_{x,max}$) (or absorption maximum when $\lambda_{x,max}$ was not available) were taken from the literature for almost all dyes (some were measured by the author). The references can be located in the section identified in the column "Ref." (Table D.1). The sections of Chapter 4 were split into small sub-sections to facilitated the location of these references.

Molecular structures together with some photophysical parameters are shown in Chapter 4. Chapter 4 also provides a detailed list of references where the photophysical parameters were taken. Panoramic views of the photophysical properties and classes are given in Chapter 5 using scattergrams. Methods to measure photostability and techniques to improve photostability are discussed in Chapter 7. The photostability (relative photobleaching mean-lifetime) of some dyes is also shown in Chapter 7.

Acronyms used in Table D.1: A, acetate; AC, acetic acid; ACN, acetonitrile; BODIPY, borondipyrromethene; BSA, bovine serum albumin; C, carbonate; CA, carboxylic acid; FBS, fetal bovine serum; Et, ethanol; nd, non-disclosed; ni, not informed; P, phosphate; pb, probably; pH >7, basic medium; pH < 7, acidic medium (when the exact value of the pH is not given in the literature); SA, sulfonic acid; and TX-100, triton X-100.

1 Abberior STAR is a trademark of Abberior GmbH (Göttingen, Germany). AlexaFluor, BODIPY, Cascade Blue, FAM, JOE, TET, HEX, Marina Blue, Oregon Green, Pacific Blue, ROX, TAMRA, and Texas Red are trademarks of Thermo Fisher Scientific (Waltham, MA, USA). ATTO is a trademark of ATTO-Tec GmbH (Siegen, Germany). CF Dye is a registered trademark of Biotium, Inc. (Fremont, CA, USA). Cy3, Cy3.5, Cy5, and Cy5.5 are trademarks of GE Healthcare (Chicago, IL, USA). Dy is a trademark of Dyomics GmbH (Jena, Germany). Seta, SeTau, and Square dyes are trademarks of SETA BioMedicals (Urbana, IL, USA).

Fluorescent Dye Labels and Stains: A Database of Photophysical Properties, First Edition. Tarso B. Ledur Kist.

Table D.1 Photophysical properties of fluorescent labels and stains in aqueous solutions ranked by Stokes shift (Δ_S).

Ent.	Short Names	Aqueous Solutions	$\lambda_{x,-0.25}$ nm	$\lambda_{x,-0.50}$ nm	$\lambda_{x,max}$ nm	$\lambda_{x,+0.50}$ nm	$\lambda_{e,-0.25}$ nm	$\lambda_{e,-0.50}$ nm	$\lambda_{e,max}$ nm	$\lambda_{e,+0.50}$ nm	$\lambda_{e,+0.25}$ nm	$\lambda_{e,+0.10}$ nm	Δ_S	ε_{max} mM^{-1}cm^{-1}	Φ_f	B_{max} mM^{-1}cm^{-1}	Class	Ref.
1	Fluorene PP6	H_2O			373				620				247	16.5	0.07	1.1	Fluorene	4.39.2
2	Fluorene PP11	H_2O			377				621				244	36.0	0.05	1.8	Fluorene	4.39.2
3	Carbostyril 7	H_2O			366				580				214	9.0	0.07	0.6	Quinoline	4.29
4	(HO)-Luciferyl adenylate	H_2O pH 5			335				548				213	15.1	0.11	1.6	Benzothiazole	4.30
5	Fluorene PP12	H_2O			345				555				210	24.0	0.16	3.8	Fluorene	4.39.2
6	(HO)-Luciferin	H_2O pH 5			328				537				209	18.2	0.25	4.5	Benzothiazole	4.30
7	(HO)-Dehydroluciferin	H_2O pH 4.5			348				555				207	18.2	0.26	4.7	Benzothiazole	4.30
8	Si-9-Iminopyronine 2f	~1 mM NaOH, 1% DMSO pH 11	375	389	401	408	583	590	606	629	643	675	205	17.8	0.35	6.2	Si(CH3)2-Pyron	4.5
9	PONy C7	PBS pH 7.4			419				613				194	22.0	0.04	0.8	Hybrid-Bonded	4.40.2
10	Si-9-Iminopyronine 2c	~1 mM NaOH, 1% DMSO pH 11	390	405	418	426	584	591	607	630	645	676	189	12.9	0.46	5.9	Si(CH3)2-Pyron	4.5
11	Edans	H_2O	281	300	336	370	472	487	520	557	574	601	184	6.2	0.27	1.6	Naphthalene	4.34
12	Acedan 8	H_2O	278	296	332	365	468	482	515	551	568	595	183	12.6	0.23	2.9	Naphthalene	4.34
13	Acridizinium I-L 5b	H_2O			369				551				182	11.0	0.02	0.2	Acridizinium	4.39.1
14	Acedan 2	H_2O	282	301	338	372	468	482	515	551	568	595	177	14.6	0.30	4.3	Naphthalene	4.34
15	Seta-400-carboxylate	P pH 7.4			399				576				177	24.0	0.34	8.1	nd	4.41
16	Prodan	H_2O	300	322	364	404	481	496	531	569	587	616	167	14.5	0.10	1.4	Naphthalene	4.34
17	Atto 490LS carboxylate	PBS pH 7.4,22 °C			495				658				163	40.0	0.30	12.0	nd (pb Coum.)	4.41
18	Si-9-Iminopyronine 2g	PBS pH 7.4	434	452	469	479	606	613	631	656	672	706	162	11	0.43	4.7	Si(CH3)2-Pyron	4.5
19	THQ-rosol	50 mM P pH 7.4	500	523	550	565	631	653	710	802	854	925	160	29.4	0.03	0.8	Rosol	4.19
20	($^-$O)-Dehydroluciferin	H_2O pH 11			396				555				159	18.2	0.62	11.2	Benzothiazole	4.30
21	Carbostyril 23b	H_2O pH 12			397				556				159	17.3	0.26	4.5	Quinoline	4.29
22	3-(4-OH-Ph)-6-MeO-chromone	100 mM Tris pH 8			311				468				157	12.6	0.05	0.6	Chromone	4.31
23	8-Aminonaphthalene-1,3,6-triSA	100 mM P pH 7	295	315	356	394	466	479	512	547	564	591	156	15.0	0.20	3.0	Naphthalene	4.34
24	Carbostyril 6	H_2O			389				544				155	7.0	0.09	0.6	Quinoline	4.29

Ent.	Short Names	Aqueous Solutions	$\lambda_{x,-0.25}$ nm	$\lambda_{x,-0.50}$ nm	$\lambda_{x,max}$ nm	$\lambda_{x,+0.50}$ nm	$\lambda_{e,-0.25}$ nm	$\lambda_{e,-0.50}$ nm	$\lambda_{e,max}$ nm	$\lambda_{e,+0.50}$ nm	$\lambda_{e,+0.25}$ nm	$\lambda_{e,+0.10}$ nm	Δ_S	ε_{max} mM^{-1}cm^{-1}	Φ_f	B_{max} mM^{-1}cm^{-1}	Class	Ref.
25	Acedan	H_2O	299	320	362	402	468	482	515	551	568	595	153	11.2	0.20	2.2	Naphthalene	4.34
26	(⁻O)-Luciferin	H_2O pH 11			384				537				153	18.5	0.62	11.4	Benzothiazole	4.30
27	Carbostyril 4	H_2O			379				531				152	8.7	0.07	0.6	Quinoline	4.29
28	Abberior Star 470SXP	PBS pH 7.4	416	434	472	499	583	594	624	681	722	767	152	29.0	0.12	3.4	Coumarin 7-Am	4.26.3
29	Acyman	H_2O	302	324	367	408	470	484	517	553	570	597	150	9.1	0.10	0.9	Naphthalene	4.34
30	Pacific orange	C pH 9.5			404				553				149	25.0	0.62	15.5	Benzothiazole	4.30
31	Acedan 4	H_2O	294	315	355	393	457	470	501	535	551	576	146	13.5	0.32	4.3	Naphthalene	4.34
32	Acedan 5	H_2O	295	316	357	396	458	472	503	537	553	579	146	17.4	0.40	6.9	Naphthalene	4.34
33	Acedan 6	H_2O	295	316	357	396	458	472	503	537	553	579	146	15.4	0.39	6.0	Naphthalene	4.34
34	Cascade yellow	H_2O	329	356	404	440	491	505	549	603	657	663	145	24.0	0.05	1.2	Hybrid-Bonded	4.40.2
35	Acedan 7	H_2O	291	311	351	388	452	465	495	528	544	568	144	11.0	0.31	3.4	Naphthalene	4.34
36	Acedan 3	H_2O	295	316	357	396	457	470	501	535	551	576	144	14.5	0.31	4.5	Naphthalene	4.34
37	BMeS-*p*-A	H_2O	322	341	377	404	458	472	517	576	610	643	140	3.8	0.67	2.5	Benzene	4.38
38	Carbostyril 13	H_2O pH 7			420				556				136	17.3	0.26	4.5	Quinoline	4.29
39	BMeS-*p*-A-COOH	H_2O	333	353	392	421	466	481	527	589	624	659	135	3.9	0.60	2.3	Benzene	4.38
40	trans-PO-PB460	PBS pH 7.4			459				592				133	9.0	0.25	2.2	Phosphole	4.39.3
41	Coumarin 152	H_2O	316	334	365	389	457	467	497	539	564	593	132	14.3	0.37	5.2	Coumarin 7-Am	4.26.2
42	BMeS-*p*-A-2COOH	H_2O	341	362	403	434	473	487	535	599	635	671	132	3.7	0.56	2.0	Benzene	4.38
43	Carbostyril 23b	H_2O pH 7			424				556				132	17.3	0.26	4.5	Quinoline	4.29
44	7-Hydroxy-4-methylcoumarin	H_2O pH 5	276	295	320	341	415	424	450	483	501	523	130	14.0	0.78	10.9	Coumarin 7-OH	4.26.1
45	Coumarin P-R 12	PBS pH 7.5	282	301	327	348	421	430	457	491	510	533	130	7.9	0.76	5.9	Coumarin 7-OH	4.26.1
46	HCDN	PBS pH 7.4	564	603	625	641	725	735	755	778	789	802	130	195.0	0.28	54.5	Cyanine-Hepta	4.13.3
47	5-Methoxysalicylic acid	H_2O pH 7	285	295	319	339	403	414	447	489	511	526	128	3.0	0.19	0.5	Benzene	4.38
48	Seta-385-carboxylate	P pH 7.4			384				512				128	25.0	0.78	19.5	nd	4.41
49	Isoquinoline 8	PBS pH 7.4			406				534				128	16.8	0.37	6.2	Isoquinoline	4.37
50	cis-PO-PB430	PBS pH 7.4			421				548				127	10.0	0.71	7.1	Phosphole	4.39.3
51	Epicocconone P-F 17a	H_2O, 0.1% SDS pH 2.5			391				517				126	8.8	0.35	3.0	Epicocconone	4.39.2

(Continued)

Ent.	Short Names	Aqueous Solutions	$\lambda_{x,-0.25}$ nm	$\lambda_{x,-0.50}$ nm	$\lambda_{x,max}$ nm	$\lambda_{x,+0.50}$ nm	$\lambda_{e,-0.25}$ nm	$\lambda_{e,-0.50}$ nm	$\lambda_{e,max}$ nm	$\lambda_{e,+0.50}$ nm	$\lambda_{e,+0.25}$ nm	$\lambda_{e,+0.10}$ nm	Δ_S	ε_{max} mM^{-1}cm^{-1}	Φ_f	B_{max} mM^{-1}cm^{-1}	Class	Ref.
52	Epicocconone P-F 17d	H_2O, 0.1% SDS pH 2.5			392				518				126	13.2	0.40	5.2	Epicocconone	4.39.2
53	trans-PO-PB430	PBS pH 7.4			422				545				123	10.3	0.74	7.6	Phosphole	4.39.3
54	3,6,7,8,10-PentaMe-isoalloxazine	H_2O pH 7			448				571				123	8.3	0.10	0.8	Pteridine	4.36
55	Isoquinoline 7f	PBS pH 7.4			406				528				122	16.3	0.16	2.6	Isoquinoline	4.37
56	Isoquinoline 7a	PBS pH 7.4			407				528				121	14.1	0.34	4.7	Isoquinoline	4.37
57	3-Buthyl-7-hydroxychromone	100 mM Tris pH 8			334				454				120	12.7	0.21	2.6	Chromone	4.31
58	Fluorene I-6	H_2O			353				473				120	30.5	0.13	3.9	Fluorene	4.39.2
59	Acridizinium I-L 5a	H_2O			387				507				120	17.4	0.12	2.0	Acridizinium	4.39.1
60	Isoquinoline 7b	PBS pH 7.4			406				526				120	14.4	0.46	6.6	Isoquinoline	4.37
61	Coumarin P-R 5	PBS pH 7.5	296	317	346	370	428	437	465	500	520	544	119	10.3	0.79	8.1	Coumarin 7-Am	4.26.2
62	Isoquinoline 3h	PBS pH 7.4			406				525				119	13.7	0.38	5.2	Isoquinoline	4.37
63	3,7,9,10-TetraMe-isoalloxazine	H_2O pH 7			452				571				119	6.8	0.10	0.6	Pteridine	4.36
64	Sulfonated coumarin N-H 7-H	PBS pH 7.4	453	474	520	552	596	607	639	699	742	789	119	40.0	0.06	2.4	Coumarin 7-Am	4.26.3
65	Stefani HB-7	H_2O			331				449				118	42.0	0.09	3.6	Benzothiazole	4.30
66	Isoquinoline 3m	PBS pH 7.4			408				526				118	18.1	0.36	6.5	Isoquinoline	4.37
67	Isoquinoline 7d	PBS pH 7.4			408				526				118	15.0	0.14	2.1	Isoquinoline	4.37
68	Lucigenin	H_2O			368				485				117	37.6	0.50	18.8	Hybrid-Bonded	4.40.2
69	Epicocconone P-F 4d	H_2O, 0.1% SDS pH 2.5			417				534				117	14.5	0.37	5.3	Epicocconone	4.39.2
70	Epicocconone P-F 4c	H_2O, 0.1% SDS pH 2.5			425				542				117	40.3	0.40	16.2	Epicocconone	4.39.2
71	Coumarin P-R 21	PBS pH 7.5	295	316	345	369	425	434	461	495	515	538	116	13.2	0.79	10.4	Coumarin 7-Am	4.26.2
72	Coumarin 307	H_2O	335	355	392	417	471	481	508	550	576	606	116	9.0	0.33	2.9	Coumarin 7-Am	4.26.2
73	Isoquinoline 3i	PBS pH 7.4			405				521				116	15.4	0.40	6.1	Isoquinoline	4.37
74	Isoquinoline 3e	PBS pH 7.4			408				524				116	15.8	0.41	6.4	Isoquinoline	4.37

Ent.	Short Names	Aqueous Solutions	$\lambda_{x,-0.25}$ nm	$\lambda_{x,-0.50}$ nm	$\lambda_{x,max}$ nm	$\lambda_{x,+0.50}$ nm	$\lambda_{e,-0.25}$ nm	$\lambda_{e,-0.50}$ nm	$\lambda_{e,max}$ nm	$\lambda_{e,+0.50}$ nm	$\lambda_{e,+0.25}$ nm	$\lambda_{e,+0.10}$ nm	Δ_S	ε_{max} mM^{-1}cm^{-1}	Φ_f	B_{max} mM^{-1}cm^{-1}	Class	Ref.
75	Isoquinoline 7c	PBS pH 7.4			408				524				116	15.8	0.26	4.1	Isoquinoline	4.37
76	Isoquinoline 3j	PBS pH 7.4			409				525				116	18.8	0.45	8.4	Isoquinoline	4.37
77	Isoquinoline 10b	PBS pH 7.4			410				526				116	15.5	0.17	2.6	Isoquinoline	4.37
78	PB430	PBS pH 7.4			426				542				116	10.4	0.66	6.8	Phosphole	4.39.3
79	Naphthalimide B-K 1	10 mM P pH 7.0	380	398	436	470	508	520	552	594	615	639	116	13.2	0.08	1.0	Naphthalimide	4.32
80	Acrylodan	H_2O	315	339	387	433	457	471	502	536	552	577	115	12.9	0.18	2.3	Naphthalene	4.34
81	Isoquinoline 3f	PBS pH 7.4			405				519				114	15.4	0.39	6.0	Isoquinoline	4.37
82	Isoquinoline 4b	PBS pH 7.4			436				550				114	15.5	0.36	5.5	Isoquinoline	4.37
83	5-Methylsalicylic acid	H_2O pH 7	268	281	305	323	379	388	418	456	477	497	113	3.0	0.16	0.4	Benzene	4.38
84	7-Azetidinyl-4-Me-coumarin	10 mM HEPES pH 7.3	308	325	354	377	432	441	467	504	525	551	113	15.0	0.96	14.4	Coumarin 7-Am	4.26.2
85	SeTau-405-carboxylate	P pH 7.4			405				518				113	13.8	0.80	11.0	nd	4.41
86	Isoquinoline 3l	PBS pH 7.4			410				522				112	16.8	0.42	7.0	Isoquinoline	4.37
87	Isoquinoline 3k	PBS pH 7.4			411				523				112	18.6	0.11	2.0	Isoquinoline	4.37
88	Isoquinoline 3o	PBS pH 7.4			410				521				111	19.2	0.09	1.7	Isoquinoline	4.37
89	Lucifer yellow CH	H_2O pH 7	375	392	429	462	498	509	540	580	600	623	111	11.3	0.21	2.3	Naphthalimide	4.32
90	Chromenoquinoline L-S 4e	H_2O			502				613				111	33.16	0.07	2.2	Hybrid-Fused	4.40.1
91	Isoquinoline 3n	PBS pH 7.4			411				521				110	16.6	0.41	6.8	Isoquinoline	4.37
92	Alexa Fluor 430	PBS pH 7.4	373	393	431	456	508	517	541	584	613	645	110	16.0	0.55	8.8	Coumarin 7-Am	4.26.3
93	Abberior Star 520SXP	PBS pH 7.4	454	475	522	555	590	601	632	691	733	779	110	42.5	0.15	6.3	Coumarin 7-Am	4.26.3
94	Atto 430LS carboxylate	PBS pH 7.4, 22 °C			436				545				109	32.0	0.65	20.8	nd (pb Coum.)	4.41
95	Isoquinoline 3g	PBS pH 7.4			414				522				108	17.5	0.41	7.1	Isoquinoline	4.37
96	Chromenoquinoline L-S 4d	H_2O			500				608				108	36.06	0.08	2.8	Hybrid-Fused	4.40.1
97	Salicylic acid	H_2O pH 7	272	281	300	316	370	377	407	441	461	482	107	3.0	0.16	0.4	Benzene	4.38
98	4-Methylsalicylic acid	H_2O pH 7	252	272	300	317	370	379	407	446	467	491	107	3.0	0.11	0.3	Benzene	4.38
99	AI3	H_2O pH 7.4	387	405	445	481	508	520	552	594	615	639	107	15.0	0.21	3.1	Naphthalimide	4.32
100	8-Trifluoromethyl-BODIPY 12	100 mM P pH 7.4	439	456	477	485	568	574	584	601	624	645	107	41.8	0.70	29.2	Bodipy-Small	4.14.1

(Continued)

Ent.	Short Names	Aqueous Solutions	$\lambda_{x,-0.25}$ nm	$\lambda_{x,-0.50}$ nm	$\lambda_{x,max}$ nm	$\lambda_{x,+0.50}$ nm	$\lambda_{e,-0.25}$ nm	$\lambda_{e,-0.50}$ nm	$\lambda_{e,max}$ nm	$\lambda_{e,+0.50}$ nm	$\lambda_{e,+0.25}$ nm	$\lambda_{e,+0.10}$ nm	Δ_S	ε_{max} mM^{-1}cm^{-1}	Φ_f	B_{max} mM^{-1}cm^{-1}	Class	Ref.
101	Coumarin P-R 13	PBS pH 7.5	319	337	368	393	438	447	474	512	534	560	106	15.0	0.63	9.4	Coumarin 7-OH	4.26.1
102	BC-H	100 mM P, 20% Et pH 7.4			365				470				105	27.2	0.09	2.4	Hybrid-Bonded	4.40.2
103	Luminarine	H_2O pH 7			428				533				105	9.0	0.47	4.2	Pteridine	4.36
104	Lucifer yellow VS	H_2O pH 7	376	393	430	463	493	505	535	574	594	616	105	11.3	0.24	2.7	Naphthalimide	4.32
105	Quinine	100 mM $HClO_4$			347				451				104	5.7	0.60	3.4	Quinoline	4.29
106	Quinine	50 mM H_2SO_4			347				451				104	5.7	0.55	3.1	Quinoline	4.29
107	Seta-375-carboxylate	P pH 7.4			372				476				104	24.5	0.86	21.0	nd	4.41
108	Stefani HB-9	H_2O pH 6			320				423				103	35.9	0.12	4.2	Hybrid-Bonded	4.40.2
109	Tri-*O*-acetyl-P-luminoarosin	H_2O pH 7			425				528				103	10.8	0.62	6.7	Pteridine	4.36
110	Epicocconone P-F 4b	H_2O, 0.1% SDS pH 2.5			435				538				103	19.8	0.20	3.9	Epicocconone	4.39.2
111	3-Bt-6-MeO-7-hydroxychromone	100 mM Tris pH 8			345				447				102	16.4	0.48	7.8	Chromone	4.31
112	Coumarin P-R 3	PBS pH 7.5	297	319	348	372	415	424	450	483	501	523	102	12.5	0.82	10.2	Coumarin 7-Am	4.26.2
113	BC-BA	100 mM P, 20% Et pH 7.4			371				473				102	25.5	0.17	4.3	Hybrid-Bonded	4.40.2
114	Spirobifluorene 1	H_2O			378				480				102	51	0.50	25.6	Spirobifluorene	4.39.3
115	Isoquinoline 3d	PBS pH 7.4			418				520				102	14.5	0.43	6.2	Isoquinoline	4.37
116	Chromenoquinoline L-S 4b	H_2O			477				579				102	36.05	0.15	5.2	Hybrid-Fused	4.40.1
117	7-Am-4-Me-2-hydroxyquinoline	H_2O pH 5-9			340				440				100	16.0	0.97	15.5	Quinoline	4.29
118	Acridizinium J-N 1e	H_2O			355				455				100	6.7	0.20	1.3	Acridizinium	4.39.1
119	3-OH-4-nitroso-2,7-naphth-diSA	A pH 4	304	326	370	412	431	442	470	500	514	535	100	19.0	0.22	4.1	Naphthalene	4.34
120	3-OH-4-nitroso-2,7-naphth-diSA	T pH 9	339	367	423	478	475	489	523	560	578	605	100	20.0	0.28	5.6	Naphthalene	4.34
121	Alexa Fluor 350	H_2O pH 7	296	317	346	370	411	420	445	477	495	516	99	19.0	0.89	16.9	Coumarin 7-Am	4.26.3
122	3-AB-6-MeO-7-OH-chromone	100 mM Tris pH 8			346				445				99	16.3	0.41	6.6	Chromone	4.31

Ent.	Short Names	Aqueous Solutions	$\lambda_{x,-0.25}$ nm	$\lambda_{x,-0.50}$ nm	$\lambda_{x,max}$ nm	$\lambda_{x,+0.50}$ nm	$\lambda_{e,-0.25}$ nm	$\lambda_{e,-0.50}$ nm	$\lambda_{e,max}$ nm	$\lambda_{e,+0.50}$ nm	$\lambda_{e,+0.25}$ nm	$\lambda_{e,+0.10}$ nm	Δ_S	ε_{max} $mM^{-1}cm^{-1}$	Φ_f	B_{max} $mM^{-1}cm^{-1}$	Class	Ref.
123	Coumarin 120	H_2O	293	314	342	366	407	415	440	471	489	510	98	16.0	0.94	15.0	Coumarin 7-Am	4.26.2
124	7-Dimethylamino-4-Me-coumarin	10 mм HEPES pH 7.3	322	340	372	397	434	443	470	507	529	555	98	18.0	0.19	3.4	Coumarin 7-Am	4.26.2
125	Coumarin P-R 22	PBS pH 7.5	307	324	353	376	417	425	450	484	504	527	97	13.4	0.75	10.0	Coumarin 7-Am	4.26.2
126	Coumarin P-R 6	PBS pH 7.5	309	326	355	378	419	427	452	486	507	530	97	12.6	0.71	8.9	Coumarin 7-Am	4.26.2
127	Marina Blue	100 mм P pH 10/8	311	328	358	381	421	430	455	490	510	534	97	18.7	0.89	16.6	Coumarin 7-OH	4.26.1
128	Naphtahlimide 10b	H_2O	389	408	448	484	502	514	545	586	606	630	97	12.4	0.21	2.6	Naphthalimide	4.32
129	Chromenoquinoline L-S 4a	H_2O			473				570				97	36.19	0.15	5.4	Hybrid-Fused	4.40.1
130	Coumarin 102	H_2O	336	356	393	418	455	464	489	528	552	579	96	21.8	0.66	14.3	Coumarin 7-Am	4.26.2
131	7-Hydroxy-4-Me-coumarin-3-CA	H_2O pH 5	313	330	360	384	421	430	455	490	510	534	95	18.0	0.82	14.7	Coumarin 7-OH	4.26.1
132	Carbostyril 23b	H2O pH 2.8			380				475				95	18.0	0.41	7.3	Quinoline	4.29
133	COUPY 11	H_2O			567				662				95	13.0	0.02	0.2	Hybrid-Bridged	4.40.3
134	nor-Harmane (β-carboline)	50 mм H_2SO_4	333	340	360	381	418	425	454	492	514	538	94	6.3	0.58	3.6	Indole	4.33
135	Coumarin 2	H_2O	314	332	362	386	422	431	456	491	512	535	94	16.9	0.95	16.0	Coumarin 7-Am	4.26.2
136	3-Amino-9-ethylcarbazole 2	H_2O			364				458				94	1.8	0.76	1.3	Carbazole	4.39.1
137	Harmine	H_2O pH 9	309	315	332	350	393	400	425	458	477	498	93	11.0	0.38	4.1	Indole	4.33
138	DMPT	H_2O pH 7			382				475				93	12.4	0.06	0.7	Pteridine	4.36
139	Phosphorilated coum. N-H 22c	PBS pH 7.4	348	370	410	437	467	477	503	545	570	599	93	14.0	0.44	6.1	Coumarin 7-Am	4.26.3
140	3,6,10-Trimethyliso-alloxazine	H_2O pH 7			439				532				93	8.5	0.08	0.6	Pteridine	4.36
141	Chromenoquinoline L-S 4f	H_2O			529				622				93	40.64	0.05	1.9	Hybrid-Fused	4.40.1
142	7-(4-Hydroxy)cyclohexyl coum.	H_2O	319	337	368	393	426	434	460	496	517	541	92	16.5	0.59	9.7	Coumarin 7-Am	4.26.2
143	Harmaline	5 mм H_2SO_4	352	360	383	407	436	444	475	517	541	568	92	28.6	0.32	9.1	Indole	4.33
144	DPP 7	PBS			456				548				92	14.7	0.76	11.1	Diketopyrrolop	4.39.1
145	DY-510XL carboxylate	H_2O	432	451	493	522	549	558	585	635	670	709	92	40.0		0.0	Coumarin 7-Am	4.26.3

(Continued)

Ent.	Short Names	Aqueous Solutions	$\lambda_{x,-0.25}$ nm	$\lambda_{x,-0.50}$ nm	$\lambda_{x,max}$ nm	$\lambda_{x,+0.50}$ nm	$\lambda_{e,-0.25}$ nm	$\lambda_{e,-0.50}$ nm	$\lambda_{e,max}$ nm	$\lambda_{e,+0.50}$ nm	$\lambda_{e,+0.25}$ nm	$\lambda_{e,+0.10}$ nm	Δ_S	ε_{max} mм$^{-1}$cm^{-1}	Φ_f	B_{max} mм$^{-1}$cm^{-1}	Class	Ref.
146	Epicocconone	H_2O, 0.1% SDS pH 2.5			444				535				91	68.6	0.07	5.0	Epicocconone	4.39.2
147	Chromenoquinoline L-S 4c	H_2O			500				591				91	33.16	0.03	1.1	Hybrid-Fused	4.40.1
148	3-Aminopyridine	H_2O			300				390				90	2.8	0.32	0.9	Pyridine	4.39.1
149	Harmane	50 mм H_2SO_4	324	331	350	370	405	412	440	474	494	517	90	6.3	0.83	5.1	Indole	4.33
150	7-Ethylamino-coumarin	H_2O	319	337	368	393	424	432	458	493	514	538	90	11.9	0.51	6.0	Coumarin 7-Am	4.26.2
151	NBQ	H_2O pH 2-5			380				470				90	35.0	0.35	12.2	Quinoline	4.29
152	Coumarin 1	H_2O	329	349	382	409	436	445	472	510	532	558	90	20.1	0.07	1.4	Coumarin 7-Am	4.26.2
153	HBQ	H_2O pH 2-5			390				480				90	35.0	0.15	5.2	Quinoline	4.29
154	Briliant sulfaflavine	H_2O	376	393	430	463	481	491	520	557	575	596	90	16.0	0.27	4.3	Naphthalimide	4.32
155	DPP 9	PBS			451				541				90	13.9	0.90	12.5	Diketopyrrolop	4.39.1
156	Naphtahlimide 10c	H_2O	393	412	453	490	500	512	543	583	603	627	90	17.3	0.30	5.1	Naphthalimide	4.32
157	7-Hydroxy-4-methylcoumarin	H_2O pH 10	313	330	360	384	416	424	449	483	503	526	89	18.0	0.86	15.4	Coumarin 7-OH	4.26.1
158	6-Azetidinyl naphthalimide	10 mм HEPES pH 7.3	401	421	464	503	509	521	553	595	616	640	89	18.0	0.28	5.0	Naphthalimide	4.32
159	3-Amino-9-ethylcarbazole 1	H_2O			364				452				88	1.8	0.63	1.1	Carbazole	4.39.1
160	Coumarin P-R 14	PBS pH 7.5	322	341	373	398	426	435	461	497	518	542	88	17.7	0.62	10.9	Coumarin 7-OH	4.26.1
161	CF450	H_2O			450				538				88	40.0	0.56	22.3	nd (Cyanine)	4.41
162	COUPY 12	H_2O			595				683				88	6.9	0.05	0.3	Hybrid-Bridged	4.40.3
163	Carbostyril 5	H_2O			353				440				87	6.6	0.16	1.0	Quinoline	4.29
164	7-Diethylamino-coumarin	H_2O	333	354	388	416	438	448	475	513	536	562	87	18.3	0.05	0.9	Coumarin 7-Am	4.26.2
165	Harmine	50 mм H_2SO_4	304	309	326	343	383	389	412	445	464	483	86	11.9	0.45	5.3	Indole	4.33
166	Atto 390 carboxylate	PBS pH 7.4, 22 °C	343	357	390	415	441	450	476	516	540	568	86	24.0	0.90	21.6	Coumarin 7-Am	4.26.3
167	3,7,10-Trimethyliso-alloxazine	H_2O pH 7			446				532				86	9.8	0.32	3.1	Pteridine	4.36
168	SeTau-380-carboxylate	P pH 7.3			383				468				85	23.8	0.56	13.3	nd	4.41
169	NOBQ	H_2O pH 2-5			385				470				85	35.0	0.41	14.3	Quinoline	4.29
170	OBQ	H_2O pH 2-5			385				470				85	35.0	0.43	15.0	Quinoline	4.29

Ent.	Short Names	Aqueous Solutions	$\lambda_{x,-0.25}$ nm	$\lambda_{x,-0.50}$ nm	$\lambda_{x,max}$ nm	$\lambda_{x,+0.50}$ nm	$\lambda_{e,-0.25}$ nm	$\lambda_{e,-0.50}$ nm	$\lambda_{e,max}$ nm	$\lambda_{e,+0.50}$ nm	$\lambda_{e,+0.25}$ nm	$\lambda_{e,+0.10}$ nm	Δ_S	ε_{max} mM^{-1}cm^{-1}	Φ_f	B_{max} mM^{-1}cm^{-1}	Class	Ref.
171	3,8,10-Trimethyliso-alloxazine	H_2O pH 7			433				518				85	11.5	0.12	1.3	Pteridine	4.36
172	3,10-Dimethylisoalloxazine	H_2O pH 7			435				520				85	9.5	0.15	1.4	Pteridine	4.36
173	3,7,8,10-TetraMe-isoalloxazine	H_2O pH 7			444				529				85	12.4	0.25	3.1	Pteridine	4.36
174	2-Methyl harmine	50 mM H_2SO_4	324	331	350	370	399	406	433	466	485	509	83	6.3	0.45	2.8	Indole	4.33
175	7-Azetidinyl-3-CA coumarin	10 mM HEPES pH 7.3	333	353	387	414	434	443	470	507	529	555	83	24.0	0.84	20.1	Coumarin 7-Am	4.26.2
176	LS482	Buffer pH 8	632	682	710	731	760	771	793	819	830	845	83	84.0	0.10	8.4	Cyanine-Hepta	4.13.3
177	10-Ethylacridine	P pH 7	376	390	412	423	467	474	494	534	560	589	82	3.5	0.98	3.4	Acridine	4.23.1
178	8-Aminopyrene-1,3,6-triSA	H_2O	363	380	424	433	482	487	505	552	581	607	81	20.6	0.90	18.5	Pyrene	4.28
179	Phosphorilated coum. N-H 21c	PBS pH 7.4	374	394	432	457	482	490	512	550	576	604	80	20.4	0.81	16.5	Coumarin 7-Am	4.26.3
180	Flavin mononucleotide (FMN)	H_2O			450				530				80	12.2	0.25	3.0	Pteridine	4.36
181	DY-350 carboxylate	PBS pH 7.4	301	323	353	378	400	408	432	462	479	499	79	19.0	0.84	15.9	Coumarin 7-Am	4.26.3
182	Abberior Star 440SXP	PBS pH 7.4	377	398	436	461	485	493	515	554	580	608	79	22.7	0.68	15.4	Coumarin 7-Am	4.26.3
183	Sulfonated coumarin N-H 6-H	PBS pH 7.4	377	398	436	461	485	493	515	554	580	608	79	31.0	0.57	17.6	Coumarin 7-Am	4.26.3
184	DY-485XL carboxylate	H_2O	424	442	482	510	527	535	560	605	638	672	78	48.0		0.0	Coumarin 7-Am	4.26.3
185	4,10-Dibromonaphtho-fluorescein	C pH 9.5	549	578	605	624	646	655	682	728	768	811	77	32.0	0.05	1.6	Fluores-analog	4.8.4
186	Carbostyril 124	H_2O			341				417				76	15.8	0.46	7.2	Quinoline	4.29
187	Isoquinoline 4c	PBS pH 7.4			453				529				76	22.1	0.24	5.3	Isoquinoline	4.37
188	Carbostyril 1	H_2O			358				432				74	11.6	0.24	2.7	Quinoline	4.29
189	8-Trifluoromethyl-BODIPY 10	100 mM P pH 7.4	432	449	469	476	529	533	542	557	576	594	73	17.1	0.74	12.6	Bodipy-Small	4.14.1
190	PONy CP1	PBS pH 7.4	640	681	710	728	737	754	783	817	843	870	73	17.0	0.19	3.2	C(CH3)2-Pyron	4.5
191	CF430	H_2O			426				498				72	40.0	0.79	31.6	nd (Cyanine)	4.41
192	DPP 10	PBS			523				594				71	17.1	0.61	10.4	Diketopyrrolop	4.39.1

(Continued)

Ent.	Short Names	Aqueous Solutions	$\lambda_{x,-0.25}$ nm	$\lambda_{x,-0.50}$ nm	$\lambda_{x,max}$ nm	$\lambda_{x,+0.50}$ nm	$\lambda_{e,-0.25}$ nm	$\lambda_{e,-0.50}$ nm	$\lambda_{e,max}$ nm	$\lambda_{e,+0.50}$ nm	$\lambda_{e,+0.25}$ nm	$\lambda_{e,+0.10}$ nm	Δ_S	ε_{max} mM^{-1}cm^{-1}	Φ_f	B_{max} mM^{-1}cm^{-1}	Class	Ref.
193	Carbostyril 2	H_2O			360				430				70	6.7	0.14	0.9	Quinoline	4.29
194	Safranin T	H_2O			517				586				69	27.0	0.11	2.9	Phenazine	4.39.1
195	COUPY 10	H_2O			562				631				69	33.0	0.03	0.8	Hybrid-Bridged	4.40.3
196	DPP 8	PBS			527				594				67	18.1	0.56	10.1	Diketopyrrolop	4.39.1
197	Acridizinium J-N 1c	H_2O			354				420				66	8.4	0.30	2.5	Acridizinium	4.39.1
198	6-Chloro-harmine	H_2O pH 4	335	341	362	383	396	402	428	462	481	502	66	9.0	0.16	1.4	Indole	4.33
199	3,6-Diaminoacridine (Proflavine)	H_2O pH 7	404	420	445	457	482	490	511	554	582	614	66	38.9	0.34	13.2	Acridine	4.23.1
200	Naphthofluorescein	C pH 9.5	541	568	595	613	626	635	660	703	740	780	65	44.0	0.14	6.1	Fluores-analog	4.8.4
201	2-Aminopyridine	H_2O			300				363				63	4.1	0.71	2.9	Pyridine	4.39.1
202	8-Chloro-harmine	H_2O pH 4	334	341	361	382	392	398	423	456	474	495	62	5.9	0.11	0.6	Indole	4.33
203	COUPY 9	H_2O			543				605				62	31.0	0.15	4.6	Hybrid-Bridged	4.40.3
204	7-Diethylamino-3-CA coumarin	10 mM HEPES pH 7.3	357	376	410	432	446	452	471	503	525	548	61	35.0	0.03	1.0	Coumarin 7-Am	4.26.2
205	6,8-Dichloroharmine (6,8-DiCL-Ha)	H_2O pH 4	341	348	369	391	397	403	429	463	482	503	60	6.3	0.20	1.2	Indole	4.33
206	Benzo[h]quinoline (BQ)	H_2O pH 2-5			370				430				60	35.0	0.32	11.2	Quinoline	4.29
207	7-Hydroxy-4-Me-coumarin-3-CA	H_2O pH 10	332	352	386	413	414	422	446	479	499	522	60	33.0	0.85	28.0	Coumarin 7-OH	4.26.1
208	DPP 5	PBS			455				515				60	22.7	0.57	12.9	Diketopyrrolop	4.39.1
209	PONy A1	PBS pH 7.4	476	499	535	553	556	566	595	654	693	739	60	38.0	0.10	3.8	Acridine	4.23.1
210	Harmine	H_2O pH 4	330	337	357	378	387	393	416	449	465	484	59	7.3	0.49	3.6	Indole	4.33
211	SNARF-1	P pH 10	519	543	573	590	594	606	631	667	703	744	58	44.0	0.09	4.0	Rhodol-analog	4.15.3
212	DPP 6	PBS			530				586				56	30.7	0.49	15.0	Diketopyrrolop	4.39.1
213	Vita Blue	C pH 9.5	560	586	609	625	636	643	665	731	758	793	56	36.0	0.15	5.4	S-Fluore-analog	4.10
214	DY-431 carboxylate	PBS pH 7.4	388	405	442	467	470	477	496	529	550	573	54	35.0	0.75	26.2	Coumarin 7-Am	4.26.3
215	8-Hydroxypyrene-1,3, 6-triSA	H_2O pH 10	389	408	460	470	490	495	514	563	593	620	54	24.0	0.95	22.8	Pyrene	4.28
216	Distyryl-BODIPY 11	PBS pH 7.5	546	572	601	617	626	632	655	703	739	784	54	17.9	0.04	0.7	Bodipy-Med.	4.14.2

Ent.	Short Names	Aqueous Solutions	$\lambda_{x,-0.25}$ nm	$\lambda_{x,-0.50}$ nm	$\lambda_{x,max}$ nm	$\lambda_{x,+0.50}$ nm	$\lambda_{e,-0.25}$ nm	$\lambda_{e,-0.50}$ nm	$\lambda_{e,max}$ nm	$\lambda_{e,+0.50}$ nm	$\lambda_{e,+0.25}$ nm	$\lambda_{e,+0.10}$ nm	Δ_S	ε_{max} mм$^{-1}$cm^{-1}	Φ_f	B_{max} mм$^{-1}$cm^{-1}	Class	Ref.
217	Atto 465 carboxylate	PBS pH 7.4, 22 °C	404	423	453	467	471	480	506	551	579	611	53	75.0	0.75	56.2	Acridine	4.23.1
218	NCR	H_2O			565				618				53	25.8	0.00	0.0	Hybrid-Fused	4.40.1
219	Acridizinium J-N 1d	H_2O			356				408				52	5.4	0.20	1.0	Acridizinium	4.39.1
220	Quinine	H_2O pH 7			331				382				51	5.7	0.50	2.8	Quinoline	4.29
221	Seta-405-carboxylate	P pH 7.4			401				452				51	30.0	0.70	21.0	nd	4.41
222	ECXi	H_2O			887				938				51	93.3	0.01	1.3	Pentacene	4.39.2
223	*meso*-CaboxyBODIPY J-K 1	10 mм PBS pH 7.4	463	482	505	514	540	545	554	569	589	608	49	62.0	0.78	48.3	Bodipy-Small	4.14.1
224	Stefani HB-6	H_2O pH 6			317				365				48	35.9	0.03	0.9	Hybrid-Bonded	4.40.2
225	Atto 425 carboxylate	PBS pH 7.4, 22 °C	395	409	439	460	460	466	485	521	548	576	46	45.0	0.90	40.5	Coumarin 7-Am	4.26.3
226	BC-OH	100 mм P, 20% Et pH 7.4			442				488				46	38.1	0.85	32.3	Hybrid-Bonded	4.40.2
227	Pacific Blue	PBS pH 7.2, 22 °C	339	366	410	437	428	436	455	486	505	526	45	29.5	0.75	22.1	Coumarin 7-OH	4.26.1
228	Cyan 500 carboxylate	PBS pH 7.4, 22 °C			450				495				45	45.0	0.85	38.2	nd (pb Coum.)	4.41
229	CF405M	H_2O			408				452				44	41.0	0.76	31.1	nd (Coumarin)	4.41
230	PONy P5	PBS pH 7.4	592	627	660	679	673	682	704	735	755	799	44	38.0	0.10	3.8	Pyronine	4.4
231	DY-430 carboxylate	PBS pH 7.4	394	410	444	467	462	468	487	521	545	570	43	37.0	0.72	26.6	Coumarin 7-Am	4.26.3
232	Carbostyril 9	H_2O			340				382				42	6.0	0.17	0.9	Quinoline	4.29
233	2-OH-6-DiEtAm-benzo[c]-xanthen	H_2O	523	547	578	595	584	596	620	654	689	728	42	38.1	0.18	6.8	Rhodol-analog	4.15.3
234	DMAC-SR101	PBS			605				647				42	80.0	0.09	7.2	Hybrid-Fused	4.40.1
235	SO_2R5	10 mм PBS, 0.25% ACN pH 7.4	639	684	710	732	715	726	752	790	817	848	42	129.0	0.07	8.7	SO2-Rosamine	4.18.3
236	Bpe-H	H_2O pH 7.4	567	601	640	657	657	665	681	706	726	769	41	48.1	0.23	11.0	Phenoxazine	4.25
237	8-Hydroxypyrene-1,3, 6-triSA	H_2O pH 2	348	363	403	411	425	429	443	479	501	519	40	20.0	0.95	19.0	Pyrene	4.28
238	Acridine orange	H_2O	439	458	488	503	497	505	528	574	604	639	40	50.0	0.12	6.0	Acridine	4.23.1
239	Cresyl violet (CV)	H_2O	523	553	585	599	605	611	625	646	663	698	40	33.1	0.36	11.9	Phenoxazine	4.25

(Continued)

Ent.	Short Names	Aqueous Solutions	$\lambda_{x,-0.25}$ nm	$\lambda_{x,-0.50}$ nm	$\lambda_{x,max}$ nm	$\lambda_{x,+0.50}$ nm	$\lambda_{e,-0.25}$ nm	$\lambda_{e,-0.50}$ nm	$\lambda_{e,max}$ nm	$\lambda_{e,+0.50}$ nm	$\lambda_{e,+0.25}$ nm	$\lambda_{e,+0.10}$ nm	Δ_S	ε_{max} mм$^{-1}$cm^{-1}	Φ_f	B_{max} mм$^{-1}$cm^{-1}	Class	Ref.
240	PONy P2	PBS pH 7.4	569	602	632	649	644	652	672	700	718	758	40	51.0	0.16	8.1	Pyronine	4.4
241	SO_2R3	10 mм PBS, 0.25% ACN pH 7.4	636	681	707	728	711	721	747	784	811	842	40	69	0.05	3.4	SO2-Rosamine	4.18.3
242	3-Methoxymethyl ether rhodol	50 mм P, <0.1% ACN pH 8	439	457	477	488	492	500	516	539	563	587	39	58.0	0.85	49.3	Rhodol	4.15.2
243	3,6-Di(azetidin-1-yl)-acridine	10 mм HEPES pH 7.3	442	461	492	507	500	508	531	578	608	643	39	47.0	0.52	24.4	Acridine	4.23.1
244	*meso*-CaboxyBODIPY J-K 3a	10 mм PBS pH 7.4	477	497	519	531	537	541	558	592	618	649	39	77.0	0.67	51.5	Bodipy-Small	4.14.1
245	NCRMe	H_2O			568				607				39	22.8	0.01	0.1	Hybrid-Fused	4.40.1
246	SR101-NaphtOH	PBS	481	509	580	607	580	588	619	676	700	735	39	80.0	0.15	12.0	Sulforhodol-an	4.27
247	PONy P10	PBS pH 7.4	567	599	629	646	640	648	668	696	714	753	39	52.0	0.12	6.2	Pyronine	4.4
248	Bzm-H	H_2O pH 7.4	563	597	635	652	651	658	674	698	718	760	39	29.9	0.14	4.1	Phenoxazine	4.25
249	SO_2R2	10 mм PBS, 0.25% ACN pH 7.4	633	678	703	724	706	717	742	779	805	836	39	95	0.07	6.9	SO2-Rosamine	4.18.3
250	JF479	10 mм HEPES pH 7.3	432	450	479	493	488	495	517	561	589	623	38	47.9	0.62	29.7	Acridine	4.23.1
251	3-MOMO-6-propylamino-rhodol	50 mм P, <0.1% ACN pH 8	453	472	494	506	507	515	532	557	582	608	38	43.0	0.35	15.0	Rhodol	4.15.2
252	Pip-rosol	50 mм P pH 7.4	479	500	525	538	537	544	563	599	630	669	38	42.5	0.05	2.0	Rosol	4.19
253	SO_2R4	10 mм PBS, 0.25% ACN pH 7.4	634	679	704	725	706	717	742	779	805	836	38	97.0	0.10	9.4	SO2-Rosamine	4.18.3
254	Me-rosol	50 mм P pH 7.4	475	496	520	533	536	542	557	591	624	653	37	46.1	0.18	8.4	Rosol	4.19
255	SR101-NaphțNH_2	PBS	493	526	589	610	593	602	626	666	691	722	37	80.0	0.19	15.2	Srhod.-analog	4.6
256	PONy P1	PBS pH 7.4	535	564	590	605	602	610	627	652	667	701	37	59.0	0.33	19.4	Pyronine	4.4
257	PONy P9	PBS pH 7.4	541	570	597	613	608	615	633	658	674	709	36	66.0	0.22	14.5	Pyronine	4.4
258	PONy P13	PBS pH 7.4	568	600	630	647	638	646	666	694	712	750	36	52.0	0.17	8.8	Pyronine	4.4
259	6-Benzamidorhodol	50 mм P, <0.1% ACN pH 8	450	469	490	502	501	508	525	549	573	599	35	30.1	0.16	4.8	Rhodol	4.15.2

Ent.	Short Names	Aqueous Solutions	$\lambda_{x,-0.25}$ nm	$\lambda_{x,-0.50}$ nm	$\lambda_{x,max}$ nm	$\lambda_{x,+0.50}$ nm	$\lambda_{e,-0.25}$ nm	$\lambda_{e,-0.50}$ nm	$\lambda_{e,max}$ nm	$\lambda_{e,+0.50}$ nm	$\lambda_{e,+0.25}$ nm	$\lambda_{e,+0.10}$ nm	Δ_S	ε_{max} mм$^{-1}$cm^{-1}	Φ_f	B_{max} mм$^{-1}$cm^{-1}	Class	Ref.
260	Acridine orange	HEPES pH 7.3	443	462	493	508	497	505	528	574	604	639	35	50.0	0.21	10.5	Acridine	4.23.1
261	PONy P6	PBS pH 7.4	546	576	603	619	613	620	638	663	680	715	35	69.0	0.09	6.2	Pyronine	4.4
262	NIR1	H_2O	672	729	761	785	763	773	796	822	834	849	35	268.0	0.23	61.6	Cyanine-Hepta	4.13.3
263	IRDye78-CA	10 mм HEPES pH 7.4	680	738	771	796	772	783	806	832	845	860	35	145.7	0.14	20.6	Cyanine-Hepta	4.13.3
264	6-Acrylamidorhodol	50 mм P, <0.1% ACN pH 8	450	469	490	502	500	507	524	548	572	598	34	28.5	0.19	5.4	Rhodol	4.15.2
265	Et-rosol	50 mм P pH 7.4	478	500	524	537	538	544	558	589	622	649	34	42.6	0.10	4.1	Rosol	4.19
266	Acridine red	H_2O	484	507	528	540	542	548	562	582	594	621	34	100.0	0.51	51.0	Pyronine	4.4
267	Jul-rosol	50 mм P pH 7.4	492	514	540	554	548	554	574	614	648	686	34	30.4	0.41	12.5	Rosol	4.19
268	Alexa Fluor 546	PBS pH 7.2, 22 °C	506	537	546	567	553	558	580	592	621	648	34	112.0	0.79	88.4	Rhodam Hi Ch	4.2.5
269	DY-635 carboxylate	PBS pH 7.4	550	583	635	658	630	641	669	729	766	808	34	175.0	0.05	8.7	Merocyanine	4.24
270	Bza-OH	H_2O pH 7.4	551	584	620	636	631	638	653	676	694	733	33	12.8	0.37	4.7	Phenoxazine	4.25
271	SO_2R1	10 mм PBS, 0.25% ACN pH 7.4	633	678	703	724	701	711	736	772	798	828	33	39	0.07	2.6	SO2-Rosamine	4.18.3
272	Mercaptofluorescein MK-61	1 mм NaOH	457	477	495	508	505	511	527	554	577	601	32	38.0	0.24	9.1	Fluorescein-SH	4.8.3
273	CF514	H_2O			516				548				32	105.0	0.38	39.8	nd (Rhodam.)	4.41
274	6-Piperidinorhodol	50 mм P, <0.1% ACN pH 8	478	499	523	536	528	536	555	582	609	638	32	62.5	0.10	6.2	Rhodol	4.15.2
275	1,2,7,8-Dibenzo-4,5-Br-fluorescein	C pH 9.5	503	526	549	565	554	562	581	614	642	672	32	24.0	0.05	1.2	Fluorescein-ha	4.8.2
276	HMDSeR	100 mм P pH 7.4	516	540	558	571	568	574	590	631	656	680	32	92.0	0.01	0.5	Se-Rhodol	4.16
277	PONy P1-Halo	PBS pH 7.4	546	576	603	619	610	617	635	660	676	711	32	18.0	0.26	4.6	Pyronine	4.4
278	DEAC-SR101	PBS			610				642				32	80.0	0.10	8.0	Hybrid-Fused	4.40.1
279	Stefani HB-11	H_2O pH 6			329				360				31	38.1	0.11	4.0	Hybrid-Bonded	4.40.2
280	Mercaptofluorescein MK-43	1 mм NaOH	457	477	495	508	504	510	526	553	576	600	31	34.0	0.22	7.4	Fluorescein-SH	4.8.3
281	Mercaptofluorescein MK-69	1 mм NaOH	457	477	495	508	504	510	526	553	576	600	31	25.0	0.20	5.0	Fluorescein-SH	4.8.3

(Continued)

Ent.	Short Names	Aqueous Solutions	$\lambda_{x,-0.25}$ nm	$\lambda_{x,-0.50}$ nm	$\lambda_{x,max}$ nm	$\lambda_{x,+0.50}$ nm	$\lambda_{e,-0.25}$ nm	$\lambda_{e,-0.50}$ nm	$\lambda_{e,max}$ nm	$\lambda_{e,+0.50}$ nm	$\lambda_{e,+0.25}$ nm	$\lambda_{e,+0.10}$ nm	Δ_S	ε_{max} mM^{-1}cm^{-1}	Φ_f	B_{max} mM^{-1}cm^{-1}	Class	Ref.
282	JF502	10 mM HEPES pH 7.3	450	470	502	518	502	510	533	580	610	646	31	57.8	0.71	41.0	Acridine	4.23.1
283	Mercaptofluorescein MK-75	1 mM NaOH	463	483	502	515	511	517	533	561	584	609	31	12.0	0.16	1.9	Fluorescein-SH	4.8.3
284	6-Pyrrolidinorhodol	50 mM P, <0.1% ACN pH 8	474	495	519	532	523	532	550	577	603	632	31	63.0	0.12	7.5	Rhodol	4.15.2
285	6-Morpholinorhodol	50 mM P, <0.1% ACN pH 8	477	498	522	535	526	534	553	580	607	636	31	61.0	0.13	7.9	Rhodol	4.15.2
286	CF532	H_2O			527				558				31	96.0	0.73	69.8	nd (Rhodam.)	4.41
287	1,2,7,8-Dibenzofluorescein	C pH 9.5	490	513	534	549	540	547	565	596	623	651	31	81.0	0.24	19.4	Fluorescein-nh	4.8.1
288	2-PrAc-6-quinolinorhodol	50 mM P pH 8-9	493	515	541	555	543	552	572	601	630	661	31	58.2	0.49	28.5	Rhodol	4.15.1
289	2-Ethyl-6-quinolino-carboxyrhodol	50 mM P pH 8-9	494	517	543	557	545	554	574	603	632	664	31	79.4	0.49	38.9	Rhodol	4.15.1
290	Bzh-Cl	H_2O pH 7.4	551	584	620	636	629	636	651	674	692	731	31	36.1	0.31	11.1	Phenoxazine	4.25
291	Bzh-H	H_2O pH 7.4	551	584	620	636	629	636	651	674	692	731	31	11.3	0.44	4.9	Phenoxazine	4.25
292	MetACSO3H	H_2O pH 5.5	364	376	403	411	418	423	433	470	487	495	30	5.6	0.76	4.2	Acridones	4.23.2
293	*meso*-CaboxyBODIPY J-K 3b	10 mM PBS pH 7.4	477	497	519	531	528	532	549	582	607	637	30	78.0	0.83	64.7	Bodipy-Small	4.14.1
294	6-Quinolinocarboxyrhodol	50 mM P pH 8-9	492	514	540	554	541	550	570	599	627	658	30	66.8	0.43	28.7	Rhodol	4.15.1
295	TAMRA	H_2O pH 7	501	525	546	561	554	560	576	602	624	651	30	90.0	0.20	18.0	Rhoda 500-550	4.2.2
296	DAOTA 2	H_2O	493	515	564	583	569	576	594	627	654	678	30	50.0	0.27	13.5	Triangulenium	4.22
297	Oxazine 170	H_2O pH 2-7	539	560	620	643	624	632	650	668	679	692	30	48.0	0.24	11.5	Phenoxazine	4.25
298	HC6H-SR101	PBS, 5% BSA			624				654				30	80.0	0.15	12.0	Hybrid-Fused	4.40.1
299	Y-A 3	PBS pH 7.2	684	744	777	802	773	784	807	834	846	861	30	215.0	0.03	6.4	Cyanine-Hepta	4.13.3
300	Seta-780-carboxylate	P pH 7.4			777				807				30	215.0	0.01	2.5	nd	4.41
301	Y-A 9	PBS pH 7.2	686	745	779	804	775	786	809	836	848	864	30	187.0	0.03	5.6	Cyanine-Hepta	4.13.3
302	ICG	PBS	687	746	780	805	776	787	810	837	849	865	30	211.2	0.08	16.9	Cyanine-Hepta	4.13.3
303	1,3,6,8-Pyrene tetrasulfonic acid	H_2O pH 7	335	350	374	378	391	394	403	430	451	465	29	51.0	0.56	28.5	Pyrene	4.28
304	NBDY	H_2O	462	481	504	513	520	524	533	547	566	583	29	74.4	0.33	24.5	Bodipy-Small	4.14.1

Ent.	Short Names	Aqueous Solutions	$\lambda_{x,-0.25}$ nm	$\lambda_{x,-0.50}$ nm	$\lambda_{x,max}$ nm	$\lambda_{x,+0.50}$ nm	$\lambda_{e,-0.25}$ nm	$\lambda_{e,-0.50}$ nm	$\lambda_{e,max}$ nm	$\lambda_{e,+0.50}$ nm	$\lambda_{e,+0.25}$ nm	$\lambda_{e,+0.10}$ nm	Δ_S	ε_{max} mм$^{-1}$cm^{-1}	Φ_f	B_{max} mм$^{-1}$cm^{-1}	Class	Ref.
305	Fluorinated rhodamine 1r	Buffer pH 7	483	506	525	538	533	539	554	578	598	623	29	31.0	0.28	8.6	Rhoda 500-550	4.2.2
306	DAOTA 1	H_2O	489	511	558	578	563	570	587	621	650	674	29	50.0	0.31	15.5	Triangulenium	4.22
307	DAOTA 3	H_2O	449	511	565	583	571	577	594	626	654	681	29	50.0	0.28	14.0	Triangulenium	4.22
308	570CPH	PBS pH 7.4	580	602	571	633	617	623	600	673	707	740	29	79.0	0.71	56.0	C(CH3)2-Rhoda	4.3.1
309	Atto 590 carboxylate	PBS pH 7.4, 22 °C	535	569	593	609	594	601	622	651	687	718	29	120.0	0.80	96.0	Rhoda 550-600	4.2.3
310	POF	PBS pH 7.4	561	591	632	652	635	643	661	687	714	750	29	54.5	0.24	13.0	POPh-Fluorone	4.12
311	Pyrodone	100 mм P pH 7.4	436	454	472	481	486	490	500	513	523	539	28	60.0	0.37	22.2	Pyrodone	4.21.2
312	2,6-Sulfon.-1,3,5,7,8-Me-BODIPY	H_2O	451	469	491	499	507	511	519	532	550	567	28	99.0	0.83	82.1	Bodipy-Small	4.14.1
313	Atto 495 carboxylate	PBS pH 7.4, 22 °C	455	471	498	512	503	509	526	569	597	631	28	80.0	0.20	16.0	Acridine	4.23.1
314	4,7-DiMe-6-EtAm-carboxyrhodol	50 mм P pH 8-9	472	492	516	529	518	526	544	570	596	624	28	56.9	0.43	24.4	Rhodol	4.15.1
315	6-Dimethylaminorhodol	10 mм HEPES pH 7.3	473	494	518	531	520	528	546	572	599	626	28	60.0	0.21	12.6	Rhodol	4.15.2
316	JOE	H_2O pH 9	478	500	520	534	524	531	548	577	602	628	28	73.0	0.60	43.8	Fluorescein-ha	4.8.2
317	6-Dibutylaminorhodol	50 mм P, <0.1% ACN pH 8	479	500	525	538	526	534	553	580	607	636	28	63.0	0.08	5.0	Rhodol	4.15.2
318	DAOTA 5	H_2O	479	505	548	567	551	558	576	606	630	656	28	50.0	0.56	28.0	Triangulenium	4.22
319	Sulforhodamine B	H_2O pH 7	508	537	554	567	561	568	582	599	609	628	28	84.0	0.39	32.7	Sulforhodam.	4.6
320	Bza-Oet	H_2O pH 7.4	555	588	625	641	631	638	653	676	694	733	28	33.5	0.26	8.7	Phenoxazine	4.25
321	nor-SqB1	P pH 7.4	589	619	638	654	644	651	666	688	706	748	28	88.0	0.20	17.6	Squaraine	4.35
322	nor-SqB2	P pH 7.4	589	619	638	654	644	651	666	688	706	748	28	114.0	0.22	25.0	Squaraine	4.35
323	PORL	PBS, 1% DMSO pH 7.4	609	643	670	686	671	679	698	724	744	776	28	66.1	0.11	7.2	POPh-Rosol	4.20
324	Seta-750-carboxylate	P pH 7.4			751				779				28	235.0	0.21	49.3	nd	4.41
325	Y-A 2	PBS pH 7.2	685	744	778	803	772	783	806	832	845	860	28	190.0	0.03	5.7	Cyanine-Hepta	4.13.3
326	Y-A 10	PBS pH 7.2	687	746	780	805	774	785	808	835	847	862	28	219.0	0.03	6.5	Cyanine-Hepta	4.13.3
327	IR-783	PBS	688	748	782	807	776	787	810	837	849	865	28	261.0	0.08	21.9	Cyanine-Hepta	4.13.3
328	Acridizinium J-N 1a	H_2O			377				404				27	8.9	0.33	2.9	Acridizinium	4.39.1

(Continued)

Ent.	Short Names	Aqueous Solutions	$\lambda_{x,-0.25}$ nm	$\lambda_{x,-0.50}$ nm	$\lambda_{x,max}$ nm	$\lambda_{x,+0.50}$ nm	$\lambda_{e,-0.25}$ nm	$\lambda_{e,-0.50}$ nm	$\lambda_{e,max}$ nm	$\lambda_{e,+0.50}$ nm	$\lambda_{e,+0.25}$ nm	$\lambda_{e,+0.10}$ nm	Δ_S	ε_{max} mM^{-1}cm^{-1}	Φ_f	B_{max} mM^{-1}cm^{-1}	Class	Ref.
329	Acridizinium J-N 1b	H_2O			381				408				27	6.7	0.34	2.2	Acridizinium	4.39.1
330	Cascade Blue	50 mM P pH 8	348	363	403	411	413	417	430	464	484	502	27	28.0	0.90	25.2	Pyrene	4.28
331	CF405S	H_2O			404				431				27	33.0	0.70	23.0	nd (Pyrene)	4.41
332	4-Benzamidofluorescein	100 mM P pH 7.5	453	472	490	502	496	502	517	543	565	588	27	80.0	0.79	63.2	Fluorescein-nh	4.8.1
333	DY-495 carboxylate	PBS pH 7.4	456	476	494	507	499	505	521	547	570	593	27	78.0	0.68	53.2	Fluorescein-nh	4.8.1
334	6-Allylaminorhodol	50 mM P, <0.1% ACN pH 8	464	484	507	519	509	517	534	559	584	611	27	61.5	0.92	56.5	Rhodol	4.15.2
335	2-Et-7-Me-6-EtAm-carboxyrhodol	50 mM P pH 8-9	468	488	511	524	512	520	538	563	589	616	27	73.0	0.84	61.3	Rhodol	4.15.1
336	6-Azetidinylaminorhodol	10 mM HEPES pH 7.3	474	495	519	532	520	528	546	572	599	626	27	59.0	0.85	50.1	Rhodol	4.15.2
337	6-Diethylaminorhodol	50 mM P, <0.1% ACN pH 8	477	498	522	535	522	531	549	576	602	630	27	61.0	0.15	9.1	Rhodol	4.15.2
338	6-Dipropylaminorhodol	50 mM P, <0.1% ACN pH 8	479	500	525	538	525	533	552	579	606	634	27	59.0	0.10	5.9	Rhodol	4.15.2
339	M-S 10b	H_2O, 0.1% TX-100	501	523	547	560	551	556	574	610	638	671	27	64.0	0.78	49.9	Bodipy-Small	4.14.1
340	TAMRA	50 mM PBS pH 8-9	502	526	547	562	552	558	574	599	622	648	27	77.0	0.35	26.9	Rhoda 500-550	4.2.2
341	2-Cloro-6-quinolino-carboxyrhodol	50 mM P pH 8-9	498	521	548	563	546	555	575	604	634	665	27	74.5	0.42	31.2	Rhodol	4.15.1
342	DAOTA 4	H_2O	481	502	549	567	553	559	576	607	632	655	27	50.0	0.56	28.0	Triangulenium	4.22
343	560CP	PBS pH 7.4	522	541	561	576	565	572	588	625	646	669	27	61.0	0.76	46.3	C(CH3)2-Rhoda	4.3.1
344	W-B 15	100 mM P pH 7.4	518	549	567	582	571	578	594	620	651	681	27	76.5	0.13	9.9	Rhoda 550-600	4.2.3
345	Carborhodol S-H 13-H	PBS pH 7.4	539	567	586	601	589	595	613	657	685	711	27	59.0	0.32	18.8	C(CH3)2-Rhod	4.16
346	Alexa Fluor 594	PBS pH 7.2/7.0, 20/22 °C	537	570	590	606	593	599	617	646	679	712	27	92.0	0.66	60.7	Rhodam Hi Ch	4.2.5
347	Alexa Fluor 660	PBS pH 7.2/7.5, 20/22 °C			663				690				27	132.0	0.37	48.8	nd	4.41
348	NIR3	H_2O	663	719	750	773	745	756	777	802	813	827	27	275.0	0.28	77.0	Cyanine-Hepta	4.13.3
349	Y-A 1 (Cypate)	PBS pH 7.2	685	744	778	803	771	782	805	831	844	859	27	224.0	0.03	6.7	Cyanine-Hepta	4.13.3
350	Y-A 4	PBS pH 7.2	685	744	778	803	771	782	805	831	844	859	27	201.0	0.03	6.0	Cyanine-Hepta	4.13.3

Ent.	Short Names	Aqueous Solutions	$\lambda_{x,-0.25}$ nm	$\lambda_{x,-0.50}$ nm	$\lambda_{x,max}$ nm	$\lambda_{x,+0.50}$ nm	$\lambda_{e,-0.25}$ nm	$\lambda_{e,-0.50}$ nm	$\lambda_{e,max}$ nm	$\lambda_{e,+0.50}$ nm	$\lambda_{e,+0.25}$ nm	$\lambda_{e,+0.10}$ nm	Δ_S	ε_{max} mм$^{-1}$cm^{-1}	Φ_f	B_{max} mм$^{-1}$cm^{-1}	Class	Ref.
351	Y-A 6	PBS pH 7.2	685	744	778	803	771	782	805	831	844	859	27	224.0	0.03	6.7	Cyanine-Hepta	4.13.3
352	Y-A 11	PBS pH 7.2	687	746	780	805	773	784	807	834	846	861	27	199.0	0.03	5.9	Cyanine-Hepta	4.13.3
353	HCDS	PBS pH 7.4	695	756	791	817	783	794	818	845	858	874	27	218.8	0.39	85.3	Cyanine-Hepta	4.13.3
354	Sulfobetaine pyrene 21	H_2O	314	327	359	365	371	375	385	412	428	441	26	30.6	0.32	9.7	Pyrene	4.28
355	Fluorescein	100 mм NaOH in D_2O	453	472	489.7	502	495	501	516.3	542	564	587	26	67.0	0.98	65.6	Fluorescein-nh	4.8.1
356	Fluorescein	100 mм NaOH in H_2O	453	472	490.2	503	495	501	516.3	542	564	587	26	67.0	0.92	61.6	Fluorescein-nh	4.8.1
357	5(6)-Carboxyrhodol	50 mм P pH 8-9	451	472	494	503	495	501	520	539	562	584	26	59.9	0.89	53.3	Rhodol	4.15.1
358	Fluorinated rhodamine 1j	Buffer pH 7	458	482	496	507	504	509	522	542	566	588	26	53.0	0.89	47.1	Rhodam <500	4.2.1
359	Rhodamine 123	H_2O pH <7	464	485	502	514	509	514	528	549	568	590	26	75.0	0.90	67.5	Rhoda 500-550	4.2.2
360	6-Benzylaminorhodol	50 mм P, <0.1% ACN pH 8	461	480	503	515	504	512	529	554	578	604	26	71.0	0.95	67.4	Rhodol	4.15.2
361	7-Me-6-EtAm-carboxyrhodol	50 mм P pH 8-9	465	485	508	521	509	517	534	559	584	611	26	78.1	0.82	64.0	Rhodol	4.15.1
362	2-PrAc-7-Me-6-EtAm-rhodol	50 mм P pH 8-9	466	486	509	522	510	518	535	560	585	612	26	79.4	0.92	73.0	Rhodol	4.15.1
363	2,7-DiMe-6-EtAm-carboxyrhodol	50 mм P pH 8-9	467	487	510	522	511	519	536	561	586	612	26	66.3	0.89	59.0	Rhodol	4.15.1
364	6-Dimethylaminorhodol	50 mм P, <0.1% ACN pH 8	473	494	518	531	518	526	544	570	596	624	26	64.0	0.20	12.8	Rhodol	4.15.2
365	Rhodamine 6G	D_2O pH <7	484	506	525.5	539	531	537	552.0	575	596	620	26	105.0	0.98	102.9	Rhoda 500-550	4.2.2
366	Rhodamine 6G	H_2O pH <7	485	507	526.3	540	532	537	552.3	576	596	621	26	105.0	0.93	97.6	Rhoda 500-550	4.2.2
367	1,2,7,8-Dibenzosulfo-fluorescein	C pH 9.5	502	523	541	554	546	551	567	614	633	658	26	81.0	0.48	38.8	Sulfofluoresc.	4.10
368	Rosamine W-B 5h	100 mм P pH 7.4	495	524	542	557	546	552	568	587	603	634	26	71.5	0.14	10.0	Rosamine	4.17
369	2′,7′-Difluorocarbo-fluorescein	100 mм NaOH	514	536	555	569	559	564	581	631	651	677	26	90.9	0.40	36.3	C(CH3)2-Fluore	4.9
370	DAOTA 6	H_2O	489	509	558	575	560	567	584	614	640	660	26	50.0	0.36	18.0	Triangulenium	4.22
371	Atto 565 carboxylate	PBS pH 7.4, 22 °C	513	543	564	578	565	572	590	617	648	676	26	120.0	0.90	108.0	Rhoda 550-600	4.2.3
372	DY-636 carboxylate	PBS pH 7.4	558	591	645	669	632	643	671	732	769	811	26	190.0	0.09	16.3	Merocyanine	4.24

(Continued)

Ent.	Short Names	Aqueous Solutions	$\lambda_{x,-0.25}$ nm	$\lambda_{x,-0.50}$ nm	$\lambda_{x,max}$ nm	$\lambda_{x,+0.50}$ nm	$\lambda_{e,-0.25}$ nm	$\lambda_{e,-0.50}$ nm	$\lambda_{e,max}$ nm	$\lambda_{e,+0.50}$ nm	$\lambda_{e,+0.25}$ nm	$\lambda_{e,+0.10}$ nm	Δ_S	ε_{max} mᴍ$^{-1}$cm^{-1}	Φ_f	B_{max} mᴍ$^{-1}$cm^{-1}	Class	Ref.
373	B-Z 19	H_2O	587	617	651	670	646	652	677	728	767	816	26	118.5	0.23	27.2	Bodipy-Large	4.14.3
374	Alexa Fluor 750	PBS/H_2O pH 7.2, 20/22 °C			749				775				26	290.0	0.12	34.8	nd	4.41
375	Y-A 5	PBS pH 7.2	685	744	778	803	770	781	804	830	842	858	26	185.0	0.03	5.5	Cyanine-Hepta	4.13.3
376	Y-A 7	PBS pH 7.2	686	745	779	804	771	782	805	831	844	859	26	224.0	0.03	6.7	Cyanine-Hepta	4.13.3
377	Y-A 8	PBS pH 7.2	687	746	780	805	772	783	806	832	845	860	26	197.0	0.03	5.9	Cyanine-Hepta	4.13.3
378	Pyrene-1-sulfonate	H_2O	327	340	375	382	385	389	400	429	446	461	25	30.6	0.56	17.1	Pyrene	4.28
379	PetACSO3H	H_2O pH 5.5	367	379	406	414	417	421	431	467	480	495	25	6.2	0.92	5.6	Acridones	4.23.2
380	BOPAHY H-D 18bb′	H_2O	397	411	428	434	444	447	453	463	476	489	25	51	0.49	24.9	Bodipy-analog	4.14.5
381	5(6)-Carboxyfluorescein (FAM)	100 mᴍ PBS pH 9	457	477	495	508	499	504	520	546	569	592	25	83.0	0.90	74.7	Fluorescein-nh	4.8.1
382	Mercaptofluorescein MK-67	1 mᴍ NaOH	463	483	502	515	505	511	527	554	577	601	25	27.0	0.17	4.5	Fluorescein-SH	4.8.3
383	DY-505 carboxylate	PBS pH 7.4	459	478	505	515	506	512	530	552	570	590	25	85.0	0.74	63.2	Rhoda 500-550	4.2.2
384	Fluorinated rhodamine 1k	Buffer pH 7	467	488	506	518	512	517	531	553	572	594	25	54.0	0.66	35.6	Rhoda 500-550	4.2.2
385	6-Butylaminorhodol	50 mᴍ P, <0.1% ACN pH 8	465	485	508	521	508	516	533	558	583	609	25	60.0	0.93	55.8	Rhodol	4.15.2
386	2-Cl-7-Me-6-EtAm-carboxyrhodol	50 mᴍ P pH 8-9	479	495	514	538	514	525	539	573	600	634	25	80.5	0.82	66.0	Rhodol	4.15.1
387	6-Piperidinocarboxyrhodol	50 mᴍ P pH 8-9	472	492	516	529	515	523	541	567	593	620	25	80.9	0.79	63.9	Rhodol	4.15.1
388	Rhodamine RBzET	H_2O	485	508	527	541	531	537	552	575	596	620	25	82.0	0.87	71.3	Rhoda 500-550	4.2.2
389	Carborhodamine	10 mᴍ HEPES pH 7.3	515	537	552	564	558	564	577	605	628	654	25	65.0	0.64	41.6	C(CH3)2-Rhoda	4.3.1
390	DY-555 carboxylate	PBS pH 7.4	508	537	555	569	558	564	580	605	635	663	25	100.0	0.32	31.7	Rhoda 550-600	4.2.3
391	5(6)-Carboxy-X-rhodamine (ROX)	H_2O pH <7	525	557	576	591	578	584	601	628	660	691	25	82.0	0.70	57.4	Rhoda 550-600	4.2.3
392	Alexa Fluor 568	PBS pH 7.2/7.0, 20/22 °C	527	559	578	593	580	586	603	630	662	693	25	88.0	0.69	60.7	Rhodam Hi Ch	4.2.5
393	DY-651 carboxylate	PBS pH 7.4	599	631	653	672	652	659	678	699	711	731	25	160.0	0.15	24.1	Merocyanine	4.24
394	Cy5.204	PBS pH 7.4	615	658	682	698	683	689	707	726	739	772	25	197.0	0.10	19.7	Cyanine-Penta	4.13.2
395	Zy7 carboxylate	P pH 7.4	663	718	749	772	743	753	774	798	810	824	25	200.0	0.13	26.0	Cyanine-Hepta	4.13.3
396	Alexa Fluor 405	PBS pH 7.2, 22 °C	346	360	400	408	408	411	424	457	476	494	24	35.0	0.50	17.5	Pyrene	4.28

Ent.	Short Names	Aqueous Solutions	$\lambda_{x,-0.25}$ nm	$\lambda_{x,-0.50}$ nm	$\lambda_{x,max}$ nm	$\lambda_{x,+0.50}$ nm	$\lambda_{e,-0.25}$ nm	$\lambda_{e,-0.50}$ nm	$\lambda_{e,max}$ nm	$\lambda_{e,+0.50}$ nm	$\lambda_{e,+0.25}$ nm	$\lambda_{e,+0.10}$ nm	Δ_S	ε_{max} mM^{-1}cm^{-1}	Φ_f	B_{max} mM^{-1}cm^{-1}	Class	Ref.
397	BOPAHY H-D 18ab′	H_2O	398	412	429	435	444	447	453	463	476	489	24	51	0.55	28.0	Bodipy-analog	4.14.5
398	Fluorescein (S-H 1)	P pH 9	453	472	490	502	493	499	514	540	561	584	24	90.0	0.92	82.8	Fluorescein-nh	4.8.1
399	Fluorinated fluorescein S-H 17	P pH 9	453	472	490	502	493	499	514	540	561	584	24	82.4	0.97	79.9	Fluorescein-ha	4.8.2
400	Oregon Green 488	H_2O pH 9	453	472	490	502	493	499	514	540	561	584	24	76.0	0.97	73.7	Fluorescein-ha	4.8.2
401	DY-490 carboxylate	PBS pH 7.4	452	468	491	504	495	500	515	536	553	572	24	73.0	0.90	65.4	Rhodam Hi Ch	4.2.5
402	5-Carboxyfluorescein (S-H 2)	P pH 9	454	474	492	504	495	501	516	542	564	587	24	84.7	0.92	77.9	Fluorescein-nh	4.8.1
403	Fluorinated fluorescein S-H 14a,b	P pH 9	454	474	492	504	495	501	516	542	564	587	24	85.9	0.92	79.0	Fluorescein-ha	4.8.2
404	Rhodamine 110	H_2O pH <7	458	482	496	507	503	507	520	540	564	586	24	83.0	0.89	73.8	Rhodam <500	4.2.1
405	Fluorinated rhodam. 1f-5′-COOH	Buffer pH 7	463	484	501	513	506	512	525	546	565	586	24	85.0	0.95	80.7	Rhoda 500-550	4.2.2
406	6-Propylaminorhodol	50 mM P, <0.1% ACN pH 8	464	484	507	519	506	514	531	556	581	607	24	61.0	0.94	57.3	Rhodol	4.15.2
407	Fluorinated fluorescein S-H 15a,b	P pH 9	470	490	510	523	511	518	534	562	585	610	24	84.2	0.43	36.2	Fluorescein-ha	4.8.2
408	Fluorinated fluorescein S-H 16a,b	P pH 9	470	490	510	523	511	518	534	562	585	610	24	78.1	0.59	46.0	Fluorescein-ha	4.8.2
409	2-Cl-6-Piperidinocarboxy-rhodol	50 mM P pH 8-9	478	500	524	537	522	530	548	574	601	629	24	77.4	0.72	55.7	Rhodol	4.15.1
410	JF525	10 mM HEPES pH 7.3	483	506	525	538	529	534	549	572	593	617	24	94.0	0.91	85.5	Rhoda 500-550	4.2.2
411	JF526	10 mM HEPES pH 7.3	484	507	526	539	530	535	550	573	594	618	24	19.0	0.87	16.5	Rhoda 500-550	4.2.2
412	Atto Rho6G carboxylate	PBS pH 7.4, 22 °C	489	513	533	546	534	540	557	583	609	636	24	115.0	0.94	108.1	Rhoda 500-550	4.2.2
413	M-S 10c	H_2O, 0.1% TX-100	489	510	533	546	536	540	557	591	617	648	24	54.0	0.61	32.9	Bodipy-Small	4.14.1
414	HEX	H_2O pH 9	491	513	535	550	534	541	559	589	616	643	24	98.0	0.70	68.6	Fluorescein-ha	4.8.2
415	6-Quinoline-2-pentanoic-pyrodol	50 mM P pH 8-9	492	515	538	550	544	549	562	579	591	611	24	64.1	0.70	44.8	Pyrodol	4.21.1
416	6-Quinoline-2-propanoic-pyrodol	50 mM P pH 8-9	492	515	538	550	544	549	562	579	591	611	24	65.9	0.70	46.1	Pyrodol	4.21.1

(Continued)

Ent.	Short Names	Aqueous Solutions	$\lambda_{x,-0.25}$ nm	$\lambda_{x,-0.50}$ nm	$\lambda_{x,max}$ nm	$\lambda_{x,+0.50}$ nm	$\lambda_{e,-0.25}$ nm	$\lambda_{e,-0.50}$ nm	$\lambda_{e,max}$ nm	$\lambda_{e,+0.50}$ nm	$\lambda_{e,+0.25}$ nm	$\lambda_{e,+0.10}$ nm	Δ_S	ε_{max} mM^{-1}cm^{-1}	Φ_f	B_{max} mM^{-1}cm^{-1}	Class	Ref.
417	Tetramethylrhodamine	10 mM HEPES pH 7.3	503	527	548	563	550	556	572	597	619	646	24	78.0	0.45	35.1	Rhoda 500-550	4.2.2
418	SR101-OH	H_2O	493	516	548	564	550	556	572	595	628	652	24	80.0	0.35	28.0	Sulforhodol	4.27
419	Seta-580-carboxylate	H_2O			578				602				24	90.0	0.55	49.5	nd	4.41
420	JF585	10 mM HEPES pH 7.3	544	568	585	598	588	594	609	640	666	696	24	1.5	0.78	1.1	C(CH3)2-Rhoda	4.3.1
421	Atto Rho13 carboxylate	PBS pH 7.4, 22 °C	547	584	603	618	602	609	627	654	690	723	24	120.0	0.80	96.0	Rhodam >600	4.2.4
422	Atto 635 carboxylate	PBS pH 7.4, 22 °C	579	612	635	649	626	638	659	683	701	720	24	130.0	0.25	32.5	C(CH3)2-Pyron	4.5
423	NR638	10 mM PBS pH 7.4	583	613	638	652	637	645	662	686	703	732	24	55.6	0.21	11.6	POPh-Rosol	4.20
424	Carborhodamine 21b	H_2O	591	619	640	656	639	646	664	701	732	768	24	68.0	0.56	38.0	C(CH3)2-Rhoda	4.3.1
425	Carborhodamine 22	H_2O	591	619	640	656	639	646	664	701	732	768	24	67.0	0.54	36.1	C(CH3)2-Rhoda	4.3.1
426	W-L 8a	PBS	583	621	643	657	645	651	667	684	696	725	24	87.0	0.12	10.4	Cyanine-Penta	4.13.2
427	W-L 8b	PBS	584	622	644	659	646	652	668	685	697	726	24	189.0	0.15	28.3	Cyanine-Penta	4.13.2
428	Cy5.12-OH	H_2O/PBS	589	628	650	665	652	658	674	692	703	733	24	215.0	0.13	27.9	Cyanine-Penta	4.13.2
429	DY-654 carboxylate	PBS pH 7.4	594	627	653	673	651	659	677	699	711	732	24	220.0	0.16	35.2	Merocyanine	4.24
430	JF724	10 mM HEPES pH 7.3	647	688	724	750	711	721	748	792	831	879	24	6.6	0.05	0.3	SO2-Rhodam	4.3.3
431	DY-405 carboxylate	PBS pH 7.4	336	349	400	410	404	408	423	458	474	492	23	32.0	0.97	30.9	Pyrene	4.28
432	DMAX-EP	100 mM P pH 7.5	454	474	492	504	494	500	515	541	563	585	23	71.0	0.81	57.5	Fluorescein-nh	4.8.1
433	6-Aminorhodol (Rhodol)	50 mM P, <0.1% ACN pH 8	452	471	493	505	492	500	516	539	563	587	23	70.0	0.98	68.6	Rhodol	4.15.2
434	Alexa Fluor 488	PBS pH 7.2, 22 °C	457	477	494	506	499	504	517	538	555	576	23	73.0	0.92	67.1	Rhodam Hi Ch	4.2.5
435	Fluorinated rhodamine 1e	Buffer pH 7	461	485	499	510	504	509	522	542	566	588	23	81.0	0.88	71.2	Rhodam <500	4.2.1
436	2-Br-7-Me-6-EtAm-carboxyrhodol	50 mM P pH 8-9	472	492	516	529	513	521	539	565	590	618	23	80.9	0.61	49.3	Rhodol	4.15.1
437	*meso*-CaboxyBODIPY J-K 3c	10 mM PBS pH 7.4	480	501	517	527	519	523	540	577	598	629	23	76.0	0.97	73.7	Bodipy-Small	4.14.1
438	6-Dimethylaminorhodol	C pH 9.5	477	498	522	535	519	527	545	571	597	625	23	57.0	0.20	11.4	Rhodol	4.15.1
439	Rhodamine 6G +2	H_2O pH <7	487	511	530	544	533	539	553	571	583	603	23	101.0	0.93	93.9	Rhoda 500-550	4.2.2
440	Alexa Fluor 532	PBS pH 7.2, 22 °C	486	514	531	545	533	539	554	572	587	616	23	81.0	0.61	49.4	Rosamine	4.17

Ent.	Short Names	Aqueous Solutions	$\lambda_{x,-0.25}$ nm	$\lambda_{x,-0.50}$ nm	$\lambda_{x,max}$ nm	$\lambda_{x,+0.50}$ nm	$\lambda_{e,-0.25}$ nm	$\lambda_{e,-0.50}$ nm	$\lambda_{e,max}$ nm	$\lambda_{e,+0.50}$ nm	$\lambda_{e,+0.25}$ nm	$\lambda_{e,+0.10}$ nm	Δ_S	ε_{max} mм$^{-1}$cm^{-1}	Φ_f	B_{max} mм$^{-1}$cm^{-1}	Class	Ref.
441	Carbofluorescein	100 mм NaOH	505	526	544	557	546	551	567	614	633	658	23	108.0	0.62	66.9	C(CH3)2-Fluore	4.9
442	Pyronine Y (OP)	PBS pH 7.4	500	525	546	560	519	526	569	557	567	586	23	77.0	0.36	27.7	Pyronine	4.4
443	JF552	10 mм HEPES pH 7.3	505	535	552	566	554	560	575	600	629	657	23	95.2	0.83	79.0	Rhoda 550-600	4.2.3
444	Tetraethylrhodamine	H_2O pH 7	510	539	557	571	558	564	580	605	635	663	23	105.0	0.24	25.2	Rhoda 550-600	4.2.3
445	Atto Rho3B carboxylate	PBS pH 7.4, 22 °C	517	548	566	579	566	572	589	613	645	675	23	120.0	0.50	60.0	Rhoda 550-600	4.2.3
446	JF570	10 mм HEPES pH 7.3	521	548	570	586	569	576	593	620	644	673	23	83.6	0.63	52.6	S-Rhodamine	4.3.3
447	Atto Rho11 carboxylate	PBS pH 7.4, 22 °C	522	553	572	585	571	577	595	619	650	680	23	120.0	0.80	96.0	Rhoda 550-600	4.2.3
448	575RH	PBS pH 7.4	524	555	574	589	574	580	597	624	655	685	23	55.0	0.74	40.7	Rhoda 550-600	4.2.3
449	Atto Rho12 carboxylate	PBS pH 7.4, 22 °C	526	557	577	590	577	583	600	627	659	689	23	120.0	0.80	96.0	Rhoda 550-600	4.2.3
450	Atto 594 carboxylate	PBS pH 7.4, 22 °C	547	583	603	617	601	608	626	652	688	718	23	120.0	0.85	102.0	Rhodam >600	4.2.4
451	NR604	PBS, 1% DMSO pH 7.4	554	584	604	618	604	610	627	654	672	697	23	45.7	0.48	21.9	POOH-Fluoron	4.12
452	JF608	10 mм HEPES pH 7.3	564	589	608	623	609	615	631	664	692	725	23	99.0	0.67	66.3	C(CH3)2-Rhoda	4.3.1
453	SqB1	P pH 7.4	580	609	627	642	629	636	650	671	688	728	23	150.0	0.05	7.5	Squaraine	4.35
454	NR632	10 mм PBS pH 7.4	578	608	632	646	631	638	655	678	695	723	23	64.0	0.26	16.6	POPh-Rosol	4.20
455	KK1119	H_2O	577	618	637	654	634	641	660	687	721	763	23	90.0	0.55	49.5	Rhodam >600	4.2.4
456	2-Me SiR640	100 mм P, 1% DMSO pH 7.4	509	548	637	658	635	643	660	683	707	733	23	54.0	0.18	9.7	Si(CH3)2-Rosa	4.18.1
457	KK114	H_2O	580	619	637	652	635	642	660	685	713	762	23	92.0	0.55	50.6	Rhodam >600	4.2.4
458	Carborhodamine 21a	H_2O	591	619	640	656	638	646	663	700	731	767	23	70.0	0.56	39.2	C(CH3)2-Rhoda	4.3.1
459	W-L 8c	PBS	583	621	643	657	645	650	666	683	695	724	23	107.0	0.12	12.8	Cyanine-Penta	4.13.2
460	W-L 8d	PBS	583	621	643	657	645	650	666	683	695	724	23	169.0	0.13	21.9	Cyanine-Penta	4.13.2
461	nor-SqCN1	P pH 7.4	616	650	670	687	669	677	693	717	737	782	23	123.0	0.19	23.3	Squaraine	4.35
462	nor-SqCN2	P pH 7.4	616	650	670	687	669	677	693	717	737	782	23	167.0	0.19	31.7	Squaraine	4.35
463	2-Me SiR675	100 mм P, 1% DMSO pH 7.4	592	635	675	694	672	679	698	721	742	772	23	88.0	0.07	6.1	Si(CH3)2-Rosa	4.18.1
464	Alexa Fluor 680	PBS pH 7.2/7.5, 20/22 °C	613	655	679	695	678	685	702	721	734	766	23	183.0	0.36	65.8	Cyanine-Penta	4.13.2

(Continued)

Ent.	Short Names	Aqueous Solutions	$\lambda_{x,-0.25}$ nm	$\lambda_{x,-0.50}$ nm	$\lambda_{x,max}$ nm	$\lambda_{x,+0.50}$ nm	$\lambda_{e,-0.25}$ nm	$\lambda_{e,-0.50}$ nm	$\lambda_{e,max}$ nm	$\lambda_{e,+0.50}$ nm	$\lambda_{e,+0.25}$ nm	$\lambda_{e,+0.10}$ nm	Δ_S	ε_{max} mм$^{-1}$cm^{-1}	Φ_f	B_{max} mм$^{-1}$cm^{-1}	Class	Ref.
465	Atto 725 carboxylate	PBS pH 7.4, 22 °C			728				751				23	120.0	0.10	12.0	nd	4.41
466	DY-731 carboxylate	PBS pH 7.4	678	709	736	761	725	735	759	783	794	811	23	225.0	0.11	23.6	Merocyanine	4.24
467	Fluorescein (G-C 1)	H_2O pH 9.18, 25 °C	454	474	492	504	493	499	514	540	561	584	22	78.4	0.90	70.5	Fluorescein-nh	4.8.1
468	2-Me RG	PBS pH 7.4	458	483	498	510	502	507	520	536	549	574	22	82.0	0.84	68.8	Rosamine	4.17
469	Fluorinated rhodam. 1f-6′-COOH	Buffer pH 7	463	484	501	513	505	510	523	544	562	584	22	88.0	0.93	81.8	Rhoda 500-550	4.2.2
470	BCECF	100 mм NaOH	466	486	505	518	505	511	527	554	577	601	22	67.0	0.84	56.2	Fluorescein-nh	4.8.1
471	BCPCF	100 mм NaOH	466	486	505	518	505	511	527	554	577	601	22	67.0	0.83	55.6	Fluorescein-nh	4.8.1
472	2,4,5,7-Tetra-F-sulfo-fluorescein	P pH 9.18	480	499	515	527	518	523	537	579	596	618	22	63.2	0.80	50.5	Sulfofluoresc.	4.10
473	2,4,5,7-Tetra-Br-sulfo-fluorescein	100 mм NaOH pH 12	488	507	524	536	526	531	546	590	607	630	22	87.0	0.35	30.4	Sulfofluoresc.	4.10
474	Rhodamine 6G +3	H_2O pH <7	486	510	529	544	532	537	551	570	582	602	22	101.0	0.91	91.9	Rhodam Hi Ch	4.2.5
475	JF549	10 mм HEPES pH 7.3	504	528	549	564	549	555	571	596	618	644	22	101	0.88	88.8	Rhoda 500-550	4.2.2
476	Neo-Cy3	PBS	488	509	551	562	555	560	573	588	622	640	22	98.0	0.02	1.9	Cyanine-Trimet	4.13.1
477	Atto 550 carboxylate	PBS pH 7.4, 22 °C	507	535	554	567	554	560	576	598	626	654	22	120.0	0.80	96.0	Rhoda 550-600	4.2.3
478	SR101-110	PBS	485	507	554	570	555	560	576	600	638	661	22	80.0	0.55	44.0	Sulforhodam.	4.6
479	Atto Rho101 carboxylate	PBS pH 7.4, 22 °C	535	569	587	601	586	592	609	635	664	698	22	120.0	0.80	96.0	Rhoda 550-600	4.2.3
480	2′,7′-Difluoro-Si-Fluorescein	100 mм NaOH	549	574	596	612	593	599	618	675	698	727	22	60.9	0.42	25.5	Si(CH3)2-Fluore	4.9
481	CF620R	H_2O			617				639				22	115.0	0.30	34.0	nd (Rhodam.)	4.41
482	Atto 620 carboxylate	PBS pH 7.4, 22 °C	580	602	620	633	617	623	642	673	707	740	22	120.0	0.50	60.0	C(CH3)2-Rhoda	4.3.1
483	Square-655-carboxylate	P pH 7.4	576	604	622	638	623	630	644	664	682	720	22	177.0	0.09	15.9	Squaraine	4.35
484	KK9046	H_2O	573	613	632	648	629	636	654	681	714	755	22	90.0	0.45	40.5	Rhodam >600	4.2.4
485	Carborhodamine 1b	H_2O	590	619	639	655	636	644	661	698	728	764	22	80.0	0.58	46.4	C(CH3)2-Rhoda	4.3.1
486	Carborhodamine 1c	H_2O	591	619	640	656	637	645	662	699	730	766	22	89.0	0.53	47.1	C(CH3)2-Rhoda	4.3.1

Ent.	Short Names	Aqueous Solutions	$\lambda_{x,-0.25}$ nm	$\lambda_{x,-0.50}$ nm	$\lambda_{x,max}$ nm	$\lambda_{x,+0.50}$ nm	$\lambda_{e,-0.25}$ nm	$\lambda_{e,-0.50}$ nm	$\lambda_{e,max}$ nm	$\lambda_{e,+0.50}$ nm	$\lambda_{e,+0.25}$ nm	$\lambda_{e,+0.10}$ nm	Δ_S	ε_{max} mм$^{-1}$cm^{-1}	Φ_f	B_{max} mм$^{-1}$cm^{-1}	Class	Ref.
487	Carborhodamine 20b	H_2O	592	620	641	657	638	646	663	700	731	767	22	80.0	0.53	42.4	C(CH3)2-Rhoda	4.3.1
488	W-L 8e	PBS	583	621	643	657	644	649	665	682	694	722	22	195.0	0.09	17.5	Cyanine-Penta	4.13.2
489	AF647	PBS pH 7.2, 20/22 °C	589	628	650	665	650	656	672	690	701	731	22	270.0	0.33	89.1	Cyanine-Penta	4.13.2
490	AFD647	PBS pH 7.2, 20/22 °C	589	628	650	665	650	656	672	690	701	731	22	270.0	0.33	89.1	Cyanine-Penta	4.13.2
491	Alexa Fluor 647	PBS pH 7.2, 20/22 °C	589	628	650	665	650	656	672	690	701	731	22	270.0	0.33	89.1	Cyanine-Penta	4.13.2
492	nor-SqB3	P pH 7.4	610	642	662	679	660	668	684	707	727	770	22	102.0	0.18	18.3	Squaraine	4.35
493	NIR2	H_2O	599	639	662	677	661	667	684	702	714	745	22	250.0	0.34	85.0	Cyanine-Penta	4.13.2
494	Seta-670-pH-di-carboxylate	P pH 6			672				694				22	93.0	0.19	17.6	nd	4.41
495	Cy5.203	PBS pH 7.4	614	656	680	696	678	685	702	721	734	766	22	190.0	0.10	19.0	Cyanine-Penta	4.13.2
496	NR700	PBS, 1% DMSO pH 7.4	630	674	700	720	685	696	722	752	772	808	22	71.0	0.11	7.8	PO2R-Rosam.	4.18.2
497	DY-747P1	PBS pH 7.4	661	716	747	770	738	748	769	793	804	818	22	240.0	0.12	28.5	Cyanine-Hepta	4.13.3
498	$ACSO_3H$	H_2O pH 5.5	353	366	396	404	405	408	417	449	458	472	21	4.4	0.82	3.6	Acridones	4.23.2
499	2-OMe TokyoGreen	100 mм NaOH	456	478	494	506	497	502	515	554	569	589	21	85.0	0.86	73.1	Fluorone	4.11
500	4-Carboxy Pennsylvania Green	PBS pH 7.4	456	478	494	506	497	502	515	554	569	589	21	62.0	0.89	55.1	Fluorone	4.11
501	Abberior Star 488 carboxylate	PBS pH 7.4	465	485	503	515	505	511	524	545	564	585	21	64.5	0.89	57.4	Rhodam Hi Ch	4.2.5
502	Fluorinated rhodamine 3-H	Buffer pH 7	471	492	510	523	512	517	531	553	572	594	21	57.0	0.86	49.0	Rhoda 500-550	4.2.2
503	Atto 514 carboxylate	PBS pH 7.4, 22 °C	474	494	511	522	512	517	532	553	575	599	21	115.0	0.85	97.7	Rhoda 500-550	4.2.2
504	2,7-DiCl-6-methylamino rhodol	50 mм P pH 8-9	469	488	512	525	507	515	533	558	583	611	21	71.0	0.67	47.5	Rhodol	4.15.1
505	Atto 520 carboxylate	PBS pH 7.4, 22 °C	474	497	517	529	520	525	538	556	567	592	21	110.0	0.90	99.0	Pyronine	4.4
506	AR116	100 mм P pH 7.5	483	506	527	539	529	535	548	567	578	604	21	64.8	0.07	4.5	Pyronine	4.4
507	530RH	PBS pH 7.4	491	514	532	547	533	538	553	574	592	618	21	56.0	0.89	49.8	Rhoda 500-550	4.2.2
508	DY-530 carboxylate	PBS pH 7.4	490	513	533	547	533	539	554	578	598	623	21	100.0	0.81	80.6	Rhoda 500-550	4.2.2
509	SR101-6G	PBS	508	540	565	579	563	571	586	608	638	666	21	80.0	0.68	54.4	Sulforhodam.	4.6
510	Z-L G	0.5 м PBS pH 7.4	526	550	577	592	574	578	598	637	667	704	21	50	0.21	10.6	Bodipy-Large	4.14.3

(Continued)

Ent.	Short Names	Aqueous Solutions	$\lambda_{x,-0.25}$ nm	$\lambda_{x,-0.50}$ nm	$\lambda_{x,max}$ nm	$\lambda_{x,+0.50}$ nm	$\lambda_{e,-0.25}$ nm	$\lambda_{e,-0.50}$ nm	$\lambda_{e,max}$ nm	$\lambda_{e,+0.50}$ nm	$\lambda_{e,+0.25}$ nm	$\lambda_{e,+0.10}$ nm	Δ_S	ε_{max} mм$^{-1}$cm^{-1}	Φ_f	B_{max} mм$^{-1}$cm^{-1}	Class	Ref.
511	CF594	H_2O			593				614				21	115.0	0.45	51.7	nd (Rhodam.)	4.41
512	2′COOH-Si-rhodol	100 mм P pH 7.4	559	588	609	625	605	611	630	677	706	734	21	70.0	0.22	15.4	Si(CH3)2-Rhod	4.16
513	Atto 633 carboxylate	PBS pH 7.4, 22 °C	575	610	630	646	627	635	651	687	724	756	21	130.0	0.64	83.2	C(CH3)2-Rhoda	4.3.1
514	TMDHS or SiP-Me	50 mм PBS pH 7.4	570	603	633	651	627	635	654	681	698	735	21	64.2	0.41	26.3	Si(CH3)2-Pyron	4.5
515	GeR	PBS pH 7.4	601	621	634	658	635	643	655	681	697	734	21	97.0	0.43	41.7	Ge(CH3)2-Rhoda	4.3.3
516	Carborhodamine 16a	H_2O	588	616	636	652	633	640	657	693	723	759	21	114.0	0.62	70.6	C(CH3)2-Rhoda	4.3.1
517	Carborhodamine 20a	H_2O	591	619	640	656	636	644	661	698	728	764	21	70.0	0.53	37.1	C(CH3)2-Rhoda	4.3.1
518	640SiRH	PBS pH 7.4	584	519	641	658	639	646	662	681	694	726	21	51.0	0.42	21.4	Si(CH3)2-Rhoda	4.3.2
519	Cy5.11-OH	H_2O/PBS	586	624	646	661	645	651	667	684	696	725	21	215.0	0.06	12.9	Cyanine-Penta	4.13.2
520	Cy5.13-OH	H_2O/PBS	588	626	648	663	647	653	669	686	698	727	21	215.0	0.12	25.8	Cyanine-Penta	4.13.2
521	NIR4	H_2O	589	628	650	665	649	655	671	688	700	729	21	260.0	0.43	111.8	Cyanine-Penta	4.13.2
522	Seta-670-carboxylate	P pH 7.4			667				688				21	180.0	0.07	12.6	nd	4.41
523	CF680R	H_2O			680				701				21	140	0.03	4.7	nd (Rhodam.)	4.41
524	2-Me SiR700	H_2O	617	665	691	710	686	695	712	732	746	764	21	100.0	0.12	12.0	Si(CH3)2-Rosa	4.18.1
525	nor-SqCN3	P pH 7.4	636	671	693	712	688	697	714	739	761	809	21	113.0	0.15	16.9	Squaraine	4.35
526	Alexa Fluor 700	PBS/H_2O pH 7.2, 20/22 °C			702				723				21	205.0	0.25	51.2	nd	4.41
527	JF711	10 mм HEPES pH 7.3	642	618	711	732	702	711	732	756	774	815	21	12.4	0.17	2.1	POPh-Rhodam	4.3.3
528	JF722	10 mм HEPES pH 7.3	651	626	722	743	712	721	743	768	786	829	21	87.2	0.11	9.5	POPh-Rhodam	4.3.3
529	9-Aminoacridine	P pH 7	366	379	400	410	400	405	420	449	467	487	20	9.3	0.97	9.0	Acridine	4.23.1
530	N-Methylacridone	H_2O	369	382	411	419	417	421	431	467	480	493	20	8.8	0.99	8.7	Acridones	4.23.2
531	NX	100 mм P pH 7.5	454	474	492	504	491	497	512	537	559	582	20	69.0	0.83	57.2	Fluorescein-nh	4.8.1
532	2-OMe-5-Me TokyoGreen	100 mм NaOH	456	478	494	506	497	501	514	553	568	587	20	85.0	0.50	42.5	Fluorone	4.11
533	Pennsylvania Green	PBS pH 7.4	456	478	494	506	497	501	514	553	568	587	20	82.0	0.91	74.6	Fluorone	4.11
534	6-Sulfonato-1,3,5,7-Me-BODIPY	H_2O	454	473	495	503	503	507	515	528	545	562	20	52.5	0.85	44.6	Bodipy-Small	4.14.1

Ent.	Short Names	Aqueous Solutions	$\lambda_{x,-0.25}$ nm	$\lambda_{x,-0.50}$ nm	$\lambda_{x,max}$ nm	$\lambda_{x,+0.50}$ nm	$\lambda_{e,-0.25}$ nm	$\lambda_{e,-0.50}$ nm	$\lambda_{e,max}$ nm	$\lambda_{e,+0.50}$ nm	$\lambda_{e,+0.25}$ nm	$\lambda_{e,+0.10}$ nm	Δ_S	ε_{max} mM^{-1}cm^{-1}	Φ_f	B_{max} mM^{-1}cm^{-1}	Class	Ref.
535	Sulfofluorescein	C pH 9.5	464	482	497	508	499	504	517	556	572	591	20	78.0	0.92	71.7	Sulfofluoresc.	4.10
536	Atto 488 carboxylate	PBS pH 7.4, 22 °C	461	486	500	511	503	507	520	540	564	586	20	90.0	0.80	72.0	Rhodam <500	4.2.1
537	6-Azido-4,5,7-trifluorofluorescein	100 mM NaOH	467	488	507	520	505	511	527	554	577	601	20	79.9	0.82	65.5	Fluorescein-ha	4.8.2
538	Atto 532 carboxylate	PBS pH 7.4, 22 °C	491	515	532	544	533	538	552	575	600	625	20	115.0	0.90	103.5	Rhoda 500-550	4.2.2
539	Atto 542 carboxylate	PBS pH 7.4, 22 °C			542				562				20	120.0	0.93	111.6	nd (pb Rhoda)	4.41
540	Z-L J	0.5 M PBS pH 7.4	498	519	543	556	541	546	563	598	624	656	20	50	0.34	16.8	Bodipy-Large	4.14.3
541	Pyronine Y (OP)	H_2O	498	523	546	559	546	550	566	584	598	631	20	77.0	0.22	16.9	Pyronine	4.4
542	Tetramethylrosamine	10 mM PBS pH 7.4	501	532	550	565	548	554	570	589	605	636	20	74.5	0.31	23.1	Rosamine	4.17
543	Cy3	H_2O	487	508	550	561	553	557	570	585	618	636	20	150.0	0.06	8.2	Cyanine-Trimet	4.13.1
544	Pyronine B	H_2O	504	529	552	565	551	557	572	592	605	633	20	94.1	0.18	16.9	Pyronine	4.4
545	JF559	10 mM HEPES pH 7.3	511	541	559	573	557	563	579	604	633	662	20	106.0	0.85	90.1	Rhoda 550-600	4.2.3
546	Pyronine 630	H_2O, 25 °C	509	535	559	573	558	564	579	600	613	642	20	100.0	0.92	92.0	Pyronine	4.4
547	SR101-Me_2	PBS	517	553	573	587	572	578	593	613	632	668	20	80.0	0.66	52.8	Sulforhodam.	4.6
548	JF574	10 mM HEPES pH 7.3	524	555	574	589	571	578	594	620	651	681	20	103	0.78	80.3	Rhoda 550-600	4.2.3
549	SR101-B	PBS	522	557	575	590	574	580	595	615	631	665	20	80.0	0.65	52.0	Sulforhodam.	4.6
550	Si-Fluorescein	100 mM NaOH	535	559	579	594	575	581	599	652	674	701	20	93.0	0.53	49.2	Si(CH3)2-Fluore	4.9
551	Abberior Star 580 carboxylate	PBS pH 7.4	535	567	587	603	583	590	607	635	667	699	20	85.0	0.90	76.5	Rhoda 550-600	4.2.3
552	2-Me SiR600	H_2O	542	574	593	607	593	599	613	633	650	674	20	91.0	0.38	34.5	Si(CH3)2-Rosa	4.18.1
553	Tetramethycarbo-rhodamine	10 mM HEPES pH 7.3	562	588	606	620	604	610	626	659	686	718	20	121.0	0.52	62.9	C(CH3)2-Rhoda	4.3.1
554	2′COOH-4CH_2OH-Si-rhodol	100 mM P pH 7.4	560	589	610	626	605	611	630	677	706	734	20	70.0	0.21	14.7	Si(CH3)2-Rhod	4.16
555	JS-R	10 mM HEPES pH 7.4	558	589	611	625	609	617	631	648	659	671	20	130.0	0.59	76.7	B(OH)2-Pyron	4.5
556	Pyrrolidinylcarbo-rhodamine	10 mM HEPES pH 7.3	568	594	613	628	610	617	633	666	694	727	20	87.0	0.54	46.9	C(CH3)2-Rhoda	4.3.1

(Continued)

Ent.	Short Names	Aqueous Solutions	$\lambda_{x,-0.25}$ nm	$\lambda_{x,-0.50}$ nm	$\lambda_{x,max}$ nm	$\lambda_{x,+0.50}$ nm	$\lambda_{e,-0.25}$ nm	$\lambda_{e,-0.50}$ nm	$\lambda_{e,max}$ nm	$\lambda_{e,+0.50}$ nm	$\lambda_{e,+0.25}$ nm	$\lambda_{e,+0.10}$ nm	Δ_S	ε_{max} mм$^{-1}$cm^{-1}	Φ_f	B_{max} mм$^{-1}$cm^{-1}	Class	Ref.
557	Atto Rho14 carboxylate	PBS pH 7.4, 22 °C	564	606	626	639	622	628	646	668	693	738	20	140.0	0.80	112.0	Rhodam >600	4.2.4
558	630GeRH	PBS pH 7.4	582	612	631	647	629	635	651	671	683	722	20	61.0	0.60	36.6	Ge(CH3)2-Rhoda	4.3.3
559	SiP-Pr	50 mм PBS pH 7.4	579	613	632	647	607	619	652	661	675	691	20	64.2	0.21	13.4	C(CH3)2-Pyron	4.5
560	Abberior Star 635 carboxylate	PBS pH 7.4	574	615	634	650	629	636	654	681	714	755	20	63.0	0.51	32.1	Rhodam >600	4.2.4
561	Abberior Star 635P carboxylate	PBS pH 7.4	575	616	635	652	630	637	655	682	715	756	20	80.0	0.55	44.0	Rhodam >600	4.2.4
562	Atto 647 carboxylate	PBS pH 7.4, 22 °C	609	624	647	659	639	646	667	700	734	767	20	120.0	0.20	24.0	C(CH3)2-Pyron	4.5
563	Seta-650-carboxylate	P pH 7.4			651				671				20	200.0	0.28	56.0	nd	4.41
564	SiRF 13a	H_2O	598	577	657	675	651	659	677	698	712	747	20	53.0	0.63	33.3	Si(CH3)2-Rhoda	4.3.2
565	NR669	10 mм PBS pH 7.4	619	653	669	684	664	672	689	712	726	757	20	77.2	0.30	23.1	PO2H-Rosam.	4.18.2
566	Cy5.205	PBS pH 7.4	609	650	674	690	671	677	694	713	725	757	20	190.0	0.23	43.7	Cyanine-Penta	4.13.2
567	Cy5.5	PBS	609	650	674	690	671	677	694	713	725	757	20	190.0	0.23	43.7	Cyanine-Penta	4.13.2
568	NR675	10 mм PBS pH 7.4	625	657	675	689	668	675	695	718	733	762	20	73.2	0.24	17.5	PO2H-Rosam.	4.18.2
569	Seta-680-carboxylate	P pH 7.4			678				698				20	200.0	0.24	48.0	nd	4.41
570	DY-682 carboxylate	PBS pH 7.4	617	654	690	708	671	685	710	733	745	764	20	120.0	0.41	48.6	Merocyanine	4.24
571	Atto 740 carboxylate	PBS pH 7.4, 22 °C			743				763				20	120.0	0.10	12.0	nd	4.41
572	NR744	PBS, 1% DMSO pH 7.4	672	719	744	763	725	737	764	795	812	841	20	80.0	0.16	12.8	PO2R-Rosam.	4.18.2
573	BOPAHY H-D 19ba′	H_2O	344	355	367	372	379	381	386	393	403	412	19	51	0.03	1.5	Bodipy-analog	4.14.5
574	BOPAHY H-D 19aa′	H_2O	345	355	368	373	380	382	387	394	404	413	19	51	0.02	0.7	Bodipy-analog	4.14.5
575	BOPAHY H-D 19ab′	H_2O	387	400	416	422	426	429	435	444	457	468	19	51	0.20	10.2	Bodipy-analog	4.14.5
576	BOPAHY H-D 20bb′	H_2O	387	400	416	422	426	429	435	444	457	468	19	51	0.17	8.6	Bodipy-analog	4.14.5
577	2,4-DiMe TokyoGreen	100 mм NaOH	453	475	491	503	493	497	510	548	563	582	19	85.0	0.87	73.5	Fluorone	4.11
578	2,5-DiMe TokyoGreen	100 mм NaOH	453	475	491	503	493	497	510	548	563	582	19	85.0	0.89	75.4	Fluorone	4.11
579	2-Me TokyoGreen	H_2O pH 9	453	475	491	503	493	497	510	548	563	582	19	85.0	0.85	72.2	Fluorone	4.11

Ent.	Short Names	Aqueous Solutions	$\lambda_{x,-0.25}$ nm	$\lambda_{x,-0.50}$ nm	$\lambda_{x,max}$ nm	$\lambda_{x,+0.50}$ nm	$\lambda_{e,-0.25}$ nm	$\lambda_{e,-0.50}$ nm	$\lambda_{e,max}$ nm	$\lambda_{e,+0.50}$ nm	$\lambda_{e,+0.25}$ nm	$\lambda_{e,+0.10}$ nm	Δ_S	ε_{max} mм$^{-1}$cm^{-1}	Φ_f	B_{max} mм$^{-1}$cm^{-1}	Class	Ref.
580	2,6-Sulfon.-1,3,5,7-Me-BODIPY	H_2O	451	469	491	499	498	502	510	523	540	556	19	60.0	0.85	51.0	Bodipy-Small	4.14.1
581	2,4-DiOMe TokyoGreen	100 mм NaOH	456	478	494	506	496	500	513	551	567	586	19	85.0	0.20	17.0	Fluorone	4.11
582	Chlorinated fluorescein G-C 2(3)	H_2O pH 9.18, 25 °C	460	480	499	512	497	503	518	544	566	589	19	48.4	0.93	45.0	Fluorescein-ha	4.8.2
583	Chlorinated fluorescein G-C 1	H_2O pH 9.18, 25 °C	464	484	503	516	500	506	522	548	571	595	19	103.0	0.86	88.5	Fluorescein-ha	4.8.2
584	6-CA-MeThio-4,5,7-F-fluorescein	100 mм NaOH	467	488	507	520	504	510	526	553	576	600	19	63.5	0.88	55.8	Fluorescein-ha	4.8.2
585	Fluorinated fluorescein S-H 18	P pH 9	468	489	508	521	505	511	527	554	577	601	19	85.6	0.85	72.7	Fluorescein-ha	4.8.2
586	Fluorinated fluorescein S-H 19	P pH 9	468	489	508	521	505	511	527	554	577	601	19	81.2	0.96	77.9	Fluorescein-ha	4.8.2
587	Oregon Green 514	PBS pH 7.2, 22 °C	471	491	511	524	508	514	530	557	581	605	19	88.0	0.96	84.4	Fluorescein-ha	4.8.2
588	Abberior Star 512 carboxylate	PBS pH 7.4	472	493	511	524	511	516	530	552	571	593	19	84.0	0.82	68.8	Rhoda 500-550	4.2.2
589	Niu 25	PBS pH 7.4	479	499	521	533	520	524	540	572	596	625	19	58.8	0.82	48.2	Bodipy-Med.	4.14.2
590	Yakima Yellow carboxylate	100 mм P/Tris pH 7.0/7.4	487	509	530	545	525	532	549	578	603	630	19	84.0	0.96	80.6	Fluorescein-ha	4.8.2
591	Z-L D	0.5 м PBS pH 7.4	492	513	536	549	534	538	555	589	614	645	19	50	0.02	0.9	Bodipy-Large	4.14.3
592	Coumarin-pyronine 9	100 mм P pH 7.5			560				579				19	132.5	0.13	17.2	Hybrid-Bonded	4.40.2
593	JF571	10 mм HEPES pH 7.3	521	552	571	586	568	574	590	616	647	676	19	101	0.78	78.7	Rhoda 550-600	4.2.3
594	Sulforhodamine 101	H_2O pH 7	534	569	586	602	584	590	605	622	632	647	19	116.0	0.90	104.4	Sulforhodam.	4.6
595	JF593	10 mм HEPES pH 7.3	541	569	593	610	587	594	612	641	667	697	19	90.3	0.55	49.6	S-Rhodamine	4.3.3
596	NR600	PBS, 1% DMSO pH 7.4	551	580	600	613	597	603	619	646	665	689	19	67.6	0.68	45.9	POOH-Fluoron	4.12
597	Abberior Star 600 carboxylate	PBS pH 7.4	550	587	604	619	600	606	623	647	677	714	19	93.0	0.80	74.4	Rhodam >600	4.2.4
598	TMDHS or SiP-Me	H_2O	571	604	634	652	626	634	653	680	697	734	19	64.2	0.18	11.5	Si(CH3)2-Pyron	4.5

(Continued)

Ent.	Short Names	Aqueous Solutions	$\lambda_{x,-0.25}$ nm	$\lambda_{x,-0.50}$ nm	$\lambda_{x,max}$ nm	$\lambda_{x,+0.50}$ nm	$\lambda_{e,-0.25}$ nm	$\lambda_{e,-0.50}$ nm	$\lambda_{e,max}$ nm	$\lambda_{e,+0.50}$ nm	$\lambda_{e,+0.25}$ nm	$\lambda_{e,+0.10}$ nm	Δ_S	ε_{max} mм$^{-1}$cm^{-1}	Φ_f	B_{max} mм$^{-1}$cm^{-1}	Class	Ref.
599	SiTMR	10 mм HEPES pH 7.3	586	566	643	660	638	645	662	682	696	729	19	28.2	0.41	11.5	Si(CH3)2-Rhoda	4.3.2
600	CypHer5	A pH 4.6	583	621	643	657	641	646	662	679	690	719	19	100.0	0.19	19.0	Cyanine-Penta	4.13.2
601	Cy5.10-OH	H_2O/PBS	588	626	648	663	645	651	667	684	696	725	19	215.0	0.09	19.3	Cyanine-Penta	4.13.2
602	SiRF 15a	H_2O	598	577	657	675	651	658	676	697	711	746	19	40.0	0.40	16.0	Si(CH3)2-Rhoda	4.3.2
603	Square-660-carboxylate	P pH 7.4	600	635	657	675	654	660	676	702	730	769	19	182.0	0.03	5.4	Squaraine	4.35
604	Squaraine T-P 4d	67 mм P pH 7.4	606	638	658	675	654	662	677	700	719	762	19	182.0	0.03	5.4	Squaraine	4.35
605	Atto Oxa12 carboxylate	PBS pH 7.4, 22 °C	584	621	662	680	657	665	681	706	726	769	19	125.0	0.30	37.5	Phenoxazine	4.25
606	CF660R	H_2O			663				682				19	100	0.08	7.6	nd (Rhodam.)	4.41
607	NR666	PBS, 1% DMSO pH 7.4	608	646	666	683	653	661	685	719	741	778	19	165.0	0.38	62.7	PO2H-Rosam.	4.18.2
608	JF668	10 mм HEPES pH 7.3	607	585	668	686	661	668	687	708	724	760	19	26.7	0.34	9.0	POOH-Rhoda	4.3.3
609	2-Me SiR720	H_2O	627	675	721	742	710	719	740	766	790	824	19	160.0	0.05	8.0	Si(CH3)2-Rosa	4.18.1
610	HCDO	PBS pH 7.4	678	736	769	793	755	766	788	813	825	840	19	218.8	0.30	65.6	Cyanine-Hepta	4.13.3
611	BOPAHY H-D 19bb′	H_2O	387	400	416	422	425	428	434	443	455	467	18	51	0.17	8.6	Bodipy-analog	4.14.5
612	2,5-DiOMe TokyoGreen	100 mм NaOH	456	478	494	506	495	499	512	550	566	585	18	85.0	0.01	0.8	Fluorone	4.11
613	BODIPY	H_2O	457	475	498	506	504	508	516	529	547	563	18	90.0	0.90	81.2	Bodipy-Small	4.14.1
614	Niu 27	H_2O	474	494	510	520	508	512	528	563	583	612	18	63.3	0.81	51.2	Bodipy-Med.	4.14.2
615	Fluorinated rhodamine 5-SO_3H	Buffer pH 7	472	494	512	525	511	516	530	552	571	593	18	52.0	0.86	44.7	Rhoda 500-550	4.2.2
616	2′,7′-Cl-4,5,6,7-F-fluorescein	100 mм NaOH	478	500	520	534	515	521	538	566	590	615	18	88.5	0.80	70.8	Fluorescein-ha	4.8.2
617	HMDER	PBS pH 7.4	479	500	525	538	517	525	543	569	595	622	18	70.0	0.14	9.8	Rhodol	4.15.2
618	Fluorinated fluorescein S-H 20	P pH 9	491	513	535	550	529	535	553	583	608	635	18	83.4	0.47	39.2	Fluorescein-ha	4.8.2
619	Cy3.10-OH	H_2O/PBS	489	510	552	563	553	557	570	585	618	636	18	130.0	0.03	3.9	Cyanine-Trimet	4.13.1
620	Cy3.12-OH	H_2O/PBS	492	513	556	567	556	561	574	590	623	641	18	130.0	0.05	6.5	Cyanine-Trimet	4.13.1
621	2-COOH TM	100 mм P, 1% DMSO pH 10	536	560	580	595	575	580	598	651	672	700	18	110.0	0.38	41.8	Si(CH3)2-Fluore	4.9
622	2′Me-Si-rhosoL	100 mм P pH 7.4	553	582	612	627	608	614	630	651	667	693	18	70.0	0.28	19.6	Si(CH3)2-Rosol	4.20

Ent.	Short Names	Aqueous Solutions	$\lambda_{x,-0.25}$ nm	$\lambda_{x,-0.50}$ nm	$\lambda_{x,max}$ nm	$\lambda_{x,+0.50}$ nm	$\lambda_{e,-0.25}$ nm	$\lambda_{e,-0.50}$ nm	$\lambda_{e,max}$ nm	$\lambda_{e,+0.50}$ nm	$\lambda_{e,+0.25}$ nm	$\lambda_{e,+0.10}$ nm	Δ_S	ε_{max} mM^{-1}cm^{-1}	Φ_f	B_{max} mM^{-1}cm^{-1}	Class	Ref.
623	2′Me-4CH_2OH-Si-rosol	100 mM P pH 7.4	554	583	613	628	609	615	631	652	668	694	18	70.0	0.30	21.0	Si(CH3)2-Rosol	4.20
624	2′,4,5,6,7,7′-HexaF-Si-fluorescein	100 mM NaOH	567	594	617	634	609	615	635	695	720	751	18	67.9	0.32	21.7	Si(CH3)2-Fluore	4.9
625	2-Me SiR620	100 mM P, 1% DMSO pH 7.4	550	586	620	636	616	622	638	657	675	699	18	96.0	0.21	20.1	Si(CH3)2-Rosa	4.18.1
626	nor-SqO2	P pH 7.4	590	620	639	655	635	642	657	678	696	736	18	183.0	0.33	60.3	Squaraine	4.35
627	Cy5′	H_2O	583	621	642	656	639	645	660	677	688	716	18	138.0	0.17	23.4	Cyanine-Penta	4.13.2
628	JF646	10 mM HEPES pH 7.3	589	568	646	663	639	646	664	684	698	732	18	5.6	0.54	3.0	Si(CH3)2-Rhoda	4.3.2
629	Atto 647N carboxylate	PBS pH 7.4, 22 °C	592	626	646	661	642	648	664	690	706	753	18	150.0	0.65	97.5	C(CH3)2-Rhoda	4.3.1
630	Zy5	P pH 7.4	587	625	647	662	644	649	665	682	694	722	18	250.0	0.27	67.5	Cyanine-Penta	4.13.2
631	SMSiR	H_2O	584	626	652	671	643	651	670	695	717	751	18	120.0	0.31	37.2	Si(CH3)2-Srhod	4.7
632	Square-650-carboxylate	P pH 5.6	598	632	653	668	651	658	671	694	712	765	18	135.0	0.16	21.6	Squaraine	4.35
633	2-Me SiR654	100 mM P, 1% DMSO pH 7.4	576	616	654	671	647	655	672	693	713	740	18	80.0	0.24	19.2	Si(CH3)2-Rosa	4.18.1
634	SiRF 1a-H	H_2O	602	581	662	680	654	662	680	701	716	751	18	118.0	0.66	77.8	Si(CH3)2-Rhoda	4.3.2
635	Atto 665 carboxylate	PBS pH 7.4, 22 °C			662				680				18	160.0	0.60	96.0	nd (pb Pyron.)	4.41
636	$SiOH_2R$	10 mM PBS pH 7.4	599	639	663	679	656	662	681	700	713	738	18	105.0	0.43	45.1	Si(OH)2-Rosa	4.18.1
637	Si-9-Iminopyronine 2f	~1 mM HCl, 1% DMSO pH 3	595	631	664	683	653	661	682	711	730	770	18	123	0.26	31.9	Si(CH3)2-Pyron	4.5
638	2-Me SiR666	100 mM P, 1% DMSO pH 7.4	585	627	666	684	659	666	684	706	726	755	18	80.0	0.25	20.0	Si(CH3)2-Rosa	4.18.1
639	SqCN1	P pH 7.4	614	647	667	685	661	669	685	708	728	772	18	188.0	0.07	13.1	Squaraine	4.35
640	Squaraine T-P 5b	67 mM P pH 7.4	614	647	667	685	661	669	685	708	728	772	18	188.0	0.07	13.1	Squaraine	4.35
641	CF660C	H_2O			667				685				18	200	0.06	11.1	nd (Cyanine)	4.41
642	Zy5.5	P pH 7.4	610	652	676	692	671	677	694	713	725	757	18	190.0	0.23	43.7	Cyanine-Penta	4.13.2
643	DY-679P1	PBS pH 7.4	612	655	679	696	673	680	697	715	726	745	18	200.0	0.23	46.6	Cyanine-Penta	4.13.2
644	Me-PR	PBS, 1% DMSO pH 7.4	632	672	694	711	682	690	712	740	757	789	18	92.0	0.11	10.1	POR-Rosam.	4.18.2

(Continued)

Ent.	Short Names	Aqueous Solutions	$\lambda_{x,-0.25}$ nm	$\lambda_{x,-0.50}$ nm	$\lambda_{x,max}$ nm	$\lambda_{x,+0.50}$ nm	$\lambda_{e,-0.25}$ nm	$\lambda_{e,-0.50}$ nm	$\lambda_{e,max}$ nm	$\lambda_{e,+0.50}$ nm	$\lambda_{e,+0.25}$ nm	$\lambda_{e,+0.10}$ nm	Δ_S	ε_{max} mᴍ$^{-1}$cm^{-1}	Φ_f	B_{max} mᴍ$^{-1}$cm^{-1}	Class	Ref.
645	PR	PBS, 1% DMSO pH 7.4	632	672	694	711	682	690	712	740	757	789	18	65.0	0.06	3.9	POR-Rosam.	4.18.2
646	DY-780 carboxylate	PBS pH 7.4	657	705	782	817	745	760	800	888	943	1007	18	102.0	0.03	3.2	Merocyanine	4.24
647	2-Me-4-OMe TokyoGreen	100 mᴍ NaOH	454	476	492	504	492	496	509	547	562	581	17	85.0	0.84	71.4	Fluorone	4.11
648	Z-L C	0.5 ᴍ PBS pH 7.4	489	510	533	546	529	533	550	583	608	638	17	50	0.01	0.6	Bodipy-Large	4.14.3
649	AF555	PBS pH 7.2, 22 °C	491	512	555	566	555	559	572	587	620	639	17	158.0	0.09	14.2	Cyanine-Trimet	4.13.1
650	Alexa Fluor 555	PBS pH 7.2, 22 °C	491	512	555	566	555	559	572	587	620	639	17	155.0	0.10	15.5	Cyanine-Trimet	4.13.1
651	Atto 610 carboxylate	PBS pH 7.4, 22 °C	529	554	616	581	568	573	633	608	634	659	17	150.0	0.70	105.0	C(CH3)2-Pyron	4.5
652	BODIPY 630/650	H_2O	569	597	629	647	618	623	646	692	728	772	17	97.0	0.50	48.5	Bodipy-Small	4.14.1
653	nor-SqO1	P pH 7.4	588	618	636	652	631	639	653	674	692	731	17	149.0	0.32	47.6	Squaraine	4.35
654	B-Z 14	H_2O	576	605	638	656	626	632	655	703	739	784	17	102.0	0.42	42.8	Bodipy-Large	4.14.3
655	Abberior Star Red carboxylate	PBS pH 7.4	578	619	638	655	630	637	655	682	715	756	17	100.0	0.66	66.0	Rhodam >600	4.2.4
656	Neo-Cy5	PBS pH 7.4	583	621	643	657	639	645	660	677	688	716	17	170.0	0.10	17.0	Cyanine-Penta	4.13.2
657	3-Methyl-JF646	10 mᴍ HEPES pH 7.3	572	612	649	666	642	649	666	687	706	733	17	118	0.47	55.4	Si(CH3)2-Rosa	4.18.1
658	EVOblue 30	H_2O	575	610	650	667	644	651	667	691	710	751	17	100.0	0.10	10.0	Phenoxazine	4.25
659	Cy5.18.OH	PBS	590	628	650	664	646	651	667	685	698	739	17	250.0	0.27	67.5	Cyanine-Penta	4.13.2
660	Squaraine T-P 4b	67 mᴍ P pH 7.4	607	639	659	676	653	661	676	698	718	760	17	207.0	0.02	4.1	Squaraine	4.35
661	Phosphorilated oxazine 2-H	H_2O pH ~7	583	620	661	679	654	662	678	703	723	765	17	60.0	0.28	16.8	Phenoxazine	4.25
662	nor-SqO3	P pH 7.4	610	642	662	679	656	663	679	702	721	764	17	124.0	0.21	26.0	Squaraine	4.35
663	SiRF 1b-H	H_2O	603	581	663	681	654	662	680	701	716	751	17	90.0	0.70	63.0	Si(CH3)2-Rhoda	4.3.2
664	Atto 655 carboxylate	PBS pH 7.4, 22 °C	587	629	663	679	656	664	680	707	733	784	17	125.0	0.30	37.5	Phenoxazine	4.25
665	2-Me TeOR	PBS pH 7.4	599	642	669	692	654	663	686	714	737	784	17	120.0	0.18	21.6	TeO-Rosamine	4.18.3
666	2-Me SiR670	100 mᴍ P, 1% DMSO pH 7.4	588	630	670	688	661	669	687	709	730	759	17	90.0	0.27	24.3	Si(CH3)2-Rosa	4.18.1
667	Atto 680 carboxylate	PBS pH 7.4, 22 °C	603	648	681	696	675	682	698	727	749	807	17	125.0	0.30	37.5	Phenoxazine	4.25
668	CF680	H_2O			681				698				17	210	0.04	8.2	nd (Cyanine)	4.41
669	JF690	10 mᴍ HEPES pH 7.3	625	602	690	710	679	687	707	730	746	784	17	150.0	0.24	36.0	POOH-Rhoda	4.3.3

Ent.	Short Names	Aqueous Solutions	$\lambda_{x,-0.25}$ nm	$\lambda_{x,-0.50}$ nm	$\lambda_{x,max}$ nm	$\lambda_{x,+0.50}$ nm	$\lambda_{e,-0.25}$ nm	$\lambda_{e,-0.50}$ nm	$\lambda_{e,max}$ nm	$\lambda_{e,+0.50}$ nm	$\lambda_{e,+0.25}$ nm	$\lambda_{e,+0.10}$ nm	Δ_S	ε_{max} mм$^{-1}$cm^{-1}	Φ_f	B_{max} mм$^{-1}$cm^{-1}	Class	Ref.
670	DY-681 carboxylate	PBS pH 7.4	592	630	691	718	665	676	708	776	818	866	17	125.0	0.09	11.2	Merocyanine	4.24
671	tMe-PR	PBS, 1% DMSO pH 7.4	634	674	696	714	682	691	713	741	759	790	17	68.0	0.15	10.2	POR-Rosam.	4.18.2
672	Chlorinated fluorescein G-C 5	H_2O pH 9.18, 25 °C	471	491	511	524	505	511	527	554	577	601	16	91.6	0.84	76.9	Fluorescein-ha	4.8.2
673	Chlorinated fluorescein G-C 4	H_2O pH 9.18, 25 °C	471	492	512	526	506	512	528	555	578	602	16	93.8	0.87	81.6	Fluorescein-ha	4.8.2
674	DAMBO-P^{H}-T	100 mм P pH 7.4	479	499	521	533	517	521	537	569	592	621	16	56.0	0.74	41.4	Bodipy-Small	4.14.1
675	Z-L E	0.5 м PBS pH 7.4	490	511	534	547	529	533	550	583	608	638	16	50	0.09	4.4	Bodipy-Large	4.14.3
676	Cy3.13-OH	H_2O/PBS	487	508	550	561	549	553	566	581	613	631	16	130.0	0.03	3.9	Cyanine-Trimet	4.13.1
677	Redmond Red	100 mм P pH 7.1	518	547	579	593	577	583	595	614	629	661	16	52.3	0.84	43.9	Phenoxazine	4.25
678	2-Me Tokyo Magenta	100 mм P pH 9	532	563	582	595	578	583	598	617	633	655	16	110.0	0.42	46.2	Si(CH3)2-Fluoron	4.12
679	2-COOH DCTM	100 mм P, 1% DMSO pH 10	545	570	591	606	583	589	607	662	684	712	16	120.0	0.48	57.6	Si(CH3)2-Fluore	4.9
680	2′,4′,5′,7′-TetraF-Si-fluorescein	100 mм NaOH	550	575	597	613	588	594	613	669	691	720	16	65.0	0.67	43.5	Si(CH3)2-Fluore	4.9
681	4,5,6,7-TetraF-Si-fluorescein	100 mм NaOH	551	576	598	614	589	595	614	670	693	722	16	84.1	0.50	42.0	Si(CH3)2-Fluore	4.9
682	RF620	10 mм PBS pH 7.4	569	600	620	635	614	620	636	659	674	701	16	109.0	0.36	39.2	B(OH)2-Rosa	4.18.3
683	Seta-635-pH-di-carboxylate	P pH 9			640				656				16	188.0	0.33	62.0	nd	4.41
684	Distyryl-BODIPY 13	PBS pH 7.5	579	608	641	659	628	633	657	705	742	787	16	58.0	0.25	14.5	Bodipy-Med.	4.14.2
685	3,6-Azetid.-9-CA.-Si-pyronine	10 mм HEPES pH 7.3	577	610	641	659	630	638	657	684	701	739	16	120.0	0.26	31.2	Si(CH3)2-Pyron	4.5
686	SiR-COOH	H_2O	588	567	645	662	637	644	661	681	695	728	16	100.0	0.39	39.0	Si(CH3)2-Rhoda	4.3.2
687	SiR	PBS pH 7.4	589	623	645	663	640	647	661	687	702	737	16	93.0	0.41	38.1	Si(CH3)2-Rhoda	4.3.2
688	Sulfo-cyanine5	H_2O	586	624	646	661	641	646	662	679	690	719	16	271.0	0.20	54.2	Cyanine-Penta	4.13.2
689	Cy5	H_2O	587	625	647	662	642	647	663	680	691	720	16	250.0	0.27	67.5	Cyanine-Penta	4.13.2
690	SiP-Ph	50 mм PBS pH 7.4	588	624	649	665	613	630	665	672	683	698	16	80.0	0.23	18.4	C(CH3)2-Pyron	4.5

(Continued)

Ent.	Short Names	Aqueous Solutions	$\lambda_{x,-0.25}$ nm	$\lambda_{x,-0.50}$ nm	$\lambda_{x,max}$ nm	$\lambda_{x,+0.50}$ nm	$\lambda_{e,-0.25}$ nm	$\lambda_{e,-0.50}$ nm	$\lambda_{e,max}$ nm	$\lambda_{e,+0.50}$ nm	$\lambda_{e,+0.25}$ nm	$\lambda_{e,+0.10}$ nm	Δ_S	ε_{max} mм$^{-1}$cm^{-1}	Φ_f	B_{max} mм$^{-1}$cm^{-1}	Class	Ref.
691	Cy5.29.OH	PBS	589	628	650	665	645	650	666	683	695	724	16	250.0	0.20	50.0	Cyanine-Penta	4.13.2
692	Thienyl-JF646	10 mм HEPES pH 7.3	593	572	651	668	642	649	667	687	701	735	16	113	0.40	45.2	Si(CH3)2-Rhoda	4.3.2
693	S-JF646	10 mм HEPES pH 7.3	586	620	652	671	640	648	668	696	714	753	16	21.2	0.25	5.3	Si(CH3)2-Pyron	4.5
694	Square-680-carboxylate	P pH 7.4	603	634	654	671	647	655	670	692	711	753	16	138.0	0.04	5.1	Squaraine	4.35
695	Z-L H	0.5 м PBS pH 7.4	594	625	660	679	645	651	676	727	766	815	16	50	0.01	0.6	Bodipy-Large	4.14.3
696	2-Me SiR671	100 mм P, 1% DMSO pH 7.4	589	631	671	689	661	669	687	709	730	759	16	99.0	0.29	28.7	Si(CH3)2-Rosa	4.18.1
697	Seta-700-carboxylate	P pH 7.4			687				703				16	177.0	0.06	10.6	nd	4.41
698	Seta-700-di-carboxylate	P pH 7.4			688				704				16	180.0	0.11	19.8	nd	4.41
699	Atto 700 carboxylate	PBS pH 7.4, 22 °C	622	666	700	717	693	701	716	744	768	828	16	120.0	0.25	30.0	Phenoxazine	4.25
700	ZW800-1	FBS	680	739	772	797	755	766	788	813	825	840	16	249.0	0.15	37.6	Cyanine-Hepta	4.13.3
701	ZW800-3a	FBS	682	741	774	799	757	768	790	815	827	842	16	309.0	0.16	49.7	Cyanine-Hepta	4.13.3
702	RS800	FBS	690	750	784	809	766	777	800	826	838	853	16	240.0	0.17	40.5	Cyanine-Hepta	4.13.3
703	Niu 18	H_2O	455	474	496	504	499	503	511	524	541	557	15	75.5	0.20	15.1	Bodipy-Med.	4.14.2
704	2,7-DiCl-6-pyrrolidino sulforhodol	PBS pH 7.4	470	491	520	534	516	521	535	555	584	604	15	83.0	0.27	22.4	Sulforhodol	4.27
705	TET	H_2O pH 9	479	501	521	535	513	519	536	564	588	613	15	73.0	0.70	51.1	Fluorescein-ha	4.8.2
706	Chlorinated fluorescein G-C 6	H_2O pH 9.18, 25 °C	480	501	522	536	514	520	537	565	589	614	15	65.8	0.83	54.6	Fluorescein-ha	4.8.2
707	Cy3′	H_2O	483	504	545	556	543	547	560	575	606	624	15	120.0	0.04	4.8	Cyanine-Trimet	4.13.1
708	Sulfo-cyanine3	H_2O	486	506	548	559	546	550	563	578	610	628	15	162.0	0.10	16.2	Cyanine-Trimet	4.13.1
709	Cy3.18.OH	PBS	487	508	550	561	548	552	565	580	612	630	15	150.0	0.04	6.0	Cyanine-Trimet	4.13.1
710	Seta-555-carboxylate	P pH 7.4			555				570				15	155.0	0.07	10.8	nd	4.41
711	8-Trifluoromethyl-BODIPY 11	100 mм P pH 7.4	519	543	569	583	561	565	584	622	650	685	15	57.0	0.95	54.1	Bodipy-Small	4.14.1
712	Cy3.205	PBS pH 7.4	509	532	578	590	574	579	593	610	645	665	15	120.0	0.14	16.8	Cyanine-Trimet	4.13.1
713	2-COOH DFTM	100 mм P, 1% DMSO pH 10	537	560	581	596	573	578	596	649	670	697	15	120.0	0.54	64.8	Si(CH3)2-Fluore	4.9

Ent.	Short Names	Aqueous Solutions	$\lambda_{x,-0.25}$ nm	$\lambda_{x,-0.50}$ nm	$\lambda_{x,max}$ nm	$\lambda_{x,+0.50}$ nm	$\lambda_{e,-0.25}$ nm	$\lambda_{e,-0.50}$ nm	$\lambda_{e,max}$ nm	$\lambda_{e,+0.50}$ nm	$\lambda_{e,+0.25}$ nm	$\lambda_{e,+0.10}$ nm	Δ_S	ε_{max} mM^{-1}cm^{-1}	Φ_f	B_{max} mM^{-1}cm^{-1}	Class	Ref.
714	Cy3.5	H_2O	511	534	581	593	577	582	596	613	649	669	15	120.0	0.14	16.8	Cyanine-Trimet	4.13.1
715	Alexa Fluor 633	PBS/H_2O pH 7.2, 20/22 °C	568	608	632	650	621	629	647	670	691	722	15	100.0	0.64	64.0	Sulforhodam.	4.6
716	Distyryl-BODIPY 12	PBS pH 7.5	579	609	642	660	628	633	657	705	742	787	15	55.1	0.22	12.1	Bodipy-Med.	4.14.2
717	3-Methyl-5-Methoxy-JF646	10 mM HEPES pH 7.3	572	612	649	666	640	647	664	685	704	731	15	110	0.49	53.9	Si(CH3)2-Rosa	4.18.1
718	CF647	H_2O			650				665				15	240.0	0.12	27.9	nd (Cyanine)	4.41
719	3-Hydroxy-JF646	10 mM HEPES pH 7.3	574	614	651	668	642	649	666	687	706	733	15	136	0.48	65.2	Si(CH3)2-Rosa	4.18.1
720	Sulfo-JF646	10 mM HEPES pH 7.3	586	629	655	674	643	651	670	695	717	751	15	124.0	0.51	63.2	Si(CH3)2-Srhod	4.7
721	Si-9-Iminopyronine 2c	~1 mM HCl, 1% DMSO pH 3	597	633	666	685	652	660	681	710	729	769	15	83.2	0.28	23.3	Si(CH3)2-Pyron	4.5
722	4,5,6,7-Tetrafluoro-SiTMR	10 mM HEPES pH 7.3	606	584	667	685	656	664	682	703	718	754	15	132.0	0.31	40.9	Si(CH3)2-Rhoda	4.3.2
723	2-Me SiR680	H_2O	592	634	674	693	663	671	689	711	732	761	15	130.0	0.35	45.5	Si(CH3)2-Rosa	4.18.1
724	LS479	Buffer pH 8	661	716	747	770	731	741	762	786	796	810	15	105.0	0.04	4.4	Cyanine-Hepta	4.13.3
725	ICG	FBS	708	771	807	834	787	798	822	850	862	878	15	121.0	0.09	11.2	Cyanine-Hepta	4.13.3
726	L-B 2a	H_2O	473	493	509	519	503	507	523	557	577	606	14	72.4	0.27	19.5	Bodipy-Small	4.14.1
727	Cy3B	H_2O	493	515	558	569	555	559	572	587	620	639	14	130.0	0.67	87.1	Cyanine-Trimet	4.13.1
728	M-S 10a	H_2O, 0.1% TX-100	516	539	565	579	556	561	579	616	644	678	14	69.0	0.65	44.8	Bodipy-Small	4.14.1
729	M-S 10e	H_2O, 0.1% TX-100	532	556	584	599	574	578	598	637	667	704	14	124.0	0.41	50.8	Bodipy-Small	4.14.1
730	M-S 10f	H_2O, 0.1% TX-100	536	561	589	604	578	583	603	643	674	711	14	74.0	0.18	13.3	Bodipy-Small	4.14.1
731	Squaraine T-P 4a	67 mM P pH 7.4	581	610	628	644	621	628	642	662	679	718	14	163.0	0.03	4.8	Squaraine	4.35
732	SiP-F	50 mM PBS pH 7.4	577	611	637	653	602	612	651	655	667	679	14	64.2	0.20	12.8	C(CH3)2-Pyron	4.5
733	2-Me SiR650	H_2O	591	628	646	661	636	642	660	681	707	749	14	110.0	0.31	34.1	Si(CH3)2-Rosa	4.18.1
734	3,7-Di(azetidin)phenoxazine	10 mM HEPES pH 7.3	572	607	647	664	638	646	661	684	703	743	14	99.0	0.24	23.7	Phenoxazine	4.25

(Continued)

Ent.	Short Names	Aqueous Solutions	$\lambda_{x,-0.25}$ nm	$\lambda_{x,-0.50}$ nm	$\lambda_{x,max}$ nm	$\lambda_{x,+0.50}$ nm	$\lambda_{e,-0.25}$ nm	$\lambda_{e,-0.50}$ nm	$\lambda_{e,max}$ nm	$\lambda_{e,+0.50}$ nm	$\lambda_{e,+0.25}$ nm	$\lambda_{e,+0.10}$ nm	Δ_S	ε_{max} mM^{-1}cm^{-1}	Φ_f	B_{max} mM^{-1}cm^{-1}	Class	Ref.
735	Des-carboxy-JF646	10 mM HEPES pH 7.3	571	611	648	665	638	645	662	683	701	728	14	116	0.20	23.2	Si(CH3)2-Rosa	4.18.1
736	SiR-methyl	H_2O	571	611	648	665	638	645	662	683	701	728	14	100.0	0.39	39.0	Si(CH3)2-Rosa	4.18.1
737	3,7-Di(ethylamino)-phenoxazine	10 mM HEPES pH 7.3	578	615	655	673	646	653	669	693	712	754	14	111.0	0.07	7.7	Phenoxazine	4.25
738	5-Aza-3-methyl-JF646	10 mM HEPES pH 7.3	578	618	656	674	646	653	670	691	710	738	14	133	0.48	63.8	Si(CH3)2-Rosa	4.18.1
739	JF669-6-SCH2CO2H	10 mM HEPES pH 7.3	606	584	667	685	655	663	681	702	717	752	14	130.0	0.38	49.4	Si(CH3)2-Rhoda	4.3.2
740	NR698	PBS, 1% DMSO pH 7.4	631	676	698	715	686	693	712	738	751	773	14	25.9	0.32	8.2	PO2H-Rosam.	4.18.2
741	CW800	FBS	691	752	786	812	766	777	800	826	838	853	14	237.0	0.14	33.6	Cyanine-Hepta	4.13.3
742	L-B 1a	H_2O	453	472	494	502	495	499	507	520	537	552	13	100.0	0.47	47.0	Bodipy-Small	4.14.1
743	L-B 4a	H_2O	453	472	494	502	495	499	507	520	537	552	13	69.2	0.34	23.5	Bodipy-Small	4.14.1
744	L-B 5	H_2O	457	475	498	506	499	503	511	524	541	557	13	79.4	0.49	38.9	Bodipy-Small	4.14.1
745	G-L 14	H_2O	472	492	508	518	501	505	521	555	575	603	13	59.0	0.77	45.4	Bodipy-Med.	4.14.2
746	G-L 13b	H_2O	473	493	509	519	502	506	522	556	576	604	13	56.0	0.67	37.5	Bodipy-Med.	4.14.2
747	B-Z 7b	H_2O	479	500	516	526	509	513	529	564	584	614	13	60.5	0.68	41.1	Bodipy-Large	4.14.3
748	Distyryl-BODIPY 18	H_2O	568	596	628	646	613	619	641	687	721	764	13	71.0	0.38	26.9	Bodipy-Med.	4.14.2
749	Distyryl-BODIPY 17	H_2O	570	598	630	648	615	620	643	689	724	767	13	76.6	0.30	22.9	Bodipy-Med.	4.14.2
750	3,6-Azetid.-Si-pyronine	10 mM HEPES pH 7.3	573	606	636	654	623	630	649	675	692	729	13	97.1	0.62	60.2	Si(CH3)2-Pyron	4.5
751	3,7-Dimethyl-JF646	10 mM HEPES pH 7.3	574	614	651	668	640	647	664	685	704	731	13	89.6	0.51	45.7	Si(CH3)2-Rosa	4.18.1
752	JF669-6-SCH2CO2Et	10 mM HEPES pH 7.3	607	585	668	686	655	663	681	702	717	752	13	119.0	0.38	45.2	Si(CH3)2-Rhoda	4.3.2
753	JF669	10 mM HEPES pH 7.3	608	586	669	687	656	664	682	703	718	754	13	112.0	0.37	41.4	Si(CH3)2-Rhoda	4.3.2
754	G-L 13a	H_2O	463	482	505	514	505	509	517	530	548	564	12	35.0	0.47	16.4	Bodipy-Med.	4.14.2
755	L-B 2b	H_2O	476	496	512	522	504	508	524	559	578	607	12	77.6	0.41	31.8	Bodipy-Small	4.14.1
756	Square-635-carboxylate	P pH 5.6	591	617	634	648	624	633	646	665	679	725	12	135.0	0.16	21.6	Squaraine	4.35

Ent.	Short Names	Aqueous Solutions	$\lambda_{x,-0.25}$ nm	$\lambda_{x,-0.50}$ nm	$\lambda_{x,max}$ nm	$\lambda_{x,+0.50}$ nm	$\lambda_{e,-0.25}$ nm	$\lambda_{e,-0.50}$ nm	$\lambda_{e,max}$ nm	$\lambda_{e,+0.50}$ nm	$\lambda_{e,+0.25}$ nm	$\lambda_{e,+0.10}$ nm	Δ_S	ε_{max} mM^{-1}cm^{-1}	Φ_f	B_{max} mM^{-1}cm^{-1}	Class	Ref.
757	Squaraine T-P 5a	67 mM P pH 7.4	592	620	636	653	624	631	648	666	680	705	12	216.0	0.06	12.9	Squaraine	4.35
758	TG170	H_2O	695	756	791	817	769	780	803	829	841	857	12	300.0	0.02	7.4	Cyanine-Hepta	4.13.3
759	Z-L B	0.5 M PBS pH 7.4	455	474	496	504	495	499	507	520	537	552	11	50	0.36	17.8	Bodipy-Large	4.14.3
760	L-B 1b	H_2O	457	475	498	506	497	501	509	522	539	555	11	100.0	0.37	37.0	Bodipy-Small	4.14.1
761	L-B 4b	H_2O	457	475	498	506	497	501	509	522	539	555	11	77.6	0.15	11.6	Bodipy-Small	4.14.1
762	Z-L A	0.5 M PBS pH 7.4	457	476	499	507	498	502	510	523	540	556	11	50	0.04	2.1	Bodipy-Large	4.14.3
763	Niu 29	H_2O	477	497	519	531	511	515	530	561	584	612	11	28.0	0.61	17.0	Bodipy-Med.	4.14.2
764	Squaraine T-P 4c	67 mM P pH 7.4	577	606	624	639	615	621	635	655	672	709	11	245.0	0.02	4.9	Squaraine	4.35
765	Seta-633-carboxylate	P pH 7.4			633				644				11	250.0	0.07	17.5	nd	4.41
766	Seta-660-di-carboxylate	P pH 7.4			661				672				11	220.0	0.11	24.2	nd	4.41
767	Z-L I	0.5 M PBS pH 7.4	451	469	491	499	489	493	501	513	530	545	10	50	0.68	34.0	Bodipy-Large	4.14.3
768	Z-L F	0.5 M PBS pH 7.4	459	478	501	509	499	503	511	524	541	557	10	50	0.31	15.6	Bodipy-Large	4.14.3
769	2-Me DCTM	100 mM P pH 9	544	577	597	611	586	592	607	627	643	666	10	130.0	0.48	62.4	Si(CH3)2-Fluoron	4.12
770	SqO1	P pH 7.4	584	614	632	648	621	628	642	662	679	718	10	265.0	0.07	18.5	Squaraine	4.35
771	Squaraine T-P 5c	67 mM P pH 7.4	584	614	632	648	621	628	642	662	679	718	10	265.0	0.06	15.9	Squaraine	4.35
772	Seta-646-carboxylate	P pH 7.4			646				656				10	207.0	0.10	20.7	nd	4.41
773	DAMBO-T	100 mM P pH 7.4	457	475	498	506	495	499	507	520	537	552	9	52.0	0.40	20.8	Bodipy-Small	4.14.1
774	B-Z 7a	H_2O	458	477	500	508	497	501	509	522	539	555	9	64.5	0.61	39.3	Bodipy-Large	4.14.3
775	BODIPY FL	H_2O	460	479	502	511	499	503	511	524	541	557	9	92.0	0.97	89.2	Bodipy-Small	4.14.1
776	Seta-632-carboxylate	P pH 7.4			632				641				9	280.0	0.06	16.8	nd	4.41
777	M-S 10d	H_2O, 0.1% TX-100	523	547	574	589	559	563	582	619	648	682	8	118.0	0.57	67.2	Bodipy-Small	4.14.1

Appendix E

E1. Ranked by Brightness

Appendix E contains the database of over 700 flourescent dyes identified by their short names and ranked by descending brightness values. Many short names are registered trademarks and are listed in footnote[1]. Maximum brightness (B_{max}) is one of the most important photophysical parameters of a fluorescent dye and is defined as the product of the molar absorption coefficient (ε_{max}) and fluorescence quantum yield (ϕ_f). Both molar absorption coefficient and fluorescence quantum yield were measured in aqueous solutions such as plain water, phosphate-buffered saline, or any aqueous buffer (Table E.1). Fluorescent dyes whose photophysical parameters were measured in aqueous buffers with more than 1% (V/V) organic modifier were not included in this database (with only a few exceptions). Most of the values of ε_{max} and Φ_f were taken from the literature (articles, thesis, patents, catalogs, and books), and only a few were measured by the author (~5%). The locations of the references are indicated in the column "Ref." Important properties of the dyes are revealed in Chapter 5 through informative scattergrams.

The molecular structures together with some photophysical parameters are shown in Chapter 4. Chapter 4 also provides a detailed list of references where the photophysical parameters were taken. Important properties of the dyes are also revealed in Chapter 5. Panoramic views of the photophysical properties and classes are given in Chapter 5 using scattergrams. Methods to measure photostability and techniques to improve photostability are discussed in Chapter 7. The photostability (relative photobleaching mean-lifetime) of some dyes is also shown in Chapter 7.

Acronyms used in Table E.1: A, acetate; AC, acetic acid; ACN, acetonitrile; BODIPY, borondipyrromethene; BSA, bovine serum albumin; C, carbonate; CA, carboxylic acid; FBS, fetal bovine serum; Et, ethanol; nd, non-disclosed; ni, not informed; P, phosphate; pb, probably; pH >7, basic medium; pH < 7, acidic medium (when the exact value of the pH is not given in the literature); SA, sulfonic acid; and TX-100, triton X-100.

1 Abberior STAR is a trademark of Abberior GmbH (Göttingen, Germany). AlexaFluor, BODIPY, Cascade Blue, FAM, JOE, TET, HEX, Marina Blue, Oregon Green, Pacific Blue, ROX, TAMRA, and Texas Red are trademarks of Thermo Fisher Scientific (Waltham, MA, USA). ATTO is a trademark of ATTO-Tec GmbH (Siegen, Germany). CF Dye is a registered trademark of Biotium, Inc. (Fremont, CA, USA). Cy3, Cy3.5, Cy5, and Cy5.5 are trademarks of GE Healthcare (Chicago, IL, USA). Dy is a trademark of Dyomics GmbH (Jena, Germany). Seta, SeTau, and Square dyes are trademarks of SETA BioMedicals (Urbana, IL, USA).

Fluorescent Dye Labels and Stains: A Database of Photophysical Properties, First Edition. Tarso B. Ledur Kist.

Table E.1 Photophysical properties of fluorescent labels and stains in aqueous solutions ranked by brightnesses (B_{max}).

Ent.	Short names	Aqueous solution	$\lambda_{x,-0.25}$ nm	$\lambda_{x,-0.50}$ nm	$\lambda_{x,max}$ nm	$\lambda_{x,+0.50}$ nm	$\lambda_{e,-0.25}$ nm	$\lambda_{e,-0.50}$ nm	$\lambda_{e,max}$ nm	$\lambda_{e,+0.50}$ nm	$\lambda_{e,+0.25}$ nm	$\lambda_{e,+0.10}$ nm	Δ_S nm	ε_{max} mM^{-1}cm^{-1}	Φ_f	B_{max} mM^{-1}cm^{-1}	Class	Ref.
1	Atto Rho14 carboxylate	PBS pH 7.4, 22 °C	564	606	626	639	622	628	646	668	693	738	20	140.0	0.80	112.0	Rhodam >600	4.2.4
2	NIR4	H_2O	589	628	650	665	649	655	671	688	700	729	21	260.0	0.43	111.8	Cyanine-Penta	4.13.2
3	Atto 542 carboxylate	PBS pH 7.4, 22 °C			542				562				20	120.0	0.93	111.6	nd (pb Rhoda)	4.41
4	Atto Rho6G carboxylate	PBS pH 7.4, 22 °C	489	513	533	546	534	540	557	583	609	636	24	115.0	0.94	108.1	Rhoda 500-550	4.2.2
5	Atto 565 carboxylate	PBS pH 7.4, 22 °C	513	543	564	578	565	572	590	617	648	676	26	120.0	0.90	108.0	Rhoda 550-600	4.2.3
6	Atto 610 carboxylate	PBS pH 7.4, 22 °C	529	554	616	581	568	573	633	608	634	659	17	150.0	0.70	105.0	C(CH3)2-Pyron	4.5
7	Sulforhodamine 101	H_2O pH 7	534	569	586	602	584	590	605	622	632	647	19	116.0	0.90	104.4	Sulforhodam.	4.6
8	Atto 532 carboxylate	PBS pH 7.4, 22 °C	491	515	532	544	533	538	552	575	600	625	20	115.0	0.90	103.5	Rhoda 500-550	4.2.2
9	Rhodamine 6G	D_2O pH <7	484	506	525.5	539	531	537	552.0	575	596	620	26	105.0	0.98	102.9	Rhoda 500-550	4.2.2
10	Atto 594 carboxylate	PBS pH 7.4, 22 °C	547	583	603	617	601	608	626	652	688	718	23	120.0	0.85	102.0	Rhodam >600	4.2.4
11	Atto 520 carboxylate	PBS pH 7.4, 22 °C	474	497	517	529	520	525	538	556	567	592	21	110.0	0.90	99.0	Pyronine	4.4
12	Atto 514 carboxylate	PBS pH 7.4, 22 °C	474	494	511	522	512	517	532	553	575	599	21	115.0	0.85	97.7	Rhoda 500-550	4.2.2
13	Rhodamine 6G	H_2O pH <7	485	507	526.3	540	532	537	552.3	576	596	621	26	105.0	0.93	97.6	Rhoda 500-550	4.2.2
14	Atto 647N carboxylate	PBS pH 7.4, 22 °C	592	626	646	661	642	648	664	690	706	753	18	150.0	0.65	97.5	C(CH3)2-Rhoda	4.3.1
15	Atto 550 carboxylate	PBS pH 7.4, 22 °C	507	535	554	567	554	560	576	598	626	654	22	120.0	0.80	96.0	Rhoda 550-600	4.2.3
16	Atto Rho11 carboxylate	PBS pH 7.4, 22 °C	522	553	572	585	571	577	595	619	650	680	23	120.0	0.80	96.0	Rhoda 550-600	4.2.3
17	Atto Rho12 carboxylate	PBS pH 7.4, 22 °C	526	557	577	590	577	583	600	627	659	689	23	120.0	0.80	96.0	Rhoda 550-600	4.2.3

Ent.	Short names	Aqueous solution	$\lambda_{x,-0.25}$ nm	$\lambda_{x,-0.50}$ nm	$\lambda_{x,max}$ nm	$\lambda_{x,+0.50}$ nm	$\lambda_{e,-0.25}$ nm	$\lambda_{e,-0.50}$ nm	$\lambda_{e,max}$ nm	$\lambda_{e,+0.50}$ nm	$\lambda_{e,+0.25}$ nm	$\lambda_{e,+0.10}$ nm	Δ_S nm	ε_{max} mм$^{-1}$cm^{-1}	Φ_f	B_{max} mм$^{-1}$cm^{-1}	Class	Ref.
18	Atto Rho101 carboxylate	PBS pH 7.4, 22 °C	535	569	587	601	586	592	609	635	664	698	22	120.0	0.80	96.0	Rhoda 550-600	4.2.3
19	Atto 590 carboxylate	PBS pH 7.4, 22 °C	535	569	593	609	594	601	622	651	687	718	29	120.0	0.80	96.0	Rhoda 550-600	4.2.3
20	Atto Rho13 carboxylate	PBS pH 7.4, 22 °C	547	584	603	618	602	609	627	654	690	723	24	120.0	0.80	96.0	Rhodam >600	4.2.4
21	Atto 665 carboxylate	PBS pH 7.4, 22 °C			662				680				18	160.0	0.60	96.0	nd (pb Pyron.)	4.41
22	Rhodamine 6G +2	H_2O pH <7	487	511	530	544	533	539	553	571	583	603	23	101.0	0.93	93.9	Rhoda 500-550	4.2.2
23	Pyronine 630	H_2O, 25 °C	509	535	559	573	558	564	579	600	613	642	20	100.0	0.92	92.0	Pyronine	4.4
24	Rhodamine 6G +3	H_2O pH <7	486	510	529	544	532	537	551	570	582	602	22	101.0	0.91	91.9	Rhodam Hi Ch	4.2.5
25	JF559	10 mм HEPES pH 7.3	511	541	559	573	557	563	579	604	633	662	20	106.0	0.85	90.1	Rhoda 550-600	4.2.3
26	BODIPY FL	H_2O	460	479	502	511	499	503	511	524	541	557	9	92.0	0.97	89.2	Bodipy-Small	4.14.1
27	AF647	PBS pH 7.2, 20/22 °C	589	628	650	665	650	656	672	690	701	731	22	270.0	0.33	89.1	Cyanine-Penta	4.13.2
28	AFD647	PBS pH 7.2, 20/22 °C	589	628	650	665	650	656	672	690	701	731	22	270.0	0.33	89.1	Cyanine-Penta	4.13.2
29	Alexa Fluor 647	PBS pH 7.2, 20/22 °C	589	628	650	665	650	656	672	690	701	731	22	270.0	0.33	89.1	Cyanine-Penta	4.13.2
30	JF549	10 mм HEPES pH 7.3	504	528	549	564	549	555	571	596	618	644	22	101	0.88	88.8	Rhoda 500-550	4.2.2
31	Chlorinated fluorescein G-C 1	H_2O pH 9.18, 25 °C	464	484	503	516	500	506	522	548	571	595	19	103.0	0.86	88.5	Fluorescein-ha	4.8.2
32	Alexa Fluor 546	PBS pH 7.2, 22 °C	506	537	546	567	553	558	580	592	621	648	34	112.0	0.79	88.4	Rhodam Hi Ch	4.2.5

(Continued)

Ent.	Short names	Aqueous solution	$\lambda_{x,-0.25}$ nm	$\lambda_{x,-0.50}$ nm	$\lambda_{x,max}$ nm	$\lambda_{x,+0.50}$ nm	$\lambda_{e,-0.25}$ nm	$\lambda_{e,-0.50}$ nm	$\lambda_{e,max}$ nm	$\lambda_{e,+0.50}$ nm	$\lambda_{e,+0.25}$ nm	$\lambda_{e,+0.10}$ nm	Δ_S nm	ε_{max} mм$^{-1}$cm^{-1}	Φ_f	B_{max} mм$^{-1}$cm^{-1}	Class	Ref.
33	Cy3B	H_2O	493	515	558	569	555	559	572	587	620	639	14	130.0	0.67	87.1	Cyanine-Trimet	4.13.1
34	JF525	10 mм HEPES pH 7.3	483	506	525	538	529	534	549	572	593	617	24	94.0	0.91	85.5	Rhoda 500-550	4.2.2
35	HCDS	PBS pH 7.4	695	756	791	817	783	794	818	845	858	874	27	218.8	0.39	85.3	Cyanine-Hepta	4.13.3
36	NIR2	H_2O	599	639	662	677	661	667	684	702	714	745	22	250.0	0.34	85.0	Cyanine-Penta	4.13.2
37	Oregon Green 514	PBS pH 7.2, 22 °C	471	491	511	524	508	514	530	557	581	605	19	88.0	0.96	84.4	Fluorescein-ha	4.8.2
38	Atto 633 carboxylate	PBS pH 7.4, 22 °C	575	610	630	646	627	635	651	687	724	756	21	130.0	0.64	83.2	C(CH3)2-Rhoda	4.3.1
39	Fluorescein (S-H 1)	P pH 9	453	472	490	502	493	499	514	540	561	584	24	90.0	0.92	82.8	Fluorescein-nh	4.8.1
40	2,6-Sulfon.-1,3,5,7,8-Me-BODIPY	H_2O	451	469	491	499	507	511	519	532	550	567	28	99.0	0.83	82.1	Bodipy-Small	4.14.1
41	Fluorinated rhodam. 1f-6′-COOH	Buffer pH 7	463	484	501	513	505	510	523	544	562	584	22	88.0	0.93	81.8	Rhoda 500-550	4.2.2
42	Chlorinated fluorescein G-C 4	H_2O pH 9.18, 25 °C	471	492	512	526	506	512	528	555	578	602	16	93.8	0.87	81.6	Fluorescein-ha	4.8.2
43	BODIPY	H_2O	457	475	498	506	504	508	516	529	547	563	18	90.0	0.90	81.2	Bodipy-Small	4.14.1
44	Fluorinated rhodam. 1f-5′-COOH	Buffer pH 7	463	484	501	513	506	512	525	546	565	586	24	85.0	0.95	80.7	Rhoda 500-550	4.2.2
45	Yakima Yellow carboxylate	100 mм P/Tris pH 7.0/7.4	487	509	530	545	525	532	549	578	603	630	19	84.0	0.96	80.6	Fluorescein-ha	4.8.2
46	DY-530 carboxylate	PBS pH 7.4	490	513	533	547	533	539	554	578	598	623	21	100.0	0.81	80.6	Rhoda 500-550	4.2.2
47	JF574	10 mм HEPES pH 7.3	524	555	574	589	571	578	594	620	651	681	20	103	0.78	80.3	Rhoda 550-600	4.2.3
48	Fluorinated fluorescein S-H 17	P pH 9	453	472	490	502	493	499	514	540	561	584	24	82.4	0.97	79.9	Fluorescein-ha	4.8.2

Ent.	Short names	Aqueous solution	$\lambda_{x,-0.25}$ nm	$\lambda_{x,-0.50}$ nm	$\lambda_{x,max}$ nm	$\lambda_{x,+0.50}$ nm	$\lambda_{e,-0.25}$ nm	$\lambda_{e,-0.50}$ nm	$\lambda_{e,max}$ nm	$\lambda_{e,+0.50}$ nm	$\lambda_{e,+0.25}$ nm	$\lambda_{e,+0.10}$ nm	Δ_S nm	ε_{max} mM^{-1}cm^{-1}	Φ_f	B_{max} mM^{-1}cm^{-1}	Class	Ref.
49	Fluorinated fluorescein S-H 14a,b	P pH 9	454	474	492	504	495	501	516	542	564	587	24	85.9	0.92	79.0	Fluorescein-ha	4.8.2
50	JF552	10 mM HEPES pH 7.3	505	535	552	566	554	560	575	600	629	657	23	95.2	0.83	79.0	Rhoda 550-600	4.2.3
51	JF571	10 mM HEPES pH 7.3	521	552	571	586	568	574	590	616	647	676	19	101	0.78	78.7	Rhoda 550-600	4.2.3
52	5-Carboxyfluorescein (S-H 2)	P pH 9	454	474	492	504	495	501	516	542	564	587	24	84.7	0.92	77.9	Fluorescein-nh	4.8.1
53	Fluorinated fluorescein S-H 19	P pH 9	468	489	508	521	505	511	527	554	577	601	19	81.2	0.96	77.9	Fluorescein-ha	4.8.2
54	SiRF 1a-H	H_2O	602	581	662	680	654	662	680	701	716	751	18	118.0	0.66	77.8	Si(CH3)2-Rhoda	4.3.2
55	NIR3	H_2O	663	719	750	773	745	756	777	802	813	827	27	275.0	0.28	77.0	Cyanine-Hepta	4.13.3
56	Chlorinated fluorescein G-C 5	H_2O pH 9.18, 25 °C	471	491	511	524	505	511	527	554	577	601	16	91.6	0.84	76.9	Fluorescein-ha	4.8.2
57	JS-R	10 mM HEPES pH 7.4	558	589	611	625	609	617	631	648	659	671	20	130.0	0.59	76.7	B(OH)2-Pyron	4.5
58	Abberior Star 580 carboxylate	PBS pH 7.4	535	567	587	603	583	590	607	635	667	699	20	85.0	0.90	76.5	Rhoda 550-600	4.2.3
59	2,5-DiMe TokyoGreen	100 mM NaOH	453	475	491	503	493	497	510	548	563	582	19	85.0	0.89	75.4	Fluorone	4.11
60	5(6)-Carboxyfluorescein (FAM)	100 mM PBS pH 9	457	477	495	508	499	504	520	546	569	592	25	83.0	0.90	74.7	Fluorescein-nh	4.8.1
61	Pennsylvania Green	PBS pH 7.4	456	478	494	506	497	501	514	553	568	587	20	82.0	0.91	74.6	Fluorone	4.11
62	Abberior Star 600 carboxylate	PBS pH 7.4	550	587	604	619	600	606	623	647	677	714	19	93.0	0.80	74.4	Rhodam >600	4.2.4
63	Rhodamine 110	H_2O pH <7	458	482	496	507	503	507	520	540	564	586	24	83.0	0.89	73.8	Rhodam <500	4.2.1

(Continued)

Ent.	Short names	Aqueous solution	$\lambda_{x,-0.25}$ nm	$\lambda_{x,-0.50}$ nm	$\lambda_{x,max}$ nm	$\lambda_{x,+0.50}$ nm	$\lambda_{e,-0.25}$ nm	$\lambda_{e,-0.50}$ nm	$\lambda_{e,max}$ nm	$\lambda_{e,+0.50}$ nm	$\lambda_{e,+0.25}$ nm	$\lambda_{e,+0.10}$ nm	Δ_S nm	ε_{max} mм$^{-1}$cm^{-1}	Φ_f	B_{max} mм$^{-1}$cm^{-1}	Class	Ref.
64	Oregon Green 488	H_2O pH 9	453	472	490	502	493	499	514	540	561	584	24	76.0	0.97	73.7	Fluorescein-ha	4.8.2
65	*meso*-CaboxyBODIPY J-K 3c	10 mм PBS pH 7.4	480	501	517	527	519	523	540	577	598	629	23	76.0	0.97	73.7	Bodipy-Small	4.14.1
66	2,4-DiMe TokyoGreen	100 mм NaOH	453	475	491	503	493	497	510	548	563	582	19	85.0	0.87	73.5	Fluorone	4.11
67	2-OMe TokyoGreen	100 mм NaOH	456	478	494	506	497	502	515	554	569	589	21	85.0	0.86	73.1	Fluorone	4.11
68	2-PrAc-7-Me-6-EtAm-rhodol	50 mм P pH 8-9	466	486	509	522	510	518	535	560	585	612	26	79.4	0.92	73.0	Rhodol	4.15.1
69	Fluorinated fluorescein S-H 18	P pH 9	468	489	508	521	505	511	527	554	577	601	19	85.6	0.85	72.7	Fluorescein-ha	4.8.2
70	2-Me TokyoGreen	H_2O pH 9	453	475	491	503	493	497	510	548	563	582	19	85.0	0.85	72.2	Fluorone	4.11
71	Atto 488 carboxylate	PBS pH 7.4, 22 °C	461	486	500	511	503	507	520	540	564	586	20	90.0	0.80	72.0	Rhodam <500	4.2.1
72	Sulfofluorescein	C pH 9.5	464	482	497	508	499	504	517	556	572	591	20	78.0	0.92	71.7	Sulfofluoresc.	4.10
73	2-Me-4-OMe TokyoGreen	100 mм NaOH	454	476	492	504	492	496	509	547	562	581	17	85.0	0.84	71.4	Fluorone	4.11
74	Rhodamine RBzET	H_2O	485	508	527	541	531	537	552	575	596	620	25	82.0	0.87	71.3	Rhoda 500-550	4.2.2
75	Fluorinated rhodamine 1e	Buffer pH 7	461	485	499	510	504	509	522	542	566	588	23	81.0	0.88	71.2	Rhodam <500	4.2.1
76	2′,7′-Cl-4,5,6,7-F-fluorescein	100 mм NaOH	478	500	520	534	515	521	538	566	590	615	18	88.5	0.80	70.8	Fluorescein-ha	4.8.2
77	Carborhodamine 16a	H_2O	588	616	636	652	633	640	657	693	723	759	21	114.0	0.62	70.6	C(CH3)2-Rhoda	4.3.1
78	Fluorescein (G-C 1)	H_2O pH 9.18, 25 °C	454	474	492	504	493	499	514	540	561	584	22	78.4	0.90	70.5	Fluorescein-nh	4.8.1
79	CF532	H_2O			527				558				31	96.0	0.73	69.8	nd (Rhodam.)	4.41
80	2-Me RG	PBS pH 7.4	458	483	498	510	502	507	520	536	549	574	22	82.0	0.84	68.8	Rosamine	4.17

Ent.	Short names	Aqueous solution	$\lambda_{x,-0.25}$ nm	$\lambda_{x,-0.50}$ nm	$\lambda_{x,max}$ nm	$\lambda_{x,+0.50}$ nm	$\lambda_{e,-0.25}$ nm	$\lambda_{e,-0.50}$ nm	$\lambda_{e,max}$ nm	$\lambda_{e,+0.50}$ nm	$\lambda_{e,+0.25}$ nm	$\lambda_{e,+0.10}$ nm	Δ_S nm	ε_{max} mм$^{-1}$cm^{-1}	Φ_f	B_{max} mм$^{-1}$cm^{-1}	Class	Ref.
81	Abberior Star 512 carboxylate	PBS pH 7.4	472	493	511	524	511	516	530	552	571	593	19	84.0	0.82	68.8	Rhoda 500-550	4.2.2
82	6-Aminorhodol (Rhodol)	50 mм P, <0.1% ACN pH 8	452	471	493	505	492	500	516	539	563	587	23	70.0	0.98	68.6	Rhodol	4.15.2
83	HEX	H_2O pH 9	491	513	535	550	534	541	559	589	616	643	24	98.0	0.70	68.6	Fluorescein-ha	4.8.2
84	Rhodamine 123	H_2O pH <7	464	485	502	514	509	514	528	549	568	590	26	75.0	0.90	67.5	Rhoda 500-550	4.2.2
85	Cy5	H_2O	587	625	647	662	642	647	663	680	691	720	16	250.0	0.27	67.5	Cyanine-Penta	4.13.2
86	Zy5	P pH 7.4	587	625	647	662	644	649	665	682	694	722	18	250.0	0.27	67.5	Cyanine-Penta	4.13.2
87	Cy5.18.OH	PBS	590	628	650	664	646	651	667	685	698	739	17	250.0	0.27	67.5	Cyanine-Penta	4.13.2
88	6-Benzylaminorhodol	50 mм P, <0.1% ACN pH 8	461	480	503	515	504	512	529	554	578	604	26	71.0	0.95	67.4	Rhodol	4.15.2
89	M-S 10d	H_2O, 0.1% TX-100	523	547	574	589	559	563	582	619	648	682	8	118.0	0.57	67.2	Bodipy-Small	4.14.1
90	Alexa Fluor 488	PBS pH 7.2, 22 °C	457	477	494	506	499	504	517	538	555	576	23	73.0	0.92	67.1	Rhodam Hi Ch	4.2.5
91	Carbofluorescein	100 mм NaOH	505	526	544	557	546	551	567	614	633	658	23	108.0	0.62	66.9	C(CH3)2-Fluore	4.9
92	JF608	10 mм HEPES pH 7.3	564	589	608	623	609	615	631	664	692	725	23	99.0	0.67	66.3	C(CH3)2-Rhoda	4.3.1
93	2-Cl-7-Me-6-EtAm-carboxyrhodol	50 mм P pH 8-9	479	495	514	538	514	525	539	573	600	634	25	80.5	0.82	66.0	Rhodol	4.15.1
94	Abberior Star Red carboxylate	PBS pH 7.4	578	619	638	655	630	637	655	682	715	756	17	100.0	0.66	66.0	Rhodam >600	4.2.4
95	Alexa Fluor 680	PBS pH 7.2/7.5, 20/22 °C	613	655	679	695	678	685	702	721	734	766	23	183.0	0.36	65.8	Cyanine-Penta	4.13.2

(Continued)

Ent.	Short names	Aqueous solution	$\lambda_{x,-0.25}$ nm	$\lambda_{x,-0.50}$ nm	$\lambda_{x,max}$ nm	$\lambda_{x,+0.50}$ nm	$\lambda_{e,-0.25}$ nm	$\lambda_{e,-0.50}$ nm	$\lambda_{e,max}$ nm	$\lambda_{e,+0.50}$ nm	$\lambda_{e,+0.25}$ nm	$\lambda_{e,+0.10}$ nm	Δ_S nm	ε_{max} mM^{-1}cm^{-1}	Φ_f	B_{max} mM^{-1}cm^{-1}	Class	Ref.
96	Fluorescein	100 mM NaOH in D_2O	453	472	489.7	502	495	501	516.3	542	564	587	26	67.0	0.98	65.6	Fluorescein-nh	4.8.1
97	HCDO	PBS pH 7.4	678	736	769	793	755	766	788	813	825	840	19	218.8	0.30	65.6	Cyanine-Hepta	4.13.3
98	6-Azido-4,5,7-trifluorofluorescein	100 mM NaOH	467	488	507	520	505	511	527	554	577	601	20	79.9	0.82	65.5	Fluorescein-ha	4.8.2
99	DY-490 carboxylate	PBS pH 7.4	452	468	491	504	495	500	515	536	553	572	24	73.0	0.90	65.4	Rhodam Hi Ch	4.2.5
100	3-Hydroxy-JF646	10 mM HEPES pH 7.3	574	614	651	668	642	649	666	687	706	733	15	136	0.48	65.2	Si(CH3)2-Rosa	4.18.1
101	2-COOH DFTM	100 mM P, 1% DMSO pH 10	537	560	581	596	573	578	596	649	670	697	15	120.0	0.54	64.8	Si(CH3)2-Fluore	4.9
102	*meso*-CaboxyBODIPY J-K 3b	10 mM PBS pH 7.4	477	497	519	531	528	532	549	582	607	637	30	78.0	0.83	64.7	Bodipy-Small	4.14.1
103	7-Me-6-EtAm-carboxyrhodol	50 mM P pH 8-9	465	485	508	521	509	517	534	559	584	611	26	78.1	0.82	64.0	Rhodol	4.15.1
104	Alexa Fluor 633	PBS/H_2O pH 7.2, 20/22 °C	568	608	632	650	621	629	647	670	691	722	15	100.0	0.64	64.0	Sulforhodam.	4.6
105	6-Piperidinocarboxyrhodol	50 mM P pH 8-9	472	492	516	529	515	523	541	567	593	620	25	80.9	0.79	63.9	Rhodol	4.15.1
106	5-Aza-3-methyl-JF646	10 mM HEPES pH 7.3	578	618	656	674	646	653	670	691	710	738	14	133	0.48	63.8	Si(CH3)2-Rosa	4.18.1
107	4-Benzamidofluorescein	100 mM P pH 7.5	453	472	490	502	496	502	517	543	565	588	27	80.0	0.79	63.2	Fluorescein-nh	4.8.1
108	DY-505 carboxylate	PBS pH 7.4	459	478	505	515	506	512	530	552	570	590	25	85.0	0.74	63.2	Rhoda 500-550	4.2.2

Ent.	Short names	Aqueous solution	$\lambda_{x,-0.25}$ nm	$\lambda_{x,-0.50}$ nm	$\lambda_{x,max}$ nm	$\lambda_{x,+0.50}$ nm	$\lambda_{e,-0.25}$ nm	$\lambda_{e,-0.50}$ nm	$\lambda_{e,max}$ nm	$\lambda_{e,+0.50}$ nm	$\lambda_{e,+0.25}$ nm	$\lambda_{e,+0.10}$ nm	Δ_S nm	ε_{max} $mM^{-1}cm^{-1}$	Φ_f	B_{max} $mM^{-1}cm^{-1}$	Class	Ref.
109	Sulfo-JF646	10 mM HEPES pH 7.3	586	629	655	674	643	651	670	695	717	751	15	124.0	0.51	63.2	Si(CH3)2-Srhod	4.7
110	SiRF 1b-H	H_2O	603	581	663	681	654	662	680	701	716	751	17	90.0	0.70	63.0	Si(CH3)2-Rhoda	4.3.2
111	Tetramethycarborhodamine	10 mM HEPES pH 7.3	562	588	606	620	604	610	626	659	686	718	20	121.0	0.52	62.9	C(CH3)2-Rhoda	4.3.1
112	NR666	PBS, 1% DMSO pH 7.4	608	646	666	683	653	661	685	719	741	778	19	165.0	0.38	62.7	PO2H-Rosam.	4.18.2
113	2-Me DCTM	100 mM P pH 9	544	577	597	611	586	592	607	627	643	666	10	130.0	0.48	62.4	Si(CH3)2-Fluoron	4.12
114	Seta-635-pH-di-carboxylate	P pH 9			640				656				16	188.0	0.33	62.0	nd	4.41
115	Fluorescein	100 mM NaOH in H_2O	453	472	490.2	503	495	501	516.3	542	564	587	26	67.0	0.92	61.6	Fluorescein-nh	4.8.1
116	NIR1	H_2O	672	729	761	785	763	773	796	822	834	849	35	268.0	0.23	61.6	Cyanine-Hepta	4.13.3
117	2-Et-7-Me-6-EtAm-carboxyrhodol	50 mM P pH 8-9	468	488	511	524	512	520	538	563	589	616	27	73.0	0.84	61.3	Rhodol	4.15.1
118	Alexa Fluor 568	PBS pH 7.2/7.0, 20/22 °C	527	559	578	593	580	586	603	630	662	693	25	88.0	0.69	60.7	Rhodam Hi Ch	4.2.5
119	Alexa Fluor 594	PBS pH 7.2/7.0, 20/22 °C	537	570	590	606	593	599	617	646	679	712	27	92.0	0.66	60.7	Rhodam Hi Ch	4.2.5
120	nor-SqO2	P pH 7.4	590	620	639	655	635	642	657	678	696	736	18	183.0	0.33	60.3	Squaraine	4.35
121	3,6-Azetid.-Si-pyronine	10 mM HEPES pH 7.3	573	606	636	654	623	630	649	675	692	729	13	97.1	0.62	60.2	Si(CH3)2-Pyron	4.5
122	Atto Rho3B carboxylate	PBS pH 7.4, 22 °C	517	548	566	579	566	572	589	613	645	675	23	120.0	0.50	60.0	Rhoda 550-600	4.2.3
123	Atto 620 carboxylate	PBS pH 7.4, 22 °C	580	602	620	633	617	623	642	673	707	740	22	120.0	0.50	60.0	C(CH3)2-Rhoda	4.3.1

(Continued)

Ent.	Short names	Aqueous solution	$\lambda_{x,-0.25}$ nm	$\lambda_{x,-0.50}$ nm	$\lambda_{x,max}$ nm	$\lambda_{x,+0.50}$ nm	$\lambda_{e,-0.25}$ nm	$\lambda_{e,-0.50}$ nm	$\lambda_{e,max}$ nm	$\lambda_{e,+0.50}$ nm	$\lambda_{e,+0.25}$ nm	$\lambda_{e,+0.10}$ nm	Δ_S nm	ε_{max} mм$^{-1}$cm^{-1}	Φ_f	B_{max} mм$^{-1}$cm^{-1}	Class	Ref.
124	2,7-DiMe-6-EtAm-carboxyrhodol	50 mм P pH 8-9	467	487	510	522	511	519	536	561	586	612	26	66.3	0.89	59.0	Rhodol	4.15.1
125	2-COOH DCTM	100 mм P, 1% DMSO pH 10	545	570	591	606	583	589	607	662	684	712	16	120.0	0.48	57.6	Si(CH3)2-Fluore	4.9
126	DMAX-EP	100 mм P pH 7.5	454	474	492	504	494	500	515	541	563	585	23	71.0	0.81	57.5	Fluorescein-nh	4.8.1
127	Abberior Star 488 carboxylate	PBS pH 7.4	465	485	503	515	505	511	524	545	564	585	21	64.5	0.89	57.4	Rhodam Hi Ch	4.2.5
128	5(6)-Carboxy-X-rhodamine (ROX)	H_2O pH <7	525	557	576	591	578	584	601	628	660	691	25	82.0	0.70	57.4	Rhoda 550-600	4.2.3
129	6-Propylaminorhodol	50 mм P, <0.1% ACN pH 8	464	484	507	519	506	514	531	556	581	607	24	61.0	0.94	57.3	Rhodol	4.15.2
130	NX	100 mм P pH 7.5	454	474	492	504	491	497	512	537	559	582	20	69.0	0.83	57.2	Fluorescein-nh	4.8.1
131	6-Allylaminorhodol	50 mм P, <0.1% ACN pH 8	464	484	507	519	509	517	534	559	584	611	27	61.5	0.92	56.5	Rhodol	4.15.2
132	Atto 465 carboxylate	PBS pH 7.4, 22 °C	404	423	453	467	471	480	506	551	579	611	53	75.0	0.75	56.2	Acridine	4.23.1
133	BCECF	100 mм NaOH	466	486	505	518	505	511	527	554	577	601	22	67.0	0.84	56.2	Fluorescein-nh	4.8.1
134	570CPH	PBS pH 7.4	580	602	571	633	617	623	600	673	707	740	29	79.0	0.71	56.0	C(CH3)2-Rhoda	4.3.1
135	Seta-650-carboxylate	P pH 7.4			651				671				20	200.0	0.28	56.0	nd	4.41
136	6-CA-MeThio-4,5,7-F-fluorescein	100 mм NaOH	467	488	507	520	504	510	526	553	576	600	19	63.5	0.88	55.8	Fluorescein-ha	4.8.2
137	6-Butylaminorhodol	50 mм P, <0.1% ACN pH 8	465	485	508	521	508	516	533	558	583	609	25	60.0	0.93	55.8	Rhodol	4.15.2

Ent.	Short names	Aqueous solution	$\lambda_{x,-0.25}$ nm	$\lambda_{x,-0.50}$ nm	$\lambda_{x,max}$ nm	$\lambda_{x,+0.50}$ nm	$\lambda_{e,-0.25}$ nm	$\lambda_{e,-0.50}$ nm	$\lambda_{e,max}$ nm	$\lambda_{e,+0.50}$ nm	$\lambda_{e,+0.25}$ nm	$\lambda_{e,+0.10}$ nm	Δ_S nm	ε_{max} $mM^{-1}cm^{-1}$	Φ_f	B_{max} $mM^{-1}cm^{-1}$	Class	Ref.
138	2-Cl-6-Piperidinocarboxy-rhodol	50 mM P pH 8-9	478	500	524	537	522	530	548	574	601	629	24	77.4	0.72	55.7	Rhodol	4.15.1
139	BCPCF	100 mM NaOH	466	486	505	518	505	511	527	554	577	601	22	67.0	0.83	55.6	Fluorescein-nh	4.8.1
140	3-Methyl-JF646	10 mM HEPES pH 7.3	572	612	649	666	642	649	666	687	706	733	17	118	0.47	55.4	Si(CH3)2-Rosa	4.18.1
141	4-Carboxy Pennsylvania Green	PBS pH 7.4	456	478	494	506	497	502	515	554	569	589	21	62.0	0.89	55.1	Fluorone	4.11
142	Chlorinated fluorescein G-C 6	H_2O pH 9.18, 25 °C	480	501	522	536	514	520	537	565	589	614	15	65.8	0.83	54.6	Fluorescein-ha	4.8.2
143	HCDN	PBS pH 7.4	564	603	625	641	725	735	755	778	789	802	130	195.0	0.28	54.5	Cyanine-Hepta	4.13.3
144	SR101-6G	PBS	508	540	565	579	563	571	586	608	638	666	21	80.0	0.68	54.4	Sulforhodam.	4.6
145	Sulfo-cyanine5	H_2O	586	624	646	661	641	646	662	679	690	719	16	271.0	0.20	54.2	Cyanine-Penta	4.13.2
146	8-Trifluoromethyl-BODIPY 11	100 mM P pH 7.4	519	543	569	583	561	565	584	622	650	685	15	57.0	0.95	54.1	Bodipy-Small	4.14.1
147	3-Methyl-5-Methoxy-JF646	10 mM HEPES pH 7.3	572	612	649	666	640	647	664	685	704	731	15	110	0.49	53.9	Si(CH3)2-Rosa	4.18.1
148	5(6)-Carboxyrhodol	50 mM P pH 8-9	451	472	494	503	495	501	520	539	562	584	26	59.9	0.89	53.3	Rhodol	4.15.1
149	DY-495 carboxylate	PBS pH 7.4	456	476	494	507	499	505	521	547	570	593	27	78.0	0.68	53.2	Fluorescein-nh	4.8.1
150	SR101-Me_2	PBS	517	553	573	587	572	578	593	613	632	668	20	80.0	0.66	52.8	Sulforhodam.	4.6
151	JF570	10 mM HEPES pH 7.3	521	548	570	586	569	576	593	620	644	673	23	83.6	0.63	52.6	S-Rhodamine	4.3.3
152	SR101-B	PBS	522	557	575	590	574	580	595	615	631	665	20	80.0	0.65	52.0	Sulforhodam.	4.6
153	CF594	H_2O			593				614				21	115.0	0.45	51.7	nd (Rhodam.)	4.41
154	*meso*-CaboxyBODIPY J-K 3a	10 mM PBS pH 7.4	477	497	519	531	537	541	558	592	618	649	39	77.0	0.67	51.5	Bodipy-Small	4.14.1

(Continued)

Ent.	Short names	Aqueous solution	$\lambda_{x,-0.25}$ nm	$\lambda_{x,-0.50}$ nm	$\lambda_{x,max}$ nm	$\lambda_{x,+0.50}$ nm	$\lambda_{e,-0.25}$ nm	$\lambda_{e,-0.50}$ nm	$\lambda_{e,max}$ nm	$\lambda_{e,+0.50}$ nm	$\lambda_{e,+0.25}$ nm	$\lambda_{e,+0.10}$ nm	Δ_S nm	ε_{max} mм$^{-1}$cm^{-1}	Φ_f	B_{max} mм$^{-1}$cm^{-1}	Class	Ref.
155	Niu 27	H_2O	474	494	510	520	508	512	528	563	583	612	18	63.3	0.81	51.2	Bodipy-Med.	4.14.2
156	Alexa Fluor 700	PBS/H_2O pH 7.2, 20/22 °C			702				723				21	205.0	0.25	51.2	nd	4.41
157	TET	H_2O pH 9	479	501	521	535	513	519	536	564	588	613	15	73.0	0.70	51.1	Fluorescein-ha	4.8.2
158	2,6-Sulfon.-1,3,5,7-Me-BODIPY	H_2O	451	469	491	499	498	502	510	523	540	556	19	60.0	0.85	51.0	Bodipy-Small	4.14.1
159	Acridine red	H_2O	484	507	528	540	542	548	562	582	594	621	34	100.0	0.51	51.0	Pyronine	4.4
160	M-S 10e	H_2O, 0.1% TX-100	532	556	584	599	574	578	598	637	667	704	14	124.0	0.41	50.8	Bodipy-Small	4.14.1
161	KK114	H_2O	580	619	637	652	635	642	660	685	713	762	23	92.0	0.55	50.6	Rhodam >600	4.2.4
162	2,4,5,7-Tetra-F-sulfofluorescein	P pH 9.18	480	499	515	527	518	523	537	579	596	618	22	63.2	0.80	50.5	Sulfofluoresc.	4.10
163	6-Azetidinylaminorhodol	10 mм HEPES pH 7.3	474	495	519	532	520	528	546	572	599	626	27	59.0	0.85	50.1	Rhodol	4.15.2
164	Cy5.29.OH	PBS	589	628	650	665	645	650	666	683	695	724	16	250.0	0.20	50.0	Cyanine-Penta	4.13.2
165	M-S 10b	H_2O, 0.1% TX-100	501	523	547	560	551	556	574	610	638	671	27	64.0	0.78	49.9	Bodipy-Small	4.14.1
166	530RH	PBS pH 7.4	491	514	532	547	533	538	553	574	592	618	21	56.0	0.89	49.8	Rhoda 500-550	4.2.2
167	ZW800-3a	FBS	682	741	774	799	757	768	790	815	827	842	16	309.0	0.16	49.7	Cyanine-Hepta	4.13.3
168	JF593	10 mм HEPES pH 7.3	541	569	593	610	587	594	612	641	667	697	19	90.3	0.55	49.6	S-Rhodamine	4.3.3
169	Seta-580-carboxylate	H_2O			578				602				24	90.0	0.55	49.5	nd	4.41
170	KK1119	H_2O	577	618	637	654	634	641	660	687	721	763	23	90.0	0.55	49.5	Rhodam >600	4.2.4
171	Alexa Fluor 532	PBS pH 7.2, 22 °C	486	514	531	545	533	539	554	572	587	616	23	81.0	0.61	49.4	Rosamine	4.17

Ent.	Short names	Aqueous solution	$\lambda_{x,-0.25}$ nm	$\lambda_{x,-0.50}$ nm	$\lambda_{x,max}$ nm	$\lambda_{x,+0.50}$ nm	$\lambda_{e,-0.25}$ nm	$\lambda_{e,-0.50}$ nm	$\lambda_{e,max}$ nm	$\lambda_{e,+0.50}$ nm	$\lambda_{e,+0.25}$ nm	$\lambda_{e,+0.10}$ nm	Δ_S nm	ε_{max} mM^{-1}cm^{-1}	Φ_f	B_{max} mM^{-1}cm^{-1}	Class	Ref.
172	JF669-6-SCH2CO2H	10 mM HEPES pH 7.3	606	584	667	685	655	663	681	702	717	752	14	130.0	0.38	49.4	Si(CH3)2-Rhoda	4.3.2
173	3-Methoxymethyl ether rhodol	50 mM P, <0.1% ACN pH 8	439	457	477	488	492	500	516	539	563	587	39	58.0	0.85	49.3	Rhodol	4.15.2
174	2-Br-7-Me-6-EtAm-carboxyrhodol	50 mM P pH 8-9	472	492	516	529	513	521	539	565	590	618	23	80.9	0.61	49.3	Rhodol	4.15.1
175	Seta-750-carboxylate	P pH 7.4			751				779				28	235.0	0.21	49.3	nd	4.41
176	Si-Fluorescein	100 mM NaOH	535	559	579	594	575	581	599	652	674	701	20	93.0	0.53	49.2	Si(CH3)2-Fluore	4.9
177	Fluorinated rhodamine 3-H	Buffer pH 7	471	492	510	523	512	517	531	553	572	594	21	57.0	0.86	49.0	Rhoda 500-550	4.2.2
178	Alexa Fluor 660	PBS pH 7.2/7.5, 20/22 °C			663				690				27	132.0	0.37	48.8	nd	4.41
179	DY-682 carboxylate	PBS pH 7.4	617	654	690	708	671	685	710	733	745	764	20	120.0	0.41	48.6	Merocyanine	4.24
180	BODIPY 630/650	H_2O	569	597	629	647	618	623	646	692	728	772	17	97.0	0.50	48.5	Bodipy-Small	4.14.1
181	*meso*-CaboxyBODIPY J-K 1	10 mM PBS pH 7.4	463	482	505	514	540	545	554	569	589	608	49	62.0	0.78	48.3	Bodipy-Small	4.14.1
182	Niu 25	PBS pH 7.4	479	499	521	533	520	524	540	572	596	625	19	58.8	0.82	48.2	Bodipy-Med.	4.14.2
183	Seta-680-carboxylate	P pH 7.4			678				698				20	200.0	0.24	48.0	nd	4.41
184	nor-SqO1	P pH 7.4	588	618	636	652	631	639	653	674	692	731	17	149.0	0.32	47.6	Squaraine	4.35
185	2,7-DiCl-6-methylamino rhodol	50 mM P pH 8-9	469	488	512	525	507	515	533	558	583	611	21	71.0	0.67	47.5	Rhodol	4.15.1
186	Fluorinated rhodamine 1j	Buffer pH 7	458	482	496	507	504	509	522	542	566	588	26	53.0	0.89	47.1	Rhodam <500	4.2.1
187	Carborhodamine 1c	H_2O	591	619	640	656	637	645	662	699	730	766	22	89.0	0.53	47.1	C(CH3)2-Rhoda	4.3.1
188	L-B 1a	H_2O	453	472	494	502	495	499	507	520	537	552	13	100.0	0.47	47.0	Bodipy-Small	4.14.1

(Continued)

Ent.	Short names	Aqueous solution	$\lambda_{x,-0.25}$ nm	$\lambda_{x,-0.50}$ nm	$\lambda_{x,max}$ nm	$\lambda_{x,+0.50}$ nm	$\lambda_{e,-0.25}$ nm	$\lambda_{e,-0.50}$ nm	$\lambda_{e,max}$ nm	$\lambda_{e,+0.50}$ nm	$\lambda_{e,+0.25}$ nm	$\lambda_{e,+0.10}$ nm	Δ_S nm	ε_{max} $mM^{-1}cm^{-1}$	Φ_f	B_{max} $mM^{-1}cm^{-1}$	Class	Ref.
189	Pyrrolidinylcarborhodamine	10 mM HEPES pH 7.3	568	594	613	628	610	617	633	666	694	727	20	87.0	0.54	46.9	C(CH3)2-Rhoda	4.3.1
190	DY-679P1	PBS pH 7.4	612	655	679	696	673	680	697	715	726	745	18	200.0	0.23	46.6	Cyanine-Penta	4.13.2
191	Carborhodamine 1b	H_2O	590	619	639	655	636	644	661	698	728	764	22	80.0	0.58	46.4	C(CH3)2-Rhoda	4.3.1
192	560CP	PBS pH 7.4	522	541	561	576	565	572	588	625	646	669	27	61.0	0.76	46.3	C(CH3)2-Rhoda	4.3.1
193	2-Me Tokyo Magenta	100 mM P pH 9	532	563	582	595	578	583	598	617	633	655	16	110.0	0.42	46.2	Si(CH3)2-Fluoron	4.12
194	6-Quinoline-2-propanoic-pyrodol	50 mM P pH 8-9	492	515	538	550	544	549	562	579	591	611	24	65.9	0.70	46.1	Pyrodol	4.21.1
195	Fluorinated fluorescein S-H 16a,b	P pH 9	470	490	510	523	511	518	534	562	585	610	24	78.1	0.59	46.0	Fluorescein-ha	4.8.2
196	NR600	PBS, 1% DMSO pH 7.4	551	580	600	613	597	603	619	646	665	689	19	67.6	0.68	45.9	POOH-Fluoron	4.12
197	3,7-Dimethyl-JF646	10 mM HEPES pH 7.3	574	614	651	668	640	647	664	685	704	731	13	89.6	0.51	45.7	Si(CH3)2-Rosa	4.18.1
198	2-Me SiR680	H_2O	592	634	674	693	663	671	689	711	732	761	15	130.0	0.35	45.5	Si(CH3)2-Rosa	4.18.1
199	G-L 14	H_2O	472	492	508	518	501	505	521	555	575	603	13	59.0	0.77	45.4	Bodipy-Med.	4.14.2
200	Thienyl-JF646	10 mM HEPES pH 7.3	593	572	651	668	642	649	667	687	701	735	16	113	0.40	45.2	Si(CH3)2-Rhoda	4.3.2
201	JF669-6-SCH2CO2Et	10 mM HEPES pH 7.3	607	585	668	686	655	663	681	702	717	752	13	119.0	0.38	45.2	Si(CH3)2-Rhoda	4.3.2
202	$SiOH_2R$	10 mM PBS pH 7.4	599	639	663	679	656	662	681	700	713	738	18	105.0	0.43	45.1	Si(OH)2-Rosa	4.18.1
203	Chlorinated fluorescein G-C 2(3)	H_2O pH 9.18, 25 °C	460	480	499	512	497	503	518	544	566	589	19	48.4	0.93	45.0	Fluorescein-ha	4.8.2

Ent.	Short names	Aqueous solution	$\lambda_{x,-0.25}$ nm	$\lambda_{x,-0.50}$ nm	$\lambda_{x,max}$ nm	$\lambda_{x,+0.50}$ nm	$\lambda_{e,-0.25}$ nm	$\lambda_{e,-0.50}$ nm	$\lambda_{e,max}$ nm	$\lambda_{e,+0.50}$ nm	$\lambda_{e,+0.25}$ nm	$\lambda_{e,+0.10}$ nm	Δ_S nm	ε_{max} mм$^{-1}$cm^{-1}	Φ_f	B_{max} mм$^{-1}$cm^{-1}	Class	Ref.
204	6-Quinoline-2-pentanoic-pyrodol	50 mм P pH 8-9	492	515	538	550	544	549	562	579	591	611	24	64.1	0.70	44.8	Pyrodol	4.21.1
205	M-S 10a	H_2O, 0.1% TX-100	516	539	565	579	556	561	579	616	644	678	14	69.0	0.65	44.8	Bodipy-Small	4.14.1
206	Fluorinated rhodamine 5-SO_3H	Buffer pH 7	472	494	512	525	511	516	530	552	571	593	18	52.0	0.86	44.7	Rhoda 500-550	4.2.2
207	6-Sulfonato-1,3,5,7-Me-BODIPY	H_2O	454	473	495	503	503	507	515	528	545	562	20	52.5	0.85	44.6	Bodipy-Small	4.14.1
208	SR101-110	PBS	485	507	554	570	555	560	576	600	638	661	22	80.0	0.55	44.0	Sulforhodam.	4.6
209	Abberior Star 635P carboxylate	PBS pH 7.4	575	616	635	652	630	637	655	682	715	756	20	80.0	0.55	44.0	Rhodam >600	4.2.4
210	Redmond Red	100 mм P pH 7.1	518	547	579	593	577	583	595	614	629	661	16	52.3	0.84	43.9	Phenoxazine	4.25
211	JOE	H_2O pH 9	478	500	520	534	524	531	548	577	602	628	28	73.0	0.60	43.8	Fluorescein-ha	4.8.2
212	Cy5.205	PBS pH 7.4	609	650	674	690	671	677	694	713	725	757	20	190.0	0.23	43.7	Cyanine-Penta	4.13.2
213	Cy5.5	PBS	609	650	674	690	671	677	694	713	725	757	20	190.0	0.23	43.7	Cyanine-Penta	4.13.2
214	Zy5.5	P pH 7.4	610	652	676	692	671	677	694	713	725	757	18	190.0	0.23	43.7	Cyanine-Penta	4.13.2
215	2′,4′,5′,7′-TetraF-Si-fluorescein	100 mм NaOH	550	575	597	613	588	594	613	669	691	720	16	65.0	0.67	43.5	Si(CH3)2-Fluore	4.9
216	B-Z 14	H_2O	576	605	638	656	626	632	655	703	739	784	17	102.0	0.42	42.8	Bodipy-Large	4.14.3
217	2-OMe-5-Me TokyoGreen	100 mм NaOH	456	478	494	506	497	501	514	553	568	587	20	85.0	0.50	42.5	Fluorone	4.11
218	Carborhodamine 20b	H_2O	592	620	641	657	638	646	663	700	731	767	22	80.0	0.53	42.4	C(CH3)2-Rhoda	4.3.1
219	4,5,6,7-TetraF-Si-fluorescein	100 mм NaOH	551	576	598	614	589	595	614	670	693	722	16	84.1	0.50	42.0	Si(CH3)2-Fluore	4.9
220	2-COOH TM	100 mм P, 1% DMSO pH 10	536	560	580	595	575	580	598	651	672	700	18	110.0	0.38	41.8	Si(CH3)2-Fluore	4.9

(Continued)

Ent.	Short names	Aqueous solution	$\lambda_{x,-0.25}$ nm	$\lambda_{x,-0.50}$ nm	$\lambda_{x,max}$ nm	$\lambda_{x,+0.50}$ nm	$\lambda_{e,-0.25}$ nm	$\lambda_{e,-0.50}$ nm	$\lambda_{e,max}$ nm	$\lambda_{e,+0.50}$ nm	$\lambda_{e,+0.25}$ nm	$\lambda_{e,+0.10}$ nm	Δ_S nm	ε_{max} mм$^{-1}$cm^{-1}	Φ_f	B_{max} mм$^{-1}$cm^{-1}	Class	Ref.
221	GeR	PBS pH 7.4	601	621	634	658	635	643	655	681	697	734	21	97.0	0.43	41.7	Ge(CH3)2-Rhoda	4.3.3
222	Carborhodamine	10 mм HEPES pH 7.3	515	537	552	564	558	564	577	605	628	654	25	65.0	0.64	41.6	C(CH3)2-Rhoda	4.3.1
223	DAMBO-P^H-T	100 mм P pH 7.4	479	499	521	533	517	521	537	569	592	621	16	56.0	0.74	41.4	Bodipy-Small	4.14.1
224	JF669	10 mм HEPES pH 7.3	608	586	669	687	656	664	682	703	718	754	13	112.0	0.37	41.4	Si(CH3)2-Rhoda	4.3.2
225	B-Z 7b	H_2O	479	500	516	526	509	513	529	564	584	614	13	60.5	0.68	41.1	Bodipy-Large	4.14.3
226	JF502	10 mм HEPES pH 7.3	450	470	502	518	502	510	533	580	610	646	31	57.8	0.71	41.0	Acridine	4.23.1
227	4,5,6,7-Tetrafluoro-SiTMR	10 mм HEPES pH 7.3	606	584	667	685	656	664	682	703	718	754	15	132.0	0.31	40.9	Si(CH3)2-Rhoda	4.3.2
228	575RH	PBS pH 7.4	524	555	574	589	574	580	597	624	655	685	23	55.0	0.74	40.7	Rhoda 550-600	4.2.3
229	Atto 425 carboxylate	PBS pH 7.4, 22 °C	395	409	439	460	460	466	485	521	548	576	46	45.0	0.90	40.5	Coumarin 7-Am	4.26.3
230	KK9046	H_2O	573	613	632	648	629	636	654	681	714	755	22	90.0	0.45	40.5	Rhodam >600	4.2.4
231	RS800	FBS	690	750	784	809	766	777	800	826	838	853	16	240.0	0.17	40.5	Cyanine-Hepta	4.13.3
232	CF514	H_2O			516				548				32	105.0	0.38	39.8	nd (Rhodam.)	4.41
233	B-Z 7a	H_2O	458	477	500	508	497	501	509	522	539	555	9	64.5	0.61	39.3	Bodipy-Large	4.14.3
234	Fluorinated fluorescein S-H 20	P pH 9	491	513	535	550	529	535	553	583	608	635	18	83.4	0.47	39.2	Fluorescein-ha	4.8.2
235	RF620	10 mм PBS pH 7.4	569	600	620	635	614	620	636	659	674	701	16	109.0	0.36	39.2	B(OH)2-Rosa	4.18.3
236	Carborhodamine 21a	H_2O	591	619	640	656	638	646	663	700	731	767	23	70.0	0.56	39.2	C(CH3)2-Rhoda	4.3.1
237	SiR-COOH	H_2O	588	567	645	662	637	644	661	681	695	728	16	100.0	0.39	39.0	Si(CH3)2-Rhoda	4.3.2

Ent.	Short names	Aqueous solution	$\lambda_{x,-0.25}$ nm	$\lambda_{x,-0.50}$ nm	$\lambda_{x,max}$ nm	$\lambda_{x,+0.50}$ nm	$\lambda_{e,-0.25}$ nm	$\lambda_{e,-0.50}$ nm	$\lambda_{e,max}$ nm	$\lambda_{e,+0.50}$ nm	$\lambda_{e,+0.25}$ nm	$\lambda_{e,+0.10}$ nm	Δ_S nm	ε_{max} mM^{-1}cm^{-1}	Φ_f	B_{max} mM^{-1}cm^{-1}	Class	Ref.
238	SiR-methyl	H_2O	571	611	648	665	638	645	662	683	701	728	14	100.0	0.39	39.0	Si(CH3)2-Rosa	4.18.1
239	L-B 5	H_2O	457	475	498	506	499	503	511	524	541	557	13	79.4	0.49	38.9	Bodipy-Small	4.14.1
240	2-Ethyl-6-quinolinocarboxyrhodol	50 mM P pH 8-9	494	517	543	557	545	554	574	603	632	664	31	79.4	0.49	38.9	Rhodol	4.15.1
241	1,2,7,8-Dibenzosulfofluorescein	C pH 9.5	502	523	541	554	546	551	567	614	633	658	26	81.0	0.48	38.8	Sulfofluoresc.	4.10
242	Cyan 500 carboxylate	PBS pH 7.4, 22 °C			450				495				45	45.0	0.85	38.2	nd (pb Coum.)	4.41
243	SiR	PBS pH 7.4	589	623	645	663	640	647	661	687	702	737	16	93.0	0.41	38.1	Si(CH3)2-Rhoda	4.3.2
244	Carborhodamine 21b	H_2O	591	619	640	656	639	646	664	701	732	768	24	68.0	0.56	38.0	C(CH3)2-Rhoda	4.3.1
245	ZW800-1	FBS	680	739	772	797	755	766	788	813	825	840	16	249.0	0.15	37.6	Cyanine-Hepta	4.13.3
246	G-L 13b	H_2O	473	493	509	519	502	506	522	556	576	604	13	56.0	0.67	37.5	Bodipy-Med.	4.14.2
247	Atto Oxa12 carboxylate	PBS pH 7.4, 22 °C	584	621	662	680	657	665	681	706	726	769	19	125.0	0.30	37.5	Phenoxazine	4.25
248	Atto 655 carboxylate	PBS pH 7.4, 22 °C	587	629	663	679	656	664	680	707	733	784	17	125.0	0.30	37.5	Phenoxazine	4.25
249	Atto 680 carboxylate	PBS pH 7.4, 22 °C	603	648	681	696	675	682	698	727	749	807	17	125.0	0.30	37.5	Phenoxazine	4.25
250	SMSiR	H_2O	584	626	652	671	643	651	670	695	717	751	18	120.0	0.31	37.2	Si(CH3)2-Srhod	4.7
251	Carborhodamine 20a	H_2O	591	619	640	656	636	644	661	698	728	764	21	70.0	0.53	37.1	C(CH3)2-Rhoda	4.3.1
252	L-B 1b	H_2O	457	475	498	506	497	501	509	522	539	555	11	100.0	0.37	37.0	Bodipy-Small	4.14.1
253	630GeRH	PBS pH 7.4	582	612	631	647	629	635	651	671	683	722	20	61.0	0.60	36.6	Ge(CH3)2-Rhoda	4.3.3
254	2′,7′-Difluorocarbo-fluorescein	100 mM NaOH	514	536	555	569	559	564	581	631	651	677	26	90.9	0.40	36.3	C(CH3)2-Fluore	4.9
255	Fluorinated fluorescein S-H 15a,b	P pH 9	470	490	510	523	511	518	534	562	585	610	24	84.2	0.43	36.2	Fluorescein-ha	4.8.2
256	Carborhodamine 22	H_2O	591	619	640	656	639	646	664	701	732	768	24	67.0	0.54	36.1	C(CH3)2-Rhoda	4.3.1

(Continued)

Ent.	Short names	Aqueous solution	$\lambda_{x,-0.25}$ nm	$\lambda_{x,-0.50}$ nm	$\lambda_{x,max}$ nm	$\lambda_{x,+0.50}$ nm	$\lambda_{e,-0.25}$ nm	$\lambda_{e,-0.50}$ nm	$\lambda_{e,max}$ nm	$\lambda_{e,+0.50}$ nm	$\lambda_{e,+0.25}$ nm	$\lambda_{e,+0.10}$ nm	Δ_S nm	ε_{max} mM^{-1}cm^{-1}	Φ_f	B_{max} mM^{-1}cm^{-1}	Class	Ref.
257	JF690	10 mM HEPES pH 7.3	625	602	690	710	679	687	707	730	746	784	17	150.0	0.24	36.0	POOH-Rhoda	4.3.3
258	Fluorinated rhodamine 1k	Buffer pH 7	467	488	506	518	512	517	531	553	572	594	25	54.0	0.66	35.6	Rhoda 500-550	4.2.2
259	DY-654 carboxylate	PBS pH 7.4	594	627	653	673	651	659	677	699	711	732	24	220.0	0.16	35.2	Merocyanine	4.24
260	Tetramethylrhodamine	10 mM HEPES pH 7.3	503	527	548	563	550	556	572	597	619	646	24	78.0	0.45	35.1	Rhoda 500-550	4.2.2
261	Alexa Fluor 750	PBS/H_2O pH 7.2, 20/22 °C			749				775				26	290.0	0.12	34.8	nd	4.41
262	2-Me SiR600	H_2O	542	574	593	607	593	599	613	633	650	674	20	91.0	0.38	34.5	Si(CH3)2-Rosa	4.18.1
263	2-Me SiR650	H_2O	591	628	646	661	636	642	660	681	707	749	14	110.0	0.31	34.1	Si(CH3)2-Rosa	4.18.1
264	Z-L I	0.5 M PBS pH 7.4	451	469	491	499	489	493	501	513	530	545	10	50	0.68	34.0	Bodipy-Large	4.14.3
265	CF620R	H_2O			617				639				22	115.0	0.30	34.0	nd (Rhodam.)	4.41
266	CW800	FBS	691	752	786	812	766	777	800	826	838	853	14	237.0	0.14	33.6	Cyanine-Hepta	4.13.3
267	SiRF 13a	H_2O	598	577	657	675	651	659	677	698	712	747	20	53.0	0.63	33.3	Si(CH3)2-Rhoda	4.3.2
268	M-S 10c	H_2O, 0.1% TX-100	489	510	533	546	536	540	557	591	617	648	24	54.0	0.61	32.9	Bodipy-Small	4.14.1
269	Sulforhodamine B	H_2O pH 7	508	537	554	567	561	568	582	599	609	628	28	84.0	0.39	32.7	Sulforhodam.	4.6
270	Atto 635 carboxylate	PBS pH 7.4, 22 °C	579	612	635	649	626	638	659	683	701	720	24	130.0	0.25	32.5	C(CH3)2-Pyron	4.5
271	BC-OH	100 mM P, 20% Et pH 7.4			442				488				46	38.1	0.85	32.3	Hybrid-Bonded	4.40.2
272	Abberior Star 635 carboxylate	PBS pH 7.4	574	615	634	650	629	636	654	681	714	755	20	63.0	0.51	32.1	Rhodam >600	4.2.4

Ent.	Short names	Aqueous solution	$\lambda_{x,-0.25}$ nm	$\lambda_{x,-0.50}$ nm	$\lambda_{x,max}$ nm	$\lambda_{x,+0.50}$ nm	$\lambda_{e,-0.25}$ nm	$\lambda_{e,-0.50}$ nm	$\lambda_{e,max}$ nm	$\lambda_{e,+0.50}$ nm	$\lambda_{e,+0.25}$ nm	$\lambda_{e,+0.10}$ nm	Δ_S nm	ε_{max} mм$^{-1}$cm^{-1}	Φ_f	B_{max} mм$^{-1}$cm^{-1}	Class	Ref.
273	Si-9-Iminopyronine 2f	~1 mм HCl, 1% DMSO pH 3	595	631	664	683	653	661	682	711	730	770	18	123	0.26	31.9	Si(CH3)2-Pyron	4.5
274	L-B 2b	H_2O	476	496	512	522	504	508	524	559	578	607	12	77.6	0.41	31.8	Bodipy-Small	4.14.1
275	DY-555 carboxylate	PBS pH 7.4	508	537	555	569	558	564	580	605	635	663	25	100.0	0.32	31.7	Rhoda 550-600	4.2.3
276	nor-SqCN2	P pH 7.4	616	650	670	687	669	677	693	717	737	782	23	167.0	0.19	31.7	Squaraine	4.35
277	CF430	H_2O			426				498				72	40.0	0.79	31.6	nd (Cyanine)	4.41
278	2-Cloro-6-quinolinocarboxy-rhodol	50 mм P pH 8-9	498	521	548	563	546	555	575	604	634	665	27	74.5	0.42	31.2	Rhodol	4.15.1
279	3,6-Azetid.-9-CA.-Si-pyronine	10 mм HEPES pH 7.3	577	610	641	659	630	638	657	684	701	739	16	120.0	0.26	31.2	Si(CH3)2-Pyron	4.5
280	CF405M	H_2O			408				452				44	41.0	0.76	31.1	nd (Coumarin)	4.41
281	DY-405 carboxylate	PBS pH 7.4	336	349	400	410	404	408	423	458	474	492	23	32.0	0.97	30.9	Pyrene	4.28
282	2,4,5,7-Tetra-Br-sulfofluorescein	100 mм NaOH pH 12	488	507	524	536	526	531	546	590	607	630	22	87.0	0.35	30.4	Sulfofluoresc.	4.10
283	Atto 700 carboxylate	PBS pH 7.4, 22 °C	622	666	700	717	693	701	716	744	768	828	16	120.0	0.25	30.0	Phenoxazine	4.25
284	JF479	10 mм HEPES pH 7.3	432	450	479	493	488	495	517	561	589	623	38	47.9	0.62	29.7	Acridine	4.23.1
285	8-Trifluoromethyl-BODIPY 12	100 mм P pH 7.4	439	456	477	485	568	574	584	601	624	645	107	41.8	0.70	29.2	Bodipy-Small	4.14.1
286	6-Quinolinocarboxyrhodol	50 mм P pH 8-9	492	514	540	554	541	550	570	599	627	658	30	66.8	0.43	28.7	Rhodol	4.15.1
287	2-Me SiR671	100 mм P, 1% DMSO pH 7.4	589	631	671	689	661	669	687	709	730	759	16	99.0	0.29	28.7	Si(CH3)2-Rosa	4.18.1

(Continued)

Ent.	Short names	Aqueous solution	$\lambda_{x,-0.25}$ nm	$\lambda_{x,-0.50}$ nm	$\lambda_{x,max}$ nm	$\lambda_{x,+0.50}$ nm	$\lambda_{e,-0.25}$ nm	$\lambda_{e,-0.50}$ nm	$\lambda_{e,max}$ nm	$\lambda_{e,+0.50}$ nm	$\lambda_{e,+0.25}$ nm	$\lambda_{e,+0.10}$ nm	Δ_S nm	ε_{max} mM^{-1}cm^{-1}	Φ_f	B_{max} mM^{-1}cm^{-1}	Class	Ref.
288	1,3,6,8-Pyrenetetrasulfonic acid	H_2O pH 7	335	350	374	378	391	394	403	430	451	465	29	51.0	0.56	28.5	Pyrene	4.28
289	2-PrAc-6-quinolinorhodol	50 mM P pH 8-9	493	515	541	555	543	552	572	601	630	661	31	58.2	0.49	28.5	Rhodol	4.15.1
290	DY-747P1	PBS pH 7.4	661	716	747	770	738	748	769	793	804	818	22	240.0	0.12	28.5	Cyanine-Hepta	4.13.3
291	W-L 8b	PBS	584	622	644	659	646	652	668	685	697	726	24	189.0	0.15	28.3	Cyanine-Penta	4.13.2
292	7-Hydroxy-4-Me-coumarin-3-CA	H_2O pH 10	332	352	386	413	414	422	446	479	499	522	60	33.0	0.85	28.0	Coumarin 7-OH	4.26.1
293	BOPAHY H-D 18ab′	H_2O	398	412	429	435	444	447	453	463	476	489	24	51	0.55	28.0	Bodipy-analog	4.14.5
294	SR101-OH	H_2O	493	516	548	564	550	556	572	595	628	652	24	80.0	0.35	28.0	Sulforhodol	4.27
295	DAOTA 5	H_2O	479	505	548	567	551	558	576	606	630	656	28	50.0	0.56	28.0	Triangulenium	4.22
296	DAOTA 4	H_2O	481	502	549	567	553	559	576	607	632	655	27	50.0	0.56	28.0	Triangulenium	4.22
297	CF647	H_2O			650				665				15	240.0	0.12	27.9	nd (Cyanine)	4.41
298	Cy5.12-OH	H_2O/PBS	589	628	650	665	652	658	674	692	703	733	24	215.0	0.13	27.9	Cyanine-Penta	4.13.2
299	Pyronine Y (OP)	PBS pH 7.4	500	525	546	560	519	526	569	557	567	586	23	77.0	0.36	27.7	Pyronine	4.4
300	B-Z 19	H_2O	587	617	651	670	646	652	677	728	767	816	26	118.5	0.23	27.2	Bodipy-Large	4.14.3
301	TAMRA	50 mM PBS pH 8-9	502	526	547	562	552	558	574	599	622	648	27	77.0	0.35	26.9	Rhoda 500-550	4.2.2
302	Distyryl-BODIPY 18	H_2O	568	596	628	646	613	619	641	687	721	764	13	71.0	0.38	26.9	Bodipy-Med.	4.14.2
303	DY-430 carboxylate	PBS pH 7.4	394	410	444	467	462	468	487	521	545	570	43	37.0	0.72	26.6	Coumarin 7-Am	4.26.3
304	TMDHS or SiP-Me	50 mM PBS pH 7.4	570	603	633	651	627	635	654	681	698	735	21	64.2	0.41	26.3	Si(CH3)2-Pyron	4.5
305	DY-431 carboxylate	PBS pH 7.4	388	405	442	467	470	477	496	529	550	573	54	35.0	0.75	26.2	Coumarin 7-Am	4.26.3
306	nor-SqO3	P pH 7.4	610	642	662	679	656	663	679	702	721	764	17	124.0	0.21	26.0	Squaraine	4.35
307	Zy7 carboxylate	P pH 7.4	663	718	749	772	743	753	774	798	810	824	25	200.0	0.13	26.0	Cyanine-Hepta	4.13.3
308	Cy5.13-OH	H_2O/PBS	588	626	648	663	647	653	669	686	698	727	21	215.0	0.12	25.8	Cyanine-Penta	4.13.2

Ent.	Short names	Aqueous solution	$\lambda_{x,-0.25}$ nm	$\lambda_{x,-0.50}$ nm	$\lambda_{x,max}$ nm	$\lambda_{x,+0.50}$ nm	$\lambda_{e,-0.25}$ nm	$\lambda_{e,-0.50}$ nm	$\lambda_{e,max}$ nm	$\lambda_{e,+0.50}$ nm	$\lambda_{e,+0.25}$ nm	$\lambda_{e,+0.10}$ nm	Δ_S nm	ε_{max} mм$^{-1}$cm^{-1}	Φ_f	B_{max} mм$^{-1}$cm^{-1}	Class	Ref.
309	Spirobifluorene 1	H_2O			378				480				102	51	0.50	25.6	Spirobifluorene	4.39.3
310	2′,7′-Difluoro-Si-Fluorescein	100 mм NaOH	549	574	596	612	593	599	618	675	698	727	22	60.9	0.42	25.5	Si(CH3)2-Fluore	4.9
311	Cascade Blue	50 mм P pH 8	348	363	403	411	413	417	430	464	484	502	27	28.0	0.90	25.2	Pyrene	4.28
312	Tetraethylrhodamine	H_2O pH 7	510	539	557	571	558	564	580	605	635	663	23	105.0	0.24	25.2	Rhoda 550-600	4.2.3
313	nor-SqB2	P pH 7.4	589	619	638	654	644	651	666	688	706	748	28	114.0	0.22	25.0	Squaraine	4.35
314	BOPAHY H-D 18bb′	H_2O	397	411	428	434	444	447	453	463	476	489	25	51	0.49	24.9	Bodipy-analog	4.14.5
315	NBDY	H_2O	462	481	504	513	520	524	533	547	566	583	29	74.4	0.33	24.5	Bodipy-Small	4.14.1
316	3,6-Di(azetidin-1-yl)acridine	10 mм HEPES pH 7.3	442	461	492	507	500	508	531	578	608	643	39	47.0	0.52	24.4	Acridine	4.23.1
317	4,7-DiMe-6-EtAm-carboxyrhodol	50 mм P pH 8-9	472	492	516	529	518	526	544	570	596	624	28	56.9	0.43	24.4	Rhodol	4.15.1
318	2-Me SiR670	100 mм P, 1% DMSO pH 7.4	588	630	670	688	661	669	687	709	730	759	17	90.0	0.27	24.3	Si(CH3)2-Rosa	4.18.1
319	Seta-660-di-carboxylate	P pH 7.4			661				672				11	220.0	0.11	24.2	nd	4.41
320	DY-651 carboxylate	PBS pH 7.4	599	631	653	672	652	659	678	699	711	731	25	160.0	0.15	24.1	Merocyanine	4.24
321	Atto 647 carboxylate	PBS pH 7.4, 22 °C	609	624	647	659	639	646	667	700	734	767	20	120.0	0.20	24.0	C(CH3)2-Pyron	4.5
322	3,7-Di(azetidin)phenoxazine	10 mм HEPES pH 7.3	572	607	647	664	638	646	661	684	703	743	14	99.0	0.24	23.7	Phenoxazine	4.25
323	DY-731 carboxylate	PBS pH 7.4	678	709	736	761	725	735	759	783	794	811	23	225.0	0.11	23.6	Merocyanine	4.24
324	L-B 4a	H_2O	453	472	494	502	495	499	507	520	537	552	13	69.2	0.34	23.5	Bodipy-Small	4.14.1
325	Cy5′	H_2O	583	621	642	656	639	645	660	677	688	716	18	138.0	0.17	23.4	Cyanine-Penta	4.13.2

(Continued)

Ent.	Short names	Aqueous solution	$\lambda_{x,-0.25}$ nm	$\lambda_{x,-0.50}$ nm	$\lambda_{x,max}$ nm	$\lambda_{x,+0.50}$ nm	$\lambda_{e,-0.25}$ nm	$\lambda_{e,-0.50}$ nm	$\lambda_{e,max}$ nm	$\lambda_{e,+0.50}$ nm	$\lambda_{e,+0.25}$ nm	$\lambda_{e,+0.10}$ nm	Δ_S nm	ε_{max} mM^{-1}cm^{-1}	Φ_f	B_{max} mM^{-1}cm^{-1}	Class	Ref.
326	Si-9-Iminopyronine 2c	~1 mM HCl, 1% DMSO pH 3	597	633	666	685	652	660	681	710	729	769	15	83.2	0.28	23.3	Si(CH3)2-Pyron	4.5
327	nor-SqCN1	P pH 7.4	616	650	670	687	669	677	693	717	737	782	23	123.0	0.19	23.3	Squaraine	4.35
328	Des-carboxy-JF646	10 mM HEPES pH 7.3	571	611	648	665	638	645	662	683	701	728	14	116	0.20	23.2	Si(CH3)2-Rosa	4.18.1
329	Tetramethylrosamine	10 mM PBS pH 7.4	501	532	550	565	548	554	570	589	605	636	20	74.5	0.31	23.1	Rosamine	4.17
330	NR669	10 mM PBS pH 7.4	619	653	669	684	664	672	689	712	726	757	20	77.2	0.30	23.1	PO2H-Rosam.	4.18.2
331	CF405S	H_2O			404				431				27	33.0	0.70	23.0	nd (Pyrene)	4.41
332	Distyryl-BODIPY 17	H_2O	570	598	630	648	615	620	643	689	724	767	13	76.6	0.30	22.9	Bodipy-Med.	4.14.2
333	8-Hydroxypyrene-1,3,6-triSA	H_2O pH 10	389	408	460	470	490	495	514	563	593	620	54	24.0	0.95	22.8	Pyrene	4.28
334	2,7-DiCl-6-pyrrolidino sulforhodol	PBS pH 7.4	470	491	520	534	516	521	535	555	584	604	15	83.0	0.27	22.4	Sulforhodol	4.27
335	CF450	H_2O			450				538				88	40.0	0.56	22.3	nd (Cyanine)	4.41
336	Pyrodone	100 mM P pH 7.4	436	454	472	481	486	490	500	513	523	539	28	60.0	0.37	22.2	Pyrodone	4.21.2
337	Pacific Blue	PBS pH 7.2, 22 °C	339	366	410	437	428	436	455	486	505	526	45	29.5	0.75	22.1	Coumarin 7-OH	4.26.1
338	NR604	PBS, 1% DMSO pH 7.4	554	584	604	618	604	610	627	654	672	697	23	45.7	0.48	21.9	POOH-Fluoron	4.12
339	W-L 8d	PBS	583	621	643	657	645	650	666	683	695	724	23	169.0	0.13	21.9	Cyanine-Penta	4.13.2
340	IR-783	PBS	688	748	782	807	776	787	810	837	849	865	28	261.0	0.08	21.9	Cyanine-Hepta	4.13.3
341	2′,4,5,6,7,7′-HexaF-Si-fluorescein	100 mM NaOH	567	594	617	634	609	615	635	695	720	751	18	67.9	0.32	21.7	Si(CH3)2-Fluore	4.9
342	Atto 390 carboxylate	PBS pH 7.4, 22 °C	343	357	390	415	441	450	476	516	540	568	86	24.0	0.90	21.6	Coumarin 7-Am	4.26.3

Ent.	Short names	Aqueous solution	$\lambda_{x,-0.25}$ nm	$\lambda_{x,-0.50}$ nm	$\lambda_{x,max}$ nm	$\lambda_{x,+0.50}$ nm	$\lambda_{e,-0.25}$ nm	$\lambda_{e,-0.50}$ nm	$\lambda_{e,max}$ nm	$\lambda_{e,+0.50}$ nm	$\lambda_{e,+0.25}$ nm	$\lambda_{e,+0.10}$ nm	Δ_S nm	ε_{max} mм$^{-1}$cm^{-1}	Φ_f	B_{max} mм$^{-1}$cm^{-1}	Class	Ref.
343	Square-635-carboxylate	P pH 5.6	591	617	634	648	624	633	646	665	679	725	12	135.0	0.16	21.6	Squaraine	4.35
344	Square-650-carboxylate	P pH 5.6	598	632	653	668	651	658	671	694	712	765	18	135.0	0.16	21.6	Squaraine	4.35
345	2-Me TeOR	PBS pH 7.4	599	642	669	692	654	663	686	714	737	784	17	120.0	0.18	21.6	TeO-Rosamine	4.18.3
346	640SiRH	PBS pH 7.4	584	519	641	658	639	646	662	681	694	726	21	51.0	0.42	21.4	Si(CH3)2-Rhoda	4.3.2
347	Seta-375-carboxylate	P pH 7.4			372				476				104	24.5	0.86	21.0	nd	4.41
348	Seta-405-carboxylate	P pH 7.4			401				452				51	30.0	0.70	21.0	nd	4.41
349	2′Me-4CH_2OH-Si-rosol	100 mм P pH 7.4	554	583	613	628	609	615	631	652	668	694	18	70.0	0.30	21.0	Si(CH3)2-Rosol	4.20
350	Atto 430LS carboxylate	PBS pH 7.4, 22 °C			436				545				109	32.0	0.65	20.8	nd (pb Coum.)	4.41
351	DAMBO-T	100 mм P pH 7.4	457	475	498	506	495	499	507	520	537	552	9	52.0	0.40	20.8	Bodipy-Small	4.14.1
352	Seta-646-carboxylate	P pH 7.4			646				656				10	207.0	0.10	20.7	nd	4.41
353	IRDye78-CA	10 mм HEPES pH 7.4	680	738	771	796	772	783	806	832	845	860	35	145.7	0.14	20.6	Cyanine-Hepta	4.13.3
354	7-Azetidinyl-3-CA coumarin	10 mм HEPES pH 7.3	333	353	387	414	434	443	470	507	529	555	83	24.0	0.84	20.1	Coumarin 7-Am	4.26.2
355	2-Me SiR620	100 mм P, 1% DMSO pH 7.4	550	586	620	636	616	622	638	657	675	699	18	96.0	0.21	20.1	Si(CH3)2-Rosa	4.18.1
356	2-Me SiR666	100 mм P, 1% DMSO pH 7.4	585	627	666	684	659	666	684	706	726	755	18	80.0	0.25	20.0	Si(CH3)2-Rosa	4.18.1
357	Seta-700-di-carboxylate	P pH 7.4			688				704				16	180.0	0.11	19.8	nd	4.41
358	Cy5.204	PBS pH 7.4	615	658	682	698	683	689	707	726	739	772	25	197.0	0.10	19.7	Cyanine-Penta	4.13.2
359	2′Me-Si-rhosoL	100 mм P pH 7.4	553	582	612	627	608	614	630	651	667	693	18	70.0	0.28	19.6	Si(CH3)2-Rosol	4.20

(Continued)

Ent.	Short names	Aqueous solution	$\lambda_{x,-0.25}$ nm	$\lambda_{x,-0.50}$ nm	$\lambda_{x,max}$ nm	$\lambda_{x,+0.50}$ nm	$\lambda_{e,-0.25}$ nm	$\lambda_{e,-0.50}$ nm	$\lambda_{e,max}$ nm	$\lambda_{e,+0.50}$ nm	$\lambda_{e,+0.25}$ nm	$\lambda_{e,+0.10}$ nm	Δ_S nm	ε_{max} $mM^{-1}cm^{-1}$	Φ_f	B_{max} $mM^{-1}cm^{-1}$	Class	Ref.
360	Seta-385-carboxylate	P pH 7.4			384				512				128	25.0	0.78	19.5	nd	4.41
361	L-B 2a	H_2O	473	493	509	519	503	507	523	557	577	606	14	72.4	0.27	19.5	Bodipy-Small	4.14.1
362	1,2,7,8-Dibenzofluorescein	C pH 9.5	490	513	534	549	540	547	565	596	623	651	31	81.0	0.24	19.4	Fluorescein-nh	4.8.1
363	PONy P1	PBS pH 7.4	535	564	590	605	602	610	627	652	667	701	37	59.0	0.33	19.4	Pyronine	4.4
364	Cy5.10-OH	H_2O/PBS	588	626	648	663	645	651	667	684	696	725	19	215.0	0.09	19.3	Cyanine-Penta	4.13.2
365	2-Me SiR654	100 mM P, 1% DMSO pH 7.4	576	616	654	671	647	655	672	693	713	740	18	80.0	0.24	19.2	Si(CH3)2-Rosa	4.18.1
366	8-Hydroxypyrene-1,3,6-triSA	H_2O pH 2	348	363	403	411	425	429	443	479	501	519	40	20.0	0.95	19.0	Pyrene	4.28
367	CypHer5	A pH 4.6	583	621	643	657	641	646	662	679	690	719	19	100.0	0.19	19.0	Cyanine-Penta	4.13.2
368	Cy5.203	PBS pH 7.4	614	656	680	696	678	685	702	721	734	766	22	190.0	0.10	19.0	Cyanine-Penta	4.13.2
369	Lucigenin	H_2O			368				485				117	37.6	0.50	18.8	Hybrid-Bonded	4.40.2
370	Carborhodol S-H 13-H	PBS pH 7.4	539	567	586	601	589	595	613	657	685	711	27	59.0	0.32	18.8	C(CH3)2-Rhod	4.16
371	8-Aminopyrene-1,3,6-triSA	H_2O	363	380	424	433	482	487	505	552	581	607	81	20.6	0.90	18.5	Pyrene	4.28
372	SqO1	P pH 7.4	584	614	632	648	621	628	642	662	679	718	10	265.0	0.07	18.5	Squaraine	4.35
373	SiP-Ph	50 mM PBS pH 7.4	588	624	649	665	613	630	665	672	683	698	16	80.0	0.23	18.4	C(CH3)2-Pyron	4.5
374	nor-SqB3	P pH 7.4	610	642	662	679	660	668	684	707	727	770	22	102.0	0.18	18.3	Squaraine	4.35
375	TAMRA	H_2O pH 7	501	525	546	561	554	560	576	602	624	651	30	90.0	0.20	18.0	Rhoda 500-550	4.2.2
376	DAOTA 6	H_2O	489	509	558	575	560	567	584	614	640	660	26	50.0	0.36	18.0	Triangulenium	4.22
377	Z-L B	0.5 M PBS pH 7.4	455	474	496	504	495	499	507	520	537	552	11	50	0.36	17.8	Bodipy-Large	4.14.3
378	Sulfonated coumarin N-H 6-H	PBS pH 7.4	377	398	436	461	485	493	515	554	580	608	79	31.0	0.57	17.6	Coumarin 7-Am	4.26.3

Ent.	Short names	Aqueous solution	$\lambda_{x,-0.25}$ nm	$\lambda_{x,-0.50}$ nm	$\lambda_{x,max}$ nm	$\lambda_{x,+0.50}$ nm	$\lambda_{e,-0.25}$ nm	$\lambda_{e,-0.50}$ nm	$\lambda_{e,max}$ nm	$\lambda_{e,+0.50}$ nm	$\lambda_{e,+0.25}$ nm	$\lambda_{e,+0.10}$ nm	Δ_S nm	ε_{max} mм$^{-1}$cm^{-1}	Φ_f	B_{max} mм$^{-1}$cm^{-1}	Class	Ref.
379	nor-SqB1	P pH 7.4	589	619	638	654	644	651	666	688	706	748	28	88.0	0.20	17.6	Squaraine	4.35
380	Seta-670-pH-di-carboxylate	P pH 6			672				694				22	93.0	0.19	17.6	nd	4.41
381	Alexa Fluor 405	PBS pH 7.2, 22 °C	346	360	400	408	408	411	424	457	476	494	24	35.0	0.50	17.5	Pyrene	4.28
382	Seta-633-carboxylate	P pH 7.4			633				644				11	250.0	0.07	17.5	nd	4.41
383	W-L 8e	PBS	583	621	643	657	644	649	665	682	694	722	22	195.0	0.09	17.5	Cyanine-Penta	4.13.2
384	NR675	10 mм PBS pH 7.4	625	657	675	689	668	675	695	718	733	762	20	73.2	0.24	17.5	PO2H-Rosam.	4.18.2
385	Coumarin-pyronine 9	100 mм P pH 7.5			560				579				19	132.5	0.13	17.2	Hybrid-Bonded	4.40.2
386	Pyrene-1-sulfonate	H_2O	327	340	375	382	385	389	400	429	446	461	25	30.6	0.56	17.1	Pyrene	4.28
387	2,4-DiOMe TokyoGreen	100 mм NaOH	456	478	494	506	496	500	513	551	567	586	19	85.0	0.20	17.0	Fluorone	4.11
388	Niu 29	H_2O	477	497	519	531	511	515	530	561	584	612	11	28.0	0.61	17.0	Bodipy-Med.	4.14.2
389	Neo-Cy5	PBS pH 7.4	583	621	643	657	639	645	660	677	688	716	17	170.0	0.10	17.0	Cyanine-Penta	4.13.2
390	Alexa Fluor 350	H_2O pH 7	296	317	346	370	411	420	445	477	495	516	99	19.0	0.89	16.9	Coumarin 7-Am	4.26.3
391	Pyronine Y (OP)	H_2O	498	523	546	559	546	550	566	584	598	631	20	77.0	0.22	16.9	Pyronine	4.4
392	Pyronine B	H_2O	504	529	552	565	551	557	572	592	605	633	20	94.1	0.18	16.9	Pyronine	4.4
393	nor-SqCN3	P pH 7.4	636	671	693	712	688	697	714	739	761	809	21	113.0	0.15	16.9	Squaraine	4.35
394	ICG	PBS	687	746	780	805	776	787	810	837	849	865	30	211.2	0.08	16.9	Cyanine-Hepta	4.13.3
395	Z-L J	0.5 м PBS pH 7.4	498	519	543	556	541	546	563	598	624	656	20	50	0.34	16.8	Bodipy-Large	4.14.3
396	Cy3.205	PBS pH 7.4	509	532	578	590	574	579	593	610	645	665	15	120.0	0.14	16.8	Cyanine-Trimet	4.13.1
397	Cy3.5	H_2O	511	534	581	593	577	582	596	613	649	669	15	120.0	0.14	16.8	Cyanine-Trimet	4.13.1
398	Seta-632-carboxylate	P pH 7.4			632				641				9	280.0	0.06	16.8	nd	4.41
399	Phosphorilated oxazine 2-H	H_2O pH ~7	583	620	661	679	654	662	678	703	723	765	17	60.0	0.28	16.8	Phenoxazine	4.25

(Continued)

Ent.	Short names	Aqueous solution	$\lambda_{x,-0.25}$ nm	$\lambda_{x,-0.50}$ nm	$\lambda_{x,max}$ nm	$\lambda_{x,+0.50}$ nm	$\lambda_{e,-0.25}$ nm	$\lambda_{e,-0.50}$ nm	$\lambda_{e,max}$ nm	$\lambda_{e,+0.50}$ nm	$\lambda_{e,+0.25}$ nm	$\lambda_{e,+0.10}$ nm	Δ_S nm	ε_{max} mM^{-1}cm^{-1}	Φ_f	B_{max} mM^{-1}cm^{-1}	Class	Ref.
400	Marina Blue	100 mM P pH 10/8	311	328	358	381	421	430	455	490	510	534	97	18.7	0.89	16.6	Coumarin 7-OH	4.26.1
401	NR632	10 mM PBS pH 7.4	578	608	632	646	631	638	655	678	695	723	23	64.0	0.26	16.6	POPh-Rosol	4.20
402	Phosphorilated coum. N-H 21c	PBS pH 7.4	374	394	432	457	482	490	512	550	576	604	80	20.4	0.81	16.5	Coumarin 7-Am	4.26.3
403	JF526	10 mM HEPES pH 7.3	484	507	526	539	530	535	550	573	594	618	24	19.0	0.87	16.5	Rhoda 500-550	4.2.2
404	G-L 13a	H_2O	463	482	505	514	505	509	517	530	548	564	12	35.0	0.47	16.4	Bodipy-Med.	4.14.2
405	DY-636 carboxylate	PBS pH 7.4	558	591	645	669	632	643	671	732	769	811	26	190.0	0.09	16.3	Merocyanine	4.24
406	Epicocconone P-F 4c	H_2O, 0.1% SDS pH 2.5			425				542				117	40.3	0.40	16.2	Epicocconone	4.39.2
407	Sulfo-cyanine3	H_2O	486	506	548	559	546	550	563	578	610	628	15	162.0	0.10	16.2	Cyanine-Trimet	4.13.1
408	Coumarin 2	H_2O	314	332	362	386	422	431	456	491	512	535	94	16.9	0.95	16.0	Coumarin 7-Am	4.26.2
409	Atto 495 carboxylate	PBS pH 7.4, 22 °C	455	471	498	512	503	509	526	569	597	631	28	80.0	0.20	16.0	Acridine	4.23.1
410	SiRF 15a	H_2O	598	577	657	675	651	658	676	697	711	746	19	40.0	0.40	16.0	Si(CH3)2-Rhoda	4.3.2
411	DY-350 carboxylate	PBS pH 7.4	301	323	353	378	400	408	432	462	479	499	79	19.0	0.84	15.9	Coumarin 7-Am	4.26.3
412	Square-655-carboxylate	P pH 7.4	576	604	622	638	623	630	644	664	682	720	22	177.0	0.09	15.9	Squaraine	4.35
413	Squaraine T-P 5c	67 mM P pH 7.4	584	614	632	648	621	628	642	662	679	718	10	265.0	0.06	15.9	Squaraine	4.35
414	Z-L F	0.5 M PBS pH 7.4	459	478	501	509	499	503	511	524	541	557	10	50	0.31	15.6	Bodipy-Large	4.14.3
415	7-Am-4-Me-2-hydroxyquinoline	H_2O pH 5-9			340				440				100	16.0	0.97	15.5	Quinoline	4.29
416	Pacific orange	C pH 9.5			404				553				149	25.0	0.62	15.5	Benzothiazole	4.30

Ent.	Short names	Aqueous solution	$\lambda_{x,-0.25}$ nm	$\lambda_{x,-0.50}$ nm	$\lambda_{x,max}$ nm	$\lambda_{x,+0.50}$ nm	$\lambda_{e,-0.25}$ nm	$\lambda_{e,-0.50}$ nm	$\lambda_{e,max}$ nm	$\lambda_{e,+0.50}$ nm	$\lambda_{e,+0.25}$ nm	$\lambda_{e,+0.10}$ nm	Δ_S nm	ε_{max} mм$^{-1}$cm^{-1}	Φ_f	B_{max} mм$^{-1}$cm^{-1}	Class	Ref.
417	Alexa Fluor 555	PBS pH 7.2, 22 °C	491	512	555	566	555	559	572	587	620	639	17	155.0	0.10	15.5	Cyanine-Trimet	4.13.1
418	DAOTA 1	H_2O	489	511	558	578	563	570	587	621	650	674	29	50.0	0.31	15.5	Triangulenium	4.22
419	7-Hydroxy-4-methylcoumarin	H_2O pH 10	313	330	360	384	416	424	449	483	503	526	89	18.0	0.86	15.4	Coumarin 7-OH	4.26.1
420	Abberior Star 440SXP	PBS pH 7.4	377	398	436	461	485	493	515	554	580	608	79	22.7	0.68	15.4	Coumarin 7-Am	4.26.3
421	2′COOH-Si-rhodol	100 mм P pH 7.4	559	588	609	625	605	611	630	677	706	734	21	70.0	0.22	15.4	Si(CH3)2-Rhod	4.16
422	SR101-NaphtNH$_2$	PBS	493	526	589	610	593	602	626	666	691	722	37	80.0	0.19	15.2	Srhod.-analog	4.6
423	Niu 18	H_2O	455	474	496	504	499	503	511	524	541	557	15	75.5	0.20	15.1	Bodipy-Med.	4.14.2
424	Coumarin 120	H_2O	293	314	342	366	407	415	440	471	489	510	98	16.0	0.94	15.0	Coumarin 7-Am	4.26.2
425	OBQ	H_2O pH 2-5			385				470				85	35.0	0.43	15.0	Quinoline	4.29
426	3-MOMO-6-propylaminor-hodol	50 mм P, <0.1% ACN pH 8	453	472	494	506	507	515	532	557	582	608	38	43.0	0.35	15.0	Rhodol	4.15.2
427	DPP 6	PBS			530				586				56	30.7	0.49	15.0	Diketopyrrolop	4.39.1
428	7-Hydroxy-4-Me-coumarin-3-CA	H_2O pH 5	313	330	360	384	421	430	455	490	510	534	95	18.0	0.82	14.7	Coumarin 7-OH	4.26.1
429	2′COOH-4CH$_2$OH-Si-rhodol	100 mм P pH 7.4	560	589	610	626	605	611	630	677	706	734	20	70.0	0.21	14.7	Si(CH3)2-Rhod	4.16
430	PONy P9	PBS pH 7.4	541	570	597	613	608	615	633	658	674	709	36	66.0	0.22	14.5	Pyronine	4.4
431	Distyryl-BODIPY 13	PBS pH 7.5	579	608	641	659	628	633	657	705	742	787	16	58.0	0.25	14.5	Bodipy-Med.	4.14.2
432	7-Azetidinyl-4-Me-coumarin	10 mм HEPES pH 7.3	308	325	354	377	432	441	467	504	525	551	113	15.0	0.96	14.4	Coumarin 7-Am	4.26.2
433	NOBQ	H_2O pH 2-5			385				470				85	35.0	0.41	14.3	Quinoline	4.29

(Continued)

Ent.	Short names	Aqueous solution	$\lambda_{x,-0.25}$ nm	$\lambda_{x,-0.50}$ nm	$\lambda_{x,max}$ nm	$\lambda_{x,+0.50}$ nm	$\lambda_{e,-0.25}$ nm	$\lambda_{e,-0.50}$ nm	$\lambda_{e,max}$ nm	$\lambda_{e,+0.50}$ nm	$\lambda_{e,+0.25}$ nm	$\lambda_{e,+0.10}$ nm	Δ_S nm	ε_{max} $mM^{-1}cm^{-1}$	Φ_f	B_{max} $mM^{-1}cm^{-1}$	Class	Ref.
434	Coumarin 102	H_2O	336	356	393	418	455	464	489	528	552	579	96	21.8	0.66	14.3	Coumarin 7-Am	4.26.2
435	AF555	PBS pH 7.2, 22 °C	491	512	555	566	555	559	572	587	620	639	17	158.0	0.09	14.2	Cyanine-Trimet	4.13.1
436	DAOTA 3	H_2O	449	511	565	583	571	577	594	626	654	681	29	50.0	0.28	14.0	Triangulenium	4.22
437	DAOTA 2	H_2O	493	515	564	583	569	576	594	627	654	678	30	50.0	0.27	13.5	Triangulenium	4.22
438	SiP-Pr	50 mM PBS pH 7.4	579	613	632	647	607	619	652	661	675	691	20	64.2	0.21	13.4	C(CH3)2-Pyron	4.5
439	SeTau-380-carboxylate	P pH 7.3			383				468				85	23.8	0.56	13.3	nd	4.41
440	M-S 10f	H_2O, 0.1% TX-100	536	561	589	604	578	583	603	643	674	711	14	74.0	0.18	13.3	Bodipy-Small	4.14.1
441	3,6-Diaminoacridine (Proflavine)	H_2O pH 7	404	420	445	457	482	490	511	554	582	614	66	38.9	0.34	13.2	Acridine	4.23.1
442	SqCN1	P pH 7.4	614	647	667	685	661	669	685	708	728	772	18	188.0	0.07	13.1	Squaraine	4.35
443	Squaraine T-P 5b	67 mM P pH 7.4	614	647	667	685	661	669	685	708	728	772	18	188.0	0.07	13.1	Squaraine	4.35
444	POF	PBS pH 7.4	561	591	632	652	635	643	661	687	714	750	29	54.5	0.24	13.0	POPh-Fluorone	4.12
445	DPP 5	PBS			455				515				60	22.7	0.57	12.9	Diketopyrrolop	4.39.1
446	Squaraine T-P 5a	67 mM P pH 7.4	592	620	636	653	624	631	648	666	680	705	12	216.0	0.06	12.9	Squaraine	4.35
447	Cy5.11-OH	H_2O/PBS	586	624	646	661	645	651	667	684	696	725	21	215.0	0.06	12.9	Cyanine-Penta	4.13.2
448	6-Dimethylaminorhodol	50 mM P, <0.1% ACN pH 8	473	494	518	531	518	526	544	570	596	624	26	64.0	0.20	12.8	Rhodol	4.15.2
449	SiP-F	50 mM PBS pH 7.4	577	611	637	653	602	612	651	655	667	679	14	64.2	0.20	12.8	C(CH3)2-Pyron	4.5
450	W-L 8c	PBS	583	621	643	657	645	650	666	683	695	724	23	107.0	0.12	12.8	Cyanine-Penta	4.13.2

Ent.	Short names	Aqueous solution	$\lambda_{x,-0.25}$ nm	$\lambda_{x,-0.50}$ nm	$\lambda_{x,max}$ nm	$\lambda_{x,+0.50}$ nm	$\lambda_{e,-0.25}$ nm	$\lambda_{e,-0.50}$ nm	$\lambda_{e,max}$ nm	$\lambda_{e,+0.50}$ nm	$\lambda_{e,+0.25}$ nm	$\lambda_{e,+0.10}$ nm	Δ_S nm	ε_{max} mM^{-1}cm^{-1}	Φ_f	B_{max} mM^{-1}cm^{-1}	Class	Ref.
451	NR744	PBS, 1% DMSO pH 7.4	672	719	744	763	725	737	764	795	812	841	20	80.0	0.16	12.8	PO2R-Rosam.	4.18.2
452	8-Trifluoromethyl-BODIPY 10	100 mM P pH 7.4	432	449	469	476	529	533	542	557	576	594	73	17.1	0.74	12.6	Bodipy-Small	4.14.1
453	6-Dimethylaminorhodol	10 mM HEPES pH 7.3	473	494	518	531	520	528	546	572	599	626	28	60.0	0.21	12.6	Rhodol	4.15.2
454	Seta-670-carboxylate	P pH 7.4			667				688				21	180.0	0.07	12.6	nd	4.41
455	DPP 9	PBS			451				541				90	13.9	0.90	12.5	Diketopyrrolop	4.39.1
456	Jul-rosol	50 mM P pH 7.4	492	514	540	554	548	554	574	614	648	686	34	30.4	0.41	12.5	Rosol	4.19
457	NBQ	H_2O pH 2-5			380				470				90	35.0	0.35	12.2	Quinoline	4.29
458	Distyryl-BODIPY 12	PBS pH 7.5	579	609	642	660	628	633	657	705	742	787	15	55.1	0.22	12.1	Bodipy-Med.	4.14.2
459	Atto 490LS carboxylate	PBS pH 7.4, 22 °C			495				658				163	40.0	0.30	12.0	nd (pb Coum.)	4.41
460	SR101-NaphtOH	PBS	481	509	580	607	580	588	619	676	700	735	39	80.0	0.15	12.0	Sulforhodol-an	4.27
461	HC6H-SR101	PBS, 5% BSA			624				654				30	80.0	0.15	12.0	Hybrid-Fused	4.40.1
462	2-Me SiR700	H_2O	617	665	691	710	686	695	712	732	746	764	21	100.0	0.12	12.0	Si(CH3)2-Rosa	4.18.1
463	Atto 725 carboxylate	PBS pH 7.4, 22 °C			728				751				23	120.0	0.10	12.0	nd	4.41
464	Atto 740 carboxylate	PBS pH 7.4, 22 °C			743				763				20	120.0	0.10	12.0	nd	4.41
465	Cresyl violet (CV)	H_2O	523	553	585	599	605	611	625	646	663	698	40	33.1	0.36	11.9	Phenoxazine	4.25
466	L-B 4b	H_2O	457	475	498	506	497	501	509	522	539	555	11	77.6	0.15	11.6	Bodipy-Small	4.14.1
467	NR638	10 mM PBS pH 7.4	583	613	638	652	637	645	662	686	703	732	24	55.6	0.21	11.6	POPh-Rosol	4.20
468	Oxazine 170	H_2O pH 2-7	539	560	620	643	624	632	650	668	679	692	30	48.0	0.24	11.5	Phenoxazine	4.25

(Continued)

Ent.	Short names	Aqueous solution	$\lambda_{x,-0.25}$ nm	$\lambda_{x,-0.50}$ nm	$\lambda_{x,max}$ nm	$\lambda_{x,+0.50}$ nm	$\lambda_{e,-0.25}$ nm	$\lambda_{e,-0.50}$ nm	$\lambda_{e,max}$ nm	$\lambda_{e,+0.50}$ nm	$\lambda_{e,+0.25}$ nm	$\lambda_{e,+0.10}$ nm	Δ_S nm	ε_{max} mм$^{-1}$cm^{-1}	Φ_f	B_{max} mм$^{-1}$cm^{-1}	Class	Ref.
469	TMDHS or SiP-Me	H_2O	571	604	634	652	626	634	653	680	697	734	19	64.2	0.18	11.5	Si(CH3)2-Pyron	4.5
470	SiTMR	10 mм HEPES pH 7.3	586	566	643	660	638	645	662	682	696	729	19	28.2	0.41	11.5	Si(CH3)2-Rhoda	4.3.2
471	(⁻O)-Luciferin	H_2O pH 11			384				537				153	18.5	0.62	11.4	Benzothiazole	4.30
472	6-Dimethylaminorhodol	C pH 9.5	477	498	522	535	519	527	545	571	597	625	23	57.0	0.20	11.4	Rhodol	4.15.1
473	Benzo[h]quinoline (BQ)	H_2O pH 2-5			370				430				60	35.0	0.32	11.2	Quinoline	4.29
474	(⁻O)-Dehydroluciferin	H_2O pH 11			396				555				159	18.2	0.62	11.2	Benzothiazole	4.30
475	DY-681 carboxylate	PBS pH 7.4	592	630	691	718	665	676	708	776	818	866	17	125.0	0.09	11.2	Merocyanine	4.24
476	ICG	FBS	708	771	807	834	787	798	822	850	862	878	15	121.0	0.09	11.2	Cyanine-Hepta	4.13.3
477	DPP 7	PBS			456				548				92	14.7	0.76	11.1	Diketopyrrolop	4.39.1
478	Bzh-Cl	H_2O pH 7.4	551	584	620	636	629	636	651	674	692	731	31	36.1	0.31	11.1	Phenoxazine	4.25
479	CF660C	H_2O			667				685				18	200	0.06	11.1	nd (Cyanine)	4.41
480	SeTau-405-carboxylate	P pH 7.4			405				518				113	13.8	0.80	11.0	nd	4.41
481	Bpe-H	H_2O pH 7.4	567	601	640	657	657	665	681	706	726	769	41	48.1	0.23	11.0	Phenoxazine	4.25
482	7-Hydroxy-4-methylcoumarin	H_2O pH 5	276	295	320	341	415	424	450	483	501	523	130	14.0	0.78	10.9	Coumarin 7-OH	4.26.1
483	Coumarin P-R 14	PBS pH 7.5	322	341	373	398	426	435	461	497	518	542	88	17.7	0.62	10.9	Coumarin 7-OH	4.26.1
484	Seta-555-carboxylate	P pH 7.4			555				570				15	155.0	0.07	10.8	nd	4.41
485	Z-L G	0.5 м PBS pH 7.4	526	550	577	592	574	578	598	637	667	704	21	50	0.21	10.6	Bodipy-Large	4.14.3
486	Seta-700-carboxylate	P pH 7.4			687				703				16	177.0	0.06	10.6	nd	4.41
487	Acridine orange	HEPES pH 7.3	443	462	493	508	497	505	528	574	604	639	35	50.0	0.21	10.5	Acridine	4.23.1

Ent.	Short names	Aqueous solution	$\lambda_{x,-0.25}$ nm	$\lambda_{x,-0.50}$ nm	$\lambda_{x,max}$ nm	$\lambda_{x,+0.50}$ nm	$\lambda_{e,-0.25}$ nm	$\lambda_{e,-0.50}$ nm	$\lambda_{e,max}$ nm	$\lambda_{e,+0.50}$ nm	$\lambda_{e,+0.25}$ nm	$\lambda_{e,+0.10}$ nm	Δ_S nm	ε_{max} mM^{-1}cm^{-1}	Φ_f	B_{max} mM^{-1}cm^{-1}	Class	Ref.
488	Coumarin P-R 21	PBS pH 7.5	295	316	345	369	425	434	461	495	515	538	116	13.2	0.79	10.4	Coumarin 7-Am	4.26.2
489	DPP 10	PBS			523				594				71	17.1	0.61	10.4	Diketopyrrolop	4.39.1
490	W-L 8a	PBS	583	621	643	657	645	651	667	684	696	725	24	87.0	0.12	10.4	Cyanine-Penta	4.13.2
491	Coumarin P-R 3	PBS pH 7.5	297	319	348	372	415	424	450	483	501	523	102	12.5	0.82	10.2	Coumarin 7-Am	4.26.2
492	BOPAHY H-D 19ab′	H_2O	387	400	416	422	426	429	435	444	457	468	19	51	0.20	10.2	Bodipy-analog	4.14.5
493	tMe-PR	PBS, 1% DMSO pH 7.4	634	674	696	714	682	691	713	741	759	790	17	68.0	0.15	10.2	POR-Rosam.	4.18.2
494	DPP 8	PBS			527				594				67	18.1	0.56	10.1	Diketopyrrolop	4.39.1
495	Me-PR	PBS, 1% DMSO pH 7.4	632	672	694	711	682	690	712	740	757	789	18	92.0	0.11	10.1	POR-Rosam.	4.18.2
496	Coumarin P-R 22	PBS pH 7.5	307	324	353	376	417	425	450	484	504	527	97	13.4	0.75	10.0	Coumarin 7-Am	4.26.2
497	Rosamine W-B 5h	100 mM P pH 7.4	495	524	542	557	546	552	568	587	603	634	26	71.5	0.14	10.0	Rosamine	4.17
498	EVOblue 30	H_2O	575	610	650	667	644	651	667	691	710	751	17	100.0	0.10	10.0	Phenoxazine	4.25
499	W-B 15	100 mM P pH 7.4	518	549	567	582	571	578	594	620	651	681	27	76.5	0.13	9.9	Rhoda 550-600	4.2.3
500	HMDER	PBS pH 7.4	479	500	525	538	517	525	543	569	595	622	18	70.0	0.14	9.8	Rhodol	4.15.2
501	Sulfobetaine pyrene 21	H_2O	314	327	359	365	371	375	385	412	428	441	26	30.6	0.32	9.7	Pyrene	4.28
502	7-(4-Hydroxy)cyclohexyl coum.	H_2O	319	337	368	393	426	434	460	496	517	541	92	16.5	0.59	9.7	Coumarin 7-Am	4.26.2
503	2-Me SiR640	100 mM P, 1% DMSO pH 7.4	509	548	637	658	635	643	660	683	707	733	23	54.0	0.18	9.7	Si(CH3)2-Rosa	4.18.1
504	JF722	10 mM HEPES pH 7.3	651	626	722	743	712	721	743	768	786	829	21	87.2	0.11	9.5	POPh-Rhodam	4.3.3

(Continued)

Ent.	Short names	Aqueous solution	$\lambda_{x,-0.25}$ nm	$\lambda_{x,-0.50}$ nm	$\lambda_{x,max}$ nm	$\lambda_{x,+0.50}$ nm	$\lambda_{e,-0.25}$ nm	$\lambda_{e,-0.50}$ nm	$\lambda_{e,max}$ nm	$\lambda_{e,+0.50}$ nm	$\lambda_{e,+0.25}$ nm	$\lambda_{e,+0.10}$ nm	Δ_S nm	ε_{max} mM^{-1}cm^{-1}	Φ_f	B_{max} mM^{-1}cm^{-1}	Class	Ref.
505	Coumarin P-R 13	PBS pH 7.5	319	337	368	393	438	447	474	512	534	560	106	15.0	0.63	9.4	Coumarin 7-OH	4.26.1
506	SO_2R4	10 mM PBS, 0.25% ACN pH 7.4	634	679	704	725	706	717	742	779	805	836	38	97.0	0.10	9.4	SO2-Rosamine	4.18.3
507	Harmaline	5 mM H_2SO_4	352	360	383	407	436	444	475	517	541	568	92	28.6	0.32	9.1	Indole	4.33
508	Mercaptofluorescein MK-61	1 mM NaOH	457	477	495	508	505	511	527	554	577	601	32	38.0	0.24	9.1	Fluorescein-SH	4.8.3
509	6-Diethylaminorhodol	50 mM P, <0.1% ACN pH 8	477	498	522	535	522	531	549	576	602	630	27	61.0	0.15	9.1	Rhodol	4.15.2
510	9-Aminoacridine	P pH 7	366	379	400	410	400	405	420	449	467	487	20	9.3	0.97	9.0	Acridine	4.23.1
511	JF668	10 mM HEPES pH 7.3	607	585	668	686	661	668	687	708	724	760	19	26.7	0.34	9.0	POOH-Rhoda	4.3.3
512	Coumarin P-R 6	PBS pH 7.5	309	326	355	378	419	427	452	486	507	530	97	12.6	0.71	8.9	Coumarin 7-Am	4.26.2
513	Alexa Fluor 430	PBS pH 7.4	373	393	431	456	508	517	541	584	613	645	110	16.0	0.55	8.8	Coumarin 7-Am	4.26.3
514	PONy P13	PBS pH 7.4	568	600	630	647	638	646	666	694	712	750	36	52.0	0.17	8.8	Pyronine	4.4
515	N-Methylacridone	H_2O	369	382	411	419	417	421	431	467	480	493	20	8.8	0.99	8.7	Acridones	4.23.2
516	Bza-Oet	H_2O pH 7.4	555	588	625	641	631	638	653	676	694	733	28	33.5	0.26	8.7	Phenoxazine	4.25
517	DY-635 carboxylate	PBS pH 7.4	550	583	635	658	630	641	669	729	766	808	34	175.0	0.05	8.7	Merocyanine	4.24
518	SO_2R5	10 mM PBS, 0.25% ACN pH 7.4	639	684	710	732	715	726	752	790	817	848	42	129.0	0.07	8.7	SO2-Rosamine	4.18.3

Ent.	Short names	Aqueous solution	$\lambda_{x,-0.25}$ nm	$\lambda_{x,-0.50}$ nm	$\lambda_{x,max}$ nm	$\lambda_{x,+0.50}$ nm	$\lambda_{e,-0.25}$ nm	$\lambda_{e,-0.50}$ nm	$\lambda_{e,max}$ nm	$\lambda_{e,+0.50}$ nm	$\lambda_{e,+0.25}$ nm	$\lambda_{e,+0.10}$ nm	Δ_S nm	ε_{max} mM^{-1}cm^{-1}	Φ_f	B_{max} mM^{-1}cm^{-1}	Class	Ref.
519	BOPAHY H-D 19bb′	H_2O	387	400	416	422	425	428	434	443	455	467	18	51	0.17	8.6	Bodipy-analog	4.14.5
520	BOPAHY H-D 20bb′	H_2O	387	400	416	422	426	429	435	444	457	468	19	51	0.17	8.6	Bodipy-analog	4.14.5
521	Fluorinated rhodamine 1r	Buffer pH 7	483	506	525	538	533	539	554	578	598	623	29	31.0	0.28	8.6	Rhoda 500-550	4.2.2
522	Isoquinoline 3j	PBS pH 7.4			409				525				116	18.8	0.45	8.4	Isoquinoline	4.37
523	Me-rosol	50 mM P pH 7.4	475	496	520	533	536	542	557	591	624	653	37	46.1	0.18	8.4	Rosol	4.19
524	LS482	Buffer pH 8	632	682	710	731	760	771	793	819	830	845	83	84.0	0.10	8.4	Cyanine-Hepta	4.13.3
525	Cy3	H_2O	487	508	550	561	553	557	570	585	618	636	20	150.0	0.06	8.2	Cyanine-Trimet	4.13.1
526	CF680	H_2O			681				698				17	210	0.04	8.2	nd (Cyanine)	4.41
527	NR698	PBS, 1% DMSO pH 7.4	631	676	698	715	686	693	712	738	751	773	14	25.9	0.32	8.2	PO2H-Rosam.	4.18.2
528	Coumarin P-R 5	PBS pH 7.5	296	317	346	370	428	437	465	500	520	544	119	10.3	0.79	8.1	Coumarin 7-Am	4.26.2
529	Seta-400-carboxylate	P pH 7.4			399				576				177	24.0	0.34	8.1	nd	4.41
530	PONy P2	PBS pH 7.4	569	602	632	649	644	652	672	700	718	758	40	51.0	0.16	8.1	Pyronine	4.4
531	DEAC-SR101	PBS			610				642				32	80.0	0.10	8.0	Hybrid-Fused	4.40.1
532	2-Me SiR720	H_2O	627	675	721	742	710	719	740	766	790	824	19	160.0	0.05	8.0	Si(CH3)2-Rosa	4.18.1
533	6-Morpholinorhodol	50 mM P, <0.1% ACN pH 8	477	498	522	535	526	534	553	580	607	636	31	61.0	0.13	7.9	Rhodol	4.15.2
534	3-Bt-6-MeO-7-hydroxychromone	100 mM Tris pH 8			345				447				102	16.4	0.48	7.8	Chromone	4.31
535	NR700	PBS, 1% DMSO pH 7.4	630	674	700	720	685	696	722	752	772	808	22	71.0	0.11	7.8	PO2R-Rosam.	4.18.2

(Continued)

Ent.	Short names	Aqueous solution	$\lambda_{x,-0.25}$ nm	$\lambda_{x,-0.50}$ nm	$\lambda_{x,max}$ nm	$\lambda_{x,+0.50}$ nm	$\lambda_{e,-0.25}$ nm	$\lambda_{e,-0.50}$ nm	$\lambda_{e,max}$ nm	$\lambda_{e,+0.50}$ nm	$\lambda_{e,+0.25}$ nm	$\lambda_{e,+0.10}$ nm	Δ_S nm	ε_{max} mм$^{-1}$cm^{-1}	Φ_f	B_{max} mм$^{-1}$cm^{-1}	Class	Ref.
536	3,7-Di(ethylamino) phenoxazine	10 mм HEPES pH 7.3	578	615	655	673	646	653	669	693	712	754	14	111.0	0.07	7.7	Phenoxazine	4.25
537	trans-PO-PB430	PBS pH 7.4			422				545				123	10.3	0.74	7.6	Phosphole	4.39.3
538	CF660R	H_2O			663				682				19	100	0.08	7.6	nd (Rhodam.)	4.41
539	6-Pyrrolidinorhodol	50 mм P, <0.1% ACN pH 8	474	495	519	532	523	532	550	577	603	632	31	63.0	0.12	7.5	Rhodol	4.15.2
540	SqB1	P pH 7.4	580	609	627	642	629	636	650	671	688	728	23	150.0	0.05	7.5	Squaraine	4.35
541	Mercaptofluorescein MK-43	1 mм NaOH	457	477	495	508	504	510	526	553	576	600	31	34.0	0.22	7.4	Fluorescein-SH	4.8.3
542	TG170	H_2O	695	756	791	817	769	780	803	829	841	857	12	300.0	0.02	7.4	Cyanine-Hepta	4.13.3
543	Carbostyril 23b	H2O pH 2.8			380				475				95	18.0	0.41	7.3	Quinoline	4.29
544	Carbostyril 124	H_2O			341				417				76	15.8	0.46	7.2	Quinoline	4.29
545	DMAC-SR101	PBS			605				647				42	80.0	0.09	7.2	Hybrid-Fused	4.40.1
546	PORL	PBS, 1% DMSO pH 7.4	609	643	670	686	671	679	698	724	744	776	28	66.1	0.11	7.2	POPh-Rosol	4.20
547	Isoquinoline 3g	PBS pH 7.4			414				522				108	17.5	0.41	7.1	Isoquinoline	4.37
548	cis-PO-PB430	PBS pH 7.4			421				548				127	10.0	0.71	7.1	Phosphole	4.39.3
549	Isoquinoline 3l	PBS pH 7.4			410				522				112	16.8	0.42	7.0	Isoquinoline	4.37
550	Acedan 5	H_2O	295	316	357	396	458	472	503	537	553	579	146	17.4	0.40	6.9	Naphthalene	4.34
551	SO_2R2	10 mм PBS, 0.25% ACN pH 7.4	633	678	703	724	706	717	742	779	805	836	39	95	0.07	6.9	SO2-Rosamine	4.18.3

Ent.	Short names	Aqueous solution	$\lambda_{x,-0.25}$ nm	$\lambda_{x,-0.50}$ nm	$\lambda_{x,max}$ nm	$\lambda_{x,+0.50}$ nm	$\lambda_{e,-0.25}$ nm	$\lambda_{e,-0.50}$ nm	$\lambda_{e,max}$ nm	$\lambda_{e,+0.50}$ nm	$\lambda_{e,+0.25}$ nm	$\lambda_{e,+0.10}$ nm	Δ_S nm	ε_{max} mM^{-1}cm^{-1}	Φ_f	B_{max} mM^{-1}cm^{-1}	Class	Ref.
552	Isoquinoline 3n	PBS pH 7.4			411				521				110	16.6	0.41	6.8	Isoquinoline	4.37
553	PB430	PBS pH 7.4			426				542				116	10.4	0.66	6.8	Phosphole	4.39.3
554	2-OH-6-DiEtAm-benzo[c] xanthen	H_2O	523	547	578	595	584	596	620	654	689	728	42	38.1	0.18	6.8	Rhodol-analog	4.15.3
555	Tri-*O*-acetyl-P-luminoarosin	H_2O pH 7			425				528				103	10.8	0.62	6.7	Pteridine	4.36
556	Y-A 1 (Cypate)	PBS pH 7.2	685	744	778	803	771	782	805	831	844	859	27	224.0	0.03	6.7	Cyanine-Hepta	4.13.3
557	Y-A 6	PBS pH 7.2	685	744	778	803	771	782	805	831	844	859	27	224.0	0.03	6.7	Cyanine-Hepta	4.13.3
558	Y-A 7	PBS pH 7.2	686	745	779	804	771	782	805	831	844	859	26	224.0	0.03	6.7	Cyanine-Hepta	4.13.3
559	3-AB-6-MeO-7-OH-chromone	100 mM Tris pH 8			346				445				99	16.3	0.41	6.6	Chromone	4.31
560	Isoquinoline 7b	PBS pH 7.4			406				526				120	14.4	0.46	6.6	Isoquinoline	4.37
561	Isoquinoline 3m	PBS pH 7.4			408				526				118	18.1	0.36	6.5	Isoquinoline	4.37
562	Cy3.12-OH	H_2O/PBS	492	513	556	567	556	561	574	590	623	641	18	130.0	0.05	6.5	Cyanine-Trimet	4.13.1
563	Y-A 10	PBS pH 7.2	687	746	780	805	774	785	808	835	847	862	28	219.0	0.03	6.5	Cyanine-Hepta	4.13.3
564	Isoquinoline 3e	PBS pH 7.4			408				524				116	15.8	0.41	6.4	Isoquinoline	4.37
565	Y-A 3	PBS pH 7.2	684	744	777	802	773	784	807	834	846	861	30	215.0	0.03	6.4	Cyanine-Hepta	4.13.3
566	Abberior Star 520SXP	PBS pH 7.4	454	475	522	555	590	601	632	691	733	779	110	42.5	0.15	6.3	Coumarin 7-Am	4.26.3
567	Si-9-Iminopyronine 2f	~1 mM NaOH, 1% DMSO pH 11	375	389	401	408	583	590	606	629	643	675	205	17.8	0.35	6.2	Si(CH3)2-Pyron	4.5

(Continued)

Ent.	Short names	Aqueous solution	$\lambda_{x,-0.25}$ nm	$\lambda_{x,-0.50}$ nm	$\lambda_{x,max}$ nm	$\lambda_{x,+0.50}$ nm	$\lambda_{e,-0.25}$ nm	$\lambda_{e,-0.50}$ nm	$\lambda_{e,max}$ nm	$\lambda_{e,+0.50}$ nm	$\lambda_{e,+0.25}$ nm	$\lambda_{e,+0.10}$ nm	Δ_S nm	ε_{max} mM^{-1}cm^{-1}	Φ_f	B_{max} mM^{-1}cm^{-1}	Class	Ref.
568	Isoquinoline 8	PBS pH 7.4			406				534				128	16.8	0.37	6.2	Isoquinoline	4.37
569	Isoquinoline 3d	PBS pH 7.4			418				520				102	14.5	0.43	6.2	Isoquinoline	4.37
570	6-Piperidinorhodol	50 mM P, <0.1% ACN pH 8	478	499	523	536	528	536	555	582	609	638	32	62.5	0.10	6.2	Rhodol	4.15.2
571	PONy P6	PBS pH 7.4	546	576	603	619	613	620	638	663	680	715	35	69.0	0.09	6.2	Pyronine	4.4
572	PONy P10	PBS pH 7.4	567	599	629	646	640	648	668	696	714	753	39	52.0	0.12	6.2	Pyronine	4.4
573	Isoquinoline 3i	PBS pH 7.4			405				521				116	15.4	0.40	6.1	Isoquinoline	4.37
574	Phosphorilated coum. N-H 22c	PBS pH 7.4	348	370	410	437	467	477	503	545	570	599	93	14.0	0.44	6.1	Coumarin 7-Am	4.26.3
575	Naphthofluorescein	C pH 9.5	541	568	595	613	626	635	660	703	740	780	65	44.0	0.14	6.1	Fluores-analog	4.8.4
576	2-Me SiR675	100 mM P, 1% DMSO pH 7.4	592	635	675	694	672	679	698	721	742	772	23	88.0	0.07	6.1	Si(CH3)2-Rosa	4.18.1
577	Acedan 6	H_2O	295	316	357	396	458	472	503	537	553	579	146	15.4	0.39	6.0	Naphthalene	4.34
578	7-Ethylamino-coumarin	H_2O	319	337	368	393	424	432	458	493	514	538	90	11.9	0.51	6.0	Coumarin 7-Am	4.26.2
579	Isoquinoline 3f	PBS pH 7.4			405				519				114	15.4	0.39	6.0	Isoquinoline	4.37
580	Acridine orange	H_2O	439	458	488	503	497	505	528	574	604	639	40	50.0	0.12	6.0	Acridine	4.23.1
581	Cy3.18.OH	PBS	487	508	550	561	548	552	565	580	612	630	15	150.0	0.04	6.0	Cyanine-Trimet	4.13.1
582	Y-A 4	PBS pH 7.2	685	744	778	803	771	782	805	831	844	859	27	201.0	0.03	6.0	Cyanine-Hepta	4.13.3
583	Coumarin P-R 12	PBS pH 7.5	282	301	327	348	421	430	457	491	510	533	130	7.9	0.76	5.9	Coumarin 7-OH	4.26.1

Ent.	Short names	Aqueous solution	$\lambda_{x,-0.25}$ nm	$\lambda_{x,-0.50}$ nm	$\lambda_{x,max}$ nm	$\lambda_{x,+0.50}$ nm	$\lambda_{e,-0.25}$ nm	$\lambda_{e,-0.50}$ nm	$\lambda_{e,max}$ nm	$\lambda_{e,+0.50}$ nm	$\lambda_{e,+0.25}$ nm	$\lambda_{e,+0.10}$ nm	Δ_S nm	ε_{max} mM^{-1}cm^{-1}	Φ_f	B_{max} mM^{-1}cm^{-1}	Class	Ref.
584	Si-9-Iminopyronine 2c	~1 mM NaOH, 1% DMSO pH 11	390	405	418	426	584	591	607	630	645	676	189	12.9	0.46	5.9	Si(CH3)2-Pyron	4.5
585	6-Dipropylaminorhodol	50 mM P, <0.1% ACN pH 8	479	500	525	538	525	533	552	579	606	634	27	59.0	0.10	5.9	Rhodol	4.15.2
586	Y-A 8	PBS pH 7.2	687	746	780	805	772	783	806	832	845	860	26	197.0	0.03	5.9	Cyanine-Hepta	4.13.3
587	Y-A 11	PBS pH 7.2	687	746	780	805	773	784	807	834	846	861	27	199.0	0.03	5.9	Cyanine-Hepta	4.13.3
588	Y-A 2	PBS pH 7.2	685	744	778	803	772	783	806	832	845	860	28	190.0	0.03	5.7	Cyanine-Hepta	4.13.3
589	PetACSO3H	H_2O pH 5.5	367	379	406	414	417	421	431	467	480	495	25	6.2	0.92	5.6	Acridones	4.23.2
590	3-OH-4-nitroso-2,7-naphth-diSA	T pH 9	339	367	423	478	475	489	523	560	578	605	100	20.0	0.28	5.6	Naphthalene	4.34
591	Y-A 9	PBS pH 7.2	686	745	779	804	775	786	809	836	848	864	30	187.0	0.03	5.6	Cyanine-Hepta	4.13.3
592	Isoquinoline 4b	PBS pH 7.4			436				550				114	15.5	0.36	5.5	Isoquinoline	4.37
593	Y-A 5	PBS pH 7.2	685	744	778	803	770	781	804	830	842	858	26	185.0	0.03	5.5	Cyanine-Hepta	4.13.3
594	Chromenoquinoline L-S 4a	H_2O			473				570				97	36.19	0.15	5.4	Hybrid-Fused	4.40.1
595	6-Acrylamidorhodol	50 mM P, <0.1% ACN pH 8	450	469	490	502	500	507	524	548	572	598	34	28.5	0.19	5.4	Rhodol	4.15.2
596	Vita Blue	C pH 9.5	560	586	609	625	636	643	665	731	758	793	56	36.0	0.15	5.4	S-Fluore-analog	4.10
597	Square-660-carboxylate	P pH 7.4	600	635	657	675	654	660	676	702	730	769	19	182.0	0.03	5.4	Squaraine	4.35
598	Squaraine T-P 4d	67 mM P pH 7.4	606	638	658	675	654	662	677	700	719	762	19	182.0	0.03	5.4	Squaraine	4.35

(Continued)

Ent.	Short names	Aqueous solution	$\lambda_{x,-0.25}$ nm	$\lambda_{x,-0.50}$ nm	$\lambda_{x,max}$ nm	$\lambda_{x,+0.50}$ nm	$\lambda_{e,-0.25}$ nm	$\lambda_{e,-0.50}$ nm	$\lambda_{e,max}$ nm	$\lambda_{e,+0.50}$ nm	$\lambda_{e,+0.25}$ nm	$\lambda_{e,+0.10}$ nm	Δ_S nm	ε_{max} mM^{-1}cm^{-1}	Φ_f	B_{max} mM^{-1}cm^{-1}	Class	Ref.
599	Harmine	50 mM H_2SO_4	304	309	326	343	383	389	412	445	464	483	86	11.9	0.45	5.3	Indole	4.33
600	Epicocconone P-F 4d	H_2O, 0.1% SDS pH 2.5			417				534				117	14.5	0.37	5.3	Epicocconone	4.39.2
601	Isoquinoline 4c	PBS pH 7.4			453				529				76	22.1	0.24	5.3	Isoquinoline	4.37
602	S-JF646	10 mM HEPES pH 7.3	586	620	652	671	640	648	668	696	714	753	16	21.2	0.25	5.3	Si(CH3)2-Pyron	4.5
603	Coumarin 152	H_2O	316	334	365	389	457	467	497	539	564	593	132	14.3	0.37	5.2	Coumarin 7-Am	4.26.2
604	HBQ	H_2O pH 2-5			390				480				90	35.0	0.15	5.2	Quinoline	4.29
605	Epicocconone P-F 17d	H_2O, 0.1% SDS pH 2.5			392				518				126	13.2	0.40	5.2	Epicocconone	4.39.2
606	Isoquinoline 3h	PBS pH 7.4			406				525				119	13.7	0.38	5.2	Isoquinoline	4.37
607	Chromenoquinoline L-S 4b	H_2O			477				579				102	36.05	0.15	5.2	Hybrid-Fused	4.40.1
608	Harmane	50 mM H_2SO_4	324	331	350	370	405	412	440	474	494	517	90	6.3	0.83	5.1	Indole	4.33
609	Naphtahlimide 10c	H_2O	393	412	453	490	500	512	543	583	603	627	90	17.3	0.30	5.1	Naphthalimide	4.32
610	Square-680-carboxylate	P pH 7.4	603	634	654	671	647	655	670	692	711	753	16	138.0	0.04	5.1	Squaraine	4.35
611	Epicocconone	H_2O, 0.1% SDS pH 2.5			444				535				91	68.6	0.07	5.0	Epicocconone	4.39.2
612	6-Azetidinyl naphthalimide	10 mM HEPES pH 7.3	401	421	464	503	509	521	553	595	616	640	89	18.0	0.28	5.0	Naphthalimide	4.32
613	Mercaptofluorescein MK-69	1 mM NaOH	457	477	495	508	504	510	526	553	576	600	31	25.0	0.20	5.0	Fluorescein-SH	4.8.3

Ent.	Short names	Aqueous solution	$\lambda_{x,-0.25}$ nm	$\lambda_{x,-0.50}$ nm	$\lambda_{x,max}$ nm	$\lambda_{x,+0.50}$ nm	$\lambda_{e,-0.25}$ nm	$\lambda_{e,-0.50}$ nm	$\lambda_{e,max}$ nm	$\lambda_{e,+0.50}$ nm	$\lambda_{e,+0.25}$ nm	$\lambda_{e,+0.10}$ nm	Δ_S nm	ε_{max} mм$^{-1}$cm^{-1}	Φ_f	B_{max} mм$^{-1}$cm^{-1}	Class	Ref.
614	6-Dibutylaminorhodol	50 mм P, <0.1% ACN pH 8	479	500	525	538	526	534	553	580	607	636	28	63.0	0.08	5.0	Rhodol	4.15.2
615	Bzh-H	H_2O pH 7.4	551	584	620	636	629	636	651	674	692	731	31	11.3	0.44	4.9	Phenoxazine	4.25
616	Squaraine T-P 4c	67 mм P pH 7.4	577	606	624	639	615	621	635	655	672	709	11	245.0	0.02	4.9	Squaraine	4.35
617	6-Benzamidorhodol	50 mм P, <0.1% ACN pH 8	450	469	490	502	501	508	525	549	573	599	35	30.1	0.16	4.8	Rhodol	4.15.2
618	Cy3′	H_2O	483	504	545	556	543	547	560	575	606	624	15	120.0	0.04	4.8	Cyanine-Trimet	4.13.1
619	Squaraine T-P 4a	67 mм P pH 7.4	581	610	628	644	621	628	642	662	679	718	14	163.0	0.03	4.8	Squaraine	4.35
620	(HO)-Dehydroluciferin	H_2O pH 4.5			348				555				207	18.2	0.26	4.7	Benzothiazole	4.30
621	Isoquinoline 7a	PBS pH 7.4			407				528				121	14.1	0.34	4.7	Isoquinoline	4.37
622	Si-9-Iminopyronine 2g	PBS pH 7.4	434	452	469	479	606	613	631	656	672	706	162	11	0.43	4.7	Si(CH3)2-Pyron	4.5
623	Bza-OH	H_2O pH 7.4	551	584	620	636	631	638	653	676	694	733	33	12.8	0.37	4.7	Phenoxazine	4.25
624	CF680R	H_2O			680				701				21	140	0.03	4.7	nd (Rhodam.)	4.41
625	COUPY 9	H_2O			543				605				62	31.0	0.15	4.6	Hybrid-Bridged	4.40.3
626	PONy P1-Halo	PBS pH 7.4	546	576	603	619	610	617	635	660	676	711	32	18.0	0.26	4.6	Pyronine	4.4
627	(HO)-Luciferin	H_2O pH 5			328				537				209	18.2	0.25	4.5	Benzothiazole	4.30
628	Acedan 3	H_2O	295	316	357	396	457	470	501	535	551	576	144	14.5	0.31	4.5	Naphthalene	4.34
629	Carbostyril 23b	H_2O pH 12			397				556				159	17.3	0.26	4.5	Quinoline	4.29

(Continued)

Ent.	Short names	Aqueous solution	$\lambda_{x,-0.25}$ nm	$\lambda_{x,-0.50}$ nm	$\lambda_{x,max}$ nm	$\lambda_{x,+0.50}$ nm	$\lambda_{e,-0.25}$ nm	$\lambda_{e,-0.50}$ nm	$\lambda_{e,max}$ nm	$\lambda_{e,+0.50}$ nm	$\lambda_{e,+0.25}$ nm	$\lambda_{e,+0.10}$ nm	Δ_S nm	ε_{max} mм$^{-1}$cm^{-1}	Φ_f	B_{max} mм$^{-1}$cm^{-1}	Class	Ref.
630	Carbostyril 13	H_2O pH 7			420				556				136	17.3	0.26	4.5	Quinoline	4.29
631	Carbostyril 23b	H_2O pH 7			424				556				132	17.3	0.26	4.5	Quinoline	4.29
632	Mercaptofluorescein MK-67	1 mм NaOH	463	483	502	515	505	511	527	554	577	601	25	27.0	0.17	4.5	Fluorescein-SH	4.8.3
633	AR116	100 mм P pH 7.5	483	506	527	539	529	535	548	567	578	604	21	64.8	0.07	4.5	Pyronine	4.4
634	Z-L E	0.5 м PBS pH 7.4	490	511	534	547	529	533	550	583	608	638	16	50	0.09	4.4	Bodipy-Large	4.14.3
635	LS479	Buffer pH 8	661	716	747	770	731	741	762	786	796	810	15	105.0	0.04	4.4	Cyanine-Hepta	4.13.3
636	Acedan 2	H_2O	282	301	338	372	468	482	515	551	568	595	177	14.6	0.30	4.3	Naphthalene	4.34
637	Acedan 4	H_2O	294	315	355	393	457	470	501	535	551	576	146	13.5	0.32	4.3	Naphthalene	4.34
638	BC-BA	100 mм P, 20% Et pH 7.4			371				473				102	25.5	0.17	4.3	Hybrid-Bonded	4.40.2
639	Briliant sulfaflavine	H_2O	376	393	430	463	481	491	520	557	575	596	90	16.0	0.27	4.3	Naphthalimide	4.32
640	Stefani HB-9	H_2O pH 6			320				423				103	35.9	0.12	4.2	Hybrid-Bonded	4.40.2
641	MetACSO3H	H_2O pH 5.5	364	376	403	411	418	423	433	470	487	495	30	5.6	0.76	4.2	Acridones	4.23.2
642	Luminarine	H_2O pH 7			428				533				105	9.0	0.47	4.2	Pteridine	4.36
643	Harmine	H_2O pH 9	309	315	332	350	393	400	425	458	477	498	93	11.0	0.38	4.1	Indole	4.33
644	3-OH-4-nitroso-2,7-naphth-diSA	A pH 4	304	326	370	412	431	442	470	500	514	535	100	19.0	0.22	4.1	Naphthalene	4.34
645	Isoquinoline 7c	PBS pH 7.4			408				524				116	15.8	0.26	4.1	Isoquinoline	4.37
646	Et-rosol	50 mм P pH 7.4	478	500	524	537	538	544	558	589	622	649	34	42.6	0.10	4.1	Rosol	4.19

Ent.	Short names	Aqueous solution	$\lambda_{x,-0.25}$ nm	$\lambda_{x,-0.50}$ nm	$\lambda_{x,max}$ nm	$\lambda_{x,+0.50}$ nm	$\lambda_{e,-0.25}$ nm	$\lambda_{e,-0.50}$ nm	$\lambda_{e,max}$ nm	$\lambda_{e,+0.50}$ nm	$\lambda_{e,+0.25}$ nm	$\lambda_{e,+0.10}$ nm	Δ_S nm	ε_{max} mM^{-1}cm^{-1}	Φ_f	B_{max} mM^{-1}cm^{-1}	Class	Ref.
647	Bzm-H	H_2O pH 7.4	563	597	635	652	651	658	674	698	718	760	39	29.9	0.14	4.1	Phenoxazine	4.25
648	Squaraine T-P 4b	67 mM P pH 7.4	607	639	659	676	653	661	676	698	718	760	17	207.0	0.02	4.1	Squaraine	4.35
649	Stefani HB-11	H_2O pH 6			329				360				31	38.1	0.11	4.0	Hybrid-Bonded	4.40.2
650	SNARF-1	P pH 10	519	543	573	590	594	606	631	667	703	744	58	44.0	0.09	4.0	Rhodol-analog	4.15.3
651	Fluorene I-6	H_2O			353				473				120	30.5	0.13	3.9	Fluorene	4.39.2
652	Epicocconone P-F 4b	H_2O, 0.1% SDS pH 2.5			435				538				103	19.8	0.20	3.9	Epicocconone	4.39.2
653	Cy3.13-OH	H_2O/PBS	487	508	550	561	549	553	566	581	613	631	16	130.0	0.03	3.9	Cyanine-Trimet	4.13.1
654	Cy3.10-OH	H_2O/PBS	489	510	552	563	553	557	570	585	618	636	18	130.0	0.03	3.9	Cyanine-Trimet	4.13.1
655	PR	PBS, 1% DMSO pH 7.4	632	672	694	711	682	690	712	740	757	789	18	65.0	0.06	3.9	POR-Rosam.	4.18.2
656	Fluorene PP12	H_2O			345				555				210	24.0	0.16	3.8	Fluorene	4.39.2
657	PONy A1	PBS pH 7.4	476	499	535	553	556	566	595	654	693	739	60	38.0	0.10	3.8	Acridine	4.23.1
658	PONy P5	PBS pH 7.4	592	627	660	679	673	682	704	735	755	799	44	38.0	0.10	3.8	Pyronine	4.4
659	Stefani HB-7	H_2O			331				449				118	42.0	0.09	3.6	Benzothiazole	4.30
660	Harmine	H_2O pH 4	330	337	357	378	387	393	416	449	465	484	59	7.3	0.49	3.6	Indole	4.33
661	nor-Harmane (β-carboline)	50 mM H_2SO_4	333	340	360	381	418	425	454	492	514	538	94	6.3	0.58	3.6	Indole	4.33
662	$ACSO_3H$	H_2O pH 5.5	353	366	396	404	405	408	417	449	458	472	21	4.4	0.82	3.6	Acridones	4.23.2
663	Quinine	100 mM $HClO_4$			347				451				104	5.7	0.60	3.4	Quinoline	4.29
664	Acedan 7	H_2O	291	311	351	388	452	465	495	528	544	568	144	11.0	0.31	3.4	Naphthalene	4.34

(Continued)

Ent.	Short names	Aqueous solution	$\lambda_{x,-0.25}$ nm	$\lambda_{x,-0.50}$ nm	$\lambda_{x,max}$ nm	$\lambda_{x,+0.50}$ nm	$\lambda_{e,-0.25}$ nm	$\lambda_{e,-0.50}$ nm	$\lambda_{e,max}$ nm	$\lambda_{e,+0.50}$ nm	$\lambda_{e,+0.25}$ nm	$\lambda_{e,+0.10}$ nm	Δ_S nm	ε_{max} mM^{-1}cm^{-1}	Φ_f	B_{max} mM^{-1}cm^{-1}	Class	Ref.
665	7-Dimethylamino-4-Me-coumarin	10 mM HEPES pH 7.3	322	340	372	397	434	443	470	507	529	555	98	18.0	0.19	3.4	Coumarin 7-Am	4.26.2
666	10-Ethylacridine	P pH 7	376	390	412	423	467	474	494	534	560	589	82	3.5	0.98	3.4	Acridine	4.23.1
667	Abberior Star 470SXP	PBS pH 7.4	416	434	472	499	583	594	624	681	722	767	152	29.0	0.12	3.4	Coumarin 7-Am	4.26.3
668	SO_2R3	10 mM PBS, 0.25% ACN pH 7.4	636	681	707	728	711	721	747	784	811	842	40	69	0.05	3.4	SO2-Rosamine	4.18.3
669	PONy CP1	PBS pH 7.4	640	681	710	728	737	754	783	817	843	870	73	17.0	0.19	3.2	C(CH3)2-Pyron	4.5
670	DY-780 carboxylate	PBS pH 7.4	657	705	782	817	745	760	800	888	943	1007	18	102.0	0.03	3.2	Merocyanine	4.24
671	Quinine	50 mM H_2SO_4			347				451				104	5.7	0.55	3.1	Quinoline	4.29
672	3,7,8,10-TetraMe-isoalloxazine	H_2O pH 7			444				529				85	12.4	0.25	3.1	Pteridine	4.36
673	AI3	H_2O pH 7.4	387	405	445	481	508	520	552	594	615	639	107	15.0	0.21	3.1	Naphthalimide	4.32
674	3,7,10-Trimethyliso-alloxazine	H_2O pH 7			446				532				86	9.8	0.32	3.1	Pteridine	4.36
675	8-Aminonaphthalene-1,3,6-triSA	100 mM P pH 7	295	315	356	394	466	479	512	547	564	591	156	15.0	0.20	3.0	Naphthalene	4.34
676	Epicocconone P-F 17a	H_2O, 0.1% SDS pH 2.5			391				517				126	8.8	0.35	3.0	Epicocconone	4.39.2
677	Flavin mononucleotide (FMN)	H_2O			450				530				80	12.2	0.25	3.0	Pteridine	4.36
678	JF646	10 mM HEPES pH 7.3	589	568	646	663	639	646	664	684	698	732	18	5.6	0.54	3.0	Si(CH3)2-Rhoda	4.3.2
679	2-Aminopyridine	H_2O			300				363				63	4.1	0.71	2.9	Pyridine	4.39.1

Ent.	Short names	Aqueous solution	$\lambda_{x,-0.25}$ nm	$\lambda_{x,-0.50}$ nm	$\lambda_{x,max}$ nm	$\lambda_{x,+0.50}$ nm	$\lambda_{e,-0.25}$ nm	$\lambda_{e,-0.50}$ nm	$\lambda_{e,max}$ nm	$\lambda_{e,+0.50}$ nm	$\lambda_{e,+0.25}$ nm	$\lambda_{e,+0.10}$ nm	Δ_S nm	ε_{max} $mM^{-1}cm^{-1}$	Φ_f	B_{max} $mM^{-1}cm^{-1}$	Class	Ref.
680	Acedan 8	H_2O	278	296	332	365	468	482	515	551	568	595	183	12.6	0.23	2.9	Naphthalene	4.34
681	Acridizinium J-N 1a	H_2O			377				404				27	8.9	0.33	2.9	Acridizinium	4.39.1
682	Coumarin 307	H_2O	335	355	392	417	471	481	508	550	576	606	116	9.0	0.33	2.9	Coumarin 7-Am	4.26.2
683	Safranin T	H_2O			517				586				69	27.0	0.11	2.9	Phenazine	4.39.1
684	Quinine	H_2O pH 7			331				382				51	5.7	0.50	2.8	Quinoline	4.29
685	2-Methyl harmine	50 mM H_2SO_4	324	331	350	370	399	406	433	466	485	509	83	6.3	0.45	2.8	Indole	4.33
686	Chromenoquinoline L-S 4d	H_2O			500				608				108	36.06	0.08	2.8	Hybrid-Fused	4.40.1
687	Carbostyril 1	H_2O			358				432				74	11.6	0.24	2.7	Quinoline	4.29
688	Lucifer yellow VS	H_2O pH 7	376	393	430	463	493	505	535	574	594	616	105	11.3	0.24	2.7	Naphthalimide	4.32
689	3-Buthyl-7-hydroxychromone	100 mM Tris pH 8			334				454				120	12.7	0.21	2.6	Chromone	4.31
690	Isoquinoline 7f	PBS pH 7.4			406				528				122	16.3	0.16	2.6	Isoquinoline	4.37
691	Isoquinoline 10b	PBS pH 7.4			410				526				116	15.5	0.17	2.6	Isoquinoline	4.37
692	Naphtahlimide 10b	H_2O	389	408	448	484	502	514	545	586	606	630	97	12.4	0.21	2.6	Naphthalimide	4.32
693	SO_2R1	10 mM PBS, 0.25% ACN pH 7.4	633	678	703	724	701	711	736	772	798	828	33	39	0.07	2.6	SO2-Rosamine	4.18.3
694	Acridizinium J-N 1c	H_2O			354				420				66	8.4	0.30	2.5	Acridizinium	4.39.1
695	BMeS-*p*-A	H_2O	322	341	377	404	458	472	517	576	610	643	140	3.8	0.67	2.5	Benzene	4.38
696	Seta-780-carboxylate	P pH 7.4			777				807				30	215.0	0.01	2.5	nd	4.41
697	BC-H	100 mM P, 20% Et pH 7.4			365				470				105	27.2	0.09	2.4	Hybrid-Bonded	4.40.2
698	Sulfonated coumarin N-H 7-H	PBS pH 7.4	453	474	520	552	596	607	639	699	742	789	119	40.0	0.06	2.4	Coumarin 7-Am	4.26.3

(Continued)

Ent.	Short names	Aqueous solution	$\lambda_{x,-0.25}$ nm	$\lambda_{x,-0.50}$ nm	$\lambda_{x,max}$ nm	$\lambda_{x,+0.50}$ nm	$\lambda_{e,-0.25}$ nm	$\lambda_{e,-0.50}$ nm	$\lambda_{e,max}$ nm	$\lambda_{e,+0.50}$ nm	$\lambda_{e,+0.25}$ nm	$\lambda_{e,+0.10}$ nm	Δ_S nm	ε_{max} mм$^{-1}$cm^{-1}	Φ_f	B_{max} mм$^{-1}$cm^{-1}	Class	Ref.
699	Acrylodan	H_2O	315	339	387	433	457	471	502	536	552	577	115	12.9	0.18	2.3	Naphthalene	4.34
700	BMeS-*p*-A-COOH	H_2O	333	353	392	421	466	481	527	589	624	659	135	3.9	0.60	2.3	Benzene	4.38
701	Lucifer yellow CH	H_2O pH 7	375	392	429	462	498	509	540	580	600	623	111	11.3	0.21	2.3	Naphthalimide	4.32
702	Acedan	H_2O	299	320	362	402	468	482	515	551	568	595	153	11.2	0.20	2.2	Naphthalene	4.34
703	Acridizinium J-N 1b	H_2O			381				408				27	6.7	0.34	2.2	Acridizinium	4.39.1
704	trans-PO-PB460	PBS pH 7.4			459				592				133	9.0	0.25	2.2	Phosphole	4.39.3
705	Chromenoquinoline L-S 4e	H_2O			502				613				111	33.16	0.07	2.2	Hybrid-Fused	4.40.1
706	Isoquinoline 7d	PBS pH 7.4			408				526				118	15.0	0.14	2.1	Isoquinoline	4.37
707	Z-L A	0.5 м PBS pH 7.4	457	476	499	507	498	502	510	523	540	556	11	50	0.04	2.1	Bodipy-Large	4.14.3
708	JF711	10 mм HEPES pH 7.3	642	618	711	732	702	711	732	756	774	815	21	12.4	0.17	2.1	POPh-Rhodam	4.3.3
709	Acridizinium I-L 5a	H_2O			387				507				120	17.4	0.12	2.0	Acridizinium	4.39.1
710	BMeS-*p*-A-2COOH	H_2O	341	362	403	434	473	487	535	599	635	671	132	3.7	0.56	2.0	Benzene	4.38
711	Isoquinoline 3k	PBS pH 7.4			411				523				112	18.6	0.11	2.0	Isoquinoline	4.37
712	Pip-rosol	50 mм P pH 7.4	479	500	525	538	537	544	563	599	630	669	38	42.5	0.05	2.0	Rosol	4.19
713	Mercaptofluorescein MK-75	1 mм NaOH	463	483	502	515	511	517	533	561	584	609	31	12.0	0.16	1.9	Fluorescein-SH	4.8.3
714	Chromenoquinoline L-S 4f	H_2O			529				622				93	40.64	0.05	1.9	Hybrid-Fused	4.40.1
715	Neo-Cy3	PBS	488	509	551	562	555	560	573	588	622	640	22	98.0	0.02	1.9	Cyanine-Trimet	4.13.1
716	Fluorene PP11	H_2O			377				621				244	36.0	0.05	1.8	Fluorene	4.39.2
717	Isoquinoline 3o	PBS pH 7.4			410				521				111	19.2	0.09	1.7	Isoquinoline	4.37
718	(HO)-Luciferyl adenylate	H_2O pH 5			335				548				213	15.1	0.11	1.6	Benzothiazole	4.30
719	Edans	H_2O	281	300	336	370	472	487	520	557	574	601	184	6.2	0.27	1.6	Naphthalene	4.34

Ent.	Short names	Aqueous solution	$\lambda_{x,-0.25}$ nm	$\lambda_{x,-0.50}$ nm	$\lambda_{x,max}$ nm	$\lambda_{x,+0.50}$ nm	$\lambda_{e,-0.25}$ nm	$\lambda_{e,-0.50}$ nm	$\lambda_{e,max}$ nm	$\lambda_{e,+0.50}$ nm	$\lambda_{e,+0.25}$ nm	$\lambda_{e,+0.10}$ nm	Δ_S nm	ε_{max} $mM^{-1}cm^{-1}$	Φ_f	B_{max} $mM^{-1}cm^{-1}$	Class	Ref.
720	4,10-Dibromonaphtho-fluorescein	C pH 9.5	549	578	605	624	646	655	682	728	768	811	77	32.0	0.05	1.6	Fluores-analog	4.8.4
721	BOPAHY H-D 19ba′	H_2O	344	355	367	372	379	381	386	393	403	412	19	51	0.03	1.5	Bodipy-analog	4.14.5
722	6-Chloro-harmine	H_2O pH 4	335	341	362	383	396	402	428	462	481	502	66	9.0	0.16	1.4	Indole	4.33
723	Prodan	H_2O	300	322	364	404	481	496	531	569	587	616	167	14.5	0.10	1.4	Naphthalene	4.34
724	Coumarin 1	H_2O	329	349	382	409	436	445	472	510	532	558	90	20.1	0.07	1.4	Coumarin 7-Am	4.26.2
725	3,10-Dimethylisoalloxazine	H_2O pH 7			435				520				85	9.5	0.15	1.4	Pteridine	4.36
726	Acridizinium J-N 1e	H_2O			355				455				100	6.7	0.20	1.3	Acridizinium	4.39.1
727	3-Amino-9-ethylcarbazole 2	H_2O			364				458				94	1.8	0.76	1.3	Carbazole	4.39.1
728	3,8,10-Trimethyliso-alloxazine	H_2O pH 7			433				518				85	11.5	0.12	1.3	Pteridine	4.36
729	ECXi	H_2O			887				938				51	93.3	0.01	1.3	Pentacene	4.39.2
730	6,8-Dichloroharmine (6,8-DiCL-Ha)	H_2O pH 4	341	348	369	391	397	403	429	463	482	503	60	6.3	0.20	1.2	Indole	4.33
731	Cascade yellow	H_2O	329	356	404	440	491	505	549	603	657	663	145	24.0	0.05	1.2	Hybrid-Bonded	4.40.2
732	1,2,7,8-Dibenzo-4,5-Br-fluorescein	C pH 9.5	503	526	549	565	554	562	581	614	642	672	32	24.0	0.05	1.2	Fluorescein-ha	4.8.2
733	3-Amino-9-ethylcarbazole 1	H_2O			364				452				88	1.8	0.63	1.1	Carbazole	4.39.1
734	Fluorene PP6	H_2O			373				620				247	16.5	0.07	1.1	Fluorene	4.39.2
735	Chromenoquinoline L-S 4c	H_2O			500				591				91	33.16	0.03	1.1	Hybrid-Fused	4.40.1
736	JF585	10 mM HEPES pH 7.3	544	568	585	598	588	594	609	640	666	696	24	1.5	0.78	1.1	C(CH3)2-Rhoda	4.3.1
737	Carbostyril 5	H_2O			353				440				87	6.6	0.16	1.0	Quinoline	4.29
738	Acridizinium J-N 1d	H_2O			356				408				52	5.4	0.20	1.0	Acridizinium	4.39.1
739	7-Diethylamino-3-CA coumarin	10 mM HEPES pH 7.3	357	376	410	432	446	452	471	503	525	548	61	35.0	0.03	1.0	Coumarin 7-Am	4.26.2

(Continued)

Ent.	Short names	Aqueous solution	$\lambda_{x,-0.25}$ nm	$\lambda_{x,-0.50}$ nm	$\lambda_{x,max}$ nm	$\lambda_{x,+0.50}$ nm	$\lambda_{e,-0.25}$ nm	$\lambda_{e,-0.50}$ nm	$\lambda_{e,max}$ nm	$\lambda_{e,+0.50}$ nm	$\lambda_{e,+0.25}$ nm	$\lambda_{e,+0.10}$ nm	Δ_S nm	ε_{max} mм$^{-1}$cm^{-1}	Φ_f	B_{max} mм$^{-1}$cm^{-1}	Class	Ref.
740	Naphthalimide B-K 1	10 mм P pH 7.0	380	398	436	470	508	520	552	594	615	639	116	13.2	0.08	1.0	Naphthalimide	4.32
741	3-Aminopyridine	H_2O			300				390				90	2.8	0.32	0.9	Pyridine	4.39.1
742	Stefani HB-6	H_2O pH 6			317				365				48	35.9	0.03	0.9	Hybrid-Bonded	4.40.2
743	Carbostyril 9	H_2O			340				382				42	6.0	0.17	0.9	Quinoline	4.29
744	Carbostyril 2	H_2O			360				430				70	6.7	0.14	0.9	Quinoline	4.29
745	Acyman	H_2O	302	324	367	408	470	484	517	553	570	597	150	9.1	0.10	0.9	Naphthalene	4.34
746	7-Diethylamino-coumarin	H_2O	333	354	388	416	438	448	475	513	536	562	87	18.3	0.05	0.9	Coumarin 7-Am	4.26.2
747	Z-L D	0.5 м PBS pH 7.4	492	513	536	549	534	538	555	589	614	645	19	50	0.02	0.9	Bodipy-Large	4.14.3
748	PONy C7	PBS pH 7.4			419				613				194	22.0	0.04	0.8	Hybrid-Bonded	4.40.2
749	3,6,7,8,10-PentaMe-isoalloxazine	H_2O pH 7			448				571				123	8.3	0.10	0.8	Pteridine	4.36
750	2,5-DiOMe TokyoGreen	100 mм NaOH	456	478	494	506	495	499	512	550	566	585	18	85.0	0.01	0.8	Fluorone	4.11
751	THQ-rosol	50 mм P pH 7.4	500	523	550	565	631	653	710	802	854	925	160	29.4	0.03	0.8	Rosol	4.19
752	COUPY 10	H_2O			562				631				69	33.0	0.03	0.8	Hybrid-Bridged	4.40.3
753	BOPAHY H-D 19aa′	H_2O	345	355	368	373	380	382	387	394	404	413	19	51	0.02	0.7	Bodipy-analog	4.14.5
754	DMPT	H_2O pH 7			382				475				93	12.4	0.06	0.7	Pteridine	4.36
755	Distyryl-BODIPY 11	PBS pH 7.5	546	572	601	617	626	632	655	703	739	784	54	17.9	0.04	0.7	Bodipy-Med.	4.14.2
756	3-(4-OH-Ph)-6-MeO-chromone	100 mм Tris pH 8			311				468				157	12.6	0.05	0.6	Chromone	4.31
757	8-Chloro-harmine	H_2O pH 4	334	341	361	382	392	398	423	456	474	495	62	5.9	0.11	0.6	Indole	4.33
758	Carbostyril 7	H_2O			366				580				214	9.0	0.07	0.6	Quinoline	4.29
759	Carbostyril 4	H_2O			379				531				152	8.7	0.07	0.6	Quinoline	4.29

Ent.	Short names	Aqueous solution	$\lambda_{x,-0.25}$ nm	$\lambda_{x,-0.50}$ nm	$\lambda_{x,max}$ nm	$\lambda_{x,+0.50}$ nm	$\lambda_{e,-0.25}$ nm	$\lambda_{e,-0.50}$ nm	$\lambda_{e,max}$ nm	$\lambda_{e,+0.50}$ nm	$\lambda_{e,+0.25}$ nm	$\lambda_{e,+0.10}$ nm	Δ_S nm	ε_{max} mм$^{-1}$cm^{-1}	Φ_f	B_{max} mм$^{-1}$cm^{-1}	Class	Ref.
760	Carbostyril 6	H_2O			389				544				155	7.0	0.09	0.6	Quinoline	4.29
761	3,6,10-Trimethyliso-alloxazine	H_2O pH 7			439				532				93	8.5	0.08	0.6	Pteridine	4.36
762	3,7,9,10-TetraMe-isoalloxazine	H_2O pH 7			452				571				119	6.8	0.10	0.6	Pteridine	4.36
763	Z-L C	0.5 м PBS pH 7.4	489	510	533	546	529	533	550	583	608	638	17	50	0.01	0.6	Bodipy-Large	4.14.3
764	Z-L H	0.5 м PBS pH 7.4	594	625	660	679	645	651	676	727	766	815	16	50	0.01	0.6	Bodipy-Large	4.14.3
765	5-Methoxysalicylic acid	H_2O pH 7	285	295	319	339	403	414	447	489	511	526	128	3.0	0.19	0.5	Benzene	4.38
766	HMDSeR	100 mм P pH 7.4	516	540	558	571	568	574	590	631	656	680	32	92.0	0.01	0.5	Se-Rhodol	4.16
767	Salicylic acid	H_2O pH 7	272	281	300	316	370	377	407	441	461	482	107	3.0	0.16	0.4	Benzene	4.38
768	5-Methylsalicylic acid	H_2O pH 7	268	281	305	323	379	388	418	456	477	497	113	3.0	0.16	0.4	Benzene	4.38
769	4-Methylsalicylic acid	H_2O pH 7	252	272	300	317	370	379	407	446	467	491	107	3.0	0.11	0.3	Benzene	4.38
770	COUPY 12	H_2O			595				683				88	6.9	0.05	0.3	Hybrid-Bridged	4.40.3
771	JF724	10 mм HEPES pH 7.3	647	688	724	750	711	721	748	792	831	879	24	6.6	0.05	0.3	SO2-Rhodam	4.3.3
772	Acridizinium I-L 5b	H_2O			369				551				182	11.0	0.02	0.2	Acridizinium	4.39.1
773	COUPY 11	H_2O			567				662				95	13.0	0.02	0.2	Hybrid-Bridged	4.40.3
774	NCRMe	H_2O			568				607				39	22.8	0.01	0.1	Hybrid-Fused	4.40.1

(Continued)

Appendix F

F1. Ranked by Fluorescence Mean-Lifetime

Appendix F contains the database of dyes, identified by their short names and ranked by descending fluorescence mean-lifetime values. Many short names are registered trademarks and are listed in footnote.[1] The majority of the fluorescent dyes exhibit a single predominant fluorescent decay curve. A few dyes exhibit multiple decay curves. When multiple decay curves are observed then the fluorescence mean-lifetime of the predominant curve was taken (see Table F.1). All fluorescence mean-lifetimes were measured in aqueous solutions (summarized in column 'Aqueous solution' of Table F.1). The exact constitution of these aqueous solutions can be found in the reference(s) shown in the column 'Ref.'.

The molecular structures together with some photophysical parameters are shown in Chapter 4. Chapter 4 also provides a detailed list of references where the photophysical parameters were taken. Important properties of the dyes are also revealed in Chapter 5. Panoramic views of the photophysical properties and classes are given in Chapter 5 using scattergrams. Methods to measure photostability and techniques to improve photostability are discussed in Chapter 7. The photostability (relative photobleaching mean-lifetime) of some dyes is also shown in Chapter 7.

Acronyms used in Table E.1: A, acetate; AC, acetic acid; ACN, acetonitrile; BODIPY, borondipyrromethene; BSA, bovine serum albumin; C, carbonate; CA, carboxylic acid; FBS, fetal bovine serum; Et, ethanol; nd, non-disclosed; ni, not informed; P, phosphate; pb, probably; pH >7, basic medium; pH < 7, acidic medium (when the exact value of the pH is not given in the literature); SA, sulfonic acid; and TX-100, triton X-100.

1 Abberior STAR is a trademark of Abberior GmbH (Göttingen, Germany). AlexaFluor, BODIPY, Cascade Blue, FAM, JOE, TET, HEX, Marina Blue, Oregon Green, Pacific Blue, ROX, TAMRA, and Texas Red are trademarks of Thermo Fisher Scientific (Waltham, MA, USA). ATTO is a trademark of ATTO-Tec GmbH (Siegen, Germany). CF Dye is a registered trademark of Biotium, Inc. (Fremont, CA, USA). Cy3, Cy3.5, Cy5, and Cy5.5 are trademarks of GE Healthcare (Chicago, IL, USA). Dy is a trademark of Dyomics GmbH (Jena, Germany). Seta, SeTau, and Square dyes are trademarks of SETA BioMedicals (Urbana, IL, USA).

Fluorescent Dye Labels and Stains: A Database of Photophysical Properties, First Edition. Tarso B. Ledur Kist.

Table F1 Photophysical properties of fluorescent labels and stains in aqueous solutions ranked by fluorescence mean-lifetimes (τ_f).

Ent.	Short names	Aqueous solution	$\lambda_{x,-0.25}$ nm	$\lambda_{x,-0.50}$ nm	$\lambda_{x,max}$ nm	$\lambda_{x,+0.50}$ nm	$\lambda_{e,max}$ nm	Δ_S nm	ε_{max} mM^{-1}cm^{-1}	Φ_f	B_{max} mM^{-1}cm^{-1}	τ_f ns	N_n	N_p	q	N_a	Class	Ref.
1	Pyrene-1-sulfonate	H_2O	327	340	375	382	400	25	30.6	0.56	17.1	64.0	1	0	-1	30	Pyrene	4.28
2	10-Ethylacridine	P pH 7	376	390	412	423	494	82	3.5	0.98	3.4	44.4	0	1	1	27	Acridine	4.23.1
3	Lucigenin	H_2O			368		485	117	37.6	0.50	18.8	25.3	0	2	2	52	Hybrid-Bonded	4.40.2
4	nor-Harmane (β-carboline)	50 mM H_2SO_4	333	340	360	381	454	94	6.3	0.58	3.6	21.1	0	1	1	21	Indole	4.33
5	Harmane	50 mM H_2SO_4	324	331	350	370	440	90	6.3	0.83	5.1	20.0	0	1	1	24	Indole	4.33
6	DAOTA 5	H_2O	479	505	548	567	576	28	50.0	0.56	28.0	18.6	2	3	1	93	Triangulenium	4.22
7	DAOTA 4	H_2O	481	502	549	567	576	27	50.0	0.56	28.0	18.3	0	3	3	75	Triangulenium	4.22
8	MetACSO3H	H_2O pH 5.5	364	376	403	411	433	30	5.6	0.76	4.2	16.7	2	1	-1	35	Acridones	4.23.2
9	$ACSO_3H$	H_2O pH 5.5	353	366	396	404	417	21	4.4	0.82	3.6	16.5	2	1	-1	32	Acridones	4.23.2
10	9-Aminoacridine	P pH 7	366	379	400	410	420	20	9.3	0.97	9.0	16.4	0	1	1	26	Acridine	4.23.1
11	N-Methylacridone	H_2O	369	382	411	419	431	20	8.8	0.99	8.7	15.8	0	1	1	28	Acridones	4.23.2
12	PetACSO3H	H_2O pH 5.5	367	379	406	414	431	25	6.2	0.92	5.6	15.6	2	1	-1	47	Acridones	4.23.2
13	DAOTA 1	H_2O	489	511	558	578	587	29	50.0	0.31	15.5	15.2	0	1	1	51	Triangulenium	4.22
14	DAOTA 6	H_2O	489	509	558	575	584	26	50.0	0.36	18.0	14.6	2	3	1	81	Triangulenium	4.22
15	6-Sulfonato-1,3,5,7-Me-BODIPY	H_2O	454	473	495	503	515	20	52.5	0.85	44.6	14.1	2	1	-1	36	Bodipy-Small	4.14.1
16	DAOTA 2	H_2O	493	515	564	583	594	30	50.0	0.27	13.5	13.5	1	1	0	54	Triangulenium	4.22
17	Quinine	H_2O pH 7			331		382	51	5.7	0.50	2.8	13.4	0	0	0	48	Quinoline	4.29
18	Quinine	50 mM H_2SO_4			347		451	104	5.7	0.55	3.1	13.4	0	1	1	48	Quinoline	4.29
19	1,3,6,8-Pyrenetetrasulfonic acid	H_2O pH 7	335	350	374	378	403	29	51.0	0.56	28.5	12.5	4	0	-4	38	Pyrene	4.28
20	DAOTA 3	H_2O	449	511	565	583	594	29	50.0	0.28	14.0	12.5	2	1	-1	57	Triangulenium	4.22
21	Sulfobetaine pyrene 21	H_2O	314	327	359	365	385	26	30.6	0.32	9.7	11.6	1	1	0	52	Pyrene	4.28
22	PB430	PBS pH 7.4			426		542	116	10.4	0.66	6.8	10.6	3	0	-3	115	Phosphole	4.39.3
23	trans-PO-PB430	PBS pH 7.4			422		545	123	10.3	0.74	7.6	10.3	3	0	-3	86	Phosphole	4.39.3
24	cis-PO-PB430	PBS pH 7.4			421		548	127	10.0	0.71	7.1	9.9	3	0	-3	86	Phosphole	4.39.3
25	Benzo[h]quinoline (BQ)	H_2O pH 2-5			370		430	60	35.0	0.32	11.2	9.6	0	0	0	23	Quinoline	4.29
26	SeTau-405-carboxylate	P pH 7.4			405		518	113	13.8	0.80	11.0	9.3			0	55	nd	4.41
27	Niu 27	H_2O	474	494	510	520	528	18	63.3	0.81	51.2	9.2	3	3	0	100	Bodipy-Med.	4.14.2
28	NOBQ	H_2O pH 2-5			385		470	85	35.0	0.41	14.3	9.1	0	1	1	31	Quinoline	4.29

Ent.	Short names	Aqueous solution	$\lambda_{x,-0.25}$ nm	$\lambda_{x,-0.50}$ nm	$\lambda_{x,max}$ nm	$\lambda_{x,+0.50}$ nm	$\lambda_{e,max}$ nm	Δ_S nm	ε_{max} mм$^{-1}$cm^{-1}	Φ_f	B_{max} mм$^{-1}$cm^{-1}	τ_f ns	N_n	N_p	q	N_a	Class	Ref.
29	OBQ	H_2O pH 2-5			385		470	85	35.0	0.43	15.0	8.7	0	0	0	27	Quinoline	4.29
30	NBQ	H_2O pH 2-5			380		470	90	35.0	0.35	12.2	8.5	0	1	1	28	Quinoline	4.29
31	Tri-*O*-acetyl-P-luminoarosin	H_2O pH 7			425		528	103	10.8	0.62	6.7	8.3	1	1	0	58	Pteridine	4.36
32	Acridizinium J-N 1e	H_2O			355		455	100	6.7	0.20	1.3	7.5	0	1	1	28	Acridizinium	4.39.1
33	Luminarine	H_2O pH 7			428		533	105	9.0	0.47	4.2	7.4	1	1	0	23	Pteridine	4.36
34	Harmine	H_2O pH 4	330	337	357	378	416	59	7.3	0.49	3.6	7.1	0	1	1	29	Indole	4.33
35	Niu 29	H_2O	477	497	519	531	530	11	28.0	0.61	17.0	7.0	3	3	0	107	Bodipy-Med.	4.14.2
36	5-Methoxysalicylic acid	H_2O pH 7	285	295	319	339	447	128	3.0	0.19	0.5	6.9	1	0	-1	20	Benzene	4.38
37	3,7,10-Trimethylisoalloxazine	H_2O pH 7			446		532	86	9.8	0.32	3.1	6.9	0	0	0	31	Pteridine	4.36
38	7-Hydroxy-4-methylcoumarin	H_2O pH 5	276	295	320	341	450	130	14.0	0.78	10.9	6.9	0	0	0	21	Coumarin 7-OH	4.26.1
39	Pacific Blue	PBS pH 7.2, 22 °C	339	366	410	437	455	45	29.5	0.75	22.1	6.9	1	0	-1	31	Coumarin 7-OH	4.26.1
40	7-Hydroxy-4-methylcoumarin	H_2O pH 10	313	330	360	384	449	89	18.0	0.86	15.4	6.7	1	0	-1	20	Coumarin 7-OH	4.26.1
41	DPP 9	PBS			451		541	90	13.9	0.90	12.5	6.6	4	0	-4	72	Diketopyrrolop	4.39.1
42	Harmine	50 mм H_2SO_4	304	309	326	343	412	86	11.9	0.45	5.3	6.6	0	1	1	29	Indole	4.33
43	2-Methyl harmine	50 mм H_2SO_4	324	331	350	370	433	83	6.3	0.45	2.8	6.5	0	1	1	31	Indole	4.33
44	DPP 5	PBS			455		515	60	22.7	0.57	12.9	6.5	2	0	-2	44	Diketopyrrolop	4.39.1
45	DPP 7	PBS			456		548	92	14.7	0.76	11.1	6.5	3	0	-3	58	Diketopyrrolop	4.39.1
46	7-Hydroxy-4-Me-coumarin-3-CA	H_2O pH 5	313	330	360	384	455	95	18.0	0.82	14.7	6.4	1	0	-1	26	Coumarin 7-OH	4.26.1
47	B-Z 7b	H_2O	479	500	516	526	529	13	60.5	0.68	41.1	6.4	4	1	-3	123	Bodipy-Large	4.14.3
48	Acridizinium J-N 1b	H_2O			381		408	27	6.7	0.34	2.2	6.2	0	1	1	24	Acridizinium	4.39.1
49	Coumarin 102	H_2O	336	356	393	418	489	96	21.8	0.66	14.3	6.0	0	0	0	36	Coumarin 7-Am	4.26.2
50	Acridizinium J-N 1a	H_2O			377		404	27	8.9	0.33	2.9	6.0	0	1	1	24	Acridizinium	4.39.1
51	Cascade yellow	H_2O	329	356	404	440	549	145	24.0	0.05	1.2	5.6	2	1	-1	61	Hybrid-Bonded	4.40.2
52	Acridizinium J-N 1c	H_2O			354		420	66	8.4	0.30	2.5	5.4	0	1	1	24	Acridizinium	4.39.1
53	5-Methylsalicylic acid	H_2O pH 7	268	281	305	323	418	113	3.0	0.16	0.4	5.3	1	0	-1	19	Benzene	4.38
54	8-Hydroxypyrene-1,3,6-triSA	H_2O pH 10	389	408	460	470	514	54	24.0	0.95	22.8	5.3	4	0	-4	36	Pyrene	4.28
55	Harmaline	5 mм H_2SO_4	352	360	383	407	475	92	28.6	0.32	9.1	5.3	0	1	1	30	Indole	4.33
56	G-L 14	H_2O	472	492	508	518	521	13	59.0	0.77	45.4	5.3	1	1	0	100	Bodipy-Med.	4.14.2

(Continued)

Ent.	Short names	Aqueous solution	$\lambda_{x,-0.25}$ nm	$\lambda_{x,-0.50}$ nm	$\lambda_{x,max}$ nm	$\lambda_{x,+0.50}$ nm	$\lambda_{e,max}$ nm	Δ_S nm	ε_{max} mM^{-1}cm^{-1}	Φ_f	B_{max} mM^{-1}cm^{-1}	τ_f ns	N_n	N_p	q	N_a	Class	Ref.
57	5(6)-Carboxyrhodol	50 mM P pH 8-9	451	472	494	503	520	26	59.9	0.89	53.3	5.1	2	0	-2	41	Rhodol	4.15.1
58	G-L 13b	H_2O	473	493	509	519	522	13	56.0	0.67	37.5	5.1	1	1	0	84	Bodipy-Med.	4.14.2
59	Lucifer yellow CH	H_2O pH 7	375	392	429	462	540	111	11.3	0.21	2.3	5.1	2	0	-2	38	Naphthalimide	4.32
60	Niu 25	PBS pH 7.4	479	499	521	533	540	19	58.8	0.82	48.2	5.1	3	3	0	88	Bodipy-Med.	4.14.2
61	Atto 390 carboxylate	PBS pH 7.4, 22 °C	343	357	390	415	476	86	24.0	0.90	21.6	5.0	1	0	-1	50	Coumarin 7-Am	4.26.3
62	Atto 465 carboxylate	PBS pH 7.4, 22 °C	404	423	453	467	506	53	75.0	0.75	56.2	5.0	1	1	0	39	Acridine	4.23.1
63	B-Z 7a	H_2O	458	477	500	508	509	9	64.5	0.61	39.3	5.0	4	1	-3	111	Bodipy-Large	4.14.3
64	2,7-DiMe-6-EtAm-carboxyrhodol	50 mM P pH 8-9	467	487	510	522	536	26	66.3	0.89	59.0	5.0	2	0	-2	53	Rhodol	4.15.1
65	trans-PO-PB460	PBS pH 7.4			459		592	133	9.0	0.25	2.2	4.9	3	0	-3	86	Phosphole	4.39.3
66	Coumarin 120	H_2O	293	314	342	366	440	98	16.0	0.94	15.0	4.9	0	0	0	22	Coumarin 7-Am	4.26.2
67	7-Hydroxy-4-Me-coumarin-3-CA	H_2O pH 10	332	352	386	413	446	60	33.0	0.85	28.0	4.8	2	0	-2	22	Coumarin 7-OH	4.26.1
68	8-Hydroxypyrene-1,3,6-triSA	H_2O pH 2	348	363	403	411	443	40	20.0	0.95	19.0	4.8	3	0	-3	36	Pyrene	4.28
69	DPP 10	PBS			523		594	71	17.1	0.61	10.4	4.8	4	0	-4	66	Diketopyrrolop	4.39.1
70	DPP 8	PBS			527		594	67	18.1	0.56	10.1	4.8	3	0	-3	52	Diketopyrrolop	4.39.1
71	Coumarin 2	H_2O	314	332	362	386	456	94	16.9	0.95	16.0	4.8	0	0	0	31	Coumarin 7-Am	4.26.2
72	3,10-Dimethylisoalloxazine	H_2O pH 7			435		520	85	9.5	0.15	1.4	4.5	0	0	0	28	Pteridine	4.36
73	3-Bt-6-MeO-7-hydroxychromone	100 mM Tris pH 8			345		447	102	16.4	0.48	7.8	4.5	1	0	-1	34	Chromone	4.31
74	Coumarin 152	H_2O	316	334	365	389	497	132	14.3	0.37	5.2	4.5	0	0	0	34	Coumarin 7-Am	4.26.2
75	DPP 6	PBS			530		586	56	30.7	0.49	15.0	4.4	2	0	-2	38	Diketopyrrolop	4.39.1
76	Oregon Green 514	PBS pH 7.2, 22 °C	471	491	511	524	530	19	88.0	0.96	84.4	4.4	3	0	-3	44	Fluorescein-ha	4.8.2
77	Fluorescein	100 mM NaOH in D_2O	453	472	489.7	502	516.3	26	67.0	0.98	65.6	4.4	2	0	-2	37	Fluorescein-nh	4.8.1
78	Rhodamine 6G	D_2O pH <7	484	506	525.5	539	552.0	26	105.0	0.98	102.9	4.4	0	1	1	64	Rhoda 500-550	4.2.2
79	3-AB-6-MeO-7-OH-chromone	100 mM Tris pH 8			346		445	99	16.3	0.41	6.6	4.3	1	0	-1	30	Chromone	4.31
80	575RH	PBS pH 7.4	524	555	574	589	597	23	55.0	0.74	40.7	4.3	1	1	0	71	Rhoda 550-600	4.2.3

Ent.	Short names	Aqueous solution	$\lambda_{x,-0.25}$ nm	$\lambda_{x,-0.50}$ nm	$\lambda_{x,max}$ nm	$\lambda_{x,+0.50}$ nm	$\lambda_{e,max}$ nm	Δ_S nm	ε_{max} mм$^{-1}$cm^{-1}	Φ_f	B_{max} mм$^{-1}$cm^{-1}	τ_f ns	N_n	N_p	q	N_a	Class	Ref.
81	Abberior Star 600 carboxylate	PBS pH 7.4	550	587	604	619	623	19	93.0	0.80	74.4	4.3	2	1	-1	73	Rhodam >600	4.2.4
82	Coumarin 307	H_2O	335	355	392	417	508	116	9.0	0.33	2.9	4.3	0	0	0	31	Coumarin 7-Am	4.26.2
83	Oregon Green 488	H_2O pH 9	453	472	490	502	514	24	76.0	0.97	73.7	4.3	3	0	-3	37	Fluorescein-ha	4.8.2
84	Salicylic acid	H_2O pH 7	272	281	300	316	407	107	3.0	0.16	0.4	4.3	1	0	-1	16	Benzene	4.38
85	Rhodamine 110	H_2O pH <7	458	482	496	507	520	24	83.0	0.89	73.8	4.3	1	1	0	40	Rhodam <500	4.2.1
86	AI3	H_2O pH 7.4	387	405	445	481	552	107	15.0	0.21	3.1	4.3	0	2	2	68	Naphthalimide	4.32
87	2,6-Sulfon.-1,3,5,7,8-Me-BODIPY	H_2O	451	469	491	499	519	28	99.0	0.83	82.1	4.2	3	1	-2	42	Bodipy-Small	4.14.1
88	Spirobifluorene 1	H_2O			378		480	102	51	0.50	25.6	4.23	4	0	-4	169	Spirobifluorene	4.39.3
89	5(6)-Carboxyfluorescein (FAM)	100 mм PBS pH 9	457	477	495	508	520	25	83.0	0.90	74.7	4.2	3	0	-3	40	Fluorescein-nh	4.8.1
90	Atto Rho101 carboxylate	PBS pH 7.4, 22 °C	535	569	587	601	609	22	120.0	0.80	96.0	4.2	1	1	0	84	Rhoda 550-600	4.2.3
91	Fluorinated rhodamine 3-H	Buffer pH 7	471	492	510	523	531	21	57.0	0.86	49.0	4.2	0	1	1	64	Rhoda 500-550	4.2.2
92	Rhodamine 123	H_2O pH <7	464	485	502	514	528	26	75.0	0.90	67.5	4.2	0	1	1	43	Rhoda 500-550	4.2.2
93	Carbofluorescein	100 mм NaOH	505	526	544	557	567	23	108.0	0.62	66.9	4.2	2	0	-2	46	$C(CH_3)_2$-Fluore	4.9
94	Fluorescein	100 mм NaOH in H_2O	453	472	490.2	503	516.3	26	67.0	0.92	61.6	4.2	2	0	-2	37	Fluorescein-nh	4.8.1
95	2-Ethyl-6-quinolinocarboxyrhodol	50 mм P pH 8-9	494	517	543	557	574	31	79.4	0.49	38.9	4.13	2	0	-2	61	Rhodol	4.15.1
96	Abberior Star 512 carboxylate	PBS pH 7.4	472	493	511	524	530	19	84.0	0.82	68.8	4.1	3	1	-2	75	Rhoda 500-550	4.2.2
97	Alexa Fluor 488	PBS pH 7.2, 22 °C	457	477	494	506	517	23	73.0	0.92	67.1	4.1	4	1	-3	48	Rhodam Hi Ch	4.2.5
98	Alexa Fluor 546	PBS pH 7.2, 22 °C	506	537	546	567	580	34	112.0	0.79	88.4	4.1	4	1	-3	102	Rhodam Hi Ch	4.2.5
99	Atto 488 carboxylate	PBS pH 7.4, 22 °C	461	486	500	511	520	20	90.0	0.80	72.0	4.1	3	1	-2	64	Rhodam <500	4.2.1
100	Atto Rho6G carboxylate	PBS pH 7.4, 22 °C	489	513	533	546	557	24	115.0	0.94	108.1	4.1	1	1	0	73	Rhoda 500-550	4.2.2
101	Fluorinated rhodam. 1f-5′-COOH	Buffer pH 7	463	484	501	513	525	24	85.0	0.95	80.7	4.1	2	1	-1	53	Rhoda 500-550	4.2.2
102	Fluorinated rhodamine 5-SO_3H	Buffer pH 7	472	494	512	525	530	18	52.0	0.86	44.7	4.1	2	1	-1	66	Rhoda 500-550	4.2.2
103	Rhodamine RBzET	H_2O	485	508	527	541	552	25	82.0	0.87	71.3	4.1	0	1	1	79	Rhoda 500-550	4.2.2
104	2-PrAc-6-quinolinorhodol	50 mм P pH 8-9	493	515	541	555	572	31	58.2	0.49	28.5	4.08	2	0	-2	63	Rhodol	4.15.1
105	Rhodamine 6G	H_2O pH <7	485	507	526.3	540	552.3	26	105.0	0.93	97.6	4.1	0	1	1	64	Rhoda 500-550	4.2.2

(Continued)

Ent.	Short names	Aqueous solution	$\lambda_{x,-0.25}$ nm	$\lambda_{x,-0.50}$ nm	$\lambda_{x,max}$ nm	$\lambda_{x,+0.50}$ nm	$\lambda_{e,max}$ nm	Δ_S nm	ε_{max} mM^{-1}cm^{-1}	Φ_f	B_{max} mM^{-1}cm^{-1}	τ_f ns	N_n	N_p	q	N_a	Class	Ref.
106	BCECF	100 mM NaOH	466	486	505	518	527	22	67.0	0.84	56.2	4.1	5	0	-5	53	Fluorescein-nh	4.8.1
107	2-Et-7-Me-6-EtAm-carboxyrhodol	50 mM P pH 8-9	468	488	511	524	538	27	73.0	0.84	61.3	4.0	2	0	-2	56	Rhodol	4.15.1
108	7-Me-6-EtAm-carboxyrhodol	50 mM P pH 8-9	465	485	508	521	534	26	78.1	0.82	64.0	4.0	2	0	-2	50	Rhodol	4.15.1
109	530RH	PBS pH 7.4	491	514	532	547	553	21	56.0	0.89	49.8	4.0	2	1	-1	67	Rhoda 500-550	4.2.2
110	Atto 430LS carboxylate	PBS pH 7.4, 22 °C			436		545	109	32.0	0.65	20.8	4.0			0	nd	nd (pb Coum.)	4.41
111	Atto 565 carboxylate	PBS pH 7.4, 22 °C	513	543	564	578	590	26	120.0	0.90	108.0	4.0	2	1	-1	69	Rhoda 550-600	4.2.3
112	Atto Rho11 carboxylate	PBS pH 7.4, 22 °C	522	553	572	585	595	23	120.0	0.80	96.0	4.0	1	1	0	82	Rhoda 550-600	4.2.3
113	Atto Rho12 carboxylate	PBS pH 7.4, 22 °C	526	557	577	590	600	23	120.0	0.80	96.0	4.0	1	1	0	100	Rhoda 550-600	4.2.3
114	6-Quinolinocarboxyrhodol	50 mM P pH 8-9	492	514	540	554	570	30	66.8	0.43	28.7	4.0	2	0	-2	55	Rhodol	4.15.1
115	2-Cl-7-Me-6-EtAm-carboxyrhodol	50 mM P pH 8-9	479	495	514	538	539	25	80.5	0.82	66.0	3.94	2	0	-2	50	Rhodol	4.15.1
116	3,6,7,8,10-PentaMe-isoalloxazine	H_2O pH 7			448		571	123	8.3	0.10	0.8	3.9	0	0	0	37	Pteridine	4.36
117	Alexa Fluor 594	PBS pH 7.2/7.0, 20/22 °C	537	570	590	606	617	27	92.0	0.66	60.7	3.9	4	1	-3	84	Rhodam Hi Ch	4.2.5
118	Atto 514 carboxylate	PBS pH 7.4, 22 °C	474	494	511	522	532	21	115.0	0.85	97.7	3.9	3	1	-2	76	Rhoda 500-550	4.2.2
119	Atto 594 carboxylate	PBS pH 7.4, 22 °C	547	583	603	617	626	23	120.0	0.85	102.0	3.9	3	1	-2	104	Rhodam >600	4.2.4
120	Atto Rho13 carboxylate	PBS pH 7.4, 22 °C	547	584	603	618	627	24	120.0	0.80	96.0	3.9	1	1	0	96	Rhodam >600	4.2.4
121	Distyryl-BODIPY 18	H_2O	568	596	628	646	641	13	71.0	0.38	26.9	3.9	5	5	0	175	Bodipy-Med.	4.14.2
122	Fluorinated rhodamine 1e	Buffer pH 7	461	485	499	510	522	23	81.0	0.88	71.2	3.9	1	1	0	51	Rhodam <500	4.2.1
123	2-Cl-6-Piperidinocarboxyrhodol	50 mM P pH 8-9	478	500	524	537	548	24	77.4	0.72	55.7	3.88	2	0	-2	48	Rhodol	4.15.1
124	CF405S	H_2O			404		431	27	33.0	0.70	23.0	3.9			0	nd	nd (Pyrene)	4.41
125	2-PrAc-7-Me-6-EtAm-rhodol	50 mM P pH 8-9	466	486	509	522	535	26	79.4	0.92	73.0	3.9	2	0	-2	56	Rhodol	4.15.1
126	Atto 532 carboxylate	PBS pH 7.4, 22 °C	491	515	532	544	552	20	115.0	0.90	103.5	3.8	3	1	-2	75	Rhoda 500-550	4.2.2
127	Fluorinated fluorescein S-H 14a,b	P pH 9	454	474	492	504	516	24	85.9	0.92	79.0	3.8	3	0	-3	40	Fluorescein-ha	4.8.2
128	Fluorinated rhodamine 1j	Buffer pH 7	458	482	496	507	522	26	53.0	0.89	47.1	3.8	2	1	-1	53	Rhodam <500	4.2.1
129	2-Cloro-6-quinolinocarboxyrhodol	50 mM P pH 8-9	498	521	548	563	575	27	74.5	0.42	31.2	3.77	2	0	-2	55	Rhodol	4.15.1
130	Pyronine 630	H_2O, 25 °C	509	535	559	573	579	20	100.0	0.92	92.0	3.7	0	1	1	53	Pyronine	4.4

Ent.	Short names	Aqueous solution	$\lambda_{x,-0.25}$ nm	$\lambda_{x,-0.50}$ nm	$\lambda_{x,max}$ nm	$\lambda_{x,+0.50}$ nm	$\lambda_{e,max}$ nm	Δ_S nm	ε_{max} $mM^{-1}cm^{-1}$	Φ_f	B_{max} $mM^{-1}cm^{-1}$	τ_f ns	N_n	N_p	q	N_a	Class	Ref.
131	Atto 542 carboxylate	PBS pH 7.4, 22 °C			542		562	20	120.0	0.93	111.6	3.7			0	nd	nd (pb Rhod.)	4.41
132	Atto 590 carboxylate	PBS pH 7.4, 22 °C	535	569	593	609	622	29	120.0	0.80	96.0	3.7	2	1	-1	83	Rhoda 550-600	4.2.3
133	Atto Rho14 carboxylate	PBS pH 7.4, 22 °C	564	606	626	639	646	20	140.0	0.80	112.0	3.7	1	1	0	96	Rhodam >600	4.2.4
134	KK1119	H_2O	577	618	637	654	660	23	90.0	0.55	49.5	3.7	3	1	-2	106	Rhodam >600	4.2.4
135	Abberior Star Red carboxylate	PBS pH 7.4	578	619	638	655	655	17	100.0	0.66	66.0	3.6	3	1	-2	119	Rhodam >600	4.2.4
136	Acridizinium J-N 1d	H_2O			356		408	52	5.4	0.20	1.0	3.6	0	1	1	24	Acridizinium	4.39.1
137	Alexa Fluor 568	PBS pH 7.2/7.0, 20/22 °C	527	559	578	593	603	25	88.0	0.69	60.7	3.6	4	1	-3	78	Rhodam Hi Ch	4.2.5
138	Atto 425 carboxylate	PBS pH 7.4, 22 °C	395	409	439	460	485	46	45.0	0.90	40.5	3.6	1	0	-1	56	Coumarin 7-Am	4.26.3
139	Atto 520 carboxylate	PBS pH 7.4, 22 °C	474	497	517	529	538	21	110.0	0.90	99.0	3.6	1	1	0	54	Pyronine	4.4
140	Atto 550 carboxylate	PBS pH 7.4, 22 °C	507	535	554	567	576	22	120.0	0.80	96.0	3.6	1	1	0	87	Rhoda 550-600	4.2.3
141	KK114	H_2O	580	619	637	652	660	23	92.0	0.55	50.6	3.6	3	1	-2	102	Rhodam >600	4.2.4
142	KK9046	H_2O	573	613	632	648	654	22	90.0	0.45	40.5	3.6	1	1	0	100	Rhodam >600	4.2.4
143	Pyrodone	100 mм P pH 7.4	436	454	472	481	500	28	60.0	0.37	22.2	3.6	1	0	-1	23	Pyrodone	4.21.2
144	Sulforhodamine 101	H_2O pH 7	534	569	586	602	605	19	116.0	0.90	104.4	3.6	2	1	-1	72	Sulforhodam.	4.6
145	Abberior Star 580 carboxylate	PBS pH 7.4	535	567	587	603	607	20	85.0	0.90	76.5	3.5	1	1	0	83	Rhoda 550-600	4.2.3
146	Alexa Fluor 350	H_2O pH 7	296	317	346	370	445	99	19.0	0.89	16.9	3.5	2	0	-2	95	Coumarin 7-Am	4.26.3
147	Atto 647N carboxylate	PBS pH 7.4, 22 °C	592	626	646	661	664	18	150.0	0.65	97.5	3.5	1	1	0	99	$C(CH_3)_2$-Rhoda	4.3.1
148	B-Z 14	H_2O	576	605	638	656	655	17	102.0	0.42	42.8	3.5	4	1	-3	127	Bodipy-Large	4.14.3
149	G-L 13a	H_2O	463	482	505	514	517	12	35.0	0.47	16.4	3.4	1	1	0	79	Bodipy-Med.	4.14.2
150	Acedan 7	H_2O	291	311	351	388	495	144	11.0	0.31	3.4	3.4	0	0	0	32	Naphthalene	4.34
151	Distyryl-BODIPY 17	H_2O	570	598	630	648	643	13	76.6	0.30	22.9	3.4	6	5	-1	171	Bodipy-Med.	4.14.2
152	BC-OH	100 mм P, 20% Et pH 7.4			442		488	46	38.1	0.85	32.3	3.32	1	0	-1	29	Hybrid-Bonded	4.40.2
153	Abberior Star 440SXP	PBS pH 7.4	377	398	436	461	515	79	22.7	0.68	15.4	3.3	2	1	-1	70	Coumarin 7-Am	4.26.3
154	Abberior Star 635P carboxylate	PBS pH 7.4	575	616	635	652	655	20	80.0	0.55	44.0	3.3	3	1	-2	105	Rhodam >600	4.2.4
155	Atto 633 carboxylate	PBS pH 7.4, 22 °C	575	610	630	646	651	21	130.0	0.64	83.2	3.3	1	1	0	79	$C(CH_3)_2$-Rhoda	4.3.1
156	Lucifer yellow VS	H_2O pH 7	376	393	430	463	535	105	11.3	0.24	2.7	3.3	2	0	-2	47	Naphthalimide	4.32
157	3,7,9,10-TetraMe-isoalloxazine	H_2O pH 7			452		571	119	6.8	0.10	0.6	3.2	0	0	0	34	Pteridine	4.36
158	630GeRH	PBS pH 7.4	582	612	631	647	651	20	61.0	0.60	36.6	3.2	1	1	0	64	$Ge(CH_3)_2$-Rhoda	4.3.3

(Continued)

Ent.	Short names	Aqueous solution	$\lambda_{x,-0.25}$ nm	$\lambda_{x,-0.50}$ nm	$\lambda_{x,max}$ nm	$\lambda_{x,+0.50}$ nm	$\lambda_{e,max}$ nm	Δ_S nm	ε_{max} mм$^{-1}$cm^{-1}	Φ_f	B_{max} mм$^{-1}$cm^{-1}	τ_f ns	N_n	N_p	q	N_a	Class	Ref.
159	640SiRH	PBS pH 7.4	584	519	641	658	662	21	51.0	0.42	21.4	3.2	1	1	0	64	$Si(CH_3)_2$-Rhoda	4.3.2
160	Alexa Fluor 633	PBS/H_2O pH 7.2, 20/22 °C	568	608	632	650	647	15	100.0	0.64	64.0	3.2	4	1	-3	93	Sulforhodam.	4.6
161	Atto 610 carboxylate	PBS pH 7.4, 22 °C	529	554	616	581	633	17	150.0	0.70	105.0	3.2	1	1	0	60	$C(CH_3)_2$-Pyron	4.5
162	Atto 635 carboxylate	PBS pH 7.4, 22 °C	579	612	635	649	659	24	130.0	0.25	32.5	3.2	1	1	0	68	$C(CH_3)_2$-Pyron	4.5
163	Fluorinated rhodamine 1k	Buffer pH 7	467	488	506	518	531	25	54.0	0.66	35.6	3.2	3	1	-2	57	Rhoda 500-550	4.2.2
164	Niu 18	H_2O	455	474	496	504	511	15	75.5	0.20	15.1	3.2	2	2	0	69	Bodipy-Med.	4.14.2
165	Phosphorilated coum. N-H 21c	PBS pH 7.4	374	394	432	457	512	80	20.4	0.81	16.5	3.2	2	1	-1	58	Coumarin 7-Am	4.26.3
166	Acedan 5	H_2O	295	316	357	396	503	146	17.4	0.40	6.9	3.2	0	0	0	42	Naphthalene	4.34
167	Seta-375-carboxylate	P pH 7.4			372		476	104	24.5	0.86	21.0	3.1			0	47	nd	4.41
168	2-Br-7-Me-6-EtAm-carboxyrhodol	50 mм P pH 8-9	472	492	516	529	539	23	80.9	0.61	49.3	3.05	2	0	-2	48	Rhodol	4.15.1
169	Acedan 6	H_2O	295	316	357	396	503	146	15.4	0.39	6.0	3.1	0	0	0	48	Naphthalene	4.34
170	4-Methylsalicylic acid	H_2O pH 7	252	272	300	317	407	107	3.0	0.11	0.3	3.0	1	0	-1	19	Benzene	4.38
171	4,7-DiMe-6-EtAm-carboxyrhodol	50 mм P pH 8-9	472	492	516	529	544	28	56.9	0.43	24.4	3.0	2	0	-2	53	Rhodol	4.15.1
172	3,8,10-Trimethylisoalloxazine	H_2O pH 7			433		518	85	11.5	0.12	1.3	3.0	0	0	0	31	Pteridine	4.36
173	Atto 620 carboxylate	PBS pH 7.4, 22 °C	580	602	620	633	642	22	120.0	0.50	60.0	2.9	1	1	0	76	$C(CH_3)_2$-Rhoda	4.3.1
174	Atto 665 carboxylate	PBS pH 7.4, 22 °C			662		680	18	160.0	0.60	96.0	2.9			0	nd	nd (pb Pyron.)	4.41
175	6,8-Dichloroharmine (6,8-DiCL-Ha)	H_2O pH 4	341	348	369	391	429	60	6.3	0.20	1.2	2.8	0	1	1	29	Indole	4.33
176	Abberior Star 635 carboxylate	PBS pH 7.4	574	615	634	650	654	20	63.0	0.51	32.1	2.8	2	1	-1	111	Rhodam >600	4.2.4
177	B-Z 19	H_2O	587	617	651	670	677	26	118.5	0.23	27.2	2.8	4	1	-3	139	Bodipy-Large	4.14.3
178	Cy3B	H_2O	493	515	558	569	572	14	130.0	0.67	87.1	2.8	2	1	-1	71	Cyanine-Trimet	4.13.1
179	Oxazine 170	H_2O pH 2-7	539	560	620	643	650	30	48.0	0.24	11.5	2.8	0	1	1	47	Phenoxazine	4.25
180	Si-9-Iminopyronine 2g	PBS pH 7.4	434	452	469	479	631	162	11	0.43	4.7	2.8	0	1	1	85	$Si(CH_3)_2$-Pyron	4.5
181	Distyryl-BODIPY 13	PBS pH 7.5	579	608	641	659	657	16	58.0	0.25	14.5	2.7	6	5	-1	171	Bodipy-Med.	4.14.2
182	GeR	PBS pH 7.4	601	621	634	658	655	21	97.0	0.43	41.7	2.7	1	1	0	62	$Ge(CH_3)_2$-Rhoda	4.3.3
183	Phosphorilated coum. N-H 22c	PBS pH 7.4	348	370	410	437	503	93	14.0	0.44	6.1	2.7	2	1	-1	62	Coumarin 7-Am	4.26.3
184	SiR	PBS pH 7.4	589	623	645	663	661	16	93.0	0.41	38.1	2.7	1	1	0	62	$Si(CH_3)_2$-Rhoda	4.3.2

Ent.	Short names	Aqueous solution	$\lambda_{x,-0.25}$ nm	$\lambda_{x,-0.50}$ nm	$\lambda_{x,max}$ nm	$\lambda_{x,+0.50}$ nm	$\lambda_{e,max}$ nm	Δ_S nm	ε_{max} mM^{-1}cm^{-1}	Φ_f	B_{max} mM^{-1}cm^{-1}	τ_f ns	N_n	N_p	q	N_a	Class	Ref.
185	Atto 490LS carboxylate	PBS pH 7.4, 22 °C			495		658	163	40.0	0.30	12.0	2.6			0	nd	nd (pb Coum.)	4.41
186	NBDY	H_2O	462	481	504	513	533	29	74.4	0.33	24.5	2.6	1	1	0	40	Bodipy-Small	4.14.1
187	SiR-COOH	H_2O	588	567	645	662	661	16	100.0	0.39	39.0	2.6	2	1	-1	62	$Si(CH_3)_2$-Rhoda	4.3.2
188	SiR-methyl	H_2O	571	611	648	665	662	14	100.0	0.39	39.0	2.6	1	1	0	61	Si(CH3)2-Rosa	4.18.1
189	6-Dimethylaminorhodol	C pH 9.5	477	498	522	535	545	23	57.0	0.20	11.4	2.5	1	0	-1	52	Rhodol	4.15.1
190	Alexa Fluor 532	PBS pH 7.2, 22 °C	486	514	531	545	554	23	81.0	0.61	49.4	2.5	3	1	-2	94	Rosamine	4.17
191	Carborhodol S-H 13-H	PBS pH 7.4	539	567	586	601	613	27	59.0	0.32	18.8	2.5	0	1	1	64	$C(CH_3)_2$-Rhod	4.16
192	Atto 647 carboxylate	PBS pH 7.4, 22 °C	609	624	647	659	667	20	120.0	0.20	24.0	2.4	2	1	-1	85	$C(CH_3)_2$-Pyron	4.5
193	SiRF 1a-H	H_2O	602	581	662	680	680	18	118.0	0.66	77.8	2.4	1	1	0	59	$Si(CH_3)_2$-Rhoda	4.3.2
194	SiRF 1b-H	H_2O	603	581	663	681	680	17	90.0	0.70	63.0	2.4	1	2	1	82	$Si(CH_3)_2$-Rhoda	4.3.2
195	Si-9-Iminopyronine 2c	~1 mM NaOH, 1% DMSO pH 11	390	405	418	426	607	189	12.9	0.46	5.9	2.34	0	1	1	72	$Si(CH_3)_2$-Pyron	4.5
196	Si-9-Iminopyronine 2c	~1 mM HCl, 1% DMSO pH 3	597	633	666	685	681	15	83.2	0.28	23.3	2.32	0	1	1	72	$Si(CH_3)_2$-Pyron	4.5
197	Prodan	H_2O	300	322	364	404	531	167	14.5	0.10	1.4	2.3	0	0	0	34	Naphthalene	4.34
198	Naphthalimide B-K 1	10 mM P pH 7.0	380	398	436	470	552	116	13.2	0.08	1.0	2.3	0	1	1	40	Naphthalimide	4.32
199	8-Chloro-harmine	H_2O pH 4	334	341	361	382	423	62	5.9	0.11	0.6	2.2	0	1	1	29	Indole	4.33
200	6-Chloro-harmine	H_2O pH 4	335	341	362	383	428	66	9.0	0.16	1.4	2.2	0	1	1	29	Indole	4.33
201	TAMRA	50 mM PBS pH 8-9	502	526	547	562	574	27	77.0	0.35	26.9	2.2	2	1	-1	54	Rhoda 500-550	4.2.2
202	Cresyl violet (CV)	H_2O	523	553	585	599	625	40	33.1	0.36	11.9	2.2	0	1	1	32	Phenoxazine	4.25
203	3-Buthyl-7-hydroxychromone	100 mM Tris pH 8			334		454	120	12.7	0.21	2.6	2.1	1	0	-1	30	Chromone	4.31
204	3,6,10-Trimethylisoalloxazine	H_2O pH 7			439		532	93	8.5	0.08	0.6	2.1	0	0	0	31	Pteridine	4.36
205	Si-9-Iminopyronine 2f	~1 mM HCl, 1% DMSO pH 3	595	631	664	683	682	18	123	0.26	31.9	2.1	0	2	2	87	$Si(CH_3)_2$-Pyron	4.5
206	Pyronine Y (OP)	H_2O	498	523	546	559	566	20	77.0	0.22	16.9	2.0	0	1	1	39	Pyronine	4.4
207	PONy P1-Halo	PBS pH 7.4	546	576	603	619	635	32	18.0	0.26	4.6	1.9	1	1	0	84	Pyronine	4.4
208	PONy P6	PBS pH 7.4	546	576	603	619	638	35	69.0	0.09	6.2	1.9	1	1	0	79	Pyronine	4.4
209	Si-9-Iminopyronine 2f	~1 mM NaOH, 1% DMSO pH 11	375	389	401	408	606	205	17.8	0.35	6.2	1.9	0	1	1	87	$Si(CH_3)_2$-Pyron	4.5

(Continued)

Ent.	Short names	Aqueous solution	$\lambda_{x,-0.25}$ nm	$\lambda_{x,-0.50}$ nm	$\lambda_{x,max}$ nm	$\lambda_{x,+0.50}$ nm	$\lambda_{e,max}$ nm	Δ_S nm	ε_{max} mм$^{-1}$cm^{-1}	Φ_f	B_{max} mм$^{-1}$cm^{-1}	τ_f ns	N_n	N_p	q	N_a	Class	Ref.
210	Atto 655 carboxylate	PBS pH 7.4, 22 °C	587	629	663	679	680	17	125.0	0.30	37.5	1.8	2	1	-1	70	Phenoxazine	4.25
211	Atto Oxa12 carboxylate	PBS pH 7.4, 22 °C	584	621	662	680	681	19	125.0	0.30	37.5	1.8	nd	nd		70	Phenoxazine	4.25
212	nor-SqCN1	P pH 7.4	616	650	670	687	693	23	123.0	0.19	23.3	1.8	3	1	-2	63	Squaraine	4.35
213	nor-SqO2	P pH 7.4	590	620	639	655	657	18	183.0	0.33	60.3	1.8	5	1	-4	87	Squaraine	4.35
214	Phosphorilated oxazine 2-H	H_2O pH ~7	583	620	661	679	678	17	60.0	0.28	16.8	1.8	2	1	-1	66	Phenoxazine	4.25
215	PONy P1	PBS pH 7.4	535	564	590	605	627	37	59.0	0.33	19.4	1.8	1	1	0	42	Pyronine	4.4
216	Seta-635-pH-di-carboxylate	P pH 9			640		656	16	188.0	0.33	62.0	1.8			0	92	nd	4.41
217	Sulforhodamine B	H_2O pH 7	508	537	554	567	582	28	84.0	0.39	32.7	1.8	2	1	-1	67	Sulforhodam.	4.6
218	nor-SqCN2	P pH 7.4	616	650	670	687	693	23	167.0	0.19	31.7	1.8	5	1	-4	91	Squaraine	4.35
219	3-(4-OH-Ph)-6-MeO-chromone	100 mм Tris pH 8			311		468	157	12.6	0.05	0.6	1.7	1	0	-1	32	Chromone	4.31
220	Atto 680 carboxylate	PBS pH 7.4, 22 °C	603	648	681	696	698	17	125.0	0.30	37.5	1.7	2	1	-1	67	Phenoxazine	4.25
221	nor-SqB2	P pH 7.4	589	619	638	654	666	28	114.0	0.22	25.0	1.7	5	1	-4	99	Squaraine	4.35
222	PONy P5	PBS pH 7.4	592	627	660	679	704	44	38.0	0.10	3.8	1.7	1	1	0	85	Pyronine	4.4
223	Pyronine Y (OP)	PBS pH 7.4	500	525	546	560	569	23	77.0	0.36	27.7	1.7	0	1	1	39	Pyronine	4.4
224	BC-H	100 mм P, 20% Et pH 7.4			365		470	105	27.2	0.09	2.4	1.63	0	0	0	29	Hybrid-Bonded	4.40.2
225	Acridine red	H_2O	484	507	528	540	562	34	100.0	0.51	51.0	1.6	0	1	1	33	Pyronine	4.4
226	Atto 700 carboxylate	PBS pH 7.4, 22 °C	622	666	700	717	716	16	120.0	0.25	30.0	1.6	2	1	-1	67	Phenoxazine	4.25
227	Fluorinated rhodamine 1r	Buffer pH 7	483	506	525	538	554	29	31.0	0.28	8.6	1.6	2	1	-1	58	Rhoda 500-550	4.2.2
228	nor-SqO1	P pH 7.4	588	618	636	652	653	17	149.0	0.32	47.6	1.6	3	1	-2	59	Squaraine	4.35
229	Tetraethylrhodamine	H_2O pH 7	510	539	557	571	580	23	105.0	0.24	25.2	1.5	1	1	0	64	Rhoda 550-600	4.2.3
230	Atto Rho3B carboxylate	PBS pH 7.4, 22 °C	517	548	566	579	589	23	120.0	0.50	60.0	1.5	1	1	0	80	Rhoda 550-600	4.2.3
231	nor-SqB1	P pH 7.4	589	619	638	654	666	28	88.0	0.20	17.6	1.5	3	1	-2	76	Squaraine	4.35
232	PONy A1	PBS pH 7.4	476	499	535	553	595	60	38.0	0.10	3.8	1.4	2	1	-1	54	Acridine	4.23.1
233	BC-BA	100 mм P, 20% Et pH 7.4			371		473	102	25.5	0.17	4.3	1.33	0	0	0	33	Hybrid-Bonded	4.40.2
234	Pyronine B	H_2O	504	529	552	565	572	20	94.1	0.18	16.9	1.3	0	1	1	51	Pyronine	4.4
235	Acedan	H_2O	299	320	362	402	515	153	11.2	0.20	2.2	1.3	0	0	0	31	Naphthalene	4.34
236	Acrylodan	H_2O	315	339	387	433	502	115	12.9	0.18	2.3	1.28	0	0	0	32	Naphthalene	4.34

Ent.	Short names	Aqueous solution	$\lambda_{x,-0.25}$ nm	$\lambda_{x,-0.50}$ nm	$\lambda_{x,max}$ nm	$\lambda_{x,+0.50}$ nm	$\lambda_{e,max}$ nm	Δ_S nm	ε_{max} mM^{-1}cm^{-1}	Φ_f	B_{max} mM^{-1}cm^{-1}	τ_f ns	N_n	N_p	q	N_a	Class	Ref.
237	Safranin T	H_2O			517		586	69	27.0	0.11	2.9	1.2	0	1	1	43	Phenazine	4.39.1
238	CF680	H_2O			681		698	17	210	0.04	8.2	1.23			0	nd	nd (Cyanine)	4.41
239	Cy5	H_2O	587	625	647	662	663	16	250.0	0.27	67.5	1.2	2	1	-1	72	Cyanine-Penta	4.13.2
240	nor-SqCN3	P pH 7.4	636	671	693	712	714	21	113.0	0.15	16.9	1.2	7	1	-6	109	Squaraine	4.35
241	SqO1	P pH 7.4	584	614	632	648	642	10	265.0	0.07	18.5	1.2	5	1	-4	94	Squaraine	4.35
242	CF680R	H_2O			680		701	21	140	0.03	4.7	1.22			0	nd	nd (Rhodam.)	4.41
243	Alexa Fluor 660	PBS pH 7.2/7.5, 20/22 °C			663		690	27	132.0	0.37	48.8	1.2			0	nd	nd	4.41
244	Alexa Fluor 680	PBS pH 7.2/7.5, 20/22 °C	613	655	679	695	702	23	183.0	0.36	65.8	1.2	4	1	-3	98	Cyanine-Penta	4.13.2
245	Square-650-carboxylate	P pH 5.6	598	632	653	668	671	18	135.0	0.16	21.6	1.17	nd	nd		nd	Squaraine	4.35
246	nor-SqB3	P pH 7.4	610	642	662	679	684	22	102.0	0.18	18.3	1.2	7	1	-6	117	Squaraine	4.35
247	AF647	PBS pH 7.2, 20/22 °C	589	628	650	665	672	22	270.0	0.33	89.1	1.1	5	1	-4	101	Cyanine-Penta	4.13.2
248	AFD647	PBS pH 7.2, 20/22 °C	589	628	650	665	672	22	270.0	0.33	89.1	1.1	5	1	-4	98	Cyanine-Penta	4.13.2
249	Alexa Fluor 647	PBS pH 7.2, 20/22 °C	589	628	650	665	672	22	270.0	0.33	89.1	1.1	5	1	-4	98	Cyanine-Penta	4.13.2
250	nor-SqO3	P pH 7.4	610	642	662	679	679	17	124.0	0.21	26.0	1.1	7	1	-6	105	Squaraine	4.35
251	SqB1	P pH 7.4	580	609	627	642	650	23	150.0	0.05	7.5	1.1	5	1	-4	104	Squaraine	4.35
252	CF647	H_2O			650		665	15	240.0	0.12	27.9	1.1			0	nd	nd (Cyanine)	4.41
253	Abberior Star 520SXP	PBS pH 7.4	454	475	522	555	632	110	42.5	0.15	6.3	1.0	1	1	0	81	Coumarin 7-Am	4.26.3
254	Alexa Fluor 700	PBS/H_2O pH 7.2, 20/22 °C			702		723	21	205.0	0.25	51.2	1.0			0	nd	nd	4.41
255	Atto 495 carboxylate	PBS pH 7.4, 22 °C	455	471	498	512	526	28	80.0	0.20	16.0	1.0	1	1	0	51	Acridine	4.23.1
256	Zy5	P pH 7.4	587	625	647	662	665	18	250.0	0.27	67.5	1.0	3	1	-2	83	Cyanine-Penta	4.13.2
257	Zy5.5	P pH 7.4	610	652	676	692	694	18	190.0	0.23	43.7	1.0	5	1	-4	103	Cyanine-Penta	4.13.2
258	SqCN1	P pH 7.4	614	647	667	685	685	18	188.0	0.07	13.1	0.9	5	1	-4	98	Squaraine	4.35
259	Chromenoquinoline L-S 4a	H_2O			473		570	97	36.19	0.15	5.4	0.92	0	0	0	46	Hybrid-Fused	4.40.1
260	Chromenoquinoline L-S 4b	H_2O			477		579	102	36.05	0.15	5.2	0.9	0	0	0	47	Hybrid-Fused	4.40.1
261	Seta-660-di-carboxylate	P pH 7.4			661		672	11	220.0	0.11	24.2	0.9			0	138	nd	4.41

(Continued)

Ent.	Short names	Aqueous solution	$\lambda_{x,-0.25}$ nm	$\lambda_{x,-0.50}$ nm	$\lambda_{x,max}$ nm	$\lambda_{x,+0.50}$ nm	$\lambda_{e,max}$ nm	Δ_S nm	ε_{max} mM^{-1}cm^{-1}	Φ_f	B_{max} mM^{-1}cm^{-1}	τ_f ns	N_n	N_p	q	N_a	Class	Ref.
262	Abberior Star 470SXP	PBS pH 7.4	416	434	472	499	624	152	29.0	0.12	3.4	0.8	2	1	-1	77	Coumarin 7-Am	4.26.3
263	PONy P2	PBS pH 7.4	569	602	632	649	672	40	51.0	0.16	8.1	0.8	0	1	1	50	Pyronine	4.4
264	Alexa Fluor 750	PBS/H_2O pH 7.2, 20/22 °C			749		775	26	290.0	0.12	34.8	0.7			0	nd	nd	4.41
265	CypHer5	A pH 4.6	583	621	643	657	662	19	100.0	0.19	19.0	0.70	3	1	-2	71	Cyanine-Penta	4.13.2
266	EVOblue 30	H_2O	575	610	650	667	667	17	100.0	0.10	10.0	0.6	2	1	-1	55	Phenoxazine	4.25
267	Cy5′	H_2O	583	621	642	656	660	18	138.0	0.17	23.4	0.6	2	1	-1	84	Cyanine-Penta	4.13.2
268	Chromenoquinoline L-S 4d	H_2O			500		608	108	36.06	0.08	2.8	0.62	0	0	0	48	Hybrid-Fused	4.40.1
269	Atto 740 carboxylate	PBS pH 7.4, 22 °C			743		763	20	120.0	0.10	12.0	0.6			0	nd	nd	4.41
270	Cy3.5	H_2O	511	534	581	593	596	15	120.0	0.14	16.8	0.6	4	1	-3	88	Cyanine-Trimet	4.13.1
271	Zy7 carboxylate	P pH 7.4	663	718	749	772	774	25	200.0	0.13	26.0	0.6	3	1	-2	87	Cyanine-Hepta	4.13.3
272	Chromenoquinoline L-S 4e	H_2O			502		613	111	33.16	0.07	2.2	0.56	0	0	0	49	Hybrid-Fused	4.40.1
273	Squaraine T-P 5b	67 mм P pH 7.4	614	647	667	685	685	18	188.0	0.07	13.1	0.5	4	1	-3	98	Squaraine	4.35
274	Atto 725 carboxylate	PBS pH 7.4, 22 °C			728		751	23	120.0	0.10	12.0	0.5			0	nd	nd	4.41
275	DY-635 carboxylate	PBS pH 7.4	550	583	635	658	669	34	175.0	0.05	8.7	0.5	2	1	-1	93	Merocyanine	4.24
276	Y-A 1 (Cypate)	PBS pH 7.2	685	744	778	803	805	27	224.0	0.03	6.7	0.5	2	1	-1	86	Cyanine-Hepta	4.13.3
277	Cy3′	H_2O	483	504	545	556	560	15	120.0	0.04	4.8	0.5	2	1	-1	80	Cyanine-Trimet	4.13.1
278	Harmine	H_2O pH 9	309	315	332	350	425	93	11.0	0.38	4.1	0.4	0	1	1	28	Indole	4.33
279	Chromenoquinoline L-S 4c	H_2O			500		591	91	33.16	0.03	1.1	0.43	0	0	0	44	Hybrid-Fused	4.40.1
280	Seta-670-carboxylate	P pH 7.4			667		688	21	180.0	0.07	12.6	0.42			0	103	nd	4.41
281	Chromenoquinoline L-S 4f	H_2O			529		622	93	40.64	0.05	1.9	0.41	0	0	0	46	Hybrid-Fused	4.40.1
282	Coumarin 1	H_2O	329	349	382	409	472	90	20.1	0.07	1.4	0.4	0	0	0	34	Coumarin 7-Am	4.26.2
283	Seta-646-carboxylate	P pH 7.4			646		656	10	207.0	0.10	20.7	0.38			0	87	nd	4.41
284	Squaraine T-P 5a	67 mм P pH 7.4	592	620	636	653	648	12	216.0	0.06	12.9	0.4	4	1	-3	94	Squaraine	4.35
285	Squaraine T-P 5c	67 mм P pH 7.4	584	614	632	648	642	10	265.0	0.06	15.9	0.4	4	1	-3	94	Squaraine	4.35
286	Alexa Fluor 555	PBS pH 7.2, 22 °C	491	512	555	566	572	17	155.0	0.10	15.5	0.4	5	1	-4	113	Cyanine-Trimet	4.13.1
287	AF555	PBS pH 7.2, 22 °C	491	512	555	566	572	17	158.0	0.09	14.2	0.3	5	1	-4	113	Cyanine-Trimet	4.13.1
288	Acyman	H_2O	302	324	367	408	517	150	9.1	0.10	0.9	0.27	0	0	0	40	Naphthalene	4.34
289	Square-660-carboxylate	P pH 7.4	600	635	657	675	676	19	182.0	0.03	5.4	0.27	nd	nd		nd	Squaraine	4.35

Ent.	Short names	Aqueous solution	$\lambda_{x,-0.25}$ nm	$\lambda_{x,-0.50}$ nm	$\lambda_{x,max}$ nm	$\lambda_{x,+0.50}$ nm	$\lambda_{e,max}$ nm	Δ_S nm	ε_{max} mM^{-1}cm^{-1}	Φ_f	B_{max} mM^{-1}cm^{-1}	τ_f ns	N_n	N_p	q	N_a	Class	Ref.
290	Seta-633-carboxylate	P pH 7.4			633		644	11	250.0	0.07	17.5	0.25			0	106	nd	4.41
291	Cy3	H_2O	487	508	550	561	570	20	150.0	0.06	8.2	0.2	2	1	-1	68	Cyanine-Trimet	4.13.1
292	Squaraine T-P 4d	67 mM P pH 7.4	606	638	658	675	677	19	182.0	0.03	5.4	0.2	3	1	-2	77	Squaraine	4.35
293	Squaraine T-P 4b	67 mM P pH 7.4	607	639	659	676	676	17	207.0	0.02	4.1	0.2	3	1	-2	78	Squaraine	4.35
294	Squaraine T-P 4a	67 mM P pH 7.4	581	610	628	644	642	14	163.0	0.03	4.8	0.1	3	1	-2	74	Squaraine	4.35
295	Squaraine T-P 4c	67 mM P pH 7.4	577	606	624	639	635	11	245.0	0.02	4.9	0.1	3	1	-2	74	Squaraine	4.35

Appendix G

G1. Ranked by Molecular Net Charge

Appendix G contains the database of over seven-hundred fluorescent labels and stains, identified by their short names and ranked by ascending net charge values ($q = N_p - N_n$). ($q = N_p - N_n$). The integer numbers represent the number of excess (−) or missing (+) elementary charges. Some of their names are registered and others are trademarks (see footnote[1]). The majority of the dyes are salts. When dyes occur as salts, only the charge of the fluorescent ion (anion or cation) is taken into account. Some dyes have a pK_a that is close to the pH used to measure the photophysical parameters in the literature (data in the column 'Aqueous Solutions'). In these cases, the prevalent ionic form of the dye is used to calculate the net charge. Note that the charges are related to the free acid forms of the fluorescent dyes, and that there will be one less negative charge (one less COO^-) if the dye is labeled (bioconjugated) to a target molecule (biomolecule) using, for instance, succinimidyl ester, maleimide, or another labeling moiety. The column N_a gives the number of atoms of the fluorescent structure (counter-ions of salts are not counted) of each dye in its prevalent ionic form.

About 5% of the structures of the entities of the database are non-disclosed and therefore could not be displayed in this list (Table G.1). Luckily the net charges of some of these non-disclosed structures were provided by the suppliers.

The molecular structures together with some photophysical parameters are shown in Chapter 4. Chapter 4 also provides a detailed list of references where the photophysical parameters were taken. Important properties of the dyes are also revealed in Chapter 5. Panoramic views of the photophysical properties and classes are given in Chapter 5 using scattergrams. Methods to measure photostability and techniques to improve photostability are discussed in Chapter 7. The photostability (relative photobleaching mean-lifetime) of some dyes is also shown in Chapter 7.

Acronyms used in Table G.1: A, acetate; AC, acetic acid; ACN, acetonitrile; BODIPY, borondipyrromethene; BSA, bovine serum albumin; C, carbonate; CA, carboxylic acid; FBS, fetal bovine serum; Et, ethanol; nd, nondisclosed; ni, not informed; P, phosphate; pb, probably; pH >7, basic medium; pH < 7, acidic medium (when the exact value of the pH is not given in the literature); SA, sulfonic acid; and TX-100, triton X-100.

1 Abberior STAR is a trademark of Abberior GmbH (Göttingen, Germany). Alexa Fluor, BODIPY, Cascade Blue, FAM, JOE, TET, HEX, Marina Blue, Oregon Green, Pacific Blue, ROX, TAMRA, and Texas Red are trademarks of Thermo Fisher Scientific (Waltham, MA, USA). ATTO is a trademark of ATTO-Tec GmbH (Siegen, Germany). CF Dye is a registered trademark of Biotium, Inc. (Fremont, CA, USA). Cy3, Cy3.5, Cy5, and Cy5.5 are trademarks of GE Healthcare (Chicago, IL, USA). Dy is a trademark of Dyomics GmbH (Jena, Germany). Seta, SeTau, and Square dyes are trademarks of SETA BioMedicals (Urbana, IL, USA).

Fluorescent Dye Labels and Stains: A Database of Photophysical Properties, First Edition. Tarso B. Ledur Kist.

Table G.1 Photophysical properties of fluorescent labels and stains in aqueous solutions ranked by net charges of the prevalent ionic form at the solution and pH indicated ($q = N_p - N_n$).

Ent.	Short names	Aqueous solution	$\lambda_{x,-0.25}$ nm	$\lambda_{x,-0.50}$ nm	$\lambda_{x,max}$ nm	$\lambda_{x,+0.50}$ nm	$\lambda_{e,max}$ nm	Δ_S nm	ε_{max} mM^{-1}cm^{-1}	Φ_f	B_{max} mM^{-1}cm^{-1}	τ_f ns	N_n	N_p	q	N_a	Class	Ref.
1	Y-A 5	PBS pH 7.2	685	744	778	803	804	26	185.0	0.03	5.5		8	1	−7	158	Cyanine-Hepta	4.13.3
2	nor-SqB3	P pH 7.4	610	642	662	679	684	22	102.0	0.18	18.3	1.2	7	1	−6	117	Squaraine	4.35
3	nor-SqCN3	P pH 7.4	636	671	693	712	714	21	113.0	0.15	16.9	1.2	7	1	−6	109	Squaraine	4.35
4	nor-SqO3	P pH 7.4	610	642	662	679	679	17	124.0	0.21	26.0	1.1	7	1	−6	105	Squaraine	4.35
5	BCECF	100 mM NaOH	466	486	505	518	527	22	67.0	0.84	56.2	4.1	5	0	−5	53	Fluorescein-nh	4.8.1
6	BCPCF	100 mM NaOH	466	486	505	518	527	22	67.0	0.83	55.6		5	0	−5	61	Fluorescein-nh	4.8.1
7	Cy3.205	PBS pH 7.4	509	532	578	590	593	15	120.0	0.14	16.8		6	1	−5	110	Cyanine-Trimet	4.13.1
8	Cy5.205	PBS pH 7.4	609	650	674	690	694	20	190.0	0.23	43.7		6	1	−5	114	Cyanine-Penta	4.13.2
9	Y-A 4	PBS pH 7.2	685	744	778	803	805	27	201.0	0.03	6.0		6	1	−5	134	Cyanine-Hepta	4.13.3
10	1,3,6,8-Pyrenetetrasulfonic acid	H_2O pH 7	335	350	374	378	403	29	51.0	0.56	28.5	12.5	4	0	−4	38	Pyrene	4.28
11	8-Hydroxypyrene-1,3,6-triSA	H_2O pH 10	389	408	460	470	514	54	24.0	0.95	22.8	5.3	4	0	−4	36	Pyrene	4.28
12	Alexa Fluor 405	PBS pH 7.2, 22 °C	346	360	400	408	424	24	35.0	0.50	17.5		4	0	−4	59	Pyrene	4.28
13	Cascade Blue	50 mM P pH 8	348	363	403	411	430	27	28.0	0.90	25.2		4	0	−4	40	Pyrene	4.28
14	DPP 10	PBS			523		594	71	17.1	0.61	10.4	4.8	4	0	−4	66	Diketopyrrolop	4.39.1
15	DPP 9	PBS			451		541	90	13.9	0.90	12.5	6.6	4	0	−4	72	Diketopyrrolop	4.39.1
16	DY-405 carboxylate	PBS pH 7.4	336	349	400	410	423	23	32.0	0.97	30.9		4	0	−4	61	Pyrene	4.28
17	Spirobifluorene 1	H_2O			378		480	102	51	0.50	25.6	4.23	4	0	−4	169	Spirobifluorene	4.39.3
18	AF555	PBS pH 7.2, 22 °C	491	512	555	566	572	17	158.0	0.09	14.2	0.3	5	1	−4	113	Cyanine-Trimet	4.13.1

Ent.	Short names	Aqueous solution	$\lambda_{x,-0.25}$ nm	$\lambda_{x,-0.50}$ nm	$\lambda_{x,max}$ nm	$\lambda_{x,+0.50}$ nm	$\lambda_{e,max}$ nm	Δ_S nm	ε_{max} mм$^{-1}$cm^{-1}	Φ_f	B_{max} mм$^{-1}$cm^{-1}	τ_f ns	N_n	N_p	q	N_a	Class	Ref.
19	AF647	PBS pH 7.2, 20/22 °C	589	628	650	665	672	22	270.0	0.33	89.1	1.1	5	1	−4	101	Cyanine-Penta	4.13.2
20	AFD647	PBS pH 7.2, 20/22oC	589	628	650	665	672	22	270.0	0.33	89.1	1.1	5	1	−4	98	Cyanine-Penta	4.13.2
21	Alexa Fluor 555	PBS pH 7.2, 22oC	491	512	555	566	572	17	155.0	0.10	15.5	0.4	5	1	−4	113	Cyanine-Trimet	4.13.1
22	Alexa Fluor 647	PBS pH 7.2, 20/22oC	589	628	650	665	672	22	270.0	0.33	89.1	1.1	5	1	−4	98	Cyanine-Penta	4.13.2
23	CW800	FBS	691	752	786	812	800	14	237.0	0.14	33.6		5	1	−4	117	Cyanine-Hepta	4.13.3
24	Cy5.5	PBS	609	650	674	690	694	20	190.0	0.23	43.7		5	1	−4	103	Cyanine-Penta	4.13.2
25	DY-679P1	PBS pH 7.4	612	655	679	696	697	18	200.0	0.23	46.6		5	1	−4	105	Cyanine-Penta	4.13.2
26	IRDye78-CA	10 mм HEPES pH 7.4	680	738	771	796	806	35	145.7	0.14	20.6		5	1	−4	120	Cyanine-Hepta	4.13.3
27	nor-SqB2	P pH 7.4	589	619	638	654	666	28	114.0	0.22	25.0	1.7	5	1	−4	99	Squaraine	4.35
28	nor-SqCN2	P pH 7.4	616	650	670	687	693	23	167.0	0.19	31.7	1.8	5	1	−4	91	Squaraine	4.35
29	nor-SqO2	P pH 7.4	590	620	639	655	657	18	183.0	0.33	60.3	1.8	5	1	−4	87	Squaraine	4.35
30	SqB1	P pH 7.4	580	609	627	642	650	23	150.0	0.05	7.5	1.1	5	1	−4	104	Squaraine	4.35
31	SqCN1	P pH 7.4	614	647	667	685	685	18	188.0	0.07	13.1	0.9	5	1	−4	98	Squaraine	4.35
32	SqO1	P pH 7.4	584	614	632	648	642	10	265.0	0.07	18.5	1.2	5	1	−4	94	Squaraine	4.35
33	W-L 8d	PBS	583	621	643	657	666	23	169.0	0.13	21.9		5	1	−4	108	Cyanine-Penta	4.13.2
34	Zy5.5	P pH 7.4	610	652	676	692	694	18	190.0	0.23	43.7	1.0	5	1	−4	103	Cyanine-Penta	4.13.2
35	3-OH-4-nitroso-2,7-naphth-diSA	T pH 9	339	367	423	478	523	100	20.0	0.28	5.6		3	0	−3	28	Naphthalene	4.34
36	5(6)-Carboxyfluorescein (FAM)	100 mм PBS pH 9	457	477	495	508	520	25	83.0	0.90	74.7	4.2	3	0	−3	40	Fluorescein-nh	4.8.1
37	5-Carboxyfluorescein (S-H 2)	P pH 9	454	474	492	504	516	24	84.7	0.92	77.9		3	0	−3	40	Fluorescein-nh	4.8.1

(Continued)

Ent.	Short names	Aqueous solution	$\lambda_{x,-0.25}$ nm	$\lambda_{x,-0.50}$ nm	$\lambda_{x,max}$ nm	$\lambda_{x,+0.50}$ nm	$\lambda_{e,max}$ nm	Δ_S nm	ε_{max} mм$^{-1}$cm^{-1}	Φ_f	B_{max} mм$^{-1}$cm^{-1}	τ_f ns	N_n	N_p	q	N_a	Class	Ref.
38	6-CA-MeThio-4,5,7-F-fluorescein	100 mм NaOH	467	488	507	520	526	19	63.5	0.88	55.8		3	0	−3	41	Fluorescein-ha	4.8.2
39	8-Aminonaphthalene-1,3,6-triSA	100 mм P pH 7	295	315	356	394	512	156	15.0	0.20	3.0		3	0	−3	30	Naphthalene	4.34
40	8-Aminopyrene-1,3,6-triSA	H_2O	363	380	424	433	505	81	20.6	0.90	18.5		3	0	−3	37	Pyrene	4.28
41	8-Hydroxypyrene-1,3,6-triSA	H_2O pH 2	348	363	403	411	443	40	20.0	0.95	19.0	4.8	3	0	−3	36	Pyrene	4.28
42	cis-PO-PB430	PBS pH 7.4			421		548	127	10.0	0.71	7.1	9.9	3	0	−3	86	Phosphole	4.39.3
43	DPP 7	PBS			456		548	92	14.7	0.76	11.1	6.5	3	0	−3	58	Diketopyrrolop	4.39.1
44	DPP 8	PBS			527		594	67	18.1	0.56	10.1	4.8	3	0	−3	52	Diketopyrrolop	4.39.1
45	DY-350 carboxylate	PBS pH 7.4	301	323	353	378	432	79	19.0	0.84	15.9		3	0	−3	54	Coumarin 7-Am	4.26.3
46	DY-431 carboxylate	PBS pH 7.4	388	405	442	467	496	54	35.0	0.75	26.2		3	0	−3	69	Coumarin 7-Am	4.26.3
47	DY-495 carboxylate	PBS pH 7.4	456	476	494	507	521	27	78.0	0.68	53.2		3	0	−3	59	Fluorescein-nh	4.8.1
48	Fluorinated fluorescein S-H 14a,b	P pH 9	454	474	492	504	516	24	85.9	0.92	79.0	3.8	3	0	−3	40	Fluorescein-ha	4.8.2
49	Fluorinated fluorescein S-H 15a,b	P pH 9	470	490	510	523	534	24	84.2	0.43	36.2		3	0	−3	40	Fluorescein-ha	4.8.2
50	Fluorinated fluorescein S-H 16a,b	P pH 9	470	490	510	523	534	24	78.1	0.59	46.0		3	0	−3	40	Fluorescein-ha	4.8.2
51	HEX	H_2O pH 9	491	513	535	550	559	24	98.0	0.70	68.6		3	0	−3	37	Fluorescein-ha	4.8.2
52	JOE	H_2O pH 9	478	500	520	534	548	28	73.0	0.60	43.8		3	0	−3	45	Fluorescein-ha	4.8.2
53	Oregon Green 488	H_2O pH 9	453	472	490	502	514	24	76.0	0.97	73.7	4.3	3	0	−3	37	Fluorescein-ha	4.8.2
54	Oregon Green 514	PBS pH 7.2, 22oC	471	491	511	524	530	19	88.0	0.96	84.4	4.4	3	0	−3	44	Fluorescein-ha	4.8.2
55	PB430	PBS pH 7.4			426		542	116	10.4	0.66	6.8	10.6	3	0	−3	115	Phosphole	4.39.3

Ent.	Short names	Aqueous solution	$\lambda_{x,-0.25}$ nm	$\lambda_{x,-0.50}$ nm	$\lambda_{x,max}$ nm	$\lambda_{x,+0.50}$ nm	$\lambda_{e,max}$ nm	Δ_S nm	ε_{max} mM^{-1}cm^{-1}	Φ_f	B_{max} mM^{-1}cm^{-1}	τ_f ns	N_n	N_p	q	N_a	Class	Ref.
56	TET	H_2O pH 9	479	501	521	535	536	15	73.0	0.70	51.1		3	0	−3	37	Fluorescein-ha	4.8.2
57	trans-PO-PB430	PBS pH 7.4			422		545	123	10.3	0.74	7.6	10.3	3	0	−3	86	Phosphole	4.39.3
58	trans-PO-PB460	PBS pH 7.4			459		592	133	9.0	0.25	2.2	4.9	3	0	−3	86	Phosphole	4.39.3
59	Yakima Yellow carboxylate	100 mM P/Tris pH 7.0/7.4	487	509	530	545	549	19	84.0	0.96	80.6		3	0	−3	48	Fluorescein-ha	4.8.2
60	Abberior Star 488 carboxylate	PBS pH 7.4	465	485	503	515	524	21	64.5	0.89	57.4		4	1	−3	62	Rhodam Hi Ch	4.2.5
61	Alexa Fluor 488	PBS pH 7.2, 22 °C	457	477	494	506	517	23	73.0	0.92	67.1	4.1	4	1	−3	48	Rhodam Hi Ch	4.2.5
62	Alexa Fluor 546	PBS pH 7.2, 22 °C	506	537	546	567	580	34	112.0	0.79	88.4	4.1	4	1	−3	102	Rhodam Hi Ch	4.2.5
63	Alexa Fluor 568	PBS pH 7.2/7.0, 20/22 °C	527	559	578	593	603	25	88.0	0.69	60.7	3.6	4	1	−3	78	Rhodam Hi Ch	4.2.5
64	Alexa Fluor 594	PBS pH 7.2/7.0, 20/22 °C	537	570	590	606	617	27	92.0	0.66	60.7	3.9	4	1	−3	84	Rhodam Hi Ch	4.2.5
65	Alexa Fluor 633	PBS/H_2O pH 7.2, 20/22 °C	568	608	632	650	647	15	100.0	0.64	64.0	3.2	4	1	−3	93	Sulforhodam.	4.6
66	Alexa Fluor 680	PBS pH 7.2/7.5, 20/22 °C	613	655	679	695	702	23	183.0	0.36	65.8	1.2	4	1	−3	98	Cyanine-Penta	4.13.2
67	B-Z 14	H_2O	576	605	638	656	655	17	102.0	0.42	42.8	3.5	4	1	−3	127	Bodipy-Large	4.14.3
68	B-Z 19	H_2O	587	617	651	670	677	26	118.5	0.23	27.2	2.8	4	1	−3	139	Bodipy-Large	4.14.3
69	B-Z 7a	H_2O	458	477	500	508	509	9	64.5	0.61	39.3	5.0	4	1	−3	111	Bodipy-Large	4.14.3
70	B-Z 7b	H_2O	479	500	516	526	529	13	60.5	0.68	41.1	6.4	4	1	−3	123	Bodipy-Large	4.14.3
71	Cy3.12-OH	H_2O/PBS	492	513	556	567	574	18	130.0	0.05	6.5		4	1	−3	90	Cyanine-Trimet	4.13.1

(Continued)

Ent.	Short names	Aqueous solution	$\lambda_{x,-0.25}$ nm	$\lambda_{x,-0.50}$ nm	$\lambda_{x,max}$ nm	$\lambda_{x,+0.50}$ nm	$\lambda_{e,max}$ nm	Δ_S nm	ε_{max} mм$^{-1}$cm^{-1}	Φ_f	B_{max} mм$^{-1}$cm^{-1}	τ_f ns	N_n	N_p	q	N_a	Class	Ref.
72	Cy3.18.OH	PBS	487	508	550	561	565	15	150.0	0.04	6.0		4	1	−3	90	Cyanine-Trimet	4.13.1
73	Cy3.5	H_2O	511	534	581	593	596	15	120.0	0.14	16.8	0.6	4	1	−3	88	Cyanine-Trimet	4.13.1
74	Cy5.12-OH	H_2O/PBS	589	628	650	665	674	24	215.0	0.13	27.9		4	1	−3	94	Cyanine-Penta	4.13.2
75	Cy5.18.OH	PBS	590	628	650	664	667	17	250.0	0.27	67.5		4	1	−3	94	Cyanine-Penta	4.13.2
76	Cy5.203	PBS pH 7.4	614	656	680	696	702	22	190.0	0.10	19.0		4	1	−3	106	Cyanine-Penta	4.13.2
77	Cy5.204	PBS pH 7.4	615	658	682	698	707	25	197.0	0.10	19.7		4	1	−3	106	Cyanine-Penta	4.13.2
78	DY-490 carboxylate	PBS pH 7.4	452	468	491	504	515	24	73.0	0.90	65.4		4	1	−3	61	Rhodam Hi Ch	4.2.5
79	DY-682 carboxylate	PBS pH 7.4	617	654	690	708	710	20	120.0	0.41	48.6		4	1	−3	100	Merocyanine	4.24
80	LS479	Buffer pH 8	661	716	747	770	762	15	105.0	0.04	4.4		4	1	−3	109	Cyanine-Hepta	4.13.3
81	LS482	Buffer pH 8	632	682	710	731	793	83	84.0	0.10	8.4		4	1	−3	108	Cyanine-Hepta	4.13.3
82	NIR1	H_2O	672	729	761	785	796	35	268.0	0.23	61.6		4	1	−3	93	Cyanine-Hepta	4.13.3
83	NIR2	H_2O	599	639	662	677	684	22	250.0	0.34	85.0		4	1	−3	89	Cyanine-Penta	4.13.2
84	NIR3	H_2O	663	719	750	773	777	27	275.0	0.28	77.0		4	1	−3	93	Cyanine-Hepta	4.13.3
85	NIR4	H_2O	589	628	650	665	671	21	260.0	0.43	111.8		4	1	−3	89	Cyanine-Penta	4.13.2
86	Squaraine T-P 5a	67 mм P pH 7.4	592	620	636	653	648	12	216.0	0.06	12.9	0.4	4	1	−3	94	Squaraine	4.35
87	Squaraine T-P 5b	67 mм P pH 7.4	614	647	667	685	685	18	188.0	0.07	13.1	0.5	4	1	−3	98	Squaraine	4.35
88	Squaraine T-P 5c	67 mм P pH 7.4	584	614	632	648	642	10	265.0	0.06	15.9	0.4	4	1	−3	94	Squaraine	4.35
89	Sulfo-cyanine5	H_2O	586	624	646	661	662	16	271.0	0.20	54.2		4	1	−3	94	Cyanine-Penta	4.13.2
90	Y-A 3	PBS pH 7.2	684	744	777	802	807	30	215.0	0.03	6.4		4	1	−3	110	Cyanine-Hepta	4.13.3
91	Distyryl-BODIPY 11	PBS pH 7.5	546	572	601	617	655	54	17.9	0.04	0.7		6	3	−3	145	Bodipy-Med.	4.14.2

Ent.	Short names	Aqueous solution	$\lambda_{x,-0.25}$ nm	$\lambda_{x,-0.50}$ nm	$\lambda_{x,max}$ nm	$\lambda_{x,+0.50}$ nm	$\lambda_{e,max}$ nm	Δ_S nm	ε_{max} mм$^{-1}$cm^{-1}	Φ_f	B_{max} mм$^{-1}$cm^{-1}	τ_f ns	N_n	N_p	q	N_a	Class	Ref.
92	Distyryl-BODIPY 12	PBS pH 7.5	579	609	642	660	657	15	55.1	0.22	12.1		6	3	−3	177	Bodipy-Med.	4.14.2
93	(HO)-Luciferyl adenylate	H_2O pH 5			335		548	213	15.1	0.11	1.6		2	0	−2	tl	Benzothiazole	4.30
94	($^-$O)-Dehydroluciferin	H_2O pH 11			396		555	159	18.2	0.62	11.2		2	0	−2	24	Benzothiazole	4.30
95	($^-$O)-Luciferin	H_2O pH 11			384		537	153	18.5	0.62	11.4		2	0	−2	26	Benzothiazole	4.30
96	1,2,7,8-Dibenzo-4,5-Br-fluorescein	C pH 9.5	503	526	549	565	581	32	24.0	0.05	1.2		2	0	−2	48	Fluorescein-ha	4.8.2
97	1,2,7,8-Dibenzofluorescein	C pH 9.5	490	513	534	549	565	31	81.0	0.24	19.4		2	0	−2	48	Fluorescein-nh	4.8.1
98	1,2,7,8-Dibenzosulfofluorescein	C pH 9.5	502	523	541	554	567	26	81.0	0.48	38.8		2	0	−2	48	Sulfofluoresc.	4.10
99	2′,4,5,6,7,7′-HexaF-Si-fluorescein	100 mм NaOH	567	594	617	634	635	18	67.9	0.32	21.7		2	0	−2	45	Si$(CH_3)_2$-Fluore	4.9
100	2,4,5,7-Tetra-Br-sulfofluorescein	100 mм NaOH pH 12	488	507	524	536	546	22	87.0	0.35	30.4		2	0	−2	36	Sulfofluoresc.	4.10
101	2′,4′,5′,7′-TetraF-Si-fluorescein	100 mм NaOH	550	575	597	613	613	16	65.0	0.67	43.5		2	0	−2	45	Si$(CH_3)_2$-Fluore	4.9
102	2,4,5,7-Tetra-F-sulfofluorescein	P pH 9.18	480	499	515	527	537	22	63.2	0.80	50.5		2	0	−2	36	Sulfofluoresc.	4.10
103	2′,7′-Cl-4,5,6,7-F-fluorescein	100 mм NaOH	478	500	520	534	538	18	88.5	0.80	70.8		2	0	−2	35	Fluorescein-ha	4.8.2
104	2′,7′-Difluoro-Si-Fluorescein	100 mм NaOH	549	574	596	612	618	22	60.9	0.42	25.5		2	0	−2	45	Si$(CH_3)_2$-Fluore	4.9
105	2,7-DiMe-6-EtAm-carboxyrhodol	50 mм P pH 8-9	467	487	510	522	536	26	66.3	0.89	59.0	5.0	2	0	−2	53	Rhodol	4.15.1
106	2′,7′-Difluorocarbofluorescein	100 mм NaOH	514	536	555	569	581	26	90.9	0.40	36.3		2	0	−2	45	C$(CH_3)_2$-Fluore	4.9
107	2-Br-7-Me-6-EtAm-carboxyrhodol	50 mм P pH 8-9	472	492	516	529	539	23	80.9	0.61	49.3	3.05	2	0	−2	48	Rhodol	4.15.1
108	2-Cl-6-Piperidinocarboxy-rhodol	50 mм P pH 8-9	478	500	524	537	548	24	77.4	0.72	55.7	3.88	2	0	−2	48	Rhodol	4.15.1

(Continued)

Ent.	Short names	Aqueous solution	$\lambda_{x,-0.25}$ nm	$\lambda_{x,-0.50}$ nm	$\lambda_{x,max}$ nm	$\lambda_{x,+0.50}$ nm	$\lambda_{e,max}$ nm	Δ_S nm	ε_{max} mM^{-1}cm^{-1}	Φ_f	B_{max} mM^{-1}cm^{-1}	τ_f ns	N_n	N_p	q	N_a	Class	Ref.
109	2-Cl-7-Me-6-EtAm-carboxyrhodol	50 mM P pH 8-9	479	495	514	538	539	25	80.5	0.82	66.0	3.94	2	0	−2	50	Rhodol	4.15.1
110	2-Cloro-6-quinolinocarboxy-rhodol	50 mM P pH 8-9	498	521	548	563	575	27	74.5	0.42	31.2	3.77	2	0	−2	55	Rhodol	4.15.1
111	2-COOH DCTM	100 mM P, 1% DMSO pH 10	545	570	591	606	607	16	120.0	0.48	57.6		2	0	−2	43	$Si(CH_3)_2$-Fluore	4.9
112	2-COOH DFTM	100 mM P, 1% DMSO pH 10	537	560	581	596	596	15	120.0	0.54	64.8		2	0	−2	43	$Si(CH_3)_2$-Fluore	4.9
113	2-COOH TM	100 mM P, 1% DMSO pH 10	536	560	580	595	598	18	110.0	0.38	41.8		2	0	−2	43	$Si(CH_3)_2$-Fluore	4.9
114	2-Et-7-Me-6-EtAm-carboxyrhodol	50 mM P pH 8-9	468	488	511	524	538	27	73.0	0.84	61.3	4.0	2	0	−2	56	Rhodol	4.15.1
115	2-Ethyl-6-quinolinocarboxyrhodol	50 mM P pH 8-9	494	517	543	557	574	31	79.4	0.49	38.9	4.13	2	0	−2	61	Rhodol	4.15.1
116	2-PrAc-6-quinolinorhodol	50 mM P pH 8-9	493	515	541	555	572	31	58.2	0.49	28.5	4.08	2	0	−2	63	Rhodol	4.15.1
117	2-PrAc-7-Me-6-EtAm-rhodol	50 mM P pH 8-9	466	486	509	522	535	26	79.4	0.92	73.0	3.9	2	0	−2	56	Rhodol	4.15.1
118	3-OH-4-nitroso-2,7-naphth-diSA	A pH 4	304	326	370	412	470	100	19.0	0.22	4.1		2	0	−2	26	Naphthalene	4.34
119	4,10-Dibromonaphtho-fluorescein	C pH 9.5	549	578	605	624	682	77	32.0	0.05	1.6		2	0	−2	47	Fluores-analog	4.8.4
120	4,5,6,7-TetraF-Si-fluorescein	100 mM NaOH	551	576	598	614	614	16	84.1	0.50	42.0		2	0	−2	45	$Si(CH_3)_2$-Fluore	4.9
121	4,7-DiMe-6-EtAm-carboxyrhodol	50 mM P pH 8-9	472	492	516	529	544	28	56.9	0.43	24.4	3.0	2	0	−2	53	Rhodol	4.15.1
122	4-Benzamidofluorescein	100 mM P pH 7.5	453	472	490	502	517	27	80.0	0.79	63.2		2	0	−2	49	Fluorescein-nh	4.8.1

Ent.	Short names	Aqueous solution	$\lambda_{x,-0.25}$ nm	$\lambda_{x,-0.50}$ nm	$\lambda_{x,max}$ nm	$\lambda_{x,+0.50}$ nm	$\lambda_{e,max}$ nm	Δ_S nm	ε_{max} mM^{-1}cm^{-1}	Φ_f	B_{max} mM^{-1}cm^{-1}	τ_f ns	N_n	N_p	q	N_a	Class	Ref.
123	4-Carboxy Pennsylvania Green	PBS pH 7.4	456	478	494	506	515	21	62.0	0.89	55.1		2	0	−2	36	Fluorone	4.11
124	5(6)-Carboxyrhodol	50 mM P pH 8-9	451	472	494	503	520	26	59.9	0.89	53.3	5.1	2	0	−2	41	Rhodol	4.15.1
125	6-Piperidinocarboxyrhodol	50 mM P pH 8-9	472	492	516	529	541	25	80.9	0.79	63.9		2	0	−2	48	Rhodol	4.15.1
126	6-Quinolinocarboxyrhodol	50 mM P pH 8-9	492	514	540	554	570	30	66.8	0.43	28.7	4.0	2	0	−2	55	Rhodol	4.15.1
127	7-Hydroxy-4-Me-coumarin-3-CA	H_2O pH 10	332	352	386	413	446	60	33.0	0.85	28.0	4.8	2	0	−2	22	Coumarin 7-OH	4.26.1
128	7-Me-6-EtAm-carboxyrhodol	50 mM P pH 8-9	465	485	508	521	534	26	78.1	0.82	64.0	4.0	2	0	−2	50	Rhodol	4.15.1
129	Alexa Fluor 350	H_2O pH 7	296	317	346	370	445	99	19.0	0.89	16.9	3.5	2	0	−2	95	Coumarin 7-Am	4.26.3
130	Alexa Fluor 430	PBS pH 7.4	373	393	431	456	541	110	16.0	0.55	8.8		2	0	−2	67	Coumarin 7-Am	4.26.3
131	BMeS-*p*-A-2COOH	H_2O	341	362	403	434	535	132	3.7	0.56	2.0		2	0	−2	46	Benzene	4.38
132	Carbofluorescein	100 mM NaOH	505	526	544	557	567	23	108.0	0.62	66.9	4.2	2	0	−2	46	$C(CH_3)_2$-Fluore	4.9
133	Chlorinated fluorescein G-C 1	H_2O pH 9.18, 25 °C	464	484	503	516	522	19	103.0	0.86	88.5		2	0	−2	35	Fluorescein-ha	4.8.2
134	Chlorinated fluorescein G-C 2(3)	H_2O pH 9.18, 25 °C	460	480	499	512	518	19	48.4	0.93	45.0		2	0	−2	35	Fluorescein-ha	4.8.2
135	Chlorinated fluorescein G-C 4	H_2O pH 9.18, 25 °C	471	492	512	526	528	16	93.8	0.87	81.6		2	0	−2	35	Fluorescein-ha	4.8.2
136	Chlorinated fluorescein G-C 5	H_2O pH 9.18, 25 °C	471	491	511	524	527	16	91.6	0.84	76.9		2	0	−2	35	Fluorescein-ha	4.8.2
137	Chlorinated fluorescein G-C 6	H_2O pH 9.18, 25 °C	480	501	522	536	537	15	65.8	0.83	54.6		2	0	−2	35	Fluorescein-ha	4.8.2
138	DMAX-EP	100 mM P pH 7.5	454	474	492	504	515	23	71.0	0.81	57.5		2	0	−2	57	Fluorescein-nh	4.8.1

(Continued)

Ent.	Short names	Aqueous solution	$\lambda_{x,-0.25}$ nm	$\lambda_{x,-0.50}$ nm	$\lambda_{x,max}$ nm	$\lambda_{x,+0.50}$ nm	$\lambda_{e,max}$ nm	Δ_S nm	ε_{max} mм$^{-1}$cm^{-1}	Φ_f	B_{max} mм$^{-1}$cm^{-1}	τ_f ns	N_n	N_p	q	N_a	Class	Ref.
139	DPP 5	PBS			455		515	60	22.7	0.57	12.9	6.5	2	0	−2	44	Diketopyrrolop	4.39.1
140	DPP 6	PBS			530		586	56	30.7	0.49	15.0	4.4	2	0	−2	38	Diketopyrrolop	4.39.1
141	DY-430 carboxylate	PBS pH 7.4	394	410	444	467	487	43	37.0	0.72	26.6		2	0	−2	66	Coumarin 7-Am	4.26.3
142	Fluorescein	100 mм NaOH in H_2O	453	472	490.2	503	516.3	26	67.0	0.92	61.6	4.2	2	0	−2	37	Fluorescein-nh	4.8.1
143	Fluorescein	100 mм NaOH in D_2O	453	472	489.7	502	516.3	26	67.0	0.98	65.6	4.4	2	0	−2	37	Fluorescein-nh	4.8.1
144	Fluorescein (G-C 1)	H_2O pH 9.18, 25 °C	454	474	492	504	514	22	78.4	0.90	70.5		2	0	−2	37	Fluorescein-nh	4.8.1
145	Fluorescein (S-H 1)	P pH 9	453	472	490	502	514	24	90.0	0.92	82.8		2	0	−2	37	Fluorescein-nh	4.8.1
146	Fluorinated fluorescein S-H 17	P pH 9	453	472	490	502	514	24	82.4	0.97	79.9		2	0	−2	37	Fluorescein-ha	4.8.2
147	Fluorinated fluorescein S-H 18	P pH 9	468	489	508	521	527	19	85.6	0.85	72.7		2	0	−2	35	Fluorescein-ha	4.8.2
148	Fluorinated fluorescein S-H 19	P pH 9	468	489	508	521	527	19	81.2	0.96	77.9		2	0	−2	35	Fluorescein-ha	4.8.2
149	Fluorinated fluorescein S-H 20	P pH 9	491	513	535	550	553	18	83.4	0.47	39.2		2	0	−2	37	Fluorescein-ha	4.8.2
150	Isoquinoline 3k	PBS pH 7.4			411		523	112	18.6	0.11	2.0		2	0	−2	59	Isoquinoline	4.37
151	Lucifer yellow CH	H_2O pH 7	375	392	429	462	540	111	11.3	0.21	2.3	5.1	2	0	−2	38	Naphthalimide	4.32
152	Lucifer yellow VS	H_2O pH 7	376	393	430	463	535	105	11.3	0.24	2.7	3.3	2	0	−2	47	Naphthalimide	4.32
153	Naphthofluorescein	C pH 9.5	541	568	595	613	660	65	44.0	0.14	6.1		2	0	−2	48	Fluores-analog	4.8.4
154	NR600	PBS, 1% DMSO pH 7.4	551	580	600	613	619	19	67.6	0.68	45.9		2	0	−2	41	POOH-Fluoron	4.12

Ent.	Short names	Aqueous solution	$\lambda_{x,-0.25}$ nm	$\lambda_{x,-0.50}$ nm	$\lambda_{x,max}$ nm	$\lambda_{x,+0.50}$ nm	$\lambda_{e,max}$ nm	Δ_S nm	ε_{max} mM^{-1}cm^{-1}	Φ_f	B_{max} mM^{-1}cm^{-1}	τ_f ns	N_n	N_p	q	N_a	Class	Ref.
155	NR604	PBS, 1% DMSO pH 7.4	554	584	604	618	627	23	45.7	0.48	21.9		2	0	−2	43	POOH-Fluoron	4.12
156	NX	100 mM P pH 7.5	454	474	492	504	512	20	69.0	0.83	57.2		2	0	−2	41	Fluorescein-nh	4.8.1
157	Pacific orange	C pH 9.5			404		553	149	25.0	0.62	15.5		2	0	−2	29	Benzothiazole	4.30
158	Si-Fluorescein	100 mM NaOH	535	559	579	594	599	20	93.0	0.53	49.2		2	0	−2	45	$Si(CH_3)_2$-Fluore	4.9
159	Sulfofluorescein	C pH 9.5	464	482	497	508	517	20	78.0	0.92	71.7		2	0	−2	36	Sulfofluoresc.	4.10
160	Vita Blue	C pH 9.5	560	586	609	625	665	56	36.0	0.15	5.4		2	0	−2	48	S-Fluore-analog	4.10
161	2,6-Sulfon.-1,3,5,7,8-Me-BODIPY	H_2O	451	469	491	499	519	28	99.0	0.83	82.1	4.2	3	1	−2	42	Bodipy-Small	4.14.1
162	2,6-Sulfon.-1,3,5,7-Me-BODIPY	H_2O	451	469	491	499	510	19	60.0	0.85	51.0		3	1	−2	38	Bodipy-Small	4.14.1
163	6-Azido-4,5,7-trifluorofluorescein	100 mM NaOH	467	488	507	520	527	20	79.9	0.82	65.5		3	1	−2	37	Fluorescein-ha	4.8.2
164	8-Trifluoromethyl-BODIPY 12	100 mM P pH 7.4	439	456	477	485	584	107	41.8	0.70	29.2		3	1	−2	56	Bodipy-Small	4.14.1
165	Abberior Star 512 carboxylate	PBS pH 7.4	472	493	511	524	530	19	84.0	0.82	68.8	4.1	3	1	−2	75	Rhoda 500-550	4.2.2
166	Abberior Star 635P carboxylate	PBS pH 7.4	575	616	635	652	655	20	80.0	0.55	44.0	3.3	3	1	−2	105	Rhodam >600	4.2.4
167	Abberior Star Red carboxylate	PBS pH 7.4	578	619	638	655	655	17	100.0	0.66	66.0	3.6	3	1	−2	119	Rhodam >600	4.2.4
168	Alexa Fluor 532	PBS pH 7.2, 22 °C	486	514	531	545	554	23	81.0	0.61	49.4	2.5	3	1	−2	94	Rosamine	4.17
169	Atto 488 carboxylate	PBS pH 7.4, 22 °C	461	486	500	511	520	20	90.0	0.80	72.0	4.1	3	1	−2	64	Rhodam <500	4.2.1
170	Atto 514 carboxylate	PBS pH 7.4, 22 °C	474	494	511	522	532	21	115.0	0.85	97.7	3.9	3	1	−2	76	Rhoda 500-550	4.2.2
171	Atto 532 carboxylate	PBS pH 7.4, 22 °C	491	515	532	544	552	20	115.0	0.90	103.5	3.8	3	1	−2	75	Rhoda 500-550	4.2.2

(Continued)

Ent.	Short names	Aqueous solution	$\lambda_{x,-0.25}$ nm	$\lambda_{x,-0.50}$ nm	$\lambda_{x,max}$ nm	$\lambda_{x,+0.50}$ nm	$\lambda_{e,max}$ nm	Δ_S nm	ε_{max} mM^{-1}cm^{-1}	Φ_f	B_{max} mM^{-1}cm^{-1}	τ_f ns	N_n	N_p	q	N_a	Class	Ref.
172	Atto 594 carboxylate	PBS pH 7.4, 22 °C	547	583	603	617	626	23	120.0	0.85	102.0	3.9	3	1	−2	104	Rhodam >600	4.2.4
173	Cy3.13-OH	H_2O/PBS	487	508	550	561	566	16	130.0	0.03	3.9		3	1	−2	85	Cyanine-Trimet	4.13.1
174	Cy5.13-OH	H_2O/PBS	588	626	648	663	669	21	215.0	0.12	25.8		3	1	−2	89	Cyanine-Penta	4.13.2
175	Cy5.29.OH	PBS	589	628	650	665	666	16	250.0	0.20	50.0		3	1	−2	83	Cyanine-Penta	4.13.2
176	CypHer5	A pH 4.6	583	621	643	657	662	19	100.0	0.19	19.0	0.70	3	1	−2	71	Cyanine-Penta	4.13.2
177	DAMBO-P^H-T	100 mM P pH 7.4	479	499	521	533	537	16	56.0	0.74	41.4		3	1	−2	61	Bodipy-Small	4.14.1
178	DY-530 carboxylate	PBS pH 7.4	490	513	533	547	554	21	100.0	0.81	80.6		3	1	−2	80	Rhoda 500-550	4.2.2
179	DY-636 carboxylate	PBS pH 7.4	558	591	645	669	671	26	190.0	0.09	16.3		3	1	−2	96	Merocyanine	4.24
180	DY-651 carboxylate	PBS pH 7.4	599	631	653	672	678	25	160.0	0.15	24.1		3	1	−2	102	Merocyanine	4.24
181	DY-654 carboxylate	PBS pH 7.4	594	627	653	673	677	24	220.0	0.16	35.2		3	1	−2	106	Merocyanine	4.24
182	DY-681 carboxylate	PBS pH 7.4	592	630	691	718	708	17	125.0	0.09	11.2		3	1	−2	94	Merocyanine	4.24
183	DY-731 carboxylate	PBS pH 7.4	678	709	736	761	759	23	225.0	0.11	23.6		3	1	−2	98	Merocyanine	4.24
184	DY-747P1	PBS pH 7.4	661	716	747	770	769	22	240.0	0.12	28.5		3	1	−2	90	Cyanine-Hepta	4.13.3
185	ECXi	H_2O			887		938	51	93.3	0.01	1.3		3	1	−2	112	Pentacene	4.39.2
186	Fluorinated rhodamine 1k	Buffer pH 7	467	488	506	518	531	25	54.0	0.66	35.6	3.2	3	1	−2	57	Rhoda 500-550	4.2.2
187	HCDN	PBS pH 7.4	564	603	625	641	755	130	195.0	0.28	54.5		3	1	−2	93	Cyanine-Hepta	4.13.3
188	HCDO	PBS pH 7.4	678	736	769	793	788	19	218.8	0.30	65.6		3	1	−2	106	Cyanine-Hepta	4.13.3
189	HCDS	PBS pH 7.4	695	756	791	817	818	27	218.8	0.39	85.3		3	1	−2	106	Cyanine-Hepta	4.13.3
190	KK1119	H_2O	577	618	637	654	660	23	90.0	0.55	49.5	3.7	3	1	−2	106	Rhodam >600	4.2.4
191	KK114	H_2O	580	619	637	652	660	23	92.0	0.55	50.6	3.6	3	1	−2	102	Rhodam >600	4.2.4

Ent.	Short names	Aqueous solution	$\lambda_{x,-0.25}$ nm	$\lambda_{x,-0.50}$ nm	$\lambda_{x,max}$ nm	$\lambda_{x,+0.50}$ nm	$\lambda_{e,max}$ nm	Δ_S nm	ε_{max} mм$^{-1}$cm^{-1}	Φ_f	B_{max} mм$^{-1}$cm^{-1}	τ_f ns	N_n	N_p	q	N_a	Class	Ref.
192	L-B 1b	H_2O	457	475	498	506	509	11	100.0	0.37	37.0		3	1	−2	49	Bodipy-Small	4.14.1
193	L-B 2b	H_2O	476	496	512	522	524	12	77.6	0.41	31.8		3	1	−2	37	Bodipy-Small	4.14.1
194	L-B 4b	H_2O	457	475	498	506	509	11	77.6	0.15	11.6		3	1	−2	51	Bodipy-Small	4.14.1
195	L-B 5	H_2O	457	475	498	506	511	13	79.4	0.49	38.9		3	1	−2	67	Bodipy-Small	4.14.1
196	M-S 10a	H_2O, 0.1% TX-100	516	539	565	579	579	14	69.0	0.65	44.8		3	1	−2	71	Bodipy-Small	4.14.1
197	M-S 10b	H_2O, 0.1% TX-100	501	523	547	560	574	27	64.0	0.78	49.9		3	1	−2	78	Bodipy-Small	4.14.1
198	M-S 10c	H_2O, 0.1% TX-100	489	510	533	546	557	24	54.0	0.61	32.9		3	1	−2	74	Bodipy-Small	4.14.1
199	M-S 10d	H_2O, 0.1% TX-100	523	547	574	589	582	8	118.0	0.57	67.2		3	1	−2	78	Bodipy-Small	4.14.1
200	M-S 10e	H_2O, 0.1% TX-100	532	556	584	599	598	14	124.0	0.41	50.8		3	1	−2	82	Bodipy-Small	4.14.1
201	M-S 10f	H_2O, 0.1% TX-100	536	561	589	604	603	14	74.0	0.18	13.3		3	1	−2	72	Bodipy-Small	4.14.1
202	nor-SqB1	P pH 7.4	589	619	638	654	666	28	88.0	0.20	17.6	1.5	3	1	−2	76	Squaraine	4.35
203	nor-SqCN1	P pH 7.4	616	650	670	687	693	23	123.0	0.19	23.3	1.8	3	1	−2	63	Squaraine	4.35
204	nor-SqO1	P pH 7.4	588	618	636	652	653	17	149.0	0.32	47.6	1.6	3	1	−2	59	Squaraine	4.35
205	RS800	FBS	690	750	784	809	800	16	240.0	0.17	40.5		3	1	−2	111	Cyanine-Hepta	4.13.3
206	Squaraine T-P 4a	67 mм P pH 7.4	581	610	628	644	642	14	163.0	0.03	4.8	0.1	3	1	−2	74	Squaraine	4.35
207	Squaraine T-P 4b	67 mм P pH 7.4	607	639	659	676	676	17	207.0	0.02	4.1	0.2	3	1	−2	78	Squaraine	4.35
208	Squaraine T-P 4c	67 mм P pH 7.4	577	606	624	639	635	11	245.0	0.02	4.9	0.1	3	1	−2	74	Squaraine	4.35

(Continued)

Ent.	Short names	Aqueous solution	$\lambda_{x,-0.25}$ nm	$\lambda_{x,-0.50}$ nm	$\lambda_{x,max}$ nm	$\lambda_{x,+0.50}$ nm	$\lambda_{e,max}$ nm	Δ_S nm	ε_{max} mм$^{-1}$cm^{-1}	Φ_f	B_{max} mм$^{-1}$cm^{-1}	τ_f ns	N_n	N_p	q	N_a	Class	Ref.
209	Squaraine T-P 4d	67 mм P pH 7.4	606	638	658	675	677	19	182.0	0.03	5.4	0.2	3	1	−2	77	Squaraine	4.35
210	SR101-NaphtOH	PBS	481	509	580	607	619	39	80.0	0.15	12.0		3	1	−2	62	Sulforhodol-an	4.27
211	SR101-OH	H_2O	493	516	548	564	572	24	80.0	0.35	28.0		3	1	−2	56	Sulforhodol	4.27
212	Sulfo-cyanine3	H_2O	486	506	548	559	563	15	162.0	0.10	16.2		3	1	−2	79	Cyanine-Trimet	4.13.1
213	W-L 8a	PBS	583	621	643	657	667	24	87.0	0.12	10.4		3	1	−2	158	Cyanine-Penta	4.13.2
214	W-L 8b	PBS	584	622	644	659	668	24	189.0	0.15	28.3		3	1	−2	170	Cyanine-Penta	4.13.2
215	W-L 8c	PBS	583	621	643	657	666	23	107.0	0.12	12.8		3	1	−2	172	Cyanine-Penta	4.13.2
216	W-L 8e	PBS	583	621	643	657	665	22	195.0	0.09	17.5		3	1	−2	tl	Cyanine-Penta	4.13.2
217	Y-A 2	PBS pH 7.2	685	744	778	803	806	28	190.0	0.03	5.7		3	1	−2	98	Cyanine-Hepta	4.13.3
218	Zy5	P pH 7.4	587	625	647	662	665	18	250.0	0.27	67.5	1.0	3	1	−2	83	Cyanine-Penta	4.13.2
219	Zy7 carboxylate	P pH 7.4	663	718	749	772	774	25	200.0	0.13	26.0	0.6	3	1	−2	87	Cyanine-Hepta	4.13.3
220	(HO)-Dehydroluciferin	H_2O pH 4.5			348		555	207	18.2	0.26	4.7		1	0	−1	24	Benzothiazole	4.30
221	(HO)-Luciferin	H_2O pH 5			328		537	209	18.2	0.25	4.5		1	0	−1	26	Benzothiazole	4.30
222	2,4-DiMe TokyoGreen	100 mм NaOH	453	475	491	503	510	19	85.0	0.87	73.5		1	0	−1	40	Fluorone	4.11
223	2,4-DiOMe TokyoGreen	100 mм NaOH	456	478	494	506	513	19	85.0	0.20	17.0		1	0	−1	41	Fluorone	4.11
224	2,5-DiMe TokyoGreen	100 mм NaOH	453	475	491	503	510	19	85.0	0.89	75.4		1	0	−1	40	Fluorone	4.11
225	2,5-DiOMe TokyoGreen	100 mм NaOH	456	478	494	506	512	18	85.0	0.01	0.8		1	0	−1	41	Fluorone	4.11
226	2,7-DiCl-6-methylamino rhodol	50 mм P pH 8-9	469	488	512	525	533	21	71.0	0.67	47.5		1	0	−1	41	Rhodol	4.15.1
227	2,7-DiCl-6-pyrrolidino sulforhodol	PBS pH 7.4	470	491	520	534	535	15	83.0	0.27	22.4		1	0	−1	84	Sulforhodol	4.27
228	2-Me DCTM	100 mм P pH 9	544	577	597	611	607	10	130.0	0.48	62.4		1	0	−1	46	$Si(CH_3)_2$-Fluoron	4.12

Ent.	Short names	Aqueous solution	$\lambda_{x,-0.25}$ nm	$\lambda_{x,-0.50}$ nm	$\lambda_{x,max}$ nm	$\lambda_{x,+0.50}$ nm	$\lambda_{e,max}$ nm	Δ_S nm	ε_{max} mM^{-1}cm^{-1}	Φ_f	B_{max} mM^{-1}cm^{-1}	τ_f ns	N_n	N_p	q	N_a	Class	Ref.
229	2-Me Tokyo Magenta	100 mM P pH 9	532	563	582	595	598	16	110.0	0.42	46.2		1	0	−1	44	$Si(CH_3)_2$-Fluoron	4.12
230	2-Me TokyoGreen	H_2O pH 9	453	475	491	503	510	19	85.0	0.85	72.2		1	0	−1	36	Fluorone	4.11
231	2-Me-4-OMe TokyoGreen	100 mM NaOH	454	476	492	504	509	17	85.0	0.84	71.4		1	0	−1	41	Fluorone	4.11
232	2-OMe TokyoGreen	100 mM NaOH	456	478	494	506	515	21	85.0	0.86	73.1		1	0	−1	38	Fluorone	4.11
233	2-OMe-5-Me TokyoGreen	100 mM NaOH	456	478	494	506	514	20	85.0	0.50	42.5		1	0	−1	41	Fluorone	4.11
234	3-(4-OH-Ph)-6-MeO-chromone	100 mM Tris pH 8			311		468	157	12.6	0.05	0.6	1.7	1	0	−1	32	Chromone	4.31
235	3-AB-6-MeO-7-OH-chromone	100 mM Tris pH 8			346		445	99	16.3	0.41	6.6	4.3	1	0	−1	30	Chromone	4.31
236	3-Bt-6-MeO-7-hydroxychromone	100 mM Tris pH 8			345		447	102	16.4	0.48	7.8	4.5	1	0	−1	34	Chromone	4.31
237	3-Buthyl-7-hydroxychromone	100 mM Tris pH 8			334		454	120	12.7	0.21	2.6	2.1	1	0	−1	30	Chromone	4.31
238	3-Methoxymethyl ether rhodol	50 mM P, <0.1% ACN pH 8	439	457	477	488	516	39	58.0	0.85	49.3		1	0	−1	45	Rhodol	4.15.2
239	3-MOMO-6-propylaminorhodol	50 mM P, <0.1% ACN pH 8	453	472	494	506	532	38	43.0	0.35	15.0		1	0	−1	55	Rhodol	4.15.2
240	4-Methylsalicylic acid	H_2O pH 7	252	272	300	317	407	107	3.0	0.11	0.3	3.0	1	0	−1	19	Benzene	4.38
241	5-Methoxysalicylic acid	H_2O pH 7	285	295	319	339	447	128	3.0	0.19	0.5	6.9	1	0	−1	20	Benzene	4.38
242	5-Methylsalicylic acid	H_2O pH 7	268	281	305	323	418	113	3.0	0.16	0.4	5.3	1	0	−1	19	Benzene	4.38

(Continued)

Ent.	Short names	Aqueous solution	$\lambda_{x,-0.25}$ nm	$\lambda_{x,-0.50}$ nm	$\lambda_{x,max}$ nm	$\lambda_{x,+0.50}$ nm	$\lambda_{e,max}$ nm	Δ_S nm	ε_{max} mм$^{-1}$cm^{-1}	Φ_f	B_{max} mм$^{-1}$cm^{-1}	τ_f ns	N_n	N_p	q	N_a	Class	Ref.
243	6-Acrylamidorhodol	50 mм P, <0.1% ACN pH 8	450	469	490	502	524	34	28.5	0.19	5.4		1	0	−1	43	Rhodol	4.15.2
244	6-Allylaminorhodol	50 mм P, <0.1% ACN pH 8	464	484	507	519	534	27	61.5	0.92	56.5		1	0	−1	44	Rhodol	4.15.2
245	6-Aminorhodol (Rhodol)	50 mм P, <0.1% ACN pH 8	452	471	493	505	516	23	70.0	0.98	68.6		1	0	−1	37	Rhodol	4.15.2
246	6-Azetidinyl naphthalimide	10 mм HEPES pH 7.3	401	421	464	503	553	89	18.0	0.28	5.0		1	0	−1	31	Naphthalimide	4.32
247	6-Azetidinylaminorhodol	10 mм HEPES pH 7.3	474	495	519	532	546	27	59.0	0.85	50.1		1	0	−1	46	Rhodol	4.15.2
248	6-Benzamidorhodol	50 mм P, <0.1% ACN pH 8	450	469	490	502	525	35	30.1	0.16	4.8		1	0	−1	49	Rhodol	4.15.2
249	6-Benzylaminorhodol	50 mм P, <0.1% ACN pH 8	461	480	503	515	529	26	71.0	0.95	67.4		1	0	−1	50	Rhodol	4.15.2
250	6-Butylaminorhodol	50 mм P, <0.1% ACN pH 8	465	485	508	521	533	25	60.0	0.93	55.8		1	0	−1	49	Rhodol	4.15.2
251	6-Dibutylaminorhodol	50 mм P, <0.1% ACN pH 8	479	500	525	538	553	28	63.0	0.08	5.0		1	0	−1	61	Rhodol	4.15.2

Ent.	Short names	Aqueous solution	$\lambda_{x,-0.25}$ nm	$\lambda_{x,-0.50}$ nm	$\lambda_{x,max}$ nm	$\lambda_{x,+0.50}$ nm	$\lambda_{e,max}$ nm	Δ_S nm	ε_{max} mM^{-1}cm^{-1}	Φ_f	B_{max} mM^{-1}cm^{-1}	τ_f ns	N_n	N_p	q	N_a	Class	Ref.
252	6-Diethylaminorhodol	50 mM P, <0.1% ACN pH 8	477	498	522	535	549	27	61.0	0.15	9.1		1	0	−1	49	Rhodol	4.15.2
253	6-Dimethylaminorhodol	50 mM P, <0.1% ACN pH 8	473	494	518	531	544	26	64.0	0.20	12.8		1	0	−1	43	Rhodol	4.15.2
254	6-Dimethylaminorhodol	C pH 9.5	477	498	522	535	545	23	57.0	0.20	11.4	2.5	1	0	−1	52	Rhodol	4.15.1
255	6-Dimethylaminorhodol	10 mM HEPES pH 7.3	473	494	518	531	546	28	60.0	0.21	12.6		1	0	−1	44	Rhodol	4.15.2
256	6-Dipropylaminorhodol	50 mM P, <0.1% ACN pH 8	479	500	525	538	552	27	59.0	0.10	5.9		1	0	−1	55	Rhodol	4.15.2
257	6-Morpholinorhodol	50 mM P, <0.1% ACN pH 8	477	498	522	535	553	31	61.0	0.13	7.9		1	0	−1	48	Rhodol	4.15.2
258	6-Piperidinorhodol	50 mM P, <0.1% ACN pH 8	478	499	523	536	555	32	62.5	0.10	6.2		1	0	−1	50	Rhodol	4.15.2
259	6-Propylaminorhodol	50 mM P, <0.1% ACN pH 8	464	484	507	519	531	24	61.0	0.94	57.3		1	0	−1	46	Rhodol	4.15.2
260	6-Pyrrolidinorhodol	50 mM P, <0.1% ACN pH 8	474	495	519	532	550	31	63.0	0.12	7.5		1	0	−1	47	Rhodol	4.15.2
261	6-Quinoline-2-pentanoic-pyrodol	50 mM P pH 8-9	492	515	538	550	562	24	64.1	0.70	44.8		1	0	−1	53	Pyrodol	4.21.1
262	6-Quinoline-2-propanoic-pyrodol	50 mM P pH 8-9	492	515	538	550	562	24	65.9	0.70	46.1		1	0	−1	47	Pyrodol	4.21.1

(Continued)

Ent.	Short names	Aqueous solution	$\lambda_{x,-0.25}$ nm	$\lambda_{x,-0.50}$ nm	$\lambda_{x,max}$ nm	$\lambda_{x,+0.50}$ nm	$\lambda_{e,max}$ nm	Δ_S nm	ε_{max} mM^{-1}cm^{-1}	Φ_f	B_{max} mM^{-1}cm^{-1}	τ_f ns	N_n	N_p	q	N_a	Class	Ref.
263	7-Azetidinyl-3-CA coumarin	10 mM HEPES pH 7.3	333	353	387	414	470	83	24.0	0.84	20.1		1	0	−1	29	Coumarin 7-Am	4.26.2
264	7-Hydroxy-4-Me-coumarin-3-CA	H_2O pH 5	313	330	360	384	455	95	18.0	0.82	14.7	6.4	1	0	−1	26	Coumarin 7-OH	4.26.1
265	7-Hydroxy-4-methylcoumarin	H_2O pH 10	313	330	360	384	449	89	18.0	0.86	15.4	6.7	1	0	−1	20	Coumarin 7-OH	4.26.1
266	Atto 390 carboxylate	PBS pH 7.4, 22 °C	343	357	390	415	476	86	24.0	0.90	21.6	5.0	1	0	−1	50	Coumarin 7-Am	4.26.3
267	Atto 425 carboxylate	PBS pH 7.4, 22 °C	395	409	439	460	485	46	45.0	0.90	40.5	3.6	1	0	−1	56	Coumarin 7-Am	4.26.3
268	BC-OH	100 mM P, 20% Et pH 7.4			442		488	46	38.1	0.85	32.3	3.32	1	0	−1	29	Hybrid-Bonded	4.40.2
269	BMeS-*p*-A-COOH	H_2O	333	353	392	421	527	135	3.9	0.60	2.3		1	0	−1	37	Benzene	4.38
270	Briliant sulfaflavine	H_2O	376	393	430	463	520	90	16.0	0.27	4.3		1	0	−1	40	Naphthalimide	4.32
271	Coumarin P-R 12	PBS pH 7.5	282	301	327	348	457	130	7.9	0.76	5.9		1	0	−1	22	Coumarin 7-OH	4.26.1
272	Coumarin P-R 13	PBS pH 7.5	319	337	368	393	474	106	15.0	0.63	9.4		1	0	−1	22	Coumarin 7-OH	4.26.1
273	Coumarin P-R 14	PBS pH 7.5	322	341	373	398	461	88	17.7	0.62	10.9		1	0	−1	22	Coumarin 7-OH	4.26.1
274	Coumarin P-R 3	PBS pH 7.5	297	319	348	372	450	102	12.5	0.82	10.2		1	0	−1	21	Coumarin 7-Am	4.26.2
275	Flavin mononucleotide (FMN)	H_2O			450		530	80	12.2	0.25	3.0		1	0	−1	51	Pteridine	4.36
276	Fluorene PP11	H_2O			377		621	244	36.0	0.05	1.8		1	0	−1	40	Fluorene	4.39.2
277	Isoquinoline 10b	PBS pH 7.4			410		526	116	15.5	0.17	2.6		1	0	−1	46	Isoquinoline	4.37
278	Isoquinoline 3d	PBS pH 7.4			418		520	102	14.5	0.43	6.2		1	0	−1	48	Isoquinoline	4.37
279	Isoquinoline 3e	PBS pH 7.4			408		524	116	15.8	0.41	6.4		1	0	−1	51	Isoquinoline	4.37

Ent.	Short names	Aqueous solution	$\lambda_{x,-0.25}$ nm	$\lambda_{x,-0.50}$ nm	$\lambda_{x,max}$ nm	$\lambda_{x,+0.50}$ nm	$\lambda_{e,max}$ nm	Δ_S nm	ε_{max} mM^{-1}cm^{-1}	Φ_f	B_{max} mM^{-1}cm^{-1}	τ_f ns	N_n	N_p	q	N_a	Class	Ref.
280	Isoquinoline 3f	PBS pH 7.4			405		519	114	15.4	0.39	6.0		1	0	−1	57	Isoquinoline	4.37
281	Isoquinoline 3g	PBS pH 7.4			414		522	108	17.5	0.41	7.1		1	0	−1	58	Isoquinoline	4.37
282	Isoquinoline 3h	PBS pH 7.4			406		525	119	13.7	0.38	5.2		1	0	−1	61	Isoquinoline	4.37
283	Isoquinoline 3i	PBS pH 7.4			405		521	116	15.4	0.40	6.1		1	0	−1	64	Isoquinoline	4.37
284	Isoquinoline 3j	PBS pH 7.4			409		525	116	18.8	0.45	8.4		1	0	−1	58	Isoquinoline	4.37
285	Isoquinoline 3l	PBS pH 7.4			410		522	112	16.8	0.42	7.0		1	0	−1	61	Isoquinoline	4.37
286	Isoquinoline 3m	PBS pH 7.4			408		526	118	18.1	0.36	6.5		1	0	−1	58	Isoquinoline	4.37
287	Isoquinoline 3n	PBS pH 7.4			411		521	110	16.6	0.41	6.8		1	0	−1	62	Isoquinoline	4.37
288	Isoquinoline 3o	PBS pH 7.4			410		521	111	19.2	0.09	1.7		1	0	−1	66	Isoquinoline	4.37
289	Isoquinoline 4b	PBS pH 7.4			436		550	114	15.5	0.36	5.5		1	0	−1	49	Isoquinoline	4.37
290	Isoquinoline 4c	PBS pH 7.4			453		529	76	22.1	0.24	5.3		1	0	−1	54	Isoquinoline	4.37
291	Isoquinoline 7a	PBS pH 7.4			407		528	121	14.1	0.34	4.7		1	0	−1	55	Isoquinoline	4.37
292	Isoquinoline 7b	PBS pH 7.4			406		526	120	14.4	0.46	6.6		1	0	−1	62	Isoquinoline	4.37
293	Isoquinoline 7c	PBS pH 7.4			408		524	116	15.8	0.26	4.1		1	0	−1	56	Isoquinoline	4.37
294	Isoquinoline 7d	PBS pH 7.4			408		526	118	15.0	0.14	2.1		1	0	−1	62	Isoquinoline	4.37
295	Isoquinoline 7f	PBS pH 7.4			406		528	122	16.3	0.16	2.6		1	0	−1	70	Isoquinoline	4.37
296	Isoquinoline 8	PBS pH 7.4			406		534	128	16.8	0.37	6.2		1	0	−1	50	Isoquinoline	4.37

(Continued)

Ent.	Short names	Aqueous solution	$\lambda_{x,-0.25}$ nm	$\lambda_{x,-0.50}$ nm	$\lambda_{x,max}$ nm	$\lambda_{x,+0.50}$ nm	$\lambda_{e,max}$ nm	Δ_S nm	ε_{max} mM^{-1}cm^{-1}	Φ_f	B_{max} mM^{-1}cm^{-1}	τ_f ns	N_n	N_p	q	N_a	Class	Ref.
297	Marina Blue	100 mM P pH 10/8	311	328	358	381	455	97	18.7	0.89	16.6		1	0	−1	29	Coumarin 7-OH	4.26.1
298	Mercaptofluorescein MK-43	1 mM NaOH	457	477	495	508	526	31	34.0	0.22	7.4		1	0	−1	44	Fluorescein-SH	4.8.3
299	Mercaptofluorescein MK-67	1 mM NaOH	463	483	502	515	527	25	27.0	0.17	4.5		1	0	−1	43	Fluorescein-SH	4.8.3
300	Mercaptofluorescein MK-69	1 mM NaOH	457	477	495	508	526	31	25.0	0.20	5.0		1	0	−1	44	Fluorescein-SH	4.8.3
301	NR632	10 mM PBS pH 7.4	578	608	632	646	655	23	64.0	0.26	16.6		1	0	−1	49	POPh-Rosol	4.20
302	NR638	10 mM PBS pH 7.4	583	613	638	652	662	24	55.6	0.21	11.6		1	0	−1	51	POPh-Rosol	4.20
303	Pacific Blue	PBS pH 7.2, 22 °C	339	366	410	437	455	45	29.5	0.75	22.1	6.9	1	0	−1	31	Coumarin 7-OH	4.26.1
304	Pennsylvania Green	PBS pH 7.4	456	478	494	506	514	20	82.0	0.91	74.6		1	0	−1	36	Fluorone	4.11
305	POF	PBS pH 7.4	561	591	632	652	661	29	54.5	0.24	13.0		1	0	−1	53	POPh-Fluorone	4.12
306	PONy C7	PBS pH 7.4			419		613	194	22.0	0.04	0.8		1	0	−1	47	Hybrid-Bonded	4.40.2
307	Pyrene-1-sulfonate	H_2O	327	340	375	382	400	25	30.6	0.56	17.1	64.0	1	0	−1	30	Pyrene	4.28
308	Pyrodone	100 mM P pH 7.4	436	454	472	481	500	28	60.0	0.37	22.2	3.6	1	0	−1	23	Pyrodone	4.21.2
309	Redmond Red	100 mM P pH 7.1	518	547	579	593	595	16	52.3	0.84	43.9		1	0	−1	45	Phenoxazine	4.25
310	Salicylic acid	H_2O pH 7	272	281	300	316	407	107	3.0	0.16	0.4	4.3	1	0	−1	16	Benzene	4.38
311	Stefani HB-11	H_2O pH 6			329		360	31	38.1	0.11	4.0		1	0	−1	31	Hybrid-Bonded	4.40.2
312	Stefani HB-6	H_2O pH 6			317		365	48	35.9	0.03	0.9		1	0	−1	28	Hybrid-Bonded	4.40.2

Ent.	Short names	Aqueous solution	$\lambda_{x,-0.25}$ nm	$\lambda_{x,-0.50}$ nm	$\lambda_{x,max}$ nm	$\lambda_{x,+0.50}$ nm	$\lambda_{e,max}$ nm	Δ_S nm	ε_{max} mм$^{-1}$cm^{-1}	Φ_f	B_{max} mм$^{-1}$cm^{-1}	τ_f ns	N_n	N_p	q	N_a	Class	Ref.
313	Stefani HB-7	H_2O			331		449	118	42.0	0.09	3.6		1	0	−1	30	Benzothiazole	4.30
314	Stefani HB-9	H_2O pH 6			320		423	103	35.9	0.12	4.2		1	0	−1	32	Hybrid-Bonded	4.40.2
315	2-OH-6-DiEtAm-benzo[c]-xanthen	H_2O	523	547	578	595	620	42	38.1	0.18	6.8		2	1	−1	55	Rhodol-analog	4.15.3
316	5(6)-Carboxy-X-rhodamine (ROX)	H_2O pH <7	525	557	576	591	601	25	82.0	0.70	57.4		2	1	−1	69	Rhoda 550-600	4.2.3
317	530RH	PBS pH 7.4	491	514	532	547	553	21	56.0	0.89	49.8	4.0	2	1	−1	67	Rhoda 500-550	4.2.2
318	6-Sulfonato-1,3,5,7-Me-BODIPY	H_2O	454	473	495	503	515	20	52.5	0.85	44.6	14.1	2	1	−1	36	Bodipy-Small	4.14.1
319	8-Trifluoromethyl-BODIPY 10	100 mм P pH 7.4	432	449	469	476	542	73	17.1	0.74	12.6		2	1	−1	43	Bodipy-Small	4.14.1
320	8-Trifluoromethyl-BODIPY 11	100 mм P pH 7.4	519	543	569	583	584	15	57.0	0.95	54.1		2	1	−1	40	Bodipy-Small	4.14.1
321	Abberior Star 440SXP	PBS pH 7.4	377	398	436	461	515	79	22.7	0.68	15.4	3.3	2	1	−1	70	Coumarin 7-Am	4.26.3
322	Abberior Star 470SXP	PBS pH 7.4	416	434	472	499	624	152	29.0	0.12	3.4	0.8	2	1	−1	77	Coumarin 7-Am	4.26.3
323	Abberior Star 600 carboxylate	PBS pH 7.4	550	587	604	619	623	19	93.0	0.80	74.4	4.3	2	1	−1	73	Rhodam >600	4.2.4
324	Abberior Star 635 carboxylate	PBS pH 7.4	574	615	634	650	654	20	63.0	0.51	32.1	2.8	2	1	−1	111	Rhodam >600	4.2.4
325	$ACSO_3H$	H_2O pH 5.5	353	366	396	404	417	21	4.4	0.82	3.6	16.5	2	1	−1	32	Acridones	4.23.2
326	Atto 565 carboxylate	PBS pH 7.4, 22 °C	513	543	564	578	590	26	120.0	0.90	108.0	4.0	2	1	−1	69	Rhoda 550-600	4.2.3
327	Atto 590 carboxylate	PBS pH 7.4, 22 °C	535	569	593	609	622	29	120.0	0.80	96.0	3.7	2	1	−1	83	Rhoda 550-600	4.2.3
328	Atto 647 carboxylate	PBS pH 7.4, 22 °C	609	624	647	659	667	20	120.0	0.20	24.0	2.4	2	1	−1	85	$C(CH_3)_2$-Pyron	4.5
329	Atto 655 carboxylate	PBS pH 7.4, 22 °C	587	629	663	679	680	17	125.0	0.30	37.5	1.8	2	1	−1	70	Phenoxazine	4.25

(Continued)

Ent.	Short names	Aqueous solution	$\lambda_{x,-0.25}$ nm	$\lambda_{x,-0.50}$ nm	$\lambda_{x,max}$ nm	$\lambda_{x,+0.50}$ nm	$\lambda_{e,max}$ nm	Δ_S nm	ε_{max} $mM^{-1}cm^{-1}$	Φ_f	B_{max} $mM^{-1}cm^{-1}$	τ_f ns	N_n	N_p	q	N_a	Class	Ref.
330	Atto 680 carboxylate	PBS pH 7.4, 22 °C	603	648	681	696	698	17	125.0	0.30	37.5	1.7	2	1	−1	67	Phenoxazine	4.25
331	Atto 700 carboxylate	PBS pH 7.4, 22 °C	622	666	700	717	716	16	120.0	0.25	30.0	1.6	2	1	−1	67	Phenoxazine	4.25
332	BODIPY 630/650	H_2O	569	597	629	647	646	17	97.0	0.50	48.5		2	1	−1	50	Bodipy-Small	4.14.1
333	BODIPY FL	H_2O	460	479	502	511	511	9	92.0	0.97	89.2		2	1	−1	34	Bodipy-Small	4.14.1
334	Carborhodamine 21b	H_2O	591	619	640	656	664	24	68.0	0.56	38.0		2	1	−1	107	$C(CH_3)_2$-Rhoda	4.3.1
335	Carborhodamine 22	H_2O	591	619	640	656	664	24	67.0	0.54	36.1		2	1	−1	117	$C(CH_3)_2$-Rhoda	4.3.1
336	Cascade yellow	H_2O	329	356	404	440	549	145	24.0	0.05	1.2	5.6	2	1	−1	61	Hybrid-Bonded	4.40.2
337	Cy3	H_2O	487	508	550	561	570	20	150.0	0.06	8.2	0.2	2	1	−1	68	Cyanine-Trimet	4.13.1
338	Cy3′	H_2O	483	504	545	556	560	15	120.0	0.04	4.8	0.5	2	1	−1	80	Cyanine-Trimet	4.13.1
339	Cy3.10-OH	H_2O/PBS	489	510	552	563	570	18	130.0	0.03	3.9		2	1	−1	72	Cyanine-Trimet	4.13.1
340	Cy3B	H_2O	493	515	558	569	572	14	130.0	0.67	87.1	2.8	2	1	−1	71	Cyanine-Trimet	4.13.1
341	Cy5	H_2O	587	625	647	662	663	16	250.0	0.27	67.5	1.2	2	1	−1	72	Cyanine-Penta	4.13.2
342	Cy5′	H_2O	583	621	642	656	660	18	138.0	0.17	23.4	0.6	2	1	−1	84	Cyanine-Penta	4.13.2
343	Cy5.10-OH	H_2O/PBS	588	626	648	663	667	19	215.0	0.09	19.3		2	1	−1	76	Cyanine-Penta	4.13.2
344	Cy5.11-OH	H_2O/PBS	586	624	646	661	667	21	215.0	0.06	12.9		2	1	−1	80	Cyanine-Penta	4.13.2
345	DAOTA 3	H_2O	449	511	565	583	594	29	50.0	0.28	14.0	12.5	2	1	−1	57	Triangulenium	4.22
346	DEAC-SR101	PBS			610		642	32	80.0	0.10	8.0		2	1	−1	71	Hybrid-Fused	4.40.1
347	DMAC-SR101	PBS			605		647	42	80.0	0.09	7.2		2	1	−1	65	Hybrid-Fused	4.40.1
348	DY-485XL carboxylate	H_2O	424	442	482	510	560	78	48.0		0.0		2	1	−1	69	Coumarin 7-Am	4.26.3
349	DY-505 carboxylate	PBS pH 7.4	459	478	505	515	530	25	85.0	0.74	63.2		2	1	−1	62	Rhoda 500-550	4.2.2
350	DY-510XL carboxylate	H_2O	432	451	493	522	585	92	40.0		0.0		2	1	−1	76	Coumarin 7-Am	4.26.3
351	DY-555 carboxylate	PBS pH 7.4	508	537	555	569	580	25	100.0	0.32	31.7		2	1	−1	86	Rhoda 550-600	4.2.3
352	DY-635 carboxylate	PBS pH 7.4	550	583	635	658	669	34	175.0	0.05	8.7	0.5	2	1	−1	93	Merocyanine	4.24
353	DY-780 carboxylate	PBS pH 7.4	657	705	782	817	800	18	102.0	0.03	3.2		2	1	−1	95	Merocyanine	4.24

Ent.	Short names	Aqueous solution	$\lambda_{x,-0.25}$ nm	$\lambda_{x,-0.50}$ nm	$\lambda_{x,max}$ nm	$\lambda_{x,+0.50}$ nm	$\lambda_{e,max}$ nm	Δ_S nm	ε_{max} $mM^{-1}cm^{-1}$	Φ_f	B_{max} $mM^{-1}cm^{-1}$	τ_f ns	N_n	N_p	q	N_a	Class	Ref.
354	EVOblue 30	H_2O	575	610	650	667	667	17	100.0	0.10	10.0	0.6	2	1	−1	55	Phenoxazine	4.25
355	Fluorinated rhodam. 1f-5′-COOH	Buffer pH 7	463	484	501	513	525	24	85.0	0.95	80.7	4.1	2	1	−1	53	Rhoda 500-550	4.2.2
356	Fluorinated rhodam. 1f-6′-COOH	Buffer pH 7	463	484	501	513	523	22	88.0	0.93	81.8		2	1	−1	53	Rhoda 500-550	4.2.2
357	Fluorinated rhodamine 1j	Buffer pH 7	458	482	496	507	522	26	53.0	0.89	47.1	3.8	2	1	−1	53	Rhodam <500	4.2.1
358	Fluorinated rhodamine 1r	Buffer pH 7	483	506	525	538	554	29	31.0	0.28	8.6	1.6	2	1	−1	58	Rhoda 500-550	4.2.2
359	Fluorinated rhodamine 5-SO_3H	Buffer pH 7	472	494	512	525	530	18	52.0	0.86	44.7	4.1	2	1	−1	66	Rhoda 500-550	4.2.2
360	HC6H-SR101	PBS, 5% BSA			624		654	30	80.0	0.15	12.0		2	1	−1	76	Hybrid-Fused	4.40.1
361	ICG	FBS	708	771	807	834	822	15	121.0	0.09	11.2		2	1	−1	100	Cyanine-Hepta	4.13.3
362	ICG	PBS	687	746	780	805	810	30	211.2	0.08	16.9		2	1	−1	100	Cyanine-Hepta	4.13.3
363	IR-783	PBS	688	748	782	807	810	28	261.0	0.08	21.9		2	1	−1	95	Cyanine-Hepta	4.13.3
364	JF669-6-SCH2CO2H	10 mM HEPES pH 7.3	606	584	667	685	681	14	130.0	0.38	49.4		2	1	−1	69	$Si(CH_3)_2$-Rhoda	4.3.2
365	L-B 1a	H_2O	453	472	494	502	507	13	100.0	0.47	47.0		2	1	−1	46	Bodipy-Small	4.14.1
366	L-B 2a	H_2O	473	493	509	519	523	14	72.4	0.27	19.5		2	1	−1	34	Bodipy-Small	4.14.1
367	L-B 4a	H_2O	453	472	494	502	507	13	69.2	0.34	23.5		2	1	−1	48	Bodipy-Small	4.14.1
368	Mercaptofluorescein MK-61	1 mM NaOH	457	477	495	508	527	32	38.0	0.24	9.1		2	1	−1	46	Fluorescein-SH	4.8.3
369	Mercaptofluorescein MK-75	1 mM NaOH	463	483	502	515	533	31	12.0	0.16	1.9		2	1	−1	44	Fluorescein-SH	4.8.3
370	*meso*-CaboxyBODIPY J-K 1	10 mM PBS pH 7.4	463	482	505	514	554	49	62.0	0.78	48.3		2	1	−1	29	Bodipy-Small	4.14.1
371	MetACSO3H	H_2O pH 5.5	364	376	403	411	433	30	5.6	0.76	4.2	16.7	2	1	−1	35	Acridones	4.23.2
372	Neo-Cy3	PBS	488	509	551	562	573	22	98.0	0.02	1.9		2	1	−1	75	Cyanine-Trimet	4.13.1

(Continued)

Ent.	Short names	Aqueous solution	$\lambda_{x,-0.25}$ nm	$\lambda_{x,-0.50}$ nm	$\lambda_{x,max}$ nm	$\lambda_{x,+0.50}$ nm	$\lambda_{e,max}$ nm	Δ_S nm	ε_{max} $mM^{-1}cm^{-1}$	Φ_f	B_{max} $mM^{-1}cm^{-1}$	τ_f ns	N_n	N_p	q	N_a	Class	Ref.
373	Neo-Cy5	PBS pH 7.4	583	621	643	657	660	17	170.0	0.10	17.0		2	1	−1	79	Cyanine-Penta	4.13.2
374	PetACSO3H	H_2O pH 5.5	367	379	406	414	431	25	6.2	0.92	5.6	15.6	2	1	−1	47	Acridones	4.23.2
375	Phosphorilated coum. N-H 21c	PBS pH 7.4	374	394	432	457	512	80	20.4	0.81	16.5	3.2	2	1	−1	58	Coumarin 7-Am	4.26.3
376	Phosphorilated coum. N-H 22c	PBS pH 7.4	348	370	410	437	503	93	14.0	0.44	6.1	2.7	2	1	−1	62	Coumarin 7-Am	4.26.3
377	Phosphorilated oxazine 2-H	H_2O pH ~7	583	620	661	679	678	17	60.0	0.28	16.8	1.8	2	1	−1	66	Phenoxazine	4.25
378	PONy A1	PBS pH 7.4	476	499	535	553	595	60	38.0	0.10	3.8	1.4	2	1	−1	54	Acridine	4.23.1
379	SiR-COOH	H_2O	588	567	645	662	661	16	100.0	0.39	39.0	2.6	2	1	−1	62	$Si(CH_3)_2$-Rhoda	4.3.2
380	SiRF 15a	H_2O	598	577	657	675	676	19	40.0	0.40	16.0		2	1	−1	65	$Si(CH_3)_2$-Rhoda	4.3.2
381	SNARF-1	P pH 10	519	543	573	590	631	58	44.0	0.09	4.0		2	1	−1	58	Rhodol-analog	4.15.3
382	SR101-110	PBS	485	507	554	570	576	22	80.0	0.55	44.0		2	1	−1	57	Sulforhodam.	4.6
383	SR101-6G	PBS	508	540	565	579	586	21	80.0	0.68	54.4		2	1	−1	67	Sulforhodam.	4.6
384	SR101-B	PBS	522	557	575	590	595	20	80.0	0.65	52.0		2	1	−1	69	Sulforhodam.	4.6
385	SR101-Me_2	PBS	517	553	573	587	593	20	80.0	0.66	52.8		2	1	−1	63	Sulforhodam.	4.6
386	SR101-NaphtNH_2	PBS	493	526	589	610	626	37	80.0	0.19	15.2		2	1	−1	63	Srhod.-analog	4.6
387	Sulfonated coumarin N-H 6-H	PBS pH 7.4	377	398	436	461	515	79	31.0	0.57	17.6		2	1	−1	57	Coumarin 7-Am	4.26.3
388	Sulfonated coumarin N-H 7-H	PBS pH 7.4	453	474	520	552	639	119	40.0	0.06	2.4		2	1	−1	67	Coumarin 7-Am	4.26.3
389	Sulforhodamine 101	H_2O pH 7	534	569	586	602	605	19	116.0	0.90	104.4	3.6	2	1	−1	72	Sulforhodam.	4.6
390	Sulforhodamine B	H_2O pH 7	508	537	554	567	582	28	84.0	0.39	32.7	1.8	2	1	−1	67	Sulforhodam.	4.6
391	TAMRA	H_2O pH 7	501	525	546	561	576	30	90.0	0.20	18.0		2	1	−1	54	Rhoda 500-550	4.2.2
392	TAMRA	50 mм PBS pH 8-9	502	526	547	562	574	27	77.0	0.35	26.9	2.2	2	1	−1	54	Rhoda 500-550	4.2.2

Ent.	Short names	Aqueous solution	$\lambda_{x,-0.25}$ nm	$\lambda_{x,-0.50}$ nm	$\lambda_{x,max}$ nm	$\lambda_{x,+0.50}$ nm	$\lambda_{e,max}$ nm	Δ_S nm	ε_{max} mM^{-1}cm^{-1}	Φ_f	B_{max} mM^{-1}cm^{-1}	τ_f ns	N_n	N_p	q	N_a	Class	Ref.
393	W-B 15	100 mM P pH 7.4	518	549	567	582	594	27	76.5	0.13	9.9		2	1	−1	73	Rhoda 550-600	4.2.3
394	Y-A 1 (Cypate)	PBS pH 7.2	685	744	778	803	805	27	224.0	0.03	6.7	0.5	2	1	−1	86	Cyanine-Hepta	4.13.3
395	Distyryl-BODIPY 13	PBS pH 7.5	579	608	641	659	657	16	58.0	0.25	14.5	2.7	6	5	−1	171	Bodipy-Med.	4.14.2
396	Distyryl-BODIPY 17	H_2O	570	598	630	648	643	13	76.6	0.30	22.9	3.4	6	5	−1	171	Bodipy-Med.	4.14.2
397	2′Me-4CH_2OH-Si-rosol	100 mM P pH 7.4	554	583	613	628	631	18	70.0	0.30	21.0		0	0	0	56	Si$(CH_3)_2$-Rosol	4.20
398	2′Me-Si-rhosoL	100 mM P pH 7.4	553	582	612	627	630	18	70.0	0.28	19.6		0	0	0	52	Si$(CH_3)_2$-Rosol	4.20
399	3,10-Dimethylisoalloxazine	H_2O pH 7			435		520	85	9.5	0.15	1.4	4.5	0	0	0	28	Pteridine	4.36
400	3,6,10-Trimethylisoalloxazine	H_2O pH 7			439		532	93	8.5	0.08	0.6	2.1	0	0	0	31	Pteridine	4.36
401	3,6,7,8,10-PentaMe-isoalloxazine	H_2O pH 7			448		571	123	8.3	0.10	0.8	3.9	0	0	0	37	Pteridine	4.36
402	3,7,10-Trimethylisoalloxazine	H_2O pH 7			446		532	86	9.8	0.32	3.1	6.9	0	0	0	31	Pteridine	4.36
403	3,7,8,10-TetraMe-isoalloxazine	H_2O pH 7			444		529	85	12.4	0.25	3.1		0	0	0	34	Pteridine	4.36
404	3,7,9,10-TetraMe-isoalloxazine	H_2O pH 7			452		571	119	6.8	0.10	0.6	3.2	0	0	0	34	Pteridine	4.36
405	3,8,10-Trimethylisoalloxazine	H_2O pH 7			433		518	85	11.5	0.12	1.3	3.0	0	0	0	31	Pteridine	4.36
406	3-Amino-9-ethylcarbazole 1	H_2O			364		452	88	1.8	0.63	1.1		0	0	0	30	Carbazole	4.39.1
407	7-(4-Hydroxy)cyclohexyl coum.	H_2O	319	337	368	393	460	92	16.5	0.59	9.7		0	0	0	36	Coumarin 7-Am	4.26.2
408	7-Am-4-Me-2-hydroxyquinoline	H_2O pH 5-9			340		440	100	16.0	0.97	15.5		0	0	0	25	Quinoline	4.29
409	7-Azetidinyl-4-Me-coumarin	10 mM HEPES pH 7.3	308	325	354	377	467	113	15.0	0.96	14.4		0	0	0	28	Coumarin 7-Am	4.26.2

(Continued)

Ent.	Short names	Aqueous solution	$\lambda_{x,-0.25}$ nm	$\lambda_{x,-0.50}$ nm	$\lambda_{x,max}$ nm	$\lambda_{x,+0.50}$ nm	$\lambda_{e,max}$ nm	Δ_S nm	ε_{max} mM^{-1}cm^{-1}	Φ_f	B_{max} mM^{-1}cm^{-1}	τ_f ns	N_n	N_p	q	N_a	Class	Ref.
410	7-Diethylamino-3-CA coumarin	10 mM HEPES pH 7.3	357	376	410	432	471	61	35.0	0.03	1.0		0	0	0	61	Coumarin 7-Am	4.26.2
411	7-Diethylamino-coumarin	H_2O	333	354	388	416	475	87	18.3	0.05	0.9		0	0	0	31	Coumarin 7-Am	4.26.2
412	7-Dimethylamino-4-Me-coumarin	10 mM HEPES pH 7.3	322	340	372	397	470	98	18.0	0.19	3.4		0	0	0	28	Coumarin 7-Am	4.26.2
413	7-Ethylamino-coumarin	H_2O	319	337	368	393	458	90	11.9	0.51	6.0		0	0	0	25	Coumarin 7-Am	4.26.2
414	7-Hydroxy-4-methylcoumarin	H_2O pH 5	276	295	320	341	450	130	14.0	0.78	10.9	6.9	0	0	0	21	Coumarin 7-OH	4.26.1
415	Acedan	H_2O	299	320	362	402	515	153	11.2	0.20	2.2	1.3	0	0	0	31	Naphthalene	4.34
416	Acedan 2	H_2O	282	301	338	372	515	177	14.6	0.30	4.3		0	0	0	39	Naphthalene	4.34
417	Acedan 3	H_2O	295	316	357	396	501	144	14.5	0.31	4.5		0	0	0	34	Naphthalene	4.34
418	Acedan 4	H_2O	294	315	355	393	501	146	13.5	0.32	4.3		0	0	0	28	Naphthalene	4.34
419	Acedan 5	H_2O	295	316	357	396	503	146	17.4	0.40	6.9	3.2	0	0	0	42	Naphthalene	4.34
420	Acedan 6	H_2O	295	316	357	396	503	146	15.4	0.39	6.0	3.1	0	0	0	48	Naphthalene	4.34
421	Acedan 7	H_2O	291	311	351	388	495	144	11.0	0.31	3.4	3.4	0	0	0	32	Naphthalene	4.34
422	Acedan 8	H_2O	278	296	332	365	515	183	12.6	0.23	2.9		0	0	0	36	Naphthalene	4.34
423	Acrylodan	H_2O	315	339	387	433	502	115	12.9	0.18	2.3	1.28	0	0	0	32	Naphthalene	4.34
424	Acyman	H_2O	302	324	367	408	517	150	9.1	0.10	0.9	0.27	0	0	0	40	Naphthalene	4.34
425	BC-BA	100 mM P, 20% Et pH 7.4			371		473	102	25.5	0.17	4.3	1.33	0	0	0	33	Hybrid-Bonded	4.40.2
426	BC-H	100 mM P, 20% Et pH 7.4			365		470	105	27.2	0.09	2.4	1.63	0	0	0	29	Hybrid-Bonded	4.40.2
427	Benzo[h]quinoline (BQ)	H_2O pH 2-5			370		430	60	35.0	0.32	11.2	9.6	0	0	0	23	Quinoline	4.29
428	BMeS-*p*-A	H_2O	322	341	377	404	517	140	3.8	0.67	2.5		0	0	0	28	Benzene	4.38
429	Carbostyril 1	H_2O			358		432	74	11.6	0.24	2.7		0	0	0	29	Quinoline	4.29
430	Carbostyril 124	H_2O			341		417	76	15.8	0.46	7.2		0	0	0	23	Quinoline	4.29

Ent.	Short names	Aqueous solution	$\lambda_{x,-0.25}$ nm	$\lambda_{x,-0.50}$ nm	$\lambda_{x,max}$ nm	$\lambda_{x,+0.50}$ nm	$\lambda_{e,max}$ nm	Δ_S nm	ε_{max} $mM^{-1}cm^{-1}$	Φ_f	B_{max} $mM^{-1}cm^{-1}$	τ_f ns	N_n	N_p	q	N_a	Class	Ref.
431	Carbostyril 13	H_2O pH 7			420		556	136	17.3	0.26	4.5		0	0	0	35	Quinoline	4.29
432	Carbostyril 2	H_2O			360		430	70	6.7	0.14	0.9		0	0	0	32	Quinoline	4.29
433	Carbostyril 23b	H_2O pH 12			397		556	159	17.3	0.26	4.5		0	0	0	41	Quinoline	4.29
434	Carbostyril 23b	H_2O pH 7			424		556	132	17.3	0.26	4.5		0	0	0	41	Quinoline	4.29
435	Carbostyril 23b	H2O pH 2.8			380		475	95	18.0	0.41	7.3		0	0	0	41	Quinoline	4.29
436	Carbostyril 4	H_2O			379		531	152	8.7	0.07	0.6		0	0	0	27	Quinoline	4.29
437	Carbostyril 5	H_2O			353		440	87	6.6	0.16	1.0		0	0	0	32	Quinoline	4.29
438	Carbostyril 6	H_2O			389		544	155	7.0	0.09	0.6		0	0	0	30	Quinoline	4.29
439	Carbostyril 7	H_2O			366		580	214	9.0	0.07	0.6		0	0	0	33	Quinoline	4.29
440	Carbostyril 9	H_2O			340		382	42	6.0	0.17	0.9		0	0	0	32	Quinoline	4.29
441	Chromenoquinoline L-S 4a	H_2O			473		570	97	36.19	0.15	5.4	0.92	0	0	0	46	Hybrid-Fused	4.40.1
442	Chromenoquinoline L-S 4b	H_2O			477		579	102	36.05	0.15	5.2	0.9	0	0	0	47	Hybrid-Fused	4.40.1
443	Chromenoquinoline L-S 4c	H_2O			500		591	91	33.16	0.03	1.1	0.43	0	0	0	44	Hybrid-Fused	4.40.1
444	Chromenoquinoline L-S 4d	H_2O			500		608	108	36.06	0.08	2.8	0.62	0	0	0	48	Hybrid-Fused	4.40.1
445	Chromenoquinoline L-S 4e	H_2O			502		613	111	33.16	0.07	2.2	0.56	0	0	0	49	Hybrid-Fused	4.40.1
446	Chromenoquinoline L-S 4f	H_2O			529		622	93	40.64	0.05	1.9	0.41	0	0	0	46	Hybrid-Fused	4.40.1
447	Coumarin 1	H_2O	329	349	382	409	472	90	20.1	0.07	1.4	0.4	0	0	0	34	Coumarin 7-Am	4.26.2
448	Coumarin 102	H_2O	336	356	393	418	489	96	21.8	0.66	14.3	6.0	0	0	0	36	Coumarin 7-Am	4.26.2
449	Coumarin 120	H_2O	293	314	342	366	440	98	16.0	0.94	15.0	4.9	0	0	0	22	Coumarin 7-Am	4.26.2
450	Coumarin 152	H_2O	316	334	365	389	497	132	14.3	0.37	5.2	4.5	0	0	0	34	Coumarin 7-Am	4.26.2
451	Coumarin 2	H_2O	314	332	362	386	456	94	16.9	0.95	16.0	4.8	0	0	0	31	Coumarin 7-Am	4.26.2
452	Coumarin 307	H_2O	335	355	392	417	508	116	9.0	0.33	2.9	4.3	0	0	0	31	Coumarin 7-Am	4.26.2

(Continued)

Ent.	Short names	Aqueous solution	$\lambda_{x,-0.25}$ nm	$\lambda_{x,-0.50}$ nm	$\lambda_{x,max}$ nm	$\lambda_{x,+0.50}$ nm	$\lambda_{e,max}$ nm	Δ_S nm	ε_{max} mм$^{-1}$cm^{-1}	Φ_f	B_{max} mм$^{-1}$cm^{-1}	τ_f ns	N_n	N_p	q	N_a	Class	Ref.
453	Coumarin P-R 21	PBS pH 7.5	295	316	345	369	461	116	13.2	0.79	10.4		0	0	0	45	Coumarin 7-Am	4.26.2
454	Coumarin P-R 22	PBS pH 7.5	307	324	353	376	450	97	13.4	0.75	10.0		0	0	0	45	Coumarin 7-Am	4.26.2
455	Coumarin P-R 5	PBS pH 7.5	296	317	346	370	465	119	10.3	0.79	8.1		0	0	0	24	Coumarin 7-Am	4.26.2
456	Coumarin P-R 6	PBS pH 7.5	309	326	355	378	452	97	12.6	0.71	8.9		0	0	0	24	Coumarin 7-Am	4.26.2
457	DMPT	H_2O pH 7			382		475	93	12.4	0.06	0.7		0	0	0	48	Pteridine	4.36
458	Epicocconone	H_2O, 0.1% SDS pH 2.5			444		535	91	68.6	0.07	5.0		0	0	0	52	Epicocconone	4.39.2
459	Epicocconone P-F 17a	H_2O, 0.1% SDS pH 2.5			391		517	126	8.8	0.35	3.0		0	0	0	56	Epicocconone	4.39.2
460	Epicocconone P-F 17d	H_2O, 0.1% SDS pH 2.5			392		518	126	13.2	0.40	5.2		0	0	0	58	Epicocconone	4.39.2
461	Epicocconone P-F 4b	H_2O, 0.1% SDS pH 2.5			435		538	103	19.8	0.20	3.9		0	0	0	55	Epicocconone	4.39.2
462	Epicocconone P-F 4c	H_2O, 0.1% SDS pH 2.5			425		542	117	40.3	0.40	16.2		0	0	0	48	Epicocconone	4.39.2
463	Epicocconone P-F 4d	H_2O, 0.1% SDS pH 2.5			417		534	117	14.5	0.37	5.3		0	0	0	49	Epicocconone	4.39.2
464	Et-rosol	50 mм P pH 7.4	478	500	524	537	558	34	42.6	0.10	4.1		0	0	0	47	Rosol	4.19
465	Fluorene I-6	H_2O			353		473	120	30.5	0.13	3.9		0	0	0	41	Fluorene	4.39.2
466	Fluorene PP12	H_2O			345		555	210	24.0	0.16	3.8		0	0	0	40	Fluorene	4.39.2
467	Fluorene PP6	H_2O			373		620	247	16.5	0.07	1.1		0	0	0	61	Fluorene	4.39.2
468	HMDER	PBS pH 7.4	479	500	525	538	543	18	70.0	0.14	9.8		0	0	0	51	Rhodol	4.15.2

Ent.	Short names	Aqueous solution	$\lambda_{x,-0.25}$ nm	$\lambda_{x,-0.50}$ nm	$\lambda_{x,max}$ nm	$\lambda_{x,+0.50}$ nm	$\lambda_{e,max}$ nm	Δ_S nm	ε_{max} mM^{-1}cm^{-1}	Φ_f	B_{max} mM^{-1}cm^{-1}	τ_f ns	N_n	N_p	q	N_a	Class	Ref.
469	HMDSeR	100 mM P pH 7.4	516	540	558	571	590	32	92.0	0.01	0.5		0	0	0	51	Se-Rhodol	4.16
470	Jul-rosol	50 mM P pH 7.4	492	514	540	554	574	34	30.4	0.41	12.5		0	0	0	49	Rosol	4.19
471	Me-rosol	50 mM P pH 7.4	475	496	520	533	557	37	46.1	0.18	8.4		0	0	0	41	Rosol	4.19
472	Naphtahlimide 10b	H_2O	389	408	448	484	545	97	12.4	0.21	2.6		0	0	0	34	Naphthalimide	4.32
473	Naphtahlimide 10c	H_2O	393	412	453	490	543	90	17.3	0.30	5.1		0	0	0	48	Naphthalimide	4.32
474	OBQ	H_2O pH 2-5			385		470	85	35.0	0.43	15.0	8.7	0	0	0	27	Quinoline	4.29
475	Pip-rosol	50 mM P pH 7.4	479	500	525	538	563	38	42.5	0.05	2.0		0	0	0	48	Rosol	4.19
476	PORL	PBS, 1% DMSO pH 7.4	609	643	670	686	698	28	66.1	0.11	7.2		0	0	0	67	POPh-Rosol	4.20
477	Prodan	H_2O	300	322	364	404	531	167	14.5	0.10	1.4	2.3	0	0	0	34	Naphthalene	4.34
478	Quinine	H_2O pH 7			331		382	51	5.7	0.50	2.8	13.4	0	0	0	48	Quinoline	4.29
479	THQ-rosol	50 mM P pH 7.4	500	523	550	565	710	160	29.4	0.03	0.8		0	0	0	53	Rosol	4.19
480	3,6-Azetid.-9-CA.-Si-pyronine	10 mM HEPES pH 7.3	577	610	641	659	657	16	120.0	0.26	31.2		1	1	0	52	$Si(CH_3)_2$-Pyron	4.5
481	4,5,6,7-Tetrafluoro-SiTMR	10 mM HEPES pH 7.3	606	584	667	685	682	15	132.0	0.31	40.9		1	1	0	59	$Si(CH_3)_2$-Rhoda	4.3.2
482	560CP	PBS pH 7.4	522	541	561	576	588	27	61.0	0.76	46.3		1	1	0	56	$C(CH_3)_2$-Rhoda	4.3.1
483	570CPH	PBS pH 7.4	580	602	571	633	600	29	79.0	0.71	56.0		1	1	0	66	$C(CH_3)_2$-Rhoda	4.3.1
484	575RH	PBS pH 7.4	524	555	574	589	597	23	55.0	0.74	40.7	4.3	1	1	0	71	Rhoda 550-600	4.2.3
485	630GeRH	PBS pH 7.4	582	612	631	647	651	20	61.0	0.60	36.6	3.2	1	1	0	64	$Ge(CH_3)_2$-Rhoda	4.3.3

(Continued)

Ent.	Short names	Aqueous solution	$\lambda_{x,-0.25}$ nm	$\lambda_{x,-0.50}$ nm	$\lambda_{x,max}$ nm	$\lambda_{x,+0.50}$ nm	$\lambda_{e,max}$ nm	Δ_S nm	ε_{max} $mM^{-1}cm^{-1}$	Φ_f	B_{max} $mM^{-1}cm^{-1}$	τ_f ns	N_n	N_p	q	N_a	Class	Ref.
486	640SiRH	PBS pH 7.4	584	519	641	658	662	21	51.0	0.42	21.4	3.2	1	1	0	64	$Si(CH_3)_2$-Rhoda	4.3.2
487	Abberior Star 520SXP	PBS pH 7.4	454	475	522	555	632	110	42.5	0.15	6.3	1.0	1	1	0	81	Coumarin 7-Am	4.26.3
488	Abberior Star 580 carboxylate	PBS pH 7.4	535	567	587	603	607	20	85.0	0.90	76.5	3.5	1	1	0	83	Rhoda 550-600	4.2.3
489	Atto 465 carboxylate	PBS pH 7.4, 22 °C	404	423	453	467	506	53	75.0	0.75	56.2	5.0	1	1	0	39	Acridine	4.23.1
490	Atto 495 carboxylate	PBS pH 7.4, 22 °C	455	471	498	512	526	28	80.0	0.20	16.0	1.0	1	1	0	51	Acridine	4.23.1
491	Atto 520 carboxylate	PBS pH 7.4, 22 °C	474	497	517	529	538	21	110.0	0.90	99.0	3.6	1	1	0	54	Pyronine	4.4
492	Atto 550 carboxylate	PBS pH 7.4, 22 °C	507	535	554	567	576	22	120.0	0.80	96.0	3.6	1	1	0	87	Rhoda 550-600	4.2.3
493	Atto 610 carboxylate	PBS pH 7.4, 22 °C	529	554	616	581	633	17	150.0	0.70	105.0	3.2	1	1	0	60	$C(CH_3)_2$-Pyron	4.5
494	Atto 620 carboxylate	PBS pH 7.4, 22 °C	580	602	620	633	642	22	120.0	0.50	60.0	2.9	1	1	0	76	$C(CH_3)_2$-Rhoda	4.3.1
495	Atto 633 carboxylate	PBS pH 7.4, 22 °C	575	610	630	646	651	21	130.0	0.64	83.2	3.3	1	1	0	79	$C(CH_3)_2$-Rhoda	4.3.1
496	Atto 635 carboxylate	PBS pH 7.4, 22 °C	579	612	635	649	659	24	130.0	0.25	32.5	3.2	1	1	0	68	$C(CH_3)_2$-Pyron	4.5
497	Atto 647N carboxylate	PBS pH 7.4, 22 °C	592	626	646	661	664	18	150.0	0.65	97.5	3.5	1	1	0	99	$C(CH_3)_2$-Rhoda	4.3.1
498	Atto Rho101 carboxylate	PBS pH 7.4, 22 °C	535	569	587	601	609	22	120.0	0.80	96.0	4.2	1	1	0	84	Rhoda 550-600	4.2.3
499	Atto Rho11 carboxylate	PBS pH 7.4, 22 °C	522	553	572	585	595	23	120.0	0.80	96.0	4.0	1	1	0	82	Rhoda 550-600	4.2.3
500	Atto Rho12 carboxylate	PBS pH 7.4, 22 °C	526	557	577	590	600	23	120.0	0.80	96.0	4.0	1	1	0	100	Rhoda 550-600	4.2.3
501	Atto Rho13 carboxylate	PBS pH 7.4, 22 °C	547	584	603	618	627	24	120.0	0.80	96.0	3.9	1	1	0	96	Rhodam >600	4.2.4
502	Atto Rho14 carboxylate	PBS pH 7.4, 22 °C	564	606	626	639	646	20	140.0	0.80	112.0	3.7	1	1	0	96	Rhodam >600	4.2.4

Ent.	Short names	Aqueous solution	$\lambda_{x,-0.25}$ nm	$\lambda_{x,-0.50}$ nm	$\lambda_{x,max}$ nm	$\lambda_{x,+0.50}$ nm	$\lambda_{e,max}$ nm	Δ_S nm	ε_{max} mM^{-1}cm^{-1}	Φ_f	B_{max} mM^{-1}cm^{-1}	τ_f ns	N_n	N_p	q	N_a	Class	Ref.
503	Atto Rho3B carboxylate	PBS pH 7.4, 22 °C	517	548	566	579	589	23	120.0	0.50	60.0	1.5	1	1	0	80	Rhoda 550-600	4.2.3
504	Atto Rho6G carboxylate	PBS pH 7.4, 22 °C	489	513	533	546	557	24	115.0	0.94	108.1	4.1	1	1	0	73	Rhoda 500-550	4.2.2
505	BODIPY	H_2O	457	475	498	506	516	18	90.0	0.90	81.2		1	1	0	37	Bodipy-Small	4.14.1
506	Bza-OH	H_2O pH 7.4	551	584	620	636	653	33	12.8	0.37	4.7		1	1	0	53	Phenoxazine	4.25
507	Carborhodamine	10 mM HEPES pH 7.3	515	537	552	564	577	25	65.0	0.64	41.6		1	1	0	48	$C(CH_3)_2$-Rhoda	4.3.1
508	Carborhodamine 1c	H_2O	591	619	640	656	662	22	89.0	0.53	47.1		1	1	0	94	$C(CH_3)_2$-Rhoda	4.3.1
509	Carborhodamine 20b	H_2O	592	620	641	657	663	22	80.0	0.53	42.4		1	1	0	88	$C(CH_3)_2$-Rhoda	4.3.1
510	Carborhodamine 21a	H_2O	591	619	640	656	663	23	70.0	0.56	39.2		1	1	0	111	$C(CH_3)_2$-Rhoda	4.3.1
511	Coumarin-pyronine 9	100 mM P pH 7.5			560		579	19	132.5	0.13	17.2		1	1	0	56	Hybrid-Bonded	4.40.2
512	DAMBO-T	100 mM P pH 7.4	457	475	498	506	507	9	52.0	0.40	20.8		1	1	0	45	Bodipy-Small	4.14.1
513	DAOTA 2	H_2O	493	515	564	583	594	30	50.0	0.27	13.5	13.5	1	1	0	54	Triangulenium	4.22
514	Edans	H_2O	281	300	336	370	520	184	6.2	0.27	1.6		1	1	0	32	Naphthalene	4.34
515	Fluorinated rhodamine 1e	Buffer pH 7	461	485	499	510	522	23	81.0	0.88	71.2	3.9	1	1	0	51	Rhodam <500	4.2.1
516	GeR	PBS pH 7.4	601	621	634	658	655	21	97.0	0.43	41.7	2.7	1	1	0	62	$Ge(CH_3)_2$-Rhoda	4.3.3
517	G-L 13a	H_2O	463	482	505	514	517	12	35.0	0.47	16.4	3.4	1	1	0	79	Bodipy-Med.	4.14.2
518	G-L 13b	H_2O	473	493	509	519	522	13	56.0	0.67	37.5	5.1	1	1	0	84	Bodipy-Med.	4.14.2
519	G-L 14	H_2O	472	492	508	518	521	13	59.0	0.77	45.4	5.3	1	1	0	100	Bodipy-Med.	4.14.2
520	JF479	10 mM HEPES pH 7.3	432	450	479	493	517	38	47.9	0.62	29.7		1	1	0	57	Acridine	4.23.1
521	JF502	10 mM HEPES pH 7.3	450	470	502	518	533	31	57.8	0.71	41.0		1	1	0	57	Acridine	4.23.1

(Continued)

Ent.	Short names	Aqueous solution	$\lambda_{x,-0.25}$ nm	$\lambda_{x,-0.50}$ nm	$\lambda_{x,max}$ nm	$\lambda_{x,+0.50}$ nm	$\lambda_{e,max}$ nm	Δ_S nm	ε_{max} mM^{-1}cm^{-1}	Φ_f	B_{max} mM^{-1}cm^{-1}	τ_f ns	N_n	N_p	q	N_a	Class	Ref.
522	JF525	10 mM HEPES pH 7.3	483	506	525	538	549	24	94.0	0.91	85.5		1	1	0	53	Rhoda 500-550	4.2.2
523	JF526	10 mM HEPES pH 7.3	484	507	526	539	550	24	19.0	0.87	16.5		1	1	0	53	Rhoda 500-550	4.2.2
524	JF549	10 mM HEPES pH 7.3	504	528	549	564	571	22	101	0.88	88.8		1	1	0	54	Rhoda 500-550	4.2.2
525	JF552	10 mM HEPES pH 7.3	505	535	552	566	575	23	95.2	0.83	79.0		1	1	0	54	Rhoda 550-600	4.2.3
526	JF559	10 mM HEPES pH 7.3	511	541	559	573	579	20	106.0	0.85	90.1		1	1	0	53	Rhoda 550-600	4.2.3
527	JF570	10 mM HEPES pH 7.3	521	548	570	586	593	23	83.6	0.63	52.6		1	1	0	53	S-Rhodamine	4.3.3
528	JF571	10 mM HEPES pH 7.3	521	552	571	586	590	19	101	0.78	78.7		1	1	0	54	Rhoda 550-600	4.2.3
529	JF574	10 mM HEPES pH 7.3	524	555	574	589	594	20	103	0.78	80.3		1	1	0	54	Rhoda 550-600	4.2.3
530	JF585	10 mM HEPES pH 7.3	544	568	585	598	609	24	1.5	0.78	1.1		1	1	0	61	$C(CH_3)_2$-Rhoda	4.3.1
531	JF593	10 mM HEPES pH 7.3	541	569	593	610	612	19	90.3	0.55	49.6		1	1	0	53	S-Rhodamine	4.3.3
532	JF608	10 mM HEPES pH 7.3	564	589	608	623	631	23	99.0	0.67	66.3		1	1	0	62	$C(CH_3)_2$-Rhoda	4.3.1

Ent.	Short names	Aqueous solution	$\lambda_{x,-0.25}$ nm	$\lambda_{x,-0.50}$ nm	$\lambda_{x,max}$ nm	$\lambda_{x,+0.50}$ nm	$\lambda_{e,max}$ nm	Δ_S nm	ε_{max} mM^{-1}cm^{-1}	Φ_f	B_{max} mM^{-1}cm^{-1}	τ_f ns	N_n	N_p	q	N_a	Class	Ref.
533	JF646	10 mM HEPES pH 7.3	589	568	646	663	664	18	5.6	0.54	3.0		1	1	0	61	$Si(CH_3)_2$-Rhoda	4.3.2
534	JF668	10 mM HEPES pH 7.3	607	585	668	686	687	19	26.7	0.34	9.0		1	1	0	55	POOH-Rhoda	4.3.3
535	JF669	10 mM HEPES pH 7.3	608	586	669	687	682	13	112.0	0.37	41.4		1	1	0	62	$Si(CH_3)_2$-Rhoda	4.3.2
536	JF669-6-SCH2CO2Et	10 mM HEPES pH 7.3	607	585	668	686	681	13	119.0	0.38	45.2		1	1	0	75	$Si(CH_3)_2$-Rhoda	4.3.2
537	JF690	10 mM HEPES pH 7.3	625	602	690	710	707	17	150.0	0.24	36.0		1	1	0	55	POOH-Rhoda	4.3.3
538	JF711	10 mM HEPES pH 7.3	642	618	711	732	732	21	12.4	0.17	2.1		1	1	0	65	POPh-Rhodam	4.3.3
539	JF722	10 mM HEPES pH 7.3	651	626	722	743	743	21	87.2	0.11	9.5		1	1	0	65	POPh-Rhodam	4.3.3
540	JF724	10 mM HEPES pH 7.3	647	688	724	750	748	24	6.6	0.05	0.3		1	1	0	55	SO_2-Rhodam	4.3.3
541	JS-R	10 mM HEPES pH 7.4	558	589	611	625	631	20	130.0	0.59	76.7		1	1	0	37	$B(OH)_2$-Pyron	4.5
542	KK9046	H_2O	573	613	632	648	654	22	90.0	0.45	40.5	3.6	1	1	0	100	Rhodam >600	4.2.4
543	Luminarine	H_2O pH 7			428		533	105	9.0	0.47	4.2	7.4	1	1	0	23	Pteridine	4.36
544	*meso*-CaboxyBODIPY J-K 3a	10 mM PBS pH 7.4	477	497	519	531	558	39	77.0	0.67	51.5		1	1	0	31	Bodipy-Small	4.14.1
545	*meso*-CaboxyBODIPY J-K 3b	10 mM PBS pH 7.4	477	497	519	531	549	30	78.0	0.83	64.7		1	1	0	34	Bodipy-Small	4.14.1

(Continued)

Ent.	Short names	Aqueous solution	$\lambda_{x,-0.25}$ nm	$\lambda_{x,-0.50}$ nm	$\lambda_{x,max}$ nm	$\lambda_{x,+0.50}$ nm	$\lambda_{e,max}$ nm	Δ_S nm	ε_{max} mM^{-1}cm^{-1}	Φ_f	B_{max} mM^{-1}cm^{-1}	τ_f ns	N_n	N_p	q	N_a	Class	Ref.
546	*meso*-CaboxyBODIPY J-K 3c	10 mM PBS pH 7.4	480	501	517	527	540	23	76.0	0.97	73.7		1	1	0	37	Bodipy-Small	4.14.1
547	NBDY	H_2O	462	481	504	513	533	29	74.4	0.33	24.5	2.6	1	1	0	40	Bodipy-Small	4.14.1
548	NCR	H_2O			565		618	53	25.8	0.00	0.0		1	1	0	63	Hybrid-Fused	4.40.1
549	NR666	PBS, 1% DMSO pH 7.4	608	646	666	683	685	19	165.0	0.38	62.7		1	1	0	54	PO_2H-Rosam.	4.18.2
550	NR669	10 mM PBS pH 7.4	619	653	669	684	689	20	77.2	0.30	23.1		1	1	0	57	PO_2H-Rosam.	4.18.2
551	NR675	10 mM PBS pH 7.4	625	657	675	689	695	20	73.2	0.24	17.5		1	1	0	59	PO_2H-Rosam.	4.18.2
552	NR698	PBS, 1% DMSO pH 7.4	631	676	698	715	712	14	25.9	0.32	8.2		1	1	0	70	PO_2H-Rosam.	4.18.2
553	PONy P1	PBS pH 7.4	535	564	590	605	627	37	59.0	0.33	19.4	1.8	1	1	0	42	Pyronine	4.4
554	PONy P1-Halo	PBS pH 7.4	546	576	603	619	635	32	18.0	0.26	4.6	1.9	1	1	0	84	Pyronine	4.4
555	PONy P5	PBS pH 7.4	592	627	660	679	704	44	38.0	0.10	3.8	1.7	1	1	0	85	Pyronine	4.4
556	PONy P6	PBS pH 7.4	546	576	603	619	638	35	69.0	0.09	6.2	1.9	1	1	0	79	Pyronine	4.4
557	PONy P9	PBS pH 7.4	541	570	597	613	633	36	66.0	0.22	14.5		1	1	0	46	Pyronine	4.4
558	Pyrrolidinylcarborhodamine	10 mM HEPES pH 7.3	568	594	613	628	633	20	87.0	0.54	46.9		1	1	0	68	$C(CH_3)_2$-Rhoda	4.3.1
559	Rhodamine 110	H_2O pH <7	458	482	496	507	520	24	83.0	0.89	73.8	4.3	1	1	0	40	Rhodam <500	4.2.1
560	SiR	PBS pH 7.4	589	623	645	663	661	16	93.0	0.41	38.1	2.7	1	1	0	62	$Si(CH_3)_2$-Rhoda	4.3.2
561	SiRF 13a	H_2O	598	577	657	675	677	20	53.0	0.63	33.3		1	1	0	59	$Si(CH_3)_2$-Rhoda	4.3.2

Ent.	Short names	Aqueous solution	$\lambda_{x,-0.25}$ nm	$\lambda_{x,-0.50}$ nm	$\lambda_{x,max}$ nm	$\lambda_{x,+0.50}$ nm	$\lambda_{e,max}$ nm	Δ_S nm	ε_{max} mM^{-1}cm^{-1}	Φ_f	B_{max} mM^{-1}cm^{-1}	τ_f ns	N_n	N_p	q	N_a	Class	Ref.
562	SiRF 1a-H	H_2O	602	581	662	680	680	18	118.0	0.66	77.8	2.4	1	1	0	59	Si$(CH_3)_2$-Rhoda	4.3.2
563	SiR-methyl	H_2O	571	611	648	665	662	14	100.0	0.39	39.0	2.6	1	1	0	61	Si(CH3)2-Rosa	4.18.1
564	SiTMR	10 mM HEPES pH 7.3	586	566	643	660	662	19	28.2	0.41	11.5		1	1	0	59	Si$(CH_3)_2$-Rhoda	4.3.2
565	S-JF646	10 mM HEPES pH 7.3	586	620	652	671	668	16	21.2	0.25	5.3		1	1	0	57	Si$(CH_3)_2$-Pyron	4.5
566	SMSiR	H_2O	584	626	652	671	670	18	120.0	0.31	37.2		1	1	0	63	Si$(CH_3)_2$-Srhod	4.7
567	Sulfobetaine pyrene 21	H_2O	314	327	359	365	385	26	30.6	0.32	9.7	11.6	1	1	0	52	Pyrene	4.28
568	Sulfo-JF646	10 mM HEPES pH 7.3	586	629	655	674	670	15	124.0	0.51	63.2		1	1	0	62	Si$(CH_3)_2$-Srhod	4.7
569	Tetraethylrhodamine	H_2O pH 7	510	539	557	571	580	23	105.0	0.24	25.2	1.5	1	1	0	64	Rhoda 550-600	4.2.3
570	Tetramethycarborhodamine	10 mM HEPES pH 7.3	562	588	606	620	626	20	121.0	0.52	62.9		1	1	0	60	C$(CH_3)_2$-Rhoda	4.3.1
571	Tetramethylrhodamine	10 mM HEPES pH 7.3	503	527	548	563	572	24	78.0	0.45	35.1		1	1	0	51	Rhoda 500-550	4.2.2
572	Thienyl-JF646	10 mM HEPES pH 7.3	593	572	651	668	667	16	113	0.40	45.2		1	1	0	58	Si$(CH_3)_2$-Rhoda	4.3.2
573	Tri-*O*-acetyl-P-luminoarosin	H_2O pH 7			425		528	103	10.8	0.62	6.7	8.3	1	1	0	58	Pteridine	4.36
574	Y-A 6	PBS pH 7.2	685	744	778	803	805	27	224.0	0.03	6.7		1	1	0	109	Cyanine-Hepta	4.13.3
575	Z-L A	0.5 M PBS pH 7.4	457	476	499	507	510	11	50	0.04	2.1		1	1	0	170	Bodipy-Large	4.14.3
576	Z-L B	0.5 M PBS pH 7.4	455	474	496	504	507	11	50	0.36	17.8		1	1	0	230	Bodipy-Large	4.14.3

(Continued)

Ent.	Short names	Aqueous solution	$\lambda_{x,-0.25}$ nm	$\lambda_{x,-0.50}$ nm	$\lambda_{x,max}$ nm	$\lambda_{x,+0.50}$ nm	$\lambda_{e,max}$ nm	Δ_S nm	ε_{max} $mM^{-1}cm^{-1}$	Φ_f	B_{max} $mM^{-1}cm^{-1}$	τ_f ns	N_n	N_p	q	N_a	Class	Ref.
577	Z-L C	0.5 M PBS pH 7.4	489	510	533	546	550	17	50	0.01	0.6		1	1	0	170	Bodipy-Large	4.14.3
578	Z-L D	0.5 M PBS pH 7.4	492	513	536	549	555	19	50	0.02	0.9		1	1	0	174	Bodipy-Large	4.14.3
579	Z-L E	0.5 M PBS pH 7.4	490	511	534	547	550	16	50	0.09	4.4		1	1	0	182	Bodipy-Large	4.14.3
580	Z-L F	0.5 M PBS pH 7.4	459	478	501	509	511	10	50	0.31	15.6		1	1	0	170	Bodipy-Large	4.14.3
581	Z-L G	0.5 M PBS pH 7.4	526	550	577	592	598	21	50	0.21	10.6		1	1	0	231	Bodipy-Large	4.14.3
582	Z-L H	0.5 M PBS pH 7.4	594	625	660	679	676	16	50	0.01	0.6		1	1	0	tl	Bodipy-Large	4.14.3
583	Z-L I	0.5 M PBS pH 7.4	451	469	491	499	501	10	50	0.68	34.0		1	1	0	tl	Bodipy-Large	4.14.3
584	Z-L J	0.5 M PBS pH 7.4	498	519	543	556	563	20	50	0.34	16.8		1	1	0	tl	Bodipy-Large	4.14.3
585	Niu 18	H_2O	455	474	496	504	511	15	75.5	0.20	15.1	3.2	2	2	0	69	Bodipy-Med.	4.14.2
586	Niu 25	PBS pH 7.4	479	499	521	533	540	19	58.8	0.82	48.2	5.1	3	3	0	88	Bodipy-Med.	4.14.2
587	Niu 27	H_2O	474	494	510	520	528	18	63.3	0.81	51.2	9.2	3	3	0	100	Bodipy-Med.	4.14.2
588	Niu 29	H_2O	477	497	519	531	530	11	28.0	0.61	17.0	7.0	3	3	0	107	Bodipy-Med.	4.14.2
589	ZW800-1	FBS	680	739	772	797	788	16	249.0	0.15	37.6		3	3	0	132	Cyanine-Hepta	4.13.3
590	Distyryl-BODIPY 18	H_2O	568	596	628	646	641	13	71.0	0.38	26.9	3.9	5	5	0	175	Bodipy-Med.	4.14.2
591	10-Ethylacridine	P pH 7	376	390	412	423	494	82	3.5	0.98	3.4	44.4	0	1	1	27	Acridine	4.23.1
592	2-Aminopyridine	H_2O			300		363	63	4.1	0.71	2.9		0	1	1	13	Pyridine	4.39.1

Ent.	Short names	Aqueous solution	$\lambda_{x,-0.25}$ nm	$\lambda_{x,-0.50}$ nm	$\lambda_{x,max}$ nm	$\lambda_{x,+0.50}$ nm	$\lambda_{e,max}$ nm	Δ_S nm	ε_{max} mM^{-1}cm^{-1}	Φ_f	B_{max} mM^{-1}cm^{-1}	τ_f ns	N_n	N_p	q	N_a	Class	Ref.
593	2′COOH-4CH$_2$OH-Si-rhodol	100 mM P pH 7.4	560	589	610	626	630	20	70.0	0.21	14.7		0	1	1	56	Si(CH3)2-Rhod	4.16
594	2′COOH-Si-rhodol	100 mM P pH 7.4	559	588	609	625	630	21	70.0	0.22	15.4		0	1	1	52	Si(CH3)2-Rhod	4.16
595	2-Me RG	PBS pH 7.4	458	483	498	510	520	22	82.0	0.84	68.8		0	1	1	40	Rosamine	4.17
596	2-Me SiR600	H_2O	542	574	593	607	613	20	91.0	0.38	34.5		0	1	1	48	Si(CH$_3$)$_2$-Rosa	4.18.1
597	2-Me SiR620	100 mM P, 1% DMSO pH 7.4	550	586	620	636	638	18	96.0	0.21	20.1		0	1	1	54	Si(CH$_3$)$_2$-Rosa	4.18.1
598	2-Me SiR640	100 mM P, 1% DMSO pH 7.4	509	548	637	658	660	23	54.0	0.18	9.7		0	1	1	55	Si(CH$_3$)$_2$-Rosa	4.18.1
599	2-Me SiR650	H_2O	591	628	646	661	660	14	110.0	0.31	34.1		0	1	1	60	Si(CH$_3$)$_2$-Rosa	4.18.1
600	2-Me SiR654	100 mM P, 1% DMSO pH 7.4	576	616	654	671	672	18	80.0	0.24	19.2		0	1	1	61	Si(CH$_3$)$_2$-Rosa	4.18.1
601	2-Me SiR666	100 mM P, 1% DMSO pH 7.4	585	627	666	684	684	18	80.0	0.25	20.0		0	1	1	61	Si(CH$_3$)$_2$-Rosa	4.18.1
602	2-Me SiR670	100 mM P, 1% DMSO pH 7.4	588	630	670	688	687	17	90.0	0.27	24.3		0	1	1	67	Si(CH$_3$)$_2$-Rosa	4.18.1
603	2-Me SiR671	100 mM P, 1% DMSO pH 7.4	589	631	671	689	687	16	99.0	0.29	28.7		0	1	1	59	Si(CH$_3$)$_2$-Rosa	4.18.1
604	2-Me SiR675	100 mM P, 1% DMSO pH 7.4	592	635	675	694	698	23	88.0	0.07	6.1		0	1	1	71	Si(CH$_3$)$_2$-Rosa	4.18.1

(Continued)

Ent.	Short names	Aqueous solution	$\lambda_{x,-0.25}$ nm	$\lambda_{x,-0.50}$ nm	$\lambda_{x,max}$ nm	$\lambda_{x,+0.50}$ nm	$\lambda_{e,max}$ nm	Δ_S nm	ε_{max} mM^{-1}cm^{-1}	Φ_f	B_{max} mM^{-1}cm^{-1}	τ_f ns	N_n	N_p	q	N_a	Class	Ref.
605	2-Me SiR680	H_2O	592	634	674	693	689	15	130.0	0.35	45.5		0	1	1	68	$Si(CH_3)_2$-Rosa	4.18.1
606	2-Me SiR700	H_2O	617	665	691	710	712	21	100.0	0.12	12.0		0	1	1	62	$Si(CH_3)_2$-Rosa	4.18.1
607	2-Me SiR720	H_2O	627	675	721	742	740	19	160.0	0.05	8.0		0	1	1	82	$Si(CH_3)_2$-Rosa	4.18.1
608	2-Me TeOR	PBS pH 7.4	599	642	669	692	686	17	120.0	0.18	21.6		0	1	1	53	TeO-Rosamine	4.18.3
609	2-Methyl harmine	50 mM H_2SO_4	324	331	350	370	433	83	6.3	0.45	2.8	6.5	0	1	1	31	Indole	4.33
610	3,6-Azetid.-Si-pyronine	10 mM HEPES pH 7.3	573	606	636	654	649	13	97.1	0.62	60.2		0	1	1	49	$Si(CH_3)_2$-Pyron	4.5
611	3,6-Di(azetidin-1-yl)-acridine	10 mM HEPES pH 7.3	442	461	492	507	531	39	47.0	0.52	24.4		0	1	1	41	Acridine	4.23.1
612	3,6-Diaminoacridine (Proflavine)	H_2O pH 7	404	420	445	457	511	66	38.9	0.34	13.2		0	1	1	28	Acridine	4.23.1
613	3,7-Di(azetidin)-phenoxazine	10 mM HEPES pH 7.3	572	607	647	664	661	14	99.0	0.24	23.7		0	1	1	40	Phenoxazine	4.25
614	3,7-Di(ethylamino)-phenoxazine	10 mM HEPES pH 7.3	578	615	655	673	669	14	111.0	0.07	7.7		0	1	1	50	Phenoxazine	4.25
615	3,7-Dimethyl-JF646	10 mM HEPES pH 7.3	574	614	651	668	664	13	89.6	0.51	45.7		0	1	1	65	$Si(CH_3)_2$-Rosa	4.18.1
616	3-Amino-9-ethylcarbazole 2	H_2O			364		458	94	1.8	0.76	1.3		0	1	1	47	Carbazole	4.39.1
617	3-Aminopyridine	H_2O			300		390	90	2.8	0.32	0.9		0	1	1	13	Pyridine	4.39.1
618	3-Hydroxy-JF646	10 mM HEPES pH 7.3	574	614	651	668	666	15	136	0.48	65.2		0	1	1	60	$Si(CH_3)_2$-Rosa	4.18.1
619	3-Methyl-5-Methoxy-JF646	10 mM HEPES pH 7.3	572	612	649	666	664	15	110	0.49	53.9		0	1	1	66	$Si(CH_3)_2$-Rosa	4.18.1

(Continued)

Ent.	Short names	Aqueous solution	$\lambda_{x,-0.25}$ nm	$\lambda_{x,-0.50}$ nm	$\lambda_{x,max}$ nm	$\lambda_{x,+0.50}$ nm	$\lambda_{e,max}$ nm	Δ_S nm	ε_{max} mM^{-1}cm^{-1}	Φ_f	B_{max} mM^{-1}cm^{-1}	τ_f ns	N_n	N_p	q	N_a	Class	Ref.
620	3-Methyl-JF646	10 mM HEPES pH 7.3	572	612	649	666	666	17	118	0.47	55.4		0	1	1	62	Si(CH_3)$_2$-Rosa	4.18.1
621	5-Aza-3-methyl-JF646	10 mM HEPES pH 7.3	578	618	656	674	670	14	133	0.48	63.8		0	1	1	61	Si(CH_3)$_2$-Rosa	4.18.1
622	6,8-Dichloroharmine (6,8-DiCL-Ha)	H_2O pH 4	341	348	369	391	429	60	6.3	0.20	1.2	2.8	0	1	1	29	Indole	4.33
623	6-Chloro-harmine	H_2O pH 4	335	341	362	383	428	66	9.0	0.16	1.4	2.2	0	1	1	29	Indole	4.33
624	8-Chloro-harmine	H_2O pH 4	334	341	361	382	423	62	5.9	0.11	0.6	2.2	0	1	1	29	Indole	4.33
625	9-Aminoacridine	P pH 7	366	379	400	410	420	20	9.3	0.97	9.0	16.4	0	1	1	26	Acridine	4.23.1
626	Acridine orange	H_2O	439	458	488	503	528	40	50.0	0.12	6.0		0	1	1	40	Acridine	4.23.1
627	Acridine orange	HEPES pH 7.3	443	462	493	508	528	35	50.0	0.21	10.5		0	1	1	38	Acridine	4.23.1
628	Acridine red	H_2O	484	507	528	540	562	34	100.0	0.51	51.0	1.6	0	1	1	33	Pyronine	4.4
629	Acridizinium I-L 5a	H_2O			387		507	120	17.4	0.12	2.0		0	1	1	26	Acridizinium	4.39.1
630	Acridizinium I-L 5b	H_2O			369		551	182	11.0	0.02	0.2		0	1	1	26	Acridizinium	4.39.1
631	Acridizinium J-N 1a	H_2O			377		404	27	8.9	0.33	2.9	6.0	0	1	1	24	Acridizinium	4.39.1
632	Acridizinium J-N 1b	H_2O			381		408	27	6.7	0.34	2.2	6.2	0	1	1	24	Acridizinium	4.39.1
633	Acridizinium J-N 1c	H_2O			354		420	66	8.4	0.30	2.5	5.4	0	1	1	24	Acridizinium	4.39.1
634	Acridizinium J-N 1d	H_2O			356		408	52	5.4	0.20	1.0	3.6	0	1	1	24	Acridizinium	4.39.1
635	Acridizinium J-N 1e	H_2O			355		455	100	6.7	0.20	1.3	7.5	0	1	1	28	Acridizinium	4.39.1
636	AR116	100 mM P pH 7.5	483	506	527	539	548	21	64.8	0.07	4.5		0	1	1	39	Pyronine	4.4
637	Bpe-H	H_2O pH 7.4	567	601	640	657	681	41	48.1	0.23	11.0		0	1	1	54	Phenoxazine	4.25
638	Bza-Oet	H_2O pH 7.4	555	588	625	641	653	28	33.5	0.26	8.7		0	1	1	59	Phenoxazine	4.25
639	Bzh-Cl	H_2O pH 7.4	551	584	620	636	651	31	36.1	0.31	11.1		0	1	1	51	Phenoxazine	4.25

(Continued)

Ent.	Short names	Aqueous solution	$\lambda_{x,-0.25}$ nm	$\lambda_{x,-0.50}$ nm	$\lambda_{x,max}$ nm	$\lambda_{x,+0.50}$ nm	$\lambda_{e,max}$ nm	Δ_S nm	ε_{max} mM^{-1}cm^{-1}	Φ_f	B_{max} mM^{-1}cm^{-1}	τ_f ns	N_n	N_p	q	N_a	Class	Ref.
640	Bzh-H	H_2O pH 7.4	551	584	620	636	651	31	11.3	0.44	4.9		0	1	1	50	Phenoxazine	4.25
641	Bzm-H	H_2O pH 7.4	563	597	635	652	674	39	29.9	0.14	4.1		0	1	1	48	Phenoxazine	4.25
642	Carborhodamine 1b	H_2O	590	619	639	655	661	22	80.0	0.58	46.4		0	1	1	97	$C(CH_3)_2$-Rhoda	4.3.1
643	Carborhodamine 20a	H_2O	591	619	640	656	661	21	70.0	0.53	37.1		0	1	1	91	$C(CH_3)_2$-Rhoda	4.3.1
644	Carborhodol S-H 13-H	PBS pH 7.4	539	567	586	601	613	27	59.0	0.32	18.8	2.5	0	1	1	64	$C(CH_3)_2$-Rhod	4.16
645	COUPY 10	H_2O			562		631	69	33.0	0.03	0.8		0	1	1	53	Hybrid-Bridged	4.40.3
646	COUPY 11	H_2O			567		662	95	13.0	0.02	0.2		0	1	1	44	Hybrid-Bridged	4.40.3
647	COUPY 12	H_2O			595		683	88	6.9	0.05	0.3		0	1	1	47	Hybrid-Bridged	4.40.3
648	COUPY 9	H_2O			543		605	62	31.0	0.15	4.6		0	1	1	50	Hybrid-Bridged	4.40.3
649	Cresyl violet (CV)	H_2O	523	553	585	599	625	40	33.1	0.36	11.9	2.2	0	1	1	32	Phenoxazine	4.25
650	DAOTA 1	H_2O	489	511	558	578	587	29	50.0	0.31	15.5	15.2	0	1	1	51	Triangulenium	4.22
651	Des-carboxy-JF646	10 mM HEPES pH 7.3	571	611	648	665	662	14	116	0.20	23.2		0	1	1	59	$Si(CH_3)_2$-Rosa	4.18.1
652	Fluorinated rhodamine 3-H	Buffer pH 7	471	492	510	523	531	21	57.0	0.86	49.0	4.2	0	1	1	64	Rhoda 500-550	4.2.2
653	Harmaline	5 mM H_2SO_4	352	360	383	407	475	92	28.6	0.32	9.1	5.3	0	1	1	30	Indole	4.33
654	Harmane	50 mM H_2SO_4	324	331	350	370	440	90	6.3	0.83	5.1	20.0	0	1	1	24	Indole	4.33
655	Harmine	H_2O pH 4	330	337	357	378	416	59	7.3	0.49	3.6	7.1	0	1	1	29	Indole	4.33
656	Harmine	50 mM H_2SO_4	304	309	326	343	412	86	11.9	0.45	5.3	6.6	0	1	1	29	Indole	4.33
657	Harmine	H_2O pH 9	309	315	332	350	425	93	11.0	0.38	4.1	0.4	0	1	1	28	Indole	4.33
658	HBQ	H_2O pH 2-5			390		480	90	35.0	0.15	5.2		0	1	1	24	Quinoline	4.29

Ent.	Short names	Aqueous solution	$\lambda_{x,-0.25}$ nm	$\lambda_{x,-0.50}$ nm	$\lambda_{x,max}$ nm	$\lambda_{x,+0.50}$ nm	$\lambda_{e,max}$ nm	Δ_S nm	ε_{max} mM^{-1}cm^{-1}	Φ_f	B_{max} mM^{-1}cm^{-1}	τ_f ns	N_n	N_p	q	N_a	Class	Ref.
659	Me-PR	PBS, 1% DMSO pH 7.4	632	672	694	711	712	18	92.0	0.11	10.1		0	1	1	68	POR-Rosam.	4.18.2
660	Naphthalimide B-K 1	10 mM P pH 7.0	380	398	436	470	552	116	13.2	0.08	1.0	2.3	0	1	1	40	Naphthalimide	4.32
661	NBQ	H_2O pH 2-5			380		470	90	35.0	0.35	12.2	8.5	0	1	1	28	Quinoline	4.29
662	NCRMe	H_2O			568		607	39	22.8	0.01	0.1		0	1	1	67	Hybrid-Fused	4.40.1
663	N-Methylacridone	H_2O	369	382	411	419	431	20	8.8	0.99	8.7	15.8	0	1	1	28	Acridones	4.23.2
664	NOBQ	H_2O pH 2-5			385		470	85	35.0	0.41	14.3	9.1	0	1	1	31	Quinoline	4.29
665	nor-Harmane (β-carboline)	50 mM H_2SO_4	333	340	360	381	454	94	6.3	0.58	3.6	21.1	0	1	1	21	Indole	4.33
666	NR700	PBS, 1% DMSO pH 7.4	630	674	700	720	722	22	71.0	0.11	7.8		0	1	1	61	PO_2R-Rosam.	4.18.2
667	NR744	PBS, 1% DMSO pH 7.4	672	719	744	763	764	20	80.0	0.16	12.8		0	1	1	77	PO_2R-Rosam.	4.18.2
668	Oxazine 170	H_2O pH 2-7	539	560	620	643	650	30	48.0	0.24	11.5	2.8	0	1	1	47	Phenoxazine	4.25
669	PONy CP1	PBS pH 7.4	640	681	710	728	783	73	17.0	0.19	3.2		0	1	1	78	$C(CH_3)_2$-Pyron	4.5
670	PONy P10	PBS pH 7.4	567	599	629	646	668	39	52.0	0.12	6.2		0	1	1	62	Pyronine	4.4
671	PONy P13	PBS pH 7.4	568	600	630	647	666	36	52.0	0.17	8.8		0	1	1	72	Pyronine	4.4
672	PONy P2	PBS pH 7.4	569	602	632	649	672	40	51.0	0.16	8.1	0.8	0	1	1	50	Pyronine	4.4
673	PR	PBS, 1% DMSO pH 7.4	632	672	694	711	712	18	65.0	0.06	3.9		0	1	1	66	POR-Rosam.	4.18.2
674	Pyronine 630	H_2O, 25 °C	509	535	559	573	579	20	100.0	0.92	92.0	3.7	0	1	1	53	Pyronine	4.4
675	Pyronine B	H_2O	504	529	552	565	572	20	94.1	0.18	16.9	1.3	0	1	1	51	Pyronine	4.4

(Continued)

Ent.	Short names	Aqueous solution	$\lambda_{x,-0.25}$ nm	$\lambda_{x,-0.50}$ nm	$\lambda_{x,max}$ nm	$\lambda_{x,+0.50}$ nm	$\lambda_{e,max}$ nm	Δ_S nm	ε_{max} mм$^{-1}$cm^{-1}	Φ_f	B_{max} mм$^{-1}$cm^{-1}	τ_f ns	N_n	N_p	q	N_a	Class	Ref.
676	Pyronine Y (OP)	PBS pH 7.4	500	525	546	560	569	23	77.0	0.36	27.7	1.7	0	1	1	39	Pyronine	4.4
677	Pyronine Y (OP)	H_2O	498	523	546	559	566	20	77.0	0.22	16.9	2.0	0	1	1	39	Pyronine	4.4
678	Quinine	50 mм H_2SO_4			347		451	104	5.7	0.55	3.1	13.4	0	1	1	48	Quinoline	4.29
679	Quinine	100 mм $HClO_4$			347		451	104	5.7	0.60	3.4		0	1	1	48	Quinoline	4.29
680	RF620	10 mм PBS pH 7.4	569	600	620	635	636	16	109.0	0.36	39.2		0	1	1	55	$B(OH)_2$-Rosa	4.18.3
681	Rhodamine 123	H_2O pH <7	464	485	502	514	528	26	75.0	0.90	67.5	4.2	0	1	1	43	Rhoda 500-550	4.2.2
682	Rhodamine 6G	H_2O pH <7	485	507	526.3	540	552.3	26	105.0	0.93	97.6	4.1	0	1	1	64	Rhoda 500-550	4.2.2
683	Rhodamine 6G	D_2O pH <7	484	506	525.5	539	552.0	26	105.0	0.98	102.9	4.4	0	1	1	64	Rhoda 500-550	4.2.2
684	Rhodamine RBzET	H_2O	485	508	527	541	552	25	82.0	0.87	71.3	4.1	0	1	1	79	Rhoda 500-550	4.2.2
685	Rosamine W-B 5h	100 mм P pH 7.4	495	524	542	557	568	26	71.5	0.14	10.0		0	1	1	69	Rosamine	4.17
686	Safranin T	H_2O			517		586	69	27.0	0.11	2.9	1.2	0	1	1	43	Phenazine	4.39.1
687	Si-9-Iminopyronine 2c	~1 mм HCl, 1% DMSO pH 3	597	633	666	685	681	15	83.2	0.28	23.3	2.32	0	1	1	72	$Si(CH_3)_2$-Pyron	4.5
688	Si-9-Iminopyronine 2c	~1 mм NaOH, 1% DMSO pH 11	390	405	418	426	607	189	12.9	0.46	5.9	2.34	0	1	1	72	$Si(CH_3)_2$-Pyron	4.5
689	Si-9-Iminopyronine 2f	~1 mм NaOH, 1% DMSO pH 11	375	389	401	408	606	205	17.8	0.35	6.2	1.9	0	1	1	87	$Si(CH_3)_2$-Pyron	4.5
690	Si-9-Iminopyronine 2g	PBS pH 7.4	434	452	469	479	631	162	11	0.43	4.7	2.8	0	1	1	85	$Si(CH_3)_2$-Pyron	4.5

Ent.	Short names	Aqueous solution	$\lambda_{x,-0.25}$ nm	$\lambda_{x,-0.50}$ nm	$\lambda_{x,max}$ nm	$\lambda_{x,+0.50}$ nm	$\lambda_{e,max}$ nm	Δ_S nm	ε_{max} mM^{-1}cm^{-1}	Φ_f	B_{max} mM^{-1}cm^{-1}	τ_f ns	N_n	N_p	q	N_a	Class	Ref.
691	$SiOH_2R$	10 mM PBS pH 7.4	599	639	663	679	681	18	105.0	0.43	45.1		0	1	1	56	$Si(OH)_2$-Rosa	4.18.1
692	SiP-F	50 mM PBS pH 7.4	577	611	637	653	651	14	64.2	0.20	12.8		0	1	1	53	$C(CH_3)_2$-Pyron	4.5
693	SiP-Ph	50 mM PBS pH 7.4	588	624	649	665	665	16	80.0	0.23	18.4		0	1	1	61	$C(CH_3)_2$-Pyron	4.5
694	SiP-Pr	50 mM PBS pH 7.4	579	613	632	647	652	20	64.2	0.21	13.4		0	1	1	53	$C(CH_3)_2$-Pyron	4.5
695	SO_2R1	10 mM PBS, 0.25% ACN pH 7.4	633	678	703	724	736	33	39	0.07	2.6		0	1	1	51	SO_2-Rosamine	4.18.3
696	SO_2R2	10 mM PBS, 0.25% ACN pH 7.4	633	678	703	724	742	39	95	0.07	6.9		0	1	1	54	SO_2-Rosamine	4.18.3
697	SO_2R3	10 mM PBS, 0.25% ACN pH 7.4	636	681	707	728	747	40	69	0.05	3.4		0	1	1	55	SO_2-Rosamine	4.18.3
698	SO_2R4	10 mM PBS, 0.25% ACN pH 7.4	634	679	704	725	742	38	97.0	0.10	9.4		0	1	1	57	SO_2-Rosamine	4.18.3
699	SO_2R5	10 mM PBS, 0.25% ACN pH 7.4	639	684	710	732	752	42	129.0	0.07	8.7		0	1	1	59	SO_2-Rosamine	4.18.3

(Continued)

Ent.	Short names	Aqueous solution	$\lambda_{x,-0.25}$ nm	$\lambda_{x,-0.50}$ nm	$\lambda_{x,max}$ nm	$\lambda_{x,+0.50}$ nm	$\lambda_{e,max}$ nm	Δ_S nm	ε_{max} $mM^{-1}cm^{-1}$	Φ_f	B_{max} $mM^{-1}cm^{-1}$	τ_f ns	N_n	N_p	q	N_a	Class	Ref.
700	Tetramethylrosamine	10 mM PBS pH 7.4	501	532	550	565	570	20	74.5	0.31	23.1		0	1	1	52	Rosamine	4.17
701	TG170	H_2O	695	756	791	817	803	12	300.0	0.02	7.4		0	1	1	116	Cyanine-Hepta	4.13.3
702	TMDHS or SiP-Me	50 mM PBS pH 7.4	570	603	633	651	654	21	64.2	0.41	26.3		0	1	1	47	$Si(CH_3)_2$-Pyron	4.5
703	TMDHS or SiP-Me	H_2O	571	604	634	652	653	19	64.2	0.18	11.5		0	1	1	47	$Si(CH_3)_2$-Pyron	4.5
704	tMe-PR	PBS, 1% DMSO pH 7.4	634	674	696	714	713	17	68.0	0.15	10.2		0	1	1	74	POR-Rosam.	4.18.2
705	Y-A 10	PBS pH 7.2	687	746	780	805	808	28	219.0	0.03	6.5		0	1	1	254	Cyanine-Hepta	4.13.3
706	Y-A 11	PBS pH 7.2	687	746	780	805	807	27	199.0	0.03	5.9		0	1	1	318	Cyanine-Hepta	4.13.3
707	Y-A 7	PBS pH 7.2	686	745	779	804	805	26	224.0	0.03	6.7		0	1	1	132	Cyanine-Hepta	4.13.3
708	Y-A 8	PBS pH 7.2	687	746	780	805	806	26	197.0	0.03	5.9		0	1	1	164	Cyanine-Hepta	4.13.3
709	Y-A 9	PBS pH 7.2	686	745	779	804	809	30	187.0	0.03	5.6		0	1	1	196	Cyanine-Hepta	4.13.3
710	BOPAHY H-D 18ab′	H_2O	398	412	429	435	453	24	51	0.55	28.0		1	2	1	64	Bodipy-analog	4.14.5
711	BOPAHY H-D 18bb′	H_2O	397	411	428	434	453	25	51	0.49	24.9		1	2	1	56	Bodipy-analog	4.14.5
712	BOPAHY H-D 19aa′	H_2O	345	355	368	373	387	19	51	0.02	0.7		1	2	1	58	Bodipy-analog	4.14.5
713	BOPAHY H-D 19ab′	H_2O	387	400	416	422	435	19	51	0.20	10.2		1	2	1	63	Bodipy-analog	4.14.5
714	BOPAHY H-D 19ba′	H_2O	344	355	367	372	386	19	51	0.03	1.5		1	2	1	50	Bodipy-analog	4.14.5
715	BOPAHY H-D 19bb′	H_2O	387	400	416	422	434	18	51	0.17	8.6		1	2	1	55	Bodipy-analog	4.14.5
716	BOPAHY H-D 20bb′	H_2O	387	400	416	422	435	19	51	0.17	8.6		1	2	1	55	Bodipy-analog	4.14.5
717	SiRF 1b-H	H_2O	603	581	663	681	680	17	90.0	0.70	63.0	2.4	1	2	1	82	$Si(CH_3)_2$-Rhoda	4.3.2
718	DAOTA 5	H_2O	479	505	548	567	576	28	50.0	0.56	28.0	18.6	2	3	1	93	Triangulenium	4.22
719	DAOTA 6	H_2O	489	509	558	575	584	26	50.0	0.36	18.0	14.6	2	3	1	81	Triangulenium	4.22

Ent.	Short names	Aqueous solution	$\lambda_{x,-0.25}$ nm	$\lambda_{x,-0.50}$ nm	$\lambda_{x,max}$ nm	$\lambda_{x,+0.50}$ nm	$\lambda_{e,max}$ nm	Δ_S nm	ε_{max} $mM^{-1}cm^{-1}$	Φ_f	B_{max} $mM^{-1}cm^{-1}$	τ_f ns	N_n	N_p	q	N_a	Class	Ref.
720	AI3	H_2O pH 7.4	387	405	445	481	552	107	15.0	0.21	3.1	4.3	0	2	2	68	Naphthalimide	4.32
721	Carborhodamine 16a	H_2O	588	616	636	652	657	21	114.0	0.62	70.6		0	2	2	97	$C(CH_3)_2$-Rhoda	4.3.1
722	Lucigenin	H_2O			368		485	117	37.6	0.50	18.8	25.3	0	2	2	52	Hybrid-Bonded	4.40.2
723	Rhodamine 6G +2	H_2O pH <7	487	511	530	544	553	23	101.0	0.93	93.9		0	2	2	82	Rhoda 500-550	4.2.2
724	Si-9-Iminopyronine 2f	~1 mM HCl, 1% DMSO pH 3	595	631	664	683	682	18	123	0.26	31.9	2.1	0	2	2	87	$Si(CH_3)_2$-Pyron	4.5
725	ZW800-3a	FBS	682	741	774	799	790	16	309.0	0.16	49.7		1	3	2	126	Cyanine-Hepta	4.13.3
726	DAOTA 4	H_2O	481	502	549	567	576	27	50.0	0.56	28.0	18.3	0	3	3	75	Triangulenium	4.22
727	Rhodamine 6G +3	H_2O pH <7	486	510	529	544	551	22	101.0	0.91	91.9		0	3	3	76	Rhodam Hi Ch	4.2.5

Index

Fluorescent Dye Labels and Stains: A Database of Photophysical Properties, First Edition. Tarso B. Ledur Kist.

Printed in the USA/Agawam, MA
September 16, 2026
910661.195